Systems Biology

Aleš Prokop · Béla Csukás
Editors

Systems Biology

Integrative Biology and Simulation Tools

Volume 1

Springer

Editors
Aleš Prokop
Chemical and Biological Engineering
Vanderbilt University
Nashville, TN
USA

Béla Csukás
Research Group on Process Network Engineering
Kaposvár University
Kaposvár
Hungary

ISBN 978-94-007-6802-4 ISBN 978-94-007-6803-1 (eBook)
DOI 10.1007/978-94-007-6803-1
Springer Dordrecht Heidelberg New York London

Library of Congress Control Number: 2012953557

Printed on acid-free paper

Springer is part of Springer Science+Business Media (www.springer.com)

Contents

Contributors

Hassan Ahmed European Institute for Systems Biology and Medicine, CNRS-UCBL-ENS, Université de Lyon, 50, Avenue Tony Garnier, 69366 Lyon Cedex 07, France, e-mail: hahmed@eisbm.org

Gary An Department of Surgery, University of Chicago, Chicago, USA, e-mail: docgca@gmail.com

Charles Auffray European Institute for Systems Biology and Medicine—CNRS-UCBL-ENS, Université de Lyon, 50 avenue Tony Garnier, 69007 Lyon, France, e-mail: cauffray@eisbm.org

Nitin Baliga Institute for Systems Biology, 401 Terry Avenue North, Seattle, WA 98109–5234, USA, e-mail: nbaliga@systemsbiology.org

Stéphane Ballereau European Institute for Systems Biology & Medicine, CNRS-UCBL—Université de Lyon, 50, Avenue Tony Garnier, 69007 Lyon, France, e-mail: sballereau@eisbm.org

Rudi Balling Luxembourg Centre for Systems Biomedicine, University of Luxembourg, 7 Avenue des Hauts-Fourneaux, 4362 Esch-sur-Alzette, Luxembourg, e-mail: rudi.balling@uni.lu

Amrita Basu Broad Institute of MIT and Harvard, 7 Cambridge Center, Cambridge, MA 02142, USA, e-mail: amrita@broadinstitute.org

Chiranjib Bhattacharyya Indian Institute of Science, Bangalore, India, e-mail: chiru@csa.iisc.ernet.in

Maria Biryukov Luxembourg Centre for Systems Biomedicine, University of Luxembourg, 7, Avenue des Hauts-Fourneaux, 4362 Esch-sur-Alzette, Luxembourg, e-mail: maria.biryukov@uni.lu

Dominique Boutigny Centre de Calcul de l'IN2P3, USR6402 CNRS/IN2P3, 43 Bd du 11 Novembre 1918, 69622 Villeurbanne Cedex, France, e-mail: boutigny@in2p3.fr

Amphun Chaiboonchoe European Institute for Systems Biology & Medicine, CNRS-UCBL—Université de Lyon, 50, Avenue Tony Garnier, 69007 Lyon, France, e-mail: achaiboonchoe@eisbm.org

Leonid L. Chepelev Carleton University, Ottawa, Canada, e-mail: leonid.chepelev@gmail.com

Scott Christley Department of Surgery, University of Chicago, Chicago, USA, e-mail: schristley@uchicago.edu

Lily A. Chylek Department of Chemistry and Chemical Biology, Cornell University, Ithaca, NY 14853, USA, e-mail: lily.chylek@gmail.com

Paul Clemons Broad Institute of MIT and Harvard, 7 Cambridge Center, Cambridge, MA 02142, USA, e-mail: pclemons@broadinstitute.org

Isaac Crespo Luxembourg Centre for Systems Biomedicine, University of Luxembourg, 7, Avenue des Hauts-Fourneaux, 4362 Esch-sur-Alzette, Luxembourg, e-mail: isaac.crespo@uni.lu

Béla Csukás Research Group on Process Network Engineering, Kaposvár University, Kaposvár, Hungary, e-mail: csukas.bela@ke.hu

Vlado Dančík Broad Institute of MIT and Harvard, 7 Cambridge Center, Cambridge, MA 02142, USA; Mathematical Institute, Slovak Academy of Sciences, Grešákova 6, Košice, Slovakia, e-mail: vdancik@broadinstitute.org

Matthias Dehmer UMIT, Institute for Bioinformatics and Translational Research, Eduard Wallnoefer Zentrum 1, 6060 Hall in Tyrol, Austria, e-mail: matthias.dehmer@umit.at

Antonio del Sol Luxembourg Centre for Systems Biomedicine, University of Luxembourg, 7, Avenue des Hauts-Fourneaux, 4362 Esch-sur-Alzette, Luxembourg, e-mail: antonio.delsol@uni.lu

Devdatt Dubhashi Chalmers University of Technology, Göteborg, Sweden, e-mail: dubhashi@chalmers.se

Michel Dumontier Carleton University, Ottawa, Canada, e-mail: michel.dumontier@gmail.com

Frank Emmert-Streib Computational Biology and Machine Learning Lab, Center for Cancer Research and Cell Biology, School of Medicine, Dentistry and Biomedical Sciences, Queen's University Belfast, 97 Lisburn Road, Belfast BT9 7BL, UK, e-mail: v@bio-complexity.com

László Földvári-Nagy Department of Genetics, Eötvös Loránd University, Budapest, Hungary, e-mail: foldvari-nagi@netbiol.elte.hu

Fotis Georgatos Luxembourg Centre for Systems Biomedicine, University of Luxembourg, 7 Avenue des Hauts-Fourneaux, 4362 Esch-sur-Alzette, Luxembourg, e-mail: fotis.georgatos@uni.lu

Moustafa Ghanem Department of Computing, Imperial College London, London SW7 2AZ, UK, e-mail: mmg@doc.ic.ac.uk

Enrico Glaab Luxembourg Centre for Systems Biomedicine, University of Luxembourg, 7, Avenue des Hauts-Fourneaux, 4362 Esch-sur-Alzette, Luxembourg, e-mail: enrico.glaab@uni.lu

Yi-Ke Guo Department of Computing, Imperial College London, London SW7 2AZ, UK, e-mail: y.guo@imperial.ac.uk

Hendrik Hache Max Planck Institute for Molecular Genetics, Ihnestr. 63-73, 14195 Berlin, Germany, e-mail: hache@molgen.mpg.de

Wolfgang F. Heidenreich Institut für Strahlenbiologie, Helmholtz Zentrum München, Ingolstädter Landstrasse 1, 85758 Neuherberg, Germany, e-mail: heidenreich@helmholtz-muenchen.de

Ron Henkel Department of Systems Biology and Bioinformatics, Rostock University, Rostock, Germany, e-mail: ron.henkel@uni-rostock.de

William S. Hlavacek Theoretical Biology and Biophysics Group, Theoretical Division, Los Alamos National Laboratory, Los Alamos, NM 87545, USA and Department of Biology, University of New Mexico, Albuquerque, NM 87131, USA, e-mail: wish@lanl.gov

Robert Hoehndorf Carleton University, Ottawa, Canada, e-mail: leechuck@leechuck.de

Leroy Hood Institute for Systems Biology, 401 Terry Avenue North, Seattle, WA 98109-5234, USA, e-mail: lhood@systemsbiology.org

Vinay Jethava Chalmers University of Technology, Göteborg, Sweden, e-mail: jethava@chalmers.se

Wiktor Jurkowski Luxembourg Centre for Systems Biomedicine, University of Luxembourg, 7, Avenue des Hauts-Fourneaux, 4362 Esch-sur-Alzette, Luxembourg, e-mail: wiktor.jurkowski@uni.lu

Shinichi Kikuchi The Wallace H. Coulter Department of Biomedical Engineering, Georgia Institute of Technology, 313 Ferst Drive, Atlanta, GA 30332, USA, e-mail: kikuchi@bme.gatech.edu

Alexey Kolodkin Luxembourg Centre for Systems Biomedicine, University of Luxembourg, 7, Avenue des Hauts-Fourneaux, 4362 Esch-sur-Alzette, Luxembourg; Institute for Systems Biology, 401 Terry Avenue North, Seattle, WA 98109- 5234, USA, e-mail: alexey.kolodkin@uni.lu

Tamás Korcsmáros Department of Genetics, Eötvös Loránd University, Budapest, Hungary, e-mail: korcsmaros@netbiol.elte.hu

Antony Le Béchec Luxembourg Centre for Systems Biomedicine, University of Luxembourg, 7, Avenue des Hauts-Fourneaux, 4362 Esch-sur-Alzette, Luxembourg, e-mail: anthony.lebechec@uni.lu

Katalin Lenti Faculty of Health Sciences, Department of Morphology and Physiology, Semmelweis University, Budapest, Hungary, e-mail: dr.lenti.kata@gmail.com

Guangquan Li Radiation Epidemiology Branch, National Cancer Institute, Executive Plaza South, 6120 Executive Boulevard MSC 7238, Rockville, MD 20852-7238, USA, e-mail: guang.li04@imperial.ac.uk

Mark P. Little Radiation Epidemiology Branch, National Cancer Institute, Executive Plaza South, 6120 Executive Boulevard MSC 7238, Rockville, MD 20852-7238, USA ; Department of Epidemiology and Biostatistics, School of Public Health, Imperial College of Medicine, London, W2 1PG, UK, e-mail: mark.little@nih.gov

Vincent Lotteau European Institute for Systems Biology and Medicine, CNRS-UCBL-ENS, Université de Lyon, 50, Avenue Tony Garnier, 69366 Lyon cedex 07, France, e-mail: vlotteau@eisbm.org

Laurene Meyniel European Institute for Systems Biology and Medicine, CNRS-UCBL-ENS, Université de Lyon, 50, Avenue Tony Garnier, 69366 Lyon cedex 07, France, e-mail: laurene.meyniel@inserm.fr

Dezső Módos Department of Genetics, Eötvös Loránd University, Budapest, Hungary; Faculty of Health Sciences, Department of Morphology and Physiology, Semmelweis University, Budapest, Hungary, e-mail: dezso.modos@netbiol.elte.hu

Laurin A. J. Mueller UMIT, Institute for Bioinformatics and Translational Research, Eduard Wallnoefer Zentrum 1, 6060 Hall in Tyrol, Austria, e-mail: laurin.mueller@umit.at

Máté Pálfy Department of Genetics, Eötvös Loránd University, Budapest, Hungary, e-mail: palfy.mate@gmail.com

Johann Pellet European Institute for Systems Biology and Medicine—CNRS-UCBL-ENS, Université de Lyon, 50 avenue Tony Garnier, 69007 Lyon, France, e-mail: jpellet@eisbm.org

Richard G. Posner Clinical Translational Research Division, Translational Genomics Research Institute, Phoenix, AZ 85004, USA and Department of Biological Sciences, Northern Arizona University, Flagstaff, AZ 86011, USA, e-mail: rposner@tgen.org

Nathan Price Institute for Systems Biology, 401 Terry Avenue North, Seattle, WA 98109-5234, USA, e-mail: Nathan.price@systemsbiology.org

Aleš Prokop Chemical and Biomolecular Engineering Vanderbilt University, Nashville TN 37235, USA, e-mail: ales.prokop@vanderbilt.edu

Antonio Raussel Luxembourg Centre for Systems Biomedicine, University of Luxembourg, 7, Avenue des Hauts-Fourneaux, 4362 Esch-sur-Alzette, Luxembourg, e-mail: antonio.raussel@uni.lu

Reinhard Schneider Luxembourg Centre for Systems Biomedicine, University of Luxembourg, 7 Avenue des Hauts-Fourneaux, 4362 Esch-sur-Alzette, Luxembourg, e-mail: reinhard.schneider@uni.lu

Roberto Serra Department of Physics, Computer Science and Mathematics, Modena and Reggio Emilia University, v. Campi 213b, 41125 Modena, Italy; European Centre for Living Technology, Ca'Minich, S. Marco 2940, 30124 Venezia, Italy, e-mail: rserra@unimore.it

Edward C. Stites Clinical Translational Research Division, Translational Genomics Research Institute, Phoenix, AZ 85004, USA, e-mail: edstites@gmail.com

Mónika Varga Research Group on Process Network Engineering, Kaposvár University, Kaposvár, Hungary, e-mail: varga.monika@ke.hu

Marco Villani Department of Physics, Computer Science and Mathematics, Modena and Reggio Emilia University, v. Campi 213b, 41125 Modena, Italy; European Centre for Living Technology, Ca'Minich, S. Marco 2940, 30124 Venezia, Italy, e-mail: marco.villani@unimore.it

Nikos Vlassis Luxembourg Centre for Systems Biomedicine, University of Luxembourg, 7, Avenue des Hauts-Fourneaux, 4362 Esch-sur-Alzette, Luxembourg, e-mail: nikos.vlassis@uni.lu

Eberhard O. Voit The Wallace H. Coulter Department of Biomedical Engineering, Georgia Institute of Technology, 313 Ferst Drive, Atlanta, GA 30332, USA, e-mail: eberhard.voit@bme.gatech.edu

Vitaly Volpert European Institute for Systems Biology and Medicine, CNRS-UCBL-ENS, Université de Lyon, 50, Avenue Tony Garnier, 69366 Lyon cedex 07, France, e-mail: vvolpert@eisbm.org

Dagmar Waltemath Department of Systems Biology and Bioinformatics, Rostock University, Rostock, Germany, e-mail: dagmar.waltemath@uni-rostock.de

Michael Wandling Department of Surgery, Northwestern University, Evanston, USA, e-mail: m-wandling@md.northwestern.edu

Christoph Wierling Max Planck Institute for Molecular Genetics, Ihnestr. 63-73, 14195 Berlin, Germany, e-mail: wierling@molgen.mpg.de

Felix Winter Department of Systems Biology and Bioinformatics, Rostock University, Rostock, Germany, e-mail: felix.winter@uni-rostock.de

Olaf Wolkenhauer Department of Systems Biology and Bioinformatics, Rostock University, Rostock, Germany; Stellenbosch Institute for Advanced Study (STIAS), Wallenberg Research Centre at Stellenbosch University, Marais Street, Stellenbosch 7600, South Africa, e-mail: olaf.wolkenhauer@uni-rostock.de

Editorial and Introduction

Systems biology employs a rational approach to delineate the emerging properties of biological networks and complex biotechnological schemes. It aims to explain and to predict molecular, cellular, tissue, organ, and whole-body processes, quantitatively. By using multiscale models *in silico*, SB is expected to bring numerous benefits to pharmaceutical discovery and development as the properties of a system can be studied over a wide range of length and timescales. SB can reduce the number of compounds that need to be synthesized in discovery by using refined algorithms to weed out compounds with poor pharmacokinetic and toxicology profiles. Accordingly, it will save time and money by selecting the drugs, which are more likely to succeed in clinical development.

SB features two arms: biology at health (functional/physiologic network model) and biology at disease (pathological/disease network model). In both cases, the global systems level of understanding requires two components (Gomes B Practical Applications of Systems Biology in the Pharmaceutical Industry Genomics, International Drug Discovery 5(2): 54–57, 2010).

The first is termed integrative biology. This is an attempt to meaningfully combine all the sources of information on targets, assays, and compounds. Within the pharma industry there has been an enormous expansion of information from expression profiling, proteomics, metabolomics, etc., as well as a great deal of proprietary information on compound libraries, high throughput and high content screening, cheminformatics, compound safety testing, and clinical trial information. An integration of all this information represents a challenge to integrative biology.

The second component of SB lies in the areas of computational methods, modeling, and simulation. The goal is to create simulations of metabolic processes, signaling pathways, transcription (regulatory) networks, physiological processes, or even cell- or whole organism-based models.

In terms of various methods and tools of computational modeling, there is a need for critical analysis and lack of understanding of multiscale biosystems for computational experts. The challenges and critical points are the following:

Conventional biological science has produced a vast amount of data over the last few decades so that questions arise as to how to find patterns and how to relate multi-quality data sets in the quest for underlying mechanism. It is clear that conventional, reductionism-determined research approaches must fail in understanding the mechanisms behind the complex pattern generation, the self-organization and nonlinear interaction of these multiscaled systems. This old concept was for a long time around, where one simply has to dissect the biology, investigate it in isolation, and finally put it all back together. Most conventional experimental models, however, have been developed in the reductionism era, i.e., they focus on one endpoint and emphasize one feature—with little dynamical information and lacking the possibility of studying more than one to two variables of the system at the same time. Most scientists would now admit that although this approach has led to very significant discoveries in the past it will not be able to explain the complex behavior of most biological systems. It is increasingly recognized that complex systems cannot be described in a reductionist view.

It is thus imperative to develop novel research approaches with complex system science that will become one of the grand challenges of the next decade (non-reductive experimentation and corresponding computational tools). While models must be constructed, analyzed, simulated, validated, and verified, on the experimental side, biological systems need to be identified, characterized, and tested for their reliability in cross-species comparisons, in particular between model organisms and humans. Thus, SB approaches are hypothesis-driven and involve iterative rounds of data-driven modeling, prediction, model-driven experimentation, model refinement, and development. Towards this end, tremendously increased computational power will help in the efforts ahead to analyze the immense amount of data in the biological and biomedical sciences in order to guide promising new experimental work. Such new experimental design should be based on and driven by the mathematical and computational modeling. We need to come up with such new experimental methods to test and refine the predictions made with novel theoretical models, based on 'model-aided (-driven) experimental design.' The present effort features scarcities of detailed quantitative experimental data and of computational tools designed to use such data for the development and testing of biologically meaningful models. Such experimentation will require a strong background in physics, mathematics, and computer science.

The focus will be on the challenging interface between model-based experimental design and computational simulation. We should therefore bridge the gap between experimental and computational modeling experts. This is an emerging scientific field.

The Editorial Plan for present and future volumes can be summarized as follows:

(1) Foundation of systems biology, integrative biology, and issues of biosystem complexity, redundancy, and of other network properties; setting the stage
Foundations of SB: requirements and outcomes, challenges in SB and networks, graph-theoretic analysis of networks, network modularity, disease module, network topology, clustering, redundancy, emergent (systems) properties, other structural and topological network properties: parsimony, epistasis, robustness, motifs, etc., network manipulation, network rewiring, elementary network

reconstruction tools (constraint-based, accurate mathematical reaction model), network inference (learning structure), identifying disease module, etc.

(2) Modeling of ill-defined, lack of knowledge (experimental) problems in the field of complex biosystems, including identification and validation; review of enabling technologies
Reproducibility of model-based results, identifiability of ill-defined systems, structural identifiability, standardization, network functional mapping, large-scale reaction networks, large-scale integrated data analysis (combined metabolic, signaling and regulatory networks), coarse-grained simulation tools, hybrid simulation tools, agent-based modeling methodology, etc.

(3) Critical analysis of multiscale computational methods and tools, including identification and validation; computational tools for crossing levels
Multiscale simulation tools and scale laws, crossing scale and boundaries at multiscale simulation, spatiotemporal control, Physiome systems, etc.

(4) Discussion of pros and cons of various integrative system biology approaches, methods and tools of computational modeling, including simplifications, approximations and assumptions, applied for the description of various problems; success case studies
Reconstructing cellular signaling and regulatory networks—an integrative approach, computational SB in health and disease, inferring from combined gene, signaling and regulatory networks, large-scale signaling networks, employment of computational network biology for identification of combination targets, SB and immunology, SB and cancer, SB and disease inferring, SB and cell differentiation/stem cells, SB and pharmacology, etc.

(5) Systems biology in discovery and development
Target optimization tools, lead optimization, virtual chemistry screening, lead discovery and molecular interactions, SAR/QSAR methods, *In silico* screening, quantitative pharmacology and pharmacokinetics, SB at biomarker identification, clinical model-based drug development (MBDD), personalized medicine, etc.

Note, Volume one will be based on items (1) through (3). Considering space limitations the second volume will continue to cover the same topics and starts with parts from (4) and (5). A possible third volume would cover items (4) and (5), in detail.

Recent Challenges of Computer Assisted Systems Biology: Requirements and Outcomes

Motivation

Growth in the pharmaceutical market has slowed—almost to a standstill. One reason is that governments and other payers are cutting costs in a faltering world economy. But a more fundamental problem is the failure of major companies to discover, develop, and market new drugs. Major drugs, losing patent protection or being withdrawn from the market are simply not being replaced by new therapies—the pharmaceutical market model is no longer functioning effectively and most pharmaceutical companies are failing to produce the innovation needed for success (Prokop and Michelson 2012). To document the above statement by very recent numbers, the R&D efficiency, measured in terms of the number of new drugs brought to market by the global biotechnology and pharmaceutical industries has declined steadily. According to the number of new US Food and Drug Administration approved drugs per billion US dollars of R&D spending in the drug industry has halved approximately every nine years since 1950, in inflation-adjusted terms (Scannell et al. 2012).

This multi-authored new book looks at systems biology (SB) as a vital strategy that can bring innovation to a market in need of new ideas and new products. Modeling is a significant task of systems biology. SB aims to develop and use efficient algorithms, data structures, visualization, and communication tools to orchestrate the integration of large quantities of biological data with computational modeling. It involves the simulation of biological systems, such as the networks of metabolites combining with signal transduction pathways and gene regulatory networks to analyze and visualize the complex connections of these cellular processes. SB involves a series of operational protocols used for performing research, namely a cycle composed of theory, analytic, or computational modeling to propose specific testable hypotheses about a biological system, experimental validation, and then using the newly acquired quantitative description of cells or cellular processes to refine the computational model or the theory.

Place of Systems Biology in Computational Biology

Notions of Computational systems biology, bioinformatics, and the other fields of computer-assisted biology are not defined clearly.

Bioinformatics (BI) is another relatively new discipline dealing with the computational needs of biology, which has become a highly data-intensive activity. Biology databases must deal with both variety and scale, as well as have to be able to integrate the disparate databases that are their information sources. At the same time they must provide flexible, user friendly interfaces for querying and data mining must cope with incomplete and uncertain data (van Gend and Snoep 2008). BI tools are very different from SB. However, both BI and SB work in concert and in parallel; SB and BI are different but complementary. Specifically, various BI computational methods address a broad spectrum of problems in functional genomics and cell physiology, including: analysis of sequences, (alignment, homology discovery, gene annotation), gene clustering, pattern recognition/discovery in large-scale expression data, elucidation of genetic regulatory circuits, analysis of metabolic networks, and signal transduction pathways. The underlined items may overlap with SB goals.

In our view bioinformatics is a part of computational biology, consisting of five major components, as follows:

- BI (including computational genomics);
- Structural modeling: molecular modeling and protein structures (some people consider structural tools as part of BI);
- Biophysics: molecular dynamics ("physical biology of the cell");
- Computational modeling of biosystems (CSB); and
- Biomedical informatics.

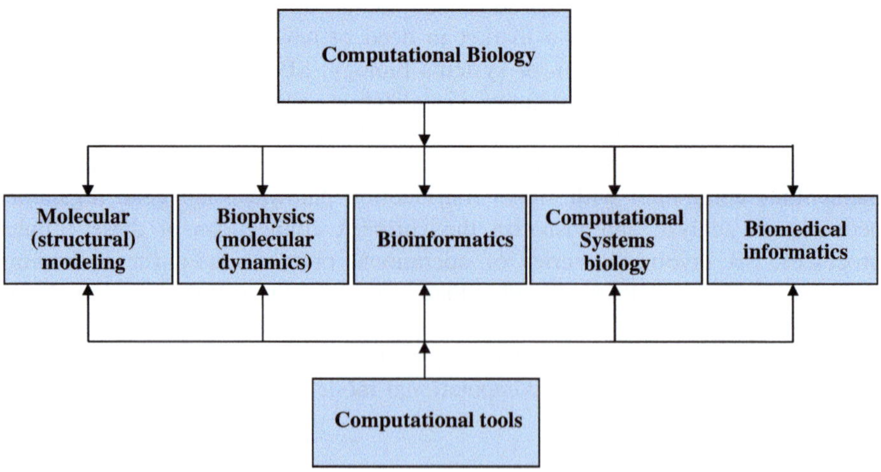

Fig. 1 Fields of Computational Biology

An overview of branches of computational biology is presented in Fig. 1 (adapted from Prokop and Michelson 2012). The question is whether there is an overlap between SB and BI. The disciplines exist side by side, in parallel. Some tools are used interchangeably between SB and BI. Biomedical informatics (BMI) is another branch of CB, like biophysics. Biophysics focuses on physical concepts and phenomena that cut across multiple biological structures and functions (Phillips et al. 2009). Three distinct biophysics approaches have been identified: (all taken from Phillips, although some approaches listed may be part of chemical engineering or SB domains now):

1. Mechanical, chemical equilibrium, entropy, statistics, and tensegrity for resting cells;
2. Statistical, chemical rate, and electrochemical tools for cell dynamics; and
3. Networking in space and time.

The SB inputs are evident in all disciplines. Although the distinction is used by NIH in their working definitions of bioinformatics and computational biology, it is clear that there is a tight coupling of developments and knowledge between the more hypothesis-driven research in computational biology and technique-driven research in bioinformatics.

Systems biology is a radically new concept resulting from the merging of two basic paradigms, reductionism and holism. It represents a combination of reductionist and holistic approaches to the relationships among the elements of a system, with the goal of identifying its emergent properties and defining molecular, cellular, tissue, organ, and whole-body processes, quantitatively. SB represents a tool for hypothesis generation about a system that is typically too large and complex to understand by simple reasoning. In our definition the SB is "quantitative, postgenomic, postproteomic, dynamic, *multiscale* physiology" (Wikswo et al. 2006).

Systems biology employs a rational approach to delineate the emerging properties of biological networks and complex biotechnological schemes. By using multiscale models *in silico*, SB is expected to bring numerous benefits to pharmaceutical discovery and development as the properties of a system can be studied over a wide range of length and timescales. SB can reduce the number of compounds that need to be synthesized in discovery by using refined algorithms to weed out compounds with poor pharmacokinetic of toxicology profiles. Also, it will save time and money by selecting the drugs, which are more likely to succeed in clinical development.

Foundation of Systems Siology, Integrative Biology, Biosystem Complexity, Redundancy, and Other Network Properties

SB features two arms: biology at health (functional/physiologic network model) and biology at disease (pathological/disease network model). In both cases, the global system level understanding requires two components (Gomes 2010). The first is termed integrative biology. This is an attempt to meaningfully combine all the sources of information on targets, assays, and compounds. Within the pharma industry there has been an enormous expansion of information from expression profiling, proteomics, metabolomics, etc., as well as a great deal of proprietary information on compound libraries, high throughput and high content screening, chemo-informatics, compound safety testing, and clinical trial information. An integration of all this information represents a challenge of integrative biology.

The second component of SB lies in the areas of computational methods, modeling, and simulation. The goal is to create simulations of metabolic processes, signaling pathways, transcription (regulatory) networks, physiological processes, or even cell- or whole organism-based models.

In terms of various methods and tools of computational modeling, there is a need for critical analysis, as well as a lack of understanding of multiscale biosystems by computational experts. As the matrix algebra is one of the basic mathematical tools employed first in SB, we stress its importance by presenting articles from this field in the first section of the book. Graph theory is an old field of mathematics where biological applications are driving new advances ("small world" networks, where most nodes are locally connected but a few have long-range links, Watts and Strogatz 1998; and "scale-free" networks, where node degree follows a power-law distribution: most nodes are connected to only a few neighbors but a few nodes are connected to many neighbors, Albert and Barabasi 2002). Perhaps the biggest open mathematical challenges, however, are in understanding the *dynamic* properties of networks that cannot be derived from static measures of their structure (Armstrong and Sorokina 2012).

Modeling of Ill-Defined, Lack of Knowledge of Experimentally Observed Problems

There is some doubt about the usefulness of SB originating from the complexity of biological systems. Simeonidis (2011) argued: in early 2010, Sydney Brenner stated that "The new science of systems biology [...] will fail because deducing models of function from the behavior of a complex system is an inverse problem that is impossible to solve." Brenner goes on to label systems biology "*anti-reductionism*," and even calls it '*low input, high throughput, no output*' biology. Nevertheless, it seems that physicists and engineers do not need to be convinced of the usefulness of holistic approaches to systems with complex, nonlinear with emerging behaviors, because they have been applying them successfully on a daily basis, for decades. Brenner's unfair criticism is an indication of the work that we still need to convince many biologists. Noble, as the creator of the Virtual Heart, has been living proof that SB works, even before the term 'systems biology' existed (his original cardiac model was published back in 1960; Noble 1960).

SB should be about understanding the parts *in terms of* the whole, and bringing everything together. While doing so, this classification allows for a precise definition of 'emergent property,' if one so desires. While the reductionist approach has identified many components of biological pathways (and key interactions), it has been less successful in describing how these interactions culminate in the emergent properties of systems. Experimentally, systems biology can be realized by perturbing a system, determining the effects of these perturbations in a rigorous and broad manner, and incorporating this information in robust computational models that may lead to testable predictions on the behavior of the system. In general, the above interactions generate rich and non-intuitive system behavior, often called emergent properties, which cannot be predicted from the properties of each intracellular component. The discovery of emergent properties is best addressed by following an SB approach. Understanding cellular networks in robust mathematical models to demonstrate emergent principles and to predict cellular function in response to perturbation remains a great challenge in biological research, both in health and disease.

Compared with the usual engineering problems, however, biological problems are ill-defined and immensely complex. The cooperation of specialists from each side is required. Models for complex biological systems may involve a large number of parameters. It may happen that some of these parameters cannot be derived from observed data via regression techniques. Such parameters are typically denoted as unidentifiable, the remaining parameters being identifiable. Closely related to this idea is that of redundancy, that a set of parameters can be expressed in terms of some smaller set. Before data are analyzed, it is critical to determine which model parameters are identifiable or redundant to avoid ill-defined and poorly convergent regression. The emphasis on identifiability is important as one naturally asks the question: how does the inaccuracy in the measurements (data noise) propagate back to errors in the inferred parameters? Therefore, it is important to consider methods that control the impact of data error on the identified parameters.

In addition, having recognized that targeting a single point often leads to a non-productive therapy, recently a focus on methods to target the networks with combinations of drugs or genetic perturbations rather than single points in the network in case of designing and implementing effective targeted therapies for cancer treatment and prevention. Recent work using omics data and integrated experimental and computational analysis to study the transient response of some signaling networks in human cancer cells revealed the effect of targeting individual molecules of signaling network and identified optimal drug combinations that inhibited cell signaling and proliferation to overcome activation of feedback loops by single agents at monotherapy (Iadevaia et al. 2010) and has led to discovery of crosstalk between signaling pathways.

Critical Analysis of Multiscale Computational Methods and Computational Tools for Crossing Levels

The emergent properties and the ways in which biological systems operate as a whole require integration and structural organization as well as the properties of the individual system components. Biological systems can be represented as networks which themselves typically contain regular (network) structures, and/or repeated occurrences of network patterns. The networks are also often organized in a multiscale manner, reflecting the physical and spatial organization of the organism, from the intracellular to the intercellular level and beyond (tissues, organs, etc). Current challenges to modeling in systems biology include those associated with issues of complexity and representing systems with multiscale attributes.

Living organisms are organized into multiple, interrelated scales so that no single one can be fully considered in isolation from the others (Martins et al. 2010). Indeed, molecular signals from the outside can elicit changes in the cell metabolism and gene expression pattern; cells acquire identity from contact with other cells and ECMs (extracellular matrix components); tissues are delineated and integrated with other tissues by specialized ECMs; and molecules carry messages from organ to organ. The timescales involved vary from seconds (for cell signaling) to years (for organism development and life span), while the spatial scales range from nanometers (for protein–DNA interactions) to meters (for nerve impulse propagation).

The aim of SB is to describe and understand biology at a global scale where biological functions are recognized as a result of complex mechanisms that happen at several scales, from the molecular to the ecosystem (Dada and Mendes 2010). Modeling and simulation are computational tools that within a single scale are invaluable for description, prediction, and understanding these mechanisms in a quantitative and integrative way. Therefore the study of biological functions is greatly aided by multiscale methods that enable the coupling and simulation of models spanning several spatial and temporal scales. Various methods have been developed for solving multiscale problems in many scientific disciplines, and they are applicable for continuum-based modeling techniques, in which the relationship between system properties is expressed with continuous mathematical equations, as well as for discrete modeling techniques that are based on individual units to model the heterogeneous microscopic elements such as individuals or cells. In this review, we survey these multiscale methods and explore their application in systems biology.

Systems biology seeks to understand not only the regulatory networks that govern cellular behavior but also the processes that govern the integration of higher level physiological structures and functions. We are now witnessing the generation of predictive models that simultaneously address such systems at multiple scales—from the genome, to the cell, to the organ—in an attempt to account for their dynamism and complexity. The development of methods that account for crossing and coupling different scales in mathematical terms is the single most important task of SB. Thus, the SB modeling with the multiscale perspective attempts to link between the scales.

Challenges of Various Integrative System Biology Approaches and Methods and Tools of Computational Modeling and Challenges

Conventional biological science has produced a vast amount of data over the last few decades so that questions arise as to how to find patterns and how to relate multi-quality data sets in the quest for underlying mechanism. It is clear that the conventional, reductionism-determined research approaches must fail in understanding the mechanisms behind the complex pattern generation, the self-organization, and nonlinear interaction of these multiscale systems. This old concept was for a long time around that one simply has to dissect the biology, investigate it in isolation, and finally put it all back together. Most conventional experimental models, however, have been developed in the reductionism era, i.e., they focus on one endpoint and emphasize one feature—with little dynamical information and lacking the possibility of studying more than one to two variables of the system at the same time. Most scientists would now admit that although this approach has led to very significant discoveries in the past, it will not be able to explain the complex behavior of most biological systems. It is increasingly recognized that complex systems cannot be described in a reductionist view.

It is thus imperative to develop novel research approaches with complex system science that will become one of the grand challenges of the next decade (non-reductive experimentation and corresponding computational tools). While models must be constructed, analyzed, simulated, validated, and verified, on the experimental side, biological systems identified, characterized, and tested for their reliability in cross-species comparisons, in particular between model organisms and humans. Thus, SB approaches are hypothesis-driven and involve iterative rounds of data-driven modeling, prediction, model-driven experimentation, model refinement, and development. Toward this end, tremendously increased computational power will help in the efforts ahead to analyze the immense amount of data in the biological and biomedical sciences in order to guide promising new experimental work. Such new experimental design should be based on and driven by the mathematical and computational modeling. We need to come up with such new experimental methods to test and refine the predictions made with novel theoretical models, based on 'model-aided (-driven) experimental design.' The present effort features scarcities of detailed quantitative experimental data and of computational tools designed to use such data for the development and testing of biologically meaningful models. Such experimentation will require a strong background in physics, mathematics, and computer science.

As pointed out by Finkelsten (Finkelstein et al. 2004) "One of systems biology's central challenges involves the tie between descriptions of experiments, observations, experimental data, interpretations derived from models, and assumptions. In short, systems biology cannot be viewed independently of an information management framework that embraces a significant part of the experimental life sciences." The focus of these Springer volumes will also be on the challenging interface between model-based experimental design and computational simulation. We should therefore bridge the gap between experimental and computational modeling experts.

We will briefly comment on the contents of this volume. Generally, we start with very fundamental (partly mathematical) approaches and then move to more applied aspects.

The first chapter by Ballereau et al., in a generalized effort, introduces systems biology, its context, aims, concepts, and strategies. They describe approaches and methods used for collection of high-dimensional structural and functional genomics data, including epigenomics, transcriptomics, proteomics, metabolomics and lipidomics, and discuss how recent technological advances in these fields have moved the bottleneck from data production to data analysis and bioinformatics. Finally, they review the most advanced mathematical and computational methods used for clustering, feature selection, prediction analysis, text mining, and pathway analysis in functional genomics and systems biology. This chapter represents a very good baseline for our effort in this volume.

The second contribution by Mueller overviews the existing methods to compare biological networks, based on approaches such as exact and inexact graph matching. Moreover, they review graph kernel-based methods and introduce an approach based on structural network measures (topological descriptors) to classify large biological networks. They introduced the superindex of topological network descriptors. The application of formal graph analysis was illustrated by two examples, classifying gene networks representing prostate cancer, and classifying metabolic networks into three domains.

The third chapter by Serra et al. emphasizes the importance of generic models of biological systems that aim at describing the features that are common to a wide class of systems, instead of studying in detail a specific subsystem in a specific cell type or organism. Among generic models of gene regulatory networks, random Boolean networks are reviewed in-depth, and it is shown that they can accurately describe some important experimental data, in particular the statistical properties of the perturbations of gene expression levels induced by the knock-out of a single gene. Boolean networks are a particular case of discrete dynamical networks, where time and states are discrete. As stated by the authors, Boolean modeling is very accessible even without a background in quantitative sciences, yet it allows life scientists to describe and explore a wide range of surprisingly complex phenomena.

Kikuchi and Voit emphasize the crosstalk, especially in signalling pathways in brain. This chapter is rather exceptional in a way that the employment of systems tools is mostly limited to microbial and mammalian cells, not specific organs. It deals with computational neuroscience and drug addiction. The authors employ three chemical reaction inspired modeling approaches. The first is a conventional ODE-based modeling method to reveal the dynamic properties of model structure. The second approach is a method to generate constraint conditions of models from the stoichiometry matrix of chemical reactions due to the lack of kinetic data. Last, the third approach is an application of complex network analysis to biological networks, focusing on *k* shortest path and *k*-cycle. In their concluding remark they emphasize the importance of the simple, transparent models.

Dančik et al. expands on network properties. Network-based approaches have a potential to significantly increase our understanding of biological systems and consequently, our understanding and treatment of human diseases. Dančik et al. address several important properties of biological networks—robustness, dynamism, modularity, and conservation. Each of these properties is an important element in establishing the 'signature' property of biological networks—emergence. They recognize that the association of diseases with network modules rather than single genes will likely impact the future of drug discovery. For treatment to have a positive impact, rather than focusing on single targets, it may be necessary to focus on the whole network associated with a disease (gene cluster).

Chaiboonchoe et al. focus on the network analysis in-depth, on the basis of graph topology. They describe several types of networks and how the combination of different analytic approaches can be used to study diseases. They provide a list of selected tools for visualization and network analysis. The use of these approaches may be extended to simulate processes on higher (cell–cell interactions) levels of organization or combined to represent multiple levels from the molecular to the organ levels, the approach of great importance in systems biology and at multiscale simulation.

Jethava et al. employ a popular assumption that the network structure features the sparsity of interaction network, i.e., each gene interacts with at most few other genes. The underlying causes govern network evolution in time-varying interaction. Authors emphasize the importance of time series data as interaction networks vary over time and in response to environmental and genetic stress during the course of the experiment. A systematic analysis of time-series data corresponding to multiple related networks allow network reconstruction, based on multiple data sources, e.g., gene interaction networks, iRNA–mRNA interactions, protein–protein interactions, as well as multiple experiments with genetic perturbations. Such approaches allow better network reconstruction by combining information from several experiments.

Sinoquet discusses the use of graph-based representation as the foundation for encoding a complete distribution over a multidimensional space. The graph is a compact representation of a set of independences that hold in the specific distribution. Typical representations are Bayesian networks and Markov networks. The purpose of this chapter is to eliminate many weak evidences from several data types that describe different biological features of genes or proteins. Probabilistic graphical models offer an appealing framework for this objective: flexibility, scalability, and ability to combine heterogeneous sources of data.

Chylek et al. employed a rule-based modeling, in which protein–protein interactions are represented at the level of functional components and thus avoiding listing of chemical species in a system, which is a necessity in traditional modeling approaches. A set of rules can be used to generate a reaction network, or to perform simulations with or without network generation and replacing systems of differential equations as formal entities. This approach serves as an excellent alternative to classical representation of chemical reaction networks typical in chemical and biological kinetics.

Waltemath et al. critically review their experience with model exchange and simulation reproducibility through standard formats gained in computational biology. For model-based results in systems biology, reproducibility requires not only a coded form of the model but also a coded form of the experimental setup to reproduce the analysis of the model. They show how sophisticated model and simulation experiment management improves the reproducibility of model-based results in systems biology. In particular, they outline the necessary steps toward reproducing a simulation result, with a particular focus on the use of standardized simulation experiment encoding. Finally, to ensure reproducibility of model-based results they recommend a "best practice" solution.

Little et al. introduce the notion of weak local identifiability and gradient weak local identifiability. These concepts are based on local properties of the likelihood, in particular the rank of the Hessian matrix. As models for complex biological systems may involve a large number of parameters, it may happen that some of these parameters cannot be derived from observed data via regression techniques. Such parameters are said to be unidentifiable, the remaining parameters being identifiable. Closely related to this idea is that of redundancy, that a set of parameters can be expressed in terms of some smaller set. Before data are analyzed, it is critical to determine which model parameters are identifiable or redundant to avoid ill-defined and poorly-convergent regression.

Durmontier et al. present a very general paper. It examines current approaches to organize systems biology knowledge and describes applications related to search, query, model similarity, integration of simulation results, and validation of model annotations. It also looks into how to formalize knowledge constructed from Semantic Web technologies and how this information could be used to build, publish, query, discover, compare, validate, and evaluate models and knowledge in systems biology.

Georgatos et al. give an overview of the infrastructure of Scientific Computing, as well as of the related network and data management, in the limelight of the necessary integration. Systems biology became an important user of computer hardware and software, as well as of the associated support. Thinking about multiscale modeling and simulation, the interaction between the computational infrastructure and the methodological development determines the capability of the applications. The authors emphasize the importance of Open Standards, as well as reproducibility, confidentiality, collaboration, and training issues. Afterwards they evaluate the advantages and the drawbacks of the available infrastructure, including HPC, computing grids, dedicated clusters, cloud computing, and desktop grids. The conclusion summarizes the challenges of Information Technology in the appropriate handling and interpretation of the ever-growing, but fragmented big datasets in the next decade. Especially anonymization issues are discussed in detail.

Wierling and Hache underline the importance of a strong interaction between wet-lab experiments, data analysis, and *in silico* modeling. Especially, the integration of experimental data and pre-existing knowledge of computational models of biological systems are of considerable importance. *In silico* simulations of model behavior under similar conditions, as in the experiment gives the possibility for model validation regarding specific experimental data. Such an integrative approach leads eventually to a more accurate and consistent description of the observed biological system. Authors then review several resources and computational tools, which support the investigation of biological networks and describe several resources and methods for integrative modeling.

An et al. view systems as aggregates of populations of interacting components, in the sense of agent-based modeling (ABMs), this represents a very suitable tool for dynamic computational systems. ABMs are particularly suited for representing the behavior of populations of cells, but have also been used to model molecular interactions, particularly when spatial and structural properties are involved as stressed by the authors. It is also a valuable tool for crossing multiple scales of biological organization and complex phenomena. Resulting is an emergence from the lower (micro) level of systems to a higher (macro) level. However, ABMs are more computationally intensive than equation-based models.

Pálfy et al. compare KEGG, Reactome, Netpath, and SignaLink pathway databases and examine their usefulness in systems-level analysis, especially in regard to signaling. They study the crosstalk as an important field in signal transduction research. To identify crosstalks and to understand their roles in development and disease, one needs to analyze signaling networks at the systems level. Biological crosstalk refers to instances in which one or more components of one signal transduction pathway affect another. This can be achieved through a number of ways with the most common form being crosstalk between proteins of signaling cascades. In these signal transduction pathways, there are often shared components that can interact with either pathway. Crosstalk between pathways provides for complex nonlinear responses to combinations of stimuli, but little is known about the density of these interactions in any specific cell. A global analysis of crosstalk suggests that many external stimuli converge on a relatively small number of interaction mechanisms to provide for context-dependent signaling.

Although this topic belongs more to network analysis (upfront Chapters) it is presented here with the emphasis on drug discovery and how the effectiveness of pathway intervention techniques (e.g., drugs, gene knockouts) changes with the presence of crosstalk.

Rider et al. provide an outline of employing the combination of multiple technologies, such as genomic, proteomic, and metabolomic data that can provide a more complete integrative picture. Much recent work has studied integrating these heterogeneous data types into single networks. They focus on describing the variety of algorithms used in integrative network inference. They concluded that any single type of data presents a one-dimensional view of a biological system. Therefore, evaluation based on a single data type may not be a baseline for the performance of an integrative method. Furthermore, different approaches tend to use different amounts and types of data, making the actual methods themselves very difficult to compare. The creation of a common body of data for evaluation and a standard for evaluation methods for integrative network approaches would allow integrative network algorithms to be truly compared.

Finally, Csukas et al. studied a simplified multiscale biosystem with a new modeling and simulation methodology. The biosystem was a consciously, but arbitrarily selected multiscale part of the p53/miR-34a related signaling process that has an important role in tumor resistance in cancer diagnostics, as well as in the therapy of various tumors. The multiscale model covered a vertical slice of the system from the change of a pathologic stage to the detailed dynamic molecular processes and vice versa. The major advantage of direct computer mapping is the unified representation of the various quantitative and qualitative sub-models, as well as the easy combination of these various models within the unified simulating environment. Regardless of the limited number of components and interactions, the investigated fictitious illustrative example demonstrated many important and interesting features of the multiscale, hybrid biosystems. The model demonstrates how the typical properties of the low-level molecular events project onto the state properties of the higher scales. These emergent properties determine typical scenarios of lower scale states and actions.

As the above chapters cover mostly items 1 and 2 of the editorial plan, the following volume will be more dedicated to items 2 and 3, mainly on problems associated with disparities between experiments and modeling and on multiscale simulation. We encourage potential contributors to submit their manuscripts without being asked.

The editors acknowledge the effort by the authors in terms of providing internal review of manuscripts to the other colleagues.

The Editors would welcome solicitation of new authors based on the previously mentioned Editorial Plan.

Aleš Prokop
Béla Csukás

References

Armstrong JD, Sorokina O (2012) Evolution of the cognitive proteome: from static to dynamic network models. Adv Exp Med Biol 736:119–34

Barabasi AL, Albert R (1999) Emergence of scaling in random networks. Science 286(5439):509–12

Brenner S (2010) Sequences and consequences. Phil Trans R Soc B 365(1537):207–212

Finkelstein A, Hetherington J, Li L et al. (2004) Computational Challenges of systems biology. IEEE Computer 37(5):26–33

Gatherer D (2010) So what do we really mean when we say that systems biology is holistic? BMC systems biology 4(22).DOI:10.1186/1752-0509-4-22

van Gend C, Snoep JL (2008) Systems biology model databases and resources. Essays Biochem 45:223–36

Gomes B (2010) Practical Applications of systems biology in the Pharmaceutical Industry. Internat Drug Discovery 5(2):54–57

Hastings A, Arzberger P, Bolker B et al. (2002) Quantitative Biology for the 21st Century, Report from a NSF funded workshops on Quantitative Environmental and Integrative Biology, the first held September 7–9, 2000 at the San Diego Supercomputer Center on the University of California, San Diego campus, and the second held 11–13 December 2002 at the University of California, San Diego, http://www.maa.org/mtc/Quant-Bio-report.pdf

Iadevaia S, Lu Y, Morales FC et al. (2010) Identification of optimal drug combinations targeting cellular networks: integrating phospho-proteomics and computational network analysis. Cancer Res 70(17):6704–6714

Noble D (1960) Cardiac action and pacemaker potentials based on the Hodgkin-Huxley equations. Nature 188:495–7

Phillips RB, Kondev J, Theriot J (2009) Physical biology of the cell. Garland Science 807

Prokop A, Michelson S (2012) systems biology in Biotech & Pharma: A Changing Paradigm. Springer Briefs in Pharmaceutical Science and Drug Development 2:145

Scannell BA, Boldon H and Warrington B (2012) Opinion: Diagnosing the decline in pharmaceutical R&D efficiency. Nature Revs Drug Discovery 11:191–200.

Simeonidis V (2011) Is systems biology doomed to fail? (NO!), http://www.pagev.net/2011/04/is-systems-biology-doomed-to-fail-no.html. Accessed 6 Apr 2011

Watts DJ, Strogatz SH (1998) Collective dynamics of 'small-world' networks. Nature 393(6684):440–2

Wikswo JP, Prokop A, Baudenbacher FJ, Cliffel D, Csukas, B and Velkovsky M (2006) The engineering challenges of BioMEMS: The integration of microfluidics, micro- and nano-devices, models, and external control for systems biology. IEE Proc Nanobiotech 153(4):81–101

Part I
Foundation of Systems Biology, Integrative Biology and Issues of Biosystem Complexity, Redundancy and of Other Network Properties

Chapter 1
Functional Genomics, Proteomics, Metabolomics and Bioinformatics for Systems Biology

Stéphane Ballereau, Enrico Glaab, Alexei Kolodkin, Amphun Chaiboonchoe, Maria Biryukov, Nikos Vlassis, Hassan Ahmed, Johann Pellet, Nitin Baliga, Leroy Hood, Reinhard Schneider, Rudi Balling and Charles Auffray

Abstract This chapter introduces Systems Biology, its context, aims, concepts and strategies, then describes approaches used in genomics, epigenomics, transcriptomics, proteomics, metabolomics and lipidomics, and how recent technological advances in these fields have moved the bottleneck from data production to data analysis. Methods for clustering, feature selection, prediction analysis, text mining and pathway analysis used to analyse and integrate the data produced are then presented.

Keywords Emergence · Holistic · Bottom-up · Top-down · Middle-out · Interactions · Data integration · Mathematical model · Functional genomics · High-throughput · Epigenomics · Transcriptomics · Proteomics · Metabolomics · Next generation sequencing · Mass spectrometry · Bioinformatics · Knowledge management · Ontology · Pathway · Network · High-dimensionality · Curse of dimensionality · Clustering · Feature selection · Prediction analysis · Text-mining

S. Ballereau (✉) · A. Chaiboonchoe · H. Ahmed · J. Pellet · C. Auffray
European Institute for Systems Biology & Medicine, CNRS-UCBL—Université de Lyon, 50 Avenue Tony Garnier, 69007 Lyon, France
e-mail: sballereau@eisbm.org, achaiboonchoe@eisbm.org, hahmed@eisbm.org, jpellet@eisbm.org, cauffray@eisbm.org

E. Glaab · A. Kolodkin · M. Biryukov · N. Vlassis · R. Schneider · R. Balling
Luxembourg Centre for Systems Biomedicine, University of Luxembourg, 7 Avenue des Hauts-Fourneaux, 4362 Esch-sur-Alzette, Luxembourg
e-mail: enrico.glaab@uni.lu, alexey.kolodkin@uni.lu, maria.biryukov@uni.lu, nikos.vlassis@uni.lu, reinhard.schneider@uni.lu, rudi.balling@uni.lu

N. Baliga · L. Hood
Institute for Systems Biology, 401 Terry Avenue North, Seattle, WA 98109–5234, USA
e-mail: nbaliga@systemsbiology.org, lhood@systemsbiology.org

A. Prokop and B. Csukás (eds.), *Systems Biology*,
DOI: 10.1007/978-94-007-6803-1_1,

Abbreviations

BASE	BioArray Software Environment
BS	BiSulphite
CATCH-IT	Covalent Attachment of Tags to Capture Histones and Identify Turnover
CFS	Correlation-based Feature Selection
CHARM	Comprehensive High-throughput Array for Relative Methylation
ChIA-PET	Chromatin Interaction Analysis by Paired-End Tag
ChIP	Chromatin ImmunoPrecipitation
CLIP	Crosslinking immunoprecipitation
DHS	DNAse I hypersensitivity
DNA	DeoxyriboNucleic Acid
EFS	Ensemble Feature Selection
ELISA	Enzyme-Linked ImmunoSorbent Assays
ENCODE	ENCyclopedia Of DNA Elements
ESI	ElectroSpray Ionisation
EWAS	Epigenome-Wide Association Studies
FAB	Fast Atom Bombardment
FAIRE	Formaldehyde-assisted isolation of regulatory elements
FDR	False Discovery Rate
FT-ICR	Fourier Transform Ion Cyclotron Resonance
FUGE	Functional Genomics Experiment data model
GAGE	Generally Applicable Gene-set Enrichment
GC	Gas Chromatography
GEO	Gene Expression Omnibus
GO	Gene Ontology
GSEA	Gene Set Enrichment Analysis
GWAS	Genome-Wide Association Studies
HITS-CLIP	HIgh-Throughput Sequencing of RNAs isolated by CrossLinking ImmunoPrecipitation
HMM	Hidden Markov Models
HPLC	High Performance Liquid Chromatography
IMS	Imaging Mass Spectrometry
IP	ImmunoPrecipitation
iTRAQ	Isobaric Tags for Relative and Absolute Quantitation
KEGG	Kyoto Encyclopedia of Genes and Genomes
kNN	k-Nearest Neighbor
LC	Liquid Chromatography
MALDI	Matrix Assisted Laser Desorption Ionisation
MBD	Methyl-CpG Binding Domain
MCAM	Multiple Clustering Analysis Methodology
MeDIP	Methylated DNA Immunoprecipitation
MGDE	Microarray Gene Expression Data
MIAME	Minimum Information About a Microarray Experiment

MIAPE	Minimum Information About a Proteomics Experiment
MINSEQE	Minimum INformation about a high-throughput SeQuencing Experiment
MMASS	Microarray-based Methylation Assessment of Single Samples
MN	Microccocal Nuclease
MRM	Multiple Reaction Monitoring
mRNA	Messenger RiboNucleic Acid
MS	Mass Spectrometry
NCBI	National Center for Biotechnology Information
NER	Named-Entity Recognition
NGS	Next Generation Sequencing
NIH	National Institutes of Health
NMR	Nuclear Magnetic Resonance
PaGE	Patterns from Gene Expression
PCR	Polymerase Chain Reaction
PRIDE	PRoteomics IDEntifications
PSM	Peptide-Spectrum Match
QMS	Quadrupole Mass Analyser
RNA	RiboNucleic Acid
RRBS	Reduced Representation Bisulphite Sequencing
RT-qPCR	Reverse Transcription quantitative PCR
SAGE	Serial Analysis of Gene Expression
SELDI	Surface Enhanced Laser Desorption Ionization
SILAC	Stable Isotope Labeling by Amino acids in Cell culture
SNP	Single Nucleotide Polymorphism
SRM	Selected Reaction Monitoring
SUMCOV	SUM of COVariances
SVM	Support Vector Machine
ToF	Time-of-Flight
UCSC	University of California, Santa Cruz
VOCs	Volatile Organic Compounds

1.1 Background

1.1.1 Context

Life in a broad scientific context can be defined as the phenomenon that emerges from particles of inorganic matter organised in molecules which interact with each other within a cell [1]. This property is systemic because it only appears in the system and not in its parts [2]. Living systems are complex, modular and hierarchical structures. Indeed, a multicellular organism consists of molecules, such as deoxyribonucleic acid (DNA), ribonucleic acid (RNA), proteins, lipids and metabolites involved in chemical reactions and structures of cells. Cells are organised in tissues forming organs with specific functions that are required for the health of the organism. Systemic properties appear at each level, for example homeostasis and response to stimuli in a single intracellular network, metabolism, growth, adaptation, reproduction in a single cell.

Information that defines an organism and its ability to react to its environment is encoded in its DNA and is expressed differentially in space and time throughout life. Typical studies in biology have until recently used the reductionist approach and addressed specific issues employing one or a few types of molecules at a small scale, each shedding light on only a small fraction of vastly complex phenomena. Some findings were remarkable, such as the discovery of the structure of DNA, and later of the way genetic information stored in DNA is transcribed in messenger RNA (mRNA) then translated in proteins, essential components of the cell machinery and the engines of life. The accumulation of such knowledge on molecules and mechanisms led to the *'bottom-up'* approach to modeling biological systems, using genes as core elements to simulate cells, organs and the whole organism. This was complementary to the *'top-down'* view of an organism as a physiological system integrating information from its various constituents and their interaction with the environment.

Major technological advances have in the last 15 years enabled biologists to eventually gather information on a larger scale in various tissues, including samples obtained with non-invasive methods, such as the collection of blood and urine. The massive increase in throughput has had several consequences. First, biologists can now study the vast majority of constituents, i.e. 'ome', of a given element, e.g. genes, of a system be it an organism, organ or cell, e.g. all genes in its genome. Second, the sheer size of data sets implies that their analysis relies increasingly on computational tools and power available to analysts. Third, because characterisation of several 'omes', e.g. genome, transcriptome, proteome and metabolome, progresses rapidly along with other disciplines such as imaging and in particular pharmaceutical research with cheminformatics, compound libraries, high throughput screening, safety and clinical data [3–5], one can now attempt to disentangle interactions between the different elements of a biological system, or 'interactome', to understand its behavior across several scales in a holistic manner, in health and disease.

1.1.2 Aims and Concepts

Systems Biology is the integrative study of complex systems in life with a *holistic* approach now based on large-scale data sets analyzed iteratively with mathematical models and simulation tools [6, 7]. Understanding each component of a complex system in isolation is not sufficient to characterise the system. Indeed, properties of the system are not only defined by the simple addition of elementary functions but also emerge from the *interactions* between the elements [7–9]. These emergent properties are studied by inferring networks of interactions between these constituents, e.g. genes, proteins and ligands, and by unraveling their regulatory mechanisms. Because of the very large number of elements in these networks, such an endeavor relies on concepts defined in the framework of the theory of complex systems [10]. Systems Biology not only aims at understanding the relationships between different levels of the expression of genetic information, via *data integration*, but also at defining the system as a whole and producing a convincing *mathematical model* of it, linking the highly complex interactions between its components to its *emergent properties* [11–14]. In this context, disease can be viewed as a shift of homeostasis from the normal range due to a large set of perturbations in the network of interacting biomolecules in the whole organism. Distinct perturbations may therefore result in a single disease phenotype, in agreement with our understanding of complex diseases. Conversely, shifting the system back to healthy homeostasis may be achieved in multiple ways and by targeting several points in the network [15, 16].

Systems Biology follows an integrative and iterative approach that relies on experimental and mathematical methods (Fig. 1.1). First, existing data relating to different hierarchical levels of the system are integrated into mathematical or graphical models to generate hypotheses towards understanding mechanisms at play and build predictions on the functions of that system. Some components of the system are then perturbed experimentally, such as in *in vitro* or *in vivo* models of a disease. The outcome is assessed in the context of the model and the initial hypotheses are revised accordingly. These revised hypotheses finally inform new perturbation experiments. The approach is repeated until the system's behaviour is faithfully simulated by the model [7]. Further complexity is added when one considers the environmental factors of the model.

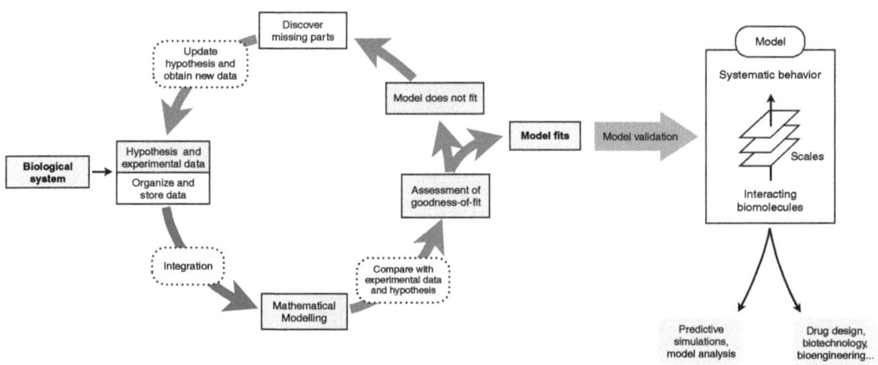

Fig. 1.1 Modeling in Systems Biology. Modeling starts with the integration of different experimental data into a single knowledge base to organize and store data. Mathematical descriptions of the interaction between model elements allow (1) simulation of the emergent behavior of the system, (2) comparison of this simulated behavior with experimental data, (3) adjustment of the model and (4) design of further experiments. When the model fits experimental data, studying the role of particular design features may help identify mechanisms at play and design principles. The model may also be used in drug design, biotechnology or bioengineering for example

1.1.3 Strategies

Three main strategies aim to build the link between the system's components and its emerging properties: 'bottom-up', 'top-down' and 'middle-out' (Fig. 1.2). The main steps of the 'bottom-up' approach are to graphically or mathematically model relationships between the components of the system, starting with those at the lowest level of the multiscale structure, hence 'bottom', e.g. genes and proteins, set model parameters using experimental values and verify the model by comparing its systemic behavior with the behavior of a real system. The term bottom-up also refers to the direction chosen: from known or assumed properties of the components one deduces system functions [17]. This molecular biology strategy has been successful in modeling biological systems with relatively low number of components, e.g. a single intracellular network or a single prokaryotic cell. It may however not be suited to reconstruction of the emergence of larger systems, e.g. the whole body physiological behavior in Mammals. In contrast, the 'top-down' or physiology approach relies on the systemic behavior. It first involves defining ways the complicated systemic function of interest varies with conditions and/or time, and then inferring hypothetical structures responsible for this function. The system behavior is perturbed and the effects studied at the level of the system components, i.e. genome, transcriptome, proteome and metabolome. This strategy is limited to an extent by the challenge of inferring DNA sequences

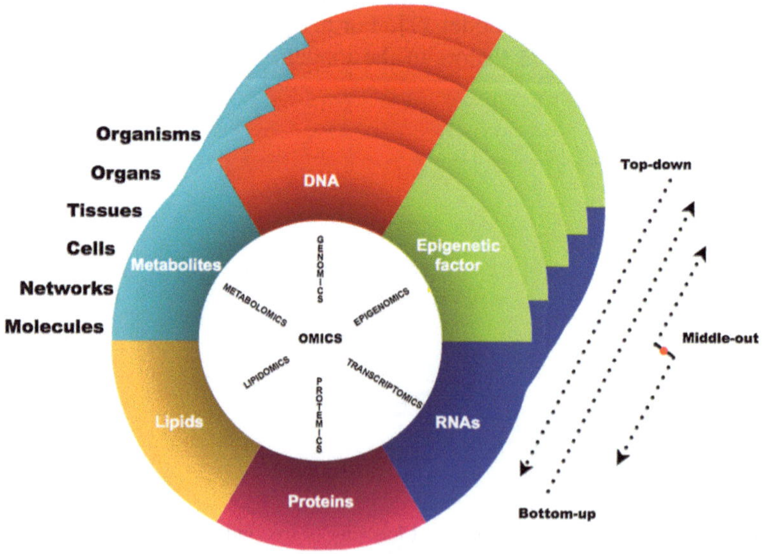

Fig. 1.2 Multiple scale strategies in Systems Biology. Starting at the molecular level, interactions between DNA, epigenetic factors, RNA, proteins, lipids and metabolites define the core biological processes required for higher order functions. These processes are defined by molecular interaction networks, which communicate with each other within a given cell, between cells in the same tissue or distinct tissues, or between organs of a complex organism

from phenotypes. Also, models built with top-down approaches must be updated with every new experiment using all existing experiments, making the analytical and computational challenges increasingly difficult. In contrast, models built with the bottom-up approach such as an *in silico* cell model comprise modules which are updated independently of each other [18]. The '*middle-out*' strategy intends to overcome the intrinsic limitations of the above approaches, taking into account that chains of causality can operate in biological systems in both directions, starting at any levels of biological organization. The behavior of a single functional system is thus modeled in terms of interactions between entities at a level sufficiently well described by experimental data ('middle'), typically of the lower levels of organization but not necessarily down to molecules. The model is then extended to higher and lower levels ('out') iteratively by combining 'bottom-up' and 'top-down' approaches. It was successfully implemented in the Physiome project [19, 20].

Systems Biology will play a crucial role in the development of personalized medicine as it will enable integration of different types of data to profile patients, identify unbiased biomarkers and produce precise disease phenotypes. It will hence help prevention, diagnosis and treatment, or Systems Medicine [21, 22].

1.2 Introduction to Functional Genomics, Proteomics, Metabolomics and Bioinformatics

Genomics is the study of the sequence, structure and content of the genome, in particular the genes and their number, structure, function and organisation along the genome. *Functional genomics* is the study of the function of genes and the regulation of their expression at the level of the cell, organ or organism, spatially and at different time points and/or health status, by deciphering the dynamics of gene transcription, translation and protein–protein interactions on a genome-wide scale using *high-throughput* technologies. The main large-scale experimental tools used to study epigenetics (*epigenomics*) and gene expression (*transcriptomics*) have so far involved microarrays and more recently next-generation sequencing. Mass spectrometry is widely used to study proteins (*proteomics*), metabolites (*metabolomics*), and more recently volatile organic compounds (VOCs) in exhaled breath condensate (breathomics). Technical advances also led to the development of computational tools to handle and analyse their output.

1.2.1 Sequencing Technologies

Whole genome sequencing started with the sequencing of a bacteriophage in 1977 using the Sanger sequencing technique. The development and maturation of

4-color automated Sanger sequencing produced the instruments that sequenced the human genome (Smith et al. 1986). Several high-throughput sequencing techniques, or *Next Generation Sequencing* (NGS), arose subsequently which were each inferior to the more established automated Sanger technique, being slower per run, less accurate, with shorter read length and more expensive, but far superior by virtue of the vastly larger number of nucleotides read [23–25]. Now 3rd generation sequencing strategies employ nanopores and single molecule reads, and promise to increase the throughput and decrease the cost of sequencing strikingly. Computational tools are being developed to process the very large amount of NGS short, low quality reads and assemble them into a genome sequence [26]. Genome sequences of over sixty pro- and eu-karyotes are annotated in online public genome browsers [27, 28]. Knowledge of whole genomes also enabled the large-scale study of gene expression and the development of functional genomics. NGS can indeed be used for DNA or RNA sequence analyses and has several advantages over microarrays: it does not require array design, enables wider scale, whole-genome studies, improved resolution, more flexibility, allele-specificity, lower cost and amount of input material. NGS now also enables routine discovery of variants in entire exomes and even large genomes [29, 30] as in Human with the 1000 Genomes Project [31], in cancer research [32, 33] and studies of allele specificity in gene expression [34]. NGS also catalyzed the massive development of metagenomics [35] and will thus help decipher host-gene-microbial interactions [36]. NGS is however not mature enough for routine use in clinical field [37]. The ever increasing speed, quality and range of applications of sequencing methods have created a huge flow of data and related challenging requirements not only for computing power, memory and storage [38–40] but also data sharing [41]. Reads mapped onto a reference genome can be displayed with other sources of annotation such as NCBI [42] with Ensembl [28] and UCSC browsers [43].

1.2.2 Mass Spectrometry

Mass spectrometry (MS) relies on deflection of charged atoms by magnetic fields in a vacuum to measure their mass/charge (m/z) ratio. A typical experiment follows five steps: (1) introduction of the sample, (2) ionisation of its particles, (3) acceleration, (4) deflection proportional to the mass and charge of the ion, and (5) detection, recorded as a spectrum showing peaks on a plot of relative quantity as a function of the m/z ratio.

Several methods for introduction, ionisation and types of spectrometers enable a wide range of analyses. Introduction methods are Gas chromatography (CG) for thermally stable mixtures, liquid chromatography (LC) for thermally labile mixtures, and solid probes. Some compounds such as large proteins and polymers must be ionized directly. Ionisation methods can be hard or soft. Hard ionisation introduces high amount of energy in the molecules that results in fragmentation and thus helps identify the compound but resulting spectra rarely contain the molecular ion. ElectroSpray Ionisation (ESI) uses high voltage to disperse and ionise macromolecules through a spray nozzle. It is soft, limits fragmentation and produces multiply charged ions, allowing detection of large compounds at lower mass/charge value, and hence increases the analyser's mass range. ESI is often coupled with LC/MS. Mixtures containing non-volatile molecules can also be analysed with Fast Atom Bombardment (FAB) and Matrix Assisted Laser Desorption Ionisation (MALDI). MALDI is used to analyse extremely large molecules, up to 200,000 Da, often coupled with time-of-flight (ToF) MS. Surface Enhanced Laser Desorption Ionization Mass Spectrometry (SELDI-MS) separates protein subsets fixed onto a surface according to specific biophysical properties, e.g. hydrophobicity. Thus, analysis of proteins, peptides and nucleotides can be performed with ESI, SELDI, MALDI, and FAB [44].

Several types of analysers exist. In a quadrupole mass analyser (QMS) ions are deflected by oscillating positive and negative electric fields. A triple-QMS contains three QMS one after the other where the first QMS enables the identification of known compounds, the second its fragmentation, and the third the identification of the fragments, thereby elucidating the compound structure. Other types of analysers include ion trap, ToF, Orbitrap, and Fourier Transform Ion Cyclotron Resonance (FT-ICR) with increasing mass resolution and accuracy. Orbitraps are cheaper, more robust and have a higher-throughput than FT-ICRs. Tandem-MS involves several steps of selection of compound using MS. MS methods mentioned above vary in throughput, robustness, sensitivity, selectivity and ease of use [44].

1.2.3 Bioinformatics

Bioinformatics comprises mathematical approaches and algorithms applied to biology and medicine using Information Technology tools, e.g. databases and mining software [45, 46]. Analysis of omics data typically follows four steps: (1) data processing and identification of molecules, (2) statistical data analysis, (3) *pathway* and *network* analysis, and (4) system modelling. Examples include *de novo* genome assembly, genome annotation, identification of co- or differentially expressed genes at the level of transcripts or proteins and the inference of protein–protein interaction networks. Bioinformatics also enables integration of heterogeneous high-throughput data sets produced by a given study and existing data sets using knowledge management, annotation and text mining tools such as the two structured vocabularies Gene Ontology (GO) for genes and associated biological processes, cellular components and molecular functions [47, 48] and Microarray Gene Expression Data (MGED) ontology [49], the PRoteomics IDEntifications (PRIDE) database [50], Functional Genomics Experiment data model (FuGE) [51], the Systems Biology Markup Language [52], the Systems Biology Graphical Notation [53], BioMART [54, 55], tranSMART [56], bioXM [57], GARUDA [58], Nexbio [59], and includes Systems Biology [23]. Identification of pathways, and network inference and analysis is covered in chapter 'Network analysis for systems biology'.

These efforts collectively aim at unraveling the molecular pathways underpinning physiology and at identifying biomarkers to describe a system with a combination of environmental, clinical, physiological measures to improve detection and monitoring of a phenomenon, such as diseases in medical research to facilitate diagnosis and therapy. Biomarker discovery relies on two types of studies: unbiased, which only depend on the technique used, and targeted, which focus on pre-defined biomarkers measured by specific methods. Experimental and bioinformatics methods and tools mentioned in the following text are listed in Tables 1.1 and 1.2.

1.3 Functional Genomics, Proteomics and Metabolomics

1.3.1 Epigenomics

Epigenomics is the genome-wide study of modifications of chromatin, i.e. DNA and associated proteins, which play an important role in gene regulation, gene-

Table 1.1 Examples of methods and tools for functional genomics, proteomics and metabolomics. This list is non exhaustive and only includes items mentioned in the text

Epigenomics methods	DNA methylation [61]: Endonucleases (MMASS, CHARM, Methyl-seq), bisulphite (BS) conversion (RRBS, MethylC-seq), and affinity (MeDIP-chip, MeDIP-seq, MDB-seq). Methylation levels can then be measured with microarrays and sequencing techniques;
	Chromatin accessibility (DNAseI-seq, FAIRE–seq, Sono-seq, 3C, 4C, 5C, ChIA-PET);
	Nucleosome positioning (CATCH-IT, MNase-se, haploChIP)
Epigenomics tools	Encyclopedia Of DNA elements (ENCODE) project [63], the NIH Roadmap Epigenomics effort [64], the Human Epigenome Project [65] and recently BLUEPRINT [67]
Transcriptomics methods	DNA microarray, SAGE, RNA-seq, ChIP-seq, CLIP-seq [108, 113, 114, 117]
Transcriptomics tools	ArrayExpress [104], GEO [106], MIAME [107], MINSEQE [119]. See [26, 120] for reviews on downstream analysis.
Proteomics methods	ELISA, 2D gel electrophoresis, NMR, MS, iTRAQ, SILAC, SRM, SELDI-ToF [126–131]
Proteomics tools	MIAPE [134], TransProteomic pipeline, protein atlas, neXProt [139–141]
Metabolomics methods	NMR [143], MS [44], IMS [144, 147]
Metabolomics tools	MetabolomeExpress [150], metaP [151], KEGG [145], human metabolome project [142]
Lipidomics methods	MS [44, 161], orbitraps [160], IMS [144, 147]
Lipidomics tools	LIPID MAPS [165], XCMS [162], MZmine2 [163]

Table 1.2 Examples of methods and tools for bioinformatics. This list is non exhaustive and only includes items mentioned in the text

Bioinformatics	Microarray gene expression data (MGED) ontology [49], the proteomics identifications (PRIDE) database [50], functional genomics experiment data model (FuGE) [51], the systems biology markup language [52], the systems biology graphical notation [53], BioMART [54, 55], tranSMART [56], bioXM [57], GARUDA [58], nexbio [59]
Clustering	Babelomics [176], BASE [177], MCAM [178]
Feature selection	Unsupervised [187], supervised [186]; filters (student's *t* test, Wilcoxon rank sum test, CFS, EFS, Markov blanket filtering) [188], wrappers (kNN [203], Naive Bayes [204], sequential forward search [205]), hybrid methods [202], mathematical programming [209], signal processing approaches [210]
Prediction analysis	Unsupervised (clustering, feature selection, dimension reduction, density estimation, and model structure learning, nonlinear dimension reduction methods) [211–213]; supervised (SVM [215], random forest [216]); semi-supervised [217]; time series (HMM [218])
Networks from literature	NER [225], iHOP [232], FActa + [221], AliBaba [233], IntAct [234], CoPub [235]
Pathway analysis	Differential expression filtering, overrepresentation statistics [236], GSEA [240], PAGE [241], GAGE [242], ontologizer [243], GeneCodis [244], elementary flux analysis [245], extreme pathways [246]

environment interactions, development and in diseases such as inflammation and cancer [60, 61]. Such modifications involve the DNA itself but not its sequence, i.e. a methylated cytosine (mC) adjacent to a guanine (CpG dinucleotides in mammals), and of chromatin proteins, i.e. methylation, acetylation and phosphorylation of histones. Epigenomics also covers chromatin accessibility, nucleosome remodelling, long-range chromatin interactions and allele-specific chromatin signatures. Technological advances are now enabling Epigenome-Wide Association Studies or EWAS, akin to Genome-Wide Association Studies or GWAS [62], and large scale studies in different cell types and tissues, as in the human ENCyclopedia Of DNA Elements (ENCODE) project [63], the NIH Roadmap Epigenomics effort [64], the Human Epigenome Project [65], [66] and recently BLUEPRINT that aims to determine the epigenome of 100 different blood cell types [67].

DNA methylation at CpG is widely studied as it mediates gene repression in a cell-specific manner by preventing the transcriptional machinery from accessing DNA. Methylated DNA can be detected with three types of DNA treatments, i.e. endonucleases, bisulphite (BS) conversion, and affinity. Methylation levels can then be measured with microarrays and sequencing techniques.

Endonucleases cleave DNA at specific sites, are sensitive to methylation and enable several DNA analyses techniques. Recent methods enable analysis of a single sample, e.g. microarray-based methylation assessment of single samples (MMASS), better statistical analyses and methods for array design, e.g. comprehensive high-throughput array for relative methylation (CHARM) [68] and the widely used NGS sequencing of DNA enriched for CpG containing regions (Methyl-seq) [61].

BS conversion modifies unmethylated cytosine in CpGs into a uracil and thus transforms an epigenetic difference into a genetic one detectable by methylation specific DNA microarrays with single-nucleotide resolution [69, 70]. Except for mC, BS treated DNA comprises only three base types and hence has reduced sequence complexity and hybridization specificity. This is overcome by enriching for CpG-containing segments as in Reduced Representation Bisulphite Sequencing (RRBS) with BS treatment and NGS. Alternatives include whole-genome BS sequencing, although that is expensive, and the widely used MethylC-seq, i.e. NGS of BS treated DNA. Throughput and coverage may increase with nanopore sequencing which can sequence mC directly, without BS treatment [71].

Genome-wide identification of DNA binding-sites and corresponding binding proteins is mainly achieved with the affinity-based approach chromatin immunoprecipitation (ChIP) whereby DNA-binding proteins, e.g. histones and transcription factors, are cross-linked *in vivo* in cells that are then lysed. DNA is fragmented by sonification, recovered by heating DNA–protein complexes and detected with microarray (ChIP-chip) or NGS (ChIP-seq) [72, 73]. Methylated DNA Immunoprecipitation (MeDIP-chip and MeDIP-seq) uses monoclonal antibody against methylated cytosine to enrich single-strand methylated DNA. Some alternatives rely instead on high affinity binding of a Methyl-CpG Binding Domain (MBD) protein complex for double-strand methylated DNA (e.g. MDB-seq) [60, 74]. Transcription factor binding sites are then predicted in the sequences identified [75]. ChIP is also widely used to study patterns of histone modifications and chromatin modifiers [63, 76]. It can be integrated to other data sets, as with Segway [77], helping development of chromatin model [78]. ChIP coupled with quantitative real-time PCR allows the study of the dynamics of DNA and proteins interactions in living cells for up to several minutes, and has now been adapted to microfluidics technology reducing the number of cells and time required [79].

Across the three types of treatment, at least 13 array- and 10 seq-based analytical methods exist, the choice of which depends on their features, the required coverage and resolution, types of bias, accuracy and reproducibility, and also on the number of samples, available DNA quality (high for affinity techniques) and quantity (high for nuclease techniques), and in particular for array-based methods: the organism. The most widely used NGS-based methods rely on BS (RRBS and MethylC-seq) or affinity (MeDIP-seq and MBD-seq) approaches [61, 80, 81].

Microarray data processing addresses imaging and scanning artefacts, background correction, batch and array normalization, and correction for GC content and CpG density. The ratio of methylated to unmethylated molecules for a given locus is a widely used metric. It is analysed with tools developed for gene expression data, potentially wrongly since they rely on assumptions violated by DNA-methylation data, e.g. independence of the number of methylated and unmethylated sites, and similarity of signal strength across samples [61, 82–84]. Processing sequencing reads involves mapping of reads to the reference genome, counting and/or analysis of bisulphite data [85, 86].

Genomic regions of chromatin accessibility, i.e. low nucleosomal content and open chromatin structure, potentially harbour regulatory sequences and can be identified with high-throughput DNAse I hypersensitivity assay (DNAseI-seq aka DHS-seq) [87], formaldehyde-assisted isolation of regulatory elements followed by sequencing (FAIRE–seq) [88] and Sono-seq [89]. And long range chromosomal interaction are identified with chromosomal conformation capture (3C) [90, 91], 3C on chip (4C) [92], 3C carbon copy (5C) [93] and coupled with NGS as in using Hi-C [94] and ChIA-PET [95]. Nucleosome positioning and remodelling is studied with CATCH-IT [96] and MNase-seq [97] while haploChIP identifies allele-specific chromatin profiles [98, 99], including SNPS that affect gene expression [100].

Methods to integrate epigenomics data are recent and currently being developed. Examples include integration with gene expression data, using an empirical Bayes model [101] and clustering of DNA methylation data followed with non-linear regression analyses [102]. Visualisation tools can display raw data genome-wide as with Circos [103] or analysis output in a similar manner to that used for GWAS, using log10 *p*-value, but on two axes: test of difference in methylation status and test of difference in gene expression [83].

1.3.2 Transcriptomics

Transcriptomics is the genome-wide identification and quantification of RNA species such as mRNAs, non-coding RNAs and small RNAs, in health and disease, and in response to external stimuli. With DNA microarrays, gene expression levels are measured as the amount of RNA in the sample that matches the set of probes fixed on the array; RNA molecules are fluorescently labelled and hybridised onto the array where the intensity of the signal measured for a given probe is assumed to be proportional to the quantity of RNA bound to it. Changes in expression levels between experimental conditions or samples with or without disease on one hand and similarity of expression pattern with a gene with known function on the other hand indicate the most likely functions of the genes. Two main public repositories for gene expression data sets exist: ArrayExpress [104, 105] and Gene Expression Omnibus (GEO) [106], both compliant with the ‘Minimum information about a microarray experiment’ (MIAME) guidelines [107]. Although microarrays are an established and very widely used technology [108], data processing and analysis methods are still being developed. For example, recent studies claim that models for background noise based on Gaussian distribution for computational efficiency may not be appropriate and non-parametric methods may harbour a lower false positive rate [109], while weighted average difference seems to be the best method to identify differentially expressed genes [110]. Two main sequencing-based alternatives exist which, unlike microarrays, do not rely on a set of pre-defined probes and are therefore considered unbiased: Serial Analysis of Gene Expression (SAGE) and genome-wide transcriptome NGS (RNA-seq).

SAGE entails sequencing tags that are unique to each gene and not defined *a priori*. SAGE was for example used to build expression profiles of long non-coding RNAs for 26 normal tissues and 19 cancers in human [111], shedding light on their poorly understood function [112]. The more recent RNA-seq provides whole transcript sequences, has very low background noise, offers a very large dynamic range, is highly accurate and reproducible, enables the discovery of novel exons, isoforms and transcripts. RNA-seq has already proved very promising but is not as mature as microarrays yet [113–115]. Rare and transient transcripts so far undetected by current methods were recently identified with targeted transcriptomics by capture on tiling array followed by NGS [116]. Currently, some experimental protocols may introduce bias due to amplification, fragmentation and ligation processes [117, 118]. Development of robust quality control standards and guidelines for microarrays occurred over a decade but should be faster for RNA-seq. Methods are being developed to describe experiments using MIAME-like 'Minimum Information about a high-throughput SeQuencing Experiment' (MINSEQE) guidelines [119], map the vast amount of short read sequences [26], assess expression levels and detect differentially expressed transcripts [120].

Estimates of expression levels of transcripts of interest must be validated by RT-qPCR and emerging techniques such as direct visualization and counting of RNA molecules [121]. These must however be standardised and applied across platforms [21]. Microarrays are still relatively cheaper than RNA-seq, their biases are known and analysis workflows are mature. They are therefore still preferred in drug discovery, though RNA-seq methods will probably replace them over the next years. Because gene expression profiles obtained with both methods correlate well, the vast amount of data acquired with microarrays is complementary to new data produced by RNA-seq [108].

Other techniques such as ChIP are also used to identify proteins binding DNA (ChIP-seq) [73] and RNA (CLIP-seq aka HITS-CLIP) [122]. These fast evolving high throughput methods are greatly improving our understanding of gene expression regulation [123, 124], at the transcriptional and post-transcriptional levels [125].

1.3.3 Proteomics

Correlation between levels of transcripts and proteins is incomplete due to variation in speed and efficiency of translation and of mRNA degradation. Many proteins undergo posttranslational modifications, e.g. phosphorylation and ubiquitination, which modulate their activity and mediate signal transduction. Proteins also play their role as part of complexes with other proteins or nucleic acids. A recent study of a human cell line identified over 10,000 proteins, with concentrations ranging over seven orders of magnitude. The human proteome has been estimated to comprise several millions distinct species which cannot currently be amplified and reflect concentrations with a very wide dynamic range [126].

Proteins can be identified using low-throughput antibody methods, Enzyme-Linked ImmunoSorbent Assays (ELISAs) and 2D gel electrophoresis. Proteomics aims at defining all of the proteins present in a cell, a tissue, or an organism (or any other biological compartment) and employs large-scale, high-throughput studies of protein content, modifications, function, structure, localisation, and interactions using high-throughput techniques. Protein microarrays capture proteins using agents fixed on their surface, which can be antibodies but also peptides, receptors, antigens, nucleic acids. Detection and quantification are often fluorescence-based and identify interactions between proteins, kinase substrates, activators of transcription factors [127]. Nanoproteomics has the potential to provide fast, high-throughput and sensitive methods using only minute amount of samples [128]. However, MS is currently the main technique for large-scale whole-proteome study with precise measurements [129, 130].

Shotgun proteomics, i.e. shotgun LC coupled with tandem MS (LC–MS/MS) is the most widely used approach. The sample of peptides resulting from the trypsin (or other enzyme) digestion of proteins is separated by High Performance Liquid Chromatography (HPLC) and peptides are identified using tandem MS: peptides are ionised and separated, producing mass spectra with peaks corresponding to peptides (first MS), which are then identified using further fragmentation and separation of resulting peptide fragments (second MS). Inclusion of labelled synthetic peptides as spike-in or labelling samples chemically (iTRAQ) or metabolically (SILAC) improves quantification [131]. Mixture complexity is addressed by fractioning the mixture. Targeted proteomics allows one to identify 100-200 proteins in a complex mixture by previously identifying the "transition peptide fragments" through the use of a triple quadrupole mass spectrometer which separates the trypsin peptide fragments, then fragments these further into "transitions" that can be quantified in the third quadrupole. One attempts to choose transitions that are unique to individual proteins and spiking in isotopically labelled transition peptides greatly improves quantification. Targeted mass spectrometry is termed Selected Reaction Monitoring (SRM) or Multiple Reaction Monitoring (MRM). SRM assays for the entire human proteome (more than 20,000 proteins) have recently been developed (R. Mortiz, personal communication).

HPLC–MS is highly sensitive, specific and fast, and thus used for bioanalysis, in particular pharmacokinetics to measure speed of drug clearance by the body, and in urine sample analysis. Drawbacks however include a bias towards identification of most abundant peptides. SELDI-ToF is more accurate than shotgun approach and is thus better suited to biomarker quantification, but may not be accurate enough for clinical diagnostics [132].

Recent techniques produce data sets of approximately one million spectra, up to 100 Gb in size, where up to 8,000 proteins can be identified [133]. Pre-processing of raw spectra entails noise filtering, baseline subtraction, peak detection, and calibration and alignment of LC/MS maps. Analysis follows four steps: (1) identification of amino-acid sequences, peptides and proteins in Peptide-Spectrum Match (PSM), and detection, quantification, annotation and alignment of features, (2) peptide and protein significance analysis, (3) class discovery and prediction, and (4) data integration and pathway analysis. Identification of amino-acid sequences mainly involves searching databases of spectra obtained experimentally or of spectra predicted from genomic sequences using *in silico* digestion, and reporting PSMs with the best scores. Statistical strength of predictions is indicated using the False Discovery Rate (FDR) computed using decoy databases, or models including the proportions of true and false identifications. Because many spectra map to many peptides and many peptides map to many proteins, identification of peptides and proteins is cumbersome and not completely solved. The issue is further complicated by post translational modifications and single amino-acid polymorphisms. Current methods identify approximately two thirds of tandem MS spectra. Proteins are reported on the basis of single-peptide match, or more stringently of match to protease specific peptides [133, 134]. Experiments are described using MIAME-like Minimum Information About a Proteomics Experiment (MIAPE) guidelines [135].

Difference in protein abundance is assessed with protein quantification (concentration estimate) and class comparison (change in abundance between conditions). The principle is to summarise all quantitative data relating to the protein by (1) spectral counting, where the number of spectra is assumed to reflect abundance with LC MS–MS, and is limited to large change for abundant proteins in low-complexity mixtures, or (2) probabilistic models incorporating all features of a protein and their variation. These models aim to address important issues, such as representation of the experimental design, treatment of missing data and control of FDR [134, 136]. Recent studies have shown convincing examples of quantitative proteomics efforts ran across different laboratories and using several experimental platforms. Currently, about two-third of human proteins predicted to exist have been detected with MS, hence the need to improve sensitivity, reproducibility of identification, and sensitivity and accuracy of quantification [133, 134, 136]. Protein–protein interactions and cell signalling cascades are mainly studied with the following approaches: yeast two-hybrid complementation, protein microarray, immunoaffinity chromatography and MS [137], and with a lower throughput by immunoprecipitation and mass spectrometry in Mammals [138, 139]. Attempts to integrate proteomics with other omics data are hindered by current drawbacks of proteomics analysis: proteome not completely sampled, uncertain identification of protein, difficulties in mapping identifiers across the different omics sources, hence the need for protein-centric knowledge bases such as TransProteomic Pipeline [140], Protein Atlas [141] and neXProt [142].

1.3.4 Metabolomics and Lipidomics

1.3.4.1 Metabolomics

Metabolomics is the high-throughput characterisation of the mixture of all metabolites in a biological system, i.e. endogenous and exogenous small molecules [143]. Metabolites are lipids, peptides, and amino, nucleic and organic acids. Metabolomics is now widely used in microbiology, nutrition, agriculture and environmental sciences, and clinical and pharmaceutical fields. Metabolites are the product of enzymatic reactions mediating complex biological processes and may therefore help understand phenotypes. They can be analysed using NMR spectroscopy although it lacks sensitivity [144] and MS (GC and LC) is usually preferred and used in targeted and untargeted approaches. Targeted strategies are specific and sensitive, allow absolute quantification and thus widely used in clinical diagnostics and drug development. Targeted approaches based on stable isotopes and models of metabolic networks allow estimation of the flux through biochemical pathways [145]. In contrast, untargeted approaches harbour a high coverage, though any metabolite identification is less specific and sensitive, and requires more intensive computational analysis. Features to use for identification are detected using univariate and multivariate analyses and then used to search databases such as Kyoto Encyclopedia of Genes and Genomes (KEGG) [146, 147]. Further experiments to distinguish isomers and characterise unidentified metabolites using tandem MS or NMR are often required. Metabolomics also include identification of substrate in *in vitro* assays of three types: (1) the protein is fixed onto a surface and ligands screened, (2) the metabolite is fixed and serve as bait for interacting proteins, or (3) activity-based protein profiling using chemical probes and beads. Last but not least, location of metabolites within cells, tissues or bodies can be studied by coupling MALDI or matrix-free MS and imaging techniques (imaging mass spectrometry, IMS) to obtain spectra by scanning the biological sample with the laser and then compiling a map of metabolite content across that sample [145, 148].

Standards for experiment description and tools for processing and analysis of metabolomics data are actively being developed [149, 150]. For example, MetabolomeExpress [151] and metaP [152] both combine tools from raw data processing, i.e. MS peak detection, to multivariate analysis.

Development of biomarkers with metabolomics and comparison between data sets depend on: (1) the characterisation of technical MS artefacts and differences in compounds discriminating samples between analysers and (2) sample type and biological variability [153]. The Human Metabolome Project quantified over 4,000 metabolites in up to 70 samples [143] out of 6,826 identified by Wishart and colleagues [154]. Another recent large-scale targeted metabolomics study quantified 122 metabolites in 377 individuals, including type 2 diabetes patients and controls, and identified 25 metabolites in plasma and 15 more in serum with different concentrations in the two groups [155].

1.3.4.2 Lipidomics

Lipids play important roles in the signalling involved in metabolism, energy storage, and cell proliferation, migration and apoptosis [156]. They are also the main components of cellular membranes, together with membrane proteins.

They thereby maintain cellular architecture and mediate membrane trafficking by enabling protein machinery assembly, as for example in dynamic clusters gathering specific proteins in lipids rafts [157]. Lipids are very diverse in their structure, physical properties and quantity. For example, signalling and structural lipids are respectively found in low and high abundance. Lipidomes, the lipids present in biological structures, are currently poorly understood [158]. The Human lipidome may contain thousands of species [159] while only 20 % of all lipids may have been detectable with existing technologies, as in 2009 [154]. Lipidomics studies aim to characterise lipids content, localisation and activity in cells and tissues [160]. The vast majority of lipids are extracted from lysed cells and tissues, and analysed with MS either directly in the shotgun method, i.e. 'top-down' lipidomics with high resolution analysers such as Orbitraps, or with LC–MS/MS 'bottom-up' lipidomics to distinguish lipids with identical charge to mass ratio [161]. Lipids have also been analysed with MALDI IMS [162]. Lipidomics MS raw data can be analysed with tools used for metabolomics, such as XCMS [163] and MZmine 2 [164].

Lipids are identified and quantified using raw data processing and statistical analysis, followed by pathway analysis and modelling [165]. Major lipidomics intiatives include the 'Lipid Metabolites And Pathways Strategy' (LIPID MAPS) which has established standards and enabled absolute rather than relative quantification [166], and the Mouse Macrophage Lipidome [167]. Absolute quantities for proteomics and lipidomics will help characterise complexes comprising both proteins and lipids [145].

Future technical advances should aim for higher accuracy better consistency, and harmonisation of protocols. Analytical developments should include: (1) automated data processing and lipid identification and mining, (2) statistical data analysis to address high-dimensionality and platform-independent computation of lipid identification false discovery rate, (3) pathway analysis to identify biochemical, signalling and regulatory processes that involve the lipids of interest characterised in a sample set, and (4) modelling in time and space within the context of physiology and systems [168].

1.4 Methods and Tools

Current high-throughput technologies produce very large data sets and have shifted the bottleneck from data production to data analysis. *Knowledge management* tools are thus very valuable to organise, store and analyse data either directly with embedded software or indirectly by exporting the data in the required format. Recent data sets also harbour very high dimensionality. Data integration aims at combining such *high-dimensionality*, large data sets differing in the type of data collected. Unsupervised integration aims to reduce the dimensionality of large data sets, without introducing a bias inherent to prior knowledge and hypotheses. It helps detect patterns within and amongst data sets and complements standard

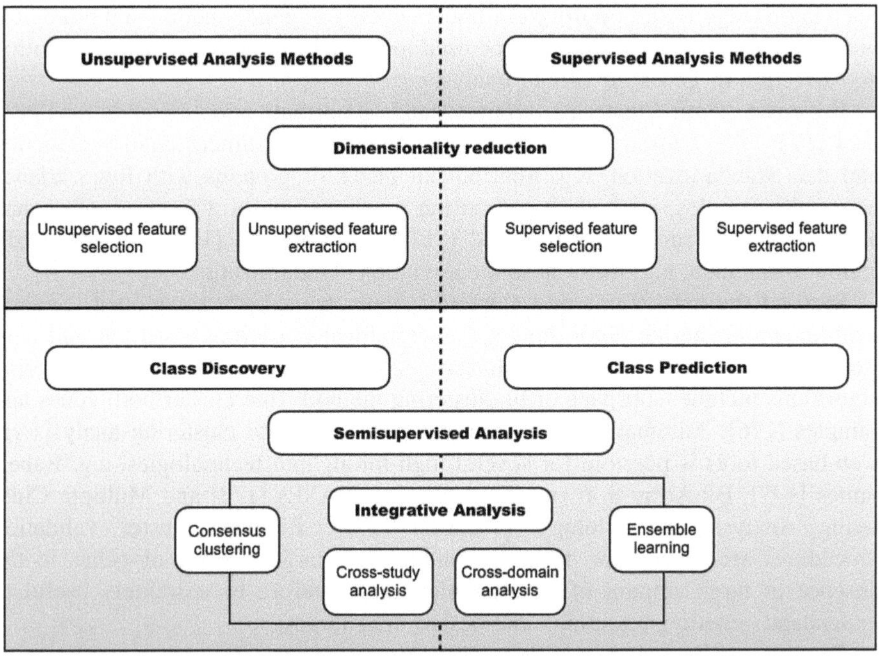

Fig. 1.3 Overview of machine learning methods. Supervised and unsupervised methods range from lower level dimensionality reduction approaches to higher-level analytical techniques and their extensions for integrative data analysis [171]

observations in building hypotheses. These are then tested analytically with supervised methods, usually only using a fraction of the available dimensions, and experimentally [58, 169, 170]. Despite its power and promises data integration is only a means to an end, not an automatic engine to generate valuable findings. Indeed, answers to the questions asked in a scientific study directly depend on the experimental design, e.g. the types of data, controls, processing and analyses, and the size of samples, within financial and time constraints. The following section describes methods for clustering, feature selection, prediction analysis, text mining and pathway analysis (Fig. 1.3).

1.4.1 Clustering

Motivation: Clustering is a data-exploration technique for multivariate analysis which divides data based on intrinsic groups without predefined labels. Clustering methods have been applied to various aspects of biomedical research, e.g. gene expression in cancer, to distinguish patients or genes subgroups based on expression levels of a set of differentially expressed genes. Clustered genes may have similar functions, be involved in the same cellular process or in similar pathways.

Such knowledge would improve our understanding of gene function and biological processes. Clustering methods can be used for visualization, hypothesis generation and selection of genes for further analysis.

Pre-processing: Clustering requires standard normalization methods for omics data [172–174]. Clustering specifically requires a prior dimensionality reduction and data standardization, e.g. filtering out genes or proteins with low variance across the samples, methods based on the maximization of a function of covariances as in the 'sum of covariances' (SUMCOV) method [175], and standardization of the data, e.g. mean absolute deviation standardization.

State-of-the-art: Numerous clustering tools have been developed. Several well-known clustering algorithms are: hierarchical clustering, partition and density-based clustering and fuzzy clustering. More recently developed clustering algorithms include: subspace or bi-clustering methods that cluster both genes and samples [176]. Automatic acquisition, pre-processing and clustering analysis via web-based tools is possible for several high-throughput technologies, e.g. Babelomics [177], BioArray Software Environment (BASE) [178] and Multiple Clustering Analysis Methodology (MCAM) [179]. Efficient cluster validation procedures are crucial for decision making with large number of genes in the absence of large amount of samples and will therefore be extremely useful to understand genetic interactions and design drug targets.

Use cases: Clustering is widely used in microarray data analysis and a wide choice of tools exists. Clustering of genes may identify a group of genes with similar functions while clustering of samples can suggest patient subgroups for stratification, response to treatments and disease subtypes or grade, e.g. childhood leukemia [180], breast cancer [181] and asthma [182, 183]. Clusters can also be integrated with pathway analysis [184].

1.4.2 Feature Selection

Motivation: Feature or attribute selection methods have a wide range of applications in Systems Biology. They enable an experimenter to identify which genes or proteins are significantly differentially expressed across different biological conditions in a cell type of interest, and which subsets of genes or proteins provide the most promising combined set of biomarkers for discriminating between these conditions (see also the section on prediction analysis). Moreover, feature selection approaches are often used to reduce the dimension of the input data before applying other higher-level statistical analysis methods. This alleviates a variety of statistical problems referred to as the *curse of dimensionality* in the literature [185]. However, in contrast to feature transformation based dimension reduction methods [186], the original features of the data are preserved, which facilitates data interpretation in subsequent analyses.

Feature selection algorithms can be grouped into *supervised* [187] and *unsupervised* approaches [188], depending on whether they incorporate information from class labels for the biological conditions. Moreover, feature selection algorithms employing prediction methods to score the informativeness of a feature subset are known as *wrappers*, whereas other univariate and combinatorial approaches to filter attributes are called *filters* [189].

Pre-processing: For most experimental platforms used in Systems Biology, several low-level pre-processing steps are required before applying feature selection methods. These include image processing [190, 191], normalisation [192] and summarisation approaches [193, 194], for microarray gene expression data [195], and raw data filtering [196], peak detection [197], peak alignment [198] and retention time normalisation methods for proteomics and metabolomics mass spectrometry data [199]. Moreover, some feature selection methods require a prior discretization of the data, e.g. if special association measures are used, such as mutual information [200].

State-of-the-art: The choice of the feature selection method depends both on the analysis goal (e.g. identifying individual biomarkers, or building a combinatorial predictive model for sample classification) and on the desired trade-off between efficiency (the run-time complexity of the algorithm) and accuracy (the predictive power of the selected features).

Among the filter approaches, simple univariate statistics like the parametric *Student's t test* and the non-parametric *Wilcoxon rank sum test* are still widely used, due to their advantages in terms of speed and the difficulty of estimating feature dependencies from noisy, high-dimensional data. More complex combinatorial methods such as *CFS* [201], *EFS* [202] and *Markov blanket filtering* [203] have recently gained influence.

Wrapper methods are becoming increasingly popular. They score feature subsets using prediction methods in combination with a search space exploration approach and their selections reach state-of-the art predictive performance in biological classification problems. Examples include combinations of fast and simple prediction methods, e.g. *kNN* [204] and Naïve Bayes [205], and search space exploration methods, e.g. sequential forward search [206]. These approaches are gradually being replaced by more complex algorithm combinations, including evolutionary algorithms [207] and kernel-based machine learning methods [208].

Finally, several recent techniques have improved the trade-off between speed and accuracy: (1) combination of filters [209], (2) combination of filters and wrappers into hybrid methods [203], (3) mathematical programming [210] and (4) signal processing approaches [211].

Use cases: Identification and prioritisation of gene, protein or metabolite biomarkers via feature selection techniques have three main aims: (1) distinguish biological conditions, e.g. presence of cancer, of viral infection, or tumor grades, (2) mediate early diagnostic, patient-tailored therapy, disease progression monitoring, and (3) help study treatment in a cell culture or animal model. However, feature selection methods are also used to filter datasets prior to the application of other higher-level data analysis methods, e.g. other machine learning methods, pathway overrepresentation analysis and network analysis. Finally, feature selection is often integrated with classification and regression techniques to decrease the complexity of machine learning models and maximize their predictive accuracy.

1.4.3 Prediction Analysis

Motivation: Prediction analysis refers to a family of methods that attempt to capture statistical dependencies and extract patterns from a set of measured data, to make predictions about future data. Such methods hold great promise in functional genomics, proteomics, metabolomics and bioinformatics, where the recent technologies provide a wealth of data such as gene and protein expression measurements, DNA and RNA sequence reads. The rate at which such data are produced makes automatic prediction analysis an indispensable tool for the biologist. Methods for prediction analysis can be unsupervised, semi-supervised, or supervised.

State-of-the-art: Unsupervised methods find regularities and hidden structure in the data. Typical approaches include clustering, feature selection, dimension reduction, density estimation, and model structure learning [212]. Classical linear dimension reduction methods are principal component analysis and independent component analysis, but recently some very powerful nonlinear dimension reduction methods have appeared [213, 214].

Supervised methods use data in the form of pairs (x, y) and estimate a function that predicts the value of y from a given input x. When y is a discrete quantity (for example a label of a number of distinct biological conditions) the method is called classification and when y is continuous the method is called regression. The key challenge is to ensure that the estimated function can generalize well to unseen situations [215]. Two methods are popular: (1) support vector machine (SVM) that estimates a discriminative function by maximizing class separation margin [216] and (2) random forest, based on tree ensembles and voting [217].

Semi-supervised methods combine ideas from supervised and unsupervised methods, to capture unsupervised structure in the data in order to boost classification performance [218].

Time series methods use data measured at different times to model and predict future values of the data, by capturing its structure and regularities and accounting for stochastic effects, e.g. with hidden Markov models (HMM) [219].

Use cases: A typical example is the classification of biological data such as gene expression data into different biological classes, e.g. disease and healthy, mostly using SVM and random forests. Prediction methods are also applied to pathway analysis, network decomposition and sequence annotation. They are often combined with a feature selection to extract the most relevant dimensions in the input data space [220].

1.4.4 Building Networks and Pathways from Literature

Motivation: Text mining joints efforts with the experimental sciences to help multifaceted disease-related research. Networks and connectivity maps are derived from text in an attempt to find connections and causal relations between components of complex biomedical systems, in order to elucidate disease mechanisms and detect co-morbidities [221, 222].

Pre-processing: Preparation of textual data consists of tokenization, removal of punctuation marks, part-of-speech tagging and sometimes syntactic parsing. Next, names of proteins, genes, chemicals, phenotypes and diseases are identified in the text. Management of biomedical terminology addresses several issues, such as appearance of new terms [221], heavy use of acronyms, abbreviations and general-purpose words that designate genes [223]. Synonymy and homonymy impose special challenges on the recognition process and complicate linking of a gene name to its unique identifier in the database [224, 225]. State-of-the-art named-entity recognition (NER) systems achieve F-measure of about 86 % [226] on biomedical corpus as opposed to 93 % on general purpose English texts [227].

State-of-the-art: Reconstruction of biological pathways from literature has evolved from undirected pairwise protein–protein co-occurrences [228] to complex biomedical events of typed and therefore directed interactions spanning multiple proteins [229–232]. The latter rely to a large extent on the richly annotated corpora, deep syntactic parsing and supervised machine learning techniques. Due to complexity of the natural language, accurate extraction of biomedical events remains a challenge. F-measure achieved by state-of-the-art systems varies from roughly 70–48 % depending largely on the event type being recognized.

Use cases: Many biomedical text-mining tools assist users at different stages of text processing, in particular for networks and pathways construction. Co-occurrence model has been successfully implemented in iHop, a hyperlinked network of genes and proteins mentioned in PubMed abstracts [233]. Facta + extends the pairwise co-occurrence model with event extraction and discovery of indirect associations between the biomedical concepts [222]. Based on PubMed abstracts, AliBaba builds networks of interacting proteins, genes—disease associations and subcellular location of proteins [234]. Networks extracted from text can be complemented with experimental data using IntAct [235] and CoPub [236].

1.4.5 Pathway Analysis

Motivation: Pathway analysis aims at identifying pathway deregulations to improve the understanding of complex phenotypes by leveraging information on known biomolecular interactions in pathways to guide the search through the space of possible functional associations. A wide range of methods exists, including enrichment analysis statistics, pathway-based disease gene prioritization methods, convex metabolic pathway analysis and *in silico* pathway prediction/reconstruction methods [237].

Pre-processing: Because experimental measurement platforms and pathway databases tend to use different identifier formats, pathway analysis usually starts with the conversion of gene/protein names into a standard format [238–240], followed by normalisation and pre-processing of the experimental data.

State-of-the-art: Several novel approaches have recently been developed to infer changes in pathway activity from high-throughput data more accurately than by the classical combination of differential expression filtering with overrepresentation statistics like the Fisher exact test (for unordered datasets) or the Kolmogorov–Smirnov test (for ranked datasets). These include parametric and nonparametric approaches that take into account unfiltered gene expression level measurements, e.g. GSEA [241], PaGE [242], GAGE [243] or exploit information from ontology graphs, e.g. Ontologizer [244] and GeneCodis [245]. For the study of metabolic pathways, two related approaches using convex analysis have become increasingly important: Elementary flux modes [246] and extreme pathways [247]. Finally, as opposed to the classical human expert-based definition of pathways, various methods for pathway prediction/reconstruction using experimental data have been proposed recently [248, 249].

Use cases: Genome-wide pathway analyses have provided new insights on the aetiology of complex diseases that cannot be obtained from classical single-locus analyses [250]. Such analyses have indeed shown that different disruptions in a pathway can cause the same disease, as in colorectal cancer [251]. Metabolic pathway analysis is used in biomedical and biotechnological applications, e.g. to increase the production yield of microorganisms by metabolic engineering, i.e. the modification of selected pathways via recombinant DNA technologies [252]. Pathway analysis can also be integrated with network analysis to identify deregulated network modules in complex diseases [253].

1.5 Conclusions

Study of individual genes and their products in model systems has shifted to high-throughput studies in laboratories and often generated by large consortia. Each type of omic data is proving very valuable and their integration promises even greater rewards. Current techniques are very diverse and can analyse complex biological samples. They harbour high sensitivity and specificity, albeit not always sufficient, as in proteomics. Ongoing developments will increase accuracy, robustness, and flexibility while reducing cost. Current technical innovations continue shifting the bottleneck from data production to data analysis. Our understanding of biology will indeed increasingly rely on data and knowledge management, and informatics infrastructure to complement advances in mathematical and computational modelling for temporal and spatial analytical techniques, which are crucial to Systems Biology.

Acknowledgments This work was supported by the CNRS, the University of Luxembourg and the ISB, and in part by the EU grants to CA in the context of the MeDALL consortium (*Mechanisms of the Development of Allergy,* Grant Agreement FP7 N°264357) and the U-BIOPRED consortium (*Unbiased Biomarkers for the PREDiction of respiratory disease outcomes*, Grant Agreement IMI 115010).

References

1. Gayon J, Malaterre C, Morange M, Raulin-Cerceau F, Tirard S (2010) Defining life: conference proceedings. Orig Life Evol Biosph 40(2):119–120
2. Westerhoff H, Hofmeyr J-H (2005) What is systems biology? From genes to function and back. In: Alberghina L, Westerhoff HV (eds) systems biology, vol 13. Springer, Berlin, pp 163–185
3. Kell DB (2006) Systems biology, metabolic modelling and metabolomics in drug discovery and development. Drug Discov Today 11(23–24):1085–1092
4. Lowe JA, Jones P, Wilson DM (2010) Network biology as a new approach to drug discovery. Curr Opin Drug Discov Devel 13(5):524–526
5. Pujol A, Mosca R, Farrés J, Aloy P (2010) Unveiling the role of network and systems biology in drug discovery. Trends Pharmacol Sci 31(3):115–123
6. Kitano H (2002) Looking beyond the details: a rise in system-oriented approaches in genetics and molecular biology. Curr Genet 41(1):1–10
7. Auffray C, Imbeaud S, Roux-Rouquié M, Hood L (2003) From functional genomics to systems biology: concepts and practices. C R Biol 326(10–11):879–892
8. Van Regenmortel MHV (2004) Reductionism and complexity in molecular biology. Scientists now have the tools to unravel biological and overcome the limitations of reductionism. EMBO Rep 5(11):1016–1020
9. Boogerd FC (2007) Systems biology: philosophical foundations. Elsevier, London, p 342
10. Wolkenhauer O (2001) Systems biology: the reincarnation of systems theory applied in biology? Brief Bioinf 2(3):258–270
11. Auffray C, Nottale L (2008) Scale relativity theory and integrative systems biology: 1. Founding principles and scale laws. Prog Biophys Mol Biol 97(1):79–114
12. Auffray C, Noble D (2009) Origins of systems biology in William Harvey's masterpiece on the movement of the heart and the blood in animals. Int J Mol Sci 10(4):1658–1669
13. Kohl P, Noble D (2009) Systems biology and the virtual physiological human. Mol Syst Biol 5:292
14. Westerhoff HV, Kolodkin A, Conradie R, Wilkinson SJ, Bruggeman FJ, Krab K, van Schuppen JH, Hardin H, Bakker BM, Moné MJ, Rybakova KN, Eijken M, van Leeuwen HJP, Snoep JL (2009) Systems biology towards life in silico: mathematics of the control of living cells. J Math Biol 58(1–2):7–34
15. Hornberg JJ, Bruggeman FJ, Westerhoff HV, Lankelma J (2006) Cancer: a systems biology disease. BioSystems 83(2–3):81–90
16. del Sol A, Balling R, Hood L, Galas D (2010) Diseases as network perturbations. Curr Opin Biotechnol 21(4):566–571
17. Westerhoff HV (2001) The silicon cell, not dead but live! Metab Eng 3(3):207–210
18. Kohl P, Crampin EJ, Quinn TA, Noble D (2010) Systems biology: an approach. Clin Pharmacol Ther 88(1):25–33
19. Hunter PJ, Borg TK (2003) Integration from proteins to organs: the physiome project. Nat Rev Mol Cell Biol 4(3):237–243
20. Hunter PJ, Crampin EJ, Nielsen PMF (2008) Bioinformatics, multiscale modeling and the IUPS physiome project. Brief Bioinf 9(4):333–343

21. Auffray C, Chen Z, Hood L (2009) Systems medicine: the future of medical genomics and healthcare. Genome Med 1(1):2
22. Hood L, Flores M (2012) A personal view on systems medicine and the emergence of proactive P4 medicine: predictive, preventive, personalized and participatory. New Biotechnol 29:613
23. Horner DS, Pavesi G, Castrignanò T, De Meo PD, Liuni S, Sammeth M, Picardi E, Pesole G (2010) Bioinformatics approaches for genomics and post genomics applications of next-generation sequencing. Brief Bioinf 11(2):181–197
24. Metzker ML (2010) Sequencing technologies—the next generation. Nat Rev Genet 11(1):31–46
25. Zhang J, Chiodini R, Badr A, Zhang G (2011) The impact of next-generation sequencing on genomics. J Genet Genomics 38(3):95–109
26. Martin JA, Wang Z (2011) Next-generation transcriptome assembly. Nat Rev Genet 12(10):671–682
27. Dreszer TR, Karolchik D, Zweig AS, Hinrichs AS, Raney BJ, Kuhn RM, Meyer LR, Wong M, Sloan CA, Rosenbloom KR, Roe G, Rhead B, Pohl A, Malladi VS, Li CH, Learned K, Kirkup V, Hsu F, Harte RA, Guruvadoo L, Goldman M, Giardine BM, Fujita PA, Diekhans M, Cline MS, Clawson H, Barber GP, Haussler D, James Kent W (2012) The UCSC Genome Browser database: extensions and updates 2011. *Nucleic Acids Res* 40(Database issue):D918–D923
28. Flicek P, Amode MR, Barrell D, Beal K, Brent S, Carvalho-Silva D, Clapham P, Coates G, Fairley S, Fitzgerald S, Gil L, Gordon L, Hendrix M, Hourlier T, Johnson N, Kähäri AK, Keefe D, Keenan S, Kinsella R, Komorowska M, Koscielny G, Kulesha E, Larsson P, Longden I, McLaren W, Muffato M, Overduin B, Pignatelli M, Pritchard B, Riat HS, Ritchie GRS, Ruffier M, Schuster M, Sobral D, Tang YA, Taylor K, Trevanion S, Vandrovcova J, White S, Wilson M, Wilder SP, Aken BL, Birney E, Cunningham F, Dunham L, Durbin R, Fernández-Suarez XM, Harrow J, Herrero J, Hubbard TJP, Parker A, Proctor G, Spudich G, Vogel J, Yates A, Zadissa A, Searle SMJ (2012) Ensembl 2012. Nucleic Acids Res 40(Database issue):D84–D90
29. Cirulli ET, Goldstein DB (2010) Uncovering the roles of rare variants in common disease through whole-genome sequencing. Nat Rev Genet 11(6):415–425
30. DePristo MA, Banks E, Poplin R, Garimella KV, Maguire JR, Hartl C, Philippakis AA, del Angel G, Rivas MA, Hanna M, McKenna A, Fennell TJ, Kernytsky AM, Sivachenko AY, Cibulskis K, Gabriel SB, Altshuler D, Daly MJ (2011) A framework for variation discovery and genotyping using next-generation DNA sequencing data. Nat Genet 43(5):491–498
31. The 1000 Genomes Project Consortium (2010) A map of human genome variation from population-scale sequencing. Nature 467(7319):1061–1073
32. Hudson TJ, Anderson W, Artez A, Barker AD, Bell C, Bernabé RR, Bhan MK, Calvo F, Eerola I, Gerhard DS, Guttmacher A, Guyer M, Hemsley FM, Jennings JL, Kerr D, Klatt P, Kolar P, Kusada J, Lane DP, Laplace F, Youyong L, Nettekoven G, Ozenberger B, Peterson J, Rao TS, Remacle J, Schafer AJ, Shibata T, Stratton MR, Vockley JG, Watanabe K, Yang H, Yuen MMF, Knoppers BM, Bobrow M, Cambon-Thomsen A, Dressler LG, Dyke SOM, Joly Y, Kato K, Kennedy KL, Nicolás P, Parker MJ, Rial-Sebbag E, Romeo-Casabona CM, Shaw KM, Wallace S, Wiesner GL, ⋯Zeps N, Lichter P, Biankin AV, Chabannon C, Chin L, Clément B, de Alava E, Degos F, Ferguson ML, Geary P, Hayes DN, Hudson TJ, Johns AL, Kasprzyk A, Nakagawa H, Penny R, Piris MA, Sarin R, Scarpa A, Shibata T, van de Vijver M, Futreal PA, Aburatani H, Bayés M, Botwell DDL, Campbell PJ, Estivill X, Gerhard DS, Grimmond SM, Gut I, Hirst M, López-Otín C, Majumder P, Marra M, McPherson JD, Nakagawa H, Ning Z, Puente XS, Ruan Y, Shibata T, Stratton MR, Stunnenberg HG, Swerdlow H, Velculescu VE, Wilson RK, Xue HH, Yang L, Spellman PT, Bader GD, Boutros PC, Campbell PJ, Flicek P, Getz G, Guigó R, Guo G, Haussler D, Heath S, Hubbard TJ, Jiang T, Jones SM, Li Q, López-Bigas N, Luo R, Muthuswamy L, Ouellette BFF, Pearson JV, Puente XS, Quesada V, Raphael BJ, Sander C, Shibata T, Speed TP, Stein LD, Stuart JM, Teague TW, Totoki Y, Tsunoda T, Valencia A, Wheeler DA, Wu H, Zhao S,

Zhou G, Stein LD, Guigó R, Hubbard TJ, Joly Y, Jones SM, Kasprzyk A, Lathrop M, López-Bigas N, Ouellette BFF, Spellman PT, Teague JW, Thomas G, Valencia A, Yoshida T, Kennedy KL, Axton M, Dyke SOM, Futreal PA, Gerhard DS, Gunter C, Guyer M, Hudson TJ, McPherson JD, Miller LJ, Ozenberger B, Shaw KM, Kasprzyk A, Stein LD, Zhang J, Haider SA, Wang J, Yung CK, Cros A, Cross A, Liang Y, Gnaneshan S, Guberman J, Hsu J, Bobrow M, Chalmers DRC, Hasel KW, Joly Y, Kaan TSH, Kennedy KL, Knoppers BM, Lowrance WW, Masui T, Nicolás P, Rial-Sebbag E, Rodriguez LL, Vergely C, Yoshida T, Grimmond SM, Biankin AV, Bowtell DDL, Cloonan N, deFazio A, Eshleman JR, Etemadmoghadam D, Gardiner BB, Gardiner BA, Kench JG, Scarpa A, Sutherland RL, Tempero MA, Waddell NJ, Wilson PJ, McPherson JD, Gallinger S, Tsao MS, Shaw PA, Petersen GM, Mukhopadhyay D, Chin L, DePinho RA, Thayer S, Muthuswamy L, Shazand K, Beck T, Sam M, Timms L, Ballin V, Lu Y, Ji J, Zhang X, Chen F, Hu X, Zhou G, Yang Q, Tian G, Zhang L, Xing X, Li X, Zhu Z, Yu Y, Yu J, Yang H, Lathrop M, Tost J, Brennan P, Holcatova I, Zaridze D, Brazma A, Egevard L, Prokhortchouk E, Banks RE, Uhlén M, Cambon-Thomsen A, Viksna J, Ponten F, Skryabin K, Stratton MR, Futreal PA, Birney E, Borg A, Børresen-Dale AL, Caldas C, Foekens JA, Martin S, Reis-Filho JS, Richardson AL, Sotiriou C, Stunnenberg HG, Thoms G, van de Vijver M, van't Veer L, Calvo F, Birnbaum D, Blanche H, Boucher P, Boyault S, Chabannon C, Gut I, Masson-Jacquemier JD, Lathrop M, Pauporté L, Pivot X, Vincent-Salomon A, Tabone E, Theillet C, Thomas G, Tost J, Treilleux I, Calvo F, Bioulac-Sage P, Clément B, Decaens T, Degos F, Franco D, Gut I, Gut M, Heath S, Lathrop M, Samuel D, Thomas G, Zucman-Rossi J, Lichter P, Eils R, Brors B, Korbel JO, Korshunov A, Landgraf P, Lehrach H, Pfister S, Radlwimmer B, Reifenberger G, Taylor MD, von Kalle C, Majumder PP, Sarin R, Rao TS, Bhan MK, Scarpa A, Pederzoli P, Lawlor RA, Delledonne M, Bardelli A, Biankin AV, Grimmond SM, Gress T, Klimstra D, Zamboni G, Shibata T, Nakamura Y, Nakagawa H, Kusada J, Tsunoda T, Miyano S, Aburatani H, Kato K, Fujimoto A, Yoshida T, Campo E, López-Otín C, Estivill X, Guigó R, de Sanjosé S, Piris MA, Montserrat E, González-Díaz M, Puente XS, Jares P, Valencia A, Himmelbauer H, Himmelbaue H, Quesada V, Bea S, Stratton MR, Futreal PA, Campbell PJ, Vincent-Salomon A, Richardson AL, Reis-Filho JS, van de Vijver M, Thomas G, Masson-Jacquemier JD, Aparicio S, Borg A, Børresen-Dale AL, Caldas C, Foekens JA, Stunnenberg HG, van't Veer L, Easton DF, Spellman PT, Martin S, Barker AD, Chin L, Collins FS, Compton CC, Ferguson ML, Gerhard DS, Getz G, Gunter C, Guttmacher A, Guyer M, Hayes DN, Lander ES, Ozenberger B, Penny R, Peterson J, Sander C, Shaw KM, Speed TP, Spellman PT, Vockley JG, Wheeler DA, Wilson RK, Hudson TJ, Chin L, Knoppers BM, Lander ES, Lichter P, Stein LD, Stratton MR, Anderson W, Barker AD, Bell C, Bobrow M, Burke W, Collins FS, Compton CC, DePinho RA, Easton DF, Futreal PA, Gerhard DS, Green AR, Guyer M, Hamilton SR, Hubbard TJ, Kallioniemi OP, Kennedy KL, Ley TJ, Liu ET, Lu Y, Majumder P, Marra M, Ozenberger B, Peterson J, Schafer AJ, Spellman PT, Stunnenberg HG, Wainwright BJ, Wilson RK, Yang H (2010) International network of cancer genome projects. Nature 464(7291):993–998

33. Meyerson M, Gabriel S, Getz G (2010) Advances in understanding cancer genomes through second-generation sequencing. Nat Rev Genet 11(10):685–696
34. Pastinen T (2010) Genome-wide allele-specific analysis: insights into regulatory variation Nat Rev Genet 11(8):533–538
35. Thomas T, Gilbert J, Meyer F (2012) Metagenomics—a guide from sampling to data analysis. Microb Inf Exp 2(1):1–12
36. Virgin HW, Todd JA (2011) Metagenomics and personalized medicine. Cell 147(1):44–56
37. Desai N, Antonopoulos D, Gilbert JA, Glass EM, Meyer F (2012) From genomics to metagenomics. Curr Opin Biotechnol 23(1):72–76
38. Langmead B, Hansen KD, Leek JT (2010) Cloud-scale RNA-sequencing differential expression analysis with Myrna. Genome Biol 11(8):R83
39. Stein LD (2010) The case for cloud computing in genome informatics. Genome Biol 11(5):207

40. Krampis K, Booth T, Chapman B, Tiwari B, Bicak M, Field D, Nelson K (2012) Cloud BioLinux: pre-configured and on-demand bioinformatics computing for the genomics community. BMC Bioinf 13(1):42
41. Field D, Sansone S-A, Collis A, Booth T, Dukes P, Gregurick SK, Kennedy K, Kolar P, Kolker E, Maxon M, Millard S, Mugabushaka A-M, Perrin N, Remacle JE, Remington K, Rocca-Serra P, Taylor CF, Thorley M, Tiwari B, Wilbanks J (2009) Megascience omics data sharing. Science 326(5950):234–236
42. Sayers EW, Barrett T, Benson DA, Bolton E, Bryant SH, Canese K, Chetvernin V, Church DM, DiCuccio M, Federhen S, Feolo M, Fingerman LM, Geer LY, Helmberg W, Kapustin Y, Landsman D, Lipman DJ, Lu Z, Madden TL, Madej T, Maglott DR, Marchler-Bauer A, Miller V, Mizrachi I, Ostell J, Panchenko A, Phan L, Pruitt KD, Schuler GD, Sequeira E, Sherry ST, Shumway M, Sirotkin K, Slotta D, Souvorov A, Starchenko G, Tatusova TA, Wagner L, Wang Y, Wilbur WJ, Yaschenko E, Ye J (2011) Database resources of the National Center for Biotechnology Information. Nucleic Acids Res 39(Database issue):D38–D51
43. Kent WJ, Sugnet CW, Furey TS, Roskin KM, Pringle TH, Zahler AM, Haussler D (2002) The human genome browser at UCSC. Genome Res 12(6):996–1006
44. Dunn WB (2011) Mass spectrometry in systems biology an introduction. Meth Enzymol 500:15–35
45. Hagen JB (2000) The origins of bioinformatics. Nat Rev Genet 1(3):231–236
46. Mount DR (2004) Bioinformatics: sequence and genome analysis. Cold Spring Harbor Laboratory Press, Second
47. Ashburner M, Ball CA, Blake JA, Botstein D, Butler H, Cherry JM, Davis AP, Dolinski K, Dwight SS, Eppig JT, Harris MA, Hill DP, Issel-Tarver L, Kasarskis A, Lewis S, Matese JC, Richardson JE, Ringwald M, Rubin GM, Sherlock G (2000) Gene ontology: tool for the unification of biology. The gene ontology consortium. Nat Genet 25(1):25–29
48. Gene Ontology Consortium (2001) Creating the gene ontology resource: design and implementation. Genome Res 11(8):425–1433
49. Whetzel PL, Parkinson H, Causton HC, Fan L, Fostel J, Fragoso G, Game L, Heiskanen M, Morrison N, Rocca-Serra P, Sansone S-A, Taylor C, White J, Stoeckert CJ Jr (2006) The MGED ontology: a resource for semantics-based description of microarray experiments. Bioinformatics 22(7):866–873
50. Csordas A, Ovelleiro D, Wang R, Foster JM, Ríos D, Vizcaíno JA, Hermjakob H (2012) PRIDE: quality control in a proteomics data repository. Database (Oxford) 2012:bas004
51. Jones AR, Miller M, Aebersold R, Apweiler R, Ball CA, Brazma A, Degreef J, Hardy N, Hermjakob H, Hubbard SJ, Hussey P, Igra M, Jenkins H, Julian RK Jr, Laursen K, Oliver SG, Paton NW, Sansone S-A, Sarkans U, Stoeckert CJ Jr, Taylor CF, Whetzel PL, White JA, Spellman P, Pizarro A (2007) The functional genomics experiment model (FuGE): an extensible framework for standards in functional genomics. Nat Biotechnol 25(10):1127–1133
52. Hucka M, Finney A, Sauro HM, Bolouri H, Doyle JC, Kitano H, Arkin AP, Bornstein BJ, Bray D, Cornish-Bowden A, Cuellar AA, Dronov S, Gilles ED, Ginkel M, Gor V, Goryanin II, Hedley WJ, Hodgman TC, Hofmeyr J-H, Hunter PJ, Juty NS, Kasberger JL, Kremling A, Kummer U, Le Novère N, Loew LM, Lucio D, Mendes P, Minch E, Mjolsness ED, Nakayama Y, Nelson MR, Nielsen PF, Sakurada T, Schaff JC, Shapiro BE, Shimizu TS, Spence HD, Stelling J, Takahashi K, Tomita M, Wagner J, Wang J (2003) The systems biology markup language (SBML): a medium for representation and exchange of biochemical network models. Bioinformatics 19(4):524–531
53. Novere NL, Hucka M, Mi H, Moodie S, Schreiber F, Sorokin A, Demir E, Wegner K, Aladjem MI, Wimalaratne SM, Bergman FT, Gauges R, Ghazal P, Kawaji H, Li L, Matsuoka Y, Villeger A, Boyd SE, Calzone L, Courtot M, Dogrusoz U, Freeman TC, Funahashi A, Ghosh S, Jouraku A, Kim S, Kolpakov F, Luna A, Sahle S, Schmidt E, Watterson S, Wu G, Goryanin I, Kell DB, Sander C, Sauro H, Snoep JL, Kohn K, Kitano H (2009) The systems biology graphical notation. Nat Biotech 27(8):735–741

54. Guberman JM, Ai J, Arnaiz O, Baran J, Blake A, Baldock R, Chelala C, Croft D, Cros A, Cutts RJ, Di Genova A, Forbes S, Fujisawa T, Gadaleta E, Goodstein DM, Gundem G, Haggarty B, Haider S, Hall M, Harris T, Haw R, Hu S, Hubbard S, Hsu J, Iyer V, Jones P, Katayama T, Kinsella R, Kong L, Lawson D, Liang Y, Lopez-Bigas N, Luo J, Lush M, Mason J, Moreews F, Ndegwa N, Oakley D, Perez-Llamas C, Primig M, Rivkin E, Rosanoff S, Shepherd R, Simon R, Skarnes B, Smedley D, Sperling L, Spooner W, Stevenson P, Stone K, Teague J, Wang J, Wang J, Whitty B, Wong DT, Wong-Erasmus M, Yao L, Youens-Clark K, Yung C, Zhang J, Kasprzyk A (2011) BioMart central portal: an open database network for the biological community. Database 2011:bar041
55. Zhang J, Haider S, Baran J, Cros A, Guberman JM, Hsu J, Liang Y, Yao L, Kasprzyk A (2011) BioMart: a data federation framework for large collaborative projects. Database (Oxford) 2011:bar038
56. Perakslis ED, Van Dam J, Szalma S (2010) How informatics can potentiate precompetitive open-source collaboration to jump-start drug discovery and development. Clin Pharmacol Ther 87(5):614–616
57. Maier D, Kalus W, Wolff M, Kalko SG, Roca J, Marin de Mas I, Turan N, Cascante M, Falciani F, Hernandez M, Villà-Freixa J, Losko S (2011) Knowledge management for systems biology a general and visually driven framework applied to translational medicine. BMC Syst Biol 5:38
58. Ghosh S, Matsuoka Y, Asai Y, Hsin K-Y, Kitano H (2011) Software for systems biology: from tools to integrated platforms. Nat Rev Genet 12(12):821–832
59. Kupershmidt I, Su QJ, Grewal A, Sundaresh S, Halperin I, Flynn J, Shekar M, Wang H, Park J, Cui W, Wall GD, Wisotzkey R, Alag S, Akhtari S, Ronaghi M (2010) Ontology-based meta-analysis of global collections of high-throughput public data. PLoS ONE 5(9):e13066
60. Huang Y-W, Huang TH-M, Wang L-S (2010) Profiling DNA methylomes from microarray to genome-scale sequencing. Technol Cancer Res Treat 9(2):139–147
61. Laird PW (2010) Principles and challenges of genome-wide DNA methylation analysis. Nat Rev Genet 11(3):191–203
62. Rakyan VK, Down TA, Balding DJ, Beck S (2011) Epigenome-wide association studies for common human diseases. Nat Rev Genet 12(8):529–541
63. Birney E, Stamatoyannopoulos JA, Dutta A, Guigó R, Gingeras TR, Margulies EH, Weng Z, Snyder M, Dermitzakis ET, Thurman RE, Kuehn MS, Taylor CM, Neph S, Koch CM, Asthana S, Malhotra A, Adzhubei I, Greenbaum JA, Andrews RM, Flicek P, Boyle PJ, Cao H, Carter NP, Clelland GK, Davis S, Day N, Dhami P, Dillon SC, Dorschner MO, Fiegler H, Giresi PG, Goldy J, Hawrylycz M, Haydock A, Humbert R, James KD, Johnson BE, Johnson EM, Frum TT, Rosenzweig ER, Karnani N, Lee K, Lefebvre GC, Navas PA, Neri F, Parker SCJ, Sabo PJ, Sandstrom R, Shafer A, Vetrie D, Weaver M, Wilcox S, Yu M, Collins FS, Dekker J, Lieb JD, Tullius TD, Crawford GE, Sunyaev S, Noble WS, Dunham I, Denoeud F, Reymond A, Kapranov P, Rozowsky J, Zheng D, Castelo R, Frankish A, Harrow J, Ghosh S, Sandelin A, Hofacker IL, Baertsch R, Keefe D, Dike S, Cheng J, Hirsch HA, Sekinger EA, Lagarde J, Abril JF, Shahab A, Flamm C, Fried C, Hackermüller J, Hertel J, Lindemeyer M, Missal K, Tanzer A, Washietl S, Korbel J, Emanuelsson O, Pedersen JS, Holroyd N, Taylor R, Swarbreck D, Matthews N, Dickson MC, Thomas DJ, Weirauch MT, Gilbert J, Drenkow J, Bell I, Zhao X, Srinivasan KG, Sung W-K, Ooi HS, Chiu KP, Foissac S, Alioto T, Brent M, Pachter L, Tress ML, Valencia A, Choo SW, Choo CY, Ucla C, Manzano C, Wyss C, Cheung E, Clark TG, Brown JB, Ganesh M, Patel S, Tammana H, Chrast J, Henrichsen CN, Kai C, Kawai J, Nagalakshmi U, Wu J, Lian Z, Lian J, Newburger P, Zhang X, Bickel P, Mattick JS, Carninci P, Hayashizaki Y, Weissman S, Hubbard T, Myers RM, Rogers J, Stadler PF, Lowe TM, Wei C-L, Ruan Y, Struhl K, Gerstein M, Antonarakis SE, Fu Y, Green ED, Karaöz U, Siepel A, Taylor J, Liefer LA, Wetterstrand KA, Good PJ, Feingold EA, Guyer MS, Cooper GM, Asimenos G, Dewey CN, Hou M, Nikolaev S, Montoya-Burgos JI, Löytynoja A, Whelan S, Pardi F, Massingham T, Huang H, Zhang NR, Holmes I, Mullikin JC, Ureta-Vidal A, Paten B, Seringhaus M, Church D,

Rosenbloom K, Kent WJ, Stone EA, Batzoglou S, Goldman N, Hardison RC, Haussler D, Miller W, Sidow A, Trinklein ND, Zhang ZD, Barrera L, Stuart R, King DC, Ameur A, Enroth S, Bieda MC, Kim J, Bhinge AA, Jiang N, Liu J, Yao F, Vega VB, Lee CWH, Ng P, Shahab A, Yang A, Moqtaderi Z, Zhu Z, Xu X, Squazzo S, Oberley MJ, Inman D, Singer MA, Richmond TA, Munn KJ, Rada-Iglesias A, Wallerman O, Komorowski J, Fowler JC, Couttet P, Bruce AW, Dovey OM, Ellis PD, Langford CF, Nix DA, Euskirchen G, Hartman S, Urban AE, Kraus P, Van Calcar S, Heintzman N, Kim TH, Wang K, Qu C, Hon G, Luna R, Glass CK, Rosenfeld MG, Aldred SF, Cooper SJ, Halees A, Lin JM, Shulha HP, Zhang X, Xu M, Haidar JNS, Yu Y, Ruan Y, Iyer VR, Green RD, Wadelius C, Farnham PJ, Ren B, Harte RA, Hinrichs AS, Trumbower H, Clawson H, Hillman-Jackson J, Zweig AS, Smith K, Thakkapallayil A, Barber G, Kuhn RM, Karolchik D, Armengol L, Bird CP, de Bakker PIW, Kern AD, Lopez-Bigas N, Martin JD, Stranger BE, Woodroffe A, Davydov E, Dimas A, Eyras E, Hallgrímsdóttir IB, Huppert J, Zody MC, Abecasis GR, Estivill X, Bouffard GG, Guan X, Hansen NF, Idol JR, Maduro VVB, Maskeri B, McDowell JC, Park M, Thomas PJ, Young AC, Blakesley RW, Muzny DM, Sodergren E, Wheeler DA, Worley KC, Jiang H, Weinstock GM, Gibbs RA, Graves T, Fulton R, Mardis ER, Wilson RK, Clamp M, Cuff J, Gnerre S, Jaffe DB, Chang JL, Lindblad-Toh K, Lander ES, Koriabine M, Nefedov M, Osoegawa K, Yoshinaga Y, Zhu B, de Jong PJ (2007) Identification and analysis of functional elements in 1% of the human genome by the ENCODE pilot project. Nature 447(7146):799–816

64. Bernstein BE, Stamatoyannopoulos JA, Costello JF, Ren B, Milosavljevic A, Meissner A, Kellis M, Marra MA, Beaudet AL, Ecker JR, Farnham PJ, Hirst M, Lander ES, Mikkelsen TS, Thomson JA (2010) The NIH roadmap epigenomics mapping consortium. Nat Biotechnol 28(10):1045–1048
65. Bradbury J (2003) Human epigenome project–up and running. PLoS Biol 1(3):E82
66. Rakyan VK, Hildmann T, Novik KL, Lewin J, Tost J, Cox AV, Andrews TD, Howe KL, Otto T, Olek A, Fischer J, Gut IG, Berlin K, Beck S (2004) DNA methylation profiling of the human major histocompatibility complex: a pilot study for the human epigenome project. PLoS Biol 2(12):e405
67. Adams D, Altucci L, Antonarakis SE, Ballesteros J, Beck S, Bird A, Bock C, Boehm B, Campo E, Caricasole A, Dahl F, Dermitzakis ET, Enver T, Esteller M, Estivill X, Ferguson-Smith A, Fitzgibbon J, Flicek P, Giehl C, Graf T, Grosveld F, Guigo R, Gut I, Helin K, Jarvius J, Kuppers R, Lehrach H, Lengauer T, Lernmark A, Leslie D, Loeffler M, Macintyre E, Mai A, Martens JH, Minucci S, Ouwehand WH, Pelicci PG, Pendeville H, Porse B, Rakyan V, Reik W, Schrappe M, Schubeler D, Seifert M, Siebert R, Simmons D, Soranzo N, Spicuglia S, Stratton M, Stunnenberg HG, Tanay A, Torrents D, Valencia A, Vellenga E, Vingron M, Walter J, Willcocks S (2012) BLUEPRINT to decode the epigenetic signature written in blood. Nat Biotech 30(3):224–226
68. Irizarry RA, Ladd-Acosta C, Carvalho B, Wu H, Brandenburg SA, Jeddeloh JA, Wen B, Feinberg AP (2008) Comprehensive high-throughput arrays for relative methylation (CHARM). Genome Res 18(5):780–790
69. Fernandez AF, Assenov Y, Martin-Subero JI, Balint B, Siebert R, Taniguchi H, Yamamoto H, Hidalgo M, Tan A-C, Galm O, Ferrer I, Sanchez-Cespedes M, Villanueva A, Carmona J, Sanchez-Mut JV, Berdasco M, Moreno V, Capella G, Monk D, Ballestar E, Ropero S, Martinez R, Sanchez-Carbayo M, Prosper F, Agirre X, Fraga MF, Graña O, Perez-Jurado L, Mora J, Puig S, Prat J, Badimon L, Puca AA, Meltzer SJ, Lengauer T, Bridgewater J, Bock C, Esteller M (2011) A DNA methylation fingerprint of 1628 human samples. Genome Res 22:407
70. Sandoval J, Heyn H, Moran S, Serra-Musach J, Pujana MA, Bibikova M, Esteller M (2011) Validation of a DNA methylation microarray for 450,000 CpG sites in the human genome. Epigenetics 6(6):692–702
71. Branton D, Deamer DW, Marziali A, Bayley H, Benner SA, Butler T, Di Ventra M, Garaj S, Hibbs A, Huang X, Jovanovich SB, Krstic PS, Lindsay S, Ling XS, Mastrangelo CH, Meller A, Oliver JS, Pershin YV, Ramsey JM, Riehn R, Soni GV, Tabard-Cossa V, Wanunu M,

Wiggin M, Schloss JA (2008) The potential and challenges of nanopore sequencing. Nat Biotechnol 26(10):1146–1153
72. Farnham PJ (2009) Insights from genomic profiling of transcription factors. Nat Rev Genet 10(9):605–616
73. Park PJ (2009) ChIP-seq: advantages and challenges of a maturing technology. Nat Rev Genet 10(10):669–680
74. Serre D, Lee BH, Ting AH (2010) MBD-isolated genome sequencing provides a high-throughput and comprehensive survey of DNA methylation in the human genome. Nucleic Acids Res 38(2):391–399
75. Macisaac KD, Fraenkel E (2010) Sequence analysis of chromatin immunoprecipitation data for transcription factors. Methods Mol Biol 674:179–193
76. Zhou VW, Goren A, Bernstein BE (2011) Charting histone modifications and the functional organization of mammalian genomes. Nat Rev Genet 12(1):7–18
77. Hoffman MM, Buske OJ, Wang J, Weng Z, Bilmes JA, Noble WS (2012) Unsupervised pattern discovery in human chromatin structure through genomic segmentation. Nature Methods 9:473
78. Korolev N, Fan Y, Lyubartsev AP, Nordenskiöld L (2012) Modelling chromatin structure and dynamics: status and prospects. Curr Opin Struct Biol 22(2):151–159
79. Geng T, Bao N, Litt MD, Glaros TG, Li L, Lu C (2011) Histone modification analysis by chromatin immunoprecipitation from a low number of cells on a microfluidic platform. Lab Chip 11(17):2842–2848
80. Bock C, Tomazou EM, Brinkman AB, Müller F, Simmer F, Gu H, Jäger N, Gnirke A, Stunnenberg HG, Meissner A (2010) Quantitative comparison of genome-wide DNA methylation mapping technologies. Nat Biotechnol 28(10):1106–1114
81. Rhee HS, Pugh BF (2011) Comprehensive genome-wide protein-DNA interactions detected at single-nucleotide resolution. Cell 147(6):1408–1419
82. Muro EM, McCann JA, Rudnicki MA, Andrade-Navarro MA (2009) Use of SNP-arrays for ChIP assays: computational aspects. Methods Mol Biol 567:145–154
83. Siegmund KD (2011) Statistical approaches for the analysis of DNA methylation microarray data. Hum Genet 129(6):585–595
84. Sun S, Huang Y-W, Yan PS, Huang TH, Lin S (2011) Preprocessing differential methylation hybridization microarray data. BioData Min 4:13
85. Huss M (2010) Introduction into the analysis of high-throughput-sequencing based epigenome data. Brief Bioinf 11(5):512–523
86. Massie CE, Mills IG (2012) Mapping protein-DNA interactions using ChIP-sequencing. Methods Mol Biol 809:157–173
87. Boyle AP, Song L, Lee B-K, London D, Keefe D, Birney E, Iyer VR, Crawford GE, Furey TS (2011) High-resolution genome-wide in vivo footprinting of diverse transcription factors in human cells. Genome Res 21(3):456–464
88. Giresi PG, Kim J, McDaniell RM, Iyer VR, Lieb JD (2007) FAIRE (formaldehyde-assisted isolation of regulatory elements) isolates active regulatory elements from human chromatin. Genome Res 17(6):877–885
89. Auerbach RK, Euskirchen G, Rozowsky J, Lamarre-Vincent N, Moqtaderi Z, Lefrançois P, Struhl K, Gerstein M, Snyder M (2009) Mapping accessible chromatin regions using sono-seq. Proc Natl Acad Sci USA 106(35):14926–14931
90. Dekker J, Rippe K, Dekker M, Kleckner N (2002) Capturing chromosome conformation. Science 295(5558):1306–1311
91. Dean A (2011) In the loop: long range chromatin interactions and gene regulation. Brief Funct Genomics 10(1):3–10
92. Simonis M, Klous P, Splinter E, Moshkin Y, Willemsen R, de Wit E, van Steensel B, de Laat W (2006) Nuclear organization of active and inactive chromatin domains uncovered by chromosome conformation capture-on-chip (4C). Nat Genet 38(11):1348–1354
93. Dostie J, Richmond TA, Arnaout RA, Selzer RR, Lee WL, Honan TA, Rubio ED, Krumm A, Lamb J, Nusbaum C, Green RD, Dekker J (2006) Chromosome conformation capture

carbon copy (5C): a massively parallel solution for mapping interactions between genomic elements. Genome Res 16(10):1299–1309
94. Lieberman-Aiden E, van Berkum NL, Williams L, Imakaev M, Ragoczy T, Telling A, Amit I, Lajoie BR, Sabo PJ, Dorschner MO, Sandstrom R, Bernstein B, Bender MA, Groudine M, Gnirke A, Stamatoyannopoulos J, Mirny LA, Lander ES, Dekker J (2009) Comprehensive mapping of long-range interactions reveals folding principles of the human genome. Science 326(5950):289–293
95. Fullwood MJ, Liu MH, Pan YF, Liu J, Xu H, Mohamed YB, Orlov YL, Velkov S, Ho A, Mei PH, Chew EGY, Huang PYH, Welboren W-J, Han Y, Ooi HS, Ariyaratne PN, Vega VB, Luo Y, Tan PY, Choy PY, Wansa KDSA, Zhao B, Lim KS, Leow SC, Yow JS, Joseph R, Li H, Desai KV, Thomsen JS, Lee YK, Karuturi RKM, Herve T, Bourque G, Stunnenberg HG, Ruan X, Cacheux-Rataboul V, Sung W-K, Liu ET, Wei C-L, Cheung E, Ruan Y (2009) An oestrogen-receptor-alpha-bound human chromatin interactome. Nature 462(7269):58–64
96. Deal RB, Henikoff JG, Henikoff S (2010) Genome-wide kinetics of nucleosome turnover determined by metabolic labeling of histones. Science 328(5982):1161–1164
97. Schones DE, Cui K, Cuddapah S, Roh T-Y, Barski A, Wang Z, Wei G, Zhao K (2008) Dynamic regulation of nucleosome positioning in the human genome. Cell 132(5):887–898
98. Knight JC, Keating BJ, Rockett KA, Kwiatkowski DP (2003) In vivo characterization of regulatory polymorphisms by allele-specific quantification of RNA polymerase loading. Nat Genet 33(4):469–475
99. McDaniell R, Lee B-K, Song L, Liu Z, Boyle AP, Erdos MR, Scott LJ, Morken MA, Kucera KS, Battenhouse A, Keefe D, Collins FS, Willard HF, Lieb JD, Furey TS, Crawford GE, Iyer VR, Birney E (2010) Heritable individual-specific and allele-specific chromatin signatures in humans. Science 328(5975):235–239
100. Kasowski M, Grubert F, Heffelfinger C, Hariharan M, Asabere A, Waszak SM, Habegger L, Rozowsky J, Shi M, Urban AE, Hong M-Y, Karczewski KJ, Huber W, Weissman SM, Gerstein MB, Korbel JO, Snyder M (2010) Variation in transcription factor binding among humans. Science 328(5975):232–235
101. Jeong J, Li L, Liu Y, Nephew KP, Huang TH-M, Shen C (2010) An empirical Bayes model for gene expression and methylation profiles in antiestrogen resistant breast cancer. BMC Med Genomics 3:55
102. Loss LA, Sadanandam A, Durinck S, Nautiyal S, Flaucher D, Carlton VEH, Moorhead M, Lu Y, Gray JW, Faham M, Spellman P, Parvin B (2010) Prediction of epigenetically regulated genes in breast cancer cell lines. BMC Bioinf 11:305
103. Krzywinski M, Schein J, Birol I, Connors J, Gascoyne R, Horsman D, Jones SJ, Marra MA (2009) Circos: an information aesthetic for comparative genomics. Genome Res 19(9):1639–1645
104. Brazma A, Parkinson H, Sarkans U, Shojatalab M, Vilo J, Abeygunawardena N, Holloway E, Kapushesky M, Kemmeren P, Lara GG, Oezcimen A, Rocca-Serra P, Sansone S-A (2003) ArrayExpress–a public repository for microarray gene expression data at the EBI. Nucleic Acids Res 31(1):68–71
105. Kapushesky M, Emam I, Holloway E, Kurnosov P, Zorin A, Malone J, Rustici G, Williams E, Parkinson H, Brazma A (2010) Gene expression atlas at the European Bioinformatics Institute. Nucleic Acids Res 38(Database issue):D690–D698
106. Barrett T, Troup DB, Wilhite SE, Ledoux P, Evangelista C, Kim LF, Tomashevsky M, Marshall KA, Phillippy KH, Sherman PM, Muertter RN, Holko M, Ayanbule O, Yefanov A, Soboleva A (2011) NCBI GEO: archive for functional genomics data sets–10 years on. Nucleic Acids Res 39(Database issue):D1005–D1010
107. Brazma A, Hingamp P, Quackenbush J, Sherlock G, Spellman P, Stoeckert C, Aach J, Ansorge W, Ball CA, Causton HC, Gaasterland T, Glenisson P, Holstege FC, Kim IF, Markowitz V, Matese JC, Parkinson H, Robinson A, Sarkans U, Schulze-Kremer S, Stewart J, Taylor R, Vilo J, Vingron M (2001) Minimum information about a microarray experiment (MIAME)-toward standards for microarray data. Nat Genet 29(4):365–371

108. Malone JH, Oliver B (2011) Microarrays, deep sequencing and the true measure of the transcriptome. BMC Biol 9:34
109. Posekany A, Felsenstein K, Sykacek P (2011) Biological assessment of robust noise models in microarray data analysis. Bioinformatics 27(6):807–814
110. Kadota K, Shimizu K (2011) Evaluating methods for ranking differentially expressed genes applied to microarray quality control data. BMC Bioinf 12:227
111. Gibb EA, Vucic EA, Enfield KSS, Stewart GL, Lonergan KM, Kennett JY, Becker-Santos DD, MacAulay CE, Lam S, Brown CJ, Lam WL (2011) Human cancer long non-coding RNA transcriptomes. PLoS ONE 6(10):e25915
112. Mercer TR, Dinger ME, Mattick JS (2009) Long non-coding RNAs: insights into functions. Nat Rev Genet 10(3):155–159
113. Wang Z, Gerstein M, Snyder M (2009) RNA-Seq: a revolutionary tool for transcriptomics. Nat Rev Genet 10(1):57–63
114. Pickrell JK, Marioni JC, Pai AA, Degner JF, Engelhardt BE, Nkadori E, Veyrieras J-B, Stephens M, Gilad Y, Pritchard JK (2010) Understanding mechanisms underlying human gene expression variation with RNA sequencing. Nature 464(7289):768–772
115. Hansen KD, Wu Z, Irizarry RA, Leek JT (2011) Sequencing technology does not eliminate biological variability. Nat Biotechnol 29(7):572–573
116. Mercer TR, Gerhardt DJ, Dinger ME, Crawford J, Trapnell C, Jeddeloh JA, Mattick JS, Rinn JL (2012) Targeted RNA sequencing reveals the deep complexity of the human transcriptome. Nat Biotechnol 30(1):99–104
117. Raz T, Kapranov P, Lipson D, Letovsky S, Milos PM, Thompson JF (2011) Protocol dependence of sequencing-based gene expression measurements. PLoS ONE 6(5):e19287
118. Schwartz S, Oren R, Ast G (2011) Detection and removal of biases in the analysis of next-generation sequencing reads. PLoS ONE 6(1):e16685
119. Brazma A (2009) Minimum information about a microarray experiment (MIAME)–successes, failures, challenges. Sci World J 9:420–423
120. Bullard JH, Purdom E, Hansen KD, Dudoit S (2010) Evaluation of statistical methods for normalization and differential expression in mRNA-Seq experiments. BMC Bioinf 11:94
121. Geiss GK, Bumgarner RE, Birditt B, Dahl T, Dowidar N, Dunaway DL, Fell HP, Ferree S, George RD, Grogan T, James JJ, Maysuria M, Mitton JD, Oliveri P, Osborn JL, Peng T, Ratcliffe AL, Webster PJ, Davidson EH, Hood L, Dimitrov K (2008) Direct multiplexed measurement of gene expression with color-coded probe pairs. Nat Biotech 26(3):317–325
122. Chi SW, Zang JB, Mele A, Darnell RB (2009) Argonaute HITS-CLIP decodes microRNA-mRNA interaction maps. Nature 460(7254):479–486
123. Jacquier A (2009) The complex eukaryotic transcriptome: unexpected pervasive transcription and novel small RNAs. Nat Rev Genet 10(12):833–844
124. Licatalosi DD, Darnell RB (2010) RNA processing and its regulation: global insights into biological networks. Nat Rev Genet 11(1):75–87
125. Vogel C, Marcotte EM (2012) Insights into the regulation of protein abundance from proteomic and transcriptomic analyses. Nat Rev Genet 13(4):227–232
126. Beck M, Schmidt A, Malmstroem A, Claassen M, Ori A, Szymborska A, Herzog F, Rinner O, Ellenberg J, Aebersold R (2011) The quantitative proteome of a human cell line. Mol Syst Biol 7:549, ISSN:1744-4292. doi:10.1038/msb.2011.82, URL:http://www.ncbi.nlm.nih.gov/pubmed/22068332. Accessed 15 Apr 2012
127. DeLuca DS, Marina O, Ray S, Zhang GL, Wu CJ, Brusic V (2011) Data processing and analysis for protein microarrays. Methods Mol Biol 723:337–347
128. Ray S, Reddy PJ, Choudhary S, Raghu D, Srivastava S (2011) Emerging nanoproteomics approaches for disease biomarker detection: a current perspective. J Proteomics 74(12):2660–2681
129. Domon B, Aebersold R (2010) Options and considerations when selecting a quantitative proteomics strategy. Nat Biotech 28(7):710–721
130. Mallick P, Kuster B (2010) Proteomics: a pragmatic perspective. Nat Biotechnol 28(7):695–709

131. Kito K, Ito T (2008) Mass spectrometry-based approaches toward absolute quantitative proteomics. Curr Genomics 9(4):263–274
132. Diao L, Clarke CH, Coombes KR, Hamilton SR, Roth J, Mao L, Czerniak B, Baggerly KA, Morris JS, Fung ET, Bast RC Jr (2011) Reproducibility of SELDI spectra across time and laboratories. Cancer Inform 10:45–64
133. Matthiesen R, Azevedo L, Amorim A, Carvalho AS (2011) Discussion on common data analysis strategies used in MS-based proteomics. Proteomics 11(4):604–619
134. Käll L, Vitek O (2011) Computational mass spectrometry-based proteomics. PLoS Comput Biol 7(12):e1002277
135. Taylor CF (2006) Minimum reporting requirements for proteomics: a MIAPE primer. Proteomics 6(Suppl 2):39–44
136. Jacob RJ (2010) Bioinformatics for LC-MS/MS-based proteomics. Methods Mol Biol 658:61–91
137. Walhout AJ, Vidal M (2001) Protein interaction maps for model organisms. Nat Rev Mol Cell Biol 2(1):55–62
138. Rinner O, Mueller LN, Hubalek M, Muller M, Gstaiger M, Aebersold R (2007) An integrated mass spectrometric and computational framework for the analysis of protein interaction networks. Nat Biotech 25(3):345–352
139. Hutchins JRA, Toyoda Y, Hegemann B, Poser I, Hériché J-K, Sykora MM, Augsburg M, Hudecz O, Buschhorn BA, Bulkescher J, Conrad C, Comartin D, Schleiffer A, Sarov M, Pozniakovsky A, Slabicki MM, Schloissnig S, Steinmacher I, Leuschner M, Ssykor A, Lawo S, Pelletier L, Stark H, Nasmyth K, Ellenberg J, Durbin R, Buchholz F, Mechtler K, Hyman AA, Peters J-M (2010) Systematic analysis of human protein complexes identifies chromosome segregation proteins. Science 328(5978):593–599
140. Deutsch EW, Shteynberg D, Lam H, Sun Z, Eng JK, Carapito C, von Haller PD, Tasman N, Mendoza L, Farrah T, Aebersold R (2010) Trans-Proteomic Pipeline supports and improves analysis of electron transfer dissociation data sets. Proteomics 10(6):1190–1195
141. Uhlen M, Oksvold P, Fagerberg L, Lundberg E, Jonasson K, Forsberg M, Zwahlen M, Kampf C, Wester K, Hober S, Wernerus H, Bjorling L, Ponten F (2010) Towards a knowledge-based human protein atlas. Nat Biotech 28(12):1248–1250
142. Lane L, Argoud-Puy G, Britan A, Cusin I, Duek PD, Evalet O, Gateau A, Gaudet P, Gleizes A, Masselot A, Zwahlen C, Bairoch A (2012) neXtProt: a knowledge platform for human proteins. Nucleic Acids Res 40(Database issue):D76–D83
143. Psychogios N, Hau DD, Peng J, Guo AC, Mandal R, Bouatra S, Sinelnikov I, Krishnamurthy R, Eisner R, Gautam B, Young N, Xia J, Knox C, Dong E, Huang P, Hollander Z, Pedersen TL, Smith SR, Bamforth F, Greiner R, McManus B, Newman JW, Goodfriend T, Wishart DS (2011) The human serum metabolome. PLoS ONE 6(2):e16957
144. Beckonert O, Keun HC, Ebbels TMD, Bundy J, Holmes E, Lindon JC, Nicholson JK (2007) Metabolic profiling, metabolomic and metabonomic procedures for NMR spectroscopy of urine, plasma, serum and tissue extracts. Nat Protoc 2(11):2692–2703
145. Lee DY, Bowen BP, Northen TR (2010) Mass spectrometry-based metabolomics, analysis of metabolite-protein interactions, and imaging. Biotechniques 49(2):557–565
146. Kanehisa M, Goto S (2000) KEGG: kyoto encyclopedia of genes and genomes. Nucleic Acids Res 28(1):27–30
147. Kotera M, Hirakawa M, Tokimatsu T, Goto S, Kanehisa M (2012) The KEGG databases and tools facilitating omics analysis: latest developments involving human diseases and pharmaceuticals. Methods Mol Biol 802:19–39
148. Sinha TK, Khatib-Shahidi S, Yankeelov TE, Mapara K, Ehtesham M, Cornett DS, Dawant BM, Caprioli RM, Gore JC (2008) Integrating spatially resolved three-dimensional MALDI IMS with in vivo magnetic resonance imaging. Nat Meth 5(1):57–59
149. Griffin JL, Steinbeck C (2010) 'So what have data standards ever done for us? The view from metabolomics. Genome Med 2(6):38

150. Sugimoto M, Kawakami M, Robert M, Soga T, Tomita M (2012) Bioinformatics tools for mass spectroscopy-based metabolomic data processing and analysis. Curr Bioinf 7(1):96–108
151. Carroll AJ, Badger MR, Harvey Millar A (2010) The metabolomeexpress project: enabling web-based processing, analysis and transparent dissemination of GC/MS metabolomics datasets. BMC Bioinf 11:376
152. Kastenmüller G, Römisch-Margl W, Wägele B, Altmaier E, Suhre K (2011) metaP-server: a web-based metabolomics data analysis tool. J Biomed Biotechnol 2011
153. Gika HG, Theodoridis GA, Earll M, Snyder RW, Sumner SJ, Wilson ID (2010) Does the mass spectrometer define the marker? A comparison of global metabolite profiling data generated simultaneously via UPLC-MS on two different mass spectrometers. Anal Chem 82(19):8226–8234
154. Wishart DS, Knox C, Guo AC, Eisner R, Young N, Gautam B, Hau DD, Psychogios N, Dong E, Bouatra S, Mandal R, Sinelnikov I, Xia J, Jia L, Cruz JA, Lim E, Sobsey CA, Shrivastava S, Huang P, Liu P, Fang L, Peng J, Fradette R, Cheng D, Tzur D, Clements M, Lewis A, De Souza A, Zuniga A, Dawe M, Xiong Y, Clive D, Greiner R, Nazyrova A, Shaykhutdinov R, Li L, Vogel HJ, Forsythe I (2009) HMDB: a knowledgebase for the human metabolome. Nucleic Acids Res 37(Database issue):D603–D610
155. Yu Z, Kastenmüller G, He Y, Belcredi P, Möller G, Prehn C, Mendes J, Wahl S, Roemisch-Margl W, Ceglarek U, Polonikov A, Dahmen N, Prokisch H, Xie L, Li Y, Wichmann H-E, Peters A, Kronenberg F, Suhre K, Adamski J, Illig T, Wang-Sattler R (2011) Differences between human plasma and serum metabolite profiles. PLoS ONE 6(7):e21230
156. Wymann MP, Schneiter R (2008) Lipid signalling in disease. Nat Rev Mol Cell Biol 9(2):162–176
157. van Meer G, Voelker DR, Feigenson GW (2008) Membrane lipids: where they are and how they behave. Nat Rev Mol Cell Biol 9(2):112–124
158. Shevchenko A. Simons K (2010) Lipidomics: coming to grips with lipid diversity. Nat Rev Mol Cell Biol 11(8):593–598
159. Quehenberger O, Armando AM, Brown AH, Milne SB, Myers DS, Merrill AH, Bandyopadhyay S, Jones KN, Kelly S, Shaner RL, Sullards CM, Wang E, Murphy RC, Barkley RM, Leiker TJ, Raetz CRH, Guan Z, Laird GM, Six DA, Russell DW, McDonald JG, Subramaniam S, Fahy E, Dennis EA (2010) Lipidomics reveals a remarkable diversity of lipids in human plasma. J Lipid Res 51(11):3299–3305
160. Han X, Gross RW (2003) Global analyses of cellular lipidomes directly from crude extracts of biological samples by ESI mass spectrometry: a bridge to lipidomics. J Lipid Res 44(6):1071–1079
161. Jung HR, Sylvänne T, Koistinen KM, Tarasov K, Kauhanen D, Ekroos K (2011) High throughput quantitative molecular lipidomics. Biochim Biophys Acta 1811(11):925–934
162. Chaurand P, Cornett DS, Angel PM, Caprioli RM (2011) From whole-body sections down to cellular level, multiscale imaging of phospholipids by MALDI mass spectrometry. Mol Cell Proteomics 10(2):O110.004259
163. Nordström A, O'Maille G, Qin C, Siuzdak G (2006) Nonlinear data alignment for UPLC-MS and HPLC-MS based metabolomics: quantitative analysis of endogenous and exogenous metabolites in human serum. Anal Chem 78(10):3289–3295
164. Pluskal T, Castillo S, Villar-Briones A, Oresic M (2010) MZmine 2: modular framework for processing, visualizing, and analyzing mass spectrometry-based molecular profile data. BMC Bioinf 11:395
165. Orešič M (2011) Informatics and computational strategies for the study of lipids. Biochim Biophys Acta 1811(11):991–999
166. Schmelzer K, Fahy E, Subramaniam S, Dennis EA (2007) The lipid maps initiative in lipidomics. Meth Enzymol 432:171–183
167. Dennis EA, Deems RA, Harkewicz R, Quehenberger O, Brown HA, Milne SB, Myers DS, Glass CK, Hardiman G, Reichart D, Merrill AH Jr, Sullards MC, Wang E, Murphy RC, Raetz CRH, Garrett TA, Guan Z, Ryan AC, Russell DW, McDonald JG, Thompson BM,

Shaw WA, Sud M, Zhao Y, Gupta S, Maurya MR, Fahy E, Subramaniam S (2010) A mouse macrophage lipidome. J Biol Chem 285(51):39976–39985
168. Niemelä PS, Castillo S, Sysi-Aho M, Oresic M (2009) Bioinformatics and computational methods for lipidomics. J Chromatogr B Analyt Technol Biomed Life Sci 877(26):2855–2862
169. Gehlenborg N, O'Donoghue SI, Baliga NS, Goesmann A, Hibbs MA, Kitano H, Kohlbacher O, Neuweger H, Schneider R, Tenenbaum D, Gavin A-C (2010) Visualization of omics data for systems biology. Nat Methods 7(3 Suppl):S56–S68
170. Hawkins RD, Hon GC, Ren B (2010) Next-generation genomics: an integrative approach. Nat Rev Genet 11(7):476–486
171. Glaab E (2011) Analysing functional genomics data using novel ensemble,consensus and data fusion techniques. University of Nottingham, Nottingham
172. Irizarry RA, Hobbs B, Collin F, Beazer-Barclay YD, Antonellis KJ, Scherf U, Speed TP (2003) Exploration, normalization, and summaries of high density oligonucleotide array probe level data. Biostatistics 4(2):249–264
173. Mueller LN, Brusniak M-Y, Mani DR, Aebersold R (2008) An assessment of software solutions for the analysis of mass spectrometry based quantitative proteomics data. J Proteome Res 7(1):51–61
174. Castillo S, Gopalacharyulu P, Yetukuri L, Orešič M (2011) Algorithms and tools for the preprocessing of LC–MS metabolomics data. Chemometrics Intell Lab Syst 108(1):23–32
175. Tritchler D, Parkhomenko E, Beyene J (2009) Filtering genes for cluster and network analysis. BMC Bioinf 10:193
176. Madeira SC, Oliveira AL (2004) Biclustering algorithms for biological data analysis: a survey. IEEE/ACM Trans Comput Biol Bioinf 1(1):24–45
177. Medina I, Carbonell J, Pulido L, Madeira SC, Goetz S, Conesa A, Tarraga J, Pascual-Montano A, Nogales-Cadenas R, Santoyo J, Garcia F, Marba M, Montaner D, Dopazo J (2010) Babelomics: an integrative platform for the analysis of transcriptomics, proteomics and genomic data with advanced functional profiling. Nucleic Acids Res 38(Web Server):W210–W213
178. Vallon-Christersson J, Nordborg N, Svensson M, Häkkinen J (2009) BASE—2nd generation software for microarray data management and analysis. BMC Bioinf 10:330
179. Naegle KM, Welsch RE, Yaffe MB, White FM, Lauffenburger DA (2011) MCAM: multiple clustering analysis methodology for deriving hypotheses and insights from high-throughput proteomic datasets. PLoS Comput Biol 7(7):e1002119
180. Chaiboonchoe A, Samarasinghe S, Kulasiri D (2009) Machine learning for childhood acute lymphoblastic leukaemia gene expression data analysis: a review. Curr Bioinform 5(2):118–133
181. Schummer M, Green A, Beatty JD, Karlan BY, Karlan S, Gross J, Thornton S, McIntosh M, Urban N (2010) Comparison of breast cancer to healthy control tissue discovers novel markers with potential for prognosis and early detection. PLoS ONE 5(2):e9122
182. Moore WC, Meyers DA, Wenzel SE, Teague WG, Li H, Li X, D'Agostino R, Castro M, Curran-Everett D, Fitzpatrick AM, Gaston B, Jarjour NN, Sorkness R, Calhoun WJ, Chung KF, Comhair SAA, Dweik RA, Israel E, Peters SP, Busse WW, Erzurum SC, Bleecker ER (2010) Identification of asthma phenotypes using cluster analysis in the severe asthma research program. Am J Respir Crit Care Med 181(4):315–323
183. Just J, Gouvis-Echraghi R, Rouve S, Wanin S, Moreau D, Annesi Maesano I (2012) Two novel severe asthma phenotypes identified during childhood using a clustering approach. Official Journal of the European Society for Clinical Respiratory Physiology, The European Respiratory Journal
184. Bjornsdottir US, Holgate ST, Reddy PS, Hill AA, McKee CM, Csimma CI, Weaver AA, Legault HM, Small CG, Ramsey RC, Ellis DK, Burke CM, Thompson PJ, Howarth PH, Wardlaw AJ, Bardin PG, Bernstein DI, Irving LB, Chupp GL, Bensch GW, Bensch GW, Stahlman JE, Karetzky M, Baker JW, Miller RL, Goodman BH, Raible DG, Goldman SJ, Miller DK, Ryan JL, Dorner AJ, Immermann FW, O'Toole M (2011) Pathways activated

during human asthma exacerbation as revealed by gene expression patterns in blood. PLoS ONE 6(7):e21902
185. Bellman R (1961) Adaptive control processes. Princeton University Press, Princeton
186. Kusiak A (2001) Feature transformation methods in data mining. IEEE Trans Electron Packaging Manuf 24(3):214–221
187. Guyon I, Elisseeff A (2003) An introduction to variable and feature selection. JMachine Learn Res 3:1157–1182
188. Dy JG (2008) Unsupervised feature selection. Comput Methods Feature Sel 2008:19–39
189. Tsamardinos I, Aliferis CF (2003) Towards principled feature selection: relevancy, filters and wrappers. In: Proceedings of the 9th international workshop on artificial intelligence and statistics, 2003
190. Bozinov D, Rahnenführer J (2002) Unsupervised technique for robust target separation and analysis of DNA microarray spots through adaptive pixel clustering. Bioinformatics 18(5):747
191. Katzer M, Kummert F, Sagerer G (2003) Methods for automatic microarray image segmentation. IEEE Transactions on NanoBiosci 2(4):202–214
192. Bolstad BM, Irizarry RA, Astrand M, Speed TP (2003) A comparison of normalization methods for high density oligonucleotide array data based on variance and bias. Bioinformatics 19(2):185–193
193. Li C, Wong WH (2001) Model-based analysis of oligonucleotide arrays: expression index computation and outlier detection. Proc Natl Acad Sci 98(1):31
194. Lazaridis EN, Sinibaldi D, Bloom G, Mane S, Jove R (2002) A simple method to improve probe set estimates from oligonucleotide arrays. Math Biosci 176(1):53–58
195. Shakya K, Ruskin H, Kerr G, Crane M, Becker J (2010) Comparison of microarray preprocessing methods. In: Arabina HR (ed) Advances in computational biology. Springer, New York, pp 139–147
196. Katajamaa M, Oresic M (2007) Data processing for mass spectrometry-based metabolomics. J Chromatogr A 1158(1–2):318–328
197. Hastings CA, Norton SM, Roy S (2002) New algorithms for processing and peak detection in liquid chromatography/mass spectrometry data. Rapid Commun Mass Spectrom 16(5):462–467
198. Smith CA, Elizabeth J, O'Maille G, Abagyan R, Siuzdak G (2006) XCMS: processing mass spectrometry data for metabolite profiling using nonlinear peak alignment, matching, and identification. Anal Chem 78(3):779–787
199. Bantscheff M, Schirle M, Sweetman G, Rick J, Kuster B (2007) Quantitative mass spectrometry in proteomics: a critical review. Anal Bioanal Chem 389(4):1017–1031
200. Hu Q, Pan W, An S, Ma P, Wei J (2010) An efficient gene selection technique for cancer recognition based on neighborhood mutual information. International Journal of Machine Learning and, Cybernetics, pp 1–12
201. Hall MA (1999) Correlation-based feature selection for machine learning. The University of Waikato, Waikato
202. Wu Y, Zhang A (2004) Feature selection for classifying high-dimensional numerical data. In: Proceedings of the 2004 IEEE computer society conference on computer vision and pattern recognition (CVPR 2004), vol. 2, p 251
203. Xing EP, Jordan MI, Karp RM, et al. (2001) Feature selection for high-dimensional genomic microarray data. In: Machine learning-international workshop then conference, 2001, pp 601–608
204. Li L, Pedersen LG, Darden TA, Weinberg CR (2002) Computational analysis of leukemia microarray expression data using the GA/KNN method. In: Methods of microarray data analysis: papers from CAMDA'00, pp 81–95
205. Blanco R, Larrañaga P, Inza I, Sierra B (2001) Selection of highly accurate genes for cancer classification by estimation of distribution algorithms. In: Workshop of Bayesian models in medicine. AIME 2001, pp 29–34

206. Inza I, Sierra B, Blanco R, Larrañaga P (2002) Gene selection by sequential search wrapper approaches in microarray cancer class prediction. J Intell Fuzzy Syst 12(1):25–33
207. Liu J, Iba H, Ishizuka M (2001) Selecting informative genes with parallel genetic algorithms in tissue classification. Genome Inform 12:14–23
208. Chen YW, Lin CJ (2006) Combining SVMs with various feature selection strategies. In: Isabella G, Andre E (eds) Feature extraction, Springer, Berlin, pp 315–324
209. Bolón-Canedo V, Sánchez-Maroño N, Alonso-Betanzos A (2011) An ensemble of filters and classifiers for microarray data classification. Pattern Recogn 45:531
210. Sun M, Xiong M (2003) A mathematical programming approach for gene selection and tissue classification. Bioinformatics 19(10):1243
211. Subramani P, Sahu R, Verma S (2006) Feature selection using haar wavelet power spectrum. BMC Bioinf 7(1):432
212. Koller D, Friedman N (2009) Probabilistic graphical models principles and techniques. MIT press, Cambridge
213. Lee JA, Verleysen M (2007) Nonlinear dimensionality reduction, 1st edn. Springer, Berlin
214. Maaten LVD, Hinton G (2008) Visualizing data using t-SNE. J Machine Learn Res 9(2579–2605):2579–2605
215. Hastie T, Tibshirani R, Friedman J (2009) The elements of statistical learning. Data mining, inference, and prediction. Springer, New York
216. Ben-Hur A, Ong CS, Sonnenburg S, Schölkopf B, Rätsch G (2008) Support vector machines and kernels for computational biology. PLoS Comput Biol 4(10):e1000173
217. Breiman L, Schapire E (2001) Random forests. Machine Learn 45:5–32
218. Chapelle O, Schölkopf B, Zien A (2010) Semi-supervised learning. MIT Press, Cambridge
219. Koski T (2002) Hidden Markov models of bioinformatics, 1st ed. Springer, Berlin
220. Bonneau R, Facciotti MT, Reiss DJ, Schmid AK, Pan M, Kaur A, Thorsson V, Shannon P, Johnson MH, Bare JC, Longabaugh W, Vuthoori M, Whitehead K, Madar A, Suzuki L, Mori T, Chang D-E, DiRuggiero J, Johnson CH, Hood L, Baliga NS (2007) A predictive model for transcriptional control of physiology in a free living cell. Cell 131(7):1354–1365
221. Krallinger M, Valencia A, Hirschman L (2008) Linking genes to literature: text mining information extraction, and retrieval applications for biology. Genome Biol 9(Suppl 2):S8
222. Tsuruoka Y, Miwa M, Hamamoto K, Tsujii J, Ananiadou S (2011) Discovering and visualizing indirect associations between biomedical concepts. Bioinformatics 27(13):i111
223. Roche M, Prince V (2010) A web-mining approach to disambiguate biomedical acronym expansions. Informatica (Slovenia) 34(2):243–253
224. Hakenberg J, Plake C, Royer L, Strobelt H, Leser U, Schroeder M (2008) Gene normalization and interaction with context and sentence motifs. Genome Biol 9(Suppl 2):S14
225. Seringhaus MB, Cayting PD, Gerstein MB (2008) Uncovering trends in gene naming. Genome Biol 9(1):401
226. Leaman R, Gonzalez G (2008) BANNER: an executable survey of advances in biomedical named entity recognition. In: Pacific symposium on biocomputing, pp 652–663
227. Marsh E, Perzanowski D (1998) MUC-7 evaluation of IE technology: overview of results. In: Proceedings of the 7th message understanding conference (MUC-7)
228. Hoffmann R, Krallinger M, Andres E, Tamames J, Blaschke C, Valencia A (2005) Text mining for metabolic pathways, signaling cascades, and protein networks. Sci STKE 2005(283):e21
229. Kim J-D, Ohta T, Pyysalo S, Kano Y, Tsujii J (2011) Extracting bio-molecular events from literature—the BioNLP'09 shared task. Comput Intell 27(4):513–540
230. Björne J, Ginter F, Pyysalo S, Tsujii J, Salakoski T (2010) Complex event extraction at PubMed scale. Bioinformatics [ISMB] 26(12):382–390
231. McClosky D, Surdeanu M, Manning CD (2011) Event extraction as dependency parsing. In: ACL 2011, pp 1626–1635
232. Riedel S, McCallum A (2011) Fast and robust joint models for biomedical event extraction. In: EMNLP 2011, pp 1–12

233. Hoffmann R, Valencia A (2005) Implementing the iHOP concept for navigation of biomedical literature. In: ECCB/JBI 2005, p 258
234. Plake C, Schiemann T, Pankalla M, Hakenberg J, Leser U (2006) ALIBABA: PubMed as a graph. Bioinformatics 22(19):2444–2445
235. Hermjakob H, Montecchi-Palazzi L, Lewington C, Mudali S, Kerrien S, Orchard S, Vingron M, Roechert B, Roepstorff P, Valencia A et al (2004) IntAct: an open source molecular interaction database. Nucleic Acids Res 32(suppl 1):D452–D455
236. Fleuren WWM, Verhoeven S, Frijters R, Heupers B, Polman J, , van Schaik R, de Vlieg J, Alkema W (2011) CoPub update: CoPub 5.0 a text mining system to answer biological questions. Nucleic Acids Res 39(Web Server):W450–W454
237. Ramanan VK, Shen L, Moore JH, Saykin AJ (2012) Pathway analysis of genomic data: concepts, methods and prospects for future development. Trends Genet 28(7): 323–332, ISSN:0168-9525. doi:10.1016/j.tig.2012.03.004
238. Dennis G Jr, Sherman BT, Hosack DA, Yang J, Gao W, Lane HC, Lempicki RA (2003) DAVID: database for annotation. Visualization, and integrated discovery. Genome Biol 4(5):P3
239. Alibés A, Yankilevich P et al (2007) IDconverter and IDClight: conversion and annotation of gene and protein IDs'. BMC Bioinf 8(1):9
240. Durinck S, Spellman PT, Birney E, Huber W (2009) Mapping identifiers for the integration of genomic datasets with the R/bioconductor package biomart. Nat Protoc 4(8):1184–1191
241. Subramanian A, Tamayo P, Mootha VK, Mukherjee S, Ebert BL, Gillette MA, Paulovich A, Pomeroy SL, Golub TR, Lander ES et al (2005) Gene set enrichment analysis: a knowledge-based approach for interpreting genome-wide expression profiles. Proc Natl Acad Sci USA 102(43):15545
242. Kim SY, Volsky D (2005) PAGE: parametric analysis of gene set enrichment. BMC Bioinf 6(1):144
243. Luo W, Friedman M, Shedden K, Hankenson K, Woolf P (2009) GAGE: generally applicable gene set enrichment for pathway analysis. BMC Bioinf 10(1):161
244. Bauer S, Grossmann S, Vingron M, Robinson PN (2008) Ontologizer 2.0—a multifunctional tool for GO term enrichment analysis and data exploration. Bioinformatics 24(14):1650
245. Carmona-Saez P, Chagoyen M, Tirado F, Carazo JM, Pascual-Montano A (2007) GENECODIS: a web-based tool for finding significant concurrent annotations in gene lists. Genome Biol 8(1):R3
246. Schuster S, Hlgetag C (1994) On elementary flux modes in biochemical reaction systems at steady state. J Biol Syst 2(2):165–182
247. Schilling CH, Letscher D, Palsson BO (2000) Theory for the systemic definition of metabolic pathways and their use in interpreting metabolic function from a pathway-oriented perspective. J Theoretical Biol 203(3):229–248
248. Kelley R, Ideker T (2005) Systematic interpretation of genetic interactions using protein networks. Nat Biotechnol 23(5):561–566
249. Ma X, Tarone AM, Li W (2008) Mapping genetically compensatory pathways from synthetic lethal interactions in yeast. PLoS ONE 3(4):e1922
250. Elbers CC, van Eijk KR, Franke L, Mulder F, van der Schouw YT, Wijmenga C, Onland-Moret NC (2009) Using genome-wide pathway analysis to unravel the etiology of complex diseases. Genet Epidemiol 33(5):419–431
251. Markowitz SD, Bertagnolli MM (2009) Molecular basis of colorectal cancer. N Engl J Med 361(25):2449–2460
252. Xu X, Cao L, Chen X (2008) Elementary flux mode analysis for optimized ethanol yield in anaerobic fermentation of glucose with Saccharomyces cerevisiae. Chin J Chem Eng 16(1):135–142
253. Cerami E, Demir E, Schultz N, Taylor BS, Sander C (2010) Automated network analysis identifies core pathways in glioblastoma. PLoS ONE 5(2):e8918

Chapter 2
Comparing Biological Networks: A Survey on Graph Classifying Techniques

Laurin A. J. Mueller, Matthias Dehmer and Frank Emmert-Streib

Abstract In order to compare biological networks numerous methods have been developed. Here, we give an overview of existing methods to compare biological networks meaningfully. Therefore we survey classical approaches of exact an inexact graph matching and discuss existing approaches to compare special types of biological networks. Moreover we review graph kernel-based methods and describe an approach based on structural network measures to classify large biological networks. The aim of this chapter is to provide a survey of techniques to compare biological networks for the interdisciplinary research community dealing with novel research questions in the field of systems biology

Keywords Graph classification · Comparative network analysis · Network biology · Exact graph matching · Gene networks · Global network alignment · Graph clustering · Graph distance · Graph Kernels · Graph matching · Graph probability distributions · Graph similarity · Inexact graph matching · Local network alignment · Metabolic networks · Network alignment · Network classification · Network clustering · Network distance · Network probability distributions · Network similarity · Structural network measures · Superindex · Systems biology · Topological network descriptors

---

L. A. J. Mueller · M. Dehmer (✉)
UMIT, Institute for Bioinformatics and Translational Research,
Eduard Wallnoefer Zentrum 1, 6060 Hall in Tyrol, Austria
e-mail: matthias.dehmer@umit.at

L. A. J. Mueller
e-mail: laurin.mueller@umit.at

F. Emmert-Streib
Computational Biology and Machine Learning Lab, Center for Cancer Research and Cell Biology, School of Medicine, Dentistry and Biomedical Sciences, Queen's University Belfast, 97 Lisburn Road, Belfast BT9 7BL, UK
e-mail: v@bio-complexity.com

A. Prokop and B. Csukás (eds.), *Systems Biology*,
DOI: 10.1007/978-94-007-6803-1_2,

Acronyms

GED	Graph Edit Distance
GEO	Gene Expression Omnibus
GO	Gene Ontology
PPI	Protein Protein Interactions
QSAR	Quantitative Structure-Activity Relationship
QSPR	Quantitative Structure-Property Relationships

2.1 Introduction

During the last decade, the amount of biological data has increased tremendously [93]. New developed technologies in molecular biology, such as DNA microarray or high-throughput sequencing, allow to create large collections of biological data and to aggregate information about different parts (i.e. genes, proteins) that are responsible for the functionality of a cell or an organism [64]. Identifying the different building blocks of a cell is the foundation of understanding the complex biological processes that take place within a cell. But to understand the mechanisms of a cell it is necessary to understand the interactions between the different components [56].

To model such interactions and the dynamics of biological processes, network-based approaches have been proven useful [3, 41, 43, 124]. A strong need for methods to compare biological networks arose in order to answer intriguing questions such as finding interactions across species, predicting functions of proteins or protein interactions, identifying functional information about poorly characterized interactions, etc. [109]. Additionally, networks-based approaches representing data from clinical studies can provide information to classify stages of complex diseases [88] or to predict tumor stages [42]. Comparing such networks can also be fruitful in order to identify biomarkers or groups of biomarkers for a better prediction and characterization of complex diseases [96].

Several methods to compare graphs for the investigation of different problems in various different scientific disciplines have been developed during the last decades [20, 24, 34, 117, 128]. Note, it is essential to understand that no general method or algorithm for comparing networks exists, as it always depends on the practical problem that has to be solved [34]. From this we see that there is a strong need to develop methods in order to compare networks. The aim of this review is to introduce existing methods to compare and classify networks to the field of system biology Additionally, this survey of graph classification techniques may help to select useful methods to shed light upon challenging problems in computational biology.

The chapter is organized as follows: in Sect. 2.2.1 we summarize classical methods for exact an inexact graph matching. A survey of different existing algorithms to align and compare biological networks is given in Sect. 2.2.2.

In Sect. 2.2.3 we discuss graph kernel-based methods for supervised machine learning and in Sect. 2.3 we describe methods based on topological network descriptors to compare and classify biological networks. Additionally we summarize two applications of the classification approach based on topological network descriptors in Sects. 2.3.3 and 2.3.4.

2.2 A Review on Graph Classification Techniques

Numerous methods and algorithms to compare graphs have been developed during the last decades [20, 24, 34, 117, 128]. It is essential to understand that there is no general approach that solves the problem of the matching of two graphs [34]. For two graphs that are structurally different, it strongly depends on the practical problem if they should be classified as similar. For example, in computational chemistry, graphs are used to represent the chemical structure of different molecules [13, 29, 118]. Comparing two molecules that are both toxic but one is a base and one an acid, it strongly depends on the research question whether the two graphs should be classified as similar (both are toxic) or not (one is a base one an acid). In general, the term classification refers to methods that identify similar objects. But in the context of machine learning, classification refers to supervised and clustering to unsupervised machine learning approaches.

In order to introduce the field of graph classification to systems biology we outline classical approaches (see Sect. 2.2.1) in this section. Moreover we give a detailed overview of recent approaches to compare biological networks (see Sect. 2.2.2). Furthermore we give an overview of kernel-based methods to classify graphs (see Sect. 2.2.3).

2.2.1 Classical Work

Methods for the comparison of graphs have been developed for the classification of structural patterns for different applications in the field of pattern recognition, such as image analysis, document processing or biometric identification [24]. In this work, we will give an overview of methods and algorithms, that can be utilized to estimate the similarity of graphs.

Exact Graph Matching: Methods to compare graphs can be categorized into two major groups: exact and inexact graph matching [19, 31]. The problem of exact graph matching is based on finding relations based on graph isomorphisms [31]. Generally speaking, two graphs $G_a = (V_a, E_a)$ and $G_b = (V_b, E_b)$ having the same number of nodes $|V_a| = |V_b|$ are isomorphic if the function $p : V_a \rightarrow V_b$ exists, such that $(g, h) \in E_a \Longleftrightarrow (p(g), p(h)) \in E_b$ [24]. Then G_a and G_b are said to be isomorphic graphs. This means that the mapping has to be edge-preserving in both directions [24]. A weaker formulation of the problem of exact graph matching is subgraph isomorphism where one graph has to be matched to a subgraph of another graph [24]. Additionally, another matching type of this category matches a subgraphs of the first graph to an isomorphic subgraph of the second one [19]. Methods that find this kind of matching are summarized under the problem of finding the maximum common subgraph [24].

Zelinka [128] was the first to suggest a measure for the similarity of graphs based on subgraph isomorphism. The measure is based on the assumption, that the larger the common induced subgraph of two graphs is the more similar the graphs are [128]. This distance measure is restricted to graphs having the same number of nodes. Sobik [117] formulated later a generalization of this method for graphs having different numbers of nodes.

The standard algorithm for graph and subgraph isomorphism was introduced by Ullmann [121]. As most algorithms for exact graph matching, this method is based on tree search using backtracking [121]. In general, the computational complexity of this algorithm is exponential, but can be reduced to polynomial time for certain graph classes [44]. In more recent work [26, 27, 74, 101] several improvements of the algorithm of Ullmann [121] have been developed. Additionally, McKay introduced an algorithm based on automorphism groups of graphs, that are used to derive a unique ordering for each equivalence class of isomorphic graphs [80]. Moreover, Messmer and Bunke [82] developed an algorithm that builds an decision tree for a whole graph library, what improves the search tremendously. Nevertheless, the complexity of constructing the decision tree has to be considered separately. Later, they further developed this method by using recursive graph decomposition for each graph in the library [83]. Although this algorithm avoids the comparison of similar subgraphs, the complexity of the matching increases with the number of graphs in the library [83].

For practical use, exact graph matching methods may not always be appropriate for several reasons [24]:

1. In the worst case, exact graph matching methods need exponential time complexity [19]. Specially, with biological networks that can consist of several thousand nodes, it is crucial to use methods that perform the process of graph matching more efficiently. Additionally, due to the inference procedure of biological networks, it is highly expectable that they do not have the same number of nodes. This makes it virtually impossible to apply exact graph matching algorithms.
2. Statistical methods for the inference of networks, particular in biology, result in non-deterministic networks [42]. The properties of such a non-deterministic network (i.e. number of nodes, edge density) is influenced by several circumstances (i.e. sample size or noise). Hence, exact graph matching may not be appropriate for such applications and the matching process has to be more tolerant [24, 42].

Inexact Graph Matching: Due to these facts, several inexact graph matching methods have been developed. Famous algorithms of this class, such as the graph edit distance [19], are based on graph edit operations to calculate the distance between two graphs [119]. Such methods are based on costs of graph transformations (i.e. insertions or deletions of nodes or edges) that are needed to transform one graph into the other [119]. Moreover, such methods have to find a mapping between two graphs, that minimizes the costs of the transformation [24]. Tsai and Fu [119] introduced a tree search-based algorithm that estimates the graph edit cost. This method only takes node and edge substitutions into account. Later [120], they extended the algorithm by taking node and edge insertions and deletions into account. Additionally, Bunke and Allermann [20] introduced the graph edit distance. The general idea of this algorithm is to define the distance between two graphs as the minimum amount of graph transformations (node and edge deletions and insertions) [21]. The main shortcoming of this approach is the computational complexity that makes it virtually impossible to apply this method to large graphs [21]. Therefore, several algorithms have been developed that calculate an approximation of the graph edit distance with acceptable computational complexity, see [21, 92, 99].

Additionally, several algorithms for inexact graph matching were developed, that include the merging or splitting of vertices, to estimate the distance between two graphs [25, 107]. Therefore, these algorithms use a defined transformation model to collapse subgraphs into one single node to estimate the graph distance [107]. Other methods that use error-tolerant graph matching based on the $A*$ search algorithm can be found in [45, 46, 104, 108, 125]. The $A*$ algorithm uses a best-first search to find a path from an initial node to a target node in order to minimize the costs on a certain path [54]. For example, Eshera et al. [45, 46] proposed a powerful method for image understanding by representing the images as attributed relational graphs. They utilize dynamic programming techniques to estimate the graph distance based on the shortest path problem. Several other algorithms have been developed in the field of pattern recognition but it would be out of the scope of this chapter to outline them here in detail. Therefore, see the extensive review of Conte [24] for an detailed overview of graph matching algorithms.

2.2.2 Recent Work on Comparative Biological Network Analysis

To model the dynamic and multidimensional nature of biological processes, network-based approaches have been proven useful [3, 43, 93]. In order to analyze the structure of biological data, Watts and Strogatz [124] and Barabási and Albert [9] introduced a network-based approach to the field of systems biology. They showed that the structure of biological networks is different from random networks [9, 124]. Representing biological data as networks has become ubiquitous in systems biology and biomedical research [41]. Therefore, several methods have been proposed to infer gene networks from biological data [4, 73, 84]. Intuitively, gene networks represent the interactions of genes or gene products [41]. From this, it is obvious that there is a strong need for methods to compare such networks. In this section we briefly summarize recent approaches to compare and classify networks.

Global and Local Alignment: The classification of networks has become an important task when analyzing protein-protein interactions (PPI). In this context, the problem of finding the best way in which nodes of one network correspond to the nodes of another network is called network alignment [69] and can be understood as a reformulation of graph matching [127].

Several local network alignment algorithms have been developed. The first method of this category is called PathBlast and was developed to compare PPI networks across species in order to identify protein pathways and complexes that have been conserved by evolution [62, 63]. The algorithm is based on the sequence aligning algorithm BLAST [5] and searches for high-scoring pathway alignments between PPI networks [62]. Kelley et al. [62] applied this algorithm to align pathways from the well studied yeast network (S. cerevisiae) to pathways of a less known bacterial PPI network (H. pylori) in order to predict protein functions. Whereas this method only allows the matching of one small protein pathways onto a larger PPI network, NetworkBLAST allows to align two PPI network in order to find all protein pathways and complexes that are common in both networks [60, 110]. Sharan et al. [110] used this method to predict physical PPIs and showed that proteins with similar sequences interact within multiple species and that such proteins occur in same conserved clusters or protein pathways. Another method based on finding the maximum induced subgraph to align PPI networks of two different species was invented by Koyutürk et al. [66]. This method is based on a model with the focus on the evolution of protein–protein interactions [65, 66]. Additionally, the alignment of two PPI networks based on this model is done by finding the maximum weighted induced subgraph [66]. Furthermore, Flannick et al. [48] proposed an local alignment algorithm based on the phylogenetic relationships of the two species in order to score possible conserved modules, called Graemlin. This method calculates the log-ratio between the probability that a module is subject to evolutionary constraints, and the probability the module is under no constraints [48].

Local network alignment seeks for single or multiple unrelated matched subgraphs of two given PPI networks and can match one protein to different local matchings. In contrast, global network alignment searches the maximum consistent matching across all vertices of the networks [127]. Therefore, Singh et al. [114] developed the first global alignment algorithm for PPI networks called IsoRank. The basic idea of this algorithm is that a protein in the first network is matched to a protein in the second network only when the neighbors of the two proteins can also be matched [127]. The calculation of a matching score of a pair of nodes is formulated as an eigenvalue problem. This score is used to identify relevant matches of corresponding nodes by using a greedy algorithm in order to extract a subgraph that represents the alignment [127]. Additionally, this method was later further developed to IsoRankN in order to distinguish the global alignment between multiple networks, see [75, 115].

Additionally Shih et al. [113] introduced an algorithm for the global alignment of multiple PPI networks. This approach calculates the similarity score based on the protein sequence in advance, and then uses clustering techniques to group similar proteins [113]. Based on a scaling parameter, similar clusters of the input networks are merged to obtain the alignment. They also showed that their method significantly outperforms IsoRankN in terms of computational complexity [113].

All previous discussed algorithms depend on information about the nodes (proteins), such as sequences of proteins. The method called GRAAL, introduced by Kuchaiev et al. [68] only considers structural features of the topology of the underlying PPI networks. In particular it matches pairs of nodes based on their graphlet degree signature similarities [97]. Graphlets are small connected non-isomorphic subgraphs of a larger network. To calculate a score for the matching of a pair of nodes GRAAL estimates how often a certain node is part of one of the 73 vertex orbits, based the automorphism groups of all graphlets with 2–5 nodes [97]. For a detailed explanation of this algorithm, we refer to [97]. Further developments of this algorithm are H-GRAAL [86] and M-GRAAL [69] that successively improved the performance in terms of computational complexity of the GRAAL algorithm. Furthermore, the MI-GRAAL algorithm improves the matching by taking protein sequences into account in order to optimize the matching of PPI networks [69]. However, a shortcoming of this approach is that the search of graphlets is accompanied by high computational costs, specially when aligning large networks.

Other Approaches: Due to the fact that the above discussed graphlet degree-based algorithms only take the topology of the underlying networks into account, it is possible not only to align PPI networks but any kind of (biological) network. Additionally, other methods have been introduced to align biological networks. Ay et al. [7] developed an algorithm to align large networks using the idea of compressing the initial networks. In particular, the method combines neighboring nodes with a low degree to so-called supernodes in order gain a compressed representation of the larger input networks. The level of compression is equal to the distance of neighboring nodes that are combined to supernodes. Subsequently, the compressed networks are matched by using a subgraph-based method called SubMAP [8]. Therefore, all possible subnetworks of size k are compared pairwise to calculate a similarity score, for details see [8]. The computational complexity of this approach depends strongly on the level of compression and also is accompanied by a loss of topological information [8]. Note, that the level of compression has to be identified for each problem individually.

Dehmer and Mehler [34] introduced a graph similarity measure to quantify the similarity between generalized trees. They utilize sequence aligning techniques based on the in- and out-degrees of the vertices to determine the similarity by transforming the generalized trees into property strings [34]. Emmert-Streib et al. [42] applied this algorithm to correlation networks inferred from microarray data, in order to classify tumor stages of cervical cancer. Each correlation network was decomposed into a set of generalized trees [42]. In particular, the proposed local decomposition algorithm generates one generalized tree for each node of the original network by using the corresponding node as root node of the tree [42]. Then, the nodes with distance k are assigned to the level k of the tree. Later, all nodes with a level greater than a predefined distance D are deleted to get the decomposed generalized tree [42]. To estimate the similarity between two networks, the obtained trees are then pairwise compared by a the above presented generalized tree-similarity-algorithm, for details see [34, 42]. They showed that by using this method it is possible to meaningfully distinguish between networks representing different tumor stages [42].

Additionally, Emmert-Streib [40] presented a network-based approach to detect differences in biological pathways. Therefore, he inferred networks based on Pearson correlations for a set of of genes corresponding certain pathways from microarray data for the chronic fatigue syndrome [40]. To compare the networks he introduced a modification of the graph edit distance. With this approach he was able to identify pathways that significantly change due to the influence of the disease [40].

In recent work Mueller et al. [88, 90] introduced a method based on topological network descriptors to classify gene networks (see Sect. 2.3.2). They were able to classify gene networks representing cancerous and benign tissue inferred from different microarray studies on prostate cancer [88, 90]. They used a network decomposition approach [90] based on the gene ontology database [53] in order to demonstrate the usefulness of their approach. Additionally, Mueller et al. applied this approach to a set of metabolic networks in order to distinguish between the three domains of life [91]. A detailed description of this approach will be presented in Sect. 2.3.2.

2.2.3 Graph Kernels for Supervised Machine Learning

The above discussed methods and algorithms focus on the matching or alignment of networks. We discussed methods for inexact and exact graph matching, algorithms that perform a local or global alignment of a pair or multiple biological networks. These methods can be used for unsupervised machine learning algorithms, such as clustering when the class label is not known. In contrast, supervised machine learning methods create a classifier based on training data where the class label is known in advance [122]. Then, this classifier can be used to assign a class label to an object where the class label is not known before.

In the following we will survey approaches that can be used to classify a set of networks corresponding to a given class label by using kernel methods such as support vector machines (SVM) [122]. The major advantage of kernel-based methods is that they access the examples via a so-called kernels [61]. In particular, only the inner product of the vector representation of the objects are used when the learning machine accesses the examples [61]. In particular, when the dimension of the vector representation of an example is very high, the dimension does not affect the task of training and classification as long as a kernel function to calculate the inner product effectively is available [61]. Kernel-based methods are often effective with high dimensional data as they map the input space into a higher dimensional space by using a kernel, where a linear separation of the examples can be performed more precisely [122]. There exist a large amount of graph kernels in order to classify graphs [17, 49, 98, 111]. In this section we briefly summarize existing approaches, in this field.

Kashima and Inokuchi [61] introduced a graph kernel based on random walks. This kernel method counts all pairs of walks of two given input graphs. In particular, walks are sequences of nodes that allow repetitions of nodes. Walks of the length k can be calculated in polynomial time by taking the adjacency matrix of a graph to the power of k [61]. Gärntner et al. [49] extended this method by defining kernel functions based on random walks for labeled and directed graphs. Additionally, Borgwardt et al. [15] developed a kernel method based on shortest paths. The advantage of this method is that it outperforms the random walk kernels in terms of computational complexity. The calculation of all paths in a graph is NP-hard, whereas determining the shortest path between all pairs of nodes can be solved in polynomial time [49], i.e. with the Floyd-Warshall algorithm [38]. Moreover, Borgwardt et al. [112] developed kernel methods based on graphlet distributions. They showed that kernels based on graphlets achieve a higher accuracy than random walk kernels by applying them to different graph sets. However, they showed that the computational complexity increases dramatically with increasing size of the used graphlets and specially for large graphs [112].

Also, Gärtner et al. [49] introduced basic graph kernels based on subgraphs comparing the neighborhoods of all pairs of nodes. Subsequently, Mathé and Vent [77] extended this method considering unbalanced subtrees. Based on this concept, Menchetti et al. introduced a subgraph kernel function based on the decomposition of the input graphs [81]. However, comparing labeled graphs using subgraph-based kernel can only be done with high computational costs and cannot be recommend for graphs with more than 100 nodes [111]. Therefore, Shervashidze and Borgwardt developed an fast subtree kernel in order to classify large labeled graphs based on the Weisfeiler-Lehman algorithm for graph isomorphism [111]. A comparison to other subgraph-based graph kernels showed that this method performs up to 5 times faster when applied to different graph data sets, see [111].

Additionally, Horvàth et al. [58] introduced a kernel method based on the number of cyclic patterns in a graph. A comparison showed that this class of graph kernels outperforms methods based on frequent subtrees patterns in terms of classification performance when applied to molecules representing drug targets for HIV [58]. However, the finding of cyclic patterns in a graph shows a higher computational complexity than kernels based on shortest path [57, 58]. For a detailed study on the classification performance of different graph kernel methods, see [17, 49, 98, 111].

Graph kernels are relatively new to the field of systems biology but have been successfully applied to several chemical and biological applications. Ralaivola et al. [98] used walk based kernels in order to predict mutagenicity and anti-cancer activity of different data sets of molecular networks representing drug targets. Additionally, they have been utilized to predict the toxicity of molecules meaningfully [77]. Borgwardt et al. [17] modified a random walk kernel in order to classify proteins based on their secondary structure into enzymes and non-enzymes. Additionally, they created a classifier based on graph kernels to predict the disease stage for PPI networks representing microarray experiments for leukemia and breast cancer [16]. Moreover, Avail et al. [1] proposed a method based on path kernels to identify protein-protein interactions from the literature. They showed that this method achieves meaningful classification performances compared to non kernel-based state-of-the art algorithms by applying them to different graph data sets [1, 2, 102]. See [105], for detailed overview of biological applications using graph-based and other kernel methods.

2.3 Comparing Networks Using Structural Measures

In order to compare and classify large biological networks we now introduce approaches based on topological network descriptors. Such descriptors have been used to quantify the complexity of networks [29, 30] and to solve problems in other disciplines [13, 78, 118, 123, 126]. A topological network descriptor is a numerical graph invariant that characterizes the structure of an underlying network quantitatively. A graph invariant is a numerical value associated with the graph G that is the same for any isomorphic graph [52]. Therefore, a large amount of topological network descriptors have been developed, but it would be out of the scope of this chapter to explain them in detail. For further investigation see the extensive overview of available network descriptors by Todeschini et al. [118]. Also, Dehmer and Mowshowitz [35] provided a recent and up to date review on information-theoretic complexity measures for graphs.

Topological network descriptors have been also used in mathematical and medical chemistry including drug design to analyze and characterize the structure of chemical compounds (QSAR/QSPR) [13, 29, 36, 118]. Basak an Magnuson [10] utilized structural graph measures to determine the structural similarity of a set of molecular networks. Additionally, Scsibrany et al. [106] used descriptors based on binary fingerprint vectors for describing molecular substructures to estimate the similarity of molecular networks.

Also, topological network descriptors can be used to indirectly improve exact graph matching techniques when comparing a large set of networks. In order to do so, descriptors showing high uniqueness [33] can can be used to reduce the amount of graphs that have to be compared with often computational expensive graph isomorphism testing algorithms [79]. For example, Dehmer et al. [33] showed that an information-theoretic measure based on degree-degree associations shows a high discriminative power on exhaustively generated sets of non-isomorphic graphs [32]. When applying such a measure to a large set of networks, isomorphism algorithms only have to be applied to networks having equal descriptor values. In particular this means the higher the uniqueness of a descriptor on a set of networks, the less the amount of graphs that have to be compared by using isomorphism testing [79].

The concept of topological network descriptors is relatively new to the field of system biology. Therefore we want to introduce approaches using unsupervised and supervised methods to classify large biological networks based on topological network descriptors.

2.3.1 Distance Measures Based on Graph Probability Distributions

As already discussed in Sect. 2.2.1 a lot of inexact graph matching methods show high computational complexity. Hence, they are not suitable for a comparative analysis of large networks. Kugler et al. [70] introduced an information-theoretic approach in order to derive a graph prototype for a set of biological networks. Therefore, they used three different probability distributions calculated form a set of gene networks. To calculate the distances between the networks, based on the derived probability distributions they used Kullback-Leibler divergence [70]. This approach motivated us to introduce a group of distance measures based on graph probability distributions. To the best knowledge of the authors, no further studies exist, that use this kind of measures for graphs in general, rather than for biological networks. However, we introduce this class of distance measures as an approach to compare large biological networks for several reasons:

1. Probability distributions of graphs, based on structural measures can be calculated in polynomial computational complexity [33].
2. They are new to the field of systems biology and have not been evaluated on networks in general, rather on biological networks.

However, the concept of graph probability distributions is not new. For instance, the degree distribution has been used in several studies in order to characterize biological networks [50, 85, 100]. It is possible to derive numerous additional different probability distributions from graphs such as i.e. the distance distribution [28] or the distribution of eigenvalues of different graph matrices [36]. Additionally, entropy based descriptors utilize different probability distribution to quantify the structural complexity of networks, see [14, 70, 87, 95].

Below in Table 2.1, we list different distance measures that we collected from the literature in order to compare discrete probability distributions. Note, that this list does not make claims of being complete. However, combining the different distance measures with different probability distributions provide a large amount of distance measures for inexact graph matching. These measures can help to answer challenging questions in systems biology. They can be used as distance

Table 2.1 Distance measures to compare probability distributions

Name	Formula	Reference
Manhattan distance	$D_L(P\|Q) = \frac{1}{2}\sum_i \lvert P(i) - Q(i)\rvert$	[67]
Euclidean distances	$D_E(P\|Q) := \sqrt{\sum_i (P(i) - Q(i))^2}$	[37]
Chebyshev distance	$D_C(P\|Q) := \sum_i \frac{\lvert P(i)-Q(i)\rvert}{\lvert P(i)\rvert + \lvert Q(i)\rvert}$	[22]
Canberra distance	$D_C(P\|Q) := \max_i(\lvert P(i) - Q(i)\rvert)$	[72]
Kullback-Leibler divergence	$D_{KL}(P\|Q) = \sum_i P(i) \ln \frac{P(i)}{Q(i)}$	[71]
ItakuraSaito distance	$D_{IS}(P\|Q) = \sum_i P(i)/Q(i) - \log \frac{P(i)}{Q(i)} - 1$	[23]
Bhattacharyya distance	$D_B(P\|Q) = -\ln\left(\sum_i \sqrt{P(i)Q(i)}\right)$	[11]

measures for different unsupervised machine learning algorithms such as clustering in order to identify common properties. Such an approach could, for example, be used to cluster PPI networks to identify functional similarities of the underlying PPIs. Additionally such measures can be used for an integrative analysis, i.e. for networks representing different microarray studies, to identify a graph prototype as a representative for all other studies, see [70].

2.3.2 Classification Using Quantitative Structural Measures

In this section we discuss an approach to classify large biological networks based on topological networks measures and supervised machine learning methods. Additionally, we discuss two applications of this approach applied to networks representing biological data. Similar approaches for determining the similarity of molecular graphs using chemical descriptors have already been applied in computational chemistry [47, 55, 55]. Feng et al. [47] used statistical methods to investigate different classes of topological network descriptors that were able to predict the toxicology of molecules. Their results show that they could not achieve this goal meaningfully but demonstrated the usefulness of the approach in general [47]. Helma et al. [55] used so-called molecular fragments to predict the mutagenicity of a set of molecular compounds. Rupp et al. [103] used chemical descriptors based on chemical substructures in order to predict the muntagenic activity of chemicals and compared different machine learning algorithms with different parameter settings. They showed that the performance of the classification strongly depends on the choice of the descriptors. For a detailed overview of the results we refer to [103]. Also, Hansen et al. [51] used several thousand descriptors from different groups to classify a benchmark set of molecular compounds in terms of mutagenicity. Later, Dehmer at al. [29] showed that comparable results could be achieved by using only seven topological network measures.

We introduced this approach into the field of systems biology for several reasons:

1. As already discussed (see Sect. 2.2) most methods of exact and inexact graph matching are not applicable to biological networks caused by their computational complexity and the typically large size of biological networks. The large size of biological networks often makes it virtually impossible to use such methods (i.e. methods based on graph isomorphism or certain graph kernels) as their computational complexity is insufficient (see Sect. 2.2.1). Moreover, when comparing a large set of networks, each network has to be compared to each other one, what results in $\frac{n(n-1)}{2}$ comparisons, where n represents the amount of networks to be compared. Whereas topological network measures have to be calculated only once for each network.
2. Some network alignment methods are only defined for special classes of biological networks (i.e. PPI networks), such as methods based on functional alignments. Whereas, other methods only take the structural characteristics of the underlying networks into account, see Sect. 2.2. In general, topological network descriptors also take only the structure of a network into account in order to quantify its structural complexity. However, there exists numerous networks measures that can be used to integrate biological information, for example by considering node or edge labels for representing biological information, see [30]. As our approach allows multiple topological network descriptors it is possible to combine measures of both classes.
3. As one network measure may be insufficient to quantify the structure of a given set of networks, the large amount of available topological network descriptors allows to combine measures that are based on different structural characteristics, such as distances, degrees, graph entropies, etc. Note, this is not possible with most of the above discussed approaches, such as graph kernels.

The idea of this approach, as illustrated in Fig. 2.1, is to classify a set of networks based on a given class label by calculating different topological network descriptors for these networks. This results in a feature vector representing the structural characteristics of each network by a single value for each applied descriptor. In order to select the features (topological network descriptors) that show an ability to distinguish between the different classes, a feature selection algorithm [76] has to be applied. The selected features are then combined to a so-called superindex that is defined as follows [12, 30]:

Definition 1 *Let $I_1, I_2, I_3 \ldots, I_n$ be topological network descriptors. The superindexSI of these measures is defined as*

$$SI := \{I_1, I_2, I_3, \ldots, I_n\}. \tag{2.1}$$

Subsequently, the resulting superindex is used as input for supervised machine learning methods, such as support vector machines [122] or random forest [18]. In order to estimate the performance of the classifier and to reduce the selection bias

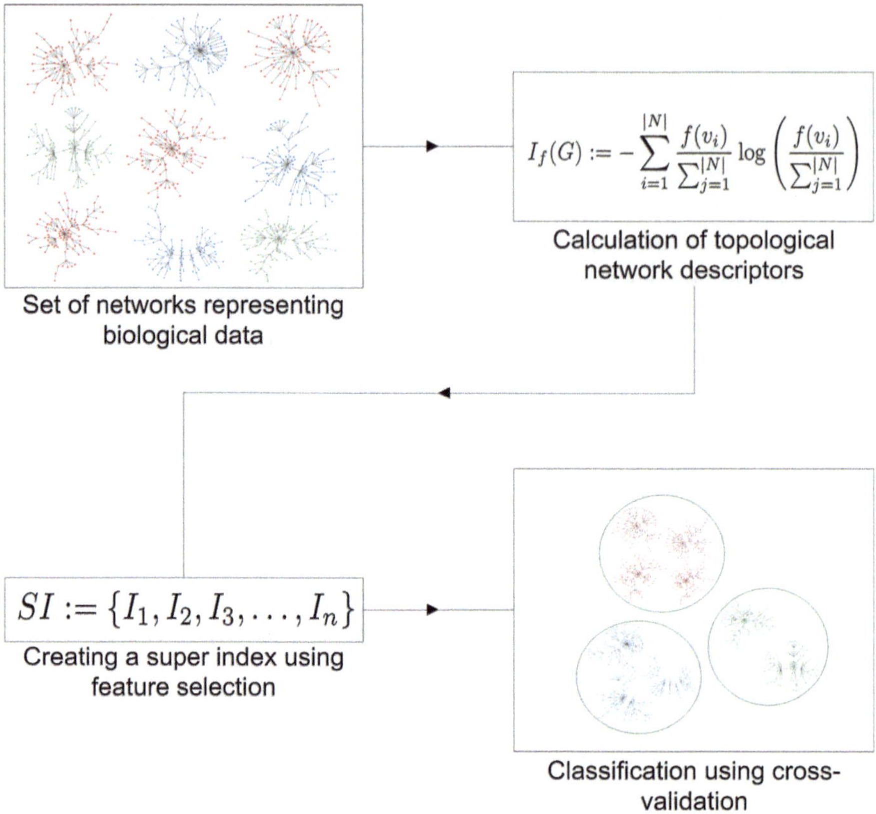

Fig. 2.1 Illustration of our approach using topological network descriptors and supervised machine learning methods to classify large biological networks

we recommend to use external cross validation [6, 116] by reperforming the feature selection within each loop of the cross-validation.

2.3.3 Classifying Gene Networks Representing Prostate Cancer

In the following we briefly survey the study due to Mueller et al. [90]. Its aim was to demonstrate the ability to classify biological networks using the recently developed R-package QuACN [89]. Currently QuACN contains about 150 different topological network measures. The focus of this application was to show the methodical perspective of our approach rather than giving a biological interpretation. We selected seven public available microarray experiments on prostate cancer form NCBI GEO [39] and EBI Arrayexpress [94]. For each study we inferred two networks using the C3NET inference algorithm [4]: One network represented benign tissue from the control group and the other one represented cancerous tissue. For each of these 14 networks we extracted subgraphs containing the genes associated with GO-terms from the gene ontology (GO) database [53]. This resulted in a total of 159 subnetworks associated with benign tissue and 108 networks associated with cancerous tissue.

In order to classify the resulting networks we calculated all descriptors available in QuACN, performed a features selection to select the 10 best features, and built a classifier using random forest. To estimate the classification performance we used external 10-fold cross-validation as suggested in Sect. 2.3.2. As a result, we achieved a classification performance with an F-Score of 0.80 and an accuracy of 0.74. These results demonstrated the usefulness of our approach, to meaningful capture class specific structural complexity by using topological network measures. Note, this was a non-trivial result as one could easily show by using other measures that the classification task would result in a random classification what would not be feasible in practice.

2.3.4 Classifying Metabolic Networks into the Three Domains of Life

Another application of our approach was to classify organism within the three domains of life using topological information of the underlying metabolic networks [91]. It has been shown that these networks share domain-independent structural similarities [59]. To get a deeper understanding of evolutionary processes we used our approach to identify domain-specific structural information of 43 metabolic networks, each representing one particular organism. These organisms can be divided in three different classes, which represent the three domains of life: Archae, Bacterium, and Eukarytote. In order to estimate the classification performance of different groups of topological network descriptors we applied our approach to the metabolic networks by using two different groups of structural measures: entropy-based and non-entropy-based descriptors. For a detailed description of the experimental setting see [91]. Our results [91] showed that the classification performance by using entropy-based descriptors is higher than with non-entropy-based descriptors. Moreover, selecting the best features combining both groups achieved the highest classification performance of an weighted F-Score and an accuracy of 0.84. Additionally we compared our results to another approach by calculating the Kullback-Leibler divergence [71] for the degree distribution of each metabolic network as already discussed in Sect. 2.3.1. We could show that this approach did not show a feasible classification of the metabolic networks within the three domains of life what additionally documented the usefulness of our approach.

2.4 Conclusion

In this chapter, we surveyed existing methods for comparative graph analysis. We put the emphasis on techniques that can be useful to answer intriguing questions in systems biology. As we have shown, numerous methods for classifying networks have been developed but their impact for the field of systems biology has not yet been explored. One reason is that different methods have been developed for a special purpose and that they only have been applied in a single discipline. Another reason is the vast amount of methods developed so far and the fact that for a special practical problem not every technique may be appropriate. We hope that the surveyed methods motivate the reader to use them in order to compare biological networks meaningfully and stimulates an interdisciplinary audience to tackle challenging problems in systems biology.

Acknowledgments This work was supported by the Tiroler Wissenschaftsfonds and the Standortagentur Tirol (formerly Tiroler Zukunftsstiftung).

References

1. Airola A, Pyysalo S, Björne J, Pahikkala T, Ginter F, Salakoski T (2008) A graph Kernel for protein-protein interaction extraction. In: Proceedings of the workshop on current trends in biomedical natural language processing, pp. 1–9. Association for, Computational Linguistics, 2008.
2. Airola A, Pyysalo S, Björne J, Pahikkala T, Ginter F, Salakoski T (2008) All-paths graph Kernel for protein-protein interaction extraction with evaluation of Cross-Corpus learning. BMC Bioinformatics 9(Suppl 11):S2
3. Alberts B, Johnson A, Lewis J, Raff M, Roberts K, Walter P (2007) Molecular biology of the cell. Garland Science, 5th edn
4. Altay G, Emmert Streib F (2010) Inferring the conservative causal core of gene regulatory networks. BMC Syst Biol 4(1):132
5. Altschul SF, Madden TL, Schäffer AA, Zhang J, Zhang Z, Miller W, Lipman DJ (1997) Gapped BLAST and PSI-BLAST: a new generation of protein database search programs. Nucleic Acids Res 25(17):3389–3402
6. Ambroise C, McLachlan GJ (2002) Selection bias in gene extraction on the basis of microarray gene-expression data. Proc Nat Acad Sci USA 99(10):6562
7. Ay F, Dang M, Kahveci T (2012) Metabolic network alignment in large scale by network compression. BMC Bioinformatics 13(Suppl 3):S2
8. Ay F, Kahveci T (2010) SubMAP: aligning metabolic pathways with subnetwork mappings. In Research in computational molecular biology. Springer, New York, pp 15–30
9. Barabási AL, Albert R (1999) Emergence of scaling in random networks. Science 206: 509–512
10. Basak SC, Magnuson VR, Niemi GJ, Regal RR (1988) Determining structural similarity of chemicals using graph-theoretic indices. Discrete Appl Math 19(1–3):17–44
11. Bhattacharyya A (1943) On a measure of divergence between two statistical populations defined by their probability distributions. Bull Calcutta Mathl Soc 35:99–109
12. Bonchev D (1983) Information theoretic indices for characterization of chemical structures. Chemometrics research studies series. Research Studies Press

13. Bonchev D, Mekenyan O, Trinajstić N (1981) Isomer discrimination by topological information approach. J Comput Chem 2(2):127–148
14. Bonchev D, Trinajstić N (1977) Information theory, distance matrix and molecular branching. J Chem Phys 67:4517–4533
15. Borgwardt KM, Kriegel HP (2005) Shortest-Path Kernels on graphs. In Data mining, Fifth IEEE International Conference on. IEEE, p 8
16. Borgwardt KM, Kriegel HP, Vishwanathan SVN, Schraudolph NN (2007) Graph Kernels for disease outcome prediction from Protein-Protein interaction networks, vol 12. In: Proceedings of Pacific symposium on biocomputing (PSB), pp 4–15
17. Borgwardt KM, Ong CS, Schönauer S, Vishwanathan SVN, Smola AJ, Kriegel HP (2005) Protein function prediction via graph Kernels. Bioinformatics 21(suppl 1):i47–i56
18. Breiman L (2001) Random Forests. Mach Learn 45(1):5–32
19. Bunke H (2000) Graph matching: theoretical foundations, algorithms, and applications. Proc Vis Interface 2000:82–88
20. Bunke H, Allermann G (1983) Inexact graph matching for structural pattern recognition. Pattern Recogn Lett 1(4):245–253
21. Bunke H, Riesen K (2009) Graph edit distance-optimal and suboptimal algorithms with applications. Anal Complex Netw pp 113–143
22. Cantrell CD (2000) Modern mathematical methods for physicists and engineers. Cambridge University Press, Cambridge
23. Chan AHS (2010) Advances in industrial engineering and operations research. Lecture notes in electrical engineering. Springer, New York
24. Conte D, Foggia P, Sansone C, Vento M (2004) Thirty years of graph matching in pattern recognition. Int J Pattern Recogn Artif Intell 18(3):265–298
25. Cordella LP, Foggia P, Sansone C, Vento M (1996) An efficient algorithm for the inexact matching of Arg Graphs using a contextual transformational model, vol 3. In: Proceedings of the 13th international conference on pattern recognition. IEEE, pp 180–184
26. Cordella LP, Foggia P, Sansone C, Vento M (2000) Fast graph matching for detecting CAD image components, vol 2. In: 15th international conference on pattern recognition. IEEE, pp 1034–1037
27. Cordella LP, Foggia P, Sansone C, Vento M (2001) An improved algorithm for matching large graphs. In: 3rd IAPR-TC15 workshop on graph-based representations in, pattern recognition, pp 149–159
28. Dankelmann P (2011) On the distance distribution of trees. Discrete Appl Math
29. Dehmer M, Barbarini N, Varmuza K, Graber A (2009) A large scale analysis of information-theoretic network complexity measures using chemical structures. PLoS ONE 4(12)
30. Dehmer M, Barbarini N, Varmuza K, Graber A (2010) Novel topological descriptors for analyzing biological networks. BMC Struct Biol 10(1):18
31. Dehmer M, Emmert Streib F (2007) Structural similarity of directed universal hierarchical graphs: a low computational complexity approach. Appl Math Comput 194(1):7–20
32. Dehmer M, Emmert Streib F, Tsoy YR, Varmuza K (2010) Quantum Frontiers of atoms and molecules, chapter quantifying structural complexity of graphs: information measures in mathematical. Nova, 2010
33. Dehmer M, Grabner M, Varmuza K (2012) Information indices with high discriminative power for graphs. PLoS ONE 7(2):e31214
34. Dehmer M, Mehler A (2007) A new method of measuring similarity for a special class of directed graphs. Tatra Mountains Math Publ 36:39–59
35. Dehmer M, Mowshowitz A (2011) A history of graph entropy measures. Inf Sci 181(1):57–78
36. Dehmer M, Sivakumar L, Varmuza K (2012) Uniquely discriminating molecular structures using novel eigenvalue based descriptors. MATCH Commun Math Comput Chem 67(1):147–172
37. Dodge CW (2004) Euclidean geometry and transformations. Dover Publications, New York

38. Dreyfus SE (1969) An appraisal of some shortest-path algorithms. Oper Res, pp 395–412
39. Edgar R, Domrachev M, Lash AE (2002) Gene expression omnibus: NCBI gene expression and hybridization array data repository. Nucleic Acids Res 30(1):207–210
40. Emmert Streib F (2007) The Chronic Fatigue syndrome: a comparative pathway analysis. J Computl Biol 14(7):961–972
41. Emmert Streib F, Dehmer M (May 2011) Networks for systems biology: conceptual connection of data and function. IET Syst Biol 5(3):185–207
42. Emmert Streib F, Dehmer M, Kilian J (2005) Classification of large graphs by a local tree decomposition. Proc DMIN 5:20–23
43. Emmert Streib F, Glazko GV (2010) Network biology: a direct approach to study biological function. Wiley Interdisciplinary Reviews. Systems biology and medicine, Dec 2010, pp 1–27
44. Eppstein D (1995). Subgraph Isomorphism in Planar Graphs and Related Problems. In Proceedings of the Sixth Annual ACM-SIAM symposium on discrete algorithms. Society for Industrial and, Applied Mathematics, pp. 632–640
45. Eshera MA, FU K (1984) A graph distance measure for image analysis. IEEE Trans Syst, Man, and Cybern 14(3):398–408
46. Eshera MA, Fu KS (1986) An image understanding system using attributed symbolic representation and inexact graph-matching. Pattern Anal Mach Intell IEEE Trans 5:604–618
47. Feng J, Lurati L, Ouyang H, Robinson T, Wang Y, Yuan S, Young SS (2003) Predictive toxicology: benchmarking molecular descriptors and statistical methods. J Chem Inf Comput Sci 43(5):1463–1470
48. Flannick J, Novak A, Srinivasan BS, McAdams HH, Batzoglou S (2006) Graemlin: General and robust alignment of multiple large interaction networks. Genome Res 16(9):1169–1181
49. Gärtner T, Flach P, Wrobel S (2003) On graph Kernels: hardness results and efficient alternatives. Learn Theory Kernel Mach, pp 129–143
50. Guzmn Vargas L, Santilln M (2008) Comparative analysis of the transcription-factor gene regulatory networks of E. Coli and S. Cerevisiae. BMC Syst Biol 2:13
51. Hansen K, Mika S, Schroeter T, Sutter A, Ter Laak A, Steger Hartmann T, Heinrich N, Muller KR (2009) Benchmark data set for in Silico prediction of Ames mutagenicity. J Chem Inf Comput Sci 49(9):2077–2081
52. Harary F (1994) Graph theory. Perseus Books, Addison-Wesley, New York, Reading
53. Harris MA, Clark J, Ireland A, Lomax J, Ashburner M, Foulger R et al (2004) The Gene Ontology (GO) database and informatics resource. Nucleic Acids Res 32(Database issue):D258
54. Hart PE, Nilsson NJ, Raphael B (1968) A formal basis for the heuristic determination of minimum cost paths. IEEE Trans Syst Sci Cybern 4(2):100–107
55. Helma C, Cramer T, Kramer S, De Raedt L (2004) Data mining and machine learning techniques for the identification of mutagenicity inducing substructures and structure activity relationships of noncongeneric compounds. J Chem Inf Comput Sci 44(4):1402–1411
56. Hood L, Heath JR, Phelps ME, Lin B (2004) Systems biology and new technologies enable predictive and preventative medicine. Science 306(5696):640
57. Horváth T (2005) Cyclic pattern Kernels revisited. Adv Knowl Discov Data Min, pp 139–140
58. Horváth T, Gärtner T, Wrobel S (2004) Cyclic pattern Kernels for predictive graph mining. In: Proceedings of the Tenth ACM SIGKDD international conference on knowledge discovery and data mining. ACM, pp 158–167
59. Jeong H, Tombor B, Albert R, Oltvai ZN, Barabsi AL (Oct 2000) The large-scale organization of metabolic networks. Nature 407(6804):651–654
60. Kalaev M, Smoot M, Ideker T, Sharan R (2008) NetworkBLAST: comparative analysis of protein networks. Bioinformatics 24(4):594–596
61. Kashima H, Inokuchi A (2002) Kernels for graph classification, vol 2002. In: ICDM workshop on active mining, p 25

62. Kelley BP, Sharan R, Karp RM, Sittler T, Root DE, Stockwell BR, Ideker T (2003) Conserved pathways within Bacteria and Yeast as revealed by global protein network alignment. Sci STKE 100(20):11394
63. Kelley BP, Yuan B, Lewitter F, Sharan R, Stockwell BR, Ideker T (2004) PathBLAST: a tool for alignment of Protein interaction networks. Nucleic Acids Res 32(suppl 2):W83–W88
64. Kitano H (2002) Systems biology: a brief overview. Science 295:1662–1664
65. Koyutürk M, Grama A, Szpankowski W (2005) Pairwise local alignment of Protein interaction networks guided by models of evolution. In: Research in computational molecular biology. Springer, New York, pp 995–995
66. Koyutürk M, Kim Y, Topkara U, Subramaniam S, Szpankowski W, Grama A (2006) Pairwise alignment of protein interaction networks. J Comput Biol 13(2):182–199
67. Krause EF (1973) Taxicab geometry. Math Teach 66(8):695–706
68. Kuchaiev O, Milenković T, Memišević V, Hayes W, Pržulj N (2010) Topological network alignment uncovers biological function and phylogeny. J Royal Soc Interf 7(50):1341–1354
69. Kuchaiev O, Pržulj N (2011) Integrative network alignment reveals large regions of global network similarity in Yeast and Human. Bioinformatics 27(10):1390–1396
70. Kugler KG, Mueller LAJ, Graber A, Dehmer M (2011) Integrative network biology: graph prototyping for co-expression cancer networks. PLoS ONE 6(7):e22843
71. Kullback S, Leibler RA (1951) On information and sufficiency. Ann Math Stat 22(1):79–86
72. Lance GN, Williams WT (1966) Computer programs for hierarchical polythetic classification (similarity analyses). Comput J 9(1):60–64
73. Langfelder P, Horvath S (January 2008) WGCNA: an R package for weighted correlation network analysis. BMC Bioinformatics 9:559
74. Larrosa J, Valiente G (2002) Constraint satisfaction algorithms for graph pattern matching. Math Struct Comput Sci 12(4):403–422
75. Liao CS, Lu K, Baym M, Singh R, Berger B (2009) IsoRankN: spectral methods for global alignment of multiple protein networks. Bioinformatics 25(12):i253–i258
76. Liu H, Motoda H (1998) Feature selection for knowledge discovery and data mining. Kluwer Academic Publishers, Dordrecht
77. Mahé P, Vert JP (2009) Graph Kernels based on tree patterns for molecules. Mach Learn 75(1):3–35
78. Mazurie A, Bonchev D, Schwikowski B, Buck GA (2008) Phylogenetic distances are encoded in networks of interacting pathways. Bioinformatics 24(22):2579
79. McKay BD Nauty. http://cs.anu.edu.au/ bdm/nauty/
80. McKay BD (1981) Practical graph isomorphism. Congressus Numerantium 30:45–87
81. Menchetti S, Costa F, Frasconi P (2005) Weighted decomposition Kernels. In: Proceedings of the 22nd international conference on machine learning. ACM, pp 585–592
82. Messmer BT, Bunke H (1999) A decision tree approach to graph and subgraph isomorphism detection. Pattern Recogn 32(12):1979–1998
83. Messmer BT, Bunke H (2000) Efficient subgraph isomorphism detection: a decomposition approach. IEEE Trans Knowl Data Eng 12(2):307–323
84. Meyer PE, Kontos K, Lafitte F, Bontempi G (2007) Information-theoretic inference of large transcriptional regulatory networks. EURASIP J Bioinform Syst Biol 8–8:2007
85. Michoel T, De Smet R, Joshi A, Van de Peer Y, Marchal K (2009) Comparative analysis of module-based versus direct methods for reverse-engineering transcriptional regulatory networks. BMC Syst Biol 3(1):49
86. Milenkoviæ T, Pržulj N (2008) Uncovering biological network function via Graphlet degree signatures. Cancer Inform 6:257
87. Mowshowitz A (1968) Entropy and the complexity of the graphs I: an index of the relative complexity of a graph. Bull Math Biophys 30:175204
88. Mueller LAJ, Kugler KG, Dander A, Graber A, Dehmer M (2010) Network-based approach to classify disease stages of prostate cancer using quantitative network measures, vol I.

Conference on bioinformatics and computational biology (BIOCOMP'10), Las Vegas/USA, pp 55–61
89. Mueller LAJ, Kugler KG, Dander A, Graber A, Dehmer M (2011) QuACN: an R package for analyzing complex biological networks quantitatively. Bioinformatics 27(1):140
90. Mueller LAJ, Kugler KG, Graber A, Emmert Streib F, Dehmer M (2011) Structural measures for network biology using QuACN. BMC Bioinformatics 12(1):492
91. Mueller LAJ, Kugler KG, Netzer M, Graber A, Dehmer M (2011) A network-based approach to classify the three domains of life. Biology Direct 6(1):53
92. Neuhaus M, Riesen K, Bunke H (2006) Fast suboptimal algorithms for the computation of graph edit distance. In: Structural, syntactic, and statistical, pattern recognition, pp 163–172
93. Palsson B (2006) Systems biology: properties of reconstructed networks. Cambridge University Press, Cambridge
94. Parkinson H, Kapushesky M, Kolesnikov N, Rustici G, Shojatalab M, Abeygunawardena N, Berube H, Dylag M, Emam I, Farne A et al (2009) ArrayExpress update from an archive of functional genomics experiments to the Atlas of gene expression. Nucleic Acids Res 37(suppl 1):D868–D872
95. Passerini F, Severini S (2009) Quantifying complexity in networks: the von Neumann entropy. Int J Agent Technol Syst (IJATS) 1(4):58–67
96. Poston KL, Eidelberg D (2009) Network biomarkers for the diagnosis and treatment of movement disorders. Neurobiol Dis 35(2):141–147
97. Pržulj N (2007) Biological network comparison using graphlet degree distribution. Bioinformatics 23(2):e177–e183
98. Ralaivola L, Swamidass SJ, Saigo H, Baldi P (2005) Graph kernels for chemical informatics. Neural Networks 18(8):1093–1110
99. Riesen K, Bunke H (2009) Approximate graph edit distance computation by means of bipartite graph matching. Image Vis Comput 27(7):950–959
100. Ruan J, Dean AK, Zhang W (2010) A general co-expression network-based approach to gene expression analysis: comparison and applications. BMC Syst Biol 4(1):8
101. Rudolf M (2000) Utilizing constraint satisfaction techniques for efficient graph pattern matching. Theory Appl Graph Transform, pp 381–394
102. Rupp M, Schneider G (2010) Graph Kernels for molecular similarity. Mol Inform 29(4):266–273
103. Rupp M, Schneider P, Schneider G (2009) Distance phenomena in high-dimensional chemical descriptor spaces: consequences for similarity-based approaches. J Comput Chem 30(14):2285–2296
104. Sanfeliu A, King Sun F (1983) A distance measure between attributed relational graphs for pattern recognition. IEEE Trans Syst Man Cybern 13(3):353–362
105. Schölkopf B, Tsuda K, Vert JP (2004) Kernel methods in computational biology. Computational molecular biology. MIT Press, Cambridge
106. Scsibrany H, Karlovits M, Demuth W, Müller F, Varmuza K (2003) Clustering and similarity of chemical structures represented by binary substructure descriptors. Chemometr Intell Lab Syst 67(2):95–108
107. Serratosa F, Alquézar R, Sanfeliu A (2003) Function-described graphs for modelling objects represented by sets of attributed graphs. Pattern Recogn 36(3):781–798
108. Shapiro LG, Haralick RM (1981) Structural descriptions and inexact graph matching. IEEE Trans Pattern Anal Mach Intell 3:504–519
109. Sharan R, Ideker T (2006) Modeling cellular machinery through biological network comparison. Nat Biotechnol 24(4):427–433
110. Sharan R, Suthram S, Kelley RM, Kuhn T, McCuine S, Uetz P, Sittler T, Karp RM, Ideker T (2005) Conserved patterns of Protein interaction in multiple species. Proc Natl Acad Sci USA 102(6):1974
111. Shervashidze N, Borgwardt KM (2009) Fast subtree Kernels on graphs. Adv Neural Inf Proc Syst 22:1660–1668

112. Shervashidze N, Vishwanathan SVN, Petri T, Mehlhorn K, Borgwardt K (2009) Efficient graphlet Kernels for large graph comparison. In: Proceedings of the international workshop on artificial intelligence and statistics. Society for Artificial Intelligence and, Statistics, 2009
113. Shih YK, Parthasarathy S (2012) Scalable global alignment for multiple biological networks. BMC Bioinformatics 13(Suppl 3):S11
114. Singh R, Xu J, Berger B (2007) Pairwise global alignment of Protein interaction networks by matching neighborhood topology. In: Research in computational molecular biology. Springer, New York, pp 16–31
115. Singh R, Xu J, Berger B (2008) Global alignment of multiple Protein interaction networks with application to functional orthology detection. Proc Natl Acad Sci 105(35):12763
116. Smialowski P, Frishman D, Kramer S (2010) Pitfalls of supervised feature selection. Bioinformatics 26(3):440
117. Sobik F (1982) Graphmetriken und Klassifikation strukturierter Objekte. ZKI-Inf, Akad Wiss DDR 2:63–122
118. Todeschini R, Mannhold R (2002) Handbook of molecular descriptors. Wiley-VCH, Weinheim, Germany
119. Tsai WH, Fu KS (1979) Error-correcting isomorphisms of attributed relational graphs for pattern analysis. IEEE Trans Syst, Man Cybern 9(12):757–768
120. Tsai WH, Fu KS (1983) Subgraph error-correcting isomorphisms for syntactic pattern recognition. IEEE Trans Syst, Man Cybern 13(1):48–62
121. Ullmann JR (1976) An algorithm for subgraph isomorphism. J ACM (JACM) 23(1):31–42
122. Vapnik VN (2000) The nature of statistical learning theory. Springer, New York
123. Wang J, Provan G (2009) Characterizing the structural complexity of real-world complex networks. Complex Sci, pp 1178–1189
124. Watts DJ, Strogatz SH (1998) Collective dynamics of 'Small-world' networks. Nature 393:440–442
125. Wong EK (1990) Three-dimensional object recognition by attributed graphs. Syntactic Struct Pattern Recogn Theory Appl, pp 381–414
126. Xia K, Fu Z, Hou L, Han JDJ (2008) Impacts of Protein-Protein interaction domains on organism and network complexity. Genome Res 18(9):1500
127. Zaslavskiy M, Bach F, Vert JP (2009) Global alignment of Protein-Protein interaction networks by graph matching methods. Bioinformatics 25(12):i259–1267
128. Zelinka A (1975) On a certain distance between isomorphism classes of graphs. Čas Pěst Mat 100:371375

Chapter 3
Emergent Properties of Gene Regulatory Networks: Models and Data

Roberto Serra and Marco Villani

Abstract We emphasize here the importance of generic models of biological systems that aim at describing the features that are common to a wide class of systems, instead of studying in detail a specific subsystem in a specific cell type or organism. Among generic models of gene regulatory networks, Random Boolean networks (RBNs) are reviewed in depth, and it is shown that they can accurately describe some important experimental data, in particular the statistical properties of the perturbations of gene expression levels induced by the knock-out of a single gene. It is also shown that this kind of study may shed light on a candidate general dynamical property of biological systems. Several biologically plausible modifications of the original model are reviewed and discussed, and it is also show how RBNs can be applied to describe cell differentiation.

Keywords Dynamical systems · Emergent properties · Generic properties · Gene regulatory networks · Random boolean networks · Modeling · Gene knock-outs · Cell differentiation

Acronyms

RBN	Random Boolean Network
DNA	Deoxyribonucleic acid
RNA	Ribonucleic acid
miRNA	micro RNA (short RNA molecule)
mRNA	messenger RNA

R. Serra (✉) · M. Villani
Department of Physics, Computer Science and Mathematics, Modena and Reggio Emilia University, v. Campi 213b, 41125 Modena, Italy
e-mail: rserra@unimore.it

M. Villani
e-mail: marco.villani@unimore.it

R. Serra · M. Villani
European Centre for Living Technology, Ca'Minich, S. Marco 2940, 30124 Venezia, Italy

A. Prokop and B. Csukás (eds.), *Systems Biology*,
DOI: 10.1007/978-94-007-6803-1_3,

cDNA	complementary DNA
SFRBN	Scale-free Random Boolean Network
TES	Threshold Ergodic Sets

3.1 Introduction

Mathematical models are a key ingredient of systems biology, so there is no real need to point out their usefulness in a book like this. However, we wish here to discuss models that are quite unfamiliar to a large part of the systems biology community, and to describe their properties and their relevance. These models are called ''generic'',[1] as they that are not tailored to specific cases but are rather meant at exploring properties that are common to several different biological systems.

This is a quite unusual approach for "classical" biologists, whose discipline has learnt in the last centuries to be skeptical towards broad generalizations. This has been a safe attitude, but it is nowadays becoming an obstacle to the growth and development of new approaches, that are required by the wealth of data and facts that have been ascertained. While in the past it has probably been wise to escape generalizations that might have been based upon a weak evidence, today we face the opposite problem, i.e. the desperate need for general concepts, to make sense out of the genomic and x-omic data deluge.

If this is the case (and indeed it is!) then one can take the risk of trying some broad generalizations. One must of course not give up with scientific rigour (this would be a disaster) but rather adopt a more daring attitude, and our scientific community should spend some time and effort in exploring the merits and the limitations of broad generalizations.

Of course, there are general theories in biology, and Darwinian evolution, with its variations, has been at the center of the stage for more than a century. There are also other broad generalizations, like the cellular theory of the living, that have gained universal validity. And there are of course other well-known facts: for example, we know that the DNA-RNA–protein mechanism is the same in almost all living creatures, that 20 aminoacids make up the proteins of the vast majority of living beings, that most organisms use ATP as an energy source, etc.

But we need to enlarge our understanding of the general laws of the biosphere, in order to be able to understand what's really going on. Since this might seem quite abstract, let us mention here a concrete example. There are regularities that have been known since a long time, that take the form of scaling laws that relate various properties of living species to their body mass. The most famous one is a relationship between oxygen consumption rate r (a proxy for power supply) and

[1] A term borrowed from condensed matter physics.

body mass M, that has been known since the seminal work of Kleiber in the 30s [1] to follow a power-law behavior:

$$r \propto M^{\frac{3}{4}} \tag{3.1}$$

This "law" (let us call it like this, although it might look just an empirical regularity) is followed by organisms of very different sizes, spanning several orders of magnitude from mouse to elephant, and even of very different types (e.g. mammals and birds). More recently the validity of the "law" has been verified down to the level of cellular organelles, i.e. mitochondria, and even to the molecular level of the respiratory complex [2]. In this way the relationship spans a huge range of body masses, about 27 orders of magnitude—very few "real" physical laws have been verified on such a wide range!

Note that Eq. 3.1 is a really systemic one: if we take the cells out of the organism and cultivate them in vitro, the oxygen consumption rate turns out to be almost independent of the mass of the animal where they come from [2]. So the question is how can one explain such an astonishing regularity, given the very different physiological organizations of the involved organisms. A possible candidate might be the surface to volume ratio, but this would give an exponent close to 2/3, while the error on the exponent 3/4 is small, and allows one to exclude this explanation.

There have been, and there still are, several attempts at deriving the value 3/4 for different organisms, but the more general one, that applies down to the molecular level, deals with the number of dimensions of the space we inhabit, i.e. 3 (excluding possible extra-dimensions that would not be perceived on a macroscopic scale, and not even at the smallest scales so far explored by our technology) [3]. Few further hypotheses are necessary to derive the 3/4 exponent: that the oxygen distribution in the interior of the organism is achieved by a fractal network that uniformly fills the available space, that the internal volume is also fractal and that evolution has been able to optimize the value of the key parameters. While it is nowadays understood that evolution cannot be regarded *in toto* as an optimization process, it is also accepted that it can perform local optimizations, like the one that is proposed here. Under these assumption the derivation is quite straightforward: the dimension of the internal surface is between 2 and 3, that of the internal volume is between 3 and 4, so, if a and b are constants, subject to the constraints $0 \le a \le 1, 0 \le b \le 1, a + b \le 1$, then the dimension of the internal surface is $2 + a$ and that of the internal volume is $(2 + a) + (1 + b) = 3 + a + b$. Therefore optimizing distribution efficiency corresponds to finding the maximum value of the ratio $r = \frac{2+a}{3+a+b}$ subject to the above constraints on a and b. Now $\frac{\partial r}{\partial b} \le 0$, so r attains it maximum when b is minimum, i.e. it takes the value 0. We are left the task of maximizing $r^* = \frac{2+a}{3+a}$; since $\frac{\partial r^*}{\partial a} = \frac{3+a-(2+a)}{(3+a)^2} = \frac{1}{(3+a)^2}$ is positive, r* attains its maximum when $a = 1$, so $r^*_{\max} = 3/4$.

This is an enlightening example of application of theoretical thinking to real data, as it not only provides an explanation of the observed regularities, but also shows a deep and unexpected relationship with the physical properties of the space we inhabit. Note also that the above property can be considered emergent, as it appears during biological evolution. Emergence is a key and yet controversial notion, and we do not want to enter here a philosophical debate, but we stress that it is a useful concept to describe some key properties of complex systems.

Another very interesting emergent property that has been proposed regards the dynamics of living beings. This is a more general claim than the above one, and it has been raised by different authors in different contexts [4–6]: in synthesis, it hypothesizes that there are general dynamical properties that should be selected by biological evolution, and that states that are intermediate "between order and chaos" should be privileged.

The qualitative argument is that chaotic states are too fragile, since a small difference can lead the system to entirely different states. Therefore survival would be quite difficult in these regimes, while on the other hand regimes that are deeply ordered could be unable to modify their behavior in response to external changes, and moreover they might not be sufficiently evolvable to cope with long-term modifications of the living environment. Therefore critical states, at the edge of chaos, should be particularly abundant in living beings, or perhaps, as Kauffmann puts it [4, 7], also ordered states close to the boundary with chaotic ones might be selected for. We will refer to this as the "criticality hypothesis".

Note that this claim is still quite vague, and requires some clarifications. We will use below the term "dynamical system" to refer either to quantitative dynamical models, described by systems of (either differential or finite-difference) equations, or to the actual physical or biological system described by that equation set. The context will avoid ambiguities. In general, a deterministic, nonlinear dynamical system can be either ordered or chaotic depending upon the value of some parameters: in the former case the attractors are fixed points or cycles, in the second one they are fractal "strange" attractors. It is therefore possible to separate regions in parameters space where the attractors are ordered from those where they are disordered: the hypersurfaces that separate these regions define the boundaries between order and chaos, i.e. the critical parameter values. In mathematical models, these hypersurfaces are zero-measure sets in the parameter space, but it has also been proposed [8] that the application of this concept to real biological systems requires to interpret the notion of critical states in a looser sense, and to include those states whose parameters belong to a small, finite volume surrounding the hypersurfaces.

So the "criticality hypothesis" states that biological evolution should be able to tune the parameters in such a way as to get the parameters of biological systems in this critical region. However, it is unlikely that this should be the case for all biological systems: indeed, the argument in favor of critical states assumes a changing environment, and it would not hold for a constant one (in this latter case an ordered state might be the optimal choice). Therefore one should look for criticality in those subsystems whose dynamics is particularly important in order to endow the whole organism with the capability to withstand environmental changes—an obvious candidate being therefore gene regulatory networks, i.e. the main topic of this chapter.

In the following sections we will address the issue of modeling these networks, and we will soon focus on a class of models where the separating surfaces between ordered and chaotic states can be clearly defined. These models are based on strong simplifications of the regulation process, that make it possible to analyze their behaviors and to simulate them with reasonable resources. But the amenability to theoretical treatment is not the only desirable feature of a model, and it will also be seen that this model is able to properly describe some quantitative experimental data on disturbances in gene expression profiles following gene knock-out. Perhaps surprisingly, this study will also allow one to get information about the dynamical regime of the biological regulatory network, and therefore to test the criticality hypothesis.

Another approach to testing this hypothesis is possible, based on the analysis of time courses of gene expression values [9]. Both approaches have provided support to the criticality hypothesis, without however yet ruling out other possibilities, on the basis of existing data.

The models considered so far are generic, i.e. they do not embed specific hypotheses about particular organisms; actually, as it will be discussed, they strongly rely on randomness of connections and transition rules for their functioning. While it is interesting and perhaps even surprising to observe how far these models can go in interpreting experimental data, it is clear that some properties have to be tuned by biological evolution[2] and that evolved networks may present properties that are different from those of random ones. Indeed, evolved networks present very interesting properties [10]; however, since most studies of their evolution have been based on criteria that do not have a sound biological base, their behavior will not be discussed here.

In the following Sect. 3.2 the topic of modeling genetic regulatory networks will be introduced, and in Sect. 3.3 we will present the model that we will be most concerned with, i.e. that of random Boolean networks (RBNs for short), deferring to Sect. 3.5 an in-depth discussion of the hypotheses behind the model, and of possible modifications in order to cope with some of its limitations. In the meantime, in Sect. 3.4 we will present evidence that RBNs, in spite of their simplifications, are able to quantitatively describe experimental data, and to shed light on the criticality hypothesis. In Sect. 3.6 we will also consider the application of RBNs to describe a fascinating phenomenon, i.e. cell differentiation. In Sect. 3.7 we will further discuss the relationship between generic models, specific models and experiments.

[2] It might then appear surprising that a random network is able to describe the data on gene knock-out: this is actually due to the fact that the initial choice of the parameters concentrated on critical networks, that are the most widely studied in the literature.

3.2 Models of Gene Regulatory Networks

Gene regulation is accomplished through several interacting processes [11], that are largely but not completely known, that include among others the control

- of transcription (from DNA to mRNA)
- of mRNA processing, transport and degradation
- of aminoacid binding to t-RNA and transport
- of ribosome activity
- of protein 3-D folding.

Transcription is usually initiated by the binding (or the leaving) of proteins to portions of the DNA chain, and the interactions between proteins and DNA can be affected by the presence of small molecules that in turn interact with the proteins, modifying their 3-D shapes. Last but not least, also miRNAs can affect the mRNA processing. In eukaryotes, one has also to consider that DNA in chromosomes is tightly packed with histones, that make most part of the genome inaccessible to binding. Therefore it is also necessary to consider the processes that control the folding/unfolding of chromatin.

Note that the above is just a simplified summary of the major interactions involved in the regulation of gene expression. Therefore a detailed description would appear to be impossible, unless perhaps one focuses on the dynamics of one or very few genes. But this is not a viable option for studying the system-level properties of gene regulatory networks. However, a possible alternative approach is suggested by some models of physical phenomena, that are able to capture interesting (qualitative and quantitative) properties in spite of seemingly drastic simplifications. An important historical example concerns the equilibrium properties of perfect gases: as it is well known, their behavior is described by the law PV = nRT (P pressure, V volume, T absolute temperature, n number of moles and R is a constant), that was derived by observations of macroscopic samples of gas. But a law with the same functional form (i.e. the product PV proportional to the number of moles and to "something else") was also obtained by considering a molecular model of the gas. Identifying the two expressions led to the conclusion that absolute temperature is proportional to the average kinetic energy of the gas molecules, a major triumph of 19th century physics.

What is interesting here is that Boltzmann and others came to the right result by considering a completely unrealistic model of the gas molecules, that were treated like rigid spheres undergoing elastic collisions. Nowadays it is well known that this is not the case, they are complex structures whose interactions are ruled by quantum mechanics. And of course this more accurate description is necessary for a proper understanding of some properties, but not for deriving the ideal gas law. So the lesson is that even a very crude model can give enlightening hints concerning the behavior of a complex system.

This way of thinking is not yet widespread as it should. It goes without saying that knowing in detail the behavior of the elements (e.g. the molecules, or the genes) can be very helpful, but it is likewise true that there are properties that are indifferent, at least within approximations, to the microscopic details, as they come out of the features of the interactions, and may hold for systems made of very different kinds of elements. Also the example quoted in the introduction, i.e. the scaling law for oxygen consumption rate, is an example of this type of indifference to the microbehavior (indeed the exponent is the same for very different kinds of animals, irrespective of their respiratory physiological differences).

There are also other examples of systems properties that do not depend upon the correctness of the interaction laws, like for example the FHP model of lattice gases [12, 13], which shows phenomena typical of fluid mechanics, like the formation of eddies of the von Karman type in the flow beyond an obstacle, although in the model the fictitious particles that represent the fluid undergo inertial motion and elastic scattering on a hexagonal grid—certainly, not a realistic description of flowing water!

Encouraged by these success stories, one can try to adopt drastic simplifications also in modeling gene regulation.

There are indeed several types of models. First of all, gene expression level means essentially the concentration of the corresponding mRNA or rather of the corresponding protein. It is this latter variable that determines the effect on other genes, but it is the former that represents the direct output of gene activation. Both concentrations are continuous variables, although in some cases the number of molecules per cell is so small that integer variables would be better suited than real variables. But there are also Boolean (i.e. two-level) models that neglect the differences in expression levels and simply choose to describe whether the gene is active or not. It has already been pointed out there is no hope of describing with some details for systems composed of several genes. This is not only, and actually not so much, a matter of computing power, but rather a matter of introducing too many parameters, therefore causing a combinatorial explosion of possible alternative sets of values.

Therefore in this review we will privilege the Boolean approach, and in particular we will consider in detail a class of models that are often called Random Boolean Networks. They have been proposed more than 40 years ago by Stuart Kauffman, and have become quite popular in the 80s, when the interest for complex dynamical systems became widespread. Indeed, together with neural networks, they were one of the few available classes of high-dimensional nonlinear dynamical systems that could be applied to interesting biological phenomena, or to artificial devices that mimicked some aspects of biology. While neural networks undergo an evolutionary process that shapes their attractors to perform useful tasks, RBNs just display a dynamical repertoire that is richer than that of (most) neural nets. It is indeed possible (see Sect. 3.3 for a more precise description) to identify ordered and disordered regions, and to precisely locate the critical boundary in parameter space. Moreover, even when subject to external influences, RBNs display a robustness of behaviors that nicely illustrated notions like autonomy and eigenbehaviors. Therefore the model became one of the favorite workhorses of complex system theorists, so its dynamical properties were studied in depth, and several variants were introduced.

At the same time Kauffman was able to show that some scaling properties of the model resembled those found in nature, thereby raising the hope that they might shed light on some properties of natural systems. However the evidence was quite vague and research in RBNs gradually faded, as the abstract properties had already been extensively analyzed. However, in the first decade of this century new results became available concerning the simultaneous expression levels of thousands of genes, so it was possible to compare the behavior of these networks with experimental data, showing that RBNs can be a useful tool for interpreting the latter. At about the same time new theorists became interested in RBNs, so we are observing in these last year a growing number of papers and a growing attention towards this model, in spite of its age.

3.3 Random Boolean Networks

Here below a synthetic description of the model main properties is presented, referring the reader to [14, 4] for a more detailed account. A classical RBN is a dynamical system composed of N genes, or nodes, which can take either the value 0 (inactive) or 1 (active). Let $x_i(t) \in \{0,1\}$ be the activation value of node *i* at time *t*, and let $X(t) = [x_1(t), x_2(t) \dots x_N(t)]$ be the vector of activation values of all the genes. As previously reported, real genes influence each other through their corresponding products and through the interaction of these products with other chemicals, by promoting or inhibiting the activation of target genes.

In the corresponding model network these relationships are represented by directed links (directed from node A to node B, if the product of gene A influences the activation of gene B) and Boolean functions, which model the response of each node to the values of its input nodes. In a classical RBN each node has the same number of incoming connections $\boldsymbol{k_{in}}$, and its $\boldsymbol{k_{in}}$ input nodes are chosen at random with uniform probability among the remaining N-1 nodes: in such a way the outgoing connectivities have a Poissonian distribution. Other widespread RBN versions release the sharp choice of a fixed number of incoming connections and allow Gaussian distributions with relatively small standard deviations: this modification yet doesn't change the Poissonian distribution of outgoing connectivities nor its dynamical behavior.

The Boolean functions can be chosen in two different ways: (1) at random for every node, by assigning to each set of input values the outcome 1 with probability π (sometimes also called the bias) or (2) at random from a predefined set of allowed transition functions with probability p. These two procedures can have similar outcomes (for example when the choice is uniform among all possible Boolean functions and $\pi = 0.5$) but could also create different nets (for example when the choice is among only a subset of the possible Boolean functions, chosen to have in average π equal to 0.5).

In the so-called *quenched* model, both the topology and the Boolean function associated to each node do not change in time. The network dynamics is discrete and synchronous, so fixed points and cycles are the only possible asymptotic states in finite networks (a single RBN can have, and usually has, more than one attractor). The model shows two main dynamical regimes, ordered and disordered, depending upon the value of the connectivity and upon the Boolean functions: typically, the average cycle length grows as a power law with the number of nodes N in the ordered region and diverges exponentially in the disordered region [4]. The dynamically disordered region also shows sensitive dependence upon the initial conditions,[3] not observed in the ordered one. RBNs temporal evolution undergoes a phase transition between order and disorder, the critical value of the connectivity k_{in_c} being given by: $k_{in\_c} = [2\pi\ (1 - \pi)]^{-1}$, [15]. This formula, which refers to an ensemble of networks rather than to a single finite realization, defines what has sometimes been referred to as the "edge of chaos" [16].

It should be mentioned that some interesting analytical results have been obtained by the so-called annealed approach, in which the topology and the Boolean functions associated to the nodes change at each step. Several results for annealed nets hold also for the corresponding ensembles of quenched networks—as for example the link between k_{in_c} and π_c. Although the annealed approximation may be useful for analytical investigations [14] in this work we will always be concerned quenched RBNs, that are closer to real gene regulatory networks.

Several works have addressed the issue of the scaling of the number of attractors in RBNs with the number of its nodes [4, 17, 18, 19]; initial claims that in critical networks the former grows as a power law of the latter (with exponent <1) were later replaced by a more accurate description, so today one knows that the scaling is higher than polynomial in the limit of infinitely large networks. Nonetheless, in simulations of large but finite networks one actually observes a power-law scaling; this is due to the inevitable undersampling of the set of initial conditions but, on the other hand, it can be argued that it is the number of attractors with a significant basin of attraction that matters for modeling finite systems—and this does actually scale as a power law.

Systems along the critical line separating ordered and disordered regions show equilibrium between robustness and adaptiveness [20]; for this reason they are supposed to be reasonable models of the living systems organization. In addition, recent results support the hypothesis that biological genetic regulatory networks operate close to the critical region (see Sect. 3.4 below) [9, 21, 22].

A very important aspect concerns how to determine and measure the RBNs' dynamical regime. The main static methods to measure the RBN dynamical regimes in fact implicitly presume ergodicity, that is, all inputs can arise with the same probability during evolution.

An example is the "sensitivity", proposed by Shmulevich and Kauffman [23] and based on the average probability of unit i to spread an incoming perturbation to its

[3] As it was observed, the asymptotic states of finite RBNs are cycles of finite length (a fixed point being a cycle of length 1), so no real chaotic dynamics is possible; however, due to the sensitive dependence upon initial conditions, the disordered region is also often termed "chaotic".

neighbors. If this average is lower than 1 the perturbation will tend to disappear (and the system is ordered), whereas if its value is higher than 1 the perturbation will tend to invade the whole system (disordered systems): the sensitivity of critical systems therefore averages to 1. In order to compute this average however Shmulevich and Kauffman suppose that all the inputs values of the Boolean functions have the same occurrence probability, a fact that is not guaranteed except that in the case of a purely random system. In fact RBN are dissipative systems, and starting from initial conditions RBNs quickly converge to their asymptotic behaviors where they remain: in this situation many input combinations are absent and therefore don't contribute to the typical dynamical behavior. It is possible to avoid this misalignment by using during the computation the correct input distributions [24, 25], explicitly introducing in such a way the dynamics into this measure.

The alternative to the static measures is that of directly exploring the dynamical behavior of the system: in particular, an interesting and well-known method directly measures the spreading of perturbations through the network. This measure involves two parallel runs of the same system, having the initial states different for only a small fraction of the units. This difference is usually measured by means of the Hamming distance h(t), defined as the number of units that have different activations on the two runs at the same time step (the measure is performed on many different initial condition realizations, so one actually considers the average value $\langle h(t) \rangle$, but we will omit below the somewhat pedantic brackets). If after a transient the two runs are likely to converge to the same state, i.e. $h(t) \rightarrow 0$, then the dynamics of the system is robust with respect to small perturbations (a signature of the ordered regime).

Therefore, a system is ordered when:

$$\lambda = \lim_{h(t) \rightarrow 0} \frac{dh(t+1)}{dh(t)} < 1 \tag{3.2}$$

whereas it is critical or disordered if λ (sometimes called the Derrida parameter) is respectively equal to or greater than 1 [14].

Following this idea, a common practice to measure the dynamical regime of a RBN is that of randomly generating a great number of pairs of initial conditions differing each other for one or more units, perform one step, measure the Hamming distance of the two resulting states, take the averages for each perturbation size, and compute the limit of the slope of the tangent of the curve as the perturbation size tends to zero (the so-called Derrida procedure [15]).

However, RBNs spend most time in their asymptotic states (the attractors): measures taken on randomly chosen states therefore do not necessarily allow a correct estimate of the effective system dynamical behavior. One can then propose the sensitivity on attractor *i* (SA_i) as the result of the Derrida procedure performed only on the states belonging to the attractor i, and the attractors sensitivity (SA) as the average of the SA_i, each SA_i being weighted according to the size of its attraction basin. This last measure indeed provides a more meaningful picture of the system dynamical regime. Near the "edge of chaos" region, static sensitivity,

Derrida parameter and attractors sensitivity tend to coincide; note however that in single realizations the three measures can differ, and that their difference becomes significant (also on average) when distant from this area [25].

In the classical model of RBNs, Boolean functions are either chosen at random among all those which are possible or generated by the random procedure described above with bias π. However, a detailed study of tens of actual genetic control circuits that have been analyzed and interpreted according to a Boolean logic [26] has shown that in real biological systems only canalizing functions are found in the great majority of the cases: a function is said to be canalizing if there is at least one value of one of its inputs that uniquely determines the output [26].

Other biologically plausible sets of Boolean functions are

(a) separable functions, whose output is determined by comparing a weighted sum of their inputs with a threshold [27]
(b) coherent functions, in the sense that a single gene either favors or opposes the expression of another gene located downstream—acts as activator or inhibitor—but not play both roles for the same downstream gene [28].

It is possible to observe that the three sets of canalizing, coherent and separable functions coincide in the case of two input connections, where only two out of 16 functions are excluded (namely the XOR and its negation). In the case $\boldsymbol{k_{in}} = 3$ one finds that the sets of coherent and separable functions are identical, while that of canalyzing functions is different. However, for all $\boldsymbol{k_{in}} > 3$ the three sets don't coincide [28]. The network dynamics is therefore affected by the choice of the set of Boolean functions, and consequently it is important to precisely specify how they are chosen.

3.4 Interpretation of Experimental Data

A bold hypothesis was put forth by Kauffman since his first papers, namely that the different attractors of genetic network models should be associated to different cell types [4, 7, 29]. While this might be surprising at first sight, it turns out to be fully reasonable if one takes into account that cell types (excluding those involved in sexual reproduction) share the same genome but differ in the expression levels of their genes, so that both dynamical attractors and cell types describe different coherent patterns of expression of the same genome.

On the basis of this identification, he noticed a similarity between the power-law scaling of the number of attractors with the number of genes in critical networks on one side,[4] and that of the number of cell types with respect to the total DNA content of different animal species on the other. In both cases an

[4] Remember that, as it was observed in Sect. 3.3, the number of attractors with a significant attraction basin grows in this way.

approximate power-law increase with an exponent close to 1/2 was observed. However, more recent discoveries about the genome have questioned the supposed proportionality of the number of genes to the total genome size, so this argument has lost some of its appeal.

While technological progress, having led to a more complex view of the genome, has cast doubts on the first attempt to relate the properties of RBNs to those of real biological organisms, technology has also opened new ways to test the suitability of the model to describe experimental results. In particular, DNA-microarrays allow to simultaneously determine the expression pattern of a huge number of genes, and comparison of the statistical features of these expression patterns with those of mathematical models becomes possible.

A particularly useful set of data comes from cDNA microarray measurements of gene expression profiles of *Saccharomyces cerevisiae* subject to the knock-out of genes, one at a time [30]. In a typical experiment, one compares the expression levels of all the genes in cells with a knocked-out gene, to those in normal ("wild type") cells. In this way, all the experimental data can be cast in matrix form E_{ij}, where E_{ij} is the ratio of the expression of gene i in experiment j to the expression of gene i in the wild type cell. In the case examined, there were more than 6,300 genes and about 230 experiments, i.e. a remarkably wide data set.

In order to compare these measured continuous data with those obtained in simulations of Boolean models it is necessary to "binarize" the experimental data, i.e. to ascertain whether a gene has changed its expression or not. Microarray data are noisy, therefore statistical methods, like e.g. p-values, are usually applied to determine whether the change in expression level is to be considered meaningful, with respect to a "null hypothesis" that the two levels are the same. This is a sensible method to find out those genes that are good candidates for being associated to the differences, but it tends to privilege the null hypothesis: two genes are regarded as "different" only if they pass a severe scrutiny. But this is not a good method to determine how many genes have been perturbed, in this case one should not favor one alternative with respect to the other. Therefore one can simply define a threshold level, such that the difference is regarded as "meaningful" if the ratio of expression levels is greater than the threshold θ (or smaller than $1/\theta$) and neglected otherwise. The choice of the threshold is essentially the only free parameter involved in this study.

Let Y be the Boolean matrix which can be obtained by E by posing $y_{ij} = 1$ if $E_{ij} > \theta$, or $E_{ij} < 1/\theta$; $y_{ij} = 0$ otherwise ($y_{ij} = 1$ therefore means that the modification of the expression level of gene i in experiment j is accepted as "meaningful"). In order to describe the global features of these experiments, two important aggregate variables are (a) the avalanche, defined as the size of the perturbation induced by a particular experiment (in experiment j, $V_j = \Sigma_i y_{ij}$), and (b) its dual quantity, the susceptibility (the susceptibility of gene i, S_i, is equal to the number of experiments where that gene has been significantly affected: $S_i = \Sigma_j y_{ij}$).

It is possible to compare the experimental data concerning the matrix V with those obtained in model RBNs, where the original random network (RBNw) corresponds to the wild type cell, while the knocked-out cell is simulated by a network (RBNk) that is identical to the previous one, except for the fact that a single gene of RBNw is permanently silenced. The reader is referred for details to the original papers [31, 32]. The initial simulations were performed with a "classical" RBNw with exactly two input links for each node, coming from two genes chosen at random with uniform probability among the remaining $N - 1$ genes. All the 16 Boolean functions of two inputs were allowed and they were chosen at random with the same probability. The number of nodes was equal to that of the matrix of experimental data, larger than 6,000.

The choice of the parameters ($k_{in} = 2$, all the Boolean functions) had no particular reason, it was just the most widely studied model. Therefore it turned out particularly surprising to find out that the model was able to reproduce the statistical features of the experiment quite well, in particular the frequency of avalanches of various sizes was quite similar for synthetic RBNs and for actual genetic networks of *S. cerevisiae.*

As usual, critical networks displayed a high variance, so some experiments were also performed limiting the set of Boolean functions to those that are canalyzing, on the ground of their higher biological plausibility (see Sect. 3.3): in this case the variance of the simulated data was smaller and the agreement was also very good. The distribution of susceptibilities are also well approximated (see Fig. 3.1).

These results therefore provide support to the claim that strongly simplified models like RBNs can provide useful information on the behavior of real organisms. But they also open up a new deep question: why does all this seem to work? In biological modeling one almost never hits the target on the first strike, moreover the model had exactly two input links per gene, and it is well known that the

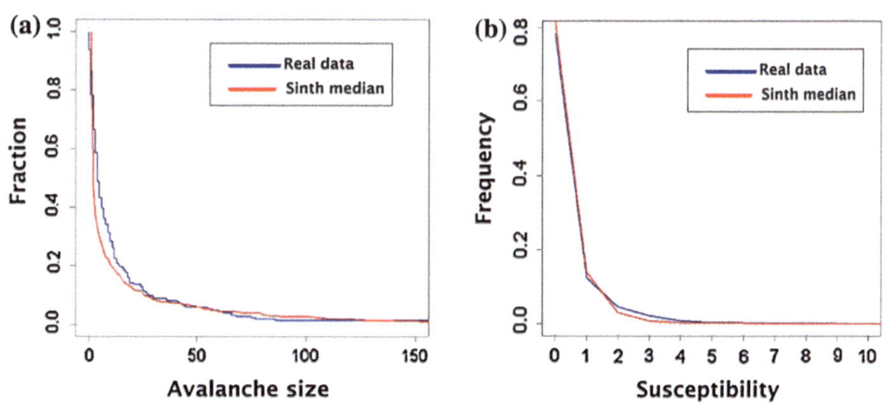

Fig. 3.1 Comparison between avalanches in *S. cerevisiae* (a gene being involved if its expression changes by a factor higher than 6—or lower than 1/6) and the median distribution of 10 synthetic networks with only canalizing boolean functions: **a** avalanche size and **b** gene susceptibilities (the number of experiments where each gene has been significantly affected by an avalanche)

distribution of links in *S. cerevisiae* is not so regular [33]. It is therefore interesting to try to understand the reasons for such an agreement.

Let us anticipate that such a good agreement is ultimately due to the fact that the synthetic networks used are critical, or close to criticality. And, if the Kauffman hypothesis holds, also real biological organisms have been driven by evolution close to a critical state.

That criticality explains the agreement can be shown by an approximate analytical calculation. One can indeed compare what happens in the RBNw and RBNk: at the beginning, a single node (that is, the knocked-out one, that will also be called the root of the perturbation) will differ in the two cases, so the size of the initial avalanche will be 1. If, at the next time step, no one of the nodes that receive input from the root changes its value, then the avalanche stops there and it will turn out to be of size 1.

Therefore one can compute p_1, i.e. the probability that an avalanche has size 1. Let q be the probability that a node chosen at random changes its value if one (and only one) of its inputs changes its value; p_1 is then the probability that all the output nodes of the root do not change, and if there are k outgoing connections, this probability is q^k; therefore, integrating over the outgoing distribution:

$$p_1 = \sum_{k=0}^{N-1} p_{out}(k) q^k \tag{3.3}$$

where $p_{out}(k)$ is the probability that a node chosen at random has *k* outgoing connections.

As far as larger avalanches are concerned, one can limit the study to the case of large sparse networks with (on average) a few connections per node, as is the case for *S. cerevisiae*; therefore the probability that an output node of the root is also one of its input nodes is negligible. In this case the probability that an avalanche has size 2 equals the probability that only one of the output nodes of the root (i.e. a node at level 1) changes its value, and that the perturbation does not propagate downwards to level 2 (i.e. that nodes which receive connections from the affected node do not change their value).

Therefore:

$$p_2 = \sum_{k=0}^{N-1} k p_{out}(k) q^{k-1}(1-q) \sum_{m=0}^{N-k-1} p_{out}(m) q^m \tag{3.4}$$

By applying the same reasoning, one can continue and compute the probability of avalanches of increasing size. Of course, calculations become more and more cumbersome, as the same size can be achieved in different ways (for example, an avalanche of size 3 may be composed by the root and by two nodes at level 2, none at level 3, or by the root, one node at level 2 and one at level 3).

It is however possible to show that every p_m can be written as a function of the moment generating function defined as

$$F = \sum_{k=0}^{N-1} q^m p_{out}(m) \tag{3.5}$$

and of its derivatives. Indeed p_1 directly coincides with F (see Eq. 3.3); noting that $\frac{\partial F}{\partial q} = \sum_{k=0}^{N-1} p_{out}(k) k q^{k-1}$, one can show that p_2 (Eq. 3.4) can be written as

$$p_2 = (1-q) F \frac{\partial F}{\partial q} \tag{3.6}$$

In the same way it can be shown [22] that also the higher order terms can be expressed as functions of F and its derivatives.

This fact has an immediate consequence: under the assumptions made, the moment generating function, that determines the distribution of avalanches, depends only upon the outdegree distribution: that is why the assumption of exactly 2 inputs per node does not affect the validity of the agreement with experimental data.

One can move one step further by taking into account the fact that the outdegree distribution in the (classical) model networks is Poissonian:

$$p_{out}(k) = e^{-A} \frac{A^k}{k!} \tag{3.7}$$

where A = <k> (note that the average of the number of ingoing connections equals that of the outgoing connections, so there is no need to specify). In this case Eq. 3.5 becomes

$$F = \sum_{k=0}^{N-1} q^k e^{-A} \frac{A^k}{k!} \cong \sum_{k=0}^{\infty} q^k e^{-A} \frac{A^k}{k!} = e^{-A} e^{qA} \tag{3.8}$$

therefore, introducing the variable $\lambda = \ln(1/F)$:

$$\begin{aligned} \lambda &= (1-q)A \\ F &= e^{-\lambda} \\ P_n &= B_n \lambda^{n-1} e^{n\lambda} \end{aligned} \tag{3.9}$$

From Eq. 3.9 one can observe that F, and therefore the avalanche distribution (the coefficient B_n depending only on the average branching of avalanches), depends only upon the parameter λ that is the product of two terms, i.e. [probability that a node changes value if one of its input changes] $\times$ [average number of connections per node].

Suppose now that one performs a transient flip on a node of the network RBNw. Under the above assumptions (i.e. that a node has a negligible probability to be perturbed by nodes placed lower levels in the avalanche tree) the situation is the same as that of the avalanche experiment. And indeed λ coincides with the Derrida exponent.

This is perhaps the result that has the deepest meaning: the avalanche distribution turns out to depend upon the same parameter that defines the dynamical regime, therefore one can infer information on the dynamical regime from data about the avalanche distribution. And so one can test the hypothesis that living beings are "at the edge" of chaos by estimating the dynamical parameter λ from the distribution of avalanches.

As discussed in Sect. 3.3 there are various measures of "sensitivity": static sensitivity, Derrida parameter (computed from random initial conditions) and attractor sensitivity (computed starting from attractor states). If one uses a RBN with $\boldsymbol{k_{in}} = 2$ and all the Boolean functions, these various definitions coincide and $\lambda = 1$: since the distribution of avalanches is close to the observed one, the biological network might actually be critical (although further work is needed to verify this hypothesis).

Indeed, the data from this single experiment are not conclusive: as it was mentioned, also networks with only the 14 canalyzing functions are able to reproduce the distribution of avalanches. In this case the networks are more ordered from a dynamical viewpoint, due to the absence of the XOR and EQUAL functions, while the average ratio of 0's to 1's in the truth table is still 1/2. The static sensitivity then coincides with the attractor sensitivity and turns out to be 6/7, i.e. slightly subcritical—and therefore still compatible with the extended "criticality hypothesis".

A different, independent type of data analysis on the distribution of the same avalanches has also led to estimate $\lambda = 1$ [34]. A further indication (fully compatible with the findings described above) that cells might operate in an ordered or critical state comes from the study of time-synchronized HeLa cells performed by Schmulevich et al. [9].

3.5 Beyond Classical RBNs

Let us now consider with some detail the simplifications that have been introduced in the RBN model. The most obvious one is the use of Boolean values for gene activation, that correspond to concentrations of mRNA or proteins. It is well known that different genes can display very different activation levels, and this is ignored in all the Boolean models. Moreover, the use of Boolean variables imposes the thresholding procedure on DNA microarray data described in Sect. 3.4, that is largely empirical and lacks firm theoretical grounding.

An obvious remedy could be that of using continuous [35] or multiple-valued models [36]. While these models are definitely interesting, their application to networks composed by very many genes is dubious: not only for reasons of computer resources needed for their simulations, but also because of the arbitrariness in the definition of the maximum activation levels (that are known only for a subset of genes). There is also a class of models, often referred to as Glass networks [37], that are in a sense intermediate between continuous and discrete:

here the activation of a gene is a continuous variable, but there are spikes that sometimes bring it to its maximum value, that is the same for every node, while the subsequent decrease is described by a differential equation. In its simplest form, this is a linear equation, so the decrease in activation is exponential and can be analytically computed, thus avoiding the burden of integrating a set of coupled differential equations and allowing the simulation of large networks. Glass networks also inspired more sophisticated dynamical models, where deactivation follows a nonlinear law [38, 39]; in this case however the simplifications of linear decay are lost, and only small networks can be dealt with.

It is interesting to observe that the binarized distribution of avalanches in Glass networks is essentially the same as that of the classical RBNs described in Sect. 3.4 [40]. Indeed, the behavior of Glass models turns out to be quite similar to that of RBNs in some respects, while they also allow, by using different decay constants, to describe the different degradation rates of various proteins.

Indeed, the most stringent limitation of the classical RBN model seems to be its synchronous updating, that requires that all the activations at time $t-1$ be forgotten when computing those at time $t+1$ (that depend only upon the values at time t). This updating procedure could lead to spurious synchrony, that in turn could introduce attractors not stable with other updating schemes [41]; it seems nevertheless that updating schemes don't heavily influence the RBN dynamical regimes [42, 43, 44, 45].

The major biological assumption of synchronous updating regards the fact that all the proteins synthesized two steps ago have faded, but it is known that proteins may have very different life spans. In order to assure the loss of memory of states prior to X(t) one should then define a "long" time step Δt, of the order of that of the longer living proteins, but this would mean that it becomes impossible to follow shorter term dynamics, that are likely to play a major role in cell processes.

This problem cannot be overcome by resorting to asynchronous updating, as it is usually done in condensed matter physics, because in this case the matters would become even worse, as the longest living protein would again determine the Δt. Moreover, even using Δt's close to the average life span of a protein, asynchronous updating would require unrealistic long times to update a set of some thousands of nodes.

A direct attack to the problem of the presence of different decay constants for the proteins has involved their explicit introduction in the model [46]. In this case, a Boolean model is still adopted, but there are now two kinds of nodes, that correspond to genes and proteins (or other "gene products" that may affect activation). Proteins are described by two variables: a Boolean one, that determines whether the protein is present or not, and a discrete one, that is a kind of "clock" that measures how much time has elapsed since the last moment when that protein was synthesized. Each protein has a fixed life span, chosen randomly with uniform probability up to a maximum decay time and when the clock reaches this limit value the activation is set to 0 (unless of course that protein is synthesized again at exactly that time). A gene is activated or inhibited by proteins, according as usual to a Boolean function; if the gene is activated, then the corresponding protein is synthesized and the clock is reset.

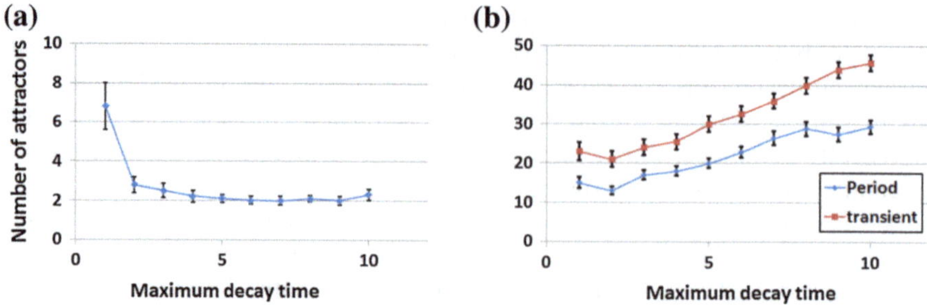

Fig. 3.2 Gene-protein model: variation of (**a**) the average number of different attractors and (**b**) the average length of the periods of the attractors and the average length of the transients for ensembles of networks characterized by a different value of the maximum decay time MDT. All the networks are designed with $\boldsymbol{k_{in}} = 2$ and $\pi = 0.5$; averages are taken over 100 different networks, and for each of them we built 10 different realizations, characterized by a different MDT, ranging from 1 to 10

Using a model like this amounts to endowing the system with a longer memory, and it is thus possible to investigate how the dynamics is affected by the duration of this memory. Several tests have been performed showing interesting effects, summarized in Fig. 3.2 and described in detail in [46, 47]. It is worthwhile to notice that while memory affects the number of attractors and their lengths, the distribution of avalanches turns out to be very close to that of the classical RBNs, and therefore also to the experimental data.

Another limitation of the classical RBN model is that of using an equal number of incoming connections for every node, a simplification probably linked to the limits of the computing resources of the heroic initial times. However it is not difficult to extend the model to include a distribution of in-degrees, and this has been done showing that the dynamics is similar to that of the model with a constant $\boldsymbol{k_{in}}$.

Another interesting variation concerns the topology of the regulatory network. At the time of the inception of the RBN model, there were basically two network topologies that were widely known and studied: random networks (of the Erdos-Renyi type) and regular networks, like e.g. cellular automata. And so the classical model is a blend of the two: as a consequence of the random procedure for determining the input nodes to a given gene, it turns out that the outdegree distribution approximates a Poissonian one (a result that was used in Sect. 3.4 to derive the role of the Derrida parameter), like in Erdos-Renyi random networks. In more recent times a great interest has been raised by different topologies, in particular those of the "small world" [48] and those of the "scale-free" types [49, 50]. A growing body of experimental data suggests that approximate realizations of the latter are widespread in biological (but also in social) systems. It is therefore interesting to consider the behavior of random Boolean networks endowed with this type of topology.

The analysis of the dynamical properties of networks with a power-law topology has been pioneered by Aldana [51], who derived the equations governing the order–disorder boundary.

The well-known formula for a scale-free distribution of outgoing links is:

$$p_{out}(k) = \frac{1}{Z}k^{-\lambda} \tag{3.10}$$

$$Z(\gamma) = \sum_{k=1}^{k_{\max}} k^{-\gamma} \tag{3.11}$$

where k can take values from 1 to a maximum possible value k_{max} (if self-coupling and multiple connections are prohibited, $k_{max} = N - 1$). Z (which coincides with the Riemann zeta function in the limit $k_{max} \to \infty$) guarantees the proper normalization. Like RBNs, also such scale-free networks show two regimes, an ordered and a disordered one, separated by a curve which determines the critical slope of the exponent γ [51]. The average value of k is:

$$\langle k \rangle \equiv \sum_{k=1}^{k_{\max}} kp(k) \cong \sum_{k=1}^{\infty} kp(k) = \frac{1}{Z(\gamma)} \sum_{k=1}^{\infty} k^{-\gamma+1} = \frac{Z(\gamma-1)}{Z(\gamma)} \tag{3.12}$$

The condition for a critical network with $\boldsymbol{k_{in}} = 0.5$ is that the last term on the r.h.s. of Eq. 3.12 be equal to 2 [51].

Comparing two different types of networks requires that the conditions for the comparison be precisely defined. To this aim, a study was performed on networks that differ only in the topology of the outgoing connections: a classical RBN and a network where all the nodes have the same indegree, but where the outgoing connections are determined by a modified Barabasi-Albert procedure, giving rise to a power-law distribution of outdegrees. In this way, the only difference is the topology, while all the other parameters (including the Boolean functions) are the same. The results are quite impressive: the scale-free networks are much more ordered, the number and length of their attractors increase with the number of nodes much less than the classical ones, and also the duration of the transients is much shorter [32] (Fig. 3.3).

One might wonder if scale-free RBNs (shortly, SFRBNs) provide different results concerning the distribution of avalanches. Again, answering to this question requires that the condition of the comparison, and therefore the way in which the SFRBN is prepared, be precisely defined.

Since, as discussed in Sect. 3.4, the distribution of avalanches does not depend on the indegree distribution, the SFRBNs can be generated keeping the number of ingoing connections equal to two for each node, exactly like the RBNs [52].

In order to compare RBNs and SFRBNs, it should be stressed that the synthetic RBNs discussed in the previous section indeed have some nodes without outgoing links (a feature likely to hold also for real genetic networks). So, in order to analyze the effects of changes of the form of the distribution of outdegrees, it is necessary to extend Eq. 3.10 to the case $k = 0$. Of course, a direct extension would lead to a meaningless divergence. The simplest generalization of Eq. 3.10 capable to include the value to $k = 0$ is then:

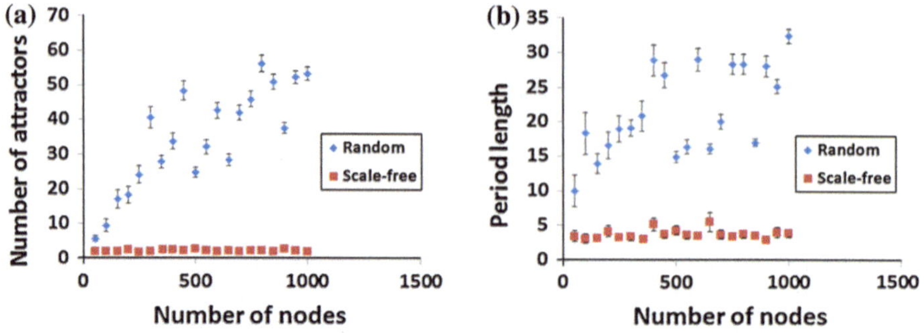

Fig. 3.3 Comparison between Boolean networks with random and scale-free topology: **a** average number of attractors and **b** average period length. Each point represent the average among 50 networks, each net being tested with 200 different initial conditions

$$\begin{cases} p_{out}(k) = \frac{1}{Z'} k^{-\gamma} & \textit{if } k \neq 0 \\ p_{out}(0) = p_0 \end{cases} \tag{3.13}$$

In the following we will refer also to the distribution given by Eq. 3.13 as a "scale-free" distribution. The normalization coefficient is now Z':

$$Z' = \frac{\sum_{k=1}^{k_{\max}} k^{-\gamma}}{1 - p_0} \tag{3.14}$$

Having said this, it is possible to analyze the distribution of avalanches in RBNs and in SFRBNs; for maintaining all the other parameters unchanged, the same threshold value (7) is used in both cases, and p_0 of Eq. 3.13 is set equal to the value of the fraction of nodes without outgoing connections in the RBN case. The results for the distribution of avalanches are slightly, but significantly different from those of the RBN case (Fig. 3.4).

The main remark concerns the presence, in the SFRBN case, of a larger fraction of smaller avalanches. This can be understood by observing that SFRBNs have some largely connected hubs: since the total number of links (2N) in the two networks is the same, this implies that in SFRBNs there will be more nodes with few connections, so the probability of getting a small avalanche increases. On the other hand, also the maximum avalanche is larger in the case of SFRBNs, because hitting hubs may lead to large avalanches. The agreement with experimental data is therefore better for RBNs, but this might be due to the way in which the comparison has been made (i.e. keeping the total number of links fixed). Moreover, when one performs the in silico knock-out only on a subset of 200–300 genes (the same number of the experiments), it is not infrequent to find values for the maximum avalanche that are much smaller than those obtained when knocking out all the 6,000 genes, and that are closer to those experimentally observed. Therefore, while RBNs appear to better describe the data, the jury is still out, waiting for further data to analyze.

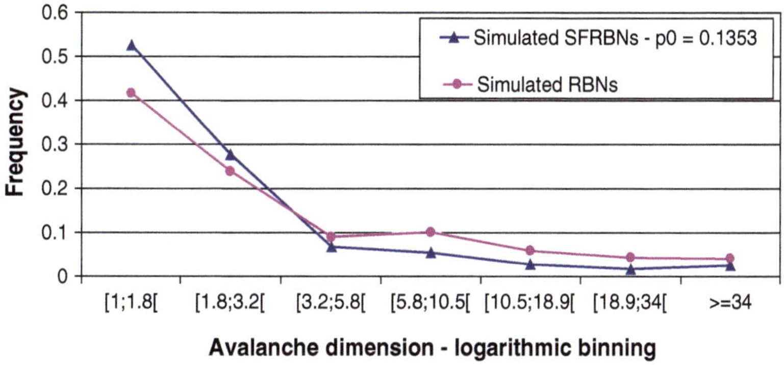

Fig. 3.4 Comparison between the distribution of the avalanches in classical RBNs (mean on 50 simulated networks) and SFRBNs with $p_{out}(0) = 0.1353$ (average taken over 10 simulated networks)—logarithmic binning

Actual living beings derive their structure (including the topology of their gene regulatory network) from a long evolutionary history: by using RBN models it is possible to duplicate some aspects of this process (for example, gene duplication and the subsequent gene divergence), in order to analyze their robustness and evolvability properties [20, 53]. Interestingly, these mechanisms lead to gene regulatory networks having scale-free topologies [54, 55]. Robustness properties are related to topological properties [56], dynamical regimes [57] and redundancy [58].

Let us finally comment on the fact that (almost) "no RBN is an island": in multicellular organisms cells interact and communicate, but also unicellular organisms present aggregated forms (e.g. colonies) that sometimes are able of very complex behaviors (like for example in the case of the slime mold [59]). Cellular communication affects the behavior of the regulatory network of single cells, so it seems particularly interesting to analyze the dynamics of a set of interacting gene network models. This can be done e.g. by placing the "model cells" on the sites a rectangular grid (allowing at most one cell per site), and allowing only interactions with neighbors [60]. Each site hosts a whole RBN, and the "genome" (topology and Boolean functions) are the same for every site. In this way one can simulate a "tissue" or even an organism undergoing development.

The interactions should simulate cell-to-cell communications, and to this aim two alternatives have been analyzed: in one case it is supposed that an RBN can feel the presence of active nodes on the neighboring sites (this feature being limited to a subset of the genes), in the other case it is supposed that the Boolean functions of some genes are affected by the activation of particular receptors that sense what has been synthesized by neighboring cells.

Extensive simulations have been carried out, leading to several interesting results [61]; the most intriguing one seems to be the non-monotonic dependency of the number of different coexisting attractors (and of other related variables) upon the interaction strength, measured as the fraction of all the nodes that are directly

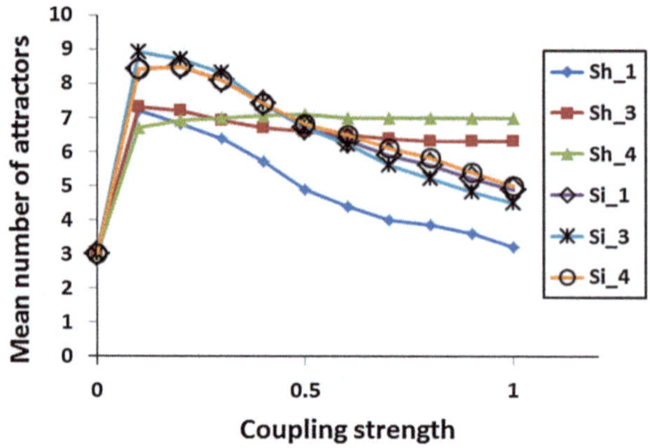

Fig. 3.5 Average number of different attractors as a function of the coupling strength. *Every curve* is a different interaction mechanism ("Sh" means that the cells can share some chemical compounds, whereas "Si" means that cells communicate by means of signaling molecules and receptors) and threshold (in order to be effective, a chemical message needs the support of at least 1, 3 or 4 neighboring cells). Note that the three signaling mechanisms show similar efficacies, whereas the three sharing mechanisms differs at high interaction strength

affected by the state of the neighboring cell. The results show that extreme values are attained at an intermediate level of connection strength, a suggestion that might be amenable to experimental tests (Fig. 3.5).

3.6 Noisy RBNs and Cell Differentiation

We will also review here some results which might shed light on a phenomenon that is both very important from the biological viewpoint and very intriguing from the modeling side. In fact it is possible to perturb attractors by means of a temporary disturbance, in which for only one time step one or more randomly chosen genes are flipped: if the perturbation is repeated with a certain frequency, we obtain a so-called noisy RBN (NRBN) [62], a system that seems to correspond to real biological networks better than simple deterministic RBNs (cells in effect are very noisy systems [63–68].

If the random perturbations are not too frequent, the network relaxes to one of the attractors of the deterministic RBN before a new perturbation takes place: under these assumptions the asymptotic dynamics of a NRBN can be properly described in terms of the attractors of the corresponding RBN.[5] However, attractors in noisy RBNs are not stable, and the effective asymptotic behavior of these systems is generated by hopping among the set of attractors that the system

[5] From now on, the term attractor will always refer to those of the deterministic RBN.

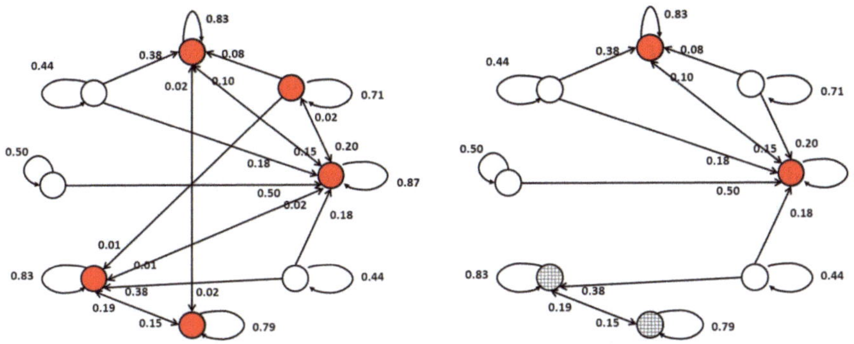

Fig. 3.6 Attractor transition graph in a particular RBN. *Circles* represent attractors; *arrows* represent transitions among attractors induced by single spin flips. The *numbers* on *each arrow* are the probability that, by flipping at random the state of a node in an attractor, that transition takes place. Homogeneously filled circles represent different TESs, whereas empty circles are unstable attractors: once *left*, the system cannot came back to these latter (semi)asymptotic states. **a** The complete attractor transition graph (only one TES_0 is present, involving 5 attractors); **b** the same graph, where links below the threshold $\theta = 0.02$ are removed (there are two $TES_{0.02}$, involving $2 + 2$ attractors)

can pass through, the so-called "ergodic set" [62]. Unfortunately, noisy RBNs typically have just a single ergodic set,[6] and therefore these systems are not suitable candidates for modeling cells in the case of multicellular organisms, where a single network should support several different asymptotic behaviors (identified with cell types). A way to avoid this difficulty [69–71], that will be summarized below, unexpectedly allows one to describe cellular differentiation, i.e. the process whereby stem cells, which can develop into different types, become more and more specialized.

The proposal is based on the observation that the main effect of (a not too high level of) noise is that of inducing transition among attractors: the transition probabilities are not all equal (for example, see in Fig. 3.6 the attractor transition graph of a particular RBN) and the smallest ones correspond to rare transitions, that are unlikely to take place during a cell lifetime; if these transitions are removed, the attractor transition graph is modified (see Fig. 3.6) and the system recovers presents different asymptotic behaviors. The direct graph composed by attractors—two attractors A and B being linked iff under noise there exists a transition that starts in A and ends in B—breaks in several disjoint groups of attractors that the system, once entered, cannot leave: they are the so-called Threshold Ergodic sets—briefly TES_θ, θ being the threshold above which the transitions are neglected. Because of their construction, TESs are robust under noise.

By tuning the level of noise (in the model, by increasing θ) is therefore possible to indirectly determine size and number of TES_θs and to modified in such a way the systems' asymptotic state, from very ample TES_θs (where noise has high level)

[6] In a few cases two ergodic sets have been observed, in networks with a few hundred nodes.

to the single attractors (situation where the noise is at minimum level). Real cells in effect can tune their internal level of noise [65], and in this vision can determine its differentiation state from stem cells wandering through a very ample portion of the state space to fully differentiated cells, confined within a smaller state space area. This process can take place according to two different modalities: (1) stochastic differentiation, in which for each lineage the proportions among the resulting different cellular types are constant, and (2) deterministic differentiation, in which the population of cells go through a particular differentiation path (a particular lineage).

The model can explain stochastic differentiation by supposing that during the noise reduction process each cell of the population remains blocked in the TES_θ that contains the particular attractor where the cell is at the moment of noise lowering.

On the other hand deterministic differentiation requires an external signal, able to drive the cells toward a particular fate. In RBNs such a signal could be modeled by permanent fixing the activities of some nodes [25, 69, 71]: indeed, one finds that in approximately one-third of critical RBNs a permanent fixing of particular genes during noise reduction can force the network towards particular final destinations (i.e. particular TES). Therefore, the repeated combinations of noise reduction and cells communication can select particular differentiation pathways, in a way that is similar to the real process [69, 71]. The nodes able to drive the transitions towards specific attractors have been termed "switch nodes" (see Fig. 3.7 for an example).

Differentiation is almost always irreversible, but there are limited exceptions under the action of appropriate signals [72, 73]: also in the model one can obtain similar effects. Moreover, in a few cases it has been possible to come back from a differentiated to a pluripotent state by forcing the expression of a node without acting on the threshold. The new pluroipotent state is similar but not identical to the original TES_0 [69, 71]: this represents the in-silico analogue of the famous experiment of Yamanaka on induced pluripotency [74, 75]. In other cases direct jumps from one differentiated cell type to another one have been observed, thereby simulating experiments as those described in Vierbuchen et al. [76].

With the same theoretical framework it is therefore possible to describe the most relevant features of cell differentiation: (a) different degrees of differentiation; (b) stochastic differentiation; (c) deterministic differentiation in well-defined lineages; (d) limited reversibility; (e) induced pluripotency and (f) induced change of cell type.

3.7 Generic Models, Specific Models and Experiments

We have described in detail the relationship between a class of generic models and the behavior of real gene regulatory networks. Of course, RBNs are just one among several types of models, so it is worthwhile to reconsider the reasons for this choice.

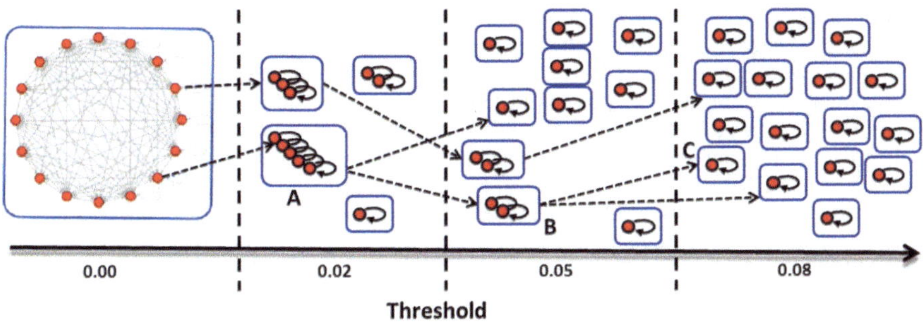

Fig. 3.7 TESs and differentiation. As the threshold increases the single TES_0 breaks into smaller disjoint TESs, corresponding to more differentiated cells, until eventually final cell types are reached. Examples of stochastic transitions are shown by *dotted lines*. By acting on particular genes at each noise reduction event, it is possible to select the particular pathway that links the TES_θ A, B and C. Note that the 16 attractors of the RBN are not always shown: the figure represents only attractors belonging to some TES, and in certain situations some attractors are not in such a condition

If one is interested in the details of gene regulation, RBNs are not particularly useful, and the same holds true if one is interested in analyzing small regulatory circuits (involving just a few genes). In this latter case RBNs could be applied, but more realistic models might be more useful. However, it must be stressed that small circuits are always embedded in larger networks, and considering a single module while ignoring the rest of the system,[7] although it is common practice, may be misleading.

In the end, when one is interested in the behavior of large networks, RBNs are still a very useful tool, because of several reasons:

- their Boolean nature makes them amenable to fast simulations
- a large body of theoretical and simulation results are already available
- the comparison with real experimental data has proven to be fruitful and able to lead to interesting insights (e.g. do systems live on the edge of chaos?)
- they can be the basis for interesting generalizations, including different topologies, the introduction of a longer memory, the interaction among neighboring cells and others
- they seem able to be the basis for analyzing cell differentiation.

Therefore, notwithstanding their age, RBNs are still an indispensable tool for analyzing vary large gene regulatory networks, although other models can prove much more accurate when one is concerned with smaller networks.

[7] In some cases the rest of the system is not ignored, but described at a very aggregated level in a crude way: the remarks in the text hold also in this case.

Acknowledgments We have developed our understanding of genetic network dynamics interacting with several scientists and collaborators. In particular, we are grateful to Stuart Kauffman who shared with us much of his insight, and of his evolving vision. Among several students and collaborators let us mention in particular Chiara Damiani, Alessandro Filisetti, Alex Graudenzi (the complete list would be too long). We also had the privilege of discussing our ideas with some biologists, in particular with Annamaria Colacci. On a more general ground, our understanding of the role of modeling such complex systems has been shaped in several fruitful discussions and joint works with David Lane and Gianni Zanarini, although of course we take full responsibility of the positions taken in this chapter. The present work has been partly supported by the Miur Project MITICA (FISR nr. 2982/Ric) and by the EU-FET projects MD (ref. 284625) and INSITE (ref. 271574) under the 7th Framework Programme.

References

1. Kleiber M (1932) Body size and metabolism. Hilgardia 6:315–351
2. West GB, Brown JH (2005) The origin of allometric scaling laws in biology from genome to ecosystems. J Exp Biol 208:1575–1592
3. West GB, Brown JH, Enquist BJ (1999) The fourth dimension of life: fractal geometry and allometric scaling of organisms. Science 284:1677–1679
4. Kauffman SA (1993) The origins of order. Oxford University Press, New York
5. Langton CG (1992) Life at the edge of chaos. In: Langton CG, Taylor C, Farmer JD, Rasmussen S (eds) Artificial life II. Addison-Wesley, Reading, pp 41–91
6. Packard NH (1988) Adaptation toward the edge of chaos. In: Kelso JAS, Mandell AJ, Shlesinger MF (eds) Dynamic patterns in complex systems. World Scientific, Singapore, pp 293–301
7. Kauffman SA (1995) At home in the universe. Oxford University Press, New York
8. Longo G, Bailly F (2008) Extended critical situations: the physical singularity of life phenomena. J Biol Syst 16(2):309–336
9. Schmulevich A, Kauffman SA, Aldana M (2005) Eukaryotic cells are dynamically ordered or critical but not chaotic. PNAS 102:13439–13444
10. Benedettini S, Villani M, Roli A, Serra R, Manfroni M, Gagliardi A, Pinciroli C, Birattari M (2012) Dynamical regimes and learning properties of evolved Boolean networks. Neurocomputing Elsevier
11. Alberts B, Johnson A, Lewis J, Raff M, Roberts K, Walter P (2007) Molecular biology of the cell. ISBN: 9780815341055
12. Frisch U, d'Humieres D, Hasslacher B, Lallemand P, Pomeau Y, Rivet JP (1987) Lattice gas hydrodynamics in two and three dimensions. Complex Syst 1:649707
13. Frisch U, Hasslacher B, Pomeau Y (1986) Lattice-gas automata for the Navier-Stokes equation. Phys Rev Lett 56(14):1505–1508
14. Aldana M, Coppersmith S, Kadanoff LP (2003) Boolean dynamics with random couplings. In: Kaplan E, Marsden JE, Sreenivasan KR (eds) Perspectives and problems in nonlinear science. Springer, New York, pp 23–89
15. Derrida B, Pomeau Y (1986) Random networks of automata: a simple annealed approximation. Europhys Lett 1:45–49
16. Langton CG (1990) Computation at the edge of chaos. Physica D 42
17. Bastolla U, Parisi G (1998) The modular structure of Kauffman networks. Physica D 115:219–233
18. Bastolla U, Parisi G (1998) Relevant elements, magnetization and dynamical properties in Kauffman networks: a numerical study. Physica D 115:203–218
19. Socolar JES, Kauffman SA (2003) Scaling in ordered and critical random Boolean networks. Phys Rev Lett 90

20. Aldana M, Balleza E, Kauffman SA, Resendiz O (2007) Robustness and evolvability in genetic regulatory networks. J Theor Biol 245:433–448
21. Balleza E, Alvarez-Buylla E, Chaos A, Kauffman SA, Shmulevich I, Aldana M (2008) Critical dynamics in genetic regulatory networks: examples from four kingdoms. PLoS ONE 3:e2456
22. Villani M, Serra R, Graudenzi A, Kauffman SA (2007) Why a simple model of genetic regulatory networks describes the distribution of avalanches in gene expression data. J Theor Biol 249:449–460
23. Shmulevich I, Kauffman SA (2004) Activities and sensitivities in Boolean network models. Phys Rev Lett 93
24. Szejka A, Mihaljev T, Drossel B (2008) The phase diagram of random threshold networks. New J Phys 10:063009
25. Villani M, Serra R (in press) Attractors perturbations in biological modeling: avalanches and cellular differentiation In: Cagnoni S, Mirolli M, Villani M (eds) Evolution, complexity and artificial life. Springer
26. Harris SE, Sawhill BK, Wuensche A, Kauffman SA (2001) A model of transcriptional regulatory networks based on biases in the observed regulation rules. Complexity 7(4):23–40
27. Raeymaekers L (2002) Dynamics of Boolean networks controlled by biologically meaningful functions. J Theor Biol 218:331–341
28. Serra R, Graudenzi A, Villani M (2009) Genetic regulatory networks and neural networks. In: Apolloni B, Bassis S, Marinaro M (eds) New directions in neural networks. IOS Press, Amsterdam, pp 109–117
29. Kauffman SA (1969) Metabolic stability and epigenesis in randomly constructed nets. J Theor Biol 22:437–467
30. Hughes TR, Marton MJ, Jones AR, Roberts CJ, Stoughton R, Armour CD, Bennett HA, Coffey E, Dai H, He YD, Kidd MJ, King AM, Meyer MR, Slade D, Lum PY, Stepaniants SB, Shoemaker DD, Gachotte D, Chakraburtty K, Simon J, Bard M, Friend SH (2000) Functional discovery via a compendium of expression profiles. Cell 102:109–126
31. Serra R, Villani M, Semeria A (2003) Robustness to damage of biological and synthetic networks. In: Banzhaf W, Christaller T, Dittrich P, Kim JT, Ziegler J (eds) Advances in artificial life. Lecture notes in artificial intelligence, 2801. Springer, Heidelberg, pp 706–715
32. Serra R, Villani M, Semeria A (2004) Genetic network models and statistical properties of gene expression data in knock-out experiments. J Theor Biol 227:149–157
33. Lee TI, Rinaldi NJ, Robert F, Odom DT et al (2002) Transcriptional regulatory networks in Saccharomyces Cerevisiae. Science 25 298(5594):799–804
34. Ramo P, Kesseli J, Yli-Harja O (2006) Perturbation avalanches and criticality in gene regulatory networks. J Theor Biol 242:164–170
35. Serra R, Villani M, Salvemini A (2001) Continuous genetic networks. Parallel Comput 27:663–683
36. Solè RV, Luque B, Kauffman SA (2011) Phase transition in random networks with multiple states working papers of Santa Fe Institute, www.santafe.edu/media/workingpapers/00-02-011.pdf
37. Kappler K, Edwards R, Glass L (2003) Dynamics in high dimensional model gene networks. Signal Process 83:789–798
38. de Jong H (2002) Modeling and simulation of genetic regulatory systems: a literature review. J Comput Biol 9(1): 67–103 (Mary Ann Liebert, Inc.)
39. Ironi L, Panzeri L, Plahte E, Simoncini V (2011) Dynamics of actively regulated gene networks. Physica D 240:779–794. doi:10.1016/j.physd.2010.12.010
40. Roli A, Vernocchi F, Serra R (2008) Continuous network models of gene expression in knock-out experiments: a preliminary study. In: Serra R, Villani M, Poli I (eds) Artificial life and evolutionary computation—Proceedings of WIVACE 2008. World Scientific Publishing
41. Klemm K, Bornholdt S (2005) Stable and unstable attractors in Boolean networks. Phys Rev E 72:055101–055104

42. Darabos C, Giacobini M, Tomassini M (2009) Generalized Boolean networks: how spatial and temporal choices influence their dynamics computational methodologies. In: Das S, Caragea D, Hsu WH, Welch SM (eds) Gene regulatory networks. Medical Information Science Reference; 1 edn. USA. ISBN: 1605666858
43. Gershenson C (2002) Classification of random Boolean networks. In: Standish RK, Abbass HA, Bedau MA (eds) Artificial life VIII. MIT Press, Cambridge, pp 1–8
44. Gershenson C (2004) Updating schemes in random Boolean networks: Do they really matter? In: Pollack J, Bedau M, Husbands P, Ikegami T, Watson RA (eds) Artificial life IX, Proceedings of the 9th international conference on the simulation and synthesis of living systems. MIT Press, pp 238–243
45. Serra R, Villani M, Agostini L (2004) On the dynamics of Boolean networks with scale-free outgoing connections. Physica A 339:665–673
46. Graudenzi A, Serra R, Villani M, Colacci A, Kauffman SA (2011) Robustness analysis of a Boolean model of gene regulatory network with memory. J Comput Biol 18(4) (Mary Ann Liebert, Inc., publishers, NY)
47. Graudenzi A, Serra R, Villani M, Damiani C, Colacci A, Kauffman SA (2011a) Dynamical properties of a Boolean model of gene regulatory network with memory. J Comput Biol 18 (Mary Ann Liebert, Inc., publishers, NY)
48. Watts DJ, Strogatz SH (1998) Collective dynamics of small world networks. Nature 393:440
49. Barabasi AL, Albert R (1999) Emergence of scaling in random networks. Science 286:509–512
50. Kitsak M, Riccaboni M, Havlin S, Pammolli F, Stanley HE (2010) Scale-free models for the structure of business firm networks. Phys Rev E 81:036117
51. Aldana M (2003) Boolean dynamics of networks with scale-free topology. Physica D 185:45–66
52. Serra R, Villani M, Graudenzi A, Colacci A, Kauffman SA (2008) The simulation of gene knock-out in scale-free random Boolean models of genetic networks. Netw Heterogen Media 3(2):333–343
53. Bornholdt S (2001) Modeling genetic networks and their evolution: a complex dynamical systems perspective. Biol Chem 382:1289–1299
54. Bhan A, Galas DJ, Dewey TG (2002) A duplication growth model of gene expression networks. Bioinformatics 18(11):1486–1493
55. Enemark J, Sneppen K (2007) Analyzing a stochastic model for evolving regulatory networks by unbiased gene duplication. JSTAT 0:P11007
56. Aldana M, Cluzel P (2003) A natural class of robust networks. PNAS 100(15):8710–8714
57. Fretter C, Drossel B (2008) Response of Boolean networks to perturbations. Eur Phys J B 62:365–371
58. Gershenson C, Kauffman SA, Shmulevich I (2006) The role of redundancy in the robustness of random Boolean networks. In: Rocha LM, Yaeger LS, Bedau MA, Floreano D, Goldstone RL, Vespignani A (eds) Artificial life X, Proceedings of the 10th international conference on the simulation and synthesis of living systems. MIT Press, pp 35–42
59. van Oss C, Panfilov AV, Hogeweg P, Siegert F, Weijer CJ (1996) Spatial pattern formation during aggregation of the slime mould Dictyostelium discoideum. J Theor Biol 181:203–213
60. Damiani C, Kauffman SA, Serra R, Villani M, Colacci A (2010) Information transfer among coupled random Boolean networks. In: Bandini S et al (eds) ACRI 2010 LNCS 6350. Springer, Berlin, pp 1–11
61. Damiani C, Serra R, Villani M, Kauffman SA, Colacci A (2011) Cell-cell interaction and diversity of emergent behaviours. IET Syst Biol 5(2):137–144. doi:10.1049/iet-syb.2010.0039
62. Ribeiro AS, Kauffman SA (2007) Noisy attractors and ergodic sets in models of gene regulatory networks. J Theor Biol 247(4):743–755
63. Blake WJ, KLrn M, Cantor CR, Collins JJ (2003) Noise in eukaryotic gene expression. Nature 422:633–637

64. Eldar A, Elowitz MB (2010) Functional roles for noise in genetic circuits. Nature 467:167–173
65. Lestas I, Paulsson J, Ross NE, Vinnicombe G (2008) Noise in gene regulatory net-works. IEEE Trans Automat Contr 53:189–200
66. McAdams HH, Arkin A (1997) Stochastic mechanisms in gene expression. PNAS 94:814–819
67. Raj A, van Oudenaarden A (2008) Nature, nurture, or chance: stochastic gene expression and its consequences. Cell 135(2):216–226
68. Swain PS, Elowitz MB, Siggia ED (2002) Intrinsic and extrinsic contributions to stochasticity in gene expression. PNAS 99:12795–12800
69. Serra R, Villani M, Barbieri A, Kauffman SA, Colacci A (2010) On the dynamics of random Boolean networks subject to noise: attractors, ergodic sets and cell types. J Theor Biol 265:185–193
70. Villani M, Barbieri A, Serra R (2011) A dynamical model of genetic networks for cell differentiation. PLoS ONE 6(3):e17703
71. Villani M, Serra R, Barbieri A, Roli A, Kauffman SA, Colacci A (2010) Noisy random Boolean networks and cell differentiation. In: Proceedings of ECCS2010—European conference on complex systems
72. Baron MH (1993) Reversibility of the differentiated state in somatic cells. Curr Opin Cell Biol 5(6):1050–1056
73. Johnson NC, Dillard ME, Baluk P, McDonald DM, Harvey NL et al (2008) Lymphatic endothelial cell identity is reversible and its maintenance requires Prox1 activity. Genes Dev 22:3282–3291
74. Takahashi K, Yamanaka S (2006) Induction of pluripotent stem cells from mouse embryonic and adult fibroblasts cultures by defined factors. Cell 126(4):663–676
75. Takahashi K, Tanabe K, Ohnuki M, Narita M, Ichisaka T et al (2007) Induction of pluripotent stem cells from adult human fibroblasts by defined factors. Cell 131(5):861–872
76. Vierbuchen T, Ostermeier A, Pang ZP, Kokubu Y, Sudhof TC et al (2010) Direct con-version of fibroblasts to functional neurons by defined factors. Nature 463:1035–1041

Chapter 4
Regulatory Crosstalk Analysis of Biochemical Networks in the Hippocampus and Nucleus Accumbens

Shinichi Kikuchi and Eberhard O. Voit

Abstract This chapter describes mathematical modeling of neuronal biochemical pathways, especially for pathological and non-pathological features of molecular and cellular mechanisms in the hippocampus and nucleus accumbens. We modeled both types of neurons with a variety of techniques: dynamic equations, constraint-based modeling, and complex network analysis. The last two approaches are called static modeling. In this chapter, we introduced these 3 methods to model the process of signal transduction, metabolism, ion fluxes, and gene regulation in a neuron, and their recent applications to the pathological characterization of the system. (1) The first one is a model of synaptic plasticity in the hippocampal CA1 neurons, which is thought to be relevant for learning and memory. We selected a constraint-based approach to model the cell, which uses constraint conditions in models from the stoichiometry matrix of chemical reactions in the absence of kinetic data. (2) The second model focuses on hippocampal signaling pathways in Alzheimer's disease, including neurite outgrowth, synaptic plasticity and neuronal death. This is an application of complex network analysis to biological networks, with a particular emphasis on the *k* shortest path and the *k*-cycle. (3) The synaptic plasticity in medium spiny neurons in the nucleus accumbens is the main topic of the third model, which is thought to be relevant for reward system. An approach to reveal the dynamic properties of the model is a conventional ordinary differential equation-based modeling and perturbation analysis. Finally, brief concluding remarks appear in Sect. 4.5.

Keywords Molecular systems neuroscience · ODE model · Stoichiometric model · Complex network analysis · Synaptic plasticity · Learning system · Reward system ·

S. Kikuchi (✉) · E. O. Voit
The Wallace H. Coulter Department of Biomedical Engineering,
Georgia Institute of Technology, 313 Ferst Drive, Atlanta, GA 30332, USA
e-mail: Skikuchi19@gatech.edu

E. O. Voit
e-mail: eberhard.voit@bme.gatech.edu

A. Prokop and B. Csukás (eds.), *Systems Biology*,
DOI: 10.1007/978-94-007-6803-1_4,

Drug addiction · Systems biology · Computational neuroscience · Chemical reactions · Dynamic model · Static model · Signal transduction · Genetic network · Hippocampus · Nucleus accumbens · Psychostimulant · Extreme pathway analysis · *k* shortest path · *k*-cycle · Sensitivity · Microarray analysis

List of Acronyms

Aβ	amyloid β
AC	adenylate cyclase
ACh	acetylcholine
AD	Alzheimer's disease
ADP	adenosine diphosphate
ADN	Alzheimer's disease network
AMPH	amphetamine
AMPT	α-methyl paratyrosine
APP	amyloid β precursor protein
ATP	adenosine triphosphate
AMPAR	alpha-amino-3-hydroxy-5-methyl-4-isoxazolepropionate receptor
BDNF	brain-derived neurotrophic factor
cAMP	cyclic adenosine monophosphate
CaM	calmodulin
CaMKII	calcium/calmodulin-dependent protein kinase type II
CaN	calcineurin
CDK5	cyclin dependent kinase 5
CN	control network
CREB	cAMP responsive element binding protein
DA	dopamine
DARPP-32	dopamine- and cAMP-regulated phosphoprotein of 32-kDa
DAT	dopamine transporter
EGF	epidermal growth factor
EP	extreme pathway
ER	endoplasmic reticulum
ES	enzyme-substrate (complex)
ESF	extreme signaling flow
FasL	Fas ligand
GABA	γ-Aminobutyric acid
GAP	GTPase-activating protein
Glu	glutamate
HFS	high frequency stimulation
ICAD	inhibitor of caspase-activated DNase
IGF1	insulin-like growth factor-1
IP_3	inositol 1, 4, 5-phosphate
I1	inhibitor 1
LFS	low frequency stimulation
LTD	long-term depression

LTP	long-term potentiation
mGluR	metabotropic glutamate receptor
MAPK	mitogen-activated protein kinase
MAO	monoamine oxidase
MDD	major depressive disorder
MINT-1	Munc18-interacting protein 1
MSNs	medium spiny neurons
NAc	nucleus accumbens
NFAT	nuclear factor of activated T cells
Ng	neurogranin
NGF	nerve growth factor
NMDAR	*N*-methyl-D-aspartate receptor
NRG	neuregulin
NT	neurotrophin
ODE	ordinary differential equation
PDE	phosphodiesterase
PKA	protein kinase A
PKC	protein kinase C
PLC_β	phospholipase C_β
PP1	protein phosphatase 1
PP2A	protein phosphatase 2A
PP2B	protein phosphatase 2B (a.k.a. calcineurin)
RRN	randomly removed network
SN	substantia nigra
TH	tyrosine hydroxylase
TNFα	tumor necrosis factor-α
VMAT2	vesicular monoamine transporter 2
VTA	ventral tegmental area

4.1 Introduction

Systems neuroscience is a multidisciplinary approach to finding the mathematical laws in neuroscience by well-defined strategies in mathematics and systems engineering [1]. It covers a wide area of studies from molecular, cellular, synapse, and circuit levels to brain function. The discovery of LTP of Glu synapses in the hippocampus launched an exciting exploration into the molecular basis of learning and memory [2, 3]. LTP and its counterpart, LTD, appear to be essential in the stabilization and elimination of synapses during the developmental fine-tuning of neural circuits in many areas of primary sensory cortex [4]. The reward circuitry in the NAc enables it to predict the future over a short period of time based on the success and failure of previous predictions, and the NAc is of interest in this respect because any drug abuse commonly causes DA release in the NAc, and behavioral observations indicate that the mesolimbic DA pathway is directly involved in rewards [5, 6]. Metabolic processes in the presynapse determine the amount of released neurotransmitters through the control of enzymatic reactions and neurotransmitter recycling between different compartments. On the postsynaptic side, the density and conductance of receptors are regulated by second messenger systems and by kinases and phosphatases. The adaptive change in the efficiency of information transmission between presynaptic and postsynaptic neurons, known as synaptic plasticity, lasts from days to weeks *in vivo* [7]. In this chapter, we discuss the molecular neurobiology of synaptic plasticity in the hippocampus and NAc using the wide range of models to examine the signal transduction mechanism as a highly complex system.

The purpose of this chapter is to introduce modeling techniques of complex biochemical networks, especially for pathological and non-pathological features of molecular and cellular mechanisms in the hippocampus and nucleus accumbens. In this chapter, we introduce these three methods to model the process of neuronal signal transduction, and their recent applications to the pathological characterization of the system. Sections 4.2 and 4.3 focus on static models for large-scale data. Section 4.2 discusses a model of synaptic plasticity in hippocampal CA1 neurons, which is thought to be relevant for learning and memory. An algebraic method is introduced to model the stoichiometric matrix from the mole ratios between reactants and products. A complex network model in Sect. 4.3 provides the characteristics of hippocampal signal transduction pathways that mediate neurite outgrowth, synaptic plasticity and neuronal death in AD, with particular emphasis on the structures of the *k* shortest path and the *k*-cycle. Section 4.4 introduces an ODE model of synaptic plasticity in MSNs in the NAc, which is thought to be relevant for the reward system. The perturbation analysis of this model reveals the dynamic properties of a model structure. Section 4.5 presents general remarks about systems biology and mathematical modeling.

4.2 Stoichiometric Analysis of Bidirectional Hippocampal Synaptic Plasticity

4.2.1 Signal Transduction Cascade and Constraint-Based Modeling

Papin and Palsson [8] used extreme pathway analysis [9], a constraints-based approach, to express reactions in a signal transduction system in terms of algebraic equations, and to define enzymes as external factors. However, it is difficult to represent large-scale signaling networks with this technique, because a cascade yielded by this method is fragmented into small pathways. Matsubara et al. [10] proposed an enhanced modeling technique for extreme pathway analysis, which applies a minimal combination of EPs to obtain a series of information flow data. This approach has been applied to analyze hippocampal neuronal plasticity in a whole postsynaptic signal transduction system. Conventional models address modeling only in terms of unidirectionality of neuronal plasticity, for example hippocampal LTP [11, 12] or cerebellar LTD [13]. We proposed an algebraic model for bidirectional synaptic plasticity in the hippocampus using the stoichiometric matrix of the signaling network (Fig. 4.1). The kinetic details and initial conditions for the differential equations and the full list of references are presented in [12].

A constraint-based model represents a group of chemical reactions described by a stoichiometric matrix. The rows and columns in this matrix correspond to network components and biochemical reactions, respectively. Each element in the matrix contains the stoichiometric coefficient of the given component in the associated reaction. A vector in the null space of the stoichiometric matrix, called the EP, indicates the minimal unique pathway of the system; the EP has been widely applied for modeling metabolic networks [14, 15], which is expressed as

$$\mathrm{EP} = \left\{ \mathbf{p} \in \mathbf{V} \mid \mathbf{S} \cdot \mathbf{p} = 0, \quad p_i^I > 0, \quad -\infty < p_j^E < \infty, \quad \forall i,j \right\}, \tag{4.1}$$

where $\mathbf{S}$ is a stoichiometric matrix, $\mathbf{V}$ a flux space, $\mathbf{p}$ an EP vector, p_i^I the ith internal flux of $\mathbf{p}$, and p_j^E the jth external flux of $\mathbf{p}$. The EP can be used for

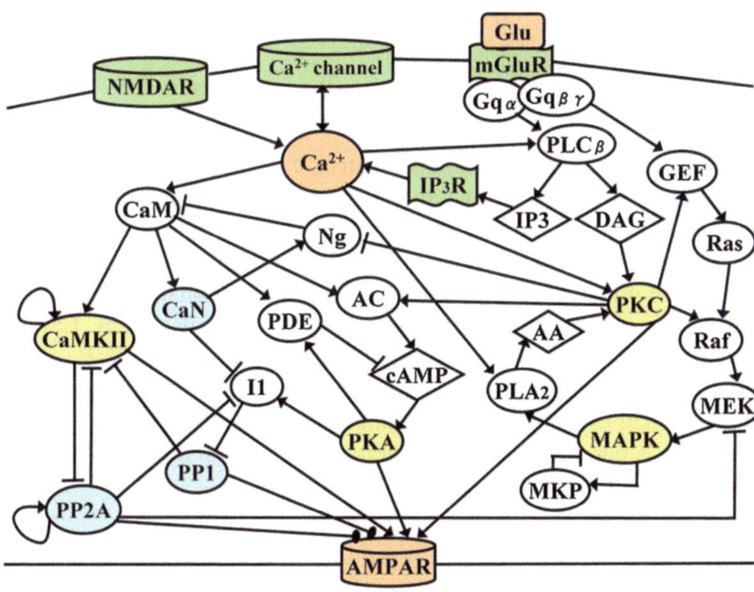

Fig. 4.1 Neuronal plasticity cascade in the hippocampal CA1 neurons. *Arrows* represent the activation (increasing) relationships, and *barred lines* the inactivation (decreasing) relationships. For the acronyms and the detailed reaction mechanisms, see Matsubara et al. [10]. *Source note*: The figure is reproduced with permission from Matsubara et al. [10]

modeling networks with catalysts of the system in the quasi-steady-state condition. Thus, only small signal transduction networks can be modeled by the EP because a molecule could be at the same time a substrate, a product, or an enzyme in different reactions.

The ESF is a minimum pathway from the input to the output signals, which creates a series of EPs of enzyme activation events and substrates depending on stoichiometric coefficients [10]. Figure 4.2a represents a typical multilevel cascade of MAPK phosphorylation, composed of two mass action and two enzymatic reactions. Initially S_i ($i = 1, 2, 3$) are activated; M_j ($j = 1, 2, 3$; intermediate ES complexes) and E_k ($k = 1, 2$) are then activated by mass conversion or enzymatic action. The signal transduction finally leads to the activation of P. In Fig. 4.2b, the enzymes in the stoichiometric matrix are classified into those before and after catalysis. The five ESFs of the network as shown in Fig. 4.2c are calculated by null-space manipulation of the converted stoichiometric matrix. Substances in the boxes (Fig. 4.2c) are defined as external substances. There are four meaningless ESFs, which are identical to EPs without external fluxes. ESF_5 connects three EPs to represent the integrated information flow from S_1 to P, thereby distinguishing the related enzymes as different states. While ESF analysis is elegant, it is difficult to infer the information flow of enzymatic reactions in this system because of the segmented EPs (Fig. 4.2d). Whereas an EP is a unique minimal unit to characterize a steady state, an ESF is a minimal functional unit for signal transduction. All ESFs are represented as non-negative linear combinations of given EPs, and are expressed as

$$\mathrm{ESF} = \left\{ \mathbf{f} \in \mathbf{V} \mid \mathbf{S}' \cdot \mathbf{f} = 0, \quad \mathbf{f} = \sum_{i=1}^{k+1} w_i \mathbf{p}_i, w_i \in \mathbf{S}', \quad \mathbf{p}_i \in \mathrm{EP}, \quad \forall i \right\}, \tag{4.2}$$

where $\mathbf{S}'$ is a converted stoichiometric matrix, $\mathbf{f}$ an ESF, w a weight coefficient yielding the stoichiometry of enzymatic reactions, $\mathbf{p}$ an EP vector, and k the number of enzymatic reactions.

In an in silico knockout analysis using EP, although EP_5 (consuming S_1 as a substrate) was eliminated, EP_8 (producing P) was not eliminated, indicating that S_1 is not necessary for the production of P. On the other hand, the ESF analysis showed that S_1 was a substrate essential for the production of P, since ESF_5 (producing P) was removed when S_1 was knocked out. This suggests that ESF is more sensitive than EP analysis in revealing the knockout influence, likely because the information is derived from the integrated information flow between the initial substances and the ultimately activated substances.

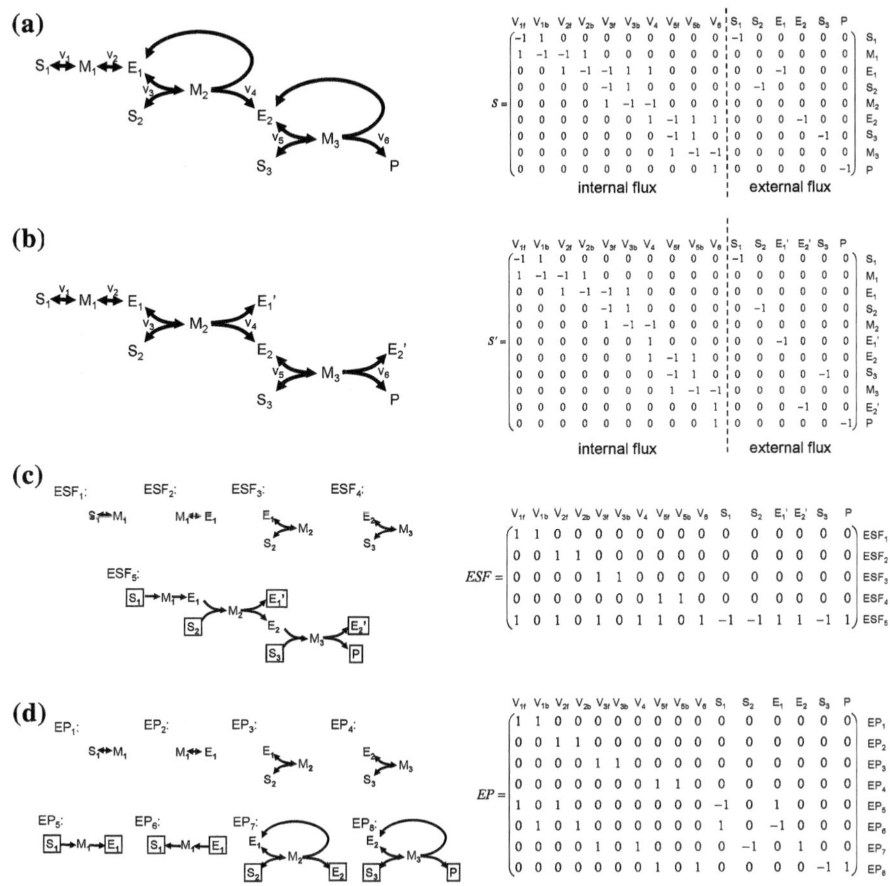

Fig. 4.2 **a** MAPK cascade as an example of a typical signal transduction cascade. It is composed of two mass action reactions ($S_1 \rightleftarrows M_1 \rightleftarrows E_1$) and two enzymatic reactions ($E_1 + S_2 \rightleftarrows M_2 \rightarrow E_1 + E_2, E_2 + S_3 \rightleftarrows M_3 \rightarrow E_2 + P$). V_{if} and V_{ib} represent the ith forward and backward reactions, respectively. The *right column* shows the stoichiometric matrix of this cascade. **b** The stoichiometric matrix is converted for ESF calculation, classifying enzymes into those before and after catalysis. It is composed of two mass action reactions and two enzymatic reactions ($E_1 + S_2 \rightleftarrows M_2 \rightarrow E_1' + E_2, E_2 + S_3 \rightleftarrows M_3 \rightarrow E_2' + P$). The *right column* shows the stoichiometric matrix of this cascade. **c** An illustration of ESFs of the above model. The ESF is represented by the null space in S', which is converted into a stoichiometric matrix. The substances in square boxes are defined as external substances. Other substances are defined as internal. **d** An illustration of EPs of the above model. The EP is represented by the null space in S. We found that ESF_5 is a combination of EP_5, EP_7, and EP_8. *Source note*: The figure is adapted with permission from Matsubara et al. [10]

4.2.2 *ESF Analysis of Signaling Cascades in the Hippocampal CA1*

4.2.2.1 Enumeration of ESFs

The ESFs of the hippocampal signaling network that underlies neuronal plasticity were calculated by using the EP algorithm [9, 16]. As a result, 13,815 ESFs were shown in this network; they were categorized into three groups: (1) The first 213 group (13,021 ESFs) induces LTP by activating CaMKII or PKC (Fig. 4.3); (2) the second one (267 ESFs) induces LTD by activating PP1 or PP2A (Fig. 4.4); (3) ESFs in the third group (527 ESFs) does not activate kinases or phosphatases that change the state of AMPARs. The number of ESFs that induce LTP was approximately 50-fold higher than the number of ESFs that induce LTD; this shows the fault-tolerance of LTP-related pathways in this cascade.

4.2.2.2 Redundancy Analysis

The redundancy of networks is evaluated by the number of identical ESFs calculated from a network (Tables 4.1 and 4.2; [8]). High redundancy is a good indicator of fault tolerance, since a specific output is generated by different inputs; low redundancy, on the other hand, indicates a high correlation between inputs and outputs. The ESFs for the induction of LTP and LTD manifest 8 and 6 output patterns, respectively. The redundancy in the LTP outputs of PKC and PKA is 98- and 44-fold higher, respectively, than the redundancy of CaMKII; the redundancy in the LTP outputs of PP2A is 24-fold higher than the redundancy of PP1. Meanwhile, CaMKII is the only kinase related to LTD induction. The redundancy in the LTD outputs of PP1 is 1.3-fold higher than the redundancy of PP2A. This suggests that PKC inherently contributes to the high redundancy of LTP, that it has no relation to LTD induction, and that the ratio of the concentrations of PP1 and CaMKII plays a role in the induction of LTD.

4.2.2.3 Reaction Participation Analysis

Reaction participation analysis scores the number of ESFs passing through a specific reaction [8, 14]. A positive correlation reportedly exists between lethality and connectivity related to a particular substance [17, 18]. This suggests that the high-scoring reactions in the reaction participation analysis indicate the factors essential to stabilize a phenotype. Figure 4.5 shows the scores of ESFs for the induction of LTP and LTD, where the thickness of the lines reflects the participation values. The scores of the Ca^{2+} exchange reactions in LTP and LTD were 99 and 78 %, respectively, consistent with the role of Ca^{2+} as a second messenger in both phenomena. The exchange reactions of ATP and ADP, not shown in the

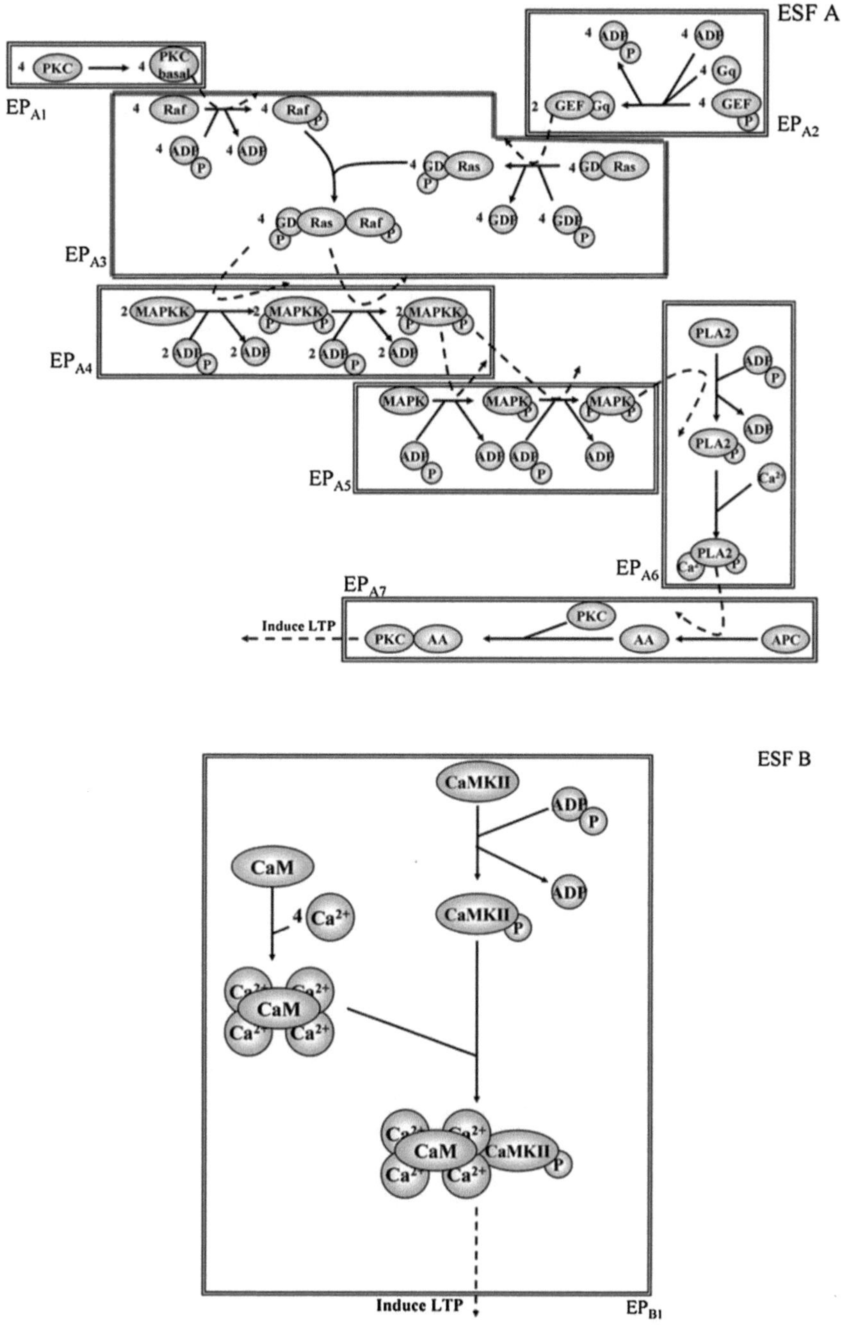

Fig. 4.3 LTP-inducing ESFs. **ESF A** (EP_{A1}–EP_{A7}) indicates activation of the MAPK cascade. **ESF B** (EP_{B1}) indicates activation of CaMKII by Ca^{2+} and CaM. *Dashed arrows* indicate enzymatic reactions. *Source note*: The figure is adapted with permission from Matsubara et al. [10]

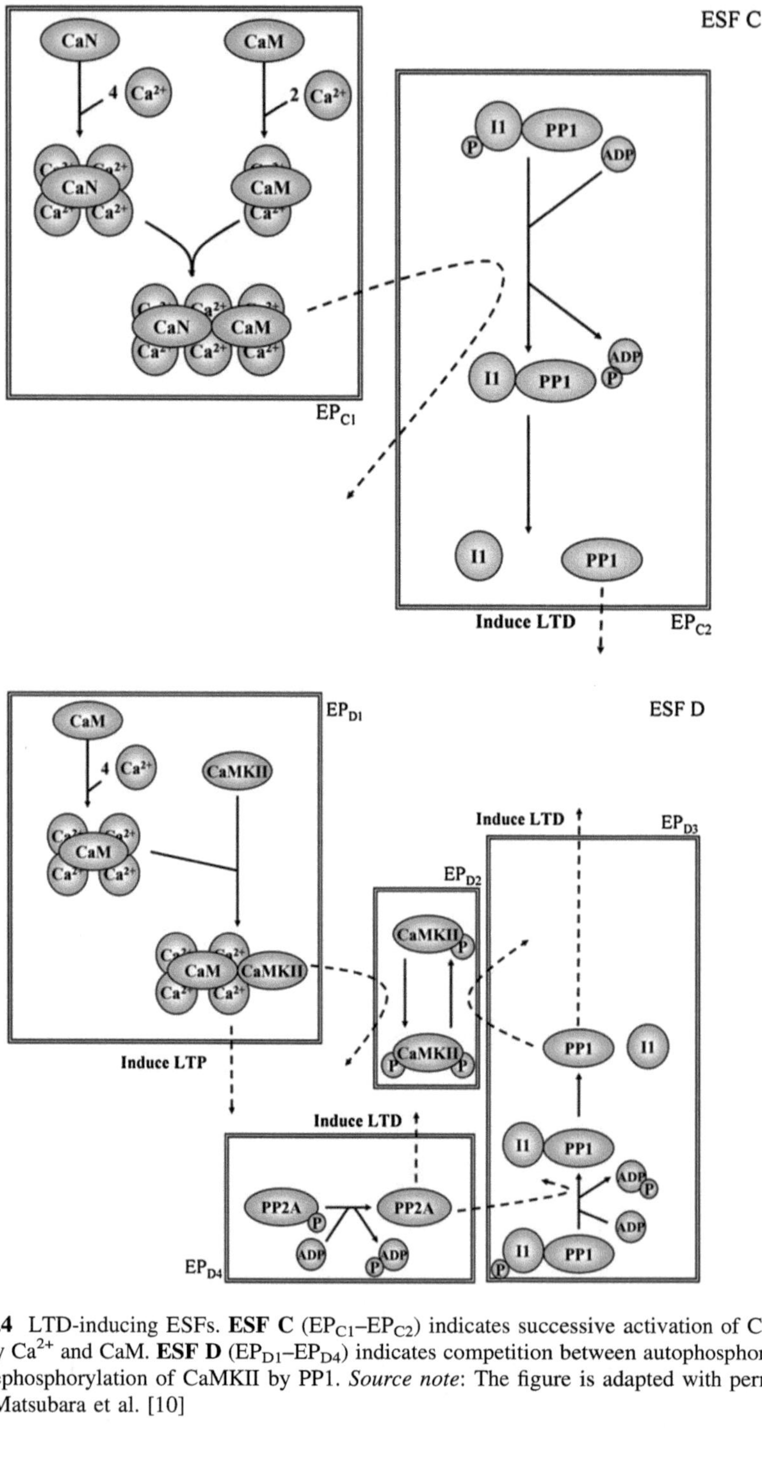

Fig. 4.4 LTD-inducing ESFs. **ESF C** (EP_{C1}–EP_{C2}) indicates successive activation of CaN and PP1 by Ca^{2+} and CaM. **ESF D** (EP_{D1}–EP_{D4}) indicates competition between autophosphorylation and dephosphorylation of CaMKII by PP1. *Source note*: The figure is adapted with permission from Matsubara et al. [10]

Table 4.1 Redundancy analysis of LTP-inducing ESFs.

Final outputs		
Kinases	Phosphatases	#ESF
PKC		5817
PKC, PKA		4896
PKC	PP2A	1361
PKC, PKA	PP2A	816
CaMKII	PPI	63
CaMKII		31
CaMKIt	PPI, PP2A	28
CaMKII	PP2A	9

#ESF is the number of ESFs. PKC and PKA have higher redundancy than CaMKII
Source note: The table is adapted with permission from Matsubara et al. [10].

Table 4.2 Redundancy analysis of LTD-inducing ESFs

Final outputs		
Kinases	Phosphatases	#ESF
CaMKII	PP1, PP2A	180
CaMKII	PPI	42
	PPI	21
	PPI, PP2A	6
CaMKII	PP2A	6
	PP2A	4

PKC and PKA are irrelevant to LTD. CaMKII and PP1 outputs are often simultaneous
Source note: The table is adapted with permission from Matsubara et al. [10].

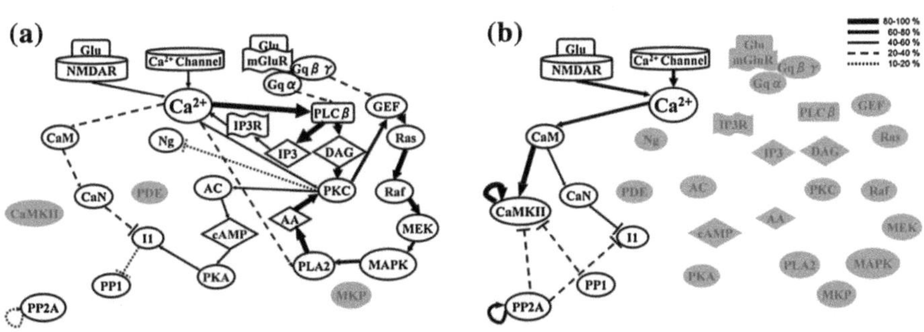

Fig. 4.5 Reaction participation analysis for: **a** LTP-inducing ESFs; **b** LTD-inducing ESFs. The *line width* reflects participation values. *Source note*: The figure is adapted with permission from Matsubara et al. [10]

cascade map, have the second-highest scores because of phosphorylation and dephosphorylation. The reactions around the positive feedback loop involving PKC and MAPK have much higher scores in LTP than the reactions that include CaMKII. Disassembly of the complex of PP1 and phosphorylated I1 has a score of

93 % in LTD. The activation of PLC_{β} induced by Ca^{2+} stimulation has a score of 88 % in LTP. The increased concentration of intracellular Ca^{2+} released from the ER, attributable to the activation of PLC_{β} and IP_3 production, plays an essential role in LTP. The successive routes of AC, cAMP, PKA, and I1 also yielded high scores in LTP. In contrast, CaM activation is the only LTD-inducing module triggered by Ca^{2+} increase.

4.2.2.4 Knockout Analysis

Table 4.3 compares the results of *in silico* knockout analysis with *in vivo* data for CA1 hippocampal region-specific gene knockout mice. ESF knockout analysis characterizes the effect of substance deletion on neuronal plasticity. If the deletion of the target substance leads to the decrease in the number of ESFs that enhance a phenomenon, it results in the suppression of the phenomenon. Conversely, the decrease in the number of suppressive ESFs indicates that the deletion results in the enhancement of the phenomenon.

In the knockout analysis shown in Table 4.3, LTP was suppressed by mGluR deletion, since mGluR is closely related to LTP-inducing PKC. The deletion of PKA suppressed LTP, but caused no change in LTD, since PKA was used by

Table 4.3 Results of *in silico* knockout analysis with *in vivo* data for CA1 hippocampal region-specific gene knockout mice

Knockout	*In vivo*		*In silico*	
	LTP	LTD	LTP	LTD
mGluR	↓[1]	↔[1]	↓	↔
PKA	↓[2,3]	↓[2,4]	↓	↔
PKC	↓[5]	↔[5]	↓	↔
CaMKII	↓[6,7,8*,9*]	↓[7]↑[8*]	↓	↑
Ras	↓[10**]		↓	↔
MAPK	↓[11,12]		↓	↔
CaN	↑[13]↔[14]	↔[13]↓[14]	↑	↓
AC	↓[15]		↓	↔
I1	↔[16]		↓	↑
Ng	↑[17]↓[18]	↓[17]↑[18]	↑	↔

↔ no effect; ↑ upregulation; ↓ downregulation. ↔ indicates that #ESFs are invariant under substance deletion. ↑ means that the number of ESFs suppressing the phenomenon is reduced by target substance deletion, resulting in the enhancement of the phenomenon. ↓ means that the number of ESFs enhancing the phenomenon is reduced by target substance deletion, resulting in the suppression of the phenomenon. [*] results of point mutations at the phosphorylation sites. [**] results of heterozygous knockouts. [1] Aiba et al. [74]; [2] Qi et al. [75]; [3] Abel et al. [76]; [4] Brandon et al. [77]; [5] Abeliovich et al. [78]; [6] Silva et al. [79]; [7] Stevens et al. [80]; [8] Giese et al. [81]; [9] Matford et al. [82]; [10] Ohno et al. [83]; [11] Mazzucchelli et al. [84]; [12] Winder et al. [85]; [13] Malleret et al. [86]; [14] Zeng et al. [87]; [15] Wong et al. [88]; [16] Allen et al. [89]; [17] Krucker et al. [90]; [18] Huang et al. [91]. *Source note*: The table is reproduced with permission from Matsubara et al. [10]

approximately half of the ESFs for the induction of LTP. LTP was suppressed by Ras, MAPK, or PKC deletion, since the activation route Ras-MAPK-PKC was used by most ESFs for the induction of LTP (not LTD). CaMKII was used by ESFs for the induction of LTD as well as LTP. LTP was suppressed, and LTD was enhanced by CaMKII deletion, since CaMKII inactivates LTD-inducing PP1. CaN deletion enhanced LTP and suppressed LTD, because it was a key phosphatase that was used by ESFs for LTP in addition to LTD. Both AC1/8 (activated by CaM) and AC2 (activated by PKC) were used by ESFs for LTP; therefore LTP was suppressed when either of them was deleted. I1 played an important role in LTD suppression due to its inhibitory binding to PP1; the deletion of I1, therefore, enhanced LTP and suppressed LTD. Ng had a constraining influence on the speed and magnitude of CaM-dependent reactions, because it interacted with high affinity with CaM in the absence of Ca^{2+}. As Ng inactivation by PKC resulted in the induction of LTP, Ng deletion resulted in LTP enhancement as well.

This is an outline of an algebraic method for signal transduction cascade. The method is especially applicable to generating multiple conditions like double knockout mice. The removal scheme helps to understand the structure of biological systems. In the next section, we introduce a method to characterize a series of chemical reactions, even without the stoichiometry of the reactions.

4.3 Complex Network Analysis of Hippocampal Signaling Pathways in AD

4.3.1 Complex Network Modeling of Signaling Pathways

Structural effects of biomolecular networks in pathological conditions have been reported in a degree analysis of cancer-related genes by using gene regulatory networks to identify the genes [19] and in various other analyses [20]. Ma'ayan et al. [21] used a directed graph of the signal transduction pathways in the human hippocampal CA1 region. The graph contains 570 nodes (signal molecules) and 1,333 edges (reactions). The edges can be categorized into three types of information defined as active, inactive, and bidirected (bidirectional activation or inactivation) reactions. The studies mentioned above assumed that the state of proteins does not change in the absence of external stimulation, and do not take into account possible intrinsic changes in gene regulation [22]. Yanashima et al. [23] extracted AD-related genes by analyzing GSE5281, a set of gene expression data in the human hippocampal CA1 region derived from patients with late-onset AD and controls [24], and proposed a complex network model of hippocampal signaling pathways using a gene expression profile of patients with AD. Complex network analysis is classified into the following three categories, outlined in Fig. 4.6.

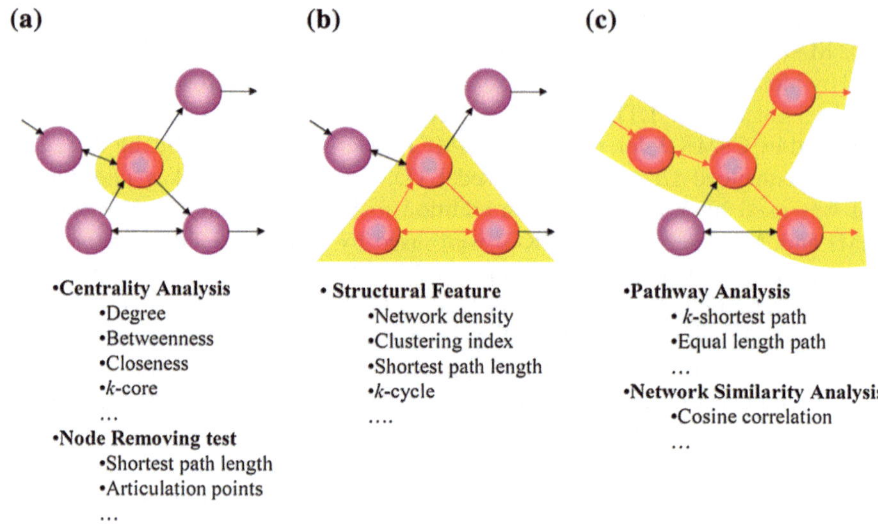

Fig. 4.6 Complex network analysis of signal transduction cascades. The *pink circles* and the *black lines* represent molecules and molecular interactions, respectively; the focus of each analysis is highlighted with *red* and *yellow colors*. The *unidirectional* and *bidirectional arrows* indicate the direction of chemical reactions. **a** Node feature analysis (centrality and changes in indicators upon node removal), **b** Structural properties, and **c** Characteristics of pathways (analysis of network similarity and pathways analysis). *Source note*: The figure is adapted with permission from Yanashima et al. [23]

4.3.1.1 Feature Analyses

Node feature analysis includes the calculation of *k*-core, betweenness centrality, closeness centrality, and degree centrality (Fig. 4.6a). The *k*-core of a graph is the maximal subgraph in which each node's degree is at least *k*. Betweenness centrality measures the importance of a node within a network, based on the enumeration of the shortest paths for all possible node pairs. Closeness centrality is defined as the number of nodes minus one, divided by the sum of all shortest path lengths from and to the given node. Degree centrality is the number of nodes to which a given node is connected. The feature analysis also includes the changes in the shortest path length for the evaluation of the small-world effect [25], and the changes in articulation points by removing nodes for the evaluation of a network connectivity.

4.3.1.2 Structural Properties

Structural index analysis is conducted by generating an ADN after removing AD-related signal molecules from the CN. A *k*-cycle structure has been used for the analysis of feedback loops in the networks (Fig. 4.6b). The network structure is defined from which duplicating nodes are removed when one node can be reached from the in-neighbors. An earlier study [21] and our pilot study showed that 90 % of all nodes can be reached within 9 steps from the input ($n = 30$). Thus, the pathways within 9 steps of each other are postulated to be important for intercellular signal transduction. Since network structure depends on the number of nodes, an RRN was generated by removing nodes from the CN to equal the number of nodes in the ADN. The network density, average clustering index, and average shortest path length change of this new CN were limited to 5 %. The *k*-cycle data can be analyzed according to Eq. 4.3:

$$C_k = \sum_{n=1}^{k} \frac{cycle_n(\mathrm{Node}_i)}{n}, \tag{4.3}$$

where C_k represents the number of *k*-cycle structures in the network. The function cycle *n* represents the number of cycle structures that can be reached from the in-neighbors.

4.3.1.3 Characteristics of Pathways

Since cellular processes are controlled by many alternate signal transduction pathways [26], it is necessary to analyze the *k*-shortest path in addition to analyzing pathway length or the shortest paths (Fig. 4.6c). The *k*-cycle of the RRN was compared with that of the ADN. Through exploration of the *k*-shortest path length, the number of pathways was determined by calculating the shortest path length between nodes and by using the depth-first iterative-deepening algorithm [27]. The *k*-shortest path with extracellular ligands ($n = 30$) as the input, and cytoskeletal proteins ($n = 24$) and transcription factors ($n = 35$) as the output were used to define 1,770 pathways. Three neuronal functions, neurite outgrowth, plasticity and death, were chosen to analyze how neuronal functions are affected in AD. Glu was set as the start point and CREB as the end point for neuronal plasticity; ACh, IGF, and Ephrin as the start points, and tubulin as the end point for neurite outgrowth; FasL and TNFα as the start points, and ICAD as the end point for neuronal death; and all inputs as the other start points, and MINT-1 and caspase 3 as the other end point for neuronal death as well. Aβ oligomers inhibit hippocampal LTP in rats *in vivo* [28]. NGF has a maintenance function in the nerve cells [29], and ACh decreases as Aβ accumulates [30]. TNFα and caspase 3 correlate positively with the accumulation of Aβ [31, 32]. An evaluation of robustness, defined in Eq. 4.4, was conducted by comparing the robustness values of all inputs and outputs of the ADN with those of the CN and RRN. The number of steps *k* used in the *k*-shortest path analysis in the *k*-cycle structure was defined as 9 steps, using the following equation:

$$R_{ij} = \frac{N_{ij} - \mathrm{mean_x}}{\mathrm{SD_x}}, \tag{4.4}$$

where R is the robustness value (R-value) of the pathway. R is the difference between the numbers of k-shortest paths obtained for all inputs to outputs in all k-shortest path sets, which is defined as X. In the pathways related to neuronal death, R is the difference in the number of k-shortest paths between node i and node j obtained in the RRN sets, which is defined as X in this case. N_{ij} is the k-shortest path number from node i to node j in the network of interest. Mean_X is the mean of all k-shortest path sets or nodes in the RRN sets. SD_X is the standard deviation of all k-shortest path sets or nodes in the RRN sets.

Equation 4.5 below shows the interpretation of network similarity by using a single value [33] for the vector space of inputs and outputs in a network and a matrix expression for the equal-length shortest path [34], which indicates pathways with equal steps. Our study analyzed the change in the entire pathway at step e [23].

$$S = \arccos\left(\frac{\overrightarrow{c^e} \cdot \overrightarrow{o^e}}{\left|\overrightarrow{c^e}\right| \cdot \left|\overrightarrow{o^e}\right|}\right), \tag{4.5}$$

where S represents network similarity between the first mode of singular value c (equal-length shortest-path matrix of CN) and o (equal-length shortest-path matrix of ADN or RRN); e represents the specific step value of the equal-length shortest-path matrix.

4.3.2 Pathological and Non-Pathological Characterization of Hippocampal Signaling Pathways

Seventy-six AD-related genes involved in downregulation of actin and β-catenin, which result in a decreased level of CaMKII [35–37], were extracted through empirical Bayes t-statistics. The number of signaling molecules (in particular kinases, adapters, receptors, transcription factors and Bcl-2 family proteins) was reduced in the ADN compared with the CN (Table 4.4). There were no significant differences between AD-related signaling molecules and other molecules in all parameters described in 3.1.1 ($P < 0.05$, Mann–Whitney U-test; Table 4.5). The ADN contained 494 nodes and 974 edges after the removal of these AD-related signal molecules. In total, 91 % of the input–output sets were connected in the CN (average path length = 5.94), and 50 % of these sets were connected in the ADN (average path length = 6.68).

Comparison of the numbers of k-cycle structures (k = 4-9) of RRN, CN, and ADN showed a decrease in the all-step k value (Fig. 4.7). However, the graph shape was similar for each RRN and for each cycle structure number corresponding to the steps in the random sampling network; the correlation coefficient between ADN/CN and RRN/CN was 0.99, indicating the rate of change in the

Table 4.4 Number of constituent signal molecules in CN and ADN

Function	Number of signal molecules in networks		
	ADN	CN	CN—ADN (%)
Adapter	89	103	14(14)
Kinase	71	86	15(17)
Receptor	39	51	12(24)
Transcriptional factor	28	35	7(20)
Ligand	30	30	0(0)
Cytoskeletal protein	21	24	3(13)
Vesicle	17	21	4(19)
Ion channel	17	20	3(15)
GEF	19	20	1(5)
Inhibitor	17	18	1(6)
GAP	13	13	0(0)
GTPase	11	13	2(15)
PDE	9	11	2(18)
G protein	9	10	1(10)
Ribosome	10	10	0(0)
Activator	8	8	0(0)
Bcl2 Family	6	8	2(25)
Protease	8	8	0(0)
Phosphatase	15	16	1(6)
Other	57	65	8(12)
	494	570	76(13)

"Other", small molecules or histones. The CN–ADN indicates the difference of the number of signal molecules between CN and ADN. The CN has the 570 nodes and 1,333 edges, and the ADN has the 494 nodes and 974 edges. 76 AD-related signal molecules were extracted, which are known to decrease actin, β-catenin, and CaMKII. This group of genes represents 13 % of the CN. According to the pathway functions of these 76 AD-related molecules, "Bcl-2 family" and "Receptor" nodes decreased at a rate greater than the network as a whole. *Source note*: The table is adapted with permission from Yanashima et al. [23]

number of k-cycle structures between ADN and CN, and between RRN and CN, respectively. This finding also demonstrates that network size, not external factors, has an effect on the cycle structure.

The k-shortest path analysis ($k = 9$) of CN, ADN, and RRN showed no notable difference in the distribution shape between all inputs and outputs. There were also no differences in the average network pathway between ADN (67 ± 216) and RRN (144 ± 342) at $k = 9$. Thus, there was no difference in the effect of AD-related signal molecules and random signal molecules on any of the inputs or outputs. The pathways associated with neuronal plasticity showed the greatest change in the robustness for each k in ADN–CN and ADN–RRN, which indicate the change in robustness between ADN and CN, and between ADN and RRN, respectively (Fig. 4.8a). Likewise, there was a large decrease in the robustness of the pathways associated with neurite outgrowth that involve NGF and ACh (Fig. 4.8b–e). The decrease was within the top 10 % of all combinations.

Table 4.5 Network feature analysis of signal molecules

	Centrality analysis				Node removal analysis	
	k-core	Between ness	Closeness	Degree	Average path length	Articulation point
AD						
ALL	0.61 ± 1.24	0.008 ± 0.013	0.21 ± 0.18	0.012 ± 0.015	5.453 ± 0.024	107.78 ± 0.75
OUT	0.66 ± 1.05	0.006 ± 0.013	0.27 ± 0.30	0.012 ± 0.015		
IN	2.62 ± 133	0.007 ± 0.016	0.24 ± 0.04	0.009 ± 0.013		
Others						
ALL	0.70 ± 1.10	0.005 ± 0.011	0.21 ± 0.17	0.010 ±0.011	5.452 ± 0.022	107.83 ± 0.64
OUT	0.76 ± 1.42	0 005 ± 0.011	0.21 ± 0.23	0.010 ±0.011		
IN	2.59 ± 1 25	0.006 ± 0.013	0.24 ± 0.04	0.008 ± 0.009		

Network feature analysis of AD-related signal molecules and other signal molecules in the network. The analysis was performed by measuring *k*-core, betweenness, closeness, degree, change in average shortest path length, and change in articulation points (mean ± SD). *IN* incoming paths; *OUT* outgoing paths; *ALL* incoming and outgoing paths. *Source note*: The table is adapted with permission from Yanashima et al. [23]

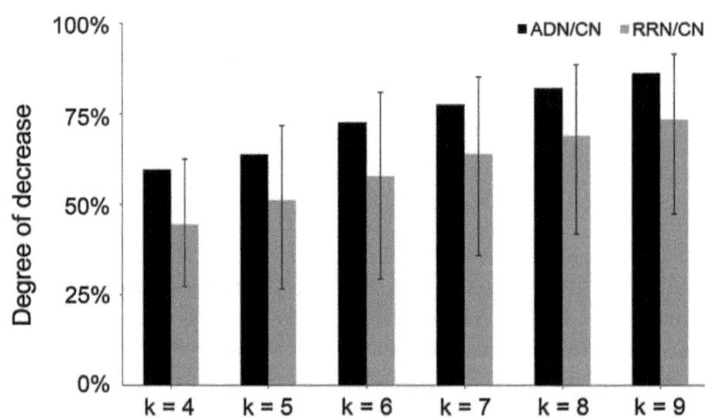

Fig. 4.7 Rate of change in the number of *k*-cycle structures between ADN and CN (ADN/CN), and between RRN and CN (RRN/CN), respectively. The *X*-axis represents step *k*, and the *Y*-axis represents the rate of decrease. The error bars represent a *top value* of 95 % and a *bottom value* of 5 %. *Source note*: The figure is adapted with permission from Yanashima et al. [23]

The largest change in robustness was found for the pathway between Glu and actin; the *R*-values were −14.9, −13.6, and −1.29 for ADN–CN, ADN–RRN, and RRN–CN, respectively (data not shown). These changes in robustness were independent of the signal molecule number, network density, the average clustering index, or the average shortest path length. The pathways associated with neuronal death that involve FasL and ICAD showed no changes in the robustness; however, CN and RRN showed increases in each step of the TNFα to ICAD and caspase 3 pathways (Fig. 4.8f–i). These results show that AD-related signal molecules have more selective effects on neural plasticity and neurite outgrowth than do random signal molecules.

Analysis of certain inputs to all outputs (Fig. 4.9) showed a large decrease in signal molecules associated with NRG, NGF, Reelin, and DA, a neuromodulator

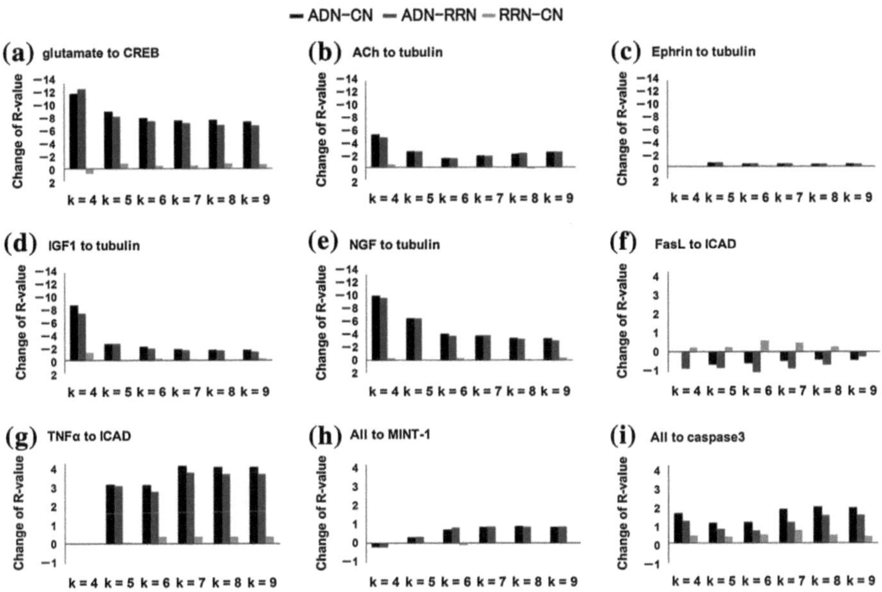

Fig. 4.8 Pathway robustness: individual input–output relationships in ADN–CN, ADN–RRN, and RRN–CN ($k = 4$–9). **a** Robustness changes in the pathways from Glu to CREB associated with neuronal plasticity. **b–e** Robustness changes in the pathways from ACh, Ephrin, IGF1, and NGF to tubulin, associated with neurite outgrowth. **f, g** Change in the number of pathways from FasL and TNFα to ICAD, associated with neural cell death. **h, i** Change in the number of pathways from all inputs to MINT-1 and caspase 3, associated with accumulation of APP. *Source note*: The figure is adapted with permission from Yanashima et al. [23]

of reward-based and motor learning (Sect. 4.4; [38, 39]). The reduction in the DA receptor level positively correlated with the severity of cognitive dysfunction in AD patients. In contrast, EGF and the NT family (including BDNF and NT4) showed an increase in associated signal molecules. The level of BDNF is increased in AD patients and in the hippocampus of a transgenic mouse model of AD [40, 41]. However, the *R*-value of inputs was between 0.8 and -1.2. This suggests that the effect of BDNF on the robustness in AD is small. Analysis of all inputs to certain outputs revealed a large decrease in signal molecules associated with CREB, actin, and tubulin (Fig. 4.9). In contrast, the transcription factor NFAT and the actin-binding proteins α-actinin and profilin showed an increase in associated signal molecules. Since the *R*-value range was between 1.2 and -4.3, the results of the comparison of input to total output imply that AD affects the expression of the output molecules more than that of the input molecules. The changes in similarity between the input and output sets in the CN, ADN, and RRN indicate that similarity is lower in the ADN than in the RRN at $e = 5$ and 9, but higher at $e = 6$, 7, and 8 (Fig. 4.10).

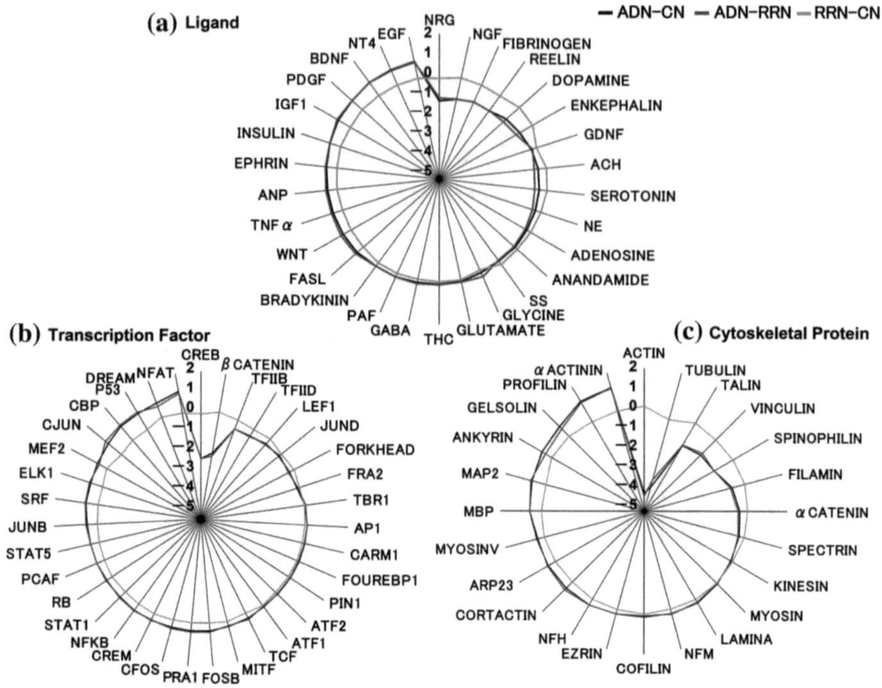

Fig. 4.9 Robustness of inputs and outputs in ADN–CN, ADN–RRN, and RRN–CN ($k = 9$), where indicate the change in robustness between ADN and CN, between ADN and RRN, and between RRN and CN, respectively. **a** Robustness analysis of the pathways from certain ligands to all outputs (transcription factors and cytoskeletal proteins). The R-value range of inputs was between 0.8 and −1.2. **b** Robustness analysis of the pathways from all ligands to certain transcription factors (R-value range, −2.1 to 1.1). **c** Robustness analysis of the pathways from all ligands to certain cytoskeletal proteins (R value range, −4.3 to 1.3). For the acronyms and the detailed reaction mechanisms, see Yanashima et al. [23]. *Source note*: The figure is adapted with permission from Yanashima et al. [23]

We introduced two static methods in the last two sections, algebraic and complex network methods. In the next section, a dynamic approach and its applications are introduced.

4.4 Perturbation Analysis of Psychostimulant-Evoked Synaptic Plasticity in the NAc

4.4.1 Why is Addiction Rewarding the Reward System?

Extracellular DA levels in the NAc are the input of the reward system. The psychostimulant promotes reverse transport of DA and blocks action of DAT, and thus causes a temporary increase in synaptic levels of DA in the NAc [42].

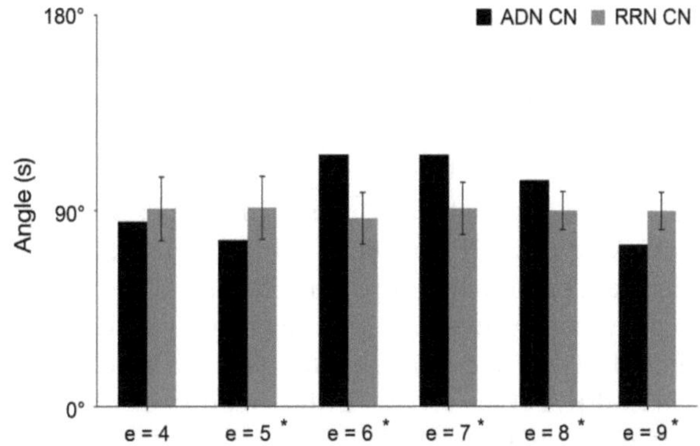

Fig. 4.10 Network similarity analysis of CN, ADN, and RRN. The *X*-axis represents step *e* and the *Y*-axis represents the angle value (*S*). Error bars, SD. *Source note*: The figure is adapted with permission from Yanashima et al. [23]

It seems to rely on synaptic plasticity of neuronal pathways involved in reward learning, and a corresponding hypothesis states that addictive drug use is a form of pathological learning [43–45]. The process of recovery from acute addiction, such as euphoria and the subsequent withdrawal, shows that the exception handler of the reward system is based on a homeostatic regulation of DA modulation [46]. Compulsive drug abuse is an exception of the exceptions for the reward system, and triggers the defense mechanism to relearn the reward prediction on the assumption of an excess of reward. The symptoms of chronic addiction are particularly challenging for patients because the regulation strategy of gene networks and the cell environment is changed in the chronic presence of stress regardless of the incorrect calculation of reward. The aim of this process (known as allostasis) is, in contrast to homeostasis, to achieve stability through change, and the cost of adaptation to cumulative stress is considered as a major cause of lifestyle diseases [47]. The pathological state of the reward system has a high risk of lethality, because the regulation of chronic addiction involves pathological learning through gene expression alterations and cytoskeletal remodeling, resulting in neuronal cell death [48].

Why is addiction so difficult to avoid? It may not seem worth the risk to have such a potentially heavy cost of the reward system, and such excess stimulation is counterintuitive. However, it seems plausible that efficient reward learning may be impossible without the “side effect” of addiction, that is, the design of the reward system may allow addiction in compensation for optimization of reward prediction. It is still an open question whether drug addiction is an acceptable trade-off as a pathological state of the reward system.

4.4.2 ODE Model of Synaptic Plasticity in the Reward Circuitry

NAc is located in the ventral striatum and receives inputs for the basal ganglia; it is a component of the important cortico-striato-pallido-thalamo-cortical loop. In addition to the DA input, NAc also receives Glu projections from the cortex that can influence the effects of DA transmission on synaptic plasticity. A model of the biochemical reactions has been developed that describes the 3 sub-modules and ultimately connects the DA and Glu signals to synaptic plasticity of GABAergic MSNs [49, 50]. Specifically, the components of the system include (1) DA metabolism and release, along with the presynaptic effects of AMPH (Fig. 4.11a; [51, 52], (2) signal transduction in the postsynapse (Fig. 4.11b; [53–56], and (3) trafficking of AMPA receptors (Fig. 4.11b; [57–59]). Color lines and dashed lines represent potentially critical mechanisms for synaptic plasticity, which are subject to perturbation investigations: the negative feedback loop PKA–PDE–cAMP–PKA (Fig. 4.11b, I: the red line); the positive feedback loop PKA–PP2A–DARPP-32-Thr75–PKA (Fig. 4.11b, II: the green line); the alternative pathways Glu–PP2B–PP1 and Glu–CaMKII–PP1 (Fig. 4.11b, III: the lavender lines); alternative pathways PKA–DARPP-32-Thr34–PP1 vs. PKA–I1–PP1 (Fig. 4.11b, IV: the blue lines). The integrative model for synaptic plasticity of MSNs in the NAc was based on ODEs and the law of mass action. All reactions were represented in the form of an enzymatic reaction (Eq. 4.6) or a simple binding reaction (Eq. 4.7):

$$S + E \underset{K_b}{\overset{K_f}{\rightleftarrows}} SE \xrightarrow{K_c} P + E, \tag{4.6}$$

$$A + B \underset{K_b}{\overset{K_f}{\rightleftarrows}} AB, \tag{4.7}$$

DA is synthesized from its precursor L-DOPA, which is produced from tyrosine. Most synthesized DA is packed into storage vesicles for later release into the synaptic cleft. DAT proteins can carry DA from the synaptic cleft back to the presynaptic terminal for recycling. In addition, DA can be enzymatically converted into other metabolites or diffuse out of the cleft. The psychostimulant AMPH

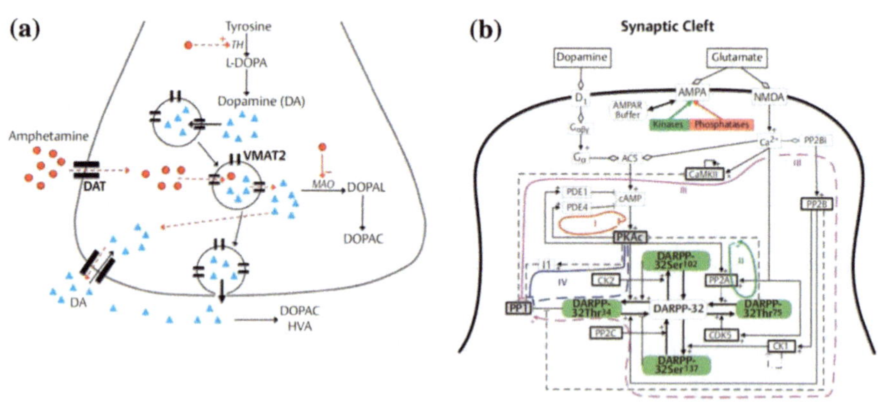

Fig. 4.11 **a** DA dynamics in the presynaptic terminal. Mechanisms that alter DA metabolism in the presynapse in response to AMPH are highlighted by *red arrows*. **b** Signal transduction in the postsynapse. For the acronyms and the detailed reaction mechanisms, see Qi et al. [49]. *Source note*: The figure is adapted with permission from Qi et al. [49]

increases release of DA from the vesicles into the cytosol through VMAT2, and to the synaptic cleft via DAT. At the same time, AMPH inhibits the enzyme MAO, which degrades excess DA, and promotes synthesis of dopamine through activation of the enzyme TH.

The effects of DA and Glu signals on synaptic plasticity depend on several important processes. DA binds to D_1 receptors and triggers production of the second-messenger cAMP, which subsequently activates PKA. The positive feedback loop PKA–PP2A–DARPP-32-Thr75–PKA is important for the effects of DA on synaptic plasticity. The negative feedback loop PKA–PDE–cAMP–PKA is also critical with respect to DA, because inhibition of this module eliminates responses of the system to DA. PKA activates PP2A, which removes the phosphate at Thr75 of DARPP-32. Since DARPP-32 phosphorylated at Thr75 inhibits PKA, these processes form a positive feedback loop (Fig. 4.12a). PKA phosphorylates DARPP-32 at Thr34 and thereby converts it into a potent inhibitor of PP1.

In contrast, Glu binds to its own receptors (AMPAR and NMDAR) and induces Ca^{2+} flux into the cell. Ca^{2+} influx activates phosphorylation of DARPP-32 by CDK5 at Thr75, which inhibits PKA activity. The elevation of Ca^{2+} activates PP2B (a.k.a. CaN), which dephosphorylates DARPP-32 and thus reduces PP1 inhibition. Thus, Glu activates PP1 through this pathway. However, the autophosphorylation of CaMKII is also activated by Glu, and leads to PP1 inhibition. Therefore, the resultant effect of Glu on synaptic plasticity can vary, depending on the relative magnitudes of PP1 activation and inhibition it causes (Fig. 4.13a).

These features regarding DA metabolism in the presynapse were taken directly from Qi et al. [51]. On the postsynaptic side, the kinetic details and initial

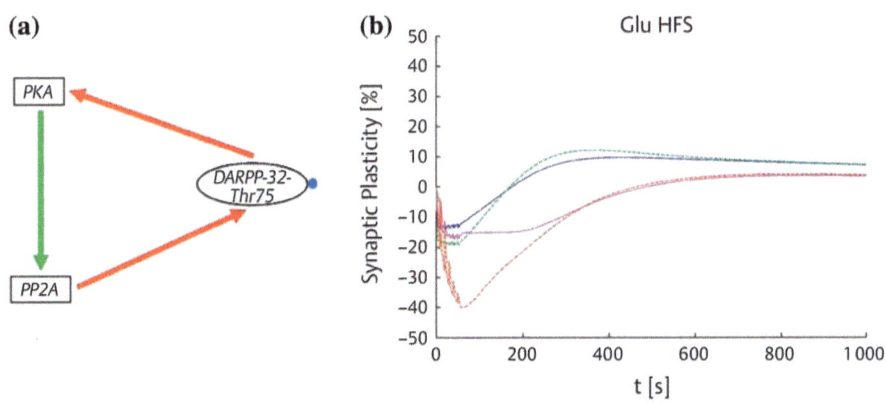

Fig. 4.12 Effect of the positive feedback loop PKA–PP2A–DARPP-32-Thr75–PKA on synaptic plasticity of MSNs. **a** Functional diagram of the positive feedback loop. *Green arrow* activation; *red arrows* inhibition. The *blue hexagon* attached to DARPP-32 indicates the phosphate group at Thr75. **b** HFS of the corticostriatal projections (Glu HFS). *Solid lines* ratios of the number of membrane-associated AMPARs after and before a stimulus. *Dashed lines* ratios of conductance of membrane-associated AMPARs after and before a stimulus. *Blue* and *green lines* represent inhibition of the loop. *Magenta* and *red lines* represent activation of the loop. *Source note*: The figure is adapted with permission from Qi et al. [49]

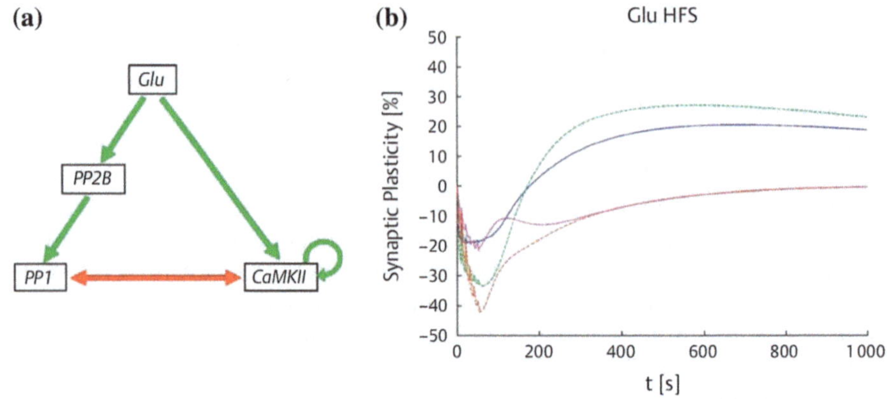

Fig. 4.13 Regulation of PP1 by Glu through CaMKII has a more significant effect on synaptic plasticity of MSNs than regulation of PP1 through PP2B. **a** Functional diagram of the regulation of PP1 by Glu through PP2B and CaMKII. *Green arrows*: activation; *red arrow*: inhibition. **b** HFS of the SN/VTA (DA HFS). *Solid lines*: ratios of membrane-associated AMPARs after and before a stimulus. *Dashed lines*: ratios of conductance of membrane-associated AMPAR after and before a stimulus. *Blue* and *green lines* represent inhibition of the Glu–CaMKII–PP1 pathway. *Magenta* and *red lines* represent activation of this pathway. *Source note*: The figure is adapted with permission from Qi et al. [49]

conditions of the model, and the full list of references are presented in [49]. Altogether, the integrated model consisted of 121 ODEs.

4.4.3 DAergic Modulation of GLUergic Synaptic Plasticity in the Presence or Absence of Psychostimulant

The published data related to the reward system reveal that (1) Corticostriatal HFS causes the release of Glu and induces synaptic depression [60–62]. (2) Simultaneous HFS of projections from the cortex and the SN/VTA causes the release of both Glu and DA, and results in synaptic potentiation [62–64]. (3) Corticostriatal HFS with simultaneous depletion of striatal DA by 6-hydroxydopamine has no effect, or results in a tendency toward synaptic depression [64, 65]. (4) HFS of the SN/VTA causes the release of DA and induces synaptic potentiation. [63]. (5) A reduction in DA release in the striatum through pretreatment with AMPT does not block the synaptic depression induced by corticostriatal HFS [64]. (6) LFS of the SN/VTA blocks synaptic depression caused by corticostriatal HFS and induces a short period of synaptic potentiation [66].

After typical diagnostics of stability and robustness (e.g., [67, 68]), the responses of the system to the above six different scenarios of HFS with or without neurotransmitter depletion or LFS were simulated (Fig. 4.14). In these simulations, the basal levels of DA and Ca^{2+} were set to 10 and 50 nM, respectively. Upon stimulation, the DA and Ca^{2+} levels peaked at 2 and 5 μM, respectively.

Subsequently, injection of the psychostimulant AMPH was simulated. Finally, the mechanisms that may critically affect synaptic plasticity were perturbed, and their effects on the performance of the system in response to various input signals were observed (Figs. 4.14 and 4.15). DA and Glu signals were considered separately as well as in combination. Overall, the results of the model simulations, as shown in Fig. 4.14, demonstrated good consistency with electrophysiological observations.

The model simulations identified 2 interesting phenomena of potential importance. First, they showed that changes in synaptic plasticity are mostly of short duration, with a typical time frame of about 10 min. This result is consistent with the observation that synaptic potentiation caused by the SN/VTA stimulation mostly lasts for 10–15 min [66]. The second interesting result is a temporary synaptic depression, which precedes synaptic potentiation in the cases of concurrent DA and Glu signals. The clinical observations of this effect have not been reported.

To study the effect of AMPH on synaptic plasticity, the mechanisms triggered by AMPH were incorporated into the model of DA metabolism, and the output of the model was plotted together with the experimental observations ([69]; Fig. 4.15). The dynamic responses of extracellular DA produced by the model were very similar to those experimentally measured. The effects of different

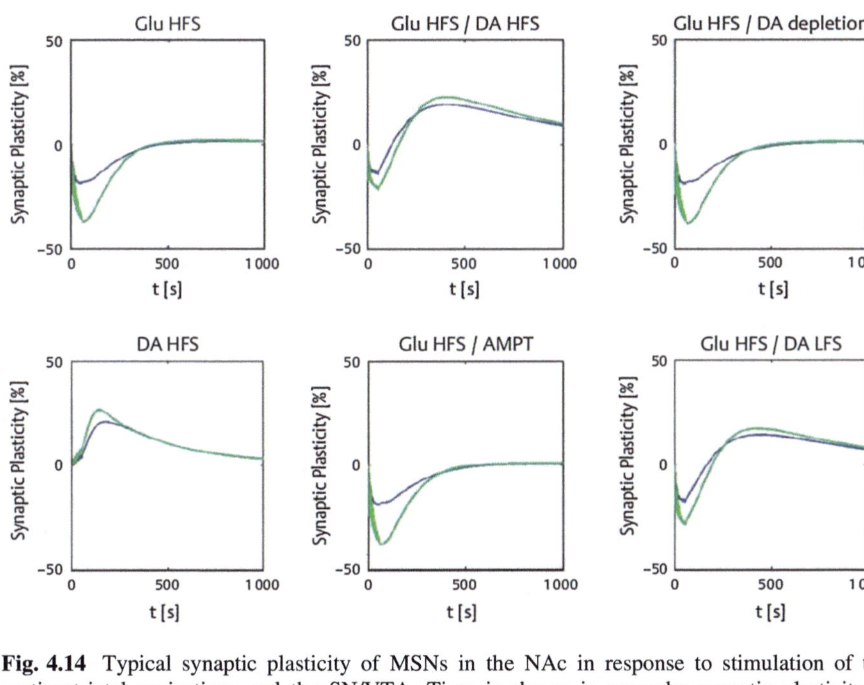

Fig. 4.14 Typical synaptic plasticity of MSNs in the NAc in response to stimulation of the corticostriatal projections and the SN/VTA. Time is shown in seconds; synaptic plasticity is expressed as the ratio of the number of membrane-associated AMPARs after and before a stimulus (*blue lines*), or as the ratio of conductance of membrane-associated AMPARs (*green lines*). *Source note*: The figure is adapted with permission from Qi et al. [49]

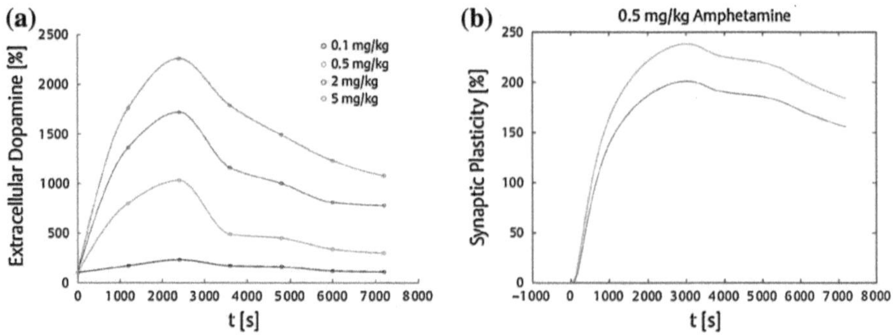

Fig. 4.15 **a** Effect of AMPH on synaptic plasticity of MSNs. Experimental measurements (connected symbols) of released DA following single injections of different doses of AMPH, namely 0.1, 0.5, 2.0, and 5.0 mg/kg. **b** Effect of AMPH on synaptic plasticity of MSNs. Synaptic plasticity caused by a single dose of 0.5 mg/kg AMPH. *Blue line*: ratio of membrane-associated AMPARs after and before the injection of AMPH. *Green line*: ratio of conductance of membrane-associated AMPARs after and before the injection of AMPH. *Source note*: The figure is adapted with permission from Qi et al. [49]

amounts of AMPH were also simulated, with dosages varying from 0.1 to 5 mg/kg, a range that corresponds to medical treatments and recreational use, respectively (Fig. 4.15a). The effective period of 0.5 mg/kg AMPH turned out to be longer than 2 h, thus requiring an increased time window for simulations. A single injection of 0.5 mg/kg AMPH can potentiate synaptic efficacy about 3-fold (Fig. 4.15b). The system behaved in a quasi-bistable way, and the synaptic potentiation lasted for over 2 h. An intervention causing reduced reward is the application of D_1 antagonists (e.g., SCH 39166), which in a model simulation caused an alteration of AMPH effects on synaptic plasticity.

4.4.4 Perturbation Analysis of Dynamic DA Modulation of Signal Transduction Pathways

We perturbed several components, which are expected to be potentially important to synaptic plasticity. Perturbations consisted of 10-fold activation and inhibition, and were implemented by multiplication of the relevant rate constants by 10 or 0.1, respectively. The simulation results show that inhibition of this loop can counteract the synaptic depression effect of Glu signals (Fig. 4.12b). For DA signals, however, activation of this loop enhanced synaptic potentiation. In response to concurrent DA and Glu signals, inhibition of this loop, rather than its activation, counteracts the synaptic depression caused by Glu. Quantitatively, this positive feedback contributes more significantly to the effects of DA than of Glu. Activation of the Glu–CaMKII–PP1 pathway enhanced the effects of both DAergic modulation and Glu signals, whereas its inhibition diminishes the normal effects of both DA and Glu, so that corticostriatal HFS induces synaptic potentiation instead

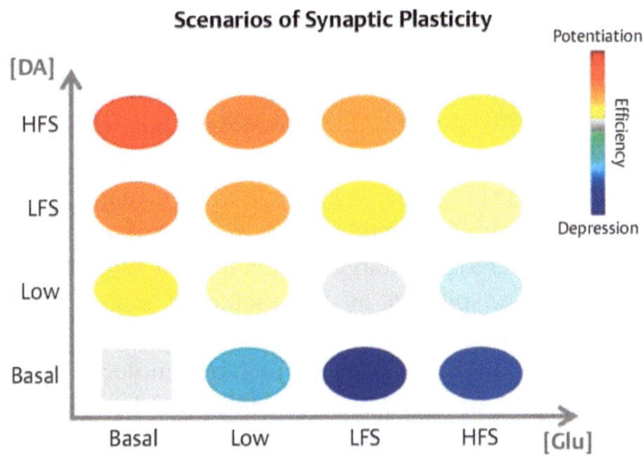

Fig. 4.16 Synaptic plasticity of MSNs in response to various stimuli. Synaptic plasticity depends on various combinations of concurrent stimuli of corticostriatal projections and the SN/VTA. *Grey areas* reflect negligible effects. *Source note*: The figure is adapted with permission from Qi et al. [49]

of synaptic depression (Fig. 4.13b). In contrast, the Glu–PP2B–PP1 pathway has a less significant impact on synaptic plasticity. For the effect of negative feedback loop PKA–PDE–cAMP–PKA on synaptic plasticity of MSNs, and the other experimental results, see Qi et al. [49].

Under normal conditions, the effects of DA and Glu last for about 10–20 min, but this time period extends to over 2 h under the influence of AMPH. Glu signals are released by neurons projected from the cortex to the striatum, while DA signals come from striatal projections of DAergic neurons. In the striatum, the MSNs typically respond to DA stimuli with synaptic potentiation, and to Glu signals with synaptic depression. Under concurrent DA and Glu signals, the synaptic plasticity varies with the input combinations as described in Fig. 4.16.

As a consequence, these two feedback loops should be further investigated, because their alteration might have the potential of reversing drug use disorders. The Glu signal from the cortex is similarly critical for drug-induced adaptations of neuronal behavior. In summary, our results suggest that Glu signaling relies more on the Glu–CaMKII–PP1 pathway than on the Glu–PP2B–PP1 mechanism with respect to the regulation of synaptic plasticity. DA and Glu signals interact with each other through multiple pathways, one of which is the inhibition of PP1 by PKA. The model simulations indicate that the indirect inhibition of PP1 through DARPP-32 phosphorylated at Thr34 might actually be more effective than its inhibition through I1.

4.5 Remarks: Systems Biology and Mathematical Modeling

Mathematical modeling is aimed at an emergent property of a system, not an entity in a system, which is often intangible, invisible, and non-measurable in nature. Nonetheless, wet-bench approaches, such as gain- and loss-of-function studies, also help us predict the emergence of complex oscillatory behavior during cell cycle, even without equations and numbers as shown in Sect. 4.2. One of the important problems in systems biology is whether molecular biosystems can be described in a simple way, even without comprehensive data. The question is not whether biosystems can be modeled, but whether models of biosystems finally become quite simple like oscillation or "$E = mc^2$", or complicated like an airplane blueprint if the model is adequate (e.g., [70]).

Systems biology models of signal transduction, and gene networks in particular, need to overcome a larger obstacle than limited data precision and lack of information. The protein as a variable in signal transduction models changes its enzymatic activity and binding affinity via posttranslational modifications (such as phosphorylation) or binding to other molecules. For instance, the dodecameric holoenzyme of CaMKII is considered to be a key molecule in postsynaptic plasticity, because its bistable activation of CaMKII acts like a memory switch, and is thought to reflect the binary history of neuronal excitability through CaMKII binding to Ca^{2+}/CaM and phosphorylation of Thr286 and Thr305/Thr306 during holoenzyme formation [71]. In addition to these key regulatory events, there are many other modifications including phosphorylation of Thr253. Assuming that the number of possible states for each average protein and the number of pertinent proteins in a cell are 20 and 200, respectively, the number of all possible states of a protein network (model variables) would be 20^{200}, not 200. Since 20^{200} precise experiments to identify kinetic parameters (model constants) are, of course, not feasible, simplification of the process of protein modification is necessary to address this basic problem in signal transduction modeling, which is called the combinatorial explosion problem [72]. Unless we have good data and a scheme of simplification, the numerical solver will not complete the numerical integration in real time. The current parameter estimation techniques cannot be compared with the vast parameter space of the model in scale [73]. On the other hand, it is helpful to think about a design of the computational roles and model granularity to understand the system, including a pathological state, for example, "why is addiction rewarding the reward system?" in the previous section.

Generally speaking, simple models are attractive for mathematical modelers, particularly in the natural sciences and engineering disciplines. In molecular systems biology, complicated models have often been selected before searching for an appropriate model granularity, because of their importance for experimental validations in molecular biology. The complexity of a model depends on the researcher's interests and the research subject. Which level of complexity is more attractive for experimental biologists oriented towards systems biology? The authors' personal choice would be a simple model. On the other hand, we also believe that if a number of simple rules emerge from complicated models, this would be a new model not only to simplify the complex phenomenon but also to overcome the fragmentation of knowledge. It is also worth noting that, in most cases, a number of complex behaviors can be captured as a part of the state space of a nonlinear system with many degrees of freedom. This is a good example to caution against overfitting to the data by providing detailed information, and to emphasize the need to focus on a system for the interpretation of the observed data.

The clarification of non-measurable characteristics is sometimes more valuable in systems biology than the accurate reproduction of the behaviors, and it might provide an important contribution to natural science by means of mathematical modeling. The authors hope that this chapter will be of some help in understanding the concepts of systems biology and the role of modeling in biology.

Acknowledgments The authors would like to thank Professor Masaru Tomita, Yoshiya Matsubara, and Ryoji Yanashima at Institute for Advanced Biosciences, Keio University, Professor Shun Ishizaki at Faculty of Environment and Information Studies, Keio University, and Dr. Zhen Qi at Department of Biomedical Engineering, Georgia Institute of Technology for helpful discussions. Part of this chapter was rewritten from collaborative papers with these coauthors.

References

1. van Hemmen JL, Sejnowski TJ. (2006) 23 Problems in systems neuroscience. Oxford University Press, Oxford
2. Lisman J, Lichtman JW, Sanes JR (2003) LTP: perils and progress. Nat Rev Neurosci 4:926–929
3. Malenka RC, Nicoll RA (1999) Long-term potentiation-a decade of progress? Science 285:1870–1874
4. Kauer JA, Malenka RC (2007) Synaptic plasticity and addiction. Nat Rev Neurosci 8:844–858
5. Di Chiara G, Imperato A (1988) Drugs abused by humans preferentially increase synaptic dopamine concentrations in the mesolimbic system of freely moving rats. Proc Natl Acad Sci USA 85:5274–5278
6. Tisch S, Silberstein P, Limousin-Dowsey P, Jahanshahi M (2004) The basal ganglia: anatomy, physiology, and pharmacology. Psychiatr Clin North Am 27:757–799
7. Cooke SF, Bliss TVP (2006) Plasticity in the human central nervous system. Brain 129:1659–1673
8. Papin JA, Palsson BO (2004) The JAK-STAT signaling network in the human B-cell: an extreme signaling pathway analysis. Biophys J 87:37–46
9. Schilling CH, Letscher D, Palsson BO (2000) Theory for the systemic definition of metabolic pathways and their use in interpreting metabolic function from a pathway-oriented perspective. J Theor Biol 203:249–283
10. Matsubara Y, Kikuchi S, Sugimoto M, Oka K, Tomita M (2008) Algebraic method for the analysis of signaling crosstalk. Artif Life 14:81–94
11. Bhalla US, Iyengar R (1999) Emergent properties of networks of biological signaling pathways. Science 283:381–387
12. Kikuchi S, Fujimoto K, Kitagawa N, Fuchikawa T, Abe M, Oka K, Takei K, Tomita M (2003) Kinetic simulation of signal transduction system in hippocampal long-term potentiation with dynamic modeling of protein phosphatase 2A. Neural Netw 16:1389–1398
13. Kuroda S, Schweighofer N, Kawato M (2001) Exploration of signal transduction pathways in cerebellar long-term depression by kinetic simulation. J Neurosci 21:5693–5702
14. Papin JA, Price ND, Palsson BO (2002) Extreme pathway lengths and reaction participation in genome-scale metabolic networks. Genome Res 12:1889–1900
15. Wiback SJ, Palsson BO (2002) Extreme pathway analysis of human red blood cell metabolism. Biophys J 83:808–818
16. Bell SL, Palsson BO (2005) ExPA: a program for calculating extreme pathways in biochemical reaction networks. Bioinformatics 21:1739–1740

17. Barabasi AL, Oltvai ZN (2004) Network biology: understanding the cell's functional organization. Nat Rev Genet 5:101–113
18. Jeong H, Mason SP, Barabasi AL, Oltvai ZN (2001) Lethality and centrality in protein networks. Nature 411:41–42
19. Futreal PA, Coin L, Marshall M, Down T, Hubbard T, Wooster R, Rahman N, Stratton MR (2004) A census of human cancer genes. Nat Rev Cancer 4:177–183
20. Ideker T, Sharan R (2008) Protein networks in disease. Genome Res 18:644–652
21. Ma'ayan A, Jenkins SL, Neves S, Hasseldine A, Grace E, Dubin-Thaler B, Eungdamrong NJ, Weng G, Ram PT, Rice JJ, Kershenbaum A, Stolovitzky GA, Blitzer RD, Iyengar R (2005) Formation of regulatory patterns during signal propagation in a mammalian cellular network. Science 309:1078–1083
22. Luscombe NM, Babu MM, Yu H, Snyder M, Teichmann SA, Gerstein M (2004) Genomic analysis of regulatory network dynamics reveals large topological changes. Nature 431:308–312
23. Yanashima R, Kitagawa N, Matsubara Y, Weatheritt R, Oka K, Kikuchi S, Tomita M, Ishizaki S (2009) Network features and pathway analyses of a signal transduction cascade. Front Neuroinform 3:13
24. Liang WS, Dunckley T, Beach TG, Grover A, Mastroeni D, Walker DG, Caselli RJ, Kukull WA, McKeel D, Morris JC, Hulette C, Schmechel D, Alexander GE, Reiman EM, Rogers J, Stephan DA (2007) Gene expression profiles in anatomically and functionally distinct regions of the normal aged human brain. Physiol Genomics 28:311–322
25. Mason O, Verwoerd M (2007) Graph theory and networks in biology. IET Syst Biol 1:89–119
26. Coulson EJ (2006) Does the p75 neurotrophin receptor mediate Aβ-induced toxicity in Alzheimer's disease? J Neurochem 98:654–660
27. Korf RE (1987) Depth-first iterative-deepening. Artif Intell 27:97–109
28. Walsh DM, Klyubin I, Fadeeva JV, Cullen WK, Anwyl R, Wolfe MS, Rowan MJ, Selkoe DJ (2002) Naturally secreted oligomers of amyloid beta protein potently inhibit hippocampal long-term potentiation in vivo. Nature 416:535–539
29. Tuszynski MH, Thal L, Pay M, Salmon DP, U HS, Bakay R, Patel P, Blesch A, Vahlsing HL, Ho G, Tong G, Potkin SG, Fallon J, Hansen L, Mufson EJ, Kordower JH, Gall C, Conner J (2005) A phase 1 clinical trial of nerve growth factor gene therapy for Alzheimer disease. Nat Med 11: 551–555
30. Hoshi M, Takashima A, Murayama M, Yasutake K, Yoshida N, Ishiguro K, Hoshino T, Imahori K (1997) Nontoxic amyloid beta peptide 1–42 suppresses acetylcholine synthesis. Possible role in cholinergic dysfunction in Alzheimer's disease. J Biol Chem 272:2038–2041
31. Cacquevel M, Lebeurrier N, Cheenne S, Vivien D (2004) Cytokines in neuroinflammation and Alzheimer's disease. Curr Drug Targets 5:529–534
32. McCusker SM, Curran MD, Dynan KB, McCullagh CD, Urquhart DD, Middleton D, Patterson CC, McIlroy SP, Passmore AP (2001) Association between polymorphism in regulatory region of gene encoding tumour necrosis factor alpha and risk of Alzheimer's disease and vascular dementia: a case-control study. Lancet 357:436–439
33. Barrett CL, Price ND, Palsson BO (2006) Network-level analysis of metabolic regulation in the human red blood cell using random sampling and singular value decomposition. BMC Bioinformatics 7:132
34. Borgwardt K, Kriegel H (2005) Shortest-path kernels on graphs. Proc IEEE Intl Conf Data Mining 5:74–81
35. Allison DW, Chervin AS, Gelfand VI, Craig AM (2000) Postsynaptic scaffolds of excitatory and inhibitory synapses in hippocampal neurons: maintenance of core components independent of actin filaments and microtubules. J Neurosci 20:4545–4554
36. Harigaya Y, Shoji M, Shirao T, Hirai S (1996) Disappearance of actin-binding protein, drebrin, from hippocampal synapses in Alzheimer's disease. J Neurosci Res 43:87–92
37. Li HL, Wang HH, Liu SJ, Deng YQ, Zhang YJ, Tian Q, Wang XC, Chen XQ, Yang Y, Zhang JY, Wang Q, Xu H, Liao FF, Wang JZ (2007) Phosphorylation of tau antagonizes apoptosis

by stabilizing β-catenin, a mechanism involved in Alzheimer's neurodegeneration. Proc Natl Acad Sci USA 104:3591–3596
38. Botella-Lopez A, Burgaya F, Gavin R, Garcia-Ayllon MS, Gomez-Tortosa E, Pena-Casanova J, Urena JM, Del Rio JA, Blesa R, Soriano E, Saez-Valero J (2006) Reelin expression and glycosylation patterns are altered in Alzheimer's disease. Proc Natl Acad Sci USA 103:5573–5578
39. Willem M, Garratt AN, Novak B, Citron M, Kaufmann S, Rittger A, DeStrooper B, Saftig P, Birchmeier C, Haass C (2006) Control of peripheral nerve myelination by the β-secretase BACE1. Science 314:664–666
40. Laske C, Stransky E, Leyhe T, Eschweiler GW, Wittorf A, Richartz E, Bartels M, Buchkremer G, Schott K (2006) Stage-dependent BDNF serum concentrations in Alzheimer's disease. J Neural Transm 113:1217–1224
41. Tang Y, Yamada K, Kanou Y, Miyazaki T, Xiong X, Kambe F, Murata Y, Seo H, Nabeshima T (2000) Spatiotemporal expression of BDNF in the hippocampus induced by the continuous intracerebroventricular infusion of β-amyloid in rats. Brain Res Mol Brain Res 80:188–197
42. Barr AM, Panenka WJ, MacEwan GW, Thornton AE, Lang DJ, Honer WG, Lecomte T (2006) The need for speed: an update on methamphetamine addiction. J Psychiatry Neurosci 31:301–313
43. Hyman SE, Malenka RC (2001) Addiction and the brain: the neurobiology of compulsion and its persistence. Nat Rev Neurosci 2:695–703
44. Hyman SE, Malenka RC, Nestler EJ (2006) Neural mechanisms of addiction: the role of reward-related learning and memory. Annu Rev Neurosci 29:565–598
45. Kauer JA (2004) Learning mechanisms in addiction: synaptic plasticity in the ventral tegmental area as a result of exposure to drugs of abuse. Annu Rev Physiol 66:447–475
46. Sulzer D, Sonders MS, Poulsen NW, Galli A (2005) Mechanisms of neurotransmitter release by amphetamines: a review. Prog Neurobiol 75:406–433
47. Schulkin J (2004) Allostasis, homeostasis, and the costs of physiological adaptation. Cambridge University Press, Cambridge
48. Yamamoto BK, Moszczynska A, Gudelsky GA (2010) Amphetamine toxicities: classical and emerging mechanisms. Ann N Y Acad Sci 1187:101–121
49. Qi Z, Kikuchi S, Tretter F, Voit EO (2011) Effects of dopamine and glutamate on synaptic plasticity: a computational modeling approach for drug abuse as comorbidity in mood disorders. Pharmacopsychiatry 44:S62–S75
50. Voit EO, Qi Z, Kikuchi S (2012) Mesoscopic models of neurotransmission as intermediates between disease simulators and tools for discovering design principles. Pharmacopsychiatry 45:22–30
51. Qi Z, Miller GW, Voit EO (2008) Computational systems analysis of dopamine metabolism. PLoS ONE 3:e2444
52. Qi Z, Miller GW, Voit EO (2008) A mathematical model of presynaptic dopamine homeostasis: implications for schizophrenia. Pharmacopsychiatry 41:S89–S98
53. Barbano PE, Spivak M, Flajolet M, Nairn AC, Greengard P, Greengard L (2007) A mathematical tool for exploring the dynamics of biological networks. Proc Natl Acad Sci USA 104:19169–19174
54. Fernandez E, Schiappa R, Girault JA, Le Novère N (2006) DARPP-32 is a robust integrator of dopamine and glutamate signals. PLoS Comput Biol 2:e176
55. Lindskog M, Kim M, Wikström MA, Blackwell KT, Kotaleski JH (2006) Transient calcium and dopamine increase PKA activity and DARPP-32 phosphorylation. PLoS Comput Biol 2:e119
56. Qi Z, Miller GW, Voit EO (2010) The internal state of medium spiny neurons varies in response to different input signals. BMC Syst Biol 4:26
57. Castellani GC, Bazzani A, Cooper LN (2009) Toward a microscopic model of bidirectional synaptic plasticity. Proc Natl Acad Sci USA 106:14091–14095
58. Hayer A, Bhalla US (2005) Molecular switches at the synapse emerge from receptor and kinase traffic. PLoS Comput Biol 1:137–154

59. Nakano T, Doi T, Yoshimoto J, Doya K (2010) A kinetic model of dopamine- and calcium-dependent striatal synaptic plasticity. PLoS Comput Biol 6:e1000670
60. Calabresi P, Centonze D, Gubellini P, Marfia GA, Bernardi G (1999) Glutamate-triggered events inducing corticostriatal long-term depression. J Neurosci 19:6102–6110
61. Calabresi P, Pisani A, Mercuri NB, Bernardi G (1992) Long-term potentiation in the striatum is unmasked by removing the voltage-dependent magnesium block of NMDA receptor channels. Eur J Neurosci 4:929–935
62. Wickens JR, Begg AJ, Arbuthnott GW (1996) Dopamine reverses the depression of rat corticostriatal synapses which normally follows high-frequency stimulation of cortex in vitro. Neuroscience 70:1–5
63. Reynolds JN, Hyland BI, Wickens JR (2001) A cellular mechanism of reward-related learning. Nature 413:67–70
64. Reynolds JN, Wickens JR (2002) Dopamine-dependent plasticity of corticostriatal synapses. Neural Netw 15:507–521
65. Tang K, Low MJ, Grandy DK, Lovinger DM (2001) Dopamine-dependent synaptic plasticity in striatum during in vivo development. Proc Natl Acad Sci USA 98:1255–1260
66. Reynolds JN, Wickens JR (2000) Substantia nigra dopamine regulates synaptic plasticity and membrane potential fluctuations in the rat neostriatum, in vivo. Neuroscience 99:199–203
67. Voit EO (2000) Computational analysis of biochemical systems: a practical guide for biochemists and molecular biologists. Cambridge University Press, Cambridge
68. Voit EO, Qi Z, Miller GW (2008) Steps of modeling complex biological systems. Pharmacopsychiatry 41:S78–S84
69. Zetterström T, Sharp T, Ungerstedt U (1986) Further evaluation of the mechanism by which AMPH reduces striatal dopamine metabolism: a brain dialysis study. Eur J Pharmacol 132:1–9
70. von Bertalanffy L. (1976) General system theory: foundations, development, applications (revised edition), George Braziller
71. Lisman J, Yasuda R, Raghavachari S (2012) Mechanisms of CaMKII action in long-term potentiation. Nat Rev Neurosci 13:169–182
72. Papin JA, Hunter T, Palsson BO, Subramaniam S (2005) Reconstruction of cellular signaling networks and analysis of their properties. Nat Rev Mol Cell Biol 6:99–111
73. Kikuchi S, Tominaga D, Arita M, Takahashi K, Tomita M (2003) Dynamic modeling of genetic networks using genetic algorithm and S-system. Bioinformatics 19:643–650
74. Aiba A, Chen C, Herrup K, Rosenmund C, Stevens CF, Tonegawa S (1994) Reduced hippocampal long-term potentiation and context-specific deficit in associative learning in mGluR1 mutant mice. Cell 79:365–375
75. Qi M, Zhuo M, Skalhegg BS, Brandon EP, Kandel ER, McKnight GS, Idzerda RL (1996) Impaired hippocampal plasticity in mice lacking the $C\beta_1$ catalytic subunit of cAMP-dependent protein kinase. Proc Natl Acad Sci USA 93:1571–1576
76. Abel T, Nguyen PV, Barad M, Deuel TAS, Kandel ER, Bourtchouladze R (1997) Genetic demonstration of a role for PKA in the late phase of LTP and in hippocampus-based long-term memory. Cell 88:615–626
77. Brandon EP, Zhuo M, Huang YY, Qi M, Gerhold KA, Burton KA, Kandel ER, McKnight GS, Idzerda RL (1995) Hippocampal long-term depression and depotentiation are defective in mice carrying a targeted disruption of the gene encoding the RI beta subunit of cAMP-dependent protein kinase. Proc Natl Acad Sci USA 92:8851–8855
78. Abeliovich A, Chen C, Goda Y, Silva AJ, Stevens CF, Tonegawa S (1993) Modified hippocampal long-term potentiation in PKC_γ-mutant mice. Cell 75:1253–1262
79. Silva AJ, Stevens CF, Tonegawa S, Wang Y (1992) Deficient hippocampal long-term potentiation in α-calcium-calmodulin kinase II mutant mice. Science 257:201–206
80. Stevens CF, Tonegawa S, Wang Y (1994) The role of calcium-calmodulin kinase II in three forms of synaptic plasticity. Curr Biol 4:687–693
81. Giese KP, Fedorov NB, Filipkowski RK, Silva AJ (1998) Autophosphorylation at Thr^{286} of the α calcium-calmodulin kinase II in LTP and learning. Science 279:870–873

82. Matford M, Kandel ER, O'Dell TJ (1995) CaMKII regulates the frequency-response function of hippocampal synapses for the production of both LTD and LTP. Cell 81:891–904
83. Ohno M, Frankland PW, Chen AP, Costa RM, Silva AJ (2001) Inducible, pharmacogenetic approaches to the study of learning and memory. Nat Neurosci 4:1238–1243
84. Mazzucchelli C, Vantaggiato C, Ciamei A, Fasano S, Pakhotin P, Krezel W, Welzl H, Wolfer DP, Pages G, Valverde O, Marowsky A, Porrazzo A, Orban PC, Maldonado R, Ehrengruber MU, Cestari V, Lipp HP, Chapman PF, Pouyssegur J, Brambilla R (2002) Knockout of ERK1 MAP kinase enhances synaptic plasticity in the striatum and facilitates striatal-mediated learning and memory. Neuron 34:807–820
85. Winder DG, Martin KC, Muzzio IA, Rohrer D, Chruscinski A, Kobilka B, Kandel ER (1999) ERK plays a regulatory role in induction of LTP by theta frequency stimulation and its modulation by β-adrenergic receptors. Neuron 24:715–726
86. Malleret G, Haditsch U, Genoux D, Jones MW, Bliss TV, Vanhoose AM, Weitlauf C, Kandel ER, Winder DG, Mansuy LM (2001) Inducible and reversible enhancement of learning, memory, and long-term potentiation by genetic inhibition of calcineurin. Cell 104:675–686
87. Zeng H, Chattarji S, Barbarosie M, Rondi-Reig L, Philpot BD, Miyakawa T, Bear MF, Tonegawa S (2001) Forebrain-specific calcineurin knockout selectively impairs bidirectional synaptic plasticity and working/episodic-like memory. Cell 107:617–629
88. Wong ST, Athos J, Figueroa XA, Pineda VV, Shaefer ML, Chavkin CC, Muglia LJ, Storm DR (1999) Calcium-stimulated adenylyl cyclase activity is critical for hippocampus-dependent long-term memory and late phase LTP. Neuron 23:787–798
89. Allen PB, Hvalby O, Jensen V, Errington ML, Ramsay M, Chaudhry FA, Bliss TVP, Stom-Mathisen J, Morris RGM, Anderson P, Greengard P (2000) Protein phosphatase-1 regulation in the induction of long-term potentiation: Heterogeneous molecular mechanisms. J Neurosci 20:3537–3543
90. Krucker T, Siggins GR, McNamara RK, Lindsley KA, Dao A, Allison DW, Lecea L, Lovenberg TW, Sutcliffe JG, Gerendasy DD (2002) Targeted disruption of RC3 reveals a calmodulin-based mechanism for regulation metaplasticity in the hippocampus. J Neurosci 22:5525–5535
91. Huang KP, Huang FL, Jager T, Li J, Reymann KG, Balschun D (2004) Neurogranin/RC3 enhances long-term potentiation and learning by promoting calcium-mediated signaling. J Neurosci 24:10660–10669

Chapter 5
Properties of Biological Networks

Vlado Dančík, Amrita Basu and Paul Clemons

Abstract Relationships in biological systems are frequently represented as networks with the goal of abstracting a system's components to nodes and connections between them. While such representations allow modeling and analysis using abstract computational methods, there are certain aspects of such modeling that are particularly important for biological networks. We explore features that are deemed necessary for living and evolving organisms and reflect the evolutionary origins of biological networks. Biological networks are robust to random alterations of their nodes and connections yet may be vulnerable to attacks targeting essential genes. Biological systems are dynamic and modular, and so are their network representations. Comparisons of biological networks across species can reveal conserved and evolved regions and shed light on evolutionary events and processes. It is important to understand networks as a whole, as significant insights might emerge from the network approach that cannot be attributed to properties of the nodes alone. Network-based approaches have a potential to significantly increase our understanding of biological systems and consequently, our understanding and treatment of human diseases.

Keywords Biological networks · Network robustness · Network dynamics · Network modularity · Network conservation · Emergent network properties · Network medicine

Acronym

(MMS)	Methyl Methanesulfonate
(ODEs)	Ordinary Differential Equations

V. Dančík (✉) · A. Basu · P. Clemons
Broad Institute of MIT and Harvard, 7 Cambridge Center, Cambridge, MA 02142, USA
e-mail: vdancik@broadinstitute.org, amrita@broadinstitute.org, pclemons@broadinstitute.org

V. Dančík
Mathematical Institute, Slovak Academy of Sciences, Grešákova 6, Košice, Slovakia

A. Prokop and B. Csukás (eds.), *Systems Biology*,
DOI: 10.1007/978-94-007-6803-1_5,

(TCA)	Tricarboxylic Acid
(PPI)	Protein-Protein Interactions
(GO)	Gene Ontology
(NP)	Nondeterministic Polynomial time
(GSEA)	Gene-Set Enrichment Analysis
(GWAS)	Genome-Wide Association Studies

5.1 Introduction

Network science has emerged as an analytical discipline with widespread applications in engineering, social sciences, and physical sciences. Network representations have been articulated for telecommunications (e.g., the Internet), social groups and interactions (e.g., transmission of ideas or diseases), and information and semantic descriptions (e.g., workflow diagrams or decision trees), among many others. Network representations also abound in the biological sciences, including food webs or evolutionary relationships among species, biological neural networks, and interactions between individual genes, proteins, or metabolites.

In all cases, the representations are made up of a set of objects, generally called nodes or vertices, and connections between pairs of objects, generally called edges or links. Nodes may be computers, species, proteins, people, or even concepts, depending on the context. Similarly, the meaning of edges will depend on the application, minimally describing the condition that there "is a connection between" two particular nodes, but possibly carrying additional information, such as direction (edges point from one node to a second node), extent (different edges have different numeric weights representing some quantity), or quality (different types of edges with different meanings are present in the same network). Because of their shared representations, the properties of networks from many disciplines can be modeled and analyzed using computational methods that are agnostic to what the particular nodes and edges represent. However, interpretation of the results of such analyses remains domain-specific and requires knowledge of the nature of the objects (nodes) and the meaning of the interactions (edges) between them.

In general, biological networks exist at a number of time and distance scales, e.g., signal transduction, gene regulation, protein interaction, metabolic, phylogenetic, and ecological [1]. In this chapter, we focus primarily on biological networks comprising elements at the molecular and cellular scales. Specifically, we will draw our examples from the area of network biology that derives from the application of bioinformatics to high-throughput biological data: molecular networks of genes or gene products (proteins) and small molecules that interact with them. In particular, we will address several important properties of biological networks—**robustness**, **dynamism**, **modularity**, and **conservation**. Each of these properties is an important element in establishing the 'signature' property of biological networks—**emergence**.

Emergence in complex adaptive systems [2] is the ability of such systems to produce phenomena that cannot easily be explained by the individual behavior of the components [3]. In other words, emergence can be considered an 'obligate' network property: it requires both the individual components (nodes, agents) and their complete system of interactions (edges, inputs). Emergence can be defined as an unexpected behavior that results from interactions among system components and between those components and their environment.

In biology, emergent phenomena are evident at several scales of organization, and indeed increasing biological organization itself is an example of emergence. Individual atoms form molecules with molecular properties not possessed by their atomic constituents. Biological macromolecules self-organize (or are chaperoned by other macromolecules) to form higher-order functional structures such as protein complexes, cellular organelles, and whole cells. Collections of cells in turn beget tissues, organs, and individual organisms. Groups of similar organisms form colonies, species, and societies, and groups of these interact to form food webs and ecosystems. At each level of biological organization, the properties of the whole depend on the components and on their complex network of interactions.

Emergent properties of biological networks depend critically on each of the other network properties discussed in this chapter. Emergence is related to robustness and dynamism as these properties respectively provide elements of stability and flexibility required to generate more complex behaviors. Emergence is related to modularity and conservation in that participating modules may function one way on their own, but change their function as contributors to larger community behavior. Thus, emergence requires network complexity sufficient to take advantage of increasing interconnectedness between individual pairwise interactions between components.

A collection of key concepts in understanding biological networks (among many other real-world networks) comes from the work of Albert-László Barabási and colleauges, who proposed a mechanism to generate the networks with scale-free degree distributions observed in the World Wide Web and genetic interaction networks [4]. Their idea was that the observed network degree distributions could be explained by two relatively simple rules: (1) that networks continually add new nodes, and (2) that new nodes are preferentially connected to nodes that already have more connections. This second rule, termed preferential attachment, was similar to a 'cumulative advantage' previously observed in networks of scientific publications [5]. Barabási and Albert comment that "the development of large networks is governed by robust self-organizing phenomena that go beyond the particulars of the individual systems". In other words, the scale-free nature of such networks is an example of an emergent property of the network itself, and networks of many different kinds may share similarities in such emergent properties. Similar studies have identified additional rules that govern real-world networks, such as the propensity of hierarchically organized systems to produce networks that are both scale-free and have a high level of clustering [6], including among highly connected modular biological units comprising several cellular components [7]. Several recent studies use modern biological network science to advance the understanding of human disease, including modeling cancer metastasis patterns [8], finding new targets in drug-resistant pathogens [9], and understanding interdependencies and disease progression in patients with multiple diseases [10]. Recent theoretical and predictive work in understanding complex systems and their emergent properties is shifting the focus from network topology to network dynamics [11] as a new frontier at which new theories and generative rules are needed.

At the end of this chapter, following our discussion of biological network properties, we present a section on network interpretation and visualization, highlighting some of the tools available to researchers studying biological networks and systems biology. Finally, we conclude the chapter with a short prospective on future possibilities for network biology to impact human disease.

5.2 Biological Networks are Robust

An important aspect of biology is the ability of an organism to thrive in dynamic, transient conditions. To achieve this, organisms must have a balance between robustness and adaptability, between resisting and permitting change in their own internal states [12]. Examples of robust biological systems are prevalent at many scales, from biochemical to ecological, as will be described in this section. At each scale, robustness can reflect the properties of individual components, or the dynamic feedback between interacting elements. For example, during temperature change, expression of a metabolic function may be robust-an enzyme maintains its shape and specificity across temperatures within a large dynamic range because a dependent network of reactions can sustain the supply of product, even when a single enzyme fails. Robustness can also be observed on a species-wide or genome-wide level. A genome may be robust because it has repair systems that minimize replication errors and is organized such that many mutations have little effect on its phenotype. For example, genetic robustness in yeast accounts for insignificant phenotypic outcomes upon deletion of many genes. This result is explained by the architecture of the genetic program in yeast, which is characterized by genes performing related functions being distributed in alternate pathways, and through gene-duplication events.

Perturbation of biological systems. How do we decide if a network is robust to genetic or environmental perturbations? Two methods used widely are measuring the essentiality of a particular node (gene) in the network, or measuring the stability of the biological system to perturbation from environmental stimuli. Insights into mutational robustness have come from large-scale synthetic lethal screens in model organisms [13, 14]. In these screens, pairs of mutations are combined systematically and the effects on viability are observed. These screens have shown that for nearly all genes, robustness to mutation depends on the continued presence of multiple additional gene products. Thus, pairs of these synthetic lethal genes can mask or buffer the effects of mutations in many other loci [15].

In addition to genetic change, organisms are also subject to changes in external conditions, such as environmental perturbations. In a normal environment there is a wide range of stochastic variation among individuals, for example in the concentrations of proteins within individual cells [13]. A frequent observation for evolved networks is a correlation in the robustness to different types of perturbation (genetic, stochastic or environmental). Historically, it had been suggested that mutational robustness may be related to the need to withstand these environmental or stochastic changes. This idea was based on Waddington's intuition that environmental change, stochastic variation, and genetic mutations are likely to have similar effects on an organism, because they act through the same underlying molecular mechanisms [16]. For example, the chaperone *Hsp*90 is known to confer robustness to both environmental (e.g., temperature) and mutational alteration. Similarly, sets of genes in *Caenorhabditis elegans* function as general buffers of genetic variation, and inactivation of these genes can result in multiple phenotypic consequences of mutation of many different genes [17]. A future challenge is to understand how dependencies among the requirements of genes for genetic, stochastic, and environmental robustness are determined [18].

Another way to determine a network's robustness is by obtaining a comprehensive understanding of the functional and dynamic changes that are caused by perturbations to the network. In a cellular network, each node may have a different biological function and therefore the effect of a perturbation cannot depend on the node's connectivity only, rather the functional identity of the node (gene product). Functional and dynamical robustness of cellular networks is supported by results that indicate that several relatively well-studied systems are robust to diverse perturbations. For example, the chemotaxis receptor module of *Escherichia coli* maintains its normal function despite significant changes in a specified set of internal or external parameters, which leaves its tumbling frequency relatively constant even under 100-fold deviations in its set of biochemical parameters [19]. Using computer simulations, the development of correct segment-polarity patterns in *Drosophila melanogaster* embryos is robust to changes in the simulated initial conditions such as starting concentrations, reaction parameters, or to the absence of several segment-polarity proteins [20, 21].

Although our understanding of network robustness is still limited, a few important themes have arisen through network property analyses. First, it is increasingly accepted that adaptation and robustness are network properties, and not the result of manipulation of an individual component's properties [19]. Second, the ability of a module to evolve intrinsically has a key role in creating or destroying robustness. Important modules that are conserved evolutionarily, and are responsible for key cellular functions, such as respiratory function, might be able to withstand uncommon errors to a lower extent. For example, orotate phosphoribosyltransferase (pyrE)-challenged *E. coli* strains have a reduced ability to tolerate additional gene inactivation, even in rich media [20, 22]. Third, scale-free networks, meaning those whose degree distributions obey a power law, are amazingly robust against accidental failures: even if a high percentage of randomly selected nodes fail, the remaining nodes still form a tight cluster with a

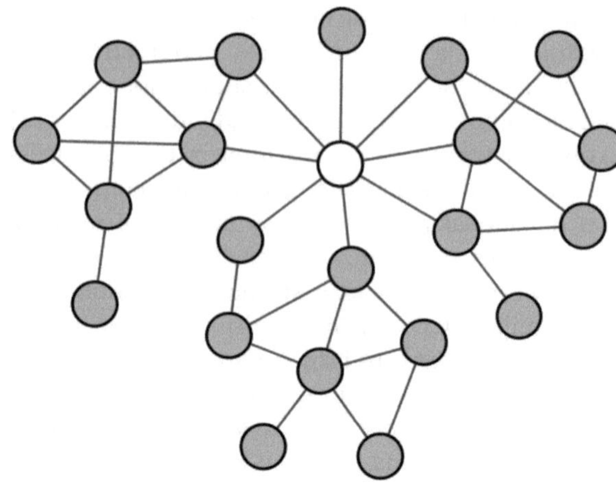

Fig. 5.1 Network robustness. Biological networks are robust with respect to disabling of a random node (*gray*), but can be affected significantly by an attack on an essential node (*white*)

short path connecting any two nodes [20]. This property occurs because random failure affects mainly the numerous lower-degree nodes, the absence of which does not disrupt the network's overall integrity. This reliance on 'hubs', on the other hand, induces an 'attack' vulnerability—the removal of a few key hubs disconnects the system into small isolated node clusters [23] (Fig. 5.1). Thus, the dual nature of scale-free networks indicates that there is a strong relationship between a protein's number of connections and its role in maintaining the viability or growth potential of a cell.

Robustness in biochemical networks. Robustness of a biochemical network is defined as the tolerance to variations in kinetic parameters with respect to the maintenance of steady state [24]. Robustness also plays an important role in discovering the mechanism of the evolutionary process of biochemical networks. Since most biochemical networks in Nature operate close to the steady state, the robustness measurement of a biochemical network is usually captured right at the steady state [24]. In addition, the sensitivity, or effect of environmental variation, is inversely related to robustness, and a biochemical network with strong robustness will be more resistant to the effects of environmental variation. Barkai and Leibler have shown in bacteria that certain key properties of biochemical networks are robust; that is, they are relatively insensitive to the exact values of biochemical parameters in a biochemical network responsible for bacterial chemotaxis [19, 25]. Bacteria such as *E. coli* are able to sense gradients of chemical ligands in their vicinity, and chemotaxis as described in a homogeneous ligand environment is insensitive to the absolute value of ligand concentration. Chemotaxis robustness therefore allows bacteria to maintain their responsiveness to chemical gradients over a wide range of attractant or repellent concentrations. Barkai et al. also find that cooperative effects among enzymatic reactions can be added without destroying the robustness of adaptation. The adaptation itself, as measured by its precision, is thus a robust property of the chemotactic network. This does not mean, however, that all the properties are equally insensitive to variations in the network parameters. For instance, the authors show that the adaptation time, which characterizes the dynamics of relaxation to the steady-state activity, displays substantial variations in the altered systems [19].

Robustness in disease and drug networks. Many times when a drug fails or produces side effects, studying drug-interaction networks may provide an explanation. A drug can be ineffective when the robustness of the cellular network of sick cells or parasites compensates for the changes caused by the drug [26]. By contrast, drug side effects can be the result of hitting an unexpected point of fragility in the perturbed network. Robustness analysis is already being used to reveal primary drug targets and methods have also been established to give a quantitative measure of changes in robustness during drug action. Other studies on drug-interaction networks include [27, 28].

Early yeast essentiality studies prompted many researchers to formulate the hypothesis that human disease genes should also have a tendency to encode hub nodes in networks. Yet, previous measurements found only a weak correlation between disease genes and hubs, resulting in an important question: what is the role of the cellular network in human diseases? Are disease genes more likely to encode hubs in the cellular network? Goh et al. [29] developed a conceptual framework to link systematically all genetic disorders (the human disease 'phenome') with the complete list of disease genes (the disease genome), resulting in a global view of the 'diseasome', the combined set of all known disorder-disease gene associations. Initial analyses appeared to support the hypothesis that disease genes, given their impact on the organism, tend to encode hubs in the interactome. In addition, disease-related proteins have a 32 % larger number of interactions with other proteins (average degree) than non-disease proteins and that high-degree proteins are more likely to be encoded by genes associated with diseases than proteins with few interactions ($P = 1.6 \times 10^{-17}$).

Other types of protein-interaction network analyses include studies of network connectivity of natural product targets compared with disease genes. Dančík et al. [30] evaluated the distributions of protein connectivities among all STRING proteins, natural product targets, and heritable disease genes. Results suggested that STRING proteins mapped from disease genes display intermediate connectivity, while STRING proteins mapped from small-molecule natural products are enriched for more highly connected proteins. This result may indicate that natural products may target proteins more essential to an organism than are disease genes.

Anvar et al. [31] further showed that the inter-species network of genes coding for the proteasome provides accurate predictions of gene-expression levels and disease phenotypes. Moreover, cross-species translation increased the stability and robustness of these networks. Unlike some other existing modeling approaches, the authors' algorithms do not require assumptions about challenging one-to-one mapping of protein orthologs or alternative transcripts, and can deal with missing data. Instead, they showed that the identified key components of the oculopharyngeal muscular dystrophy disease network can be confirmed in an unseen and independent disease model. This study presented a novel strategy in constructing inter-species disease networks that provide crucial information on regulatory relationships among genes, leading to better understanding of the disease molecular mechanisms.

Robustness in protein-interaction networks. Robustness can be measured in a protein-interaction network by measuring the integrity upon removal of the most connected nodes. Jeong et al. [32, 33] found evidence of robustness by modeling random mutations in the genome of yeast by arbitrarily removing random sets of yeast proteins. The authors identified that yeast could tolerate a deletion of a significant number of genes from its proteome, and not affect the integrity of the network. Maslov and Sneppen [34] provided more evidence of topological relationships in protein-interaction networks. These authors found that links between pairs of highly connected proteins were suppressed, whereas links between highly connected and less well-connected pairs of proteins were favored. It was suggested that this pattern increases the overall robustness of the network by localizing the effects of deleterious perturbations. As protein-interaction networks are constantly reinstated in the course of evolution, in order to integrate new proteins into the network, and to compensate for the loss of a protein, new nodes can be added to a protein interaction network by means of gene duplication [35]. This principle has been proposed as an explanation for the presence of hubs in protein-interaction networks, however later studies indicate that genes in complexes show more severe fitness effects upon deletion than other genes in protein-protein interaction networks. The authors found that this observation is not related to the number of complexes in which they are present, but that shared components between many complexes (i.e., the hubs) are not more likely to be essential than non-hub proteins [36]. Furthermore, experimentally identified protein complexes tend to be composed of uniformly essential or non-essential molecules, and thus the functional role of the whole complex is determined by the deletion phenotype of the individual proteins.

Robustness in metabolic networks. In metabolic networks, a measure of robustness is the relative change in the concentration of a metabolite to those of other metabolites. Holme [37] analyzed the average values of robustness as a function of average network modularity, and observed that robustness to global metabolic perturbations increases while the robustness to perturbations within a module remains fairly constant. The fact that their system is more robust to global than to local perturbations can be explained—a localized perturbation has a larger impact on a restricted subsystem and this subsystem cannot absorb that impact as well as the whole system would be able to do so. However, when relating robustness to modularity, one needs to specify against what kind of perturbation robustness is measured. For sudden shifts in concentration levels, more modular reaction systems are more robust and converge to a steady state faster than less modular systems.

In a different type of analysis, Barabási and coworkers presented a rigorous comparison of global properties of metabolic networks from 43 organisms, and observed that these networks had similar topological scaling properties [7, 38]. Wagner and Fell [39] and Norris and Raine [40] performed gap network analyses of *E. coli* and both groups confirmed that metabolic networks had small-world and scale-free properties. Csermly et al. [27]demonstrated the existence of a scale-free distribution of the metabolic flux of *E. coli*, and recognized the importance of the strong links, but suggested that even the weak ones may play a role in the stabilization of the system. Thus, a constrained number of hubs can control the entire metabolic network, and the small-world property of metabolic networks can ensure the stability of a network to random mutations [38, 41–43].

Our current knowledge of cellular networks and their analytical methods has arrived at a time when testing the effects of drug candidates with known cellular targets on the robustness of cellular networks is becoming increasingly possible. The more we know about disease-specific changes in cellular networks, the better we will be able to predict the efficacy of drugs *in silico* [26].

5.3 Biological Networks are Dynamic

It is well understood that biological systems are dynamic and so are the networks that represent them. Cells react to stimuli conveyed in the form of signaling molecules—typically small molecules or diffusible proteins—that originate in other cells or in the environment. For example, as a preferred nutrient becomes unavailable, cells react by activating pathways to use alternative energy sources. In addition to environmental stimuli, network dynamics also reflect signals among cells in organisms during processes like differentiation or development. Observing changes caused by perturbing the environment of cells is a main approach to explore dynamics. Large efforts have gone into studying network dynamics in the context of disease. Progress in treating diseases like cancer can benefit from using networks to understand disease mechanisms, a better understanding of network dynamics, and applying network dynamics in the development of therapies.

A recent study by Bandyopadhyay and his collaborators [44] documents the dynamic nature of biological networks. The authors selected 418 yeast genes, including most kinases, phosphatases, transcription factors, and DNA repair factors, and measured the growth of approximately 80,000 double-mutant strains in two different modes, with and without treatment by the DNA-damaging agent methyl methanesulfonate (MMS). The genetic relationship between any two mutations was determined by observing whether a strain was healthier or sicker than expected (relative to single-mutant phenotype) and collected into a genetic-interaction network. Two static networks were generated, one for the MMS-treated condition (2,297 connections) and one for the untreated condition (1,905 connections). The response to treatment was captured by a 'differential' network of 873 connections that removes 'housekeeping' interactions and reveals portions of the static networks that are sensitive to treatment. To validate their approach, the authors selected a reference set of 31 known DNA-repair genes and observed that the differential network was highly enriched for interactions among genes from the reference set. In contrast, there was no such enrichment in either of the static networks.

Constitutive and transient interactions. To explore the dynamic behavior of another network, Luscombe et al. [45] constructed a yeast network that contained 7,074 connections between 142 transcription factors and 3,420 target genes. As an illustration that some connections in networks are transient and some are constitutive, the authors determined which connections were active under each of five different conditions: cell cycle, sporulation, diauxic shift, DNA damage, and stress response. The authors observed that only 66 interactions, representing mostly regulation of housekeeping functions, were active in four or more conditions. On the other hand, half of the targets were expressed in only one of the five conditions. The authors further explored the dynamic nature of the transcription network and observed that endogenous processes (cell cycle and sporulation) progress through multiple stages while exogenous stimuli (diauxic shift, DNA damage, and stress response) exhibit rapid reaction to stimuli. In addition, some topological measures of the network (in-degrees, out-degrees, path lengths, clustering coefficients, single-input motifs, and feed-forward-loop motifs) changed considerably between endogenous and exogenous sub-networks.

Current progress in '-omics' technologies allows much more thorough investigations of network dynamics [46, 47]. A study by Nicolas et al. [48] of *Bacillus subtilis* under more than one hundred different conditions revealed a large variability of the transcriptome network. While most protein-coding sequences (85 %) were highly expressed under one or more conditions, only a few (3 %) were highly expressed under all conditions. A related study [49] focused on two conditions, nutrient shift from glucose or malate to glucose plus malate, and collected chromatin immunoprecipitation microarray [50] data, as well as mRNA, protein, and metabolite levels. These authors observed almost instant intake of malate while glucose intake was delayed, and then used integrated analysis of collected data and prior knowledge to suggest a requirement for transcriptional regulation to initiate glucose intake in *B. subtilis*.

As the previous two examples illustrate, network interactions can be divided into two types, constitutive (stable) and transient (temporary). While constitutive interactions, typically associated with protein complexes, are active all or most of the time, transient interactions only occur under certain circumstances, like reaction to outside stimuli or execution of an internal program [51]. Das et al. [52] developed a dynamic programming algorithm that uses temporal gene-expression data to identify transient interactions and explain related network dynamics.

Understanding network dynamics can provide us with a means of perturbing the network to achieve desired (or suppress undesired) phenotypes. Karlebach and Shamir [53] described an algorithm that finds the smallest possible number of perturbations required for a specific phenotypic change. They applied their method to a network of genes expressed in human glioma tumors [54, 55]. Similarly, Liu et al. [56] studied full controllability of complex networks and developed tools for identifying 'driver' nodes. Somewhat surprisingly, they observed that high-degree nodes are not likely to belong among driver nodes.

There have been many attempts to capture the dynamism of biological systems at different levels of abstraction. In principle, there are two frameworks to such modeling, quantitative or qualitative. In quantitative modeling, the variables explicitly denote values such as molecule concentrations or population sizes. Qualitative models capture systems on a more abstract level, such as whether two proteins physically interact, or whether a compound inhibits a reaction. The choice of modeling framework requires balancing model complexities that capture desired features with the availability of information needed to build good models. Here, we describe some successful approaches that were able to formalize biological systems: Boolean and Bayesian networks, Petri nets, ordinary differential equations (ODEs), and cellular automata. It should be noted that network dynamism is very active research area, and there are many additional methods to investigate network dynamics of biological systems, including ontological modeling [57], π-calculus [58], workflow modeling [59], information theory [60], graph rewriting [61], Gragner causality [62, 63], and temporal association rule mining [64], among others. For more detailed reviews of modeling formalisms see [65–69].

Boolean networks. Boolean networks, introduced by Kauffman [70], are probably the simplest models capable of capturing network dynamics. Boolean models are directed graphs in which each node, typically corresponding to a gene, has an associated Boolean value—one if a gene is expressed, and zero if it is not expressed. A collection of these values for all nodes can be seen as one state of the system. In addition, each node has an associated Boolean function (some combination of AND, OR, and NOT operators) that represents the effects of regulators of a gene on the node corresponding to that gene (Fig. 5.2). In each step of a simulation, each node is assigned a new Boolean value that is the result of applying that node's function to the current values of its input (regulator) nodes. Starting with any initial configuration, such a simulation will converge to a steady state or will cycle deterministically through a finite set of states [70].

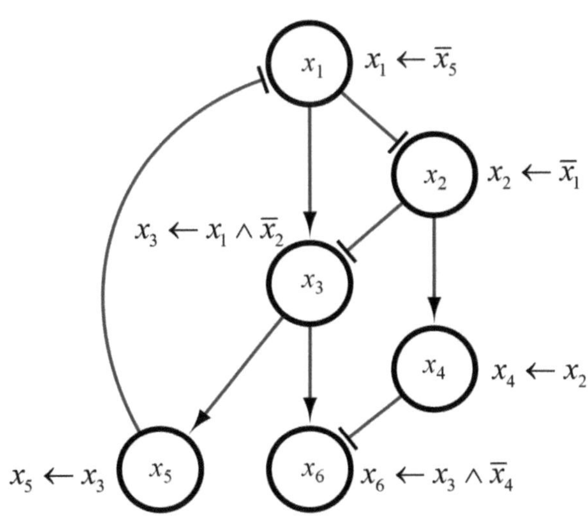

Fig. 5.2 Boolean networks. In a Boolean network each node has a Boolean variable (representing a state of the node) and a Boolean function (representing rules to change the node's state)

Despite having a simple formulation, Boolean networks are capable of capturing the complex dynamics exhibited by real biological systems like cellular states [71], feedback loops, and feed-forward loops. For example, relationships between multiple coupled feedback loops were explored by Kwon and Cho [72]. While early simulations were performed on randomly generated Boolean networks [70, 73], more recent availability of high-throughput data allowed use of Boolean networks constructed for model organisms like yeast [74]. Recent studies have explored various aspects of actual cell dynamics, including the cell cycle [75], cell proliferation [76], inflammatory signaling [77], and the FA/BRCA pathway [78].

The deterministic nature of Boolean networks can cause problems due to partial experimental evidence, stochastic components of cells, or different network rewiring due to different stimuli. Shmulevich et al. [54] addressed this problem by introducing probabilistic Boolean networks that can accommodate multiple Boolean functions per node. Recent improvements to Boolean networks include modeling of catalysts, activators, and inhibitors that is less abstract and better reflects actual molecular mechanisms [79], and use of asynchronous Boolean network ensembles to model cell-population dynamics [76].

Bayesian networks. The probabilistic nature of Bayesian networks seamlessly captures many aspects of modeling biological systems: noisy and missing measurements, limited knowledge, and natural variability. Bayesian networks are directed acyclic graphs in which every node has an associated random variable that can be discrete or continuous. The relationship among nodes is specified by probability distributions that are conditional on values of random variables associated with the input nodes (Fig. 5.3). These conditional probability distributions allow quantitative expression of relationships among related genes. Particularly popular are Gaussian Bayesian networks, in which nodes representing genes have their expression modeled by a normal distribution [80]. Friedman et al. [81] developed an algorithm for 'learning' Bayesian networks from gene-expression data and applied the method to yeast cell-cycle data. The use of a heuristic algorithm was necessary, since learning Bayesian networks is an NP-hard problem [82]. Bayesian networks were also used to detect significant sub-networks using

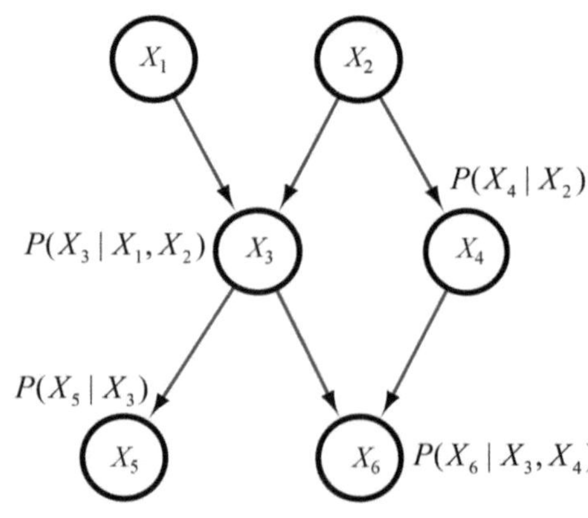

Fig. 5.3 Bayesian networks. In a Bayesian network each node has an associated random variable and the relationship between nodes is specified by a conditional probability distribution

gene-expression profiling of yeast that were genetically or environmentally perturbed [83].

The acyclic nature of the graphs underlying Bayesian networks makes it impossible to model some essential aspects of biological networks, such as feedback loops. Such limitations can be addressed by dynamic Bayesian networks, which allow cycles and introduce the concept of time, making them better-suited to model biological systems. Connections in dynamic Bayesian networks can be seen as interactions with a time delay. Dynamic Bayesian networks were used to model *E. coli* regulatory pathways in response to tryptophan stimulation or starvation [84], *E. coli* DNA repair networks [85], *Drosophila melanogaster* [86] and *Arabidopsis thaliana* [87] gene regulatory networks, yeast cell-cycle [88], and T cell activation [89, 90]. To better account for particular aspects of biological data, various extensions of dynamic Bayesian networks have been proposed, including non-parametric regression modeling [91], distribution discretization [88], and hidden state variables [89, 90].

Petri nets. Petri nets were described more than fifty years ago in Carl Adam Petri's dissertation to study the behavior of distributed systems, and soon became a popular modeling tool for other artificial systems like integrated circuits, manufacturing systems, communication networks, among many others [92]. It was only natural to extend the use of Petri nets to biological systems and especially metabolic pathways [93]. Petri nets are directed bipartite graphs consisting of two types of nodes, called 'places' and 'transitions' (corresponding to metabolites and biochemical reactions when modeling metabolic networks). The state of the system is captured by 'tokens' (corresponding to concentrations) that are assigned to places and move during simulations. The 'arcs', connecting places to transitions and transitions to places, are labeled by numbers that represent the number of tokens that 'travel' through them (corresponding to reaction stoichiometry) (Fig. 5.4). During a simulation, one of the transitions that is enabled, i.e., its input places

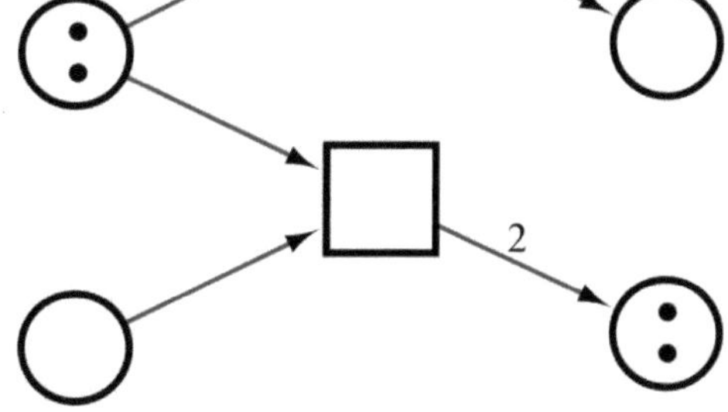

Fig. 5.4 Petri nets. An example of a Petri net with six places (metabolites), two transitions (reactions), and five tokens (concentrations). Labels on the edges represent their capacity with, by convention, labels omitted for edges with capacity equal to 1

provide a sufficient number of tokens as required by arc labels, gets fired, i.e., tokens from the input place are removed and new tokens are added to output places according to the corresponding arc labels. This relatively simple transition-firing rule is sufficient to model such complex concepts as finite state machines, nondeterministic and parallel machines, synchronization control, and priority systems. Various aspects of Petri net modeling, including reachability (can we get from one state to another?), reversibility (can we always return back to the initial state?), and 'liveness' (are there dead-lock states where no transition can fire?), can be mathematically described and investigated. For a review describing Petri net modeling and analysis in more detail see [92].

Metabolic networks modeled using Petri nets include the main glycolytic pathway and oxidative pentose phosphate pathway in erythrocytes [94], sucrose breakdown pathway in the potato tuber [95], and riboflavin production in *B. subtilis* [96]. Besides modeling metabolic networks, Petri nets have also been used to model MAPK and AKT signaling networks in breast cancer cell lines [97], the yeast mating pheromone response pathway [98], and CaMKII regulation networks [99].

Various extensions to Petri nets were proposed to capture properties pertinent to biological systems. Colored Petri nets allow for tokens and possibly transitions to be segregated into multiple groups ('colors'). Colored Petri nets were used to model the responses to EGFR and BCL2 inhibition in MCF7 breast tumor cells [100]. In stochastic Petri nets, enabled transitions fire with an exponentially distributed time delay. Stochastic Petri nets were used to model ColE1 plasmid replication [101]. One of the drawbacks of ordinary Petri nets is their discrete nature, and hybrid Petri nets address this shortcoming by introducing continuous places and continuously firing transitions. Hybrid Petri nets also use transitions with inhibitory arcs and were used to model gene regulatory networks [102]. A popular extension of hybrid Petri nets, hybrid functional Petri nets, replaces the constant speeds of continuous transitions with variable speeds that depend on values in the places. Hybrid functional Petri nets were used to model gene mechanisms for circadian rhythms and the apoptosis signaling pathway [103, 104]. For more detailed information on modeling biological systems via Petri nets see [105, 106].

Ordinary differential equations. Ordinary differential equations are a well-established approach to quantitative modeling of various dynamic systems at all levels of organization of biological systems. They can unambiguously capture the behavior of a system and correctly model responses to perturbations. Such equations are of the form

$$\frac{dx_i}{dt} = f_i(x_1, \ldots, x_n),$$

where x_i is the quantity of the i-th entity and f_i is the rate of change of x_i depending on all quantities $x_1, \ldots, x_n$ in the modeled system. Such differential equations are well-suited to capture the kinetics of individual enzymatic reactions and provide a means for modeling complex systems by combining modeled reactions [107].

Moisset et al. [108] used ordinary differential equations to model combined metabolic and gene-regulation networks in yeast. The authors' metabolic network of 39 flux reactions included glycolysis, gluconeogenesis, the TCA cycle, and fermentation reactions, and their gene-regulatory network consisted of 50 enzymes and 64 genes. The internal enzymatic reactions were modeled via first-order Michaelis-Menten kinetics, and mass-action kinetics were used for the remaining reactions. Model parameters were determined from experimental data obtained under two different conditions, one during exponential growth on glucose, and one in exponential growth on ethanol. The authors validated their model by successfully modeling the fermentation of yeast.

Numerous applications of modeling biological systems using differential equations include modeling of the toggle-switch behavior of two mutually repressive genes [109], the *E. coli* heat-shock response [110], tryptophan regulation in *E. coli* [111, 112], gene-expression [113, 114], and many others.

While differential equations were successful in precise modeling of small-to medium-sized systems, currently they are not (yet) practical for large-scale modeling. They can be too sensitive to values of model parameters and small changes in parameter values can have profound effects on the resulting models. There are a limited number of reactions with known (i.e., measured) values for kinetic parameters, and in many cases these parameters are obtained by data-fitting methods. In some cases, the functions f_i can be relatively simple, like linear differential equations, and analytically solvable. Unfortunately, much more often, the differential equations for biological systems are non-linear and therefore require computational rather than analytical solutions. To alleviate these issues, modeling via differential equations can be combined with other modeling techniques like Petri nets [107, 115], dynamic Bayesian networks [116, 117], or stochastic Bayesian networks [118].

Cellular automata. Cellular automata appeared first in the computer science literature as an abstract universal computational platform inspired by cellular population dynamics. Since that time, cellular automata were developed from a toy model to a mature modeling technique capable of realistic simulations. The main advantage of cellular automata is that, in addition to temporal dynamics, they also model spatial dynamics. Cellular automata are represented by a grid of cells (not in a biological sense) with each cell in one of a finite number of possible states (Fig. 5.5). A simulation is performed in discrete time steps according to transition rules that determine the new state of a cell depending on the states of the neighboring cells. The transition rules can be deterministic or probabilistic, creating two types of cellular automata. For reviews on modeling biological systems using cellular automata see [119–121].

Cellular automata are especially suitable for modeling systems with a significant spatial component like epidemic outbreaks. Van Ballegooijen and Boerlijst [122] studied a (spatial) susceptible-infected-resistant model of disease dynamics using probabilistic cellular automata. Each cell can be in one of the three states: susceptible, infected, or resistant. A susceptible cell can be infected by a neighboring infected cell with a probability that depends on the infection rate β and the

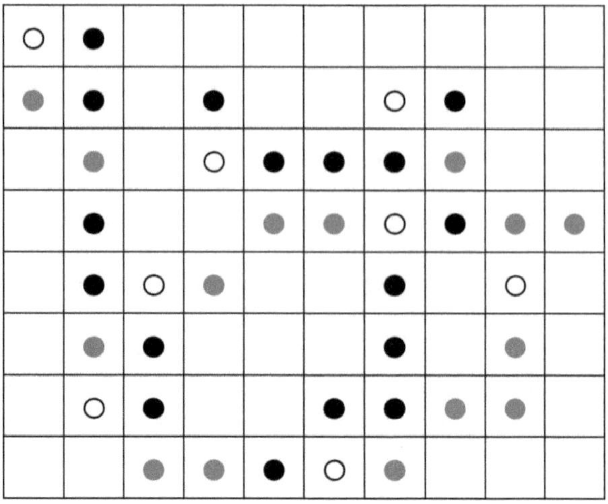

Fig. 5.5 Cellular automata. An example of a cellular automaton with cells in one of four states (*empty*, *white*, *gray*, and *black*)

number of infected neighboring cells. An infected cell becomes resistant after being infected for a time period τ_I and a resistant cell becomes susceptible after being resistant for a time period τ_R. These fairly simple transition rules can produce different patterns of behavior from localized clusters of infection to spiral or circular waves depending on the values of parameters β, τ_I, and τ_R.

Tumor modeling is another area where cellular automata were extensively used. Dorman and Deutsch [123] modeled avascular tumor growth by taking into account mitosis, apoptosis, and necrosis, including nutrient consumption and signals emitted by necrotic cells. They modeled two types of cells (tumor or necrotic) that can have five orientations (four sides + center) thus having ten discrete states. They also included two continuous variables, one for nutrients, and the other for the necrotic signals. The transition rules were probabilistic and correspond to proliferation, death, necrosis, or quiescence with appropriate probabilities depending on neighbor states and nutrient availability. The authors' cellular automata reproduce experimental results—formation of a layered pattern consisting of a central necrotic core, a rim of quiescent tumor cells, and an outer ring of proliferating cells. Additional approaches to tumor modeling with cellular automata are reviewed in [124].

Cellular automata were also used to model various biological systems including neural networks [125], thymocyte development [126], and others [119]. Vladimirov et al. [127] used cellular automata to add spatial representation of neurons when modeling very fast oscillations in the neocortex. There are various extensions to adjust cellular automata for modeling particular aspects of biological systems. Dynamic cellular automata allow movement of cell contents within the grid and more accurate modeling on molecular level [128]. Cellular automata were combined with partial differential equations to model tumor cell migration [129] and tumor growth [130].

Before the introduction of experimental high-throughput protein-interaction detection methods, the dynamic nature of biological systems was used to infer interactions, frequently involving much effort and study of one interaction at a time. At that time, rigorous modeling was limited to small, well-understood systems. As more interactions became known, these were collected into early interaction databases [131] and represented as static networks. As these networks became available to a wider community of researchers, methods using (at least partial) prior network knowledge emerged [132]. High-throughput experiments that produce large quantities of reliable data enable large-scale modeling using dynamic networks, yet the trade-off between the size of modeled systems and model accuracy remains. We expect that in the future, large-scale networks and rigorous methods will converge to well-understood, precise dynamic models of a 'virtual cell'.

5.4 Biological Networks are Modular

Biologists have observed and studied modularity at all levels, be it separation of a population into groups of individuals, or interplay of organs in an organism. Similarly, among protein-protein interactions we can observe protein complexes and functional modules [133]. Some modules can function on their own, outside of their natural cellular context. One such example is the DNA replication machinery, which was successfully isolated and is now the well-known driver of the polymerase chain reaction.

There are multiple studies confirming the modular organization of biological networks. As early yeast networks became available, Schwikowski et al. [134] explored relationships between physical protein-protein interactions (PPI) and protein functions specified by the Yeast Protein Database. The comparison revealed that proteins with related functions tend to co-localize in the interaction network, and confirmed the feasibility of using protein networks to predict functions of previously uncharacterized proteins. In another study, Tanay et al. [135] integrated yeast data from various sources, including gene expression, protein-protein interactions, phenotypic sensitivity, and transcription factors. They used log-odds ratios to quantify levels of dependencies in modules and used a search algorithm to identify 665 sub-networks with statistically significant scores. These authors used a yeast gene-ontology database to annotate detected modules and were able to predict functions for 874 uncharacterized proteins. Additional studies using yeast gene-expression data [136], yeast proteomic data [137], and yeast double mutants [138], further confirmed the modular structure of yeast networks.

Protein function prediction. With experimental evidence of modular structure in protein-protein interaction networks came use of network relationships for protein function prediction. Earlier prediction methods either relied on sequence information or used other sources of experimental data like gene expression, gene-fusion events, or phylogenetic profiling. Access to high-throughput protein-protein interaction data allowed ‘guilt-by-association’ methods to transfer putative gene annotation to other genes positioned nearby in the network. Schwikowski et al. [134] used a straightforward method in which a majority annotation of neighbors assigned annotations to 364 uncharacterized yeast proteins. The same network was analyzed by Vazquez et al. [139] who defined a scoring function that rewards connections between proteins with the same annotation and optimized it using a simulated-annealing algorithm. Hierarchical clustering is an alternative approach that was used to predict protein functions in an integrated yeast network [140] and an integrated worm network [141]. Similar approaches to protein function prediction include shortest paths in RNA co-expression networks [142], local topological weightings [143], and random walks [144].

Letovsky and Kasif [145] proposed a probabilistic approach to protein function prediction. These authors derived Bayesian likelihoods for annotations and propagated them iteratively beyond the direct neighbors from other nodes separated by two or more edges. Application of their algorithm to the yeast protein-protein interaction network yielded 320 new Gene Ontology—(GO-)term associations. In a similar approach, Deng et al. [146] developed a Markov random-field model and used Gibbs sampling [147] to estimate posterior distributions of annotations for uncharacterized proteins. The authors combined various yeast networks and applied their algorithm to the integrated network. Machine-learning approaches can be seen as extensions of probabilistic methods and various techniques of this kind have been applied, including support-vector machines [148], quadratic optimization [149], probabilistic decision trees and random forests [150, 151], Gaussian random fields [152], and multi-label kernel regularization learning [153].

Recently, Gillis and Pavlidis [154] addressed the robustness of guilt-by-association function prediction methods, both with respect to multi-functional proteins [155] and network perturbations. Multi-functional proteins, especially those with many connections, can introduce a bias in predictions that must be properly controlled. Also, gene-function information is not distributed uniformly in networks. While networks can retain much functional information even if the size is reduced by four orders of magnitude, there are a small number of connections whose removal also removes functional information. Protein-function prediction is a very active research area and new approaches are still being developed. Guilt-by-association and other methods are indispensable as only one third of human GO annotations are actually experimentally derived and the remainder rely on computational predictions [156]. The use of biological networks for protein-function prediction is reviewed in [157, 158].

Community detection. The problem of finding modules is not unique to biology and was previously studied in many different contexts. In computer science it is known as graph partitioning and was studied in connection with optimal scheduling of tasks for parallel processing or integrated circuit design. Even earlier applications were in social sciences for studying relationships within a group of people in closed environments like schools, clubs, or prisons. Motivated by sociology, module-finding algorithms are now commonly called community-detection algorithms and we will conform to this terminology as well. Community

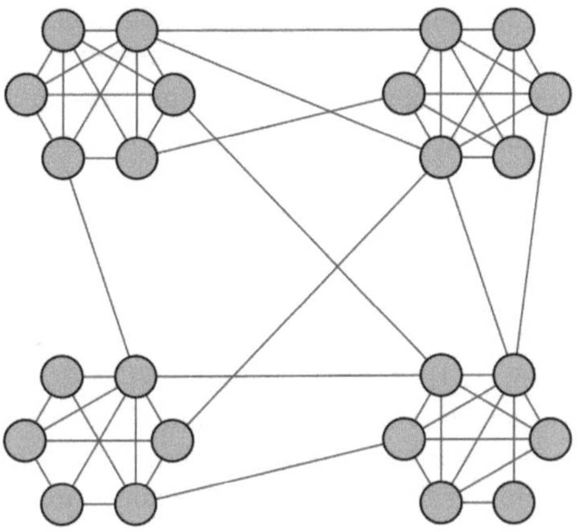

Fig. 5.6 Network modularity. A network with four distinct modules, each of which is highly connected but only lightly connected to other modules

(module, cluster) detection is a well-studied subject and there are many excellent and thorough reviews including [159–161].

Intuitively, a community is a set of nodes that have relatively dense connections with other nodes in the community and relatively sparse connections to nodes outside the community (Fig. 5.6). While there have been attempts by various authors to formalize a community definition, surprisingly there is no universally accepted measure for community strength. Probably the most popular is the modularity measure by Newman and Girvan [162], defined as the difference between the actual and expected fractions of within-community connections. Guimera and Amaral [163] used a simulated-annealing algorithm to find communities with optimal Newman-Girvan modularity scores for the metabolic networks of twelve organisms. Variants of the modularity score optimization method include randomized search [164, 165] and heuristic search [166]. The use of randomized and heuristic algorithms for finding optimal modularity scores is warranted as optimization of many modularity measures was shown to be NP-hard [167].

Divisive algorithms for community detection work by iterative removal of connections until the network is partitioned into appropriate disjoint modules. Such algorithms can also generate an overall hierarchical structure by continuing edge removal until no connections between nodes remain. Divisive algorithms differentiate themselves according to their choice for which edge or edges are to be removed. Girvan and Newman [168] define a 'betweenness' score for an edge as the number of shortest paths between pairs of nodes running through the edge (taken as the fractional number for pairs with multiple shortest paths). Connections with high edge-betweenness scores are more likely to connect different communities and are good candidates for removal by the divisive algorithm. In their algorithm, Radicchi et al. [169] removed the connection with the smallest edge-clustering coefficient, where the edge-clustering coefficient was defined as the number of triangles containing the edge divided by the maximal number of triangles that might contain the edge. By comparison to the betweenness score, which depends on overall network topology, the edge-clustering coefficient requires only knowledge of the local topology and can be computed efficiently. Variants of divisive algorithms were used to detect modules for the metabolic network of *Mycoplasma pneumoniae* [42], the network of gene co-occurrences in Medline titles and abstracts [170], the food web of marine organisms and the social network of monkeys [171], and the yeast protein-protein interaction network [172].

Modularity score optimization and divisive algorithms are examples of global methods for community detection. Global methods work with graphs as a whole and typically assign each node to one (and only one) of the communities. Other examples of global approaches to community detection include graph partitioning [173, 174], spectral methods [175–178], flow methods [179–181], hierarchical clustering [182], random walks [183, 184], and integer linear programming [185].

In contrast to global methods, local methods, starting from a single node and exploring the node's neighborhood, are capable of detecting communities without needing knowledge of the entire network structure. When starting from multiple seed nodes and applying a local search method, it is possible to generate overlapping communities. While such a structure may be a concern for some applications, it is acceptable, even desirable, for biological networks since genes can have multiple functions resulting in membership in multiple modules. Some local search algorithms can be seen as variants of clique-finding algorithms (a clique is a set of fully connected nodes), possibly followed by a clique-merging step [164, 186, 187]. Another local approach is to define some kind of community strength score and search for a set of nodes that optimizes that score [188, 189]. Random walks are also a popular approach [190, 191]. Local community detection methods are reviewed in [192].

A different approach to community detection is to group together connections rather than nodes to detect so-called 'link communities'. Naturally, such algorithms will also create overlapping node communities. Ahn et al. [193] used a hierarchical clustering algorithm to build a dendrogram with leaves corresponding to connections and branches corresponding to link communities. They applied their algorithm to detect modules in yeast protein-protein interaction networks and *E. coli* metabolic networks. The idea of connection clustering instead of node clustering was also proposed independently by Evans and Lambiotte [194] and by Gyenge et al. [195]. The algorithm of Evans and Lambiotte uses random walks on connections and was applied to social and word-association networks. The algorithm of Gyenge and collaborators, applied in machine-learning and hypertext analyses, defines an underlying Bayesian model and uses Gibbs sampling [147] to determine a model's parameters and identify the communities.

Modern approaches to community detection assume an underlying distribution according to a suitable null distribution model and express modularity in probabilistic terms. Farutin et al. [196] described an algorithm that uses random graphs with given expected degrees, in which connections are rewired in such a manner that, on average, each node's degrees are preserved [197]. Such a framework of random graphs [198] allows measures of community structure to be expressed analytically and computed efficiently. These authors used a heuristic local-search strategy to find modules with optimal community scores in human and yeast protein-protein interaction networks. The robustness of their algorithm is documented by randomly introducing false positives and false negatives and evaluating the performance of the algorithm on the perturbed networks. In similar approaches, Koyutürk et al. [199] used a piecewise degree-distribution model, and Lancichinetti et al. [200] used a null model of random graphs with an exact degree distribution. Ball et al. [201] derived a maximum-likelihood formulation of link communities and used an expectation-maximization algorithm to search for optimal solutions.

Active modules. As discussed in the section on network dynamics, it is not unusual, in response to changing conditions, that only parts of a biological network are affected. Identification of the active parts of a network can provide important insights into mechanisms of studied perturbations. Several approaches have been proposed to detect active modules, mostly using gene-expression data. Ideker et al. [202] studied the problem of detecting highly connected sub-networks with significant gene-expression differences. These authors expressed activities as z-scores calibrated against the appropriate null distribution and extended their scoring for cases of multiple conditions. Once scoring was established, the authors used a simulated-annealing algorithm with suitable heuristics to search for high-scoring modules. They used their algorithm to detect active sub-networks responsible for galactose utilization in yeast. Similar methods with alternatively defined activity scores or alternative search methods were used to identify active modules for human prostate cancer [203], the role of the immune system in melanoma [204], animal models of type 2 diabetes [205], human colorectal cancer [206], and growth hormone-treated breast cancer cells [207].

A different approach to active-module detection is to pre-process activity data into pairwise distance matrices and then combine them with network information. Hanisch et al. [208] created a combined distance measure and used a regular hierarchical clustering algorithm to generate activity clusters for yeast. Ulitsky and Shamir [209, 210] used activity data to produce log-likelihood weights on network edges and then searched the network for 'heavy' sub-networks.

Recent methods to identify active modules do not assume homogeneity of activities across the modules and adjust their approaches accordingly. An algorithm developed by Chowdhury et al. [211] captures individual combinatorial activity patterns and uses a bottom-up approach to prune the sub-network space and identify relevant modules that are then used for classification. These authors applied their method, using neural nets for classification, to predict metastasis of colorectal cancers. Dutkowski and Ideker [212] developed an algorithm that uses network-aware random forests of decision trees and applied it to determine modules responsible for human tissue differentiation during development. They also applied their method to predict clinical outcomes for breast and brain tumors.

Evolution of modules. The modularity of networks goes hand-in-hand with their hierarchical structure—small communities are grouped into larger communities and those are grouped into even larger communities [7]. While hierarchical structure is created naturally by divisive or hierarchical algorithms, any community detection method can be used to generate a hierarchy by creating new nodes from detected communities, and new edges for connected or overlapping communities, and then applying the community-detection method again on the new network of communities [196]. Information about the hierarchical structure of network modules allows more accurate functional annotation of modules by GO terms [213]. To compare hierarchy structures across various networks, Mones et al. [214] proposed a network hierarchy measure. First, they defined a local centrality score as an average inverted path length to the remaining nodes of the network, thus giving higher scores to nodes near the center of the network and lower scores to nodes on the periphery. The network hierarchy measure was then defined as the difference between the maximal and average local centrality. Intuitively, in a network with hierarchy structure, the 'root' of the hierarchy has relatively large local centrality, while in a network with no hierarchy all nodes have smaller local centralities. Alternative measures of network hierarchy, applicable to directed networks only, are defined in [215].

Modularity is indispensable for evolution, as modular systems enable modifications of individual modules while minimizing potential side effects caused by undesired interactions between modules. Also, it is easier for modular systems to react to environmental changes as they can reconfigure their modules or reuse them to acquire new functions. Although the close relationship between modularity and gene specialization has been documented, the conditions that give rise to modularity of biological systems are still the subject of intensive research [216]. Further aspects of modularity evolution are discussed in [217].

5.5 Biological Networks are Conserved

Comparison of early sequencing efforts across species provided a means of measuring levels of conservation and revealed the macromolecular basis of evolution. The availability of protein-protein interactions and other protein association networks allowed in-depth study of the dynamics of evolution and conservation. Protein interactions constrain sequence divergence as evidenced by comparing *Saccharomyces cerevisiae* and *S. pombe* orthologs [218]. Proteins involved in stable complexes have an average sequence identity of 46 %, while proteins not known to be involved in interactions have an average of 38 %. There are also examples of genes that 'rewire' during evolution. For example, the highly conserved yeast protein Puf4p regulates mitochondrial genes in the *Pezizomycotina* subphylum but targets nucleolar genes in the *Saccharomycotina* subphylum [219]. Network comparison methods are thus capable of providing valuable transfer of information between orthologs as well as distinguishing differences between taxa.

As Sharan and Ideker pointed out in a review article [220], there are three modes of network comparison: network alignment, network integration, and network query. In the sequencing world, these methods would correspond to sequence alignment, sequence assembly, and sequence database search. Network-alignment methods compare two or more networks, usually from different species, with the

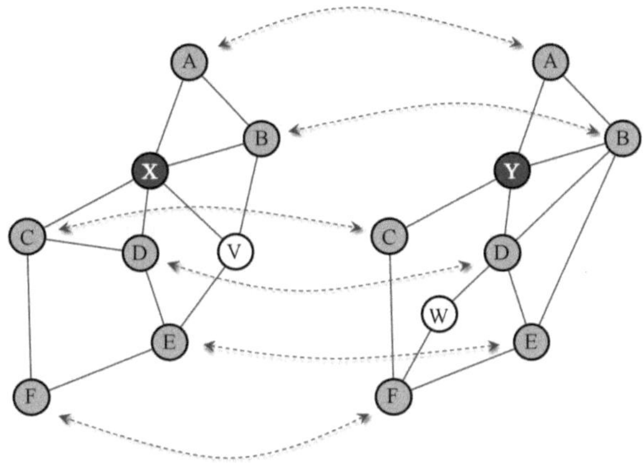

Fig. 5.7 Network alignment. An alignment of two networks with one mismatch (*dark nodes*) and two gaps (*white nodes*)

goal of identifying conserved and divergent regions (Fig. 5.7). Network alignments can be global or local, again not unlike in the case of sequence alignments. Network-integration methods use several networks obtained by different experimental methods or under different experimental conditions to build a consensus network. Network-query methods search for occurrences of a query sub-network in the whole network. Methods for network comparison are reviewed in more detail in [221, 222].

Network alignment and integration. Kelley et al. [223] were systematically searching for interaction pathways conserved between bacteria (*Helicobacter pylori*) and yeast. The authors' local network-alignment algorithm, named PATHBLAST, collects all pairs of proteins (one from each network) that have sufficient sequence similarity and uses them as nodes in a global alignment graph. Connections in the global-alignment graph are transferred from the source networks, and correspond to conserved interactions, 'gaps' (one of the source connections is indirect), and mismatches (both source connections are indirect). To assess the quality of aligned pathways, the authors proposed a score that consists of two components, one measuring the level of sequence similarity between proteins in the pathways and the other measuring the reliability of the aligned interactions. Both scores are expressed as sums of log ratios of observed and expected values. The optimal alignment is then obtained using dynamic programming on many randomized acyclic sub-networks. Other local network-alignment algorithms include NetworkBLAST [224], Graemlin [225], NetAlign [226, 227], PHUNKEE [228], match-and-split algorithm [229], and NetAligner [230].

In a global network alignment, all nodes of the source networks have to be aligned or explicitly marked as gaps. IsoRank, an algorithm of Singh et al. [231] defines a match score for pairs of nodes, one from each network, based on the stationary distribution of a random walk that is combined with a sequence similarity score. The combination is controlled by a parameter α that controls relative contributions of network topological similarity and sequence similarity. An optimal global alignment that maximizes the sum of scores of aligned nodes can be computed by a heuristic greedy algorithm, or using established methods for finding matches in bipartite graphs. These authors also extended their method to alignment of multiple networks and produced global alignments of yeast, fly, worm, mouse, and human protein-protein interaction networks [232]. Other algorithms for global network alignment include NATALIE [233], Graemlin 2.0 [234], IsoRankN [235], NetAlignBP, NetAlignMR [236], GRAAL and its variants [237, 238], PISwap [239], and an algorithm by Shih and Parthasarathy [240].

A recently introduced linear-time algorithm of Hodgkinson and Karp [241] combines network alignment with community detection. The algorithm starts by finding modules using the PageRank-Nibble method [190], then uses the detected modules to discover network conservation. To assess the performance of their algorithm, these authors investigated various evaluation measures for network alignment with a particular focus on biological networks. The alignment of human and fruit fly protein interaction networks produced by the algorithm revealed that almost ten percent of proteins belong to conserved modules. Phan and Sternberg [242] used a similar approach to align human, mouse, worm, fly, and yeast protein-protein interaction networks, and used the resulting alignments to predict protein functions. Modules were detected by a clique-percolation method and then were mapped to obtain seed-protein pairs that were, in turn, extended to produce alignments of the networks.

The presence of false positives and especially false negatives in experimental sources of protein interactions and other networks is a known problem. Integration of networks from different sources or different experimental methods can increase confidence in information supplied by the individual component networks. As an example of network integration, we describe an algorithm of Ogata et al. [243] that starts by collecting pairs of corresponding nodes from two networks, possibly creating multiple pairs for a node in cases of many-to-many correspondence. These pairs are then joined into conserved modules using single-linkage hierarchical clustering while ensuring that any two clusters can be merged only if in each of the contributing networks there is a path of limited length that connects the two clusters. The statistical significance of aligned modules is assessed by repeating the procedure on randomized networks. The authors applied their algorithm to align an *E. coli* gene co-location network to its metabolic network and identified one hundred functionally related enzyme clusters, of which 39 completely and 50 partially overlapped with known *E. coli* operons. Rhodes et al. [244] used a naïve Bayes model to integrate the protein interaction networks of yeast, worms, and flies with human protein domain data, genome-wide gene-expression data, and functional annotation data. Further integration with cancer genomics data provided networks activated in cancer. Other examples of network integration include integrated yeast networks [245], early embryogenesis in worms [141], eleven microbe species [246], mouse embryonic stem cells [247], five herpesvirus species [248], and the potato [249].

Network queries and motifs. Network queries, while similar to network alignment, do contain inherent asymmetry—a query network is usually much smaller and might be restricted to a tree or even a linear pathway. While many local network-alignment algorithms can be used to detect the presence of small subnetworks, there are approaches that specialize in network queries. Pinter et al. [250] explored querying metabolic pathways represented as directed networks in which nodes correspond to enzymes that are connected whenever a product of one enzyme is the substrate of another enzyme. A scoring function was defined so that matching similar enzymes is rewarded and gaps (unaligned nodes) are penalized. To avoid computational inefficiencies, these authors restricted queries and target networks to directed acyclic graphs. The optimal alignment for trees was then efficiently computed using dynamic programming and altered to report all suboptimal solutions with scores above a given threshold. This algorithm was used to align 113 *E. coli* pathways to 151 yeast pathways, resulting in 610 aligned pathways with statistically significant scores. In a different approach, Qian et al. [251] represented the target network as a hidden Markov model. They restricted queries to linear pathways and their algorithm allows gaps of arbitrary length. These authors evaluated their algorithm by aligning yeast pathways against networks of other model organisms. A recent algorithm of Huang et al. [252] is based on a conditional random-field model and allows both cyclic and acyclic queries and an arbitrary number of gaps. Other algorithms for network query include GenoLink [253], QPath [254], PathMatch and GraphMatch [255], SAGA [256], NetMatch [257], MetaPAT [258], QNet [259], PADA1 [260], and TORQUE [261]. Network query algorithms were recently reviewed and compared in [262, 263].

A question related to network conservation is to find conserved regions in a single network, i.e., to find network motifs. Conserved network motifs, analogous to sequence motifs, can be used, e.g., to predict protein-protein interactions [264, 265]. Ortholog prediction using network alignment produces outcomes that are improvements compared to using sequence-only methods [231, 266]. Milo et al. [267] defined a network motif as a connection pattern with significantly higher occurrence then expected by chance, and proposed an algorithm for their detection. The algorithm works by counting frequencies of all possible n-node subgraphs. The significance of n-node motifs was verified via simulation of random networks that preserve the distribution of $(n - 1)$-node subgraphs. When applied to *E. coli* [268] and yeast transcriptional regulation networks, the motif-finding algorithm found two significant motifs, a 'feed-forward loop' (a gene is regulated by two transcription factors and one of the factors also regulates the other factor) and a 'bi-fan' motif (two genes are each regulated by two different factors). These authors also applied their algorithm to a food web, a neuronal network, and a few non-biological networks. Motif distributions can be efficiently computed in analytical terms under various random-network models [269–272]. Berg and Lässig [273] extended an algorithm to approximate motif search to find motifs that are similar but not necessarily identical. The occurrence of small motifs can be combined into profiles that can be used to characterize and compare networks [274, 275], similar to an approach that has been used for decades in cheminformatics to compare chemical structures. The motif-search problem is reviewed in more detail by [276, 277].

Network evolution. Biological networks are conserved, yet they evolve. The conservation and evolution of biological networks cannot be separated from each other. As network alignment provides evidence of the existence of conserved modules, it also provides evidence of evolution. Many interactions involve various protein domains, and comparisons of conserved and evolved modules may provide insights not available from the domain arrangements revealed by protein-sequence information alone. During the course of evolution, genes get duplicated and diverge, proteins acquire new functions, and proteins change their interacting partners. Such changes, while difficult to deduce from sequence information, are reflected in the network structure and can be revealed by network comparisons.

Koyutürk et al. [278] relied on network evolution models to guide network alignment. As in previous approaches, these authors built a global alignment graph by matching orthologous interactions. To capture underlying evolutionary events, the authors defined an alignment score that, in addition to scoring orthologous interactions, also includes gene-duplication and -divergence scores. The network alignment was thus reformulated as an optimization problem, and a subgraph-growing heuristic was applied to find the optimal local network alignment. The authors applied their algorithm to budding yeast, nematode worm, and fruit fly networks, and detected 412, 146, and 83 conserved sub-networks between yeast/fly, worm/fly, and yeast/worm networks, respectively. Berg and Lässig [279] used statistical models for the evolution of nodes and connections to derive a combined model for scoring alignments. These authors applied their algorithm to align human and mouse co-expression networks. Capra et al. [280] considered a more advanced evolutionary model and integrated genome-wide comparative phylogenetic analysis with yeast functional and interaction networks. They predicted the origin (novel or duplicate) and time of creation (pre-, at, or post-whole-genome duplication) for every yeast gene and assessed various network properties for different categories of genes, observing that some properties exhibit bias. For example, younger genes are less integrated into physical interaction networks than older genes, and novel genes are less central in the network than duplicate genes of the same age. Also, proteins preferentially interact with proteins of the same age and origin. Such behaviors can be easily explained by an evolutionary model where new genes created by duplication 'inherit' interactions while *de novo* genes initially may not be fully functional.

In one step of a preferential attachment model for network evolution [4] a node is added to an existing network and is connected to older nodes randomly in such way that connections to higher-degree nodes are more likely than connections to lower-degree nodes. While such models are accurate in social networks, phylogenetic analysis of 887 co-evolving protein pairs in 15 eukaryotic species does not provide evidence for preferential attachment in protein networks [281]. Motivated by evolutionary events, gene-duplication models [282–284] add a new node to the network by randomly selecting an existing node and duplicating it and its connections (Fig. 5.8). Reflecting mutations of the duplicated genes, their connections can experience deletions and new connections can arise with certain probabilities with rewiring details varying from model to model. The duplication and rewiring

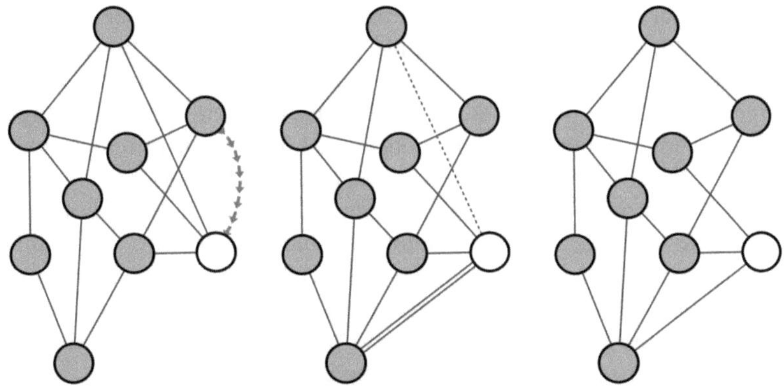

Fig. 5.8 Network evolution. A gene-duplication model for network evolution: a random node (and its interactions) is duplicated (*left*), one interaction is lost (*dashed line*) and one interaction is acquired (*double line*) due to gene mutations (*center*), resulting in an evolved network (*right*)

rates can be estimated using comparative genomics and structural information [285, 286]. It should be noted that more complex models are probably needed to precisely reflect evolutionary events, since basic gene-duplication models do not well capture the modular structure of biological networks [287]. Gibson and Goldberg [288] came to a similar conclusion and proposed a gene-duplication model enhanced by including heritable interaction sites on the surfaces of proteins. Alternative approaches, rather than duplicating one gene at time, consider whole genome duplication, motivated by such events in yeast evolution [289, 290].

Aspects of network evolution and conservation are reviewed in [291] and [35]. Importantly, there is still disagreement between estimates of the prevalence of conserved interactions, with estimates ranging from 10 to 78 % [238, 241]. Further research into network-comparison methods and more precise experimental data are needed to further our understanding of network evolution and conservation.

5.6 Biological Networks Exhibit Emergent Properties

In the Introduction, we discussed properties of networks themselves that emerge from rules or governing principles about the network or its components. Such descriptions concern emergent properties of our representation of biology as a network, allowing us to make predictions about how a biological network representation might change over time or respond to stimuli. While such work is important in understanding fundamental principles applicable to all types of networks, and understanding whether our modeling of biological systems accords with observation, some recent studies have focused on emergent properties of the biological system itself. In other words, can we build models that produce testable biological parameters or hypotheses without explicitly accounting for those parameters in our models? Studies of emergent biological properties have been reported in multiple fields and at multiple scales, even within the constraints of molecular and cellular biology, as illustrated by the following diverse approaches.

Emergent metabolic properties in adaptations to environmental change have been studied using a piecewise affine model [292]. Due to the different time-scales of change in metabolite *versus* enzyme levels, feedback from a single metabolite to transcriptional control of all enzyme levels in an un-branched metabolic pathway can give rise to multiple different phenotypes, including mono- and multi-stable ones as well as oscillations. These studies achieved simplicity of interpretation by using a formalism whose output depends, under the authors' assumptions, on only two enzymes in the pathway: the first one that initiates the pathway, and the enzyme that consumes the regulatory metabolite.

In neurobiology, computational network models that aim to model the chemical and electrical interactions between physical neuronal cells have demonstrated emergent behavior that accords with observations. For example, a neuronal network model based on the electrical spiking physiology of oxytocin cells [293] recapitulated a burst-release of oxytocin behavior that in infant mammals releases a periodic pulse of the compound into the bloodstream during suckling behavior. These emergent bursts are produced by the interactions of positive feedback and inhibitory effects between many sparsely connected cells.

Emergent properties in cellular differentiation networks, including pluripotency, de-differentiation, and trans-differentiation between cell types has been modeled using noisy random Boolean networks [294]. In this case, changing the noise thresholds in the network of interacting transcriptional regulators provided a potentially general approach to study the origins of cellular differentiation as well as the ability of scientists to reprogram cell fate. Studies such as these highlight the need for integrating theory, computational modeling, and experiment. As high-throughput data continue to grow in number and complexity, these needs to connect biological principles with sound modeling and high-quality data will become increasingly important.

In one particularly illustrative set of studies of the yeast cell cycle under different growth conditions, Barberis et al. [295] established a detailed model of cell-cycle dynamics using a system of differential equations that represent changes in concentrations over time of cell-cycle proteins. This work extended earlier models by explicitly accounting for increases in cell volume and by separating nuclear and cytoplasmic interactions between proteins. By testing model predictions under different parameters, the authors identified a critical cell size for S-phase initiation as an emergent property of their model that was dependent on nutrient source, and which agreed well with experimental observations of doubling times in glucose *versus* ethanol. Extending the fundamental model predicting critical cell size to account for additional input perturbations and additional parameters accounting for phosphorylation, allowed more precise modeling of the consequences of changing protein concentrations or phosphorylation states [296]. These studies provide a nice illustration of how a single network model, with appropriate accounting for relevant model parameters, can explain multiple emergent properties (critical cell size, cell cycle arrest, etc.) that were selected by evolution under different environmental stimuli.

The examples presented in this section are important for study of the individual systems under consideration, but also represent an important idea in network biology: that because the special 'engineering' principles of modularity, robustness, and conservation underlying biological systems reflect evolutionary processes, they are amenable to understanding using network principles [297].

5.7 Using Networks for Data Interpretation

The analysis and interpretation of network relationships between biological molecules and concepts requires tools for network visualization and interpretation, both of which are made challenging by the vast amount and heterogeneity of systems biology data. Most current network-visualization tools are applicable to a wide range of problems, but many of them reach practical limits of usability when thousands of nodes and connections have to be analyzed and visualized. We describe the functionality, and specific strengths, of these tools. Next, we describe some of the underlying methods being used to rationalize complex datasets, and that may serve as future engines of new visualization tools.

Systems biology datasets exhibit growing complexity because of numerous heterogeneous application areas and detection technologies. Integration of multiple diverse types of data is therefore gaining importance. Currently, different types of biological data, such as sequence information, protein structures and families, proteomics data, gene expression, and other experimental data are stored in distinct databases. Each existing database can be very specialized, and often stores information using specific data formats. Many of these databases also contain overlapping but not identical information with other databases, which introduces a limitation when there is a need to combine the information. Tools are currently being developed to try to simplify the interpretation of biological data by transforming raw data into visually tangible representations. The goal of most of these tools is to discover patterns and structures that remain hidden in raw datasets [298].

Network visualization. Several network visualization tools are currently available to the scientific community. Medusa [299] is based on the Fruchterman-Reingold algorithm [300], and provides 2D representations of networks of intermediate size, up to a few hundred nodes and edges. Medusa supports weighted graphs and represents the significance and importance of a connection by varying line thickness. Medusa was developed mainly to show multi-edge connections where each line represents different conceptual information. Medusa is optimized for protein-protein interaction data such as those taken from STRING [301] or protein-chemical and chemical-chemical interactions such as those taken from STITCH [302]. Cytoscape [303] is an open-source project, and mainly provides 2D representations suitable for large-scale network analysis with hundreds of thousands of nodes and edges. It can support directed, undirected, and weighted graphs, and comes with powerful visual templates that allow users to display or change the attributes of nodes or edges, zoom in or out, and browse the network. Cytoscape incorporates statistical analysis of networks and makes it easy to cluster or detect highly interconnected regions. BioLayout Express3D [304] supports both weighted and unweighted graphs together with edge annotation of pairwise relationships. It uses the Fruchterman and Reingold [300] layout algorithm for 2D and 3D graph positioning and display of the network. Other tools include Osprey [305], Pajek [306], SpectralNET [307, 308] and many others, as reviewed in [298].

Enrichment analysis. Two classes of models have been investigated by researchers to account for interactions among sets of biological entities in differential analysis of genes. The first approach is known as gene-set analysis, and its aim is to consider the joint effect of biologically related groups of genes. By performing gene-set analysis, the interaction among genes may be preserved by considering the joint effect of genes in each set [309]. The resulting inference implicitly includes interactions. While individual effects of genes may be small, the combined effects of changes in the expression of genes in a set could reveal important changes to the overall system. An examples of current methods being used widely is gene-set enrichment analysis (GSEA) [310] and its variants [311]. The second class of methods aims to incorporate information about interactions among genes and proteins into a differential analysis. Ideker et al. described an integrated genomic and proteomic analysis of perturbed networks to discover interactions among genes. Later, the same authors proposed a method to test the significance of sub-networks through permutation testing [202]. More recently, Li and Wei [312] and Wei and Pan [313] proposed Markov random field models to incorporate network information in the differential analysis of genes. In these methods, connected genes in the network are assumed to have similar expression levels and a Bayesian framework was developed using mixture models to evaluate whether each gene is differentially expressed [309]. A number of recent methods combine the advantages of incorporating network information with the strengths of enrichment analysis. Pradines et al. [314] generated connectedness profiles using random graphs with given expected degrees and used them to detect sub-networks with highly expressed genes. Sanguinetti et al. used a mixture model on graphs and a simple percolation algorithm to search for sub-networks of significant components [315]. Likewise, Shojaie et al. developed a method that incorporates network information using a mixed linear model to test whether pre-defined gene sets were differentially expressed [316].

Many of the above-mentioned methods have focused on performing single-gene analysis, or gene-set enrichment analysis for minimal experimental conditions, e.g., treatment *versus* control. Since the significance of an enrichment analysis is based on a permutation test, and the adaptation to complex experimental data including temporal correlation information is not straightforward, few methods have been developed to generalize the single-gene framework to make it more flexible. Shojaie et al. [309] analyzed arbitrary networks with directed and undirected edges, and used the flexibility of mixed linear models to develop a general inference procedure. Their method can be used to analyze changes in biological pathways and provides an inference framework for simultaneous tests of multiple hypotheses in complex experiments. Results from this method illustrate that gene-set analysis is not sensitive to small amounts of noise in network information, and provides a flexible framework used to study changes in genetic pathways.

Combining knowledge from multiple experimental sources (e.g., transcriptome [310], metabolome [317], acetylome [318]) has frequently identified functional relationships between diverse biological components. Analyses of such combined datasets can be performed at different levels: first, where each -omics dataset is analyzed separately and combining hypotheses made using each separate result, and next by integration of the datasets in advance and analyzing -omics networks of the combined data. Cavill et al. explored drug sensitivity through integration of transcriptional and metabolic data from the NCI-60 cell line panel [319]. They correlated growth inhibition with molecular profiles to identify pathways related to drug sensitivity. When combining datasets, the authors used a joint-probability estimate (to associate each pathway with each drug-sensitivity phenotype), and found 35 pathways that were significant for at least a single drug, a result they did not find when they analyzed the datasets separately. Relatively little emphasis has been placed on systematic evaluation of the extent of information overlap provided by different types of -omics data. Water et al. [320] investigated this information overlap by estimating the degree of concordance between RNA- and protein-expression changes using correlation analysis. The authors tried to reconstruct the known EGFR-regulatory network to assess whether similar biological processes are captured by each of two high-dimensional data-collection platforms by conducting gene-set enrichment analyses separately. Their results demonstrate that concordance between RNA and protein expression varies between specific functional classes of proteins, and that each data type emphasizes a specific pattern of cellular processes.

Ideally, the next generation of network-analysis methods and visualization tools should be able to better represent data from heterogeneous sources, high-throughput experiments, and text-mining applications. These tools should be able to visualize multi-edged networks across various dimensions, incorporate clustering algorithms, pattern-recognition methods, and statistical analysis methods to integrate multiple -omics datasets. Tools designed to integrate most of the aforementioned functionalities would greatly simplify large-scale research in biochemistry and molecular biology, reduce significantly time and effort spent on data processing, and provide better guidance to researchers in their experimental endeavors.

5.8 Network Medicine

Traditional approaches to disease rely on associating a patient's symptoms with pathological markers and, usually relying on a physician's experience, selecting one of many possible diagnoses and applying the appropriate treatment. As more data from various sources becomes available "the fundamental question of where function lies within a cell is slowly shifting from a singleminded [*sic*] focus on genes to the understanding that behind each cellular function there is a discernible network module consisting of genes, transcription factors, RNAs, enzymes, and metabolites" [321]. Network-based methods can increase the depth of our understanding of biological systems and processes and have far-reaching influence on various aspects of practicing medicine, including drug discovery, toxicology, biomarker discovery, and personalized medicine [322]. This viewpoint is creating a significant shift in the approach to disease treatment known by the term 'network medicine' [323, 324].

There are numerous examples of using network approaches to provide significant insight into the pathogenesis of various diseases. Ergün et al. [325] used microarray data to reverse-engineer a network that was then used to identify the androgen receptor gene among the top genetic mediators for metastatic prostate cancer. In another example, Ciriello et al. [326] developed an algorithm to identify modules that exhibit patterns of mutually exclusive genetic alterations across multiple patients and applied it to two cancers, glioblastoma multiforme and serous ovarian cancer.

A crucial first step in many drug-discovery pipelines is the selection of appropriate targets (usually proteins) and their thorough validation. It is important that target validation includes the investigation of the network neighborhood because biological networks, similarly to social and technological networks, exhibit small-world phenomena—most nodes in a network are relatively close to other nodes. Therefore, interfering with one node in the network can have pronounced effects in the node's neighborhood and across the whole network. The goal of algorithms developed by Sridhar et al. [327, 328] is to stop production of undesired compounds by a metabolic network. These authors' algorithms select enzymes to be inhibited in such a manner that the effects on remaining metabolites are minimal. Using such an approach, the authors showed that targeting arachidonate 5-lipoxygenase by benoxaprofen might not be optimal. Lee et al. [329] searched for candidate disease genes by combining data from genome-wide association studies (GWAS) with evidence from guilt-by-association predictions via an integrated human functional gene network. The algorithm, when applied to Crohn's disease and type 2 diabetes, improved the ability to detect well-validated genes. Additional network-based approaches to computational target identification are reviewed in [330–333].

In modern drug discovery, it has become increasingly difficult to develop new drugs and even more so for first-in-class drugs with novel mechanisms of action. Drug candidates face many hurdles on their way to clinical application, with undesirable on-target and off-target side effects being one of the reasons for failures. Quantitative analyses performed by Brouwers et al. [334] revealed that side-effect similarities of many drugs are caused by the network neighborhoods of the drugs' targets.

Repositioning of existing drugs is one of the approaches to alleviate difficulties facing drug candidates. While motivations for drug repositioning and the discovery of off-target effects are quite different, computationally it is essentially the same task—use a known drug-target network, possibly integrated with additional information, to predict novel drug-target interactions. Chang et al. [335] integrated structural information, obtained by protein-ligand docking, with a metabolic network, gene-expression data, and proteomic data, to build a kidney metabolic model. This model was then used to predict off-target effects for torcetrapib, a cholesteryl ester transfer protein that failed in Phase III clinical trials. Cheng et al. [336] showed that network-based predictions of drug-target interactions are superior to drug-based predictions (via 2D chemical similarity) and target-based predictions (via sequence similarity). They validated their method by repositioning five drugs to target estrogen receptors or dipeptidyl peptidase-IV.

The association of diseases with network modules rather than single genes will likely impact the future of drug discovery. For treatment to have a positive impact, rather than focusing on single targets, it may be necessary to focus on the whole network associated with a disease [337]. Network approaches will require solid understanding of the affected network, including the network's structure, dynamics, and resistance to perturbations. Future treatments will require multiple points of attack, achieved by drug combinations or development of multi-target drugs. Optimistically, there are already indications that the network medicine approach will bring successful treatment in cases where current, single-target approaches are deficient.

References

1. Junker BH, Schreiber F (2008) Analysis of biological networks. Wiley series on bioinformatics, Wiley-Interscience
2. Holland JH (1995) Hidden order: how adaptation builds complexity. Helix books, Addison-Wesley, Reading
3. Holland JH (1998) Emergence: from chaos to order. Helix books, Addison-Wesley, Reading
4. Barabasi AL, Albert R (1999) Emergence of scaling in random networks. Science 286(5439):509–512
5. Price DDS (1976) A general theory of bibliometric and other cumulative advantage processes. J Am Soc Inf Sci 27(5):292–306. doi:10.1002/asi.4630270505
6. Ravasz E, Barabasi AL (2003) Hierarchical organization in complex networks. Phys Rev E 67(2 Pt 2):026112
7. Ravasz E, Somera AL, Mongru DA, Oltvai ZN, Barabasi AL (2002) Hierarchical organization of modularity in metabolic networks. Science 297(5586):1551–1555. doi:10.1126/science.1073374
8. Chen LL, Blumm N, Christakis NA, Barabasi AL, Deisboeck TS (2009) Cancer metastasis networks and the prediction of progression patterns. Br J Cancer 101(5):749–758. doi:10.1038/sj.bjc.6605214
9. Lee DS, Burd H, Liu J, Almaas E, Wiest O, Barabasi AL, Oltvai ZN, Kapatral V (2009) Comparative genome-scale metabolic reconstruction and flux balance analysis of multiple *Staphylococcus aureus* genomes identify novel antimicrobial drug targets. J Bacteriol 191(12):4015–4024. doi:10.1128/JB.01743-08

10. Hidalgo CA, Blumm N, Barabasi AL, Christakis NA (2009) A dynamic network approach for the study of human phenotypes. PLoS Comput Biol 5(4):e1000353. doi:10.1371/journal.pcbi.1000353
11. Barabasi AL (2009) Scale-free networks: a decade and beyond. Science 325(5939):412–413. doi:10.1126/science.1173299
12. Lenski RE, Barrick JE, Ofria C (2006) Balancing robustness and evolvability. PLoS Biol 4(12):e428. doi:10.1371/journal.pbio.0040428
13. Lehner B (2010) Genes confer similar robustness to environmental, stochastic, and genetic perturbations in yeast. PLoS ONE 5(2):e9035. doi:10.1371/journal.pone.0009035
14. Dixon SJ, Andrews BJ, Boone C (2009) Exploring the conservation of synthetic lethal genetic interaction networks. Commun Integr Biol 2(2):78–81
15. Lehner B (2007) Modelling genotype-phenotype relationships and human disease with genetic interaction networks. J Exp Biol 210(Pt 9):1559–1566. doi:10.1242/jeb.002311
16. Waddington CH (1959) Canalization of development and genetic assimilation of acquired characters. Nature 183(4676):1654–1655
17. Lehner B, Crombie C, Tischler J, Fortunato A, Fraser AG (2006) Systematic mapping of genetic interactions in *Caenorhabditis* elegans identifies common modifiers of diverse signaling pathways. Nat Genet 38(8):896–903. doi:10.1038/ng1844
18. Tischler J, Lehner B, Fraser AG (2008) Evolutionary plasticity of genetic interaction networks. Nat Genet 40(4):390–391. doi:10.1038/ng.114
19. Barkai N, Leibler S (1997) Robustness in simple biochemical networks. Nature 387(6636):913–917. doi:10.1038/43199
20. Barabasi AL, Oltvai ZN (2004) Network biology: understanding the cell's functional organization. Nat Rev Genet 5(2):101–113. doi:10.1038/nrg1272
21. von Dassow G, Meir E, Munro EM, Odell GM (2000) The segment polarity network is a robust developmental module. Nature 406(6792):188–192. doi:10.1038/35018085
22. Gerdes SY, Scholle MD, Campbell JW, Balazsi G, Ravasz E, Daugherty MD, Somera AL, Kyrpides NC, Anderson I, Gelfand MS, Bhattacharya A, Kapatral V, D'Souza M, Baev MV, Grechkin Y, Mseeh F, Fonstein MY, Overbeek R, Barabasi AL, Oltvai ZN, Osterman AL (2003) Experimental determination and system level analysis of essential genes in *Escherichia coli* MG1655. J Bacteriol 185(19):5673–5684
23. Holme P, Kim BJ, Yoon CN, Han SK (2002) Attack vulnerability of complex networks. Phys Rev E 65(5 Pt 2):056109
24. Chen BS, Wang YC, Wu WS, Li WH (2005) A new measure of the robustness of biochemical networks. Bioinformatics 21(11):2698–2705. doi:10.1093/bioinformatics/bti348
25. Alon U, Surette MG, Barkai N, Leibler S (1999) Robustness in bacterial chemotaxis. Nature 397(6715):168–171. doi:10.1038/16483
26. Spiro Z, Kovacs IA, Csermely P (2008) Drug-therapy networks and the prediction of novel drug targets. J Biol 7(6):20. doi:10.1186/jbiol81
27. Csermely P, Agoston V, Pongor S (2005) The efficiency of multi-target drugs: the network approach might help drug design. Trends Pharmacol Sci 26(4):178–182. doi:10.1016/j.tips.2005.02.007
28. Ma'ayan A, Jenkins SL, Goldfarb J, Iyengar R (2007) Network analysis of FDA approved drugs and their targets. Mt Sinai J Med 74(1):27–32. doi:10.1002/msj.20002
29. Goh KI, Cusick ME, Valle D, Childs B, Vidal M, Barabasi AL (2007) The human disease network. Proc Natl Acad Sci U S A 104(21):8685–8690. doi:10.1073/pnas.0701361104
30. Dancik V, Seiler KP, Young DW, Schreiber SL, Clemons PA (2010) Distinct biological network properties between the targets of natural products and disease genes. J Am Chem Soc 132(27):9259–9261
31. Anvar SY, Tucker A, Vinciotti V, Venema A, van Ommen GJ, van der Maarel SM, Raz V, t Hoen PA (2011) Interspecies translation of disease networks increases robustness and predictive accuracy. PLoS Comput Biol 7(11):e1002258. doi:10.1371/journal.pcbi.1002258

32. Jeong H, Tombor B, Albert R, Oltvai ZN, Barabasi AL (2000) The large-scale organization of metabolic networks. Nature 407(6804):651–654
33. Jeong H, Mason SP, Barabasi AL, Oltvai ZN (2001) Lethality and centrality in protein networks. Nature 411(6833):41–42
34. Maslov S, Sneppen K (2002) Specificity and stability in topology of protein networks. Science 296(5569):910–913. doi:10.1126/science.1065103
35. Schuler A, Bornberg-Bauer E (2011) The evolution of protein interaction networks. Methods Mol Biol 696:273–289. doi:10.1007/978-1-60761-987-1_17
36. Pache RA, Babu MM, Aloy P (2009) Exploiting gene deletion fitness effects in yeast to understand the modular architecture of protein complexes under different growth conditions. BMC Syst Biol 3:74. doi:10.1186/1752-0509-3-74
37. Holme P (2011) Metabolic robustness and network modularity: a model study. PLoS ONE 6(2):e16605. doi:10.1371/journal.pone.0016605
38. Grigorov MG (2005) Global properties of biological networks. Drug Discovery Today 10(5):365–372. doi:10.1016/S1359-6446(05)03369-6
39. Wagner A, Fell DA (2001) The small world inside large metabolic networks. Proc Biol Sci 268(1478):1803–1810. doi:10.1098/rspb.2001.1711
40. Norris V, Raine D (2006) On the utility of scale-free networks. BioEssays News Rev Mol Cell Dev Biol 28(5):563–564. doi:10.1002/bies.20415
41. Ma H, Zeng AP (2003) Reconstruction of metabolic networks from genome data and analysis of their global structure for various organisms. Bioinformatics 19(2):270–277
42. Schuster S, Pfeiffer T, Moldenhauer F, Koch I, Dandekar T (2002) Exploring the pathway structure of metabolism: decomposition into subnetworks and application to *Mycoplasma pneumoniae*. Bioinformatics 18(2):351–361
43. Kaiser M, Hilgetag CC (2004) Edge vulnerability in neural and metabolic networks. Biol Cybern 90(5):311–317. doi:10.1007/s00422-004-0479-1
44. Bandyopadhyay S, Mehta M, Kuo D, Sung MK, Chuang R, Jaehnig EJ, Bodenmiller B, Licon K, Copeland W, Shales M, Fiedler D, Dutkowski J, Guenole A, van Attikum H, Shokat KM, Kolodner RD, Huh WK, Aebersold R, Keogh MC, Krogan NJ, Ideker T (2010) Rewiring of genetic networks in response to DNA damage. Science 330(6009):1385–1389. doi:10.1126/science.1195618
45. Luscombe NM, Babu MM, Yu H, Snyder M, Teichmann SA, Gerstein M (2004) Genomic analysis of regulatory network dynamics reveals large topological changes. Nature 431(7006):308–312. doi:10.1038/nature02782
46. Przytycka TM, Singh M, Slonim DK (2010) Toward the dynamic interactome: it's about time. Brief Bioinform 11(1):15–29. doi:10.1093/bib/bbp057
47. Bensimon A, Heck AJ, Aebersold R (2012) Mass spectrometry-based proteomics and network biology. Annu Rev Biochem 81:379–405. doi:10.1146/annurev-biochem-072909-100424
48. Nicolas P, Mader U, Dervyn E, Rochat T, Leduc A, Pigeonneau N, Bidnenko E, Marchadier E, Hoebeke M, Aymerich S, Becher D, Bisicchia P, Botella E, Delumeau O, Doherty G, Denham EL, Fogg MJ, Fromion V, Goelzer A, Hansen A, Hartig E, Harwood CR, Homuth G, Jarmer H, Jules M, Klipp E, Le Chat L, Lecointe F, Lewis P, Liebermeister W, March A, Mars RA, Nannapaneni P, Noone D, Pohl S, Rinn B, Rugheimer F, Sappa PK, Samson F, Schaffer M, Schwikowski B, Steil L, Stulke J, Wiegert T, Devine KM, Wilkinson AJ, van Dijl JM, Hecker M, Volker U, Bessieres P, Noirot P (2012) Condition-dependent transcriptome reveals high-level regulatory architecture in *Bacillus subtilis*. Science 335(6072):1103–1106. doi:10.1126/science.1206848
49. Buescher JM, Liebermeister W, Jules M, Uhr M, Muntel J, Botella E, Hessling B, Kleijn RJ, Le Chat L, Lecointe F, Mader U, Nicolas P, Piersma S, Rugheimer F, Becher D, Bessieres P, Bidnenko E, Denham EL, Dervyn E, Devine KM, Doherty G, Drulhe S, Felicori L, Fogg MJ, Goelzer A, Hansen A, Harwood CR, Hecker M, Hubner S, Hultschig C, Jarmer H, Klipp E, Leduc A, Lewis P, Molina F, Noirot P, Peres S, Pigeonneau N, Pohl S, Rasmussen S, Rinn B, Schaffer M, Schnidder J, Schwikowski B, Van Dijl JM, Veiga P, Walsh S,

Wilkinson AJ, Stelling J, Aymerich S, Sauer U (2012) Global network reorganization during dynamic adaptations of *Bacillus subtilis* metabolism. Science 335(6072):1099–1103. doi:10.1126/science.1206871
50. Buck MJ, Lieb JD (2004) ChIP-chip: considerations for the design, analysis, and application of genome-wide chromatin immunoprecipitation experiments. Genomics 83(3):349–360
51. Nooren IM, Thornton JM (2003) Diversity of protein-protein interactions. EMBO J 22(14):3486–3492. doi:10.1093/emboj/cdg359
52. Das J, Mohammed J, Yu H (2012) Genome-scale analysis of interaction dynamics reveals organization of biological networks. Bioinformatics 28(14):1873–1878. doi:10.1093/bioinformatics/bts283
53. Karlebach G, Shamir R (2010) Minimally perturbing a gene regulatory network to avoid a disease phenotype: the glioma network as a test case. BMC Syst Biol 4:15. doi:10.1186/1752-0509-4-15
54. Shmulevich I, Dougherty ER, Kim S, Zhang W (2002) Probabilistic Boolean networks: a rule-based uncertainty model for gene regulatory networks. Bioinformatics 18(2):261–274
55. Fuller GN, Rhee CH, Hess KR, Caskey LS, Wang R, Bruner JM, Yung WK, Zhang W (1999) Reactivation of insulin-like growth factor binding protein 2 expression in glioblastoma multiforme: a revelation by parallel gene expression profiling. Cancer Res 59(17):4228–4232
56. Liu YY, Slotine JJ, Barabasi AL (2011) Controllability of complex networks. Nature 473(7346):167–173. doi:10.1038/nature10011
57. Rzhetsky A, Koike T, Kalachikov S, Gomez SM, Krauthammer M, Kaplan SH, Kra P, Russo JJ, Friedman C (2000) A knowledge model for analysis and simulation of regulatory networks. Bioinformatics 16(12):1120–1128
58. Regev A, Silverman W, Shapiro E (2001) Representation and simulation of biochemical processes using the pi-calculus process algebra. Pac Symp Biocomput 6:459–470
59. Peleg M, Yeh I, Altman RB (2002) Modelling biological processes using workflow and Petri net models. Bioinformatics 18(6):825–837
60. Margolin AA, Nemenman I, Basso K, Wiggins C, Stolovitzky G, Dalla Favera R, Califano A (2006) ARACNE: an algorithm for the reconstruction of gene regulatory networks in a mammalian cellular context. BMC Bioinformatics 7(Suppl 1):S7. doi:10.1186/1471-2105-7-S1-S7
61. You Ch, Holder LB, Cook DJ (2009) Learning patterns in the dynamics of biological networks. In: Proceedings of the 15th ACM SIGKDD international conference on knowledge discovery and data mining. ACM, New York, pp 977–986. doi:10.1145/1557019.1557125
62. Zou C, Denby KJ, Feng J (2009) Granger causality vs. dynamic Bayesian network inference: a comparative study. BMC Bioinformatics 10:122. doi:10.1186/1471-2105-10-122
63. Zou C, Ladroue C, Guo S, Feng J (2010) Identifying interactions in the time and frequency domains in local and global networks—a Granger causality approach. BMC Bioinformatics 11:337. doi:10.1186/1471-2105-11-337
64. Nam H, Lee K, Lee D (2009) Identification of temporal association rules from time-series microarray data sets. BMC Bioinformatics 10(Suppl 3):S6. doi:10.1186/1471-2105-10-S3-S6
65. de Jong H (2002) Modeling and simulation of genetic regulatory systems: a literature review. J Comput Biol 9(1):67–103. doi:10.1089/10665270252833208
66. Fisher J, Henzinger TA (2007) Executable cell biology. Nat Biotechnol 25(11):1239–1249. doi:10.1038/nbt1356
67. Karlebach G, Shamir R (2008) Modelling and analysis of gene regulatory networks. Nat Rev Mol Cell Biol 9(10):770–780. doi:10.1038/nrm2503
68. Machado D, Costa RS, Rocha M, Ferreira EC, Tidor B, Rocha I (2011) Modeling formalisms in systems biology. AMB Express 1:45. doi:10.1186/2191-0855-1-45
69. Tenazinha N, Vinga S (2011) A survey on methods for modeling and analyzing integrated biological networks. IEEE/ACM Trans Comput Biol Bioinform 8(4):943–958. doi:10.1109/TCBB.2010.117

70. Kauffman SA (1969) Metabolic stability and epigenesis in randomly constructed genetic nets. J Theor Biol 22(3):437–467
71. Huang S (1999) Gene expression profiling, genetic networks, and cellular states: an integrating concept for tumorigenesis and drug discovery. J Mol Med (Berl) 77(6):469–480
72. Kwon YK, Cho KH (2007) Boolean dynamics of biological networks with multiple coupled feedback loops. Biophys J 92(8):2975–2981. doi:10.1529/biophysj.106.097097
73. Szallasi Z, Liang S (1998) Modeling the normal and neoplastic cell cycle with "realistic Boolean genetic networks": their application for understanding carcinogenesis and assessing therapeutic strategies. Pac Symp Biocomput 3:66–76
74. Akutsu T, Kuhara S, Maruyama O, Miyano S (1998) A system for identifying genetic networks from gene expression patterns produced by gene disruptions and overexpressions. Genome Inform Ser Workshop Genome Inform 9:151–160
75. Davidich MI, Bornholdt S (2008) Boolean network model predicts cell cycle sequence of fission yeast. PLoS ONE 3(2):e1672. doi:10.1371/journal.pone.0001672
76. Jack J, Wambaugh JF, Shah I (2011) Simulating quantitative cellular responses using asynchronous threshold Boolean network ensembles. BMC Syst Biol 5:109. doi:10.1186/1752-0509-5-109
77. Saez-Rodriguez J, Alexopoulos LG, Epperlein J, Samaga R, Lauffenburger DA, Klamt S, Sorger PK (2009) Discrete logic modelling as a means to link protein signalling networks with functional analysis of mammalian signal transduction. Mol Syst Biol 5:331. doi:10.1038/msb.2009.87
78. Rodriguez A, Sosa D, Torres L, Molina B, Frias S, Mendoza L (2012) A Boolean network model of the FA/BRCA pathway. Bioinformatics 28(6):858–866. doi:10.1093/bioinformatics/bts036
79. Handorf T, Klipp E (2012) Modeling mechanistic biological networks: an advanced Boolean approach. Bioinformatics 28(4):557–563. doi:10.1093/bioinformatics/btr697
80. Grzegorczyk M (2010) An introduction to Gaussian Bayesian networks. Methods Mol Biol 662:121–147. doi:10.1007/978-1-60761-800-3_6
81. Friedman N, Linial M, Nachman I, Pe'er D (2000) Using Bayesian networks to analyze expression data. J Comput Biol 7(3–4):601–620. doi:10.1089/106652700750050961
82. Chickering D (1996) Learning Bayesian networks is NP-complete. In: Fisher D, Lenz H (eds) Learning from data: artificial intelligence and statistics V. Springer, New York, pp 121–130
83. Pe'er D, Regev A, Elidan G, Friedman N (2001) Inferring subnetworks from perturbed expression profiles. Bioinformatics 17(Suppl 1):S215–S224
84. Ong IM, Glasner JD, Page D (2002) Modelling regulatory pathways in *E. coli* from time series expression profiles. Bioinformatics 18(Suppl 1): S241–248
85. Perrin BE, Ralaivola L, Mazurie A, Bottani S, Mallet J, d'Alche-Buc F (2003) Gene networks inference using dynamic Bayesian networks. Bioinformatics 19(Suppl 2): ii138–ii148
86. Li P, Zhang C, Perkins EJ, Gong P, Deng Y (2007) Comparison of probabilistic Boolean network and dynamic Bayesian network approaches for inferring gene regulatory networks. BMC Bioinformatics 8(Suppl 7):S13. doi:10.1186/1471-2105-8-S7-S13
87. Yan W, Zhu H, Yang Y, Chen J, Zhang Y, Shen B (2010) Effects of time point measurement on the reconstruction of gene regulatory networks. Molecules 15(8):5354–5368. doi:10.3390/molecules15085354
88. Bock M, Ogishima S, Tanaka H, Kramer S, Kaderali L (2012) Hub-centered gene network reconstruction using automatic relevance determination. PLoS ONE 7(5):e35077. doi:10.1371/journal.pone.0035077
89. Rangel C, Angus J, Ghahramani Z, Lioumi M, Sotheran E, Gaiba A, Wild DL, Falciani F (2004) Modeling T-cell activation using gene expression profiling and state-space models. Bioinformatics 20(9):1361–1372. doi:10.1093/bioinformatics/bth093

90. Beal MJ, Falciani F, Ghahramani Z, Rangel C, Wild DL (2005) A Bayesian approach to reconstructing genetic regulatory networks with hidden factors. Bioinformatics 21(3):349–356. doi:10.1093/bioinformatics/bti014
91. Kim S, Imoto S, Miyano S (2004) Dynamic Bayesian network and nonparametric regression for nonlinear modeling of gene networks from time series gene expression data. Bio Systems 75(1–3):57–65. doi:10.1016/j.biosystems.2004.03.004
92. Murata T (1989) Petri nets: properties, analysis and applications. Proc IEEE 77(4):541–580
93. Reddy VN, Mavrovouniotis ML, Liebman MN (1993) Petri net representations in metabolic pathways. Proc Int Conf Intell Syst Mol Biol 1:328–336
94. Reddy VN, Liebman MN, Mavrovouniotis ML (1996) Qualitative analysis of biochemical reaction systems. Comput Biol Med 26(1):9–24
95. Koch I, Junker BH, Heiner M (2005) Application of Petri net theory for modelling and validation of the sucrose breakdown pathway in the potato tuber. Bioinformatics 21(7):1219–1226. doi:10.1093/bioinformatics/bti145
96. D-W DING, Ln LI (2009) Modeling and analyzing the metabolism of riboflavin production using Petri nets. J Biol Syst (JBS) 17(03):479–490. doi:10.1142/S021833900900296X
97. Ruths D, Muller M, Tseng JT, Nakhleh L, Ram PT (2008) The signaling Petri net-based simulator: a non-parametric strategy for characterizing the dynamics of cell-specific signaling networks. PLoS Comput Biol 4(2):e1000005. doi:10.1371/journal.pcbi.1000005
98. Sackmann A, Heiner M, Koch I (2006) Application of Petri net based analysis techniques to signal transduction pathways. BMC Bioinformatics 7:482. doi:10.1186/1471-2105-7-482
99. Hardy S, Robillard PN (2008) Petri net-based method for the analysis of the dynamics of signal propagation in signaling pathways. Bioinformatics 24(2):209–217. doi:10.1093/bioinformatics/btm560
100. Jin G, Zhao H, Zhou X, Wong ST (2011) An enhanced Petri-net model to predict synergistic effects of pairwise drug combinations from gene microarray data. Bioinformatics 27(13):i310–i316. doi:10.1093/bioinformatics/btr202
101. Goss PJ, Peccoud J (1998) Quantitative modeling of stochastic systems in molecular biology by using stochastic Petri nets. Proc Natl Acad Sci U S A 95(12):6750–6755
102. Matsuno H, Doi A, Nagasaki M, Miyano S (2000) Hybrid Petri net representation of gene regulatory network. Pac Symp Biocomput 5:341–352
103. Matsuno H, Tanaka Y, Aoshima H, Doi A, Matsui M, Miyano S (2003) Biopathways representation and simulation on hybrid functional Petri net. In Silico Biol 3(3):389–404
104. Doi A, Fujita S, Matsuno H, Nagasaki M, Miyano S (2004) Constructing biological pathway models with hybrid functional Petri nets. In Silico Biol 4(3):271–291
105. Chaouiya C (2007) Petri net modelling of biological networks. Brief Bioinform 8(4):210–219. doi:10.1093/bib/bbm029
106. Peleg M, Rubin D, Altman RB (2005) Using Petri net tools to study properties and dynamics of biological systems. J Am Med Inform Assoc 12(2):181–199. doi:10.1197/jamia.M1637
107. Breitling R, Gilbert D, Heiner M, Orton R (2008) A structured approach for the engineering of biochemical network models, illustrated for signalling pathways. Brief Bioinform 9(5):404–421. doi:10.1093/bib/bbn026
108. Moisset P, Vaisman D, Cintolesi A, Urrutia J, Rapaport I, Andrews BA, Asenjo JA (2012) Continuous modeling of metabolic networks with gene regulation in yeast and in vivo determination of rate parameters. Biotechnol Bioeng. doi:10.1002/bit.24503
109. Zhang Y, Li P, Huang GM (2012) Quantifying dynamic stability of genetic memory circuits. IEEE/ACM Trans Comput Biol Bioinform 9(3):871–884. doi:10.1109/TCBB.2011.132
110. Liu X, Niranjan M (2012) State and parameter estimation of the heat shock response system using Kalman and particle filters. Bioinformatics 28(11):1501–1507. doi:10.1093/bioinformatics/bts161

111. Venkatesh KV, Bhartiya S, Ruhela A (2004) Multiple feedback loops are key to a robust dynamic performance of tryptophan regulation in *Escherichia coli*. FEBS Lett 563(1–3):234–240. doi:10.1016/S0014-5793(04)00310-2
112. Radde N (2012) Analyzing fixed points of intracellular regulation networks with interrelated feedback topology. BMC Syst Biol 6(1):57. doi:10.1186/1752-0509-6-57
113. Chen T, He HL, Church GM (1999) Modeling gene expression with differential equations. Pac Symp Biocomput 4:29–40
114. de Hoon MJ, Imoto S, Kobayashi K, Ogasawara N, Miyano S (2003) Inferring gene regulatory networks from time-ordered gene expression data of *Bacillus subtilis* using differential equations. Pac Symp Biocomput 8:17–28
115. Soliman S, Heiner M (2010) A unique transformation from ordinary differential equations to reaction networks. PLoS ONE 5(12):e14284. doi:10.1371/journal.pone.0014284
116. Liu B, Zhang J, Tan PY, Hsu D, Blom AM, Leong B, Sethi S, Ho B, Ding JL, Thiagarajan PS (2011) A computational and experimental study of the regulatory mechanisms of the complement system. PLoS Comput Biol 7(1):e1001059. doi:10.1371/journal.pcbi.1001059
117. Li Z, Li P, Krishnan A, Liu J (2011) Large-scale dynamic gene regulatory network inference combining differential equation models with local dynamic Bayesian network analysis. Bioinformatics 27(19):2686–2691. doi:10.1093/bioinformatics/btr454
118. Mazur J, Ritter D, Reinelt G, Kaderali L (2009) Reconstructing nonlinear dynamic models of gene regulation using stochastic sampling. BMC Bioinformatics 10:448. doi:10.1186/1471-2105-10-448
119. Ermentrout GB, Edelstein-Keshet L (1993) Cellular automata approaches to biological modeling. J Theor Biol 160(1):97–133. doi:10.1006/jtbi.1993.1007
120. Walker DC, Southgate J (2009) The virtual cell–a candidate co-ordinator for 'middle-out' modelling of biological systems. Brief Bioinform 10(4):450–461. doi:10.1093/bib/bbp010
121. Bonchev D, Thomas S, Apte A, Kier LB (2010) Cellular automata modelling of biomolecular networks dynamics. SAR QSAR Environ Res 21(1):77–102. doi:10.1080/10629360903568580
122. van Ballegooijen WM, Boerlijst MC (2004) Emergent trade-offs and selection for outbreak frequency in spatial epidemics. Proc Natl Acad Sci U S A 101(52):18246–18250. doi:10.1073/pnas.0405682101
123. Dormann S, Deutsch A (2002) Modeling of self-organized avascular tumor growth with a hybrid cellular automaton. In Silico Biol 2(3):393–406
124. Moreira J, Deutsch A (2002) Cellular automaton models of tumor development: a critical review. Advances in complex systems (ACS) 05 (02n03): 247–267. doi:10.1142/S0219525902000572
125. Goltsev AV, de Abreu FV, Dorogovtsev SN, Mendes JF (2010) Stochastic cellular automata model of neural networks. Phys Rev E 81(6 Pt 1):061921
126. Souza-e-Silva H, Savino W, Feijoo RA, Vasconcelos AT (2009) A cellular automata-based mathematical model for thymocyte development. PLoS ONE 4(12):e8233. doi:10.1371/journal.pone.0008233
127. Vladimirov N, Traub RD, Tu Y (2011) Wave speed in excitable random networks with spatially constrained connections. PLoS ONE 6(6):e20536. doi:10.1371/journal.pone.0020536
128. Wishart DS, Yang R, Arndt D, Tang P, Cruz J (2005) Dynamic cellular automata: an alternative approach to cellular simulation. In Silico Biol 5(2):139–161
129. Deroulers C, Aubert M, Badoual M, Grammaticos B (2009) Modeling tumor cell migration: from microscopic to macroscopic models. Phys Rev E 79(3 Pt 1):031917
130. Kavousanakis ME, Liu P, Boudouvis AG, Lowengrub J, Kevrekidis IG (2012) Efficient coarse simulation of a growing avascular tumor. Phys Rev E 85(3 Pt 1):031912
131. Kanehisa M, Goto S (2000) KEGG: Kyoto encyclopedia of genes and genomes. Nucleic Acids Res 28(1):27–30

132. Liao JC, Boscolo R, Yang YL, Tran LM, Sabatti C, Roychowdhury VP (2003) Network component analysis: reconstruction of regulatory signals in biological systems. Proc Natl Acad Sci U S A 100(26):15522–15527. doi:10.1073/pnas.2136632100
133. Hartwell LH, Hopfield JJ, Leibler S, Murray AW (1999) From molecular to modular cell biology. Nature 402(6761 Suppl):C47–C52. doi:10.1038/35011540
134. Schwikowski B, Uetz P, Fields S (2000) A network of protein-protein interactions in yeast. Nat Biotechnol 18(12):1257–1261. doi:10.1038/82360
135. Tanay A, Sharan R, Kupiec M, Shamir R (2004) Revealing modularity and organization in the yeast molecular network by integrated analysis of highly heterogeneous genomewide data. Proc Natl Acad Sci U S A 101(9):2981–2986. doi:10.1073/pnas.0308661100
136. Segal E, Shapira M, Regev A, Pe'er D, Botstein D, Koller D, Friedman N (2003) Module networks: identifying regulatory modules and their condition-specific regulators from gene expression data. Nat Genet 34(2):166–176. doi:10.1038/ng1165
137. Gavin AC, Aloy P, Grandi P, Krause R, Boesche M, Marzioch M, Rau C, Jensen LJ, Bastuck S, Dumpelfeld B, Edelmann A, Heurtier MA, Hoffman V, Hoefert C, Klein K, Hudak M, Michon AM, Schelder M, Schirle M, Remor M, Rudi T, Hooper S, Bauer A, Bouwmeester T, Casari G, Drewes G, Neubauer G, Rick JM, Kuster B, Bork P, Russell RB, Superti-Furga G (2006) Proteome survey reveals modularity of the yeast cell machinery. Nature 440(7084):631–636. doi:10.1038/nature04532
138. Costanzo M, Baryshnikova A, Bellay J, Kim Y, Spear ED, Sevier CS, Ding H, Koh JL, Toufighi K, Mostafavi S, Prinz J, St Onge RP, VanderSluis B, Makhnevych T, Vizeacoumar FJ, Alizadeh S, Bahr S, Brost RL, Chen Y, Cokol M, Deshpande R, Li Z, Lin ZY, Liang W, Marback M, Paw J, San Luis BJ, Shuteriqi E, Tong AH, van Dyk N, Wallace IM, Whitney JA, Weirauch MT, Zhong G, Zhu H, Houry WA, Brudno M, Ragibizadeh S, Papp B, Pal C, Roth FP, Giaever G, Nislow C, Troyanskaya OG, Bussey H, Bader GD, Gingras AC, Morris QD, Kim PM, Kaiser CA, Myers CL, Andrews BJ, Boone C (2010) The genetic landscape of a cell. Science 327(5964):425–431. doi:10.1126/science.1180823
139. Vazquez A, Flammini A, Maritan A, Vespignani A (2003) Global protein function prediction from protein-protein interaction networks. Nat Biotechnol 21(6):697–700. doi:10.1038/nbt825
140. Lee I, Date SV, Adai AT, Marcotte EM (2004) A probabilistic functional network of yeast genes. Science 306(5701):1555–1558. doi:10.1126/science.1099511
141. Gunsalus KC, Ge H, Schetter AJ, Goldberg DS, Han JD, Hao T, Berriz GF, Bertin N, Huang J, Chuang LS, Li N, Mani R, Hyman AA, Sonnichsen B, Echeverri CJ, Roth FP, Vidal M, Piano F (2005) Predictive models of molecular machines involved in *Caenorhabditis elegans* early embryogenesis. Nature 436(7052):861–865. doi:10.1038/nature03876
142. Zhou X, Kao MC, Wong WH (2002) Transitive functional annotation by shortest-path analysis of gene expression data. Proc Natl Acad Sci U S A 99(20):12783–12788. doi:10.1073/pnas.192159399
143. Chua HN, Sung WK, Wong L (2006) Exploiting indirect neighbours and topological weight to predict protein function from protein-protein interactions. Bioinformatics 22(13):1623–1630. doi:10.1093/bioinformatics/btl145
144. Yang P, Li X, Wu M, Kwoh CK, Ng SK (2011) Inferring gene-phenotype associations via global protein complex network propagation. PLoS ONE 6(7):e21502. doi:10.1371/journal.pone.0021502
145. Letovsky S, Kasif S (2003) Predicting protein function from protein/protein interaction data: a probabilistic approach. Bioinformatics 19(Suppl 1):i197–i204
146. Deng M, Chen T, Sun F (2004) An integrated probabilistic model for functional prediction of proteins. J Comput Biol 11(2–3):463–475. doi:10.1089/1066527041410346
147. Gelfand AE, Smith AFM (1990) Sampling-based approaches to calculating marginal densities. J Am Stat Assoc 85(410):398–409
148. Lanckriet GR, Deng M, Cristianini N, Jordan MI, Noble WS (2004) Kernel-based data fusion and its application to protein function prediction in yeast. Pac Symp Biocomput 9:300–311

149. Tsuda K, Shin H, Scholkopf B (2005) Fast protein classification with multiple networks. Bioinformatics 21(Suppl 2): ii59–ii65. doi:10.1093/bioinformatics/bti1110
150. Tian W, Zhang LV, Tasan M, Gibbons FD, King OD, Park J, Wunderlich Z, Cherry JM, Roth FP (2008) Combining guilt-by-association and guilt-by-profiling to predict *Saccharomyces cerevisiae* gene function. Genome Biol 9(Suppl 1):S7. doi:10.1186/gb-2008-9-s1-s7
151. Tasan M, Tian W, Hill DP, Gibbons FD, Blake JA, Roth FP (2008) An *en masse* phenotype and function prediction system for *Mus musculus*. Genome Biol 9(Suppl 1):S8. doi:10.1186/gb-2008-9-s1-s8
152. Zhang XF, Dai DQ (2012) A framework for incorporating functional interrelationships into protein function prediction algorithms. IEEE/ACM Trans Comput Biol Bioinform 9(3):740–753. doi:10.1109/TCBB.2011.148
153. Jiang JQ, McQuay LJ (2012) Predicting protein function by multi-label correlated semi-supervised learning. IEEE/ACM Trans Comput Biol Bioinform 9(4):1059–1069. doi:10.1109/TCBB.2011.156
154. Gillis J, Pavlidis P (2012) "Guilt by association" is the exception rather than the rule in gene networks. PLoS Comput Biol 8(3):e1002444. doi:10.1371/journal.pcbi.1002444
155. Gillis J, Pavlidis P (2011) The impact of multifunctional genes on "guilt by association" analysis. PLoS ONE 6(2):e17258. doi:10.1371/journal.pone.0017258
156. Tasan M, Drabkin HJ, Beaver JE, Chua HN, Dunham J, Tian W, Blake JA, Roth FP (2012) A resource of quantitative functional annotation for *Homo sapiens* genes. G3 (Bethesda) 2(2): 223–233. doi:10.1534/g3.111.000828
157. Sharan R, Ulitsky I, Shamir R (2007) Network-based prediction of protein function. Mol Syst Biol 3:88. doi:10.1038/msb4100129
158. Wang PI, Marcotte EM (2010) It's the machine that matters: predicting gene function and phenotype from protein networks. J Proteomics 73(11):2277–2289. doi:10.1016/j.jprot.2010.07.005
159. Newman MEJ (2004) Detecting community structure in networks. Eur Phys J B Condens Matter Complex Syst 38(2):321–330. doi:10.1140/epjb/e2004-00124-y
160. Schaeffer SE (2007) Graph clustering. Comput Sci Rev 1(1):27–64. doi:10.1016/j.cosrev.2007.05.001
161. Fortunato S (2010) Community detection in graphs. Phys Rep 486(3–5):75–174. doi:10.1016/j.physrep.2009.11.002
162. Newman ME, Girvan M (2004) Finding and evaluating community structure in networks. Phys Rev E 69(2 Pt 2):026113
163. Guimera R, Nunes Amaral LA (2005) Functional cartography of complex metabolic networks. Nature 433(7028):895–900. doi:10.1038/nature03288
164. Spirin V, Mirny LA (2003) Protein complexes and functional modules in molecular networks. Proc Natl Acad Sci U S A 100(21):12123–12128. doi:10.1073/pnas.2032324100
165. King AD, Przulj N, Jurisica I (2004) Protein complex prediction via cost-based clustering. Bioinformatics 20(17):3013–3020. doi:10.1093/bioinformatics/bth351
166. Duch J, Arenas A (2005) Community detection in complex networks using extremal optimization. Phys Rev E 72(2 Pt 2):027104
167. Šíma J, Schaeffer S (2006) On the NP-completeness of some graph cluster measures. In: Wiedermann J, Tel G, Pokorný J, Bieliková M, Štuller J (eds) SOFSEM 2006: theory and practice of computer science, Lecture notes in computer science, vol 3831. Springer, Berlin/Heidelberg, pp 530–537. doi:10.1007/11611257_51
168. Girvan M, Newman ME (2002) Community structure in social and biological networks. Proc Natl Acad Sci U S A 99(12):7821–7826. doi:10.1073/pnas.122653799
169. Radicchi F, Castellano C, Cecconi F, Loreto V, Parisi D (2004) Defining and identifying communities in networks. Proc Natl Acad Sci U S A 101(9):2658–2663. doi:10.1073/pnas.0400054101
170. Wilkinson DM, Huberman BA (2004) A method for finding communities of related genes. Proc Natl Acad Sci U S A 101(Suppl 1):5241–5248. doi:10.1073/pnas.0307740100

171. Fortunato S, Latora V, Marchiori M (2004) Method to find community structures based on information centrality. Phys Rev E 70(5 Pt 2):056104
172. Chen J, Yuan B (2006) Detecting functional modules in the yeast protein-protein interaction network. Bioinformatics 22(18):2283–2290. doi:10.1093/bioinformatics/btl370
173. Kernighan BW, Lin S (1970) An efficient heuristic procedure for partitioning graphs. Bell Syst Tech J 49(1):291–307
174. Pothen A (1995) Graph partitioning algorithms with applications to scientific computing. In: Keyes DE, Sameh AH, Venkatakrishnan V (eds) Parallel numerical algorithms. Kluwer Academic Press, Dordrecht
175. Fiedler M (1973) Algebraic connectivity of graphs. Czechoslovak Math J 23(98):298–305
176. Fiedler M (1975) A property of eigenvectors of nonnegative symmetric matrices and its application to graph theory. Czechoslovak Math J 25(4):619–633
177. Drineas P, Frieze A, Kannan R, Vempala S, Vinay V (2004) Clustering large graphs via the singular value decomposition. Mach Learn 56(1–3):9–33. doi:10.1023/b:mach.0000033113.59016.96
178. Coja-Oghlan A, Goerdt A, Lanka A (2006) Spectral partitioning of random graphs with given expected degrees. In: Navarro G, Bertossi L, Kohayakawa Y (eds) Fourth IFIP international conference on theoretical computer science—TCS 2006, IFIP international federation for information processing, vol 209 Springer, US, pp 271–282. doi:10.1007/978-0-387-34735-6_22
179. Flake GW, Lawrence S, Giles CL, Coetzee FM (2002) Self-organization and identification of web communities. Computer 35(3):66–71. doi:10.1109/2.989932
180. Wu F, Huberman BA (2004) Finding communities in linear time: a physics approach. Eur Phys J B 38(2):331–338
181. Weston J, Elisseeff A, Zhou D, Leslie CS, Noble WS (2004) Protein ranking: from local to global structure in the protein similarity network. Proc Natl Acad Sci U S A 101(17):6559–6563. doi:10.1073/pnas.0308067101
182. Rives AW, Galitski T (2003) Modular organization of cellular networks. Proc Natl Acad Sci U S A 100(3):1128–1133. doi:10.1073/pnas.0237338100
183. Zhou H (2003) Network landscape from a Brownian particle's perspective. Phys Rev E 67(4 Pt 1):041908
184. Zhou H (2003) Distance, dissimilarity index, and network community structure. Phys Rev E 67(6 Pt 1):061901
185. Navlakha S, Kingsford C (2010) Exploring biological network dynamics with ensembles of graph partitions. Pac Symp Biocomput 15:166–177
186. Palla G, Derenyi I, Farkas I, Vicsek T (2005) Uncovering the overlapping community structure of complex networks in nature and society. Nature 435(7043):814–818. doi:10.1038/nature03607
187. Becker E, Robisson B, Chapple CE, Guenoche A, Brun C (2012) Multifunctional proteins revealed by overlapping clustering in protein interaction network. Bioinformatics 28(1):84–90. doi:10.1093/bioinformatics/btr621
188. Clauset A (2005) Finding local community structure in networks. Phys Rev E: Stat, Nonlin, Soft Matter Phys 72(2 Pt 2):026132
189. Nepusz T, Yu H, Paccanaro A (2012) Detecting overlapping protein complexes in protein-protein interaction networks. Nat Methods 9(5):471–472. doi:10.1038/nmeth.1938
190. Andersen R, Chung F, Lang K (2007) Using pagerank to locally partition a graph. Internet Math 4(1):35–64. doi:10.1080/15427951.2007.10129139
191. Voevodski K, Teng SH, Xia Y (2009) Finding local communities in protein networks. BMC bioinformatics 10:297. doi:10.1186/1471-2105-10-297
192. Bagrow JP (2008) Evaluating local community methods in networks. J Stat Mech Theory Exp 2008(05):P05001
193. Ahn YY, Bagrow JP, Lehmann S (2010) Link communities reveal multiscale complexity in networks. Nature 466(7307):761–764. doi:10.1038/nature09182

194. Evans TS, Lambiotte R (2009) Line graphs, link partitions, and overlapping communities. Phys Rev E 80(1 Pt 2):016105
195. Gyenge A, Sinkkonen J, Benczur AA (2010) An efficient block model for clustering sparse graphs. In: Proceedings of the eighth workshop on mining and learning with graphs. ACM, New York, pp 62–69
196. Farutin V, Robison K, Lightcap E, Dancik V, Ruttenberg A, Letovsky S, Pradines J (2006) Edge-count probabilities for the identification of local protein communities and their organization. Proteins 62(3):800–818. doi:10.1002/prot.20799
197. Chung F, Lu L (2002) The average distances in random graphs with given expected degrees. Proc Natl Acad Sci U S A 99(25):15879–15882. doi:10.1073/pnas.252631999
198. Pradines JR, Farutin V, Rowley S, Dancik V (2005) Analyzing protein lists with large networks: edge-count probabilities in random graphs with given expected degrees. J Comput Bio 12(2):113–128. doi:10.1089/cmb.2005.12.113
199. Koyuturk M, Szpankowski W, Grama A (2007) Assessing significance of connectivity and conservation in protein interaction networks. J Comput Biol 14(6):747–764. doi:10.1089/cmb.2007.R014
200. Lancichinetti A, Radicchi F, Ramasco JJ, Fortunato S (2011) Finding statistically significant communities in networks. PLoS ONE 6(4):e18961. doi:10.1371/journal.pone.0018961
201. Ball B, Karrer B, Newman ME (2011) Efficient and principled method for detecting communities in networks. Phys Rev E 84(3 Pt 2):036103
202. Ideker T, Ozier O, Schwikowski B, Siegel AF (2002) Discovering regulatory and signalling circuits in molecular interaction networks. Bioinformatics 18(Suppl 1):S233–S240
203. Guo Z, Wang L, Li Y, Gong X, Yao C, Ma W, Wang D, Zhu J, Zhang M, Yang D, Rao S, Wang J (2007) Edge-based scoring and searching method for identifying condition-responsive protein-protein interaction sub-network. Bioinformatics 23(16):2121–2128. doi:10.1093/bioinformatics/btm294
204. Nacu S, Critchley-Thorne R, Lee P, Holmes S (2007) Gene expression network analysis and applications to immunology. Bioinformatics 23(7):850–858. doi:10.1093/bioinformatics/btm019
205. Liu M, Liberzon A, Kong SW, Lai WR, Park PJ, Kohane IS, Kasif S (2007) Network-based analysis of affected biological processes in type 2 diabetes models. PLoS Genet 3(6):e96. doi:10.1371/journal.pgen.0030096
206. Nibbe RK, Koyuturk M, Chance MR (2010) An integrative-omics approach to identify functional sub-networks in human colorectal cancer. PLoS Comput Biol 6(1):e1000639. doi:10.1371/journal.pcbi.1000639
207. Kim Y, Kim TK, Yoo J, You S, Lee I, Carlson G, Hood L, Choi S, Hwang D (2011) Principal network analysis: identification of subnetworks representing major dynamics using gene expression data. Bioinformatics 27(3):391–398. doi:10.1093/bioinformatics/btq670
208. Hanisch D, Zien A, Zimmer R, Lengauer T (2002) Co-clustering of biological networks and gene expression data. Bioinformatics 18(Suppl 1):S145–S154
209. Ulitsky I, Shamir R (2007) Identification of functional modules using network topology and high-throughput data. BMC Syst Biol 1:8. doi:10.1186/1752-0509-1-8
210. Ulitsky I, Shamir R (2009) Identifying functional modules using expression profiles and confidence-scored protein interactions. Bioinformatics 25(9):1158–1164. doi:10.1093/bioinformatics/btp118
211. Chowdhury SA, Nibbe RK, Chance MR, Koyuturk M (2011) Subnetwork state functions define dysregulated subnetworks in cancer. J Comput Biol 18(3):263–281. doi:10.1089/cmb.2010.0269
212. Dutkowski J, Ideker T (2011) Protein networks as logic functions in development and cancer. PLoS Comput Biol 7(9):e1002180. doi:10.1371/journal.pcbi.1002180
213. Padmanabhan K, Wang K, Samatova NF (2012) Functional annotation of hierarchical modularity. PLoS ONE 7(4):e33744. doi:10.1371/journal.pone.0033744

214. Mones E, Vicsek L, Vicsek T (2012) Hierarchy measure for complex networks. PLoS ONE 7(3):e33799. doi:10.1371/journal.pone.0033799
215. Gupte M, Shankar P, Li J, Muthukrishnan S, Iftode L (2011) Finding hierarchy in directed online social networks. In: Proceedings of the 20th international conference on World Wide Web. ACM, New York, pp 557−566
216. Espinosa-Soto C, Wagner A (2010) Specialization can drive the evolution of modularity. PLoS Comput Biol 6(3):e1000719. doi:10.1371/journal.pcbi.1000719
217. Wagner GP, Pavlicev M, Cheverud JM (2007) The road to modularity. Nat Rev Genet 8(12):921–931. doi:10.1038/nrg2267
218. Teichmann SA (2002) The constraints protein-protein interactions place on sequence divergence. J Mol Biol 324(3):399–407
219. Jiang H, Guo X, Xu L, Gu Z (2012) Rewiring of posttranscriptional RNA regulons: Puf4p in fungi as an example. Mol Biol Evol. doi:10.1093/molbev/mss085
220. Sharan R, Ideker T (2006) Modeling cellular machinery through biological network comparison. Nat Biotechnol 24(4):427–433. doi:10.1038/nbt1196
221. Kiemer L, Cesareni G (2007) Comparative interactomics: comparing apples and pears? Trends Biotechnol 25(10):448–454. doi:10.1016/j.tibtech.2007.08.002
222. Yoon BJ, Qian X, Sahraeian SME (2012) Comparative analysis of biological networks: hidden Markov model and Markov chain-based approach. IEEE Signal Process Mag 29(1): 22–34. doi:http://dx.doi.org/10.1109/MSP.2011.942819
223. Kelley BP, Sharan R, Karp RM, Sittler T, Root DE, Stockwell BR, Ideker T (2003) Conserved pathways within bacteria and yeast as revealed by global protein network alignment. Proc Natl Acad Sci U S A 100(20):11394–11399. doi:10.1073/pnas.1534710100
224. Sharan R, Suthram S, Kelley RM, Kuhn T, McCuine S, Uetz P, Sittler T, Karp RM, Ideker T (2005) Conserved patterns of protein interaction in multiple species. Proc Natl Acad Sci U S A 102(6):1974–1979. doi:10.1073/pnas.0409522102
225. Flannick J, Novak A, Srinivasan BS, McAdams HH, Batzoglou S (2006) Graemlin: general and robust alignment of multiple large interaction networks. Genome Res 16(9):1169–1181. doi:10.1101/gr.5235706
226. Liang Z, Xu M, Teng M, Niu L (2006) Comparison of protein interaction networks reveals species conservation and divergence. BMC Bioinformatics 7:457. doi:10.1186/1471-2105-7-457
227. Liang Z, Xu M, Teng M, Niu L (2006) NetAlign: a web-based tool for comparison of protein interaction networks. Bioinformatics 22(17):2175–2177. doi:10.1093/bioinformatics/btl287
228. Cootes AP, Muggleton SH, Sternberg MJ (2007) The identification of similarities between biological networks: application to the metabolome and interactome. J Mol Biol 369(4):1126–1139. doi:10.1016/j.jmb.2007.03.013
229. Narayanan M, Karp RM (2007) Comparing protein interaction networks via a graph match-and-split algorithm. J Comput Biol 14(7):892–907. doi:10.1089/cmb.2007.0025
230. Pache RA, Aloy P (2012) A novel framework for the comparative analysis of biological networks. PLoS ONE 7(2):e31220. doi:10.1371/journal.pone.0031220
231. Singh R, Xu J, Berger B (2007) Pairwise global alignment of protein interaction networks by matching neighborhood topology. In: Speed T, Huang H (eds) RECOMB 2007, Lecture notes in computer science, vol 4453. Springer, Berlin/Heidelberg, pp 16–31. doi:10.1007/978-3-540-71681-5_2
232. Singh R, Xu J, Berger B (2008) Global alignment of multiple protein interaction networks with application to functional orthology detection. Proc Natl Acad Sci U S A 105(35): 12763–12768. doi:10.1073/pnas.0806627105
233. Klau GW (2009) A new graph-based method for pairwise global network alignment. BMC Bioinformatics 10(Suppl 1):S59. doi:10.1186/1471-2105-10-S1-S59
234. Flannick J, Novak A, Do CB, Srinivasan BS, Batzoglou S (2009) Automatic parameter learning for multiple local network alignment. J Comput Biol 16(8):1001–1022. doi:10.1089/cmb.2009.0099

235. Liao CS, Lu K, Baym M, Singh R, Berger B (2009) IsoRankN: spectral methods for global alignment of multiple protein networks. Bioinformatics 25(12):i253–i258. doi:10.1093/bioinformatics/btp203
236. Bayati M, Gerritsen M, Gleich DF, Saberi A, Wang Y (2009) Algorithms for large, sparse network alignment problems. In: Proceedings of the 2009 Ninth IEEE international conference on data mining. IEEE Computer Society, Washington, pp 705–710. doi:10.1109/ICDM.2009.135
237. Kuchaiev O, Milenkovic T, Memisevic V, Hayes W, Przulj N (2010) Topological network alignment uncovers biological function and phylogeny. J R Soc Interface 7(50):1341–1354. doi:10.1098/rsif.2010.0063
238. Kuchaiev O, Przulj N (2011) Integrative network alignment reveals large regions of global network similarity in yeast and human. Bioinformatics 27(10):1390–1396. doi:10.1093/bioinformatics/btr127
239. Chindelevitch L, Liao CS, Berger B (2010) Local optimization for global alignment of protein interaction networks. Pac Symp Biocomput 15:123–132
240. Shih YK, Parthasarathy S (2012) Scalable global alignment for multiple biological networks. BMC Bioinformatics 13(Suppl 3):S11. doi:10.1186/1471-2105-13-S3-S11
241. Hodgkinson L, Karp RM (2011) Algorithms to detect multiprotein modularity conserved during evolution. IEEE/ACM Trans Comput Biol Bioinform. doi:10.1109/TCBB.2011.125
242. Phan HT, Sternberg MJ (2012) PINALOG: a novel approach to align protein interaction networks–implications for complex detection and function prediction. Bioinformatics 28(9):1239–1245. doi:10.1093/bioinformatics/bts119
243. Ogata H, Fujibuchi W, Goto S, Kanehisa M (2000) A heuristic graph comparison algorithm and its application to detect functionally related enzyme clusters. Nucleic Acids Res 28(20):4021–4028
244. Rhodes DR, Tomlins SA, Varambally S, Mahavisno V, Barrette T, Kalyana-Sundaram S, Ghosh D, Pandey A, Chinnaiyan AM (2005) Probabilistic model of the human protein-protein interaction network. Nat Biotechnol 23(8):951–959. doi:10.1038/nbt1103
245. Zhang LV, King OD, Wong SL, Goldberg DS, Tong AH, Lesage G, Andrews B, Bussey H, Boone C, Roth FP (2005) Motifs, themes and thematic maps of an integrated *Saccharomyces cerevisiae* interaction network. J Biol 4(2):6. doi:10.1186/jbiol23
246. Srinivasan B, Novak A, Flannick J, Batzoglou S, McAdams H (2006) Integrated protein interaction networks for 11 microbes. In: Apostolico A, Guerra C, Istrail S, Pevzner P, Waterman M (eds), RECOMB 2006 Lecture notes in computer science, vol 3909. Springer Berlin/Heidelberg, pp 1–14. doi:10.1007/11732990_1
247. Chen X, Xu H, Yuan P, Fang F, Huss M, Vega VB, Wong E, Orlov YL, Zhang W, Jiang J, Loh YH, Yeo HC, Yeo ZX, Narang V, Govindarajan KR, Leong B, Shahab A, Ruan Y, Bourque G, Sung WK, Clarke ND, Wei CL, Ng HH (2008) Integration of external signaling pathways with the core transcriptional network in embryonic stem cells. Cell 133(6):1106–1117. doi:10.1016/j.cell.2008.04.043
248. Fossum E, Friedel CC, Rajagopala SV, Titz B, Baiker A, Schmidt T, Kraus T, Stellberger T, Rutenberg C, Suthram S, Bandyopadhyay S, Rose D, von Brunn A, Uhlmann M, Zeretzke C, Dong YA, Boulet H, Koegl M, Bailer SM, Koszinowski U, Ideker T, Uetz P, Zimmer R, Haas J (2009) Evolutionarily conserved herpesviral protein interaction networks. PLoS Pathog 5(9):e1000570. doi:10.1371/journal.ppat.1000570
249. Acharjee A, Kloosterman B, de Vos RC, Werij JS, Bachem CW, Visser RG, Maliepaard C (2011) Data integration and network reconstruction with ~omics data using random forest regression in potato. Anal Chim Acta 705(1–2):56–63. doi:10.1016/j.aca.2011.03.050
250. Pinter RY, Rokhlenko O, Yeger-Lotem E, Ziv-Ukelson M (2005) Alignment of metabolic pathways. Bioinformatics 21(16):3401–3408. doi:10.1093/bioinformatics/bti554
251. Qian X, Sze SH, Yoon BJ (2009) Querying pathways in protein interaction networks based on hidden Markov models. J Comput Biol 16(2):145–157. doi:10.1089/cmb.2008.02TT

252. Huang Q, Wu LY, Zhang XS (2011) An efficient network querying method based on conditional random fields. Bioinformatics 27(22):3173–3178. doi:10.1093/bioinformatics/btr524
253. Durand P, Labarre L, Meil A, Divo JL, Vandenbrouck Y, Viari A, Wojcik J (2006) GenoLink: a graph-based querying and browsing system for investigating the function of genes and proteins. BMC Bioinformatics 7:21. doi:10.1186/1471-2105-7-21
254. Shlomi T, Segal D, Ruppin E, Sharan R (2006) QPath: a method for querying pathways in a protein-protein interaction network. BMC Bioinformatics 7:199. doi:10.1186/1471-2105-7-199
255. Yang Q, Sze SH (2007) Path matching and graph matching in biological networks. J Comput Biol 14(1):56–67. doi:10.1089/cmb.2006.0076
256. Tian Y, McEachin RC, Santos C, States DJ, Patel JM (2007) SAGA: a subgraph matching tool for biological graphs. Bioinformatics 23(2):232–239. doi:10.1093/bioinformatics/btl571
257. Ferro A, Giugno R, Pigola G, Pulvirenti A, Skripin D, Bader GD, Shasha D (2007) NetMatch: a Cytoscape plugin for searching biological networks. Bioinformatics 23(7):910–912. doi:10.1093/bioinformatics/btm032
258. Wernicke S, Rasche F (2007) Simple and fast alignment of metabolic pathways by exploiting local diversity. Bioinformatics 23(15):1978–1985. doi:10.1093/bioinformatics/btm279
259. Dost B, Shlomi T, Gupta N, Ruppin E, Bafna V, Sharan R (2008) QNet: a tool for querying protein interaction networks. J Comput Biol 15(7):913–925. doi:10.1089/cmb.2007.0172
260. Blin G, Sikora F, Vialette S (2010) Querying graphs in protein-protein interactions networks using feedback vertex set. IEEE/ACM Trans Comput Biol Bioinform 7(4):628–635. doi:10.1109/TCBB.2010.53
261. Bruckner S, Huffner F, Karp RM, Shamir R, Sharan R (2010) Topology-free querying of protein interaction networks. J Comput Biol 17(3):237–252. doi:10.1089/cmb.2009.0170
262. Fionda V, Palopoli L (2011) Biological network querying techniques: analysis and comparison. J Comput Biol 18(4):595–625. doi:10.1089/cmb.2009.0144
263. Zhang S, Zhang XS, Chen L (2008) Biomolecular network querying: a promising approach in systems biology. BMC Syst Biol 2:5. doi:10.1186/1752-0509-2-5
264. Albert I, Albert R (2004) Conserved network motifs allow protein-protein interaction prediction. Bioinformatics 20(18):3346–3352. doi:10.1093/bioinformatics/bth402
265. Huang TW, Lin CY, Kao CY (2007) Reconstruction of human protein interolog network using evolutionary conserved network. BMC Bioinformatics 8:152. doi:10.1186/1471-2105-8-152
266. Bandyopadhyay S, Sharan R, Ideker T (2006) Systematic identification of functional orthologs based on protein network comparison. Genome Res 16(3):428–435. doi:10.1101/gr.4526006
267. Milo R, Shen-Orr S, Itzkovitz S, Kashtan N, Chklovskii D, Alon U (2002) Network motifs: simple building blocks of complex networks. Science 298(5594):824–827. doi:10.1126/science.298.5594.824
268. Shen-Orr SS, Milo R, Mangan S, Alon U (2002) Network motifs in the transcriptional regulation network of *Escherichia coli*. Nat Genet 31(1):64–68. doi:10.1038/ng881
269. Itzkovitz S, Milo R, Kashtan N, Ziv G, Alon U (2003) Subgraphs in random networks. Phys Rev E 68(2 Pt 2):026127
270. Itzkovitz S, Alon U (2005) Subgraphs and network motifs in geometric networks. Phys Rev E 71(2 Pt 2):026117
271. Picard F, Daudin JJ, Koskas M, Schbath S, Robin S (2008) Assessing the exceptionality of network motifs. J Comput Biol 15(1):1–20. doi:10.1089/cmb.2007.0137
272. Schbath S, Lacroix V, Sagot MF (2009) Assessing the exceptionality of coloured motifs in networks. EURASIP J Bioinform Syst Biol 1:616234. doi:10.1186/1687-4153-2009-616234
273. Berg J, Lassig M (2004) Local graph alignment and motif search in biological networks. Proc Natl Acad Sci U S A 101(41):14689–14694. doi:10.1073/pnas.0305199101

274. Milo R, Itzkovitz S, Kashtan N, Levitt R, Shen-Orr S, Ayzenshtat I, Sheffer M, Alon U (2004) Superfamilies of evolved and designed networks. Science 303(5663):1538–1542. doi:10.1126/science.1089167
275. Przulj N (2007) Biological network comparison using graphlet degree distribution. Bioinformatics 23(2):e177–e183. doi:10.1093/bioinformatics/btl301
276. Alon U (2007) Network motifs: theory and experimental approaches. Nat Rev Genet 8(6):450–461. doi:10.1038/nrg2102
277. Wong E, Baur B, Quader S, Huang CH (2012) Biological network motif detection: principles and practice. Brief Bioinform 13(2):202–215. doi:10.1093/bib/bbr033
278. Koyuturk M, Kim Y, Topkara U, Subramaniam S, Szpankowski W, Grama A (2006) Pairwise alignment of protein interaction networks. J Comput Biol 13(2):182–199. doi:10.1089/cmb.2006.13.182
279. Berg J, Lassig M (2006) Cross-species analysis of biological networks by Bayesian alignment. Proc Natl Acad Sci U S A 103(29):10967–10972. doi:10.1073/pnas.0602294103
280. Capra JA, Pollard KS, Singh M (2010) Novel genes exhibit distinct patterns of function acquisition and network integration. Genome Biol 11(12):R127. doi:10.1186/gb-2010-11-12-r127
281. Pagel M, Meade A, Scott D (2007) Assembly rules for protein networks derived from phylogenetic-statistical analysis of whole genomes. BMC Evol Biol 7(Suppl 1):S16. doi:10.1186/1471-2148-7-S1-S16
282. Sole RV, Satorras P, Smith E, Kepler TB (2002) A model of large-scale proteome evolution. Adv Complex Syst 5(1):43–54
283. Vázquez A, Flammini A, Maritan A, Vespignani A (2003) Modeling of protein interaction networks. Complexus 1(1):38–44
284. Chung F, Lu L, Dewey TG, Galas DJ (2003) Duplication models for biological networks. J Comput Biol 10(5):677–687. doi:10.1089/106652703322539024
285. Beltrao P, Serrano L (2007) Specificity and evolvability in eukaryotic protein interaction networks. PLoS Comput Biol 3(2):e25. doi:10.1371/journal.pcbi.0030025
286. Berg J, Lassig M, Wagner A (2004) Structure and evolution of protein interaction networks: a statistical model for link dynamics and gene duplications. BMC Evol Biol 4:51. doi:10.1186/1471-2148-4-51
287. Emmert-Streib F (2012) Limitations of gene duplication models: evolution of modules in protein interaction networks. PLoS ONE 7(4):e35531. doi:10.1371/journal.pone.0035531
288. Gibson TA, Goldberg DS (2011) Improving evolutionary models of protein interaction networks. Bioinformatics 27(3):376–382. doi:10.1093/bioinformatics/btq623
289. Evlampiev K, Isambert H (2007) Modeling protein network evolution under genome duplication and domain shuffling. BMC Syst Biol 1:49. doi:10.1186/1752-0509-1-49
290. Gibson TA, Goldberg DS (2009) Reverse engineering the evolution of protein interaction networks. Pac Symp Biocomput 14:190–202
291. Yamada T, Bork P (2009) Evolution of biomolecular networks: lessons from metabolic and protein interactions. Nat Rev Mol Cell Biol 10(11):791–803. doi:10.1038/nrm2787
292. Oyarzun DA, Chaves M, Hoff-Hoffmeyer-Zlotnik M (2012) Multistability and oscillations in genetic control of metabolism. J Theor Biol 295:139–153. doi:10.1016/j.jtbi.2011.11.017
293. Rossoni E, Feng J, Tirozzi B, Brown D, Leng G, Moos F (2008) Emergent synchronous bursting of oxytocin neuronal network. PLoS Comput Biol 4(7):e1000123. doi:10.1371/journal.pcbi.1000123
294. Villani M, Barbieri A, Serra R (2011) A dynamical model of genetic networks for cell differentiation. PLoS ONE 6(3):e17703. doi:10.1371/journal.pone.0017703
295. Barberis M, Klipp E, Vanoni M, Alberghina L (2007) Cell size at S phase initiation: an emergent property of the G1/S network. PLoS Comput Biol 3(4):e64. doi:10.1371/journal.pcbi.0030064
296. Alberghina L, Hofer T, Vanoni M (2009) Molecular networks and system-level properties. J Biotechnol 144(3):224–233. doi:10.1016/j.jbiotec.2009.07.009

297. Alon U (2003) Biological networks: the tinkerer as an engineer. Science 301(5641):1866–1867. doi:10.1126/science.1089072
298. Pavlopoulos GA, Wegener AL, Schneider R (2008) A survey of visualization tools for biological network analysis. BioData Min 1:12. doi:10.1186/1756-0381-1-12
299. Hooper SD, Bork P (2005) Medusa: a simple tool for interaction graph analysis. Bioinformatics 21(24):4432–4433. doi:10.1093/bioinformatics/bti696
300. Fruchterman TMJ, Reingold EM (1991) Graph drawing by force-directed placement. Softw Pract Exper 21(11):1129–1164. doi:10.1002/spe.4380211102
301. Szklarczyk D, Franceschini A, Kuhn M, Simonovic M, Roth A, Minguez P, Doerks T, Stark M, Muller J, Bork P, Jensen LJ, von Mering C (2011) The STRING database in 2011: functional interaction networks of proteins, globally integrated and scored. Nucleic Acids Res 39(Database issue): D561–D568. doi:10.1093/nar/gkq973
302. Kuhn M, Szklarczyk D, Franceschini A, Campillos M, von Mering C, Jensen LJ, Beyer A, Bork P (2010) STITCH 2: an interaction network database for small molecules and proteins. Nucleic Acids Res 38(Database issue): D55–D556. doi:10.1093/nar/gkp937
303. Lopes CT, Franz M, Kazi F, Donaldson SL, Morris Q, Bader GD (2010) Cytoscape web: an interactive web-based network browser. Bioinformatics 26(18):2347–2348. doi:10.1093/bioinformatics/btq430
304. Theocharidis A, van Dongen S, Enright AJ, Freeman TC (2009) Network visualization and analysis of gene expression data using BioLayout Express(3D). Nat Protoc 4(10):1535–1550. doi:10.1038/nprot.2009.177
305. Gordon PM, Sensen CW (2004) Osprey: a comprehensive tool employing novel methods for the design of oligonucleotides for DNA sequencing and microarrays. Nucleic Acids Res 32(17):e133. doi:10.1093/nar/gnh127
306. Batagelj V, Brandenburg FJ, Didimo W, Liotta G, Palladino P, Patrignani M (2010) Visual analysis of large graphs using (X, Y)-clustering and hybrid visualizations. IEEE Trans Vis Comput Graph. doi:10.1109/TVCG.2010.265
307. Forman JJ, Clemons PA, Schreiber SL, Haggarty SJ (2005) SpectralNET–an application for spectral graph analysis and visualization. BMC Bioinformatics 6:260. doi:10.1186/1471-2105-6-260
308. Haggarty SJ, Clemons PA, Wong JC, Schreiber SL (2004) Mapping chemical space using molecular descriptors and chemical genetics: deacetylase inhibitors. Comb Chem High Throughput Screening 7(7):669–676
309. Shojaie A, Michailidis G (2010) Network enrichment analysis in complex experiments. Stat Appl Genet Mol Biol 9(1): Article22. doi:10.2202/1544-6115.1483
310. Subramanian A, Tamayo P, Mootha VK, Mukherjee S, Ebert BL, Gillette MA, Paulovich A, Pomeroy SL, Golub TR, Lander ES, Mesirov JP (2005) Gene set enrichment analysis: a knowledge-based approach for interpreting genome-wide expression profiles. Proc Natl Acad Sci U S A 102(43):15545–15550. doi:10.1073/pnas.0506580102
311. Tian L, Greenberg SA, Kong SW, Altschuler J, Kohane IS, Park PJ (2005) Discovering statistically significant pathways in expression profiling studies. Proc Natl Acad Sci U S A 102(38):13544–13549. doi:10.1073/pnas.0506577102
312. Li C, Wei Z, Li H (2010) Network-based empirical Bayes methods for linear models with applications to genomic data. J Biopharm Stat 20(2):209–222. doi:10.1080/10543400903572712
313. Wei P, Pan W (2008) Incorporating gene networks into statistical tests for genomic data via a spatially correlated mixture model. Bioinformatics 24(3):404–411. doi:10.1093/bioinformatics/btm612
314. Pradines J, Dancik V, Ruttenberg A, Farutin V (2007) Connectedness profiles in protein networks for the analysis of gene expression data. In: Speed T, Huang H (eds) RECOMB 2007, Lecture Notes in Computer Science, vol 4453. Springer, Berlin/Heidelberg, pp 296-310. doi:10.1007/978-3-540-71681-5_21

315. Sanguinetti G, Noirel J, Wright PC (2008) MMG: a probabilistic tool to identify submodules of metabolic pathways. Bioinformatics 24(8):1078–1084. doi:10.1093/bioinformatics/btn066
316. Shojaie A, Michailidis G (2009) Analysis of gene sets based on the underlying regulatory network. J Comput Biol 16(3):407–426. doi:10.1089/cmb.2008.0081
317. Oliver SG, Winson MK, Kell DB, Baganz F (1998) Systematic functional analysis of the yeast genome. Trends Biotechnol 16(9):373–378
318. Basu A, Rose KL, Zhang J, Beavis RC, Ueberheide B, Garcia BA, Chait B, Zhao Y, Hunt DF, Segal E, Allis CD, Hake SB (2009) Proteome-wide prediction of acetylation substrates. Proc Natl Acad Sci U S A 106(33):13785–13790. doi:10.1073/pnas.0906801106
319. Cavill R, Kamburov A, Ellis JK, Athersuch TJ, Blagrove MS, Herwig R, Ebbels TM, Keun HC (2011) Consensus-phenotype integration of transcriptomic and metabolomic data implies a role for metabolism in the chemosensitivity of tumour cells. PLoS Comput Biol 7(3):e1001113. doi:10.1371/journal.pcbi.1001113
320. Waters KM, Liu T, Quesenberry RD, Willse AR, Bandyopadhyay S, Kathmann LE, Weber TJ, Smith RD, Wiley HS, Thrall BD (2012) Network analysis of epidermal growth factor signaling using integrated genomic, proteomic and phosphorylation data. PLoS ONE 7(3):e34515. doi:10.1371/journal.pone.0034515
321. Barabasi AL (2007) Network medicine–from obesity to the "diseasome". New England J Med 357(4):404–407. doi:10.1056/NEJMe078114
322. Vidal M, Cusick ME, Barabasi AL (2011) Interactome networks and human disease. Cell 144(6):986–998. doi:10.1016/j.cell.2011.02.016
323. Pawson T, Linding R (2008) Network medicine. FEBS Lett 582(8):1266–1270. doi:10.1016/j.febslet.2008.02.011
324. Barabasi AL, Gulbahce N, Loscalzo J (2011) Network medicine: a network-based approach to human disease. Nat Rev Genet 12(1):56–68. doi:10.1038/nrg2918
325. Ergun A, Lawrence CA, Kohanski MA, Brennan TA, Collins JJ (2007) A network biology approach to prostate cancer. Mol Syst Biol 3:82. doi:10.1038/msb4100125
326. Ciriello G, Cerami E, Sander C, Schultz N (2012) Mutual exclusivity analysis identifies oncogenic network modules. Genome Res 22(2):398–406. doi:10.1101/gr.125567.111
327. Sridhar P, Kahveci T, Ranka S (2007) An iterative algorithm for metabolic network-based drug target identification. Pac Symp Biocomput 12:88–99
328. Sridhar P, Song B, Kahveci T, Ranka S (2008) Mining metabolic networks for optimal drug targets. Pac Symp Biocomput 13:291–302
329. Lee I, Blom UM, Wang PI, Shim JE, Marcotte EM (2011) Prioritizing candidate disease genes by network-based boosting of genome-wide association data. Genome Res 21(7):1109–1121. doi:10.1101/gr.118992.110
330. Klipp E, Wade RC, Kummer U (2010) Biochemical network-based drug-target prediction. Curr Opin Biotechnol 21(4):511–516. doi:10.1016/j.copbio.2010.05.004
331. Farkas IJ, Korcsmaros T, Kovacs IA, Mihalik A, Palotai R, Simko GI, Szalay KZ, Szalay-Beko M, Vellai T, Wang S, Csermely P (2011) Network-based tools for the identification of novel drug targets. Sci Signal 4(173): pt3. doi:10.1126/scisignal.2001950
332. Wang X, Gulbahce N, Yu H (2011) Network-based methods for human disease gene prediction. Brief Funct Genomics 10(5):280–293. doi:10.1093/bfgp/elr024
333. Piro RM, Di Cunto F (2012) Computational approaches to disease-gene prediction: rationale, classification and successes. FEBS J 279(5):678–696. doi:10.1111/j.1742-4658.2012.08471.x
334. Brouwers L, Iskar M, Zeller G, van Noort V, Bork P (2011) Network neighbors of drug targets contribute to drug side-effect similarity. PLoS ONE 6(7):e22187. doi:10.1371/journal.pone.0022187
335. Chang RL, Xie L, Bourne PE, Palsson BO (2010) Drug off-target effects predicted using structural analysis in the context of a metabolic network model. PLoS Comput Biol 6(9):e1000938. doi:10.1371/journal.pcbi.1000938

336. Cheng F, Liu C, Jiang J, Lu W, Li W, Liu G, Zhou W, Huang J, Tang Y (2012) Prediction of drug-target interactions and drug repositioning via network-based inference. PLoS Comput Biol 8(5):e1002503. doi:10.1371/journal.pcbi.1002503
337. Loscalzo J, Barabasi AL (2011) Systems biology and the future of medicine. Wiley Interdiscip Rev Syst Biol Med 3(6):619–627. doi:10.1002/wsbm.144

Part II
Modeling of Ill-Defined, Lack of Knowledge (Experimental) Problems in the Field of Complex Biosystems Including Identification and Validation: A Review of Enabling Technologies

Chapter 6
On Different Aspects of Network Analysis in Systems Biology

Amphun Chaiboonchoe, Wiktor Jurkowski, Johann Pellet, Enrico Glaab, Alexey Kolodkin, Antonio Raussel, Antony Le Béchec, Stéphane Ballereau, Laurene Meyniel, Isaac Crespo, Hassan Ahmed, Vitaly Volpert, Vincent Lotteau, Nitin Baliga, Leroy Hood, Antonio del Sol, Rudi Balling and Charles Auffray

Abstract Network analysis is an essential component of systems biology approaches toward understanding the molecular and cellular interactions underlying biological systems functionalities and their perturbations in disease. Regulatory and signalling pathways involve DNA, RNA, proteins and metabolites as key elements to coordinate most aspects of cellular functioning. Cellular processes depend on the structure and dynamics of gene regulatory networks and can be studied by employing a network representation of molecular interactions. This chapter describes several types of biological networks, how combination of different analytic approaches can be used to study diseases, and provides a list of selected tools for network visualization and analysis. It also introduces protein–protein interaction networks, gene regulatory networks, signalling networks and metabolic networks to illustrate concepts underlying network representation of cellular processes and molecular interactions. It finally discusses how the level of

An erratum to this chapter is available at 10.1007/978-94-007-6803-1_19

A. Chaiboonchoe · J. Pellet · S. Ballereau · L. Meyniel · H. Ahmed · V. Volpert · V. Lotteau · C. Auffray (✉)
European Institute for Systems Biology and Medicine, CNRS-UCBL-ENS, Université de Lyon, 50 Avenue Tony Garnier, 69366 Lyon cedex 07, France
e-mail: cauffray@eisbm.org, achaiboonchoe@eisbm.org, jpellet@eisbm.org, sballereau@eisbm.org, laurene.meyniel@inserm.fr, hahmed@eisbm.org, vvolpert@eisbm.org, vlotteau@eisbm.org

W. Jurkowski · E. Glaab · A. Kolodkin · A. Raussel · A. Le Béchec · I. Crespo · A. d. Sol · R. Balling
Luxembourg Centre for Systems Biomedicine, University of Luxembourg, 7, Avenue des Hauts-Fourneaux, 4362 Esch-sur-Alzette, Luxembourg
e-mail: wiktor.jurkowski@uni.lu, enrico.glaab@uni.lu, alexey.kolodkin@uni.lu, antonio.raussel@uni.lu, anthony.lebechec@uni.lu, isaac.crespo@uni.lu, antonio.delsol@uni.lu, rudi.balling@uni.lu

A. Kolodkin · N. Baliga · L. Hood
Institute for Systems Biology, 401 Terry Avenue North, Seattle, WA 98109-5234, USA
e-mail: nbaliga@systemsbiology.org, lhood@systemsbiology.org

A. Prokop and B. Csukás (eds.), *Systems Biology*,
DOI: 10.1007/978-94-007-6803-1_6,

accuracy in inferring functional relationships influences the choice of methods applied for the analysis of a particular biological network type.

Keywords Boolean models · Cell–cell interaction · Continuous models · Database · Discrete models · Diseasome · Gene regulatory networks · Hybrid models · Integrative disease map · Metabolic networks · Multi-scale modeling · Network analysis tools · Network modeling · Network pharmacology · Network topology · Ordinary differential equations · Protein–protein interaction · Signalling networks

Abbreviations

APID	Agile Protein Interaction DataAnalyzer
AP-MS	Affinity purification-mass spectrometry
ATP	Adenosine triphosphate
BioGRID	Biological General Repository for Interaction Datasets
CCNA2	Cyclin-A2
cMap	Connectivity map
CYP3A4	Cytochrome P_{450} 3A4
CTD	Comparative Toxicogenomics Database
DIP	Database of Interacting Proteins
DNA	Deoxyribonucleic acid
GHEN2PHEN(G2P)	Genotype-To-Phenotype
GR	Glucocorticoid receptor
GRN	Gene regulatory Network
GTP	Guanosine triphosphate
hERG	Human *Ether-à-go–go*-Related Gene
HPID	Human Protein Interaction Database
HPRD	Human Protein Reference Database
HTML	Hyper Text Markup Language
IMEx	International Molecular interaction Exchange consortium
MIMIx	Molecular Interaction eXperiment
MINT	Molecular INTeraction database
MIPS	Mammalian Protein–Protein Interaction Database
My-DTome	Myocardial infarction drug-target interactome network
NR	Nuclear Receptors
ODEs	Ordinary Differential Equations
OMIM	Online Mendelian Inheritance in Man
PDEs	Partial differential equations
PHARMGKB	Pharmacogenomics Knowledge Base
PPI	Protein-protein interaction
PSI-MI	Proteomics Standards Initiative on Molecular Interactions
RNA	Ribonucleic acid
SBML	Systems Biology Markup Language
STRING	Search Tool for the Retrieval of Interacting Genes/Proteins
XML	Extensible Markup Language
Y2H	Yeast two-hybrid

6.1 Introduction: From Pathways to Networks

In a biological pathway of interest, molecular entities such as genes or proteins very often also interact with other entities involved in distinct pathways. Since each pathway represents a specific region of a larger network in a given biological system, network analysis methods can provide additional biological insights that cannot be obtained from pathway analyses alone.

Biological networks comprise nodes that correspond to genes, proteins, metabolites or other biological entities, and edges that correspond to molecular interactions and other functional relationships between the biological entities. In general, in comparison to random networks (network elements connected by chance), biological networks of the same size and connectivity exhibit significant differences in aspects such as: wiring type or presence of topological motifs (groups of inter-connected nodes with a given structure). This affects (1) modularity i.e. the degree of division of the network into subnetworks that comprise densely connected nodes but share few edges outside the module, (2) dissortativity, i.e. the tendency of nodes to connect to other nodes in the network that are associated with different characteristics (e.g. nodes with many connections link to nodes with few connections), and (3) robustness [1] i.e. the resilience of the network to the removal of nodes or edges.

One of the most common strategies used to extract new insights from biological networks is to study the graph topology of a network, i.e. the patterns of interconnections between nodes and edges, based on a key metric: the degree or the number of connections of a node with other nodes. This led to the introduction of the concepts of scale-free networks [2], in which the node degree of connectivity distribution follows a power-law, and of small-world networks [3], in which the distance between nodes grows proportionally to the logarithm of the network size. In the latter, only few nodes act as "highly connected hubs" and the majority of the nodes are of low degree of connectivity (i.e. are engaged in only few interactions) [4, 5]. This property is believed to confer resistance to random attacks but makes scale-free network extremely susceptible to targeted perturbations [6]. Most biological networks display scale-free properties typical of small-world networks much more often than expected by chance, as is also observed for social networks [7].

These properties make network analysis an interesting approach to study biological systems in order to explain experimental observations and to formulate new hypotheses about biological functions at the molecular, cellular and higher levels of abstraction. Network analysis should however be performed with caution for two reasons. First, most of the networks are largely incomplete, that is they are missing many nodes (e.g. molecules, complexes, phenotypes) and edges (i.e. connections representing reactions, associations or influences). On the other hand, many false positive elements and interactions tend to be included in networks in the absence of contextual information on cellular or tissular localization. Second, the networks have dynamical architectures, i.e. they may change significantly in structures during biological or disease processes. There are virtually no examples where connectivity measurements have been made in a dynamical manner [8].

Despite these limitations networks have proven to be valuable tools to represent and analyse complex biological knowledge and processes. The first section of this chapter introduces the basic network types associated with gene regulation, protein–protein interactions, metabolic reactions and signalling processes. Inference of a genetic interaction network from gene co-expression data in the absence of knowledge of the underlying mechanisms, or accurate characterization of chemical reactions with known stoichiometry and kinetic parameters correspond to widely different levels of representation and require distinct computational approaches. The resulting molecular interaction networks need to be integrated with drug and phenotype networks in order to understand perturbations causing and caused by disease, and to facilitate the process of development of adequate medical interventions.

Introduction of network types (Fig. 6.1) is followed by a concise introduction to the main methods available for network topology analysis and modelling approaches: discrete and continuous models and how they can be extended to simulate cell–cell interactions.

In order to understand perturbations causing and caused by disease and to facilitate the process of development of adequate medical interventions, the

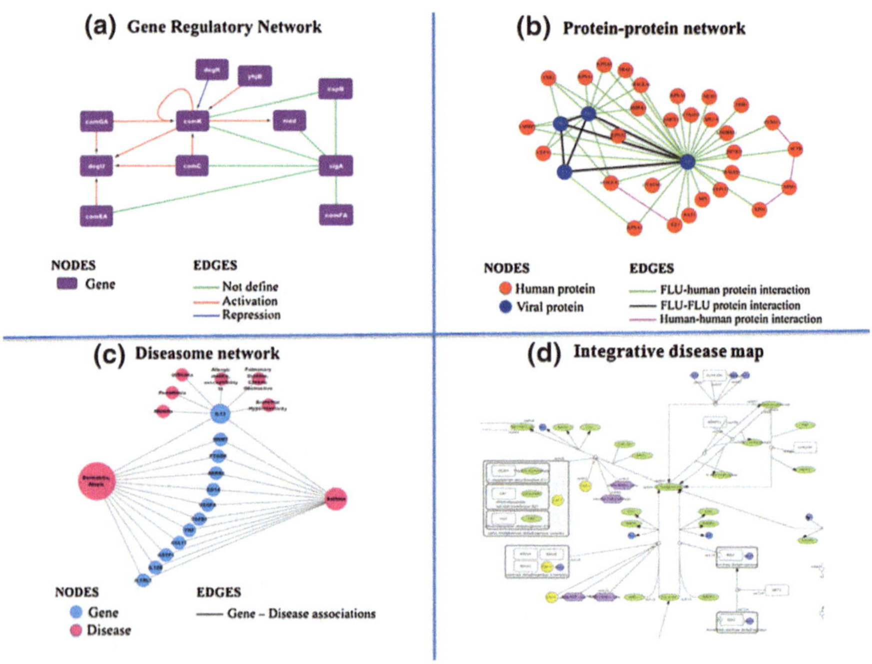

Fig. 6.1 Network types: **a** gene regulatory network, **b** protein–protein interaction network, **c** diseasome network, **d** section of an integrative network of disease model

resulting molecular interaction networks need to be integrated with drug and phenotype networks. We thus present specific examples where knowledge from multiple sources has been integrated to construct network of diseases and drug targets as well as comprehensive disease-focused cellular maps. In the last section we present a short overview of case studies applying network analysis for disease mechanism identification as an illustration of the network pharmacology trend that is now emerging in drug design and development.

6.2 Network Types

Among different types of biological networks, this section introduces the most studied molecular network types [9–11]. Although all of them can be reduced to graphs comprising nodes connected by edges, the variety of biological networks arises from the differences in annotated information or purpose for which particular biological knowledge is represented as a network. Thus, the majority of networks discussed here are directed (i.e. an arrow denotes the effect of the source node on the target node) to stress the order of the process, except the protein–protein interaction network that only combines non-causative pairwise physical interactions without implying succession in time and space. Gene regulatory networks are represented as directed graphs of proteins or small RNA molecules activating or inhibiting gene expression. Therefore a label on the edge orientation displays the regulatory effect. Signalling networks are also directed and signed, and are usually organized to separate processes by cellular localization to better illustrate mutual relationships of particular pathways. The main highlight of metabolic networks is the presence of chemical reactions that extend beyond non-covalent binding. Therefore simple edges need to be complemented with notation of additional substrates and products as well as links representing modulation e.g. activation by enzyme. Because of the complexity of mechanistic description, signalling and metabolic networks are often analysed with fine-tuned continuous models as opposed to typically simpler binary representation of gene regulatory networks.

Lastly, this section extends this simple classification of networks by presenting some examples of disease-related networks.

6.2.1 *Gene Regulatory Networks*

Construction and analysis of a gene regulatory network (GRN) play an important role in understanding the mechanisms of diseases. Recent advances in functional genomics are based on novel experimental and computational approaches that enhance the ability to comprehensively reconstruct the regulatory networks and enrich them with newly discovered components and interactions.

Gene regulatory networks can be reconstructed from the literature [12] or from experimental data using reverse engineering approaches. These approaches are based on gene expression co-variation patterns inferred from expression or on promoter region occupancy information of specific transcription factors derived from ChIP-Seq or ChIP-Chip experiments [13]. Nevertheless, networks inferred purely from experimental data and those assembled from the literature have different limitations. In the first case, a wealth of data about interactions previously described is ignored. On the other hand, literature-based networks are too disconnected from experimental data to be able to describe input–output relationships, such as cellular responses under specific biological stimuli or mechanisms that determine specific expression patterns. To bridge the gap new methods emerge that combine data from both literature and experiments and provides biological networks contextualized to certain experimental conditions [14, 15].

A GRN is commonly represented by a graph usually directed and signed showing interaction (network edges) between components (network nodes) that regulate each other. Because these regulatory components (transcription factors, cofactors, enhancers, repressors or miRNAs) have different properties, a GRN could describe mechanisms of gene regulation at multiple levels (e.g. transcription, post-transcription). Deciphering GRNs from rapidly growing microarray expression databases has been shown to be a very promising approach e.g. in cancer research [16, 17]. Many tools are emerging and have been used for constructing, inferring and analyzing such GRNs. These tools include Boolean networks, Bayesian networks and Ordinary Differential Equations (ODEs) into recently developed web-based applications [18–20]. Considering their complexity, it is often difficult to evaluate or validate the performance of the available tools. In biomedical research, GRNs are expected to improve the current understanding of development and gene interactions in complex systems [21–24].

6.2.2 Protein–Protein Interaction Networks

Protein–protein interactions (PPIs) play a vital role in mediating cellular responses in all species and interactome mapping has become an elementary aspect in all areas of systems biology as the scientific community has gathered information on thousands of protein interactions and is increasingly editing, curating and integrating these data sets.

Two complementary ways to obtain comprehensive PPI information exist. The first approach relies on high-throughput experimental methods, including yeast two-hybrid (Y2H) [25, 26], affinity purification followed by mass spectrometry (AP-MS) [27, 28], and luciferase complementation assays [29]. Although recent development of these methods aims at overcoming false-positive discoveries, experimental validation of PPIs by several methods is still crucial. The second approach is to curate all publications in the literature [30], and consult curated datasets from publicly available interaction databases [31].

The International Molecular interaction Exchange consortium (IMEx http://www.imexconsortium.org) [32] comprises eleven databases sharing data curated according to the same common standards. Interactions are reported using the Minimum Information required for reporting a Molecular Interaction eXperiment (MIMIx) [33]. The exchange of information is supported by one major data exchange format: the Proteomics Standards Initiative on Molecular Interactions (PSI-MI) [34] (see Ref. [35] for a complete review) (Table 6.1).

PPI datasets are often visualized as a network. Proteins are represented as nodes, and interactions as connections between nodes. PPI networks are dynamic: they change in time and space to adapt or switch to different physiological conditions.

Various studies have constructed molecular networks with virus proteins to identify their interactions with host proteins and reveal a host-pathogen hybrid protein-interaction network. From a systems biology perspective, a viral infection at the cell level can be viewed as a combination of molecular perturbations allowing viral components production and assembly while generating minor to massive cellular dysfunction. Thus, several large-scale studies of interactions between viral and human proteins have been performed to identify the laws governing virus-host interactomes [36–40]. Taking into account the analytical heterogeneity and the size of the interaction datasets, five corresponding virus-human interactomes were carefully and comprehensively reconstructed from the literature and uniformly analyzed using graph theory as well as structural and functional methods [41].

The systems approach to the biology of viral infection is thus beginning to unravel the global perturbations that lead to viral replication and eventually to pathogenesis. Furthermore, the list of virus-host interactors represents an invaluable resource to derive new molecules, especially for anti-viral therapy.

6.2.3 Signalling and Metabolic Networks

Signalling and metabolic networks may be built using a mechanism-based bottom-up strategy, with parameters either measured experimentally or assigned arbitrary values in the physiological range. Different questions may then be asked. For example, why is the network organized the way it is? Indeed, at first glance some features in the network may appear paradoxical. However, altering this paradoxical feature in the computer model may disclose design principles underlying the functioning of the network [42]. This approach has for example been successfully applied to design studies of nuclear receptor signalling.

Nuclear Receptors (NR) are proteins that may be activated by signalling molecules (ligands, composed of different intra- and extra-cellular metabolites such as hormones or fatty acids) and then regulate gene expression of their responsive genes. The glucocorticoid receptor (GR) is a NR with an important regulatory role in various cellular functions: gluconeogenesis and glucose uptake,

Table 6.1 Selected public interaction databases. It includes databases of protein–protein interactions (PPI), genetic interactions as well as genetic variation—disease—drug response data

Resource	URL	Data type	Comment
APID	http://bioinfow.dep.usal.es/apid	PPI	APID has been set up after the analysis of several available databases of protein–protein interactions
BioGRID	http://thebiogrid.org/	PPI, genetic interactions	Biological General Repository for Interaction Datasets
CTD	http://ctdbase.org/	Environment - disease	The Comparative Toxicogenomics Database illuminates how environmental chemicals affect human health
DIP	http://dip.doe-mbi.ucla.edu	PPI	The Database of Interacting Proteins was curated both manually by expert curators and also using computational approaches

(continued)

Table 6.1 (continued)

Resource	URL	Data type	Comment
GEN2PHEN	http://www.gen2phen.org/		Consortium for development and coordination of informatics for association studies
HPID	http://www.hpid.org	PPI	Human Protein Interaction Database
HPRD	http://www.hprd.org/		Human Protein Reference Database
IntAct	http://www.ebi.ac.uk/intact/	PPI	Open source database system and analysis tools for molecular interaction data
MIPS	http://mips.helmholtz-muenchen.de	PPI	Collection of manually curated high-quality PPI data collected from the scientific literature by expert curators
MINT	http://mint.bio.uniroma2.it/mint/	PPI	The Molecular INTeraction Database was one of the first PPI database to associate to each interaction a score estimating the reliability of the interaction

(continued)

Table 6.1 (continued)

Resource	URL	Data type	Comment
OMIM	http://www.ncbi.nlm.nih.gov/omim	Genetic variations—disease	Literature-based database of genes and diseases
PHARMGKB	http://www.pharmgkb.org/	Genetic variation—drugs	The Pharmacogenomics Knowledgebase—genetic variation on drug response data for clinicians and researchers
STRING	http://string-db.org	PPI	Known and Predicted Protein–Protein Interactions

lipolysis in adipose tissues, proteolysis in muscles, osteoblast differentiation and apoptosis [43, 44]. Ligands for GR are steroid hormones such as cortisol. GR has a high rate of nucleo-cytoplasmic shuttling and is predominantly located in the cytoplasm when unbound to a ligand. Upon ligand binding, GR changes its conformation, resulting in its increased affinity to nuclear importins and its decreased affinity to exportins. This causes translocation of the ligand-GR complex to the nucleus, where GR binds to its responsive genes and regulates their transcription [45, 46] thereby transmitting signal for gene expression. This network is also metabolic, as it involves nucleo-cytoplasmic transport of the receptor driven by GTP hydrolysis, and ATP and GTP metabolic reactions. Furthermore, concentrations of receptors, importins, exportins and ligand itself (e.g. cortisol continuously degraded by CYP3A4 enzyme) are parts of the larger metabolic network. This is the first paradoxical feature. Why is the receptor not only a receptor? Why does the receptor continuously shuttle between the nucleus and the cytoplasm? The study of GR network showed that nucleo-cytoplasmic shuttling of GR also serves as a smart shuttle for a ligand, which it pumps into the nucleus, thereby increasing the sensitivity and responsiveness of signalling [42].

This example shows that signalling and metabolic networks should not be analyzed separately, but instead be integrated together and with regulatory networks. This integration has recently become an important topic in systems biology [47].

6.2.4 Integrative Approaches Applied to Human Diseases

Disease networks can be viewed as networks of associations between disease-causing mutations and diseases or as high-resolution interaction maps integrating metabolic reactions, signalling pathways and gene regulatory networks.

The link between all genetic disorders (the human disease phenome) and the complete list of disease genes (the disease genome) results in a global view of the "diseasome", i.e. the combined set of all known disease gene associations [11, 48]. Here diseases form a network in which two diseases are connected if they share at least one gene. In the disease gene network, diseases or genes are represented as nodes and gene-disorder association as edges. In such a network representation, obesity, for example, is connected to at least seven other disorders such as diabetes, asthma, and insulin resistance because genes associated with these diseases are known to affect obesity as well [7]. In recent years several disease map projects have flourished, such as the pioneering work at the Systems Biology Institute (Okinawa, Japan) that was a hub of collaborative, community-based efforts to reconstruct a map of tuberculosis [49]. In addition, the Connectivity map (cMap) [50] developed by the Broad Institute aims to create a map connecting genes, diseases and drugs using a repository of gene expression profiles to represent different biological states including gene alterations and disease phenotypes. cMap is a web tool with preloaded data in which query results can be interpreted by strong (positive or negative) connection or absence of connection [50].

These efforts would be not possible without integrating publically available disease-related knowledge. Online Mendelian Inheritance in Man (OMIM) is a catalogue of human genes and genetic disorders and traits that has been updated continuously for several decades [51]. As of May 2012, it contained 2,795 diseases genes and 4,669 disorders for which the molecular basis is known. Other databases, including the Pharmacogenomics Knowledge Base (PHARMGKB) [52] or the Comparative Toxicogenomics Database (CTD), focus on different aspects of phenotype-genotype relationships. GEN2PHEN (G2P) is a European project aiming at gathering and curating information to build a knowledgebase of genotype-phenotype interactions [53]. This project will build a linked database from existing publicly accessible databases and integrate all available data using high-performance analytical tools.

The human cancer map project is exploiting the idea that network motifs can contribute to a network switching from one stable state to another [54]. Analyzing networks reconstructed from microarray experiments and molecular interaction maps, authors identified genes participating in bi-stable switches i.e. network motifs that can exist in two stable states and drive the change of the network states. Expression states of genes within bi-stable switches were compared between hepatocellular carcinoma or lung cancer and healthy control samples. In both cases, bi-stable switches made of differentially expressed genes were proposed to be a network mechanism for locking in disease states. Such studies have identified two important hubs: cyclins and albumin. In hepatocellular carcinoma, up-regulation of CCNA2 (cyclin A2) leads to changes in expression of downstream genes, in accordance with the general observation that perturbations of oscillations in cyclins concentrations can have a detrimental effect on cell development. For instance, ubiquitination of cyclin A1 induces apoptosis via activation of caspase-3 [55] or cyclin D1 degradation activated by a troglitazone derivative [56]. Also, the lack of phosphorylation of cyclin E, due to mutations, results in its increased stability, which has implications for breast cancer [57]. Another important hub is the up-regulated albumin gene. Albumin is a large transport molecule with an adaptable two-domain structure that can bind an array of lipids, peptides, metabolites and drugs. Allosteric modulation of albumin may change its binding and cargo transport properties, and hence directly affect downstream cellular processes. For example, electron spin resonance studies of albumin modulation by cancer-related small molecule markers revealed significant differences in binding of albumin-specific 16-doxyl-stearic acid probe. During disease progression the albumin pool saturates with cancer cellular metabolites, thus indicating an affected albumin state [58].

6.3 Network Analysis

Many software tools are available to reconstruct biological networks from experimental data and then position nodes on a graph according to a topological placement algorithm. This visual network of interconnected nodes can then be

Table 6.2 Network visualization and analysis tools

Name	Interface		License type	Network features			Other
	Graphical	Command line		Reactions	Regulatory	Meta-nodes	
Arena3D arena3d.org	+	+	Free for academic users	–	–	+	Compatibility with other tools
BioTapestry biotapestry.org	+	–	Free and open source	–	+	–	
CellDesigner celldesigner.org	+	–	Free	+	+	–	
Cytoscape cytoscape.org	+	–	GNU General Public License.	–	+	+	Multiple plugins available
COPASI copasi.org	+	+	Artistic License 2.0	+	+	–	Works with SBML files from CellDesigner
GEPHI gephi.org/	+	+	GNU General Public License	–	+	+	Customizable by plugins
Igraph igraph.sourceforge.net	–	+	GNU General Public License	–	+	+	Libraries in R and Python
Payao payaologue.org	+	–	Free	+	+	–	Java Web Application
VisANT visant.bu.edu	+	+	Free and open source	–	+	+	Java Web Application

Note All tools listed run on the main computer platforms (Linux, UNIX, Mac OSX, and Windows). Network features: Reactions—includes network representation of chemical reactions (substrates, products, reaction modifiers and reversibility); Regulatory—includes network representation of effect of regulation (activation, inhibition); Meta-node: multiple nodes can collapse into one node e.g. representing protein complex transformed into various mathematical models e.g. flux balance models, kinetic ODE models or space Partial Differential Equations (PDEs) models, which can be fitted to experimental data and used to simulate the kinetic behaviour of biological networks. Since each of the many different tools available only performs one of those tasks, designing an integrated and efficient analysis pipeline is challenging. Such difficulties prompted the development of a single unified standard language suitable for the interchange between various tools: the systems biological markup language (SBML) [59]. SBML is based on the widely used Extensible Mark-up Language (XML), allows the development of graphical interfaces and analysis frameworks to display and analyze interaction maps. It is therefore becoming a standard for the representation and annotation of biological processes. The following chapter discusses several tools and the perspectives of their future development.

6.3.1 Network Analysis Tools

Among public network management tools that currently exist to visually explore and analyse biological networks (see review in [60]) such as Arena3D [61], GEPHI [62], igraph [63] and VisANT [64] (Table 6.2), Cytoscape [65], CellDesigner [66] and Copasi [67] are the most powerful and widely used.

Cytoscape (449,030 downloads as on September 2012) is the most popular software for the visualization and analysis of interaction networks. Its functionality can be extended using the collection of plugins developed by the expanding Cytoscape community of users. Recently Cytoscape web [68] became available to embed interactive networks in an HTML page.

CellDesigner (65,105 downloads as on September 2012) is a structured diagram editor for drawing integrative maps (including gene regulatory and biochemical networks) that define reactions and interactions between various types of biochemical species (genes, proteins, small molecules) in the context of their subcellular localization and in relation to the biological or pathological processes in which they are involved.

Copasi (26,000 downloads as on January 2012) is a stand-alone program that supports models in the SBML standard and can simulate their behavior using ODEs or Gillespie's stochastic simulation algorithm.

The growing popularity of Cell Designer and Copasi stems from their compatibility and complementarity. The strength of Cell Designer is its easy-use interface for the drawing of biochemical networks while that of Copasi is its convenience for fitting the model to experimental data, metabolic control analysis and dynamic simulations. The network diagram and mathematical model created in Cell Designer can be easily transformed into Copasi and vice versa.

6.3.2 Network Topology Analysis

The discovery that many real-world biological networks exhibit scale-free and small-world properties [69, 70] has led to a surge of new topological analysis methods for biological networks. The study of global topological properties enables a general characterization of a network, e.g. providing information on its robustness to perturbations. In contrast, analysis of local topological properties can provide specific insights on single nodes (e.g. on their centrality in the network and their tendency to form dense clusters with other nodes), which can also be exploited in high-throughput data analysis applications. A comprehensive and detailed discussion of network topological properties has been compiled recently in a book dedicated to this topic [71].

Molecular interaction networks are assembled from public interaction databases like BioGRID [72], HPRD [73], IntAct [74], MIPS [75], DIP [76], HPID, [77] MINT [78], or meta-databases such as APID [79]. Several issues affect the quality of assembled networks and other integration tasks, e.g. false positives in the input data sources and incomplete lists of interactions. Commonly used pre-processing methods filter collected interactions using a combined set of criteria, e.g. the number and type of experiments that were used to verify an interaction and data source-specific confidence scores. One of the most comprehensive collections of molecular interaction data for different species is provided by the STRING database [80], which also contains different types of confidence scores for each interaction to filter the data. Since many network analysis methods require a single connected component as input, a final pre-processing step often involves removing small, disconnected components from a graph representation of the assembled interactions.

Regarding the typical applications for topological analyses, global descriptors are mainly used for the general characterization of large-scale biological networks. Since these global network properties have already been studied extensively for several biological network types and species, the corresponding analyses are only likely to provide new insights when studying a novel network type. However, new applications for employing global topological analyses as components of other algorithms have been proposed recently, e.g. to improve the generation of gene co-expression networks by analysing the scale-free property for tentative networks [81]. By contrast, local topological network characteristics have already been exploited by a wide variety of new data mining approaches recently, including methods to identify dense communities [82] of nodes [83, 84], methods to compare mapped gene and protein sets in terms of their network topological properties [85], and approaches to score distances between nodes for prioritizing disease genes [86]. Interestingly, recent studies have shown that cancer-associated genes tend to have outstanding topological characteristics [87], even when accounting for study-specific biases, and that topological information can facilitate cancer classification [88].

In summary, topological characteristics of complete biological networks, subgraphs and single nodes provide a valuable information source for the integrated analysis of functional genomics data. Network topology analyses are often combined with graph-theoretic methods to identify dense communities or clusters of nodes [83, 84], or to quantify the similarity between single nodes or node sets using different network-based distance measures [86, 89–91]. However, topological properties can also be exploited in other domains, e.g. as features in machine learning methods for clustering and prediction [92], as part of scoring criteria in de novo pathway prediction [93], and to evaluate the stability and integrity of biological networks generated from combined microarray correlation analysis and literature mining [94].

6.3.3 Network Modeling

Various approaches describe biological networks mathematically. The simplest is to build a discreet model based on graph theory. In this approach, each node (e.g. molecule) of the biological network may be present in two (e.g. 0 or 1) or several fixed states. Each state affects interactions of a node with other nodes differently and the underlying mechanisms need not be known. In contrast, continuous modeling offers alternative approaches that account for the subtle gradual changes in the concentration of species in a biological network. Three main types of continuous models exist: a continuous 'microscopic' model traces every molecule individually while a 'mesoscopic' model uses stochastic simulations with molecular concentrations described in terms of probability functions; lastly, a 'macroscopic' model neglects limitations in the diffusion of molecules on the reaction rate, considers each species of biomolecules as a single pool described with a system of ODEs. The latter is a very popular approach to model intracellular metabolic networks in which the number of molecules is rather high and could be viewed as a single pool with a certain mean concentration. Another important question relates to interactions between different cells. This requires special cell–cell interaction models. In this section we will discuss three main modeling approaches: discrete modeling, continuous modeling based on ODEs and modeling of cell–cell interactions.

6.3.3.1 Discrete Models

In the Boolean framework, the state of each node in the network is conceptually described as being active (represented by '1') or inactive (represented by '0'). Similarly, directed edges in the network are represented as activators or inhibitors. The state of each node is therefore determined by the states of the nodes that activate and/or inhibit them, following predefined logic rules inferred from experimental data and/or expert knowledge. Boolean models provide an abstraction of genetic circuits that, despite being simplistic, are able to capture important aspects of cell development [95]. Importantly, they enable the study of key dynamical properties such as steady states, defined as stable states of the network that might be stationary or oscillatory. In a Boolean model, the consequences in the global system of a given perturbation (represented as a change in the state of a node) are assessed through updating the states of the nodes in the network following the logic functions. The updating process can be done synchronously or asynchronously, depending on whether all nodes are updated simultaneously or in a step-wise manner, respectively [96, 97]. Indeed, the steady states reached following these two strategies might be different, and it is usually recommended to combine their results. Boolean networks allow integration of qualitative information into the modelling process and have been successfully applied to many relevant biological systems [6, 96, 98]. Despite being deterministic, the Boolean framework allows inclusion of stochastic components in the models, either in the states of nodes [99, 100] or in the logic functions [101]. Together with Boolean models, multiple valued logic models [97, 102] are another type of discrete logic models that allow considering more than two levels, e.g. low, medium or high expression, which might be more realistic, but is associated with a higher computational cost.

Discrete logic models have the fundamental advantage over continuous models such as ODEs that they can use qualitative data to build a gene regulatory network. The amount and availability of qualitative information is larger than the quantitative parameters required in ODEs. However, discrete logic approaches cannot model the evolution in time of the quantitative concentrations of the species in the system. To bridge the gap between ODEs and discrete logic models, a third category of approaches has been developed, where the initial discrete logic network is transformed into a system of ODEs, following different strategies with successful results [98, 103–105].

6.3.3.2 Continuous Models

ODEs are often used for dynamic modelling of regulatory networks of different levels of complexity from bacteria [106] to eukaryotes [107–110]. Changes in concentrations of each species in the network of interacting biomolecules may be expressed through balance equations and rate equations. In this framework, variations in concentrations of molecules as a function of time may be represented by differential equations establishing the stoichiometry and the reaction rates of the transformations and/or transport events [111, 112]. These equations contain the information about component properties for each molecule. Integrating all ODEs together enables reconstruction of the emergent behaviour of a whole system, e.g. simulation of its dynamics in silico.

In order to validate a dynamic model, the simulated systemic behavior is compared with the behaviour of a real system; this comparison is especially powerful if both the real object and its model are challenged with a series of perturbations that were not considered while the model was under construction nor used for parameter fitting. In many cases the predicted behaviour does not fit that of the actual biological system. This may then lead to the discovery of mechanisms missing or poorly described in the model. For example, the yeast glycolysis model built by Teusink and co-authors predicted that yeasts would invest too much of ATP in the first ATP-consuming reactions, and then die from the accumulation of these compounds and from a deficit of phosphate. The observation that yeasts are more robust in reality than in silico, provoked re-thinking of the model mechanisms and led to the discovery of an additional negative feedback loop which regulates the first phosphorylation step of glycolysis and prevents the turbo explosion in ATP-consuming reactions [113].

Once a model simulates the biological system behaviour adequately, it can be used for various goals. For example, using a in silico cell model of a metabolic network, one can design modification of the organism metabolism: e.g. the metabolism of *Escherichia coli* can be modified in such a way that *E. coli* produces polylactic acid—a biopolymer analogous to petroleum-based polymers which can be used in industry [114]; the metabolism of insects can be modified to make insects a promising source of food to meet the challenge of providing the protein supply to feed over 9 billion humans in the near future [115]. Cell models have become useful in differential network-based drug discovery: a kinetic model of the known metabolic network may help to find proper target enzymes either for correcting malfunctioning of a human cell or for killing a cancer cell [116] or a parasite [117]. For instance, comparison between glycolysis in *Trypanosoma brucei* (parasite causing African trypanosomiasis in humans), and glycolysis of human erythrocytes was used for the development of drugs killing *T. brucei* with reduced side effects [118].

The main drawback of in silico cell models is that they usually require knowledge of the mechanisms of interactions and estimates of numerous parameters for these interactions, which are only available for few systems that are very well characterized experimentally. However one may anticipate that tremendous progress in functional genomics, proteomics, metabolomics and bioinformatics should help to obtain the lacking information in the near future.

6.3.3.3 Cell-Cell Interactions and Multi-Scale Modeling

Intracellular regulatory networks are used for the analysis of a single cell or of a population of identical cells in the same state. However, different cell types usually exist in the same tissue while cells of the same type can have different concentrations of intracellular proteins due to intrinsic stochastic variations or to different cellular environments. Their interaction can influence intracellular regulatory networks e.g. through bistable switches in the network. Hence, local regulation by cell–cell interaction within the tissue and global regulation through interactions with other organs should both be taken into account in the analysis of regulatory networks. Multi-scale modeling therefore includes intracellular and molecular levels as well as interactions at the levels of cell populations and of other tissues and organs.

Intracellular regulation of individual cells and their local and global interactions can be studied with hybrid models in which cells are considered as individual objects, intracellular regulatory networks are described by ODEs or by Boolean networks, and the extracellular matrix together with nutrients, hormones and other signalling molecules by PDEs. Hybrid models can also account for natural stochastic variations of intracellular concentrations, cell motion, cell proliferation, differentiation and apoptosis. They are well adapted to the representation and analysis of various biological systems and biomedical situations. They can include the pharmacokinetics of medical treatments, prediction and optimization of their efficacy. However, they require a detailed knowledge of intracellular and extracellular regulations and sophisticated modeling tools [119, 120].

A hybrid model of erythropoiesis and leukemia treatment was described recently [121]. Erythropoiesis, or red blood cell production, occurs mainly in the bone marrow in small units called erythroblastic islands. They contain, on average, several dozens of cells in different phases of cell differentiation: erythroid progenitors, erythroblasts, reticulocytes structured around a macrophage. Normal functioning of erythropoiesis depends on the balance between proliferation, differentiation and apoptosis of erythroid progenitors. The intracellular regulatory network in erythroid progenitors, described by ODEs, determines cell fate due to bistability, where one stable equilibrium corresponds to proliferation without differentiation and another one to differentiation/apoptosis. The choice between these stable equilibria and, at the next stage, between differentiation and apoptosis is determined by two factors: local extracellular regulation, e.g. Fas-ligand produced by more mature cells and growth factors produced by macrophages; and global regulation by hormones, such as erythropoietin, with concentrations depending on the total number of erythrocytes produced by erythroblastic islands. Biochemical substances in the extracellular matrix influence intracellular regulation through the coefficients of the ODE system while their concentration is described by PDEs. This multi-level modeling simulates erythropoiesis in normal and stress situations in agreement with experimental data. It also explains the role of central macrophages in controlling erythroblastic islands. Strong perturbations of the system caused for example by mutations or dysfunction of regulatory mechanisms can result in various blood diseases such as anemia or leukemia.

6.3.4 Network Analysis for Systems Biomedicine and Pharmacology

6.3.4.1 Network Analysis of Disease

Network analysis may be also a valuable approach to study mechanisms underlying pathology and disease susceptibility. For instance studies of network dynamics can shed light on disease-related network states. Different stable steady states appear to be related to distinct phenotypic states of the cell [122].

Robustness of biological networks allows maintenance of a certain phenotypic state over a range of perturbations and may play an important role in controlling state transitions when such stable states are reached. It has been postulated that network circuits displaying bi- and multi-stability may drive network state transitions associated to disease progression and then maintain networks in diseased states [123]. In particular, bi-stable switches in protein–protein interaction or regulatory networks allow cells to enter into irreversible paths and assume different fates depending on which genes are expressed or silent [124, 125].

An analysis of pathological response to fat diet [126] or the mechanism of prion disease [127] using Boolean modelling recently showed the importance of network motifs in stabilizing the disease-related network state. Indeed circuits regulating bi-stability do not function in isolation but are assembled as an interconnected core cluster of genes that regulate one another thereby ensuring the stability of the network. In addition, differentially expressed genes involved in bi-stable switches are central to the regulatory network and can thus efficiently propagate perturbations to more distant regions of the network. These concepts are supported by previous studies focusing on network motifs to show the key role of network bi-stable feedback loops in cell fate determination and plasticity [128–130], and the implication of bi-stable circuits in the resilience and progression of human cancers, where the healthy and cancer states are considered to be the two stable states [131, 132].

6.3.4.2 Network Pharmacology

In a typical drug development approach, an active compound is optimized to act on a single protein target and other potential interactions are considered only to increase specificity of binding to a given receptor subtype while avoiding known toxic effects. This view has been recently challenged in the field of polypharmacology as it is recognized that drugs can effectively act on multiple targets, e.g. the recent discovery of the simultaneous inhibition of two families of oncogenes (tyrosine and phosphoinositide kinases) by the same effector [133]. Studies of off-target effects can lead to successful drug repurposing [134, 135] or to the prevention of adverse side effects. For instance, blocking of the hERG potassium channel is responsible for many severe drug-induced cardiac arrhythmias and is therefore included as a part of safety testing in drug development [136].

On the other hand, a single drug is unlikely to be sufficient to target the multiple facets of pathological processes. Rational drug design is now attempting to define mixtures of bioactive compounds that constitute drugs often exerting synergistic therapeutic effects, as in the long tradition of herbal medicine [137].

Systems biology approaches are used to develop tools necessary to understand complex drug-target relationships. Network pharmacology is integrating information on diseases, targets, and drugs with biological data to infer networks of drug targets, disease-related genes or drug-disease interactions [138–140]. Properties of such networks may help in understanding individual drug response due to changing genetic background [122], or contribute to the discovery of new druggable targets, therapeutic strategies to overcome adverse drug effects. In some cases, these studies are accompanied with development of resources and tools tailored for medical applications. My-DTome [141] is an example of a web-based searchable resource on drug-target interactome networks relevant to myocardial infarction.

6.4 Conclusions

Network analysis to organize and mine biological knowledge has become an inherent element of computational systems biology. By focusing on certain aspects of biochemical processes in living cells, the network may represent gene regulatory, metabolic, signalling processes and connect network elements with functional associations or, when used without imputing causality, represent physical binding of molecules. To answer specific biological questions different methodologies should be considered depending on the completeness of description that is accessible. This chapter presented an overview of approaches used to derive meaningful conclusions from graph topology, and develop simulations of network states using discrete and continuous models. The use of these approaches may be extended to simulate processes on higher (cell–cell interactions) levels of organization or combined to represent multiple levels from molecules to organs. The study of disease-related networks is increasingly impacting identification of drug targets with limited adverse effects, triggering the emergence of network pharmacology.

Acknowledgments This work was supported by the CNRS, and in part by the EU grants to CA in the context of the U-BIOPRED consortium (*Unbiased Biomarkers for the PREDiction of respiratory disease outcomes*, Grant Agreement IMI 115010) and the MeDALL consortium (*Mechanisms of the Development of Allergy,* Grant Agreement FP7 N 264357). The formation of the European Institute for Systems Biology & Medicine hosted at Claude Bernard University is supported by the Lyonbiopole competitive cluster and its academic, industrial and local authority partners, including Grand Lyon, Région Rhône-Alpes, Direction de la Recherche et de la Technologie, and the Finovi Foundation (CA).

We would like to acknowledge the support of the Luxembourg Centre for Systems Biomedicine and the University of Luxembourg (LH and RB), the NIH General Medical Sciences Center for Systems Biology GM076547 (LH) and a Department of Defense contract on Liver Toxicity W911SR-09-C-0062 (LH).

References

1. Callaway DS, Newman MEJ, Strogatz SH, Watts DJ (2000) Network robustness and fragility: percolation on random graphs. Phys Rev Lett 85:5468–5471
2. Barabási AL, Albert R (1999) Emergence of scaling in random networks. science 286, 509
3. Watts DJ, Strogatz SH (1998) Collective dynamics of 'small-world'networks. Nature 393:440–442

4. Barabási A-L, Oltvai ZN (2004) Network biology: understanding the cell's functional organization. Nat Rev Genet 5:101–113
5. Barabási A-L (2009) Scale-free networks: a decade and beyond. Science 325:412–413
6. Albert R, Jeong H, Barabasi AL (2000) Error and attack tolerance of complex networks. Nature 406, 378–382
7. Barabási A-L (2007) Network medicine–from obesity to the 'diseasome'. N Engl J Med 357:404–407
8. Ideker T, Krogan NJ (2012) Differential network biology. Mol Syst Biol 8
9. Vidal M, Cusick ME, Barabási A-L (2011) Interactome networks and human disease. Cell 144:986–998
10. Pavlopoulos G et al (2011) Using graph theory to analyze biological networks. BioData Mining 4:10
11. Barabási A-L, Gulbahce N Loscalzo J (2011) Network medicine: a network-based approach to human disease. Nat Rev Genet 12, 56–68
12. Bauer-Mehren A, Furlong LI, Sanz F (2009) Pathway databases and tools for their exploitation: benefits, current limitations and challenges. Mol Syst Biol 5:290
13. Johnson DS, Mortazavi A, Myers RM, Wold B (2007) Genome-wide mapping of in Vivo protein-DNA interactions. Science 316:1497–1502
14. Crespo I, Krishna A, Le Béchec A Del Sol A (2012) Predicting missing expression values in gene regulatory networks using a discrete logic modeling optimization guided by network stable states. Nucleic Acids Res doi:10.1093/nar/gks785
15. Saez-Rodriguez J et al (2009) Discrete logic modelling as a means to link protein signalling networks with functional analysis of mammalian signal transduction. Mol Syst Biol 5:331
16. Madhamshettiwar P, Maetschke S, Davis M, Reverter A, Ragan M (2012) Gene regulatory network inference: evaluation and application to ovarian cancer allows the prioritization of drug targets. Genome Med 4:41
17. Zhang Y, Xuan J, De los Reyes BG, Clarke R Ressom HW (2010) Reconstruction of gene regulatory modules in cancer cell cycle by multi-source data integration. PLoS ONE 5, e10268
18. Hache H, Lehrach H Herwig R (2009) Reverse engineering of gene regulatory networks: a comparative study. EURASIP J Bioinf Syst Biol, 8:1–8:12
19. Marbach D et al. (2010) Revealing strengths and weaknesses of methods for gene network inference. In: Proceedings of the national academy of sciences 107, 6286–6291
20. Haibe-Kains B et al (2011) Predictive networks: a flexible, open source, web application for integration and analysis of human gene networks. Nucleic Acids Res 40:D866–D875
21. Cooke EJ, Savage RS, Wild DL (2009) Computational approaches to the integration of gene expression, ChIP-chip and sequence data in the inference of gene regulatory networks. Semin Cell Dev Biol 20:863–868
22. Nazri A, Lio P (2012) Investigating meta-approaches for reconstructing gene networks in a mammalian cellular context. PLoS ONE 7:e28713
23. Ahmad FK, Deris S, Othman NH (2011) The inference of breast cancer metastasis through gene regulatory networks. J Biomed Inform. doi:10.1016/j.jbi.2011.11.015
24. Davidson EH (2010) Emerging properties of animal gene regulatory networks. pp 911–920
25. Fields S, Song O (1989) A novel genetic system to detect protein–protein interactions. Nature 340:245–246
26. Parrish JR, Gulyas KD, Finley RL Jr (2006) Yeast two-hybrid contributions to interactome mapping. Curr Opin Biotechnol 17:387–393
27. Rigaut G et al (1999) A generic protein purification method for protein complex characterization and proteome exploration. Nat Biotechnol 17:1030–1032
28. Köcher T, Superti-Furga G (2007) Mass spectrometry-based functional proteomics: from molecular machines to protein networks. Nat Methods 4:807–815
29. Cassonnet P et al (2011) Benchmarking a luciferase complementation assay for detecting protein complexes. Nat Methods 8:990–992
30. Roberts PM (2006) Mining literature for systems biology. Brief. Bioinformatics 7:399–406

31. Lehne B, Schlitt T (2009) Protein-protein interaction databases: keeping up with growing interactomes. Hum. Genomics 3:291–297
32. Orchard S et al (2012) Protein interaction data curation: the International Molecular Exchange (IMEx) consortium. Nat Methods 9:345–350
33. Orchard S et al (2007) The minimum information required for reporting a molecular interaction experiment (MIMIx). Nat Biotechnol 25:894–898
34. Hermjakob H et al (2004) The HUPO PSI's molecular interaction format–a community standard for the representation of protein interaction data. Nat Biotechnol 22:177–183
35. Orchard S, Kerrien S (2010) Molecular interactions and data standardisation. Methods Mol Biol 604:309–318
36. De Chassey B et al (2008) Hepatitis C virus infection protein network. Mol Syst Biol 4:230
37. Zhang L et al (2009) Analysis of vaccinia virus-host protein–protein interactions: validations of yeast two-hybrid screenings. J Proteome Res 8:4311–4318
38. Calderwood MA et al. (2007) Epstein-Barr virus and virus human protein interaction maps. Proc Natl Acad Sc. USA. 104, 7606–7611
39. Shapira SD et al (2009) A physical and regulatory map of host-influenza interactions reveals pathways in H1N1 infection. Cell 139:1255–1267
40. Khadka S et al. (2011) A physical interaction network of dengue virus and human proteins. Mol. Cell Proteomics 10, M111.012187
41. Meyniel-Schicklin L, De Chassey B, Andre P, Lotteau V (2012) Viruses and interactomes in translation. Mol Cell Proteomics: MCP. doi:10.1074/mcp.M111.014738
42. Kolodkin AN et al (2010) Design principles of nuclear receptor signaling: how complex networking improves signal transduction. Mol Syst Biol 6:446
43. Eijken M et al (2006) The essential role of glucocorticoids for proper human osteoblast differentiation and matrix mineralization. Mol Cell Endocrinol 248:87–93
44. Zhou JG, Cidlowski JA (2005) The human glucocorticoid receptor: one gene, multiple proteins and diverse responses. Steroids 70:407–417
45. Cutress ML, Whitaker HC, Mills IG, Stewart M, Neal DE (2008) Structural basis for the nuclear import of the human androgen receptor. J Cell Sci 121:957–968
46. Heitzer MD, Wolf IM, Sanchez ER, Witchel SF, DeFranco DB (2007) Glucocorticoid receptor physiology. Rev Endocr Metab Disord 8:321–330
47. Lee JM, Min Lee J, Gianchandani EP, Eddy JA, Papin JA (2008) Dynamic analysis of integrated signaling, metabolic, and regulatory networks. e1000086
48. Goh K-I et al (2007) The human disease network. Proc Natl Acad Sci USA 104:8685–8690
49. Kitano H, Ghosh S, Matsuoka Y (2011) Social engineering for virtual 'big science' in systems biology. Nat Chem Biol 7:323–326
50. Lamb J (2007) The connectivity map: a new tool for biomedical research. Nat Rev Cancer 7:54–60
51. Hamosh A, Scott AF, Amberger J, Valle D, McKusick VA (2000) Online Mendelian Inheritance in Man (OMIM). Hum Mutat 15:57–61
52. Klein TE et al (2001) Integrating genotype and phenotype information: an overview of the PharmGKB project. Pharmacogenetics research network and knowledge base. Pharmacogenomics J 1:167–170
53. Webb AJ, Thorisson GA, Brookes AJ (2011) An informatics project and online 'Knowledge Centre' supporting modern genotype-to-phenotype research. 543–550
54. Shiraishi T, Matsuyama S, Kitano H (2010) Large-scale analysis of network bistability for human cancers. PLoS Comput Biol 6:e1000851
55. Ekberg J, Persson JL (2009) Post-translational modification of cyclin A1 is associated with staurosporine and TNFalpha induced apoptosis in leukemic cells. Mol Cell Biochem 320:115–124
56. Wei S et al (2008) A novel mechanism by which thiazolidinediones facilitate the proteasomal degradation of cyclin D1 in cancer cells. J biol chem 283:26759–26770
57. Mull BB, Cox J, Bui T, Keyomarsi K (2009) Post-translational modification and stability of low molecular weight cyclin E. Oncogene 28:3167–3176

58. Gurachevsky A, Muravskaya E, Gurachevskaya T, Smirnova L, Muravsky V (2007) Cancer-associated alteration in fatty acid binding to albumin studied by spin-label electron spin resonance. Cancer Invest 25:378–383
59. Hucka M et al (2003) The systems biology markup language (SBML): a medium for representation and exchange of biochemical network models. Bioinformatics 19:524–531
60. Sanz-Pamplona R et al (2012) Tools for protein–protein interaction network analysis in cancer research. Clin Transl Oncol 14:3–14
61. Pavlopoulos GA et al (2008) Arena3D: visualization of biological networks in 3D. BMC Syst Biol 2:104
62. Bastian M, Heymann S, Jacomy Gephi M (2009) An open source software for exploring and manipulating networks
63. Csardi G, Nepusz T (2006) The igraph software package for complex network research. Int J Complex Syst Complex Sy, 1695
64. Hu Z et al (2009) VisANT 3.5: multi-scale network visualization, analysis and inference based on the gene ontology. Nucleic Acids Res 37:W115–W121
65. Shannon P et al (2003) Cytoscape: a software environment for integrated models of biomolecular interaction networks. Genome Res 13:2498–2504
66. Funahashi A et al. (2008) CellDesigner 3.5: A versatile modeling tool for biochemical networks. Proc IEEE 96, 1254–1265
67. Hoops S et al (2006) COPASI—a complex pathway simulator. Bioinformatics 22:3067–3074
68. Lopes CT et al (2010) Cytoscape web: an interactive web-based network browser. Bioinformatics 26:2347–2348
69. Wuchty S (2001) Scale-free behavior in protein domain networks. Mol Biol Evol 18:1694–1702
70. Böde C et al (2007) Network analysis of protein dynamics. FEBS Lett 581:2776–2782
71. Junker BH, Schreiber F, Ebrary I (2008) Analysis of biological networks. (Wiley online library)
72. Stark C et al (2006) BioGRID: a general repository for interaction datasets. Nucleic Acids Res 34:D535–D539
73. Peri S et al (2004) Human protein reference database as a discovery resource for proteomics. Nucleic Acids Res 32:D497–D501
74. Hermjakob H et al (2004) IntAct: an open source molecular interaction database. Nucleic Acids Res 32:D452–D455
75. Mewes HW et al (2002) MIPS: a database for genomes and protein sequences. Nucleic Acids Res 30:31–34
76. Xenarios I et al (2002) DIP, the database of interacting proteins: a research tool for studying cellular networks of protein interactions. Nucleic Acids Res 30:303–305
77. Han K, Park B, Kim H, Hong J, Park J (2004) HPID: the human protein interaction database. Bioinformatics 20:2466–2470
78. Zanzoni A et al (2002) MINT: a molecular interaction database. FEBS Lett 513:135–140
79. Prieto C, De Las Rivas J (2006) APID: agile protein interaction dataanalyzer. Nucleic Acids Res 34, W298–302
80. Szklarczyk D et al (2011) The STRING database in 2011: functional interaction networks of proteins, globally integrated and scored. Nucleic Acids Res 39:D561–D568
81. Zhang B, Horvath S (2005) A general framework for weighted gene co-expression network analysis. Stat Appli Genet Mol Biol 4:17
82. Radicchi F, Castellano C, Cecconi F, Loreto V, Parisi D (2004) Defining and identifying communities in networks. PNAS 101:2658–2663
83. Newman MEJ (2004) Fast algorithm for detecting community structure in networks. Phys Rev E 69:066133
84. Sun S, Dong X, Fu Y, Tian W (2011) An iterative network partition algorithm for accurate identification of dense network modules. Nucleic Acids Res

85. Glaab E, Baudot A, Krasnogor N, Valencia A (2010) Extending pathways and processes using molecular interaction networks to analyse cancer genome data. BMC Bioinform 11:597
86. Nitsch D et al (2011) PINTA: a web server for network-based gene prioritization from expression data. Nucleic Acids Res 39:W334
87. Jonsson PF, Bates PA (2006) Global topological features of cancer proteins in the human interactome. Bioinformatics 22:2291–2297
88. Liu CC et al (2006) Topology-based cancer classification and related pathway mining using microarray data. Nucleic Acids Res 34:4069–4080
89. Wang Q et al. (2011) A novel network-based method for measuring the functional relationship between gene sets. Bioinformatics doi:10.1093/bioinformatics/btr154
90. Alexeyenko A et al (2012) Network enrichment analysis: extension of gene-set enrichment analysis to gene networks. BMC Bioinform 13:226
91. Glaab E, Baudot A, Krasnogor N, Schneider R, Valencia A (2012) EnrichNet: network-based gene set enrichment analysis. Bioinformatics 28:i451–i457
92. Lee H, Tu Z, Deng M, Sun F, Chen T (2006) Diffusion kernel-based logistic regression models for protein function prediction. *OMICS*. J Integr Biol 10:40–55
93. Ma X, Tarone AM, Li W (2008) Mapping genetically compensatory pathways from synthetic lethal interactions in yeast. PLoS ONE 3:e1922
94. Li S, Wu L, Zhang Z (2006) Constructing biological networks through combined literature mining and microarray analysis: a LMMA approach. Bioinformatics 22:2143–2150
95. Glass L, Kauffman SA (1973) The logical analysis of continuous, non-linear biochemical control networks. J Theor Biol 39:103–129
96. Fauré A, Naldi A, Chaouiya C, Thieffry D (2006) Dynamical analysis of a generic Boolean model for the control of the mammalian cell cycle. Bioinformatics (Oxford, England) 22, e124–131
97. Garg A, Mendoza L, Xenarios I, DeMicheli G (2007) Modeling of multiple valued gene regulatory networks. In: Conference proceedings: annual international conference of the ieee engineering in medicine and biology society. IEEE engineering in medicine and biology society. Conference 2007, 1398–1404
98. Mendoza L, Xenarios I (2006) A method for the generation of standardized qualitative dynamical systems of regulatory networks. Theor Biol Med Model 3:13
99. Willadsen K, Wiles J (2007) Robustness and state-space structure of Boolean gene regulatory models. J Theor Biol 249:749–765
100. Ribeiro AS, Kauffman SA (2007) Noisy attractors and ergodic sets in models of gene regulatory networks. J Theor Biol 247:743–755
101. Garg A, Mohanram K, Di Cara A, De Micheli G Xenarios I (2009) Modeling stochasticity and robustness in gene regulatory networks. Bioinformatics (Oxford, England) 25, i101–109
102. Thomas R (1973) Boolean formalization of genetic control circuits. J Theor Biol 42:563–585
103. Sánchez-Corrales Y-E, Alvarez-Buylla ER, Mendoza L (2010) The Arabidopsis thaliana flower organ specification gene regulatory network determines a robust differentiation process. J Theor Biol 264:971–983
104. Wittmann DM et al (2009) Transforming boolean models to continuous models: methodology and application to T-cell receptor signaling. BMC Syst Biol 3:98
105. Krumsiek J, Pölsterl S, Wittmann DM, Theis FJ (2010) Odefy–from discrete to continuous models. BMC Bioinf 11:233
106. Li S, Brazhnik P, Sobral B, Tyson JJ (2008) A quantitative study of the division cycle of Caulobacter crescentus stalked cells. PLoS Comput Biol 4:e9
107. Tyson JJ, Csikasz-Nagy A, Novak B (2002) The dynamics of cell cycle regulation. BioEssays: News Rev Mol, Cell Dev Biol 24:1095–1109
108. Jaeger J et al (2004) Dynamical analysis of regulatory interactions in the gap gene system of drosophila melanogaster. Genetics 167:1721–1737

109. Chen KC et al (2004) Integrative analysis of cell cycle control in budding yeast. Mol Biol Cell 15:3841–3862
110. Locke JCW et al. (2005) Extension of a genetic network model by iterative experimentation and mathematical analysis. Mol Syst Biol 1, 2005.0013
111. Klipp E, Herwig R, Kowald A, Wierling C, Lehrach H (2005) Systems biology in practice: concepts, implementation and application. (Wiley-VCH)
112. Cao J, Qi X, Zhao H (2012) In: next generation microarray bioinformatics. Wang J, Tan AC, Tian T (eds) 802, 185–197 Humana Press
113. Teusink B, Walsh MC, Van Dam K, Westerhoff HV (1998) The danger of metabolic pathways with turbo design. Trends Biochem Sci 23:162–169
114. Jung YK, Kim TY, Park SJ, Lee SY (2010) Metabolic engineering of escherichia coli for the production of polylactic acid and its copolymers. Biotechnol Bioeng 105:161–171
115. Vogel G (2010) For more protein filet of cricket. Science 327:811
116. Hornberg JJ, Bruggeman FJ, Westerhoff HV, Lankelma J (2006) Cancer: a systems biology disease. Biosystems 83:81–90
117. Bakker BM, Westerhoff HV, Opperdoes FR, Michels PAM (2000) Metabolic control analysis of glycolysis in trypanosomes as an approach to improve selectivity and effectiveness of drugs. Mol Biochem Parasitol 106:1–10
118. Bakker BM et al. (2000) Compartmentation protects trypanosomes from the dangerous design of glycolysis. Proc Natl Acad Sci U S A 97, 2087–92
119. Alarcon T (2006) In mathematics, developmental biology and tumour growth: UIMP-RSME Lluis A. Santaló Summer School, September 11–15, 2006, Universidad Internacional Menéndez Pelayo, Santander, Spain 45–74 (American mathematical soc, 2009)
120. Osborne JM et al (2010) A hybrid approach to multi-scale modelling of cancer. Phil Trans R Soc A 368:5013–5028
121. Fischer S et al. (2012) Modeling erythroblastic islands: using a hybrid model to assess the function of central macrophage. 92–106
122. Huang S, Ernberg I, Kauffman S (2009) Cancer attractors: a systems view of tumors from a gene network dynamics and developmental perspective. Semin Cell Dev Biol 20:869–876
123. Del Sol A, Balling R, Hood L, Galas D (2010) Diseases as network perturbations. Curr Opin Biotechnol 21:566–571
124. Alon U (2007) An introduction to systems biology: design principles of biological circuits. (Chapman & Hall/CRC)
125. Materna SC, Nam J, Davidson EH (2010) High accuracy, high-resolution prevalence measurement for the majority of locally expressed regulatory genes in early sea urchin development. 177–184
126. Jurkowski W, Roomp K, Crespo I, Schneider JG, Del Sol A (2011) PPARγ population shift produces disease-related changes in molecular networks associated with metabolic syndrome. Cell Death Dis 2:e192
127. Crespo I, Roomp K, Jurkowski W, Kitano H, Del Sol A (2012) Gene regulatory network analysis supports inflammation as a key neurodegeneration process in prion disease. BMC Syst Biol 6:132
128. Huang S, Eichler G, Bar-Yam Y, Ingber DE (2005) Cell fates as high-dimensional attractor states of a complex gene regulatory network. Phys Rev Lett 94:128701
129. Maamar H, Raj A, Dubnau D (2007) Noise in gene expression determines cell fate in Bacillus subtilis. Science (New York) 317, 526–529
130. Gordon AJE et al (2009) Transcriptional infidelity promotes heritable phenotypic change in a bistable gene network. PLoS Biol 7:e44
131. Shiraishi T, Matsuyama S, Kitano H (2010) Large-scale analysis of network bistability for human cancers. PLoS Comput Biol 6:e1000851
132. Tafforeau L, Rabourdin-Combe C, Lotteau V (2012) In two hybrid technologies. Suter B, Wanker EE (eds) 812, Humana Press, pp 103–120

133. Apsel B et al (2008) Targeted polypharmacology: discovery of dual inhibitors of tyrosine and phosphoinositide kinases. Nat Chem Biol 4:691–699
134. Achenbach J, Tiikkainen P, Franke L, Proschak E (2011) Computational tools for polypharmacology and repurposing. Future med chem 3:961–968
135. Rask-Andersen M, Almén MS, Schiöth HB (2011) Trends in the exploitation of novel drug targets. Nat Rev Drug Discov 10:579–590
136. Berger SI, Iyengar R (2011) Role of systems pharmacology in understanding drug adverse events. Wiley Interdisc Rev: Syst Biol Med 3:129–135
137. Gertsch J (2011) Botanical drugs, synergy, and network pharmacology: forth and back to intelligent mixtures. Planta Med 77:1086–1098
138. Arrell DK, Terzic A (2010) Network systems biology for drug discovery. Clin Pharmacol Ther 88:120–125
139. Berger SI, Iyengar R (2009) Network analyses in systems pharmacology. Bioinformatics 25:2466–2472
140. Zhao S, Li S (2010) Network-based relating pharmacological and genomic spaces for drug target identification. doi:10.1371/journal.pone.0011764
141. Azuaje FJ, Zhang L, Devaux Y, Wagner DR (2011) Drug-target network in myocardial infarction reveals multiple side effects of unrelated drugs. Scientific reports 1

Chapter 7
Computational Approaches for Reconstruction of Time-Varying Biological Networks from Omics Data

Vinay Jethava, Chiranjib Bhattacharyya and Devdatt Dubhashi

Abstract This chapter presents a survey of recent methods for reconstruction of time-varying biological networks such as gene interaction networks based on time series node observations (e.g. gene expressions) from a modeling perspective. Time series gene expression data has been extensively used for analysis of gene interaction networks, and studying the influence of regulatory relationships on different phenotypes. Traditional correlation and regression based methods have focussed on identifying a single interaction network based on time series data. However, interaction networks vary over time and in response to environmental and genetic stress during the course of the experiment. Identifying such time-varying networks promises new insight into transient interactions and their role in the biological process. A key challenge in inferring such networks is the problem of high-dimensional data i.e. the number of unknowns p is much larger than the number of observations n. We discuss the computational aspects of this problem and examine recent methods that have addressed this problem. These methods have modeled the relationship between the latent regulatory network and the observed time series data using the framework of probabilistic graphical models. A key advantage of this approach is natural interpretability of network reconstruction results; and easy incorporation of domain knowledge into the model. We also discuss methods that have addressed the problem of inferring such time-varying regulatory networks by integrating multiple sources or experiments including time series data from multiple perturbed networks. Finally, we mention software tools that implement some of the methods discussed in this chapter. With next

V. Jethava (✉) · D. Dubhashi
Chalmers University of Technology, Göteborg, Sweden
e-mail: jethava@chalmers.se

D. Dubhashi
e-mail: dubhashi@chalmers.se

C. Bhattacharyya
Indian Institute of Science, Bangalore, India
e-mail: chiru@csa.iisc.ernet.in

A. Prokop and B. Csukás (eds.), *Systems Biology*,
DOI: 10.1007/978-94-007-6803-1_7,

generation sequencing promising yet further growth in publicly available -omics data, the potential of such methods is significant.

Keywords Time-varying networks · Dynamic interactome · Gene microarray expressions · Probabilistic graphical models · ℓ_1-regularization · Network reconstruction · Integrative analysis · Multiple networks

Acronyms

PGM	Probabilistic Graphical Model
GGM	Gaussian Graphical Model
HMM	Hidden Markov Model
BN	Bayesian Network
DBN	Dynamic Bayesian Network
MLE	Maximum Likelihood Estimate
LASSO	Least Absolute Shrinkage and Selection Operator
PML	Penalized Maximum Likelihood
GLASSO	Graphical LASSO (see LASSO)
KELLER	KErnel-reweighted Logistic Regression
TESLA	TEmporally Smoothed l1-regularized Logistic Regression
NETGEM	Network Embedded Temporal GEnerative Model for gene expression data
ERGM	Exponential Random Graph Model
PPI	Protein-Protein Interaction

7.1 Introduction

Most cellular components exert their functions through interactions with other cellular components, which can be located either in the same cell or across cells, and even across organs. In humans, the potential complexity of the resulting network—the human *interactome*—is daunting: with about 25,000 protein–coding genes, 1,000 metabolites and an undefined number of distinct proteins and functional RNA molecules, the number of cellular components that serve as the nodes of the interactome easily exceeds 100,000.

It is increasingly recognized that an understanding of a gene's network context is essential in determining the phenotypic impact of defects that affect it. To understand the behaviour of any one gene in the context of human disease, individual genes must be understood in the context of molecular networks that define the disease states. Following on from this principle, a key hypothesis is that a disease phenotype is rarely a consequence of an abnormality in a single effector gene product, but reflects various pathobiological processes that interact in a complex network. A corollary of this widely held hypothesis is that the interdependencies among a cell's molecular components lead to deep functional, molecular and causal relationships among apparently distinct phenotypes [9, 34, 55]. Analysis based on microarray expression experiments has been used extensively for exploring these interdependencies.

The past decade has seen exponential growth in publicly available datasets for analysis of gene regulatory networks, predominantly in the form of time series gene expression data. Reflecting this trend, the focus has shifted from traditional perturbation experiments (knockout/knockdown) in which a single gene or a pair of genes are inactivated and the downstream effects are studied [68, 69]; to a more holistic approach aimed at studying the influence of several regulators simultaneously based on time series gene expression data [2, 8, 14].

Initial methods for analysis of time series gene expression data focussed on correlation-based methods to identify the regulatory relationships (see Androulakis et al. [4] for an early survey focussing on correlation-based methods). Gitter et al. [27] present a recent survey discussing methods using lagged correlation and regression analysis for inferring gene regulatory networks. They also discuss methods for combining time series gene expression data with static data for reconstruction of the regulatory network.

One of the key challenges in using time series gene expression data for inference of gene regulatory networks is the relatively small number of observations compared to the number of unknown variables [28]. For example, gene expression time series data is much smaller (usually less than 8 time points, [22]) compared to the number of possible interactions between genes at different time points. Traditional methods for time-series analysis [4] fail to address this. The above problem arises in a number of domains, and is often referred to as the curse of high-dimensional data [16, 19]. This refers to the scenario when the number of unknown variables, typically represented by p is larger than the number of observations, typically represented by n i.e. large p, small n. This problem is ill-posed, and cannot be solved without additional assumptions.

Several regularization methods have been investigated for addressing this problem. One popular choice is to introduce ℓ_1 penalty on the model parameters (interaction strengths) which makes the resulting problem well-posed. The ℓ_1 penalty is known to yield sparse networks i.e. most of the interactions are absent in the resulting network. This is especially well-suited for reconstruction of gene regulatory networks, since it is known that regulatory networks are sparse i.e. only a handful of genes act as regulators for a single gene [17]. Recent advances [13, 20] in theoretical understanding of ℓ_1-based regularization have led to the development of multiple optimization methods and related applications collectively referred to using the encompassing term *compressed sensing*. See e.g. Friedman et al. [24] for a textbook introduction to ℓ_1-based regularization and related techniques, and [12] for a more advanced treatment.

The problem of high-dimensional data is exacerbated in the case of reconstruction of time-varying networks since reconstruction of the network at a time point depends on the single observation i.e. expression data at that time point. In principle, the interaction networks at different time points could be very dissimilar. However, a natural consequence of the underyling biological process is that networks at nearby time points are largely similar; except in the case of sharp changes in response to external stimula [43]. Modeling the time-varying evolution of the network is a key aspect of time-varying network reconstruction. Current methods have modeled the dynamic evolution of the network under different assumptions e.g. smooth variation i.e. the network is changing slowly over time [59], or sharp changes in network structure in response to external stimuli [1]. Other methods have used additional domain knowledge like presence or absence of certain motifs [29] and functional roles of the genes [36].

One solution to the high-dimensional data problem in the context of time-varying network reconstruction is to perform multiple measurements at each time point under similar experimental conditions. However, this is largely infeasible. A related approach is to combine data from multiple related sources or experiments. Such data and the underlying networks often exhibit commonalities such as the presence of a large common subnetwork. On the other hand, there might be significant differences in some part of the networks either to due experimental conditions or genetic perturbations. Integrating expression data for network reconstruction poses a twofold challenge, namely, modeling the common subnetwork and the variation across the networks corresponding to different data sources, and capturing the impact of network variation on the node observations. In some cases, there is additional information available such as the genetic perturbations in the networks. Recent methods [30, 36] have focussed on modeling the network variation under the assumption that there is a large common subnetwork.

Gitter et al. [27] discuss computational methods for reconstruction of regulatory networks providing a broad overview of the different approaches that have been used. In this chapter, we survey recent methods for reconstruction of time-varying networks from short time-series data from a modeling perspective. These methods use the framework of probabilistic graphical models to model the dependence between node observations and the underlying interaction network. Such methods have to make additional assumptions on network structure as well as network dynamics (how the interactions are varying with time) in order to make the inference tractable. We explore the connection between the regularization techniques and the underlying biological processes that justify these assumptions.

7.1.1 Organization

The remainder of this chapter is organized as follows: In Sect. 7.2, we discuss the framework of probabilistic graphical models in the context of network reconstruction. Section 7.3 provides a brief discussion of the ℓ_1 penalty and its usage in reconstruction of gene regulatory networks. In Sect. 7.4, we discuss recent methods for reconstruction of time-varying interaction networks. Section 7.5 presents methods for integrating time series information from multiple sources into the graphical model framework. Section 7.6 presents the relevant software tools for reconstruction of dynamic interaction networks. Finally, Sect. 7.7 presents our conclusions as well as a discussion of open problems in this area.

7.1.2 Notation

We use $\det(\cdot)$ and $\mathrm{Tr}(\cdot)$ to denote the matrix determinant and trace (sum of diagonal elements) of a matrix respectively. We use $||X||_{\ell_0}$ to denote ℓ_0-norm of a matrix X which is equal to the number of non-zero entries of X

$$||X||_{\ell_0} := \#\{X_{ij} : X_{ij} \neq 0\} \tag{7.1}$$

Similarly, we use $||X||_{\ell_1}$ to denote the ℓ_1-norm of a matrix X which is given by the sum of absolute values of entries in X

$$||X||_{\ell_1} := \sum_{i,j} |X_{ij}| \tag{7.2}$$

7.2 Background

This section describes the problem of network reconstruction from a modeling perspective focussing on time series gene expression data. Suppose gene expression levels are measured for p genes denoted by $V := \{1, \ldots, p\}$ at n different time points $T := \{1, \ldots, n\}$. We denote the gene expression levels at time t as a random variable $X^{(t)} := [X_1^{(t)}, \ldots, X_p^{(t)}]^\top$ where $X_i^{(t)} \in \mathbb{R}$ denotes gene expression level for gene i at time t.

A network-based approach models the multiple interactions among the different components in a biological system using a graph. A graph $G = (V, E)$ consists of set of nodes (or *vertices*) $V = \{1, \ldots, p\}$ representing different components in the biological systems; and set of edges $E \subseteq V \times V$ representing the dependence between the different components. In the case of a time-varying network, this is equivalent to having a different underlying graph $G^{(t)} = (V, E^{(t)})$ at each time $t \in \{1, \ldots, n\}$ wherein the set of edges varies with time.

For gene regulatory relationships, the nodes of the graph correspond to the set of genes; and the edges in the graph represent the regulatory relationships. The node observations correspond to the gene expression levels $X^{(t)}$ at each time $t \in \{1, \ldots, n\}$. It is well-known that the gene regulatory network E has an impact on the observed gene expression profile $X^{(t)}$. More precisely, the correlation between the gene expression values $X_i^{(t)}$ and $X_j^{(t)}$ measured at different times $t \in \{1, \ldots, n\}$ is a good indicator for the interaction (i,j) being present in the regulatory network. Several methods have been developed for identifying gene interaction networks which are based on the correlation of gene expression levels in gene microarray experiments. See Gitter et al. [27] for a recent survey on different methods for analysis of dynamic regulatory networks and [4] for an earlier survey.

However, random effects such as noise, measurement error, etc. may lead to false correlation between the expression profiles at two genes i and j. A probabilistic approach to model this uncertainty is to treat the gene expression profile $X^{(t)}$ at time t as a random variable drawn from a parametric distribution with unknown parameter. For static network reconstruction, the expression profiles $X^{(1)}, \ldots, X^{(n)}$ are assumed to be drawn independently and identically distributed (i.i.d.) from the same distribution. A major subtask in network reconstruction in this framework is to estimate the parameters of the underlying distribution which best fit the measured expression profiles.

More importantly, there is a strong duality between a distribution over several random variables $X = [X_1, \ldots, X_p]$ and a graph which describes the dependence between individual random variables X_i and X_j. Formally, this has been studied using the framework of *Probabilistic Graphical Models* (PGM). See e.g. Koller and Friedman [39] for a recent textbook providing a comprehensive introduction on the subject.

Several well-known models such as Hidden Markov Models (HMM), Bayesian Networks (BN), Dynamic Bayesian Networks (DBN), etc. are instances of graphical models, and have been successfully used for static network reconstruction [31, 48, 57]. However, HMMs require the number of observations (n) to be larger than the number of variables (p); and therefore, cannot be used in the case of short time-series data. DBNs also suffer from the curse of dimensionality (large p, small n), and a number of regularization methods have been investigated in order to address this [32, 37, 49, 58, 73]. More fundamentally, DBNs can only be used to identify relationships in a directed acyclic graph. In effect, while dependence between expression profiles $X_i^{(t)}$ and $X_j^{(t+1)}$ between any two genes i and j at different time points t and $(t+1)$ can be easily captured; the dependence between expression profiles $X_i^{(t)}$ and $X_j^{(t)}$ at the same time instant cannot be fully modeled as this may lead to cycles in the resulting directed graph. This problem can be addressed using undirected probabilistic graphical models, which we discuss below.

The dependence between the expression levels $X = [X_1, \ldots, X_p]^\top$ and the interaction strengths $W = \{W_{ij} : (i,j) \in E\}$ in network $G = (V, E)$ has been modeled using the conditional probability distribution [25, 44, 56, 59, 60, 66, 67]

$$P(X = x | W = w) = \frac{1}{Z(w)} \exp\left(-\frac{1}{2}\sum_{i,j \in V} w_{ij} x_i x_j\right) \tag{7.3}$$

where $Z(w)$ is a normalization constant. It has the property that whenever $w_{ij} = 0$, the node expressions for nodes i and j are *conditionally independent* given other expression levels. This has been used to construct gene association network where missing edges encode conditional independence.

Formally, the gene association network is constructed by considering the graph $G = (V, E)$ where $E \subseteq V \times V$ denotes the set of edges; with edge $(i,j) \notin E$

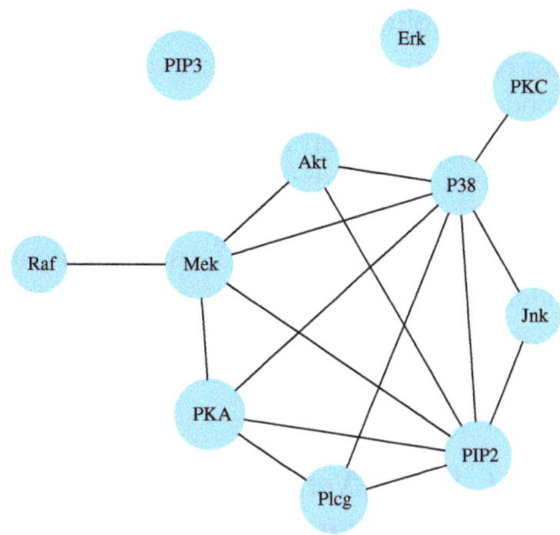

Fig. 7.1 Association network estimated from flow cytometry dataset with $p = 11$ proteins measured on $n = 7,466$ cells. A missing edge between nodes e.g. *Raf* and *PKA* means the expression levels of the two nodes is conditionally independent given the remaining expression levels. (Adapted with permission from Friedman et al. [23])

whenever genes i and j are conditionally independent ($w_{ij} = 0$). Figure 7.1 shows an association network between $p = 11$ proteins constructed using $n = 7466$ measurements [23, 54]. In Fig. 7.1, the absence of an edge between two nodes e.g. *Raf* and *PKA* indicates that these are conditionally independent given the remaining nodes.

The problem of model selection is to infer W based on i.i.d. samples $X^{(i)}$ drawn from the conditional distribution in (7.3). Static network reconstruction methods using time series data model the observations at different times $X^{(1)}, \ldots, X^{(n)}$ as being i.i.d. samples from an unknown static network W i.e.

$$P(X^{(t)} = x^{(t)} | W = w) = \frac{1}{Z(w)} \exp\left(-\frac{1}{2} \sum_{i,j \in V} w_{ij} x_i^{(t)} x_j^{(t)}\right) \tag{7.4}$$

This is not biologically consistent since it is known that the underlying network is not static during the course of the experiment. Rather, the network is varying across time and in response to external and internal stimuli [43, 50].

Therefore, one should instead consider $W^{(t)} = \{W_{ij}^{(t)} : (i,j) \in E^{(t)}\}$ as the interactions strengths in the network $G^{(t)} = (V, E^{(t)})$ at time t. The dependence between gene expression levels $X^{(t)}$ and the instantaneous interaction strengths $W^{(t)}$ is given by

$$P(X^{(t)} = x^{(t)} | W^{(t)} = w^{(t)}) = \frac{1}{Z(w^{(t)})} \exp\left(-\frac{1}{2} \sum_{i,j \in V} w_{ij}^{(t)} x_i^{(t)} x_j^{(t)}\right) \tag{7.5}$$

The problem of time-varying interaction network reconstruction is to identify the interaction strengths $W^{(t)}$ at different times $t \in \{1, \ldots, n\}$ based on the node observations $X^{(1)}, \ldots, X^{(n)}$. However, this problem is ill-posed since the number of unknowns ($W^{(t)}$) is much larger than the number of observations ($X^{(t)}$).

In the following subsection, we discuss covariance selection-a classical method for reconstructing a static network W based on real-valued node observations $X^{(t)}$ modeled as i.i.d. samples drawn from conditional distribution in (7.4). However, this method can fail if the number of observations n is smaller than the number of unknowns p. Further, the method yields a large number of false positives i.e. multiple interactions W even though it is known that underlying biological network is sparse. Section 7.3 discusses regularization method using ℓ_1 penalty which addresses the above-mentioned problems. Section 7.4 presents methods which extend these methods to inference of time-varying networks under mild assumptions on the temporal evolution of the underlying network.

7.2.1 Static Network Reconstruction Using Covariance Selection

Whenever the node observations (gene expression levels) $X_i \in \mathbb{R}$ and interaction strengths $W_{ij} \in \mathbb{R}$ are treated as real values, the conditional distribution in (7.3) is equivalent to X being drawn from a multivariate Gaussian distribution with mean 0 and covariance $\Sigma := W^{-1}$ i.e.

$$X \sim \mathcal{N}(0, \Sigma) \tag{7.6}$$

Equivalently, the conditional probability of $X^{(t)}$ conditioned on W is given by

$$P(X^{(t)}|W) = \frac{1}{\sqrt{(2\pi)^p \det(\mathrm{W}^{-1})}} \exp(-\frac{1}{2}\sum_{i,j\in V} X_i^{(t)} X_j^{(t)} W_{ij}) \tag{7.7}$$

This model is commonly referred to as the Gaussian Graphical Model [41]. We note that the normalization constant $Z(w)$ has a closed form expression given by $\sqrt{(2\pi)^p \det(\mathrm{W}^{-1})}$ for GGMs.

Construction of gene association networks requires inference of the concentration matrix W based on observations $X^{(1)}, \ldots, X^{(n)}$. This problem is known as covariance selection and was first studied by Dempster [18]. It involves computing the Maximum Likelihood Estimate (MLE) of W given observations $X^{(1)}, \ldots, X^{(n)}$. The log-likelihood of the observations is given by

$$\mathscr{L}(W) = \log P(X^{(1)}, \ldots, X^{(n)}|W) = \sum_{t=1}^{n} \log P(X^{(t)}|W) \tag{7.8}$$

$$= \frac{n}{2}\log \det W - \frac{n}{2}\mathrm{Tr}(\mathrm{SW}) - \frac{np}{2}\log(2\pi) \tag{7.9}$$

where $S \in \mathbb{R}^{p\times p}$ is the empirical covariance matrix for observations $X^{(1)}, \ldots, X^{(n)}$ given by

$$S_{ij} = \frac{1}{n}\sum_{t=1}^{n} X_i^{(t)} X_j^{(t)} \tag{7.10}$$

The Maximum Likelihood Estimate W^* is given by

$$W^* = \arg\max_{W \succ 0} \log\det W - \mathrm{Tr}(SW) \tag{7.11}$$

$$= \arg\min_{W \succ 0} \mathrm{Tr}(SW) - \log\det W \tag{7.12}$$

where $W \succ 0$ stands for positive-definiteness of W, and $Tr(SW)$ denotes the trace of the matrix. The positive definiteness constraint ensures that $W^{-1}(= \hat{\Sigma})$ is an invertible covariance matrix. Minimization of the negative log-likelihood in (7.12) is a convex optimization problem [10].

An exhaustive search for finding the non-zero elements of W^* is computationally prohibitive for moderate and large networks (more than $30 - 40$ genes). A number of methods based on greedy search have been studied for solving this problem [41, 61]. However, these are not suited whenever the number of observations n is smaller than number of genes p in the network [11, 45]. Another aspect of concern is that whenever n is much smaller than p, the solution W^* obtained by greedy methods has a large number of non-zero elements. On the other hand, it is known that the underlying interaction network is sparse [62, 70], i.e. each gene is regulated by a small number of genes.

7.2.2 Discretization of Gene Expression Levels

Microarray measurements are noisy estimates of the gene expression level. In certain applications, the qualitative level of gene expression is a better indicator of up or down regulation than the microarray measurement (which is a rough estimate of the gene expression). Therefore, the gene expression levels are sometimes quantized to a discrete set $\mathscr{X}$. For example, in the case of cDNA microarray, a choice of $\mathscr{X} = \{-1, 1\}$ corresponds to the gene being downregulated (-1) or upregulated $(+1)$. For some applications, the relative strengths of the interactions are of interest rather than the actual values. Then, the weights are quantized to some discrete set $\mathscr{W}$. For example, a choice of $\mathscr{W} = \{-1, 0, 1\}$ would correspond to the gene interaction being activator (-1), conditionally independent (0) or repressed (-1).

Notice that the convention in systems biology is to represent co-activation as $W = +1$ while $W = -1$ traditionally represents a relationship where genes mutually repress each other. This is reversed in model discussed above to due negative sign in (7.3), which is standard convention in statistical physiscs [47] where such models were first studied. Further, in case of continuous weights, the model has a natural interpretation in terms of inverse covariance (concentration) matrix of a multi-variate Gaussian distribution as discussed in 7.2.1.

A choice of $\mathscr{X} = \{-1, 1\}$ and $\mathscr{W} = \{-1, 0, 1\}$ yields the Ising model which is well-studied in statistical physics [47], wherein $W = 0$ state is modeled as the edge being absent in the network.

The normalization constant $Z(w)$ in (7.3) is given by

$$Z(w) = \sum_{x_i \in \mathcal{X}} \sum_{x_2 \in \mathcal{X}} \cdots \sum_{x_p \in \mathcal{X}} \exp\left(-\frac{1}{2}\sum_{i,j \in V} w_{ij} x_i x_j\right) \tag{7.13}$$

whenever the gene expression levels are discretized to the set $\mathcal{X}$. The network reconstruction procedure is more complicated since the log-likelihood function does not have a closed form expression as in (7.9) for Gaussian Graphical Models.

7.3 Sparse Network Reconstruction Based on ℓ_1-Regularization

As discussed in Sect. 7.2.1, traditional methods for covariance selection are computationally prohibitive and inconsistent in the high-dimensional setting i.e. whenever number of genes p is much larger than the number of microarray measurements n. Recent work [7, 23, 40, 45, 53] has addressed this by introducing an additional regularization term based on ℓ_1-norm into the optimization problem. The technique, commonly known as LASSO (least absolute shrinkage and selection operator), was first studied by Tibshirani [63] in the context of linear regression. The lasso penalty can be understood as a relaxation of the ℓ_0 norm which induces model sparsity (fewer interactions in the network). This is discussed below in the context of gene interaction network.

As mentioned earlier, the number of interactions in a gene interaction network is few compared to the total number of possible edges. In other words, a gene is regulated by few other genes. This can be ensured by introducing an additional constraint in (7.12) as follows

$$\begin{array}{ll} \arg\min_{W \succ 0} & \mathrm{Tr}(\mathrm{SW}) - \log\det W \\ & \|W\|_{\ell_0} \leq t \end{array} \tag{7.14}$$

The solution to (7.14) has at most t interactions, where t is a parameter chosen based on domain knowledge. However, the optimization in (7.14) is an instance of mixed integer programming [10], and is computationally intractable for moderate and large networks ($p \geq 30$). The lasso penalty replaces the ℓ_0-norm with ℓ_1-norm to obtain the relaxed convex optimization problem given by

$$\begin{array}{ll} \arg\min_{W \succ 0} & \mathrm{Tr}(\mathrm{SW}) - \log\det W \\ & \|W\|_{\ell_1} \leq t \end{array} \tag{7.15}$$

One can obtain the equivalent Lagrangian formulation [10] as

$$\arg\min_{W \succ 0} \mathrm{Tr}(\mathrm{SW}) - \log\det W + \lambda \|W\|_{\ell_1} \tag{7.16}$$

where λ is a user-specified parameter.

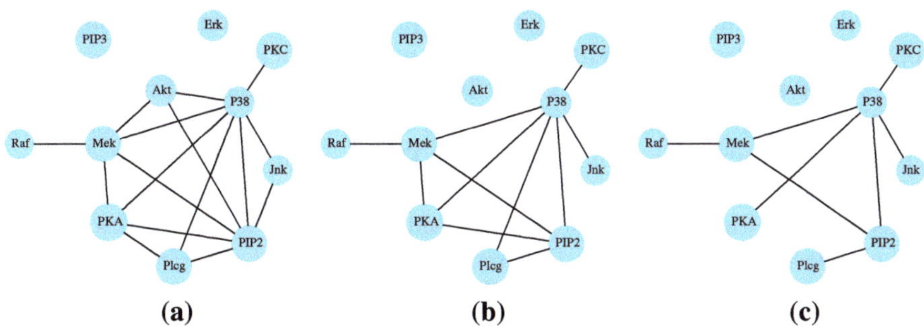

Fig. 7.2 Association network estimated from flow cytometry dataset with $p = 11$ proteins measured on $n = 7,466$ cells with different penalty parameter λ in (7.16) (Adapted with permission from Friedman et al. [23]). Increasing the parameter λ yields a network with fewer interactions

The parameter λ regulates the penalty incurred by choosing a non-sparse interaction network having many gene-gene interactions. A higher value of λ ensures higher sparsity (fewer non-zero interactions), while a choice of $\lambda = 0$ corresponds to covariance selection without any penalty. Figure 7.2 shows an association network between $p = 11$ proteins constructed using $n = 7466$ measurements with different penalty parameter λ in (7.16) [23]. As the penalty parameter increases, the number on non-zero interactions decreases. A choice of $\lambda = 0$ (no regularization) yields the fully connected network. Several methods [7, 23, 45, 71] have focussed on efficient computation of the solution in (7.16). Meinshausen and Bühlmann [45] first explored the connection to ℓ_1 regularization. They devised a neighbourhood selection procedure which used lasso regression to obtain a set of neighbours (non-zero interaction strength) for each gene based on local conditional likelihoods. In other words, for each gene, their approach chooses a small set of "most-likely neighbour" genes that have non-zero interactions, while remaining interactions are set to zero. These local neighbourhoods are used to construct the association network using either an 'AND' or an 'OR' final step procedure. This procedure corresponds to a modified penalty term in (7.16) [3].

Banerjee et al. [7] and Friedman et al. [23] improved the previous approach by casting it into the penalized maximum likelihood (PML) framework in (7.16). They used block coordinate descent to solve the resulting optimization. This means the optimization in (7.16) is solved by iteratively updating one of the rows (or columns) of W till certain convergence criteria is reached. In practice, each iteration (row or column update) requires solving a lasso problem (linear regression with ℓ_1 regularization term). This procedure is closely related to local neighbourhood search described below. Wainwright et al. [51, 65] studied network estimation in Ising models and discrete graphical models based on a similar local neighbourhood search using lasso penalty. Section 7.3.1 provides a brief description of the local neighbourhood search procedure based on logistic regression.

Ravikumar et al. and Wainwright et al. [51, 65] show from a statistical perspective that under mild conditions on sparsity of W, local neighbourhood seach recovers the interaction network correctly given a small number of measurements $n \sim d^3 \log p$ where d is the maximum number of interactions (maximum degree) of any gene with other genes in the network. This is often referred to as the *sparsistency* (sparse consistency) property (of the estimator of W). A similar condition holds for PML framework. In effect, the conditions can be understood as follows: if the underlying model was a Gaussian Graphical Model with few interactions (sparse network); then the methods will recover the interactions W precisely given enough observations n.

However, the above conditions do not apply in the context of biological networks since GGM is at best an approximation to the dependence between interactions and node observations. Instead, the reconstructed network is compared to past biological findings; and previously unknown interactions predicted by the model have to be experimentally verified.

7.3.1 Local Neighbourhood Search

We describe the local neighbourhood search procedure focussing on the special case of Ising models. This corresponds to situations where gene expression levels are discretized to the set $\mathscr{X} = \{-1, 1\}$ i.e. genes are either downregulated or upregulated.

The basic idea of the method is to iteratively estimate the concentration matrix W by updating a single row (or column) at a time. Let $W_{\setminus i} := [W_{ij}]^\top \in \mathbb{R}^{p-1}$ be a vector of length $(p-1)$ constructed by considering the ith row (or column) of W except the diagonal element. Notice that the set of non-zero elements of $W_{\setminus i}$ represent the interactions of gene i with other genes in the network i.e. the local neighbourhood of gene i given by

$$\mathscr{N}_i = \{j \in V : (W_{\setminus i})_j \neq 0\} \tag{7.17}$$

Therefore, estimating $W_{\setminus i}$ yields insight into which genes interact with the ith gene. Sparsity in $W_{\setminus i}$ is ensured by introducing an additional ℓ_1 regularization term. The resulting optimization is a convex optimization of the form

$$W^*_{\setminus i} = \arg \min_{W_{\setminus i} \in \mathbb{R}^{p-1}} \ell(W_{\setminus i}) + \lambda ||W_{\setminus i}||_{\ell_1} \tag{7.18}$$

where $\ell(W_{\setminus i})$ is the negative rescaled log likelihood $\ell(W_{\setminus i})$ is given by

$$\ell(W_{\setminus i}) := -\frac{1}{n} \sum_{t=1}^{n} \log P(X_i^{(t)} | X_{\setminus i}^{(t)}, W_{\setminus i}) \tag{7.19}$$

and $P(X_i^{(t)}|X_{\setminus i}^{(t)}, W_{\setminus i})$ is the conditional probability distribution of gene expression $X_i^{(t)}$ conditioned on the gene expressions of the other genes $X_{\setminus i}^{(t)} := \{X_j^{(t)} : j \in V\setminus i\}$ and interaction strengths $W_{\setminus i}$ is given by

$$P(X_i^{(t)}|X_{\setminus i}^{(t)}, W_{\setminus i}) = \frac{1}{1 + \exp\left(x_i^{(t)} \sum_{j \in V\setminus i} w_{ij} x_j^{(t)}\right)} \tag{7.20}$$

The optimization in (7.18) can be solved efficiently using convex solvers [21, 38, 42].

The method iteratively estimates $W_{\setminus i}$, and consequently the local neighbourhood, for different genes $i \in V$ till some convergence criteria is met, usually in terms of the size of the increments in log likelihood.

The interaction network can be constructed by either considering an "AND" configuration wherein an edge (i,j) is present in the network if i is in the local neighbourhood of j and vice versa,

$$E = \{(i,j) \in V \times V : i \in \mathcal{N}_j \, and \, j \in \mathcal{N}_i\} \tag{7.21}$$

An alternative construction is using "OR" configuration wherein an edge (i,j) is present in the network if i is in the local neighbourhood of j or vice versa,

$$E = \{(i,j) \in V \times V : i \in \mathcal{N}_j \text{ or } j \in \mathcal{N}_i\} \tag{7.22}$$

7.4 Reconstruction of Time-Varying Regulatory Networks

Sections 7.2.1 and 7.3 describe the covariance selection problem in the static setting i.e. the gene expressions $X^{(1)}, \ldots, X^{(n)}$ are modeled as independent samples from an multivariate normal distribution with fixed but unknown concentration matrix W. The zero elements of W correspond to genes whose gene expressions are conditionally independent conditioned on other genes, while non-zero elements indicate strength of interaction, either activating or repressing depending on the sign of W_{ij} *for all observation time instants.*

Nevertheless, it is known that rewiring occurs in gene interaction networks in response to environmental and genetic stress [43]. For example, genes implicated in yeast metabolism undergo significant rewiring in response to changes in nutrient availability [15]. A biologically plausible modeling of the interaction network, therefore, should incorporate interaction dynamics. This poses new challenges from a modeling perspective, which we discuss below.

The dynamics (temporal variation) of the interaction strengths should depend on the time elapsed between observation instants. For example, if we take observations at very short intervals, the interactions between nearby time instants should not differ a lot. In other words, there is sparsity in the interaction dynamics as the interaction network does not drastically change from one observation time point for the most part. Nonetheless, there might be instances at which major changes can occur in the network, often in response to environmental or genetic stress.

On the other hand, it is known that gene interactions are a vital part of functional roles performed by a gene. Indeed, one could posit that it is the changing functional requirements imposed by internal development or external stress which might drive gene rewiring. Thus, a gene interaction network may get rewired in order to satisfy a new functional role. There is a rich literature which relates the functional roles of the genes to the interaction network.

A number of recent approaches [1, 36, 59, 72] have addressed this problem. These methods can be broadly categorized into two classes: optimization-based and model-based methods. Optimization-based methods [1, 59, 72] introduce an additional term into the optimization in (7.16) which ensures that there are not many changes in the network between consecutive observation times. These methods do not incorporate additional information such as knowledge about the functional roles of the genes, network motifs, etc. into the network reconstruction. Model-based methods [26, 29, 36] incorporate additional knowledge such as information about presence or absence of specific network motifs [29], functional roles of the genes [36], etc. into network dynamics.

Inference of time-varying interactions characterizes the activity of individual genes and predicts their interactions with other genes, including unknown predictions which could serve as test candidates for experimental testing. At a broader level, it yields new insight by implicating groups of genes, often characterized by different functional roles performed by them, interacting among each other and with other groups at critical stages of a biological process [36, 59]. For example, the analysis in Song et al. [59] reveals high activity between genes related to metamorphosis, wing margin morphogenesis, wing vein morphogenesis and apposition of wing surfaces during early embryonic stage in *Drosophila Melanogaster* (fruit fly). Such interaction is typically visible during the transition from pupa stage to adult stage when wing development occurs [5]. This behaviour could potentially indicate diverse functionality of these genes. Similarly, the analysis in [36] implicates genes related to complex/cofactor binding as active during transition from glucose starvation to nitrogen starvation in *S. Cerevisiae* (baker's yeast).

7.4.1 Optimization-Based Methods for Modeling Interaction Dynamics

Zhou et al. [72] first studied estimation of time-varying Gaussian Graphical Models using ℓ_1 regularization. Their approach extends the static model in Sect. 7.2.1 by assuming the observation $X^{(t)}$ at each time t to be drawn from a Gaussian distribution

$\mathcal{N}(0, \Sigma^{(t)})$ independent of other observations. The concentration matrix at time t is given by $W^{(t)} := \Sigma^{(t)^{-1}}$. Consequently, the interaction network at time t is given by $G^{(t)} = (V, E^{(t)})$ where $E^{(t)} := \{(i,j) \in V \times V : W_{ij}^{(t)} \neq 0\}$ is the set of edges with non-zero interaction strengths. The conditional probability distribution of $X^{(t)}$ conditioned on interaction strengths $W^{(t)}$ is given by

$$P(X^{(t)}|W^{(t)}) = \frac{1}{Z(W^{(t)})}\exp\left(-\frac{1}{2}\sum_{(i,j)\in E^{(t)}} X_i^{(t)}X_j^{(t)}W_{ij}^{(t)}\right) \tag{7.23}$$

where $Z(W^{(t)})$ is a normalization constant.

Since the observation at each time is independent of other times, covariance selection procedures are not directly applicable since the empirical covariance S cannot be computed by treating different observations as i.i.d. samples. The method addresses this by constructing a weighted covariance matrix $\hat{S}^{(t)}$ at each time instant t where decreasing weights are assigned to observations with increasing time gap. At a given time t, the weight corresponding to observation at time i is defined using a symmetric non-negative kernel K as

$$w^{(t)}(i) = \frac{K(|t-i|/h)}{\sum_{i=1}^{n} K(|t-i|/h)} \tag{7.24}$$

Then, the weighted empirical covariance $\hat{S}^{(t)}$ at time t is given by

$$\hat{S}^{(t)} = \frac{1}{C}\sum_{i=1}^{n} w_i^{(t)} X^{(i)} X^{(i)^{\top}} \tag{7.25}$$

where $C := \sum_{i=1}^{n} w^{(t)}(i)$ is the scaling term. For example, if the function $K(x)$ used to assign weights to observations is given by

$$K(x) = \begin{cases} 2^{-|x|} & \text{if } x \leq 1 \\ 0 & \text{otherwise} \end{cases} \tag{7.26}$$

then measurements $X^{(t-1)}$ and $X^{(t+1)}$ made at time instances $(t-1)$ and $(t+1)$ respectively are assigned weight $1/2$ in computation of $\hat{S}^{(t)}$. The empirical covariance $\hat{S}^{(t)}$ at time t is computed as

$$\hat{S}^{(t)} = \frac{1}{2}\left(\frac{1}{2}X^{(t-1)}X^{(t-1)^{\top}} + X^{(t)}X^{(t)^{\top}} + \frac{1}{2}X^{(t+1)}X^{(t+1)^{\top}}\right) \tag{7.27}$$

This allows estimation of the concentration matrix $W^{(t)^*}$ at time t using weighted empirical covariance matrix $\hat{S}^{(t)}$ by solving the following optimization problem

$$W^{(t)^*} = \arg\max_{W \succ 0} \; Tr(\hat{S}^{(t)}W) - \log\det W + \lambda||W||_{\ell_1} \tag{7.28}$$

The optimization in (7.27) can be solved independently at each time $t \in \{1,\ldots,n\}$ using the methods described in Sect. 7.3.

Zhou et al. [72] show that the above procedure correctly recovers $W^{(t)}$ under mild smoothness conditions on $\Sigma^{(t)}$ in addition to sparsity assumption of $W^{(t)}$ (few non-zero interactions). More precisely, if $\Sigma^{(t)} = [\sigma_{ij}^{(t)}]$, where $\sigma_{ij}^{(t)} \in C^{\infty}$ is a smooth function denoting the instantaneous covariance between genes i and j at time t; then $\sigma_{ij}^{(t)}$ has bounded first and second order derivatives at all times i.e.

$$\max_{i,j} \sup_{t} |\frac{\partial}{\partial t}\sigma_{ij}^{(t)}| \leq C_1 \tag{7.29}$$

$$\max_{i,j} \sup_{t} |\frac{\partial^2}{\partial t^2}\sigma_{ij}^{(t)}| \leq C_2 \tag{7.30}$$

Put simply, each term in the covariance matrix $\Sigma^{(t)}$ *changes slowly* over time. In other words, as we make more measurements with smaller time gaps between consecutive measurements, the difference between the inferred networks at consecutive time points will decrease. This is indeed the case for most biological systems.

7.4.1.1 Inference of Dynamic Interactions Under Smooth Variation

As mentioned in Sect. 7.2.2, there are scenari where it is advantageous to discretize the measurements to some set $\mathscr{X}$. A common choice is $\mathscr{X} = \{-1, 1\}$, which corresponds to genes being downregulated or upregulated. References [1, 59] have studied inference of time-varying discrete graphical models where the gene expression are discretized to $\mathscr{X} = \{-1, 1\}$. They extended the local neighbourhood search procedure [51, 65] to handle inference of dynamic interaction networks under smoothness conditions in (7.29)–(7.30). This is achieved by introducing a weighted negative log likelihood analogous to (7.25)

$$\tilde{\ell}^{(t)}(W_{\setminus i}) = -\frac{1}{C}\sum_{k=1}^{n} w^{(t)}(k) \log P(X_i^{(k)} | X_{\setminus i}^{(k)}, W_{\setminus i}) \tag{7.31}$$

where $C = \sum_{k=1}^{n} w^{(t)}(k)$ is the scaling term, and $w^{(t)}(k)$ is given by a symmetric non-negative kernel K as described in (7.24). The modified optimization problem for the ith row at time t is given by

$$W_{\setminus i}^{(t)} = \arg \min_{W_{\setminus i} \in \mathbb{R}^{p-1}} \left(\tilde{\ell}^{(t)}(W_{\setminus i}) + \lambda ||W_{\setminus i}||_{\ell_1}\right) \tag{7.32}$$

The above optimization is same as in (7.18) except for the weighted log likelihood term $\tilde{\ell}^{(t)}(W_{\setminus i})$. Consequently, the network is constructed independently at each time t as described in 7.3.1.

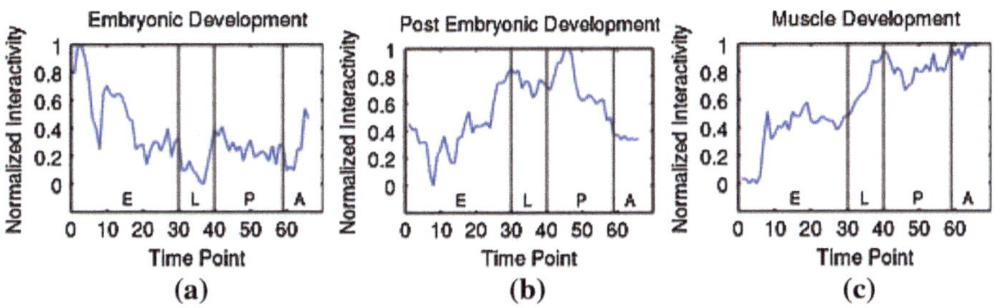

Fig. 7.3 Interactivity of three groups of genes related to (**a**) embryonic development, (**b**) post-embryonic development; and **c** muscle development. It shows the interactivity of three groups of genes during the different developmental cycles, showing that each group of genes is more active during its development stage. Thus, the time-varying networks inferred using Song et al. [59] are consistent with known biological findings. (Reprinted with permission from Song et al. [59])

They use this approach to study the evolution of gene regulatory network in *D. Melanogaster*, the common fruit fly, over its developmental cycle based on 66 gene expression measurements collected in [5]. The expression measurements can be categorized into four stages, i.e. embryonic (1–30 time point), larval (31–40 time point), pupal (41–58 time point) and adult stages (59–66 time point). In order to verify the biological findings of the method, they focus on three groups of genes consisting of 25 – 30 genes that are known to be functionally implicated in different developmental stages, namely, embryonic development, post-embryonic development and muscle development. They measure the interactivity of the group in the interaction networks found using their method, and observe that each group of genes has more non-zero interactions during their development stage. Figure 7.3 shows the interactivity of three groups of genes during the different developmental cycles, showing that each group of genes is more active during its development stage.

They also investigate the interactions between genes from different functional groups. Figure 7.4 shows the connectivity between different functional groups. This yields fresh insight into interactions between functional groups. For example, the method reveals high activity between genes related to metamorphosis, wing margin morphogenesis, wing vein morphogenesis and apposition of wing surfaces during early embryonic stage (Fig. 7.4b, c). Such interaction is typically visible during the transition from pupa stage to adult stage (Fig. 7.4r, s) when wing development occurs [5]. This behaviour could potentially indicate diverse functionality of these genes.

Ahmed and Xing [1] extended the above approach to handle sharp structural changes such as sudden rewiring of a gene network in response to a stimulus. They use the approach to study the evolution of regulatory network in *D. Melanogaster* consisting of 4,028 genes at 66 different time points over its life cycle.

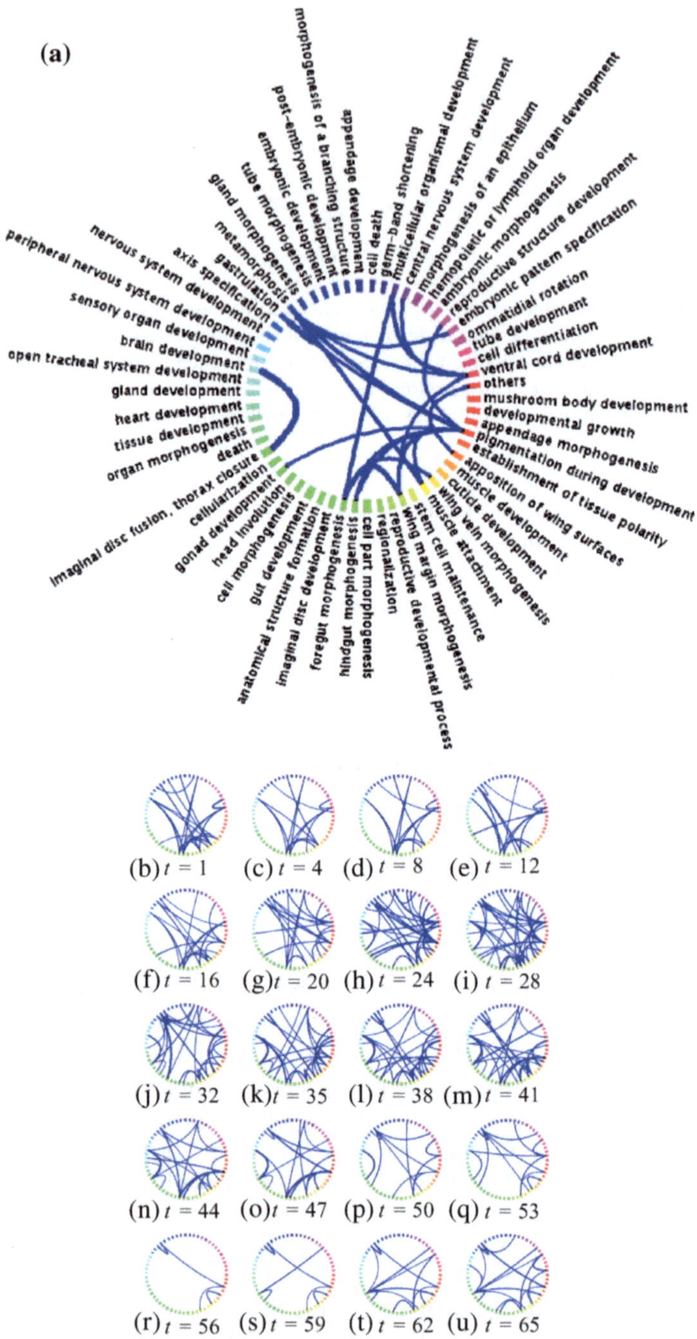

Fig. 7.4 (**a**) Average network between functional groups obtained using Song et al. [59]. Each color patch denotes an ontological group, and the position of the ontological groups remains the same from (**a**) to (**u**). The annotation in the *outer rim* indicates the function of each group. (Adapted with permission from Song et al. [59])

7.4.2 Model-Based Methods for Modeling Interaction Dynamics

Methods discussed in Sect. 7.4.1 do not make any assumptions on the structure of the network beyond sparsity (few non-zero interactions) and smoothness temporal variation (nearby time instants have similar interaction profile). However, if more information is available, it can be incorporated into the analysis. For example, the yeast (static) interaction network is known with high degree of confidence based on perturbation studies [35, 64]. It has also been observed that even though the interaction networks are highly dynamic in nature, they feature a number of building blocks i.e. motifs and subgraphs that recur over time [33]. Finally, there is extensive information available about the functional roles performed by the genes [6, 46]. Even though it is clear that this information could potentially aid in reconstruction of dynamic interaction networks; integration poses a challenge, and it is only recently that methods [29, 36] have been explored that leverage this information to aid in dynamic interaction network reconstruction.

7.4.2.1 Inference of Interaction Dynamics Based on Recurring Motifs

Guo et al. [29] studied a model-based approach that leverages information about recurring subgraphs that appear in the interaction networks at different times. Their approach considered the interaction $W_{ij}^{(t)}$ between genes i and j at time t to be either absent or present with some strength W_{ij} i.e.

$$W_{ij}^{(t)} = \begin{cases} W_{ij} & \mathit{if}\ E_{ij}^{(t)} = 1 \\ 0 & \mathit{otherwise} \end{cases} \tag{7.33}$$

They modeled the evolution of the interaction network under the Markov assumption i.e. the interaction network $E^{(t)}$ at time t depends only on the interaction network $E^{(t-1)}$ at previous time instant $(t-1)$,

$$P(E^{(1)}, E^{(2)}, \ldots, E^{(n)}) = P(E^{(1)}) \prod_{t=2}^{n} P(E^{(t)} | E^{(t-1)}) \tag{7.34}$$

The transition probability $P(E^{(t)}|E^{(t-1)})$ is specified in terms of simple features (Ψ_f) that measure some global statistic extracted from the interaction network,

$$P(E^{(t)}|E^{(t-1)}) = \frac{1}{Z(\theta, E^{(t-1)})} \exp\left(\sum_{f=1}^{F} \theta f \Psi f(E^{(t)}, E^{(t-1)}) \right) \tag{7.35}$$

Examples of simple features are "density", "stability" and "transitivity" given by

$$\Psi_1(E^{(t)}, E^{(t-1)}) = \sum_{ij} E_{ij}^{(t)} \quad \text{(density)} \tag{7.36}$$

$$\Psi_2(E^{(t)}, E^{(t-1)}) = \sum_{ij} \mathbb{I}(E_{ij}^{(t)} = E_{ij}^{(t-1)}) \quad \text{(stability)} \tag{7.37}$$

$$\Psi_3(E^{(t)}, E^{(t-1)}) = \sum_{ijk} \frac{E_{ij}^{(t)} E_{ik}^{(t-1)} E_{kj}^{(t-1)}}{\sum_{ij} E_{ik}^{(t-1)} E_{kj}^{(t-1)}} \quad \text{(transitivity)} \tag{7.38}$$

where $\mathbb{I}(\cdot)$ denotes the indicator function i.e. $\mathbb{I}(E_{ij}^{(t)} = E_{ij}^{(t-1)})$ is one if edge (i,j) is present in the interaction networks $E^{(t)}$ and $E^{(t-1)}$ at times t and $(t-1)$ respectively, and zero otherwise.

The density feature $\Psi_1(E^{(t)}, E^{(t-1)})$ counts the number of interactions in the network $E^{(t)}$ at time t. For example, if the coefficient θ_1 corresponding to the density feature in (7.35) is negative, this favors sparse interaction network $E^{(t)}$ at time t. The stability feature $\Psi_2(E^{(t)}, E^{(t-1)})$ counts the number of interactions that are preserved across the two network $E^{(t)}$ and $E^{(t-1)}$. If the coefficient θ_2 corresponding to the stability feature in (7.35) is positive, this discourages large number of changes between $E^{(t)}$ and $E^{(t-1)}$. As an extreme case, if θ_2 is extremely high, we recover a static network.

The transitivity feature $\Psi_3(E^{(t)}, E^{(t-1)})$ measures the fraction of genes i and j that are connected to a common gene k at time $(t-1)$, and also which have an edge (i,j) at time t. Figure 7.5 shows the case when $E_{ij}^{(t)} E_{ik}^{(t-1)} E_{kj}^{(t-1)} = 1$. If θ_3 corresponding to the transitivity feature in (7.35) is positive, this means a common gene interacting with two genes (which may or may not be interacting with each other) is likely to induce interactions among the two possibly non-interacting genes. In other words, this favors the triangle motif in the network.

Other motifs e.g. stars, cycles, etc. can similarly be favored or discouraged in the network structure by careful choice of features and weight function θ. This

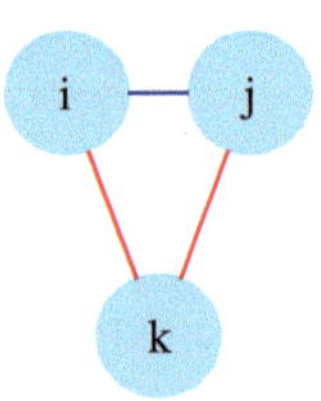

Fig. 7.5 Example of transitivity feature. The edges (i,k) and (j,k) present in $E^{(t-1)}$ at time $(t-1)$ are shown in *red*, while edge (i,j) in $E^{(t)}$ at time t are shown in *blue*. Here $E_{ij}^{(t)} E_{ik}^{(t-1)} E_{kj}^{(t-1)} = 1$

class of models are known as Exponential Random Graph Models (ERGM) (see Robins et al. [52] for an introductory survey).

Guo et al. [29] use a sampling based approach to infer the time varying networks $E^{(t)}$ based on gene expression measurements assuming that the interaction dynamics are Markovian as in (7.34). They analyze the muscle development subnetwork of *Drosophila Melanogaster* (fruit fly) consisting of 11 genes during four stages of the life cycle, including embryonic, larval, pupal and first 30 days of adulthood. Their results agree closely with known interactions, and suggest hitherto unknown linkages which can be investigated further experimentally.

7.4.2.2 Inference of Interaction Dynamics Based on Functional Roles of Genes

In a recent work, Jethava et al. [36] studied the dynamics of the interaction network in terms of the functional roles performed the genes. Their approach assumes Markovian dependence between interaction strengths $W^{(t)}$ and $W^{(t-1)}$ at times t and $(t-1)$ respectively, i.e.

$$P(W^{(t)}|W^{(t-1)},\ldots,W^{(1)}) = P(W^{(t)}|W^{(t-1)}) \tag{7.39}$$

They modeled the time-varying interaction between two genes as governed by the functions performed by the two genes. In effect, the dynamics of interaction $W_{ij}^{(t)}$ between any two genes i and j is conditionally independent of other interactions in the network conditioned on the roles. This assumption allows tractable inference in moderate and large networks (few thousand genes). A Bayesian approach is used to model network sparsity with selection of appropriate priors.

Most genes perform multiple functions at different times, each contributing to some vital requirement in the life cycle. Their model assumes that at each time, the interaction between two genes depends on the *active functional roles* of the two genes at that time. Thus, the overall network structure at each time depends on the different processes happening in the organism at that time.

The model infers the latent time-varying interactions using functional information of genes. This is used to infer the interaction network in *S. Cerevisiae* (baker's yeast) at different time points with varying nutrient availability. Figure 7.6 shows an example of the inferred network at time $t = 4.1$ h. The network changes due to external stimulus (change in nutrient availability). The model successfully captures sharp change in the network (the interaction strengths get inverted) in response to critical change in nutrient availability from Carbon-rich environment to Nitrogen-rich environment.

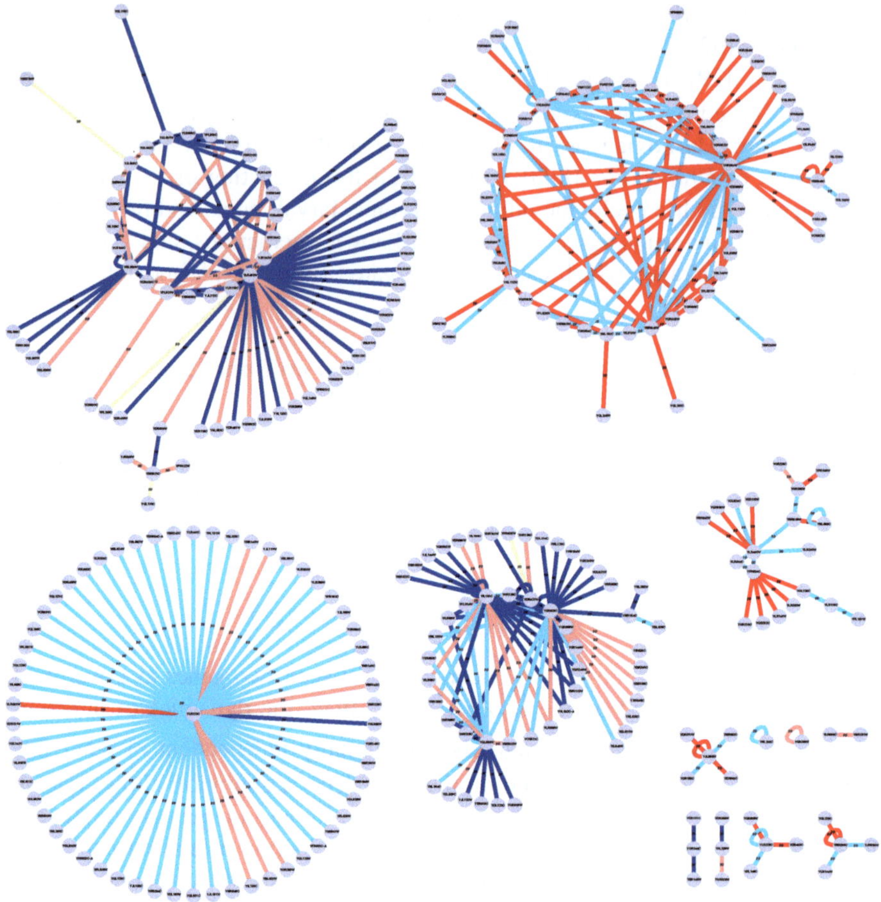

Fig. 7.6 Time varying interaction strengths between the genes at time $t = 4.1$ h in [36] in experiment with gradual change in nutrient availability. The *edge colors* denote their interaction strength, which was classified as strong repressing (*red*), low repressing (*pink*), no effect (*yellow*), low inducing (*light blue*) and strong inducing (*dark blue*). (Reprinted with permission)

7.5 Integrative Analysis from Multiple Sources

The availability of data from multiple sources such as protein-DNA binding, protein-protein interactions (PPI), miRNA-mRNA interactions, time series gene expressions under genetic perturbations etc. has led to a new challenge in network inference based on multiple sources. In many cases, the additional data belongs to a single time point under a single condition e.g. protein-DNA binding, miRNA-RNA interations, etc. (see Gitter et al. [27], Sect. 3.3 for a discussion of methods integrating static with time series data).

A new scenario has arisen where one wishes to combine several time series data to infer the related network at different time points. For example, if time series gene expressions are measured under the same experimental conditions for several different strains of an organism with minor genetic variation; the inferred networks for the different strains should be largely similar. Alternatively, if one is measuring time series expression data using different methods or under different experimental conditions, some of the interactions might be different while the remaining network structure is largely preserved.

From a computational perspective, one models such time series data as being non i.i.d. which are generated from different but closely related interaction networks. For Gaussian Graphical Models, this corresponds to multiple graphical models that share the same variables and a large part of the dependence structure. Guo et al. [30] recently investigated the joint estimation of multiple graphical models under the assumption that the underlying network structure is largely preserved across the multiple data sources. In their method, no additional assumption is made beyond the large common substructure across the different data sources. We discuss this further in Sect. 7.5.1.

In many cases, we might have further information about the variation in the network structure between different data sources. For example, if one of the sources is time series gene expression data under gene knockout; the network structure would largely vary in a close neighbourhood of the knocked out gene. Jethava et al. [36] explore this for dynamic network reconstruction in yeast based on data from several strains of yeast having genetic perturbations. This is discussed further in Sect. 7.5.2.

7.5.1 Data Integration Based on Common Network Substructure

Guo et al. [30] investigate joint estimation of network structure for multiple graphical models. Their approach assumes the underlying model structure (conditional independencies) is largely preserved across the different networks.

In order to model this, they model the interaction strength $w_{ij}^{[k]}$ between nodes i and j in kth network as a product of two terms, namely, a term θ_{ij} which is common across all networks and $\gamma_{ij}^{[k]}$ which is different across the networks arising from different data sources i.e.

$$w_{ij}^{[k]} = \theta_{ij}\gamma_{ij}^{[k]} \tag{7.40}$$

If θ_{ij} is zero, then all the networks have i and j conditionally independent i.e. no edge is present between nodes i and j across all networks. However, if θ_{ij} is not zero, some of the networks can still have $\gamma_{ij}^{[k]} = 0$ while other networks can have

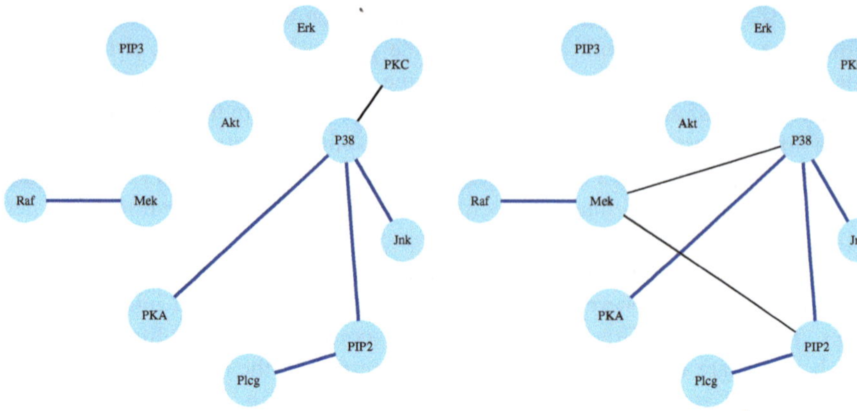

Fig. 7.7 Example of sparse networks sharing a large common substructure. The common substructure is highlighted in *blue*. Such networks can be extracted from different data sources or data from multiple experiments using approach outlined in Guo et al. [30]

$\gamma_{ij}^{(k')} \neq 0$ yielding different network structure. In order to ensure common substructure, they use ℓ_1 regularization based approach by introducing sparsity constraint on θ and $\gamma^{[k]}$ i.e.

$$\arg\min_{\Gamma^{[k]},\Theta} \sum_{k=1}^{K} (\mathrm{Tr}(\mathrm{S}^{[\mathrm{k}]}\mathrm{W}^{[\mathrm{k}]}) - \log\det W^{[k]}) + \eta_1 ||\Theta||_{\ell_1} + \eta_2 \sum_{k=1}^{K} ||^{[k]}||_{\ell_1} \tag{7.41}$$

The parameters η_1 controls the degree of commonality in the network. A high value of η_1 promotes common substructure across the different networks. The parameter η_2 controls the degree of sparsity in the networks. The resulting optimization is solved using the GLASSO software as a subroutine.

This approach allows systematic integration of data from different sources in order to obtain sparse networks with large common substructure. In order to extend this procedure to the case of dynamic networks, one can use the weighted empirical covariance $\hat{S}^{(t),[k]}$ in network k at time t based on kernel reweighting as discussed in Sect. 7.4.1. Figure 7.7 shows an example of two sparse networks with a large common substructure.

7.5.2 *Integration of Time Series Data Under Genetic Perturbations*

Jethava et al. [36] studied the problem of dynamic network reconstruction in *S. Cerevisiae* from multiple experiments with genetic perturbation i.e. where one or two genes have been knocked out. Their approach combines the network perturbation effect into dynamic network reconstruction under the assumption that the network changes drastically near the perturbations (genes knocked out); while sub-networks not related to the functional roles performed by the knocked out gene are minimally impacted.

This is modeled by considering the interaction strength $w^{(t),[k]}$ in perturbed network k at time t as a product of two terms, namely, a base interaction strength $w_{ij}^{(t)}$ and a edge damping coefficient $\gamma_{ij}^{[k]}$.

$$w_{ij}^{(t),[k]} = w_{ij}^{(t)} \gamma_{ij}^{[k]} \tag{7.42}$$

The damping coefficient $\gamma_{ij}^{[k]}$ of edge (i,j) in network k depends on the distance of nodes i and j from the genes knocked out in the perturbed network k i.e.

$$\gamma_{ij}^{[k]} = (1 - \gamma_i^{[k]})(1 - \gamma_j^{[k]}) \tag{7.43}$$

Since a base network is known for Yeast with high degree of confidence, this is used to compute the node damping $\gamma_i^{[k]}$ by diffusing the effect of the gene knockout through the network i.e. if a gene i is knocked out in network k, then $\gamma_i^{[k]}$ is 1; otherwise, $\gamma_i^{[k]}$ is computed by averaging the damping coefficients $\gamma_j^{[k]}$ for all genes j which interact with gene i.

$$\gamma_i^{[k]} = \begin{cases} 1 & \text{if gene } i \text{ is knocked out} \\ \frac{\beta}{d(i)} \sum\limits_{j \in N(i)} \gamma_j^{[k]} & \text{otherwise} \end{cases} \tag{7.44}$$

The parameter β controls the range of perturbation effects. A small value of β implies the effect of gene knockout is limited to close neighbours only; while a large value of β means the pertubation effects are long-ranged. Figure 7.8 shows an example of damping coefficients computed for a network with single gene

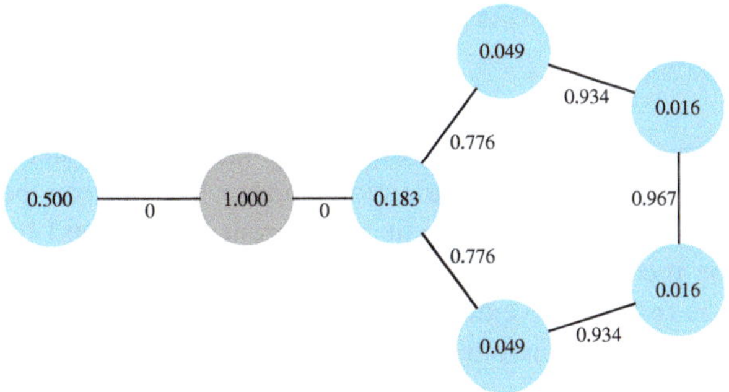

Fig. 7.8 Example of damping with one gene knocked out with damping coefficient $\beta = 0.5$. The knockout gene is indicated in *gray*; and the node and edge damping coefficients are shown. The node damping coefficients quickly decrease as one goes away from the knockout gene. Consequently, the impact of knockout decreases (edge damping close to 1) for interactions far from the knockout point

knockout. The edges next to the knockout gene are impacted the most, while interactions far away from the knockout point are treated as being the same across all networks.

This approach decouples the inference procedure from the effect of network perturbations - while allowing incorporation of data from multiple experiments into reconstrution of the dynamic networks. Consequently, one can use perturbation studies in concert with time series data.

7.6 Software

7.6.1 Glasso

Graphical lasso (GLASSO) is a popular software written in R and Matlab, for estimating sparse inverse covariance matrix using lasso (ℓ_1) penalty. This can be used to find a sparse static interaction network based on microarray expression data. The software is available at http://www-stat.stanford.edu/~tibs/glasso/.

7.6.2 Keller

KELLER is a software in Matlab for estimating time-varying regulatory networks based on time series gene expression data using ℓ_1 regularization approach. It assumes that the interaction network changes smoothly over time i.e. the network between consecutive observation times are very similar structurally. The software is available at http://cogito-b.ml.cmu.edu/keller/.

7.6.3 Tesla

TESLA is a software in Matlab for estimating time-varying networks based on node observations using ℓ_1 regularization approach. This can be used to find dynamic interaction network (different at different time-points) based on microarray expression measurements. It detects sharp changes such as sudden rewiring of the network in response to external stimulus. The software is available at http://www.sailing.cs.cmu.edu/tesla/index.html.

7.6.4 *Netgem*

NETGEM is a software in Matlab for estimating time-varying interaction network based on microarray expression data using a Bayesian approach. It models the network dynamics contingent on the functional roles performed by interacting genes. It incorporates time series data with perturbation analysis to improve network reconstuction by combining time series data from several perturbed networks. The software is available at http://www.cse.chalmers.se/~jethava/netgem.html.

7.7 Discussion

This survey discusses recent methods for reconstruction of time-varying networks based on time series gene expression data. This problem is ill-posed due to high-dimensional data i.e. number of variables p is much larger than number of observations n. Additional assumptions on the network structure as well as the temporal dynamics governing network evolution are required in order to facilitate reconstruction of time-varying interaction networks.

A popular assumption on the network structure is the sparsity of interaction network i.e. each gene interacts with at most few other genes. This agrees closely with domain knowledge and yields biologically plausible networks. This network sparsity is imposed by using ℓ_1 regularization in optimization-based methods, or an sparsity inducing prior in Bayesian approach.

The underlying causes governing network evolution in time-varying interaction networks are not well-understood. A number of simplifying assumptions have been made in order to model different aspects of the network evolution including smooth variation i.e. the underlying network changes slowly over time, piece-wise constant with sharp changes, Markovian dynamics. The reconstructions using the different methods have been shown to yield biologically plausible networks. Further, the reconstructed networks often predict transient interactions which may be experimentally verified; leading to a deeper understanding of biological processes. A clear understanding of network evolution is yet to emerge; and this is an exciting direction for future research.

We also discuss methods which allow systematic analysis of time-series data corresponding to multiple related networks. These methods allow network reconstruction based on multiple data sources e.g. gene interaction networks, miRNA-mRNA interactions, protein-protein interactions (PPI); as well as multiple experiments with genetic perturbations. Such approaches allow better network reconstruction by combining information from several experiments. This is becoming increasingly relevant with growth in publicly available -omics data due to recent advances in sequencing methods.

References

1. Ahmed A, Xing E (2009) Recovering time-varying networks of dependencies in social and biological studies. In: Proceedings of the National Academy of Sciences 106(29),11878–11883
2. Alon U (2007) An introduction to systems biology: design principles of biological circuits, vol. 10. CRC press, Boca Raton, USA
3. Ambroise C, Chiquet J, Matias C (2009) Inferring sparse Gaussian graphical models with latent structure. Electron J Stat 3:205–238
4. Androulakis I, Yang E, Almon R (2007) Analysis of time-series gene expression data: methods, challenges, and opportunities. Annu Rev Biomed Eng 9:205–228
5. Arbeitman M, Furlong E, Imam F, Johnson E, Null B, Baker B, Krasnow M, Scott M, Davis R, White K (2002) Gene expression during the life cycle of drosophila melanogaster. Science 297(5590):2270–2275
6. Ashburner M, Ball C, Blake J, Botstein D, Butler H, Cherry J, Davis A, Dolinski K, Dwight S, Eppig J et al (2000) Gene ontology: tool for the unification of biology. Nat Genet 25(1):25
7. Banerjee O, El Ghaoui L, d'Aspremont A (2008) Model selection through sparse maximum likelihood estimation for multivariate Gaussian or binary data. J Mach Learn Res 9:485–516
8. Barabási A, Oltvai Z (2004) Network biology: understanding the cell's functional organization. Nat Rev Genet 5(2):101–113
9. Barabasi L, Gulbahce N, Loscalso J (2011) Network medicine: a network-based approach to human disease. Nat Rev Genet 12:56–68
10. Boyd S, Vandenberghe L (2004) Convex optimization. Cambridge University Press, Cambridge
11. Buhl S (1993) On the existence of maximum likelihood estimators for graphical Gaussian models. Scand J Stat 20(3):263–270
12. Bühlmann P, Van De Geer S (2011) Statistics for high-dimensional data: methods theory and applications. Springer, New York Inc
13. Candes E, Tao T (2007) The dantzig selector: statistical estimation when p is much larger than n. Ann Stat 35(6):2313–2351
14. Carroll S (2005) Evolution at two levels: on genes and form. PLoS Biol 3(7):e245
15. Cipollina C, van den Brink J, Daran-Lapujade P, Pronk J, Porro D, de Winde J (2008) Saccharomyces cerevisiae sfp1: at the crossroads of central metabolism and ribosome biogenesis. Microbiology 154(6):1686–1699
16. Clarke R, Ressom H, Wang A, Xuan J, Liu M, Gehan E, Wang Y (2008) The properties of high-dimensional data spaces: implications for exploring gene and protein expression data. Nat Rev Cancer 8(1):37–49
17. Davidson E (2001) Genomic regulatory systems: development and evolution. Academic Press, London, UK
18. Dempster A (1972) Covariance selection. Biometrics, 28(1):157–175
19. Donoho D (2000) High-dimensional data analysis: the curses and blessings of dimensionality. AMS Math Challenges Lect, 1–32. http://www-stat.stanford.edu/~donoho/Lectures/AMS2000/AMS2000.html
20. Donoho D (2006) Compressed sensing. Inf Theor, IEEE Trans on 52(4):1289–1306
21. Duchi J, Shalev-Shwartz S, Singer Y, Chandra T (2008) Efficient projections onto the l 1-ball for learning in high dimensions. In: Proceedings of the 25th international conference on Machine learning, pp. 272–279. ACM
22. Ernst J, Nau G, Bar-Joseph Z (2005) Clustering short time series gene expression data. Bioinformatics 21(suppl 1):i159–i168
23. Friedman J, Hastie T, Tibshirani R (2008) Sparse inverse covariance estimation with the graphical lasso. Biostatistics 9(3):432–441
24. Hastie T, Tibshirani R, Friedman J (2008) The elements of statistical learning, 2 edn. Springer-Verlag, Springer series in statistics, 763 p

25. Friedman N, Linial M, Nachman I, Pe'er D (2000) Using Bayesian networks to analyze expression data. J Comput Biol 7(3–4):601–620
26. Fu W, Song L, Xing E (2009) Dynamic mixed membership blockmodel for evolving networks. In: Proceedings of the 26th annual international conference on machine learning, pp 329–336. ACM
27. Gitter A, Lu Y, Bar-Joseph Z (2010) Computational methods for analyzing dynamic regulatory networks. Methods in molecular biology (Clifton, NJ) 674, 419
28. Glass L, Kaplan D (1993) Time series analysis of complex dynamics in physiology and medicine. Med Progr Technol 19:115–115
29. Guo F, Hanneke S, Fu W, Xing E (2007) Recovering temporally rewiring networks: a model-based approach. In: Proceedings of the 24th international conference on Machine learning, pp 321–328. ACM
30. Guo J, Levina E, Michailidis G, Zhu J (2011) Joint estimation of multiple graphical models. Biometrika 98(1):1–15
31. Hartemink A et al (2005) Reverse engineering gene regulatory networks. Nat Biotechnol 23(5):554–555
32. de Hoon M, Imoto S, Miyano S (2002) Inferring gene regulatory networks from time-ordered gene expression data using differential equations. In: Discovery science, 283–288. Springer
33. Hu H, Yan X, Huang Y, Han J, Zhou X (2005) Mining coherent dense subgraphs across massive biological networks for functional discovery. Bioinformatics 21(suppl 1):i213–i221
34. Ideker T, Sharan R (2008) Protein networks in disease. Genome Res 18:644–652
35. Ito T, Chiba T, Ozawa R, Yoshida M, Hattori M, Sakaki Y (2001) A comprehensive two-hybrid analysis to explore the yeast protein interactome. In: Proceedings of the National Academy of Sciences 98(8):4569
36. Jethava V, Bhattacharyya C, Dubhashi D, Vemuri G (2011) Netgem:network embedded temporal generative model for gene expression data. BMC Bioinform 12(1):327
37. Kim S, Imoto S, Miyano S (2004) Dynamic Bayesian network and nonparametric regression for nonlinear modeling of gene networks from time series gene expression data. Biosystems 75(1):57–65
38. Koh K, Kim S, Boyd S (2007) An interior-point method for large-scale l1-regularized logistic regression. J Mach Learn Res 8(8):1519–1555
39. Koller D, Friedman N (2009) Probabilistic graphical models: principles and techniques. The MIT Press, Cambridge, MA
40. Lam C, Fan J (2009) Sparsistency and rates of convergence in large covariance matrix estimation. Ann Stat 37(6B), 4254
41. Lauritzen S (1996) Graphical models, vol 17. Oxford University Press, USA
42. Lin C, Weng R, Keerthi S (2008) Trust region newton method for logistic regression. J Mach Learn Res 9:627–650
43. Luscombe N, Babu M, Yu H, Snyder M, Teichmann S, Gerstein M (2004) Genomic analysis of regulatory network dynamics reveals large topological changes. Nature 431(7006):308–312
44. Ma S, Gong Q, Bohnert H (2007) An arabidopsis gene network based on the graphical Gaussian model. Genome Res 17(11):1614–1625
45. Meinshausen N, Bühlmann P (2006) High-dimensional graphs and variable selection with the lasso. Ann Stat 34(3):1436–1462
46. Mewes H, Frishman D, Gruber C, Geier B, Haase D, Kaps A, Lemcke K, Mannhaupt G, Pfeiffer F, Schüller C et al (2000) Mips: a database for genomes and protein sequences. Nucleic Acids Res 28(1):37–40
47. Parisi G, Shankar R (1988) Statistical field theory. Phys Today 41:110
48. Peer D, Regev A, Elidan G, Friedman N (2001) Inferring subnetworks from perturbed expression profiles. Bioinformatics 17(suppl 1), S215–S224
49. Perrin B, Ralaivola L, Mazurie A, Bottani S, Mallet J, dAlche Buc F (2003) Gene networks inference using dynamic Bayesian networks. Bioinformatics 19(suppl 2), ii138-ii148 .

50. Przytycka T, Singh M, Slonim D (2010) Toward the dynamic interactome: it's about time. Briefings Bioinform 11(1):15–29
51. Ravikumar P, Wainwright M, Lafferty J (2010) High-dimensional ising model selection using 1-regularized logistic regression. Ann Stat 38(3):1287–1319
52. Robins G, Pattison P, Kalish Y, Lusher D (2007) An introduction to exponential random graph p^* models for social networks. Soc Netw 29(2):173–191
53. Rothman A, Bickel P, Levina E, Zhu J (2008) Sparse permutation invariant covariance estimation. Electron J Stat 2:494–515
54. Sachs K, Perez O, Pe'er D, Lauffenburger D, Nolan G (2005) Causal protein-signaling networks derived from multiparameter single-cell data. Science's STKE 308(5721), 523
55. Schadt E (2009) Molecular networks as sensors and drivers of common human diseases. Nature 416:218–223
56. Schäfer J, Strimmer K (2005) An empirical Bayes approach to inferring large-scale gene association networks. Bioinformatics 21(6):754–764
57. Schliep A, Schönhuth A, Steinhoff C (2003) Using hidden Markov models to analyze gene expression time course data. Bioinformatics 19(suppl 1):i255–i263
58. Shermin A, Orgun M (2009) Using dynamic bayesian networks to infer gene regulatory networks from expression profiles. In: Proceedings of the 2009 ACM symposium on applied computing, 799–803. ACM
59. Song L, Kolar M, Xing E (2009) Keller: estimating time-varying interactions between genes. Bioinformatics 25(12):i128–i136
60. Soranzo N, Bianconi G, Altafini C (2007) Comparing association network algorithms for reverse engineering of large-scale gene regulatory networks: synthetic versus real data. Bioinformatics 23(13):1640–1647
61. Speed T, Kiiveri H (1986) Gaussian Markov distributions over finite graphs. Ann Stat 14(1):138–150
62. Tegner J, Yeung M, Hasty J, Collins J (2003) Reverse engineering gene networks: integrating genetic perturbations with dynamical modeling. In: Proceedings of the National Academy of Sciences 100(10):5944
63. Tibshirani R (1996) Regression shrinkage and selection via the lasso. J Roy Stat Soc. Series B (Methodological) 58(1):267–288
64. Uetz P, Giot L, Cagney G, Mansfield T, Judson R, Knight J, Lockshon D, Narayan V, Srinivasan M, Pochart P et al (2000) A comprehensive analysis of protein-protein interactions in saccharomyces cerevisiae. Nature 403(6770):623–627
65. Wainwright M, Ravikumar P, Lafferty J (2007) High-dimensional graphical model selection using 1˜ 1-regularized logistic regression. Advances in neural information processing systems 19:1465
66. Werhli A, Grzegorczyk M, Husmeier D (2006) Comparative evaluation of reverse engineering gene regulatory networks with relevance networks, graphical Gaussian models and Bayesian networks. Bioinformatics 22(20):2523–2531
67. Wille A, Zimmermann P, Vranová E, Fürholz A, Laule O, Bleuler S, Hennig L, Prelic A, Von Rohr P, Thiele L et al (2004) Sparse graphical Gaussian modeling of the isoprenoid gene network in arabidopsis thaliana. Genome Biol 5(11):R92
68. Workman C, Mak H, McCuine S, Tagne J, Agarwal M, Ozier O, Begley T, Samson L, Ideker T (2006) A systems approach to mapping dna damage response pathways. Science's STKE 312(5776):1054
69. Yeang C, Mak H, McCuine S, Workman C, Jaakkola T, Ideker T (2005) Validation and refinement of gene-regulatory pathways on a network of physical interactions. Genome Biol 6(7):R62
70. Yeung M, Tegnér J, Collins J (2002) Reverse engineering gene networks using singular value decomposition and robust regression. In: Proceedings of the National Academy of Sciences 99(9):6163
71. Yuan M, Lin Y (2007) Model selection and estimation in the Gaussian graphical model. Biometrika 94(1):19–35

72. Zhou S, Lafferty J, Wasserman L (2010) Time varying undirected graphs. Mach Learn 80(2):295–319
73. Zou M, Conzen S (2005) A new dynamic Bayesian network (dbn) approach for identifying gene regulatory networks from time course microarray data. Bioinformatics 21(1):71–79

Chapter 8
Probabilistic Graphical Modeling in Systems Biology: A Framework for Integrative Approaches

Christine Sinoquet

Abstract Systems biology may be defined as a discipline aiming at integrating various sources of heterogeneous data, with the objective to describe and predict the function of biological systems. The purpose is to cross many (possibly weak) evidences from several data types that describe different biological features of genes or proteins. Probabilistic graphical models offer an appealing framework for this objective. Through the thorough review of five selected examples, this chapter highlights how probabilistic graphical models can contribute to build the bridge between biology and computational modeling. In this methodological framework, the five cases illustrate three features of these models, which we discuss: flexibility, scalability and ability to combine heterogeneous sources of data. The applications covered address genetic association studies, identification of protein–protein interactions, identification of the target genes of transcription factors, inference of causal phenotype networks and protein function prediction.

Keywords Systems biology · Integrative approach · Integration of omics data · Heterogeneous sources of data · Computational modeling · Machine learning · Probabilistic framework · Probabilistic graphical model · Bayesian network · Markov random field

List of Acronyms

BN	Bayesian network
ChIP-chip	Chromatin immunoprecipitation on chip
ChIP-seq	Chromatin immunoprecipitation followed by sequencing
CPN	Causal phenotype network
DDI	Domain-domain interaction
DNA	Deoxyribonucleic acid

C. Sinoquet (✉)
LINA, UMR CNRS 6241, Université de Nantes, 2 rue de la Houssinière,
BP 92208 44322 Nantes Cedex, France
e-mail: christine.sinoquet@univ-nantes.fr

A. Prokop and B. Csukás (eds.), *Systems Biology*,
DOI: 10.1007/978-94-007-6803-1_8,

GA	Genetic architecture
GO	Gene ontology
GOS	GO sub-ontology
GWAS	Genome wide association study
MCMC	Monte Carlo Markov chain
MRF	Markov random field
MRF-MJM	MRF mixture joint model
PGM	Probabilistic graphical model
PPI	Protein–protein interaction
QTL	Quantitative trait loci
RNA	Ribonucleic acid
RNAi	RNA interference
ROC curve	Receiver operating characteristic curve
SMM	Standard mixture model
TF	Transcription factor

8.1 Introduction

In the machine learning domain, probabilistic graphical models provide a unified framework to both represent dependences between variables and model uncertain knowledge about the quantitative dependences between these variables. In the post-genomic era, the provision of voluminous and complex heterogeneous data by high-throughput omics technologies has brought increased attention to these models. Notably, their flexibility, scalability and ability to combine heterogeneous sources of data are expected to enhance the gain in biological and biomedical discoveries. Data integration is intended to make useful connections that could lead to novel biological knowledge.

Besides, if there is one area where transdisciplinarity is the daily lot, designing new computational methods based on advanced models devoted to applications in systems biology is this area. A constructive cooperation with a domain specialist requires ability to hold productive dialogue, which therefore demands a good understanding of the models by the non expert. Bridging the gap between biology, statistics and computer science is a condition to achieve progress in systems biology. Albeit dedicated to specific applications, the five models presented in this chapter remain general enough to help foster reflections about addressing other applications in systems biology, in an integrative framework.

Methods based on probabilistic graphical models (PGMs) may be complex and thus might be disconcerting to scientists non familiar with them, which is likely to hamper the dissemination of such methods. Thus there was a challenge in attempting to demystify the concepts and mechanisms behind such models in the perspective of using them for systems biology. To this aim, this chapter was conceived as a thorough review of five illustrative approaches of the use of probabilistic graphical modeling as an integrative framework in systems biology. We first provide an intuitive presentation of the concept of conditional independence, which is the fundamental principle PGMs all rely on; then we introduce the Bayesian networks and the Markov random fields, which are the two classes of PGMs addressed in this chapter. Subsequently we present the five illustrations selected to cover major application fields in systems biology: (1) enhancement of genome-wide association studies with knowledge on biological pathways, (2) identification of protein–protein interactions, (3) identification of the regulatory target genes of transcription factors, (4) inference of causal relationships among phenotypes *via* integration of QTL genotypes, (5) prediction of protein function through ontology-enriched networks connecting multiple related species. A brief insight about the performance of each method is provided on the fly. We conclude this chapter highlighting the pros and cons of this modeling framework, when used for integration purpose in systems biology and we indicate some directions for future work.

The order of presentation for the contributions is not incidental: it puts forward an increasing gradient in the heterogeneity of the data sources integrated in the probabilistic framework. For example, approaches (2) and (3) both integrate information coming from gene ontologies but such information is used similarly to that coming from the other data sources. In contrast, accounting for this ontological knowledge thoroughly impacts the probabilistic inference scheme in the last approach. At the opposite extremity of the data integration spectrum, it is worth mentioning that PGMs provide the ability to integrate *meta-knowledge* about a *single* data source, at the genome-wide scale. An enlightening example is the modeling of genetic data, where the so-called linkage disequilibrium encompasses short-range, long-range and chromosome-wide dependences within these data [24–25]. Such meta-knowledge integration in a genetic association study aims at enhancing power and accuracy in identifying the causal factors of a disease [20, 35, 37]. In this book chapter, we focus on the integration of *multiple* data sources.

8.2 Preliminaries

In the present section, the concepts indispensable for further understanding are introduced in an intuitive manner. Besides, we highlight why probabilistic graphical models are appealing to model biological data in an integrative framework. Within the scope of this section, we suppose that the data available are as follows: p data samples are each described by n variables $X = \{X_1, \ldots, X_n\}$. In a general probabilistic framework, computing the joint probability distribution for large data is generally not tractable as, by virtue of the so-called product-rule, the only formalization applicable is

$$\mathbb{P}(X) = \mathbb{P}(X_1)\, \mathbb{P}(X_2 \mid X_1)\, \mathbb{P}(X_3 \mid X_1,\, X_2) \ldots \mathbb{P}(X_n \mid X_1,\, X_2,\, \ldots,\, X_{n-1}). \quad (8.1)$$

If we denote by x_i a value in the domain of the possible values for the random variable X_i, it should be noted that computing the probability distribution $\mathbb{P}(X)$ means that for any possible instantiation $x = (x_1, x_2, \ldots, x_n)$ of X, we know how to calculate $\mathbb{P}(x)$. It should also be kept in mind that from now on, symbols in lower cases will denote values taken by the variables. Third, the expression "probability distribution" will be reserved to discrete random variables whereas the expression "density probability" will be used for continuous random variables. In the above formula, $X_1, \ldots, X_r$, to be understood as the *event* $(X_1 = x_1, \ldots, X_r = x_r)$, denotes the joint observation of values $x_1, \ldots, x_r$. The product rule in Eq. 8.1 involves *conditional probabilities*.

The conditional probability[1] of event D_1 given event D_2, $\mathbb{P}(D_1 \mid D_2)$, is the probability of D_1 with the additional information that D_2 has already occurred. It is defined as:

$$\mathbb{P}(D_1 \mid D_2) = \frac{\mathbb{P}(D_1, D_2)}{\mathbb{P}(D_2)}, \text{ with } \mathbb{P}(D_2) \neq 0.$$

For instance, if D_1 and D_2 are two diseases, such that D_2 is observed with probability 0.05, and D_1 and D_2 are simultaneously observed with probability 0.001, then the onset probability for D_1, when D_2 is present, is 0.02.

Probabilistic graphical models are appealing models because they rely on *conditional independence*, to offer the immense advantage of a factorized formulation of probability distributions. Let us first introduce the concept of conditional independence. In the above example, suppose we calculate that the *prior probability* $\mathbb{P}(D_1)$ is equal to the *posterior probability* $\mathbb{P}(D_1 \mid D_2)$. Intuitively, this means that knowing whether D_2 occurs ($\mathbb{P}(D_1 \mid D_2)$) does not refine our knowledge about whether D_1 occurs. The diseases D_1 and D_2 are therefore *independent*: $D_1 \perp\!\!\!\perp D_2$. Conditional independence is a little bit more complex:

Definition 1 (*Conditional independence*) Given three variables A, B and C, conditional independence between A and B given the state of C ($A \perp\!\!\!\perp B \mid C$) is defined as: $\mathbb{P}(A \mid B,\, C) = \mathbb{P}(A \mid C)$ (with $\mathbb{P}(C) > 0$). The concept of conditional independence given a unique variable is easily extended to conditional independence given a set of variables.

Intuitively, A and B are conditionally independent given C ($A \perp\!\!\!\perp B \mid C$) if and only if, given any value of C, the probability distribution of A remains the same for all values of B: $\mathbb{P}(A \mid B = b_1,\, C = c) = \mathbb{P}(A \mid B = b_2,\, C = c) = \mathbb{P}(A \mid C = c)$. Suppose now that a third variable E measures the effects of the disease D_1, and that these effects cause the disease D_2 (symbolized through $D_1 \rightarrow E \rightarrow D_2$); indoubtedly, D_1 and D_2 are dependent; however, D_1 and D_2 are conditionally independent

[1] Depending on the context, the *conditional probability* of D_1 given D_2, $\mathbb{P}(D_1 \mid D_2)$, is also called the *posterior probability* of D_1 conditional on D_2.

given E (see Table 8.1). Intuitively, this means that the status of D_1 can be inferred from the status of E. Therefore, when dependences exist within data, conditional independence shields a given variable from the remaining variables, given some set of variables. Biological data are often described by a network, if not several networks, in the integrative framework. The conditional independence property is known as the *Markov property*. The Markov property is the corner stone for simplifying probability distributions, thus directly achieving tractability or making easier approximations to further obtain tractability. Intrinsically, all five models illustrated in this chapter rely on the Markov property, to infer knowledge from one or several biological networks.

Probabilistic graphical models (PGMs) provide a powerful framework for representing and reasoning with uncertainty and dependences. The qualitative part of a PGM is a graph $\mathcal{G}$ that encodes dependences (and independences) between the variables, represented by nodes in the graph. Uncertain knowledge about the qualitative dependences between the variables is formalized with the aid of probability distributions. Besides differences in their graphs, we now briefly show the variants of Markov property for the two kinds of probabilistic graphical model (PGMs) addressed in this chapter. One of the most popular kinds of PGMs is the Bayesian network (BN).

Table 8.1 Conditional independence of two variables D_1 and D_2 given a third-variable E

		D_2E	$D_2\bar{E}$	$\bar{D}_2E$	$\bar{D}_2\bar{E}$
D_1	0	120	40	40	20
	1	180	160	60	80

(a) Counts

		D_2E	$D_2\bar{E}$	$\bar{D}_2E$	$\bar{D}_2\bar{E}$
D_1	0	0.171	0.057	0.057	0.029
	1	0.257	0.229	0.086	0.114

(b) Joint distribution $\mathbb{P}(D1, D2, E)$

		D_2 0	D_2 1
D_1	1	0.086	0.229
	0	0.200	0.485

(c) Marginal distribution $\mathbb{P}(D1, D2)$

		E = 0 D_2 0	**E = 0** D_2 1
D_1	1	0.2	0.2
	0	0.8	0.8

		E = 1 D_2 0	**E = 1** D_2 1
D_1	1	0.4	0.4
	0	0.6	0.6

Conditional distributions

(d) $\mathbb{P}(D1 \mid D2 = i, \mathbf{E} = \mathbf{0})$ **(e)** $\mathbb{P}(D1 \mid D2 = i, \mathbf{E} = \mathbf{1})$

c Marginal probabilities are obtained through summing ("marginalizing") probabilities over the domain of E; $\mathbb{P}(D_1 \mid D_2) = \frac{\mathbb{P}(D_1, D_2)}{\mathbb{P}(D_2)} = \frac{0.485}{0.679} = 0.714 \neq \mathbb{P}(D_1) = 0.685$, thus D_1 and D_2 are dependent variables. **d** and **e** D_1 and D_2 are conditionally independent given E ($D_1 \perp\!\!\!\perp D_2 \mid E$) since the columns are identical within each table.

Definition 2 (*Bayesian network*) In a BN, the qualitative component is a directed acyclic graph (acyclic because no directed path $X_{i_1} \rightarrow X_{i_2} \rightarrow \cdots \rightarrow X_{i_r}$, where $i_1 = i_r$, is allowed). Conditional distributions are defined for each variable X_i: $\theta_i = [\mathbb{P}(X_i/Pa_{X_i})]$ where Pa_{X_i} denotes node i's parents. The local Markov property states that each variable is conditionally independent of its non-descendants given a known state of its parent variables: $X_i \perp\!\!\!\perp X \setminus desc(X_i) \mid Pa_{X_i}$, where the notation $X \setminus Y$ stands for the set $\{X_i \in X \text{ and } X_i \notin Y\}$ and $desc(X_i)$ is the set of descendants of X_i. The local Markov property entails that the joint distribution writes as a product of local distributions conditional on the parent variables:

$$\mathbb{P}(X) = \prod_{i \in \{1,\ldots,n\}} \theta_i.$$

Figure 8.1a shows a Bayesian network. Another widely used model is the Markov random field.

Definition 3 (*Markov random field*) In Markov random fields (MRFs), the qualitative component $\mathcal{G}$ is an undirected graph which may have cycles (that is (undirected) cycles $X_{i_1} - X_{i_2} - \cdots - X_{i_r}$, where $i_1 = i_r$, are allowed). The joint distribution is factorized over cliques "covering" the set X. A clique is defined by any set of pairwise connected nodes, such as $\{X_1, X_2\}$ or $\{X_1, X_2, X_4\}$ in Fig. 8.1b. A set of random variables X is an MRF if there exist so-called function potentials such that the joint distribution writes:

$$\mathbb{P}(X = x) = \frac{1}{Z}\, b(x)$$

$$\mathbb{P}(X = x) = \frac{1}{Z} \prod_{C \in cliques(\mathcal{G})} \varphi_C(x_C).$$

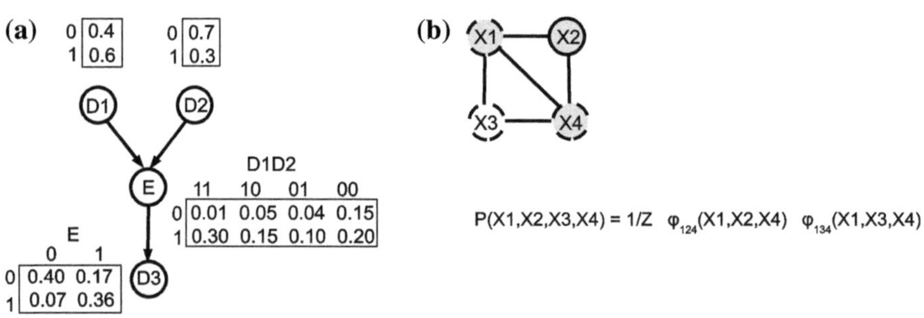

Fig. 8.1 Probabilistic graphical models. **a** Bayesian network. **b** Markov random field. **a** The prior probability distributions $\mathbb{P}(D_1)$ and $\mathbb{P}(D_2)$, and the conditional distributions $\mathbb{P}(E \mid D_1, D_2)$ and $\mathbb{P}(D_3 \mid E)$ are shown. The node E has two parents (D_1 and D_2). The node D_1 has one child (E) and two descendants (E and D_3). **b** The factorization of the joint distribution $\mathbb{P}(X_1, X_2, X_3, X_4)$ involves the potentials relative to the two cliques (X_1, X_2, X_4) and (X_1, X_3, X_4). The node X_1 has two neighbors: X_2 and X_3

There, x_C denotes some possible instantiation for the variables encompassed by clique C. Function φ_C is called a clique potential. Z is the normalizing function used to ensure that $\mathbb{P}$ be a probability distribution ($Z = \sum_x b(x)$ guarantees that $\sum_x \mathbb{P}(X = x) = 1$). In the case of the MRF, the local Markov property states that a variable is conditionally independent of all other variables given its set of neighbours N_i: $\mathbb{P}(X_i \mid X_{-i}) = \mathbb{P}(X_i \mid N_i)$, where X_{-i} designates the set X deprived of variable X_i.

Figure 8.1b shows a Markov random field. In particular, this chapter will refer to pairwise MRFs, which consider cliques of size 2 and whose joint distribution writes:

$$\mathbb{P}(X = x) = \frac{1}{Z} \prod_{i=1}^{n} \varphi_i(x_i) \prod_{(i,j) \in \mathcal{G}} \varphi_{i,j}(x_i, x_j). \tag{8.2}$$

Finally, we recall some additional notions to non specialists. Given a model M and the observed data D, according to Bayes theorem,[2] the relation between *posterior distribution*, *prior distribution* and *likelihood* writes: $\mathbb{P}(M \mid D) \propto \mathbb{P}(D \mid M)\, \mathbb{P}(M)$. The proportionality is explained by the fact that the probability to observe the data, $\mathbb{P}(D)$, is a constant. Model learning consists in evaluating how a candidate M fits the data D. Maximizing the likelihood $\mathbb{P}(D \mid M)$ is a standard procedure to achieve this purpose. Due to additional knowledge (D), the prior distribution $\mathbb{P}(M)$ is refined into the posterior distribution $\mathbb{P}(M \mid D)$. The reader is also reminded that $\mathcal{U}(a, b)$ designates the uniform probability distribution over interval $[a, b]$ and that $\mathcal{N}(\mu, \sigma^2)$ represents the normal (or Gaussian) probability distribution with mean μ and variance σ^2. The multivariate normal distribution is a generalization of the latter distribution to higher dimensions; then the normal distribution is summarized by a mean vector and a covariance matrix. To denote that a random variable A follows a given distribution, say $\mathcal{N}(\mu, \sigma^2)$, we will write: $A \sim \mathcal{N}(\mu, \sigma^2)$.

8.3 Enhancement of Genome-Wide Association Studies with Knowledge on Biological Pathways

To decipher the genetic causes of diseases, genome-wide association studies (GWASs) compare the genomes of affected people to those of unaffected. The aim is to identify associations between genetic variants and the disease. GWASs pose a formidable challenge since most of the time the effects from individual genetic variants are weak and the sample size is not large enough to guarantee sufficient power. To overcome this issue, various strategies have been proposed. Multilocus

---

[2] $\mathbb{P}(M \mid D)\, \mathbb{P}(D) = \mathbb{P}(D \mid M)\, \mathbb{P}(M)$.

association tests benefit from linkage disequilibrium—that is dependences existing within genetic data—by considering sets of correlated markers instead of single markers. An alternative lead lies in integrating evidences from external data sources, in the single locus approach. Various approaches based on the integration of prior biological knowledge were designed to prioritize candidate disease genes (see [16] for a survey). In GWASs, evidence from the gene level is recognized as the most promising. In particular, incorporating prior biological knowledge about pathways has a role to play [31, 40]: as genes interact with each other in biological pathways, they are likely to jointly affect disease susceptibility. However, so far, no GWAS approach had taken into account knowledge about *regulatory relationships* between genes of a given pathway. Not surprisingly, in this domain, the pioneering approach of Chen and collaborators takes full advantage of probabilistic graphical modeling [5].

In the following, we denote $S = \{S_1, \ldots, S_n\}$ the set of gene labels to be predicted based on the observed association data and the knowledge on the pathway topology. $S_i = 1$ states that gene i is associated with the disease; otherwise, the label is $S_i = 0$. Typically, the association data are p-values $P_1, \ldots, P_n$ resulting from n single-locus association tests. Usually, given some significant threshold P^*, $P_i < P^*$ (respectively $P_i \geq P^*$) indicates that S_i should be set to 1 (respectively 0). The probabilistic framework adopted by Chen and collaborators aims at improving the reliability in predicting the labels: the ultimate goal is thus to estimate the posterior distribution of S conditional on the data P, that is $\mathbb{P}(S \mid P)$. By virtue of the Bayes theorem, $\mathbb{P}(S \mid P) \propto \mathbb{P}(S)\ \mathbb{P}(P \mid S)$. The key to the prediction improvement by Chen et al. lies in the integration of knowledge on the pathway topology in the model: such knowledge is incorporated in the prior distribution $\mathbb{P}(S)$.

8.3.1 Exploiting Knowledge from the Gene Pathway

In the following, N_i denotes the set of the n_i neighbors of gene i in the pathway of concern; $\mathcal{G}$ denotes the pathway topology. To capture the idea that two neighbor genes i and j tend to share a common association status ($S_i = S_j$), Chen et al. adjust a nearest neighbor Gibbs measure [15] as follows:

$$\begin{aligned}\mathbb{P}(S = s \mid \theta_0) = &\frac{1}{Z} exp \\ &[h + \sum_i I_1(S_i) + \tau_0 \sum_{(i,j)\in\mathcal{G}} (w_i + w_j)\ I_0(S_i)\ I_0(S_j) \\ &+ \tau_1 \sum_{(i,j)\in\mathcal{G}} (w_i + w_j)\ I_1(S_i)\ I_1(S_j)].\end{aligned} \tag{8.3}$$

The symbol $s = (s_1, \ldots, s_n)$ denotes one label assignment (amongst the 2^n possible assignments), for instance $(0, 1, 1, \ldots, 1, 0)$. $\theta = (h, \tau_0, \tau_1)$ denotes hyperparameters fixed by the user. I_0 and I_1 are indicator functions, meaning that $I_1(S_i) = 1$

if $S_i = 1$ and $I_1(S_i) = 0$ otherwise, and, symmetrically $I_0(S_i) = 1$ if $S_i = 0$ and $I_0(S_i) = 0$ otherwise. Finally, the model in Eq. 8.3 also reflects the fact that genes showing many interactions in a pathway are likely to play a prominent role in a biological process; thus they are likely to exert a large influence. Consequently, weights w_is are incorporated in the model, that depend on the neighborhood sizes: $w_i = \sqrt{n_i}$ is an increasing function of the number n_i of neighbors of gene i.

Equation 8.3 formalizes the joint probability for S so that genes connected with each other tend to have the same labels, that is the same association status. The third term concerns all edges connecting neighbors sharing the common label 0. The fourth term concerns neighbors that share the label 1. Besides, τ_0 and τ_1 assign weights to such edges, depending on the shared labels. Positive parameters τ_0 and τ_1 will favor assignments s of S in which neighbor genes share the same label.

A property of nearest neighbor Gibbs measures is that they always define a Markov random field. In this case, the conditional independence assumption entails: $\mathbb{P}(S_i \mid S_{-i}, \theta_0) = \mathbb{P}(S_i \mid S_{N_i}, \theta_0)$, where we recall that $S_{-i} = (S_1, \ldots, S_{i-1}, S_{i+1}, \ldots, S_n)$. Besides, using Eq. 8.3, Chen et al. show that the conditional distribution $\mathbb{P}(S \mid S_{N_i}, \theta_0)$ has a logistic regression form. A standard linear regression model is not convenient to represent a binary (0/1) variable B as $B = a_0 + a_1 A_1 + a_2 A_2 + \cdots + a_k A_k$, since the predictors A_i are unconstrained. Instead, one deals with $p = \mathbb{P}(B = 1) \in [0, 1]$ and a logit transformation is therefore required to apply a linear regression model to $logit(p) = log\left(\frac{p}{1-p}\right) \in]-\infty, +\infty[$. In the case illustrated, the logistic form is:

$$logit(\mathbb{P}(S_i \mid S_{N_i}, \theta_0)) = h + \tau_1\left(w_i\ J_i^1 + \sum_{k \in N_i} w_k\ I_1(S_k)\right) - \tau_0\left(w_i\ J_i^0 + \sum_{k \in N_i} w_k\ I_0(S_k)\right), \tag{8.4}$$

where $J_i^0 = \sum_{k \in N_i} I_0(S_k)$ and, similarly, $J_i^1 = \sum_{k \in N_i} I_1(S_k)$.

In the configuration where τ_0 and τ_1 are both null, all genes are interpreted as independent; the so-called intercept h then determines the posterior probability $\mathbb{P}(S_i \mid h, \tau_0 = \tau_1 = 0) = \frac{exp(h)}{1+exp(h)}$.

To recapitulate, the prior acknowledging for the pathway topology is the conditional distribution $\mathbb{P}(S \mid S_{N_i}, \theta_0)$. This prior has the logistic regression form:

$$\begin{aligned}
&logit(\mathbb{P}(S_i \mid S_{N_i}, \theta_0)) = \beta_{i0} + \beta_{i1}\ S_1 + \cdots + \beta_{in}\ S_n \\
&\text{with} \\
&\beta_{i0} = h \\
&\beta_{ij} = 0\ \textit{if}\ i = j\ \textit{or}\ j \notin N_i \\
&\beta_{ij} = (w_i + w_j)\ (\tau_1\ I_1(S_j) - \tau_0\ I_0(S_j))\ \textit{otherwise}.
\end{aligned}$$

In the following, for concision, we will omit the references to S_{N_i} and θ_0 and the joint prior distribution will merely be denoted $\mathbb{P}(S)$ as in the end of the introductory paragraph of Sect. 8.3.

8.3.2 Posterior Distribution of Association Status

The posterior distribution integrates the knowledge about the pathway topology (from the prior) and the evidence from the observed association data (i.e. the p-values):

$$\mathbb{P}(S \mid P) \propto \mathbb{P}(S)\ \mathbb{P}(P \mid S). \tag{8.5}$$

A model remains to be defined for $\mathbb{P}(S \mid P)$. Chen and collaborators model instead $\mathbb{P}(S \mid Y)$, where any p-value P_i is converted into $Y_i = \Phi^{-1}(1 - P_i/2)$. Therein, Φ is the cumulative distribution function of $\mathcal{N}(0, 1)$. The justification for this conversion is simplification in further algebraic derivations. Then, the joint density of Y readily writes:

$$f(Y \mid S) = \prod_{i:S_i=0} f_0(Y_i) \prod_{i:S_i=1} f_1(Y_i),$$

where f_0 and f_1 respectively denote the distributions of Y_i under the null hypothesis and the hypothesis of association, that is: $f_0(Y_i) = \mathbb{P}(Y_i \mid S_i = 0)$ and $f_1(Y_i) = \mathbb{P}(Y_i \mid S_i = 1)$. Under the null hypothesis (no association, $S_i = 0$), any value in $[0, 1]$ is acceptable for the p-value (probability) P_i. P_i is therefore modeled to follow the uniform distribution $\mathcal{U}(0, 1)$. This setting entails that $f_0(Y_i)$ follows the Gaussian distribution $\mathcal{N}(0, 1)$. On the other hand, the unknown distribution of Y_i under the hypothesis of association is assumed to follow a Gaussian distribution: $f_1(Y_i) \sim \mathcal{N}(\mu_i, \sigma_i^2)$.

Under these settings, the algebraic derivation of the posterior distribution (see Eq. 8.5)

$$\mathbb{P}(S \mid Y) \propto \mathbb{P}(S)\mathbb{P}(Y \mid S)$$

shows that, similarly to the prior $\mathbb{P}(S)$, the posterior distribution $\mathbb{P}(S \mid Y)$ has a logistic regression form. The regression forms are identical in all points except for the intercept, which is now $h + log\ LR(Y_i)$ where $LR(Y_i) = \frac{f_1(Y_i)}{f_0(Y_i)}$ is the usual likelihood ratio. Importantly, the conditional independence assumption of the prior distribution is kept: $\mathbb{P}(S_i \mid S_{-i}, \ldots) = \mathbb{P}(S_i \mid S_{N_i}, \ldots)$.

Finally, the assignment of labels to the genes is performed by running an MCMC. The MCMC starts from some initial value $s^{(0)}$ assigned (at random) to S. Then step k sequentially updates the labels of the genes according to the following scheme:

$$logit(\mathbb{P}(s_i^{(k)} \mid Y, s_1^{(k)}, \ldots, s_{i-1}^{(k)}, s_{i+1}^{(k-1)}, \ldots, s_n^{(k-1)}) = \beta'_{i0} + \beta'_{i1} s_1 + \cdots + \beta'_{in} s_n.$$

An important point is that the conditional independence assumption in Eq. 8.4 holds for the posterior distribution, which is therefore also a Markov random field. The practical consequence is that the computation involved in the sampling of s_i only requires values s_j where j belongs to the neighborhood N_i: otherwise, the β'_{ij} coefficient is null if genes i and j are not neighbors in the pathway.

8.3.3 Performances

Incorporating prior biological knowledge to enhance GWASs is not new (see for instance Prioritizer [10], CANDID [13], CIPHER [42]). However, such approaches do not consider the functional relationships existing among genes. In contrast, the approach of Chen et al. takes advantage of knowledge on known associations to infer novel association knowledge on other genes related to the former through pathways.

The relevance of the model of Chen and collaborators was supported by a preliminary study. These authors considered 3,735 genes over 350 pathways. On the other hand, association results from a GWAS on Crohn's disease were available. In each of the pathways, the number of edges N_{++} with both extremities associated with the disease was observed. Over the 350 pathways, an overwhelming proportion of counts N_{++} showed exceptionally large values. This clearly confirms the hypothesis: in a given pathway, most often, genes that are associated with the disease are neighbors.

The approach of Chen et al. was then evaluated based on 289 pathways and GWAS data relative to Crohn's disease. Thirty-two genes associated with the disease were known (target genes). It was shown that ranking the genes according to their posterior probabilities is more faithful to the reality than ranking them based on their p-values. Finally, as expected, it was verified that compared to other genes in the pathway, the genes with an improved rank are more densely connected to target genes; besides, such genes are also more densily connected with each other.

8.4 Identification of Protein–Protein Interactions

Protein–protein interactions (PPIs) provide invaluable clues to help elucidate biological processes or cellular functions. Wetlab technologies such as co-affinity purification followed by mass spectrometry [12] may only provide PPI data with both low coverage and accuracy. *In silico* prediction of PPI networks falls into three categories: high-throughput data-based, sequence-based and ortholog-based methods. In the first category, for instance, correlation between mRNA expressions may suggest the existence of a PPI [7]. Sequence-based methods examine for example protein/domain structures [27], gene neighborhoods [21] and gene fusion events [9].[3] In ortholog-based methods, annotation transfer between genomes is the key to detect conserved PPIs—or interologs—*via* gene orthologs [46].

To face the ever-growing accumulation of high-dimensional data, combined with the apparition of new types of data, Xia and collaborators designed a flexible model, able to integrate up to 27 data sets of various data types. This model is

[3] Gene fusion is likely to detect a PPI since two proteins interacting in the genome of one species are more likely to be fused into one single protein in the genome of another species.

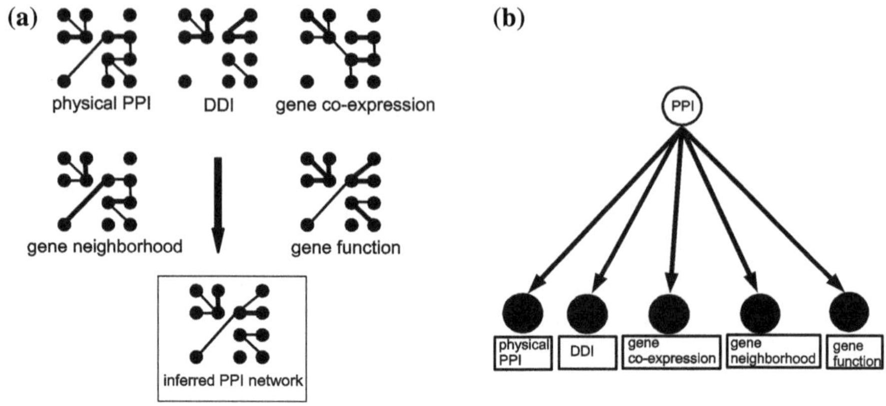

Fig. 8.2 Integration scheme in the approach of Xia et al. [43]. **a** The various data type evidences for the prediction of protein–protein interactions (PPI). **b** The naive Bayes classifier combining these data types. The variable class is binary: *PPI*/no *PPI*. *DDI:* domain-domain interaction.

based on a naive Bayes classifier [43]. A naive Bayes classifier is a Bayesian network whose elementary tree structure consists of a single parent (the class variable) and of children (here, the data types) that are independent of one another, given the class variable (see Fig. 8.2).

8.4.1 Coping with a Vast Spectrum of Heterogeneous Data Types

Before we explain thoroughly how the designed classifier, IntNetDB, integrates various types of data to find PPIs in the human genome, we emphasize the wide heterogeneity of the data types used by this system. In the most recent instance of IntNetDB, Xia and co-workers integrated 27 heterogeneous genomic, proteomic and functional data sets, encompassing 7 data types. For some model organisms (Yeast, Worm, Fruitfly...), annotations about **physical protein–protein interactions** are available. Mapping each such genome interactome to human genome through protein orthologs evidences interologs. **Domain-domain interactions** (DDIs) are known to mediate many protein–protein interactions. Structural domain information databases exist, that provide DDI scores. A DDI score is assigned to the pair of proteins that respectively harbour the two domains. **Gene co-expression** is often a reliable indicator for PPI.

In addition to data describing **gene fusion** and **gene neighborhood**, another data type also depicts gene contexts: **gene co-occurrence** is often indicative of a PPI; provided that the organism genome is fully sequenced, it is recognized that two interacting proteins are likely to be either both absent or both present in this genome [38]. On the other hand, proteins sharing the same biological function are often involved in a PPI [33]. Thus, the **Gene Ontology** (GO) [36] provides supplementary evidence for PPI. Mapping human genes to orthologs in four model organisms (Yeast, Worm, Fruitfly and Mouse) was considered in [43].

Xia and co-workers also integrated two novel data types to help predict PPIs: **phenotypic distances** and **genetic interactions**. RNAi phenotype data have been used to predict PPIs for model organisms: under knock-out experiments, the respective phenotype profiles of interacting proteins tend to be similar. To transfer these phenotype data to human, Xia and collaborators mapped to their human orthologs the genes in the model organisms. Then similarity between the mapped phenotypes was assessed for the pair of genes in human. On the other hand, synthetic genetic analysis is a technology that was used in *Saccharomyces cerevisiae* to provide a global map of genetic interactions. Genetic interactions are recognized as high reliable indicators of PPIs. Xia and co-workers mapped the genetical interaction network of the Yeast model to human interologs.

8.4.2 Heterogeneous Data Integration by Naive Bayes Classifier

In the integrative model, each of the T data types used for the integration contributes an evidence e_i $(1 \leq i \leq T)$ for some given pair of proteins. To assess PPI for this pair of proteins, the likelihood ratio is

$$LR(e_1, \ldots, e_T) = \frac{\mathbb{P}(e_1, \ldots, e_T \mid PPI)}{\mathbb{P}(e_1, \ldots, e_T \mid \neg PPI)},$$

where $\mathbb{P}(e_1, \ldots, e_T \mid H)$ represents the probability that the evidence $(e_1, \ldots, e_T)$ has been observed under hypothesis H. The two alternative hypotheses we are interested in are PPI, the existence of a protein–protein interaction, and $\neg PPI$, the absence of such an interaction. Thus, if the numerator is significantly higher than the denominator, PPI will be assessed. Symmetrically, a low likelihood ratio will support the $\neg PPI$ hypothesis.

Under the assumption that the data sources are independent, the likelihood ratio writes as a product:

$$LR(e_1, \ldots, e_T) = \prod_{i=1}^{T} LR(e_i) = \prod_{i=1}^{T} \frac{\mathbb{P}(e_i \mid PPI)}{\mathbb{P}(e_i \mid \neg PPI)}.$$

The likelihood ratio for data type i providing evidence e_i is calculated from a set of assessed PPIs (positive set) and assessed counter-examples (negative set). The Human Protein Reference Database (HPRD) was used as the positive set; it references 19,438 experimentally verified PPIs for 5,983 proteins [32] (at the time of the integration by Xia et al.). The negative set was generated by Rhodes and coworkers [33]: it spans all pairwise combinations between two sets of proteins located in two different subcellular compartments [plasma membrane (1,397 proteins) and nucleus (2,224 proteins) respectively]. Evaluating both positive and negative sets for each data type provides reference evidences, which allows to compute the desired likelihoods. Discretization into intervals is used for the purpose. Suppose we have to assess $\mathbb{P}(e_i \mid PPI)$ for some pair of proteins, where e_i is the evidence observed for this pair. The positive PPI reference set does not necessarily exhibit a protein pair showing the *exact* evidence e_i. Therefore, the value $\mathbb{P}(e_i \mid PPI)$ is replaced with $\mathbb{P}(I_{positive\ set}(e_i) \mid PPI)$, where $I_{positive\ set}(e_i)$ is an interval around e_i. This interval was obtained from the discretization into intervals of the evidences observed for the protein pairs of the positive PPI reference set. $\mathbb{P}(e_i \mid \neg PPI)$ is calculated similarly.

Care is required when several data sets contributing to the same data type are integrated. In this case, to avoid the bias due to dependence, the maximal likelihood (over the data sets) is retained for the data type.

8.4.3 *Performances*

The literature on alternative methods is vast. Machine learning methods addressing PPI prediction encompass Bayesian classifiers, decision trees, random forests, logistic regression and support vector machines. The reader is referred to [44] (for instance) for a recent overview of existing computational methods.

Two variants of the IntNetDB method were run. The two executions differed by the HPRD version (more than 10,000 newly annotated PPIs), the integration of three novel data types (phenotypic, genetic, gene context) in addition to PPI, GO, gene expression, DDI, and the incorporation of fourteen extra data sets. The comparison showed a drastical gain in coverage, for a similar ratio of true positives to false positives: the reinforced integration increased prediction coverage by five-fold (38,379 PPIs for 5,791 proteins versus 180,010 PPIs for 9,901 proteins). Besides, not only is the depicted probabilistic approach a simple yet efficient system to standardize the contributions of heterogeneous data types *via* likelihoods, it is also a flexible method: the combined likelihood easily supports the integration of any novel type of data.

8.5 Identification of the Regulatory Target Genes of Transcription Factors

A transcription factor (TF) is a protein that controls the expression of its target gene by binding to some specific DNA site located in the regulatory region of the gene. ChIP-chip and ChIP-seq techniques (Chromatin Immuno-Precipitation respectively followed by microarray gene expression measurements and by massively parallel DNA sequencing) provide the genome-wide list of the physical *binding* sites, for a given TF. Exploiting *sequence* similarity to a consensus obtained for already known binding sites is also likely to pinpoint putative binding sites for the TF of interest. Another source of evidence lies in the variation in gene *expression* induced by knock-out or mutation of the gene coding for the TF. However, none of the above data types alone can achieve accurate and complete identification. First, high-throughput data are prone to present high noise level. Besides, ChIP-chip and ChIP-seq technologies only inform about physical DNA-TF interactions. Third, putative binding sites predicted based on sequence similarity with a canonical motif might actually not be bound by the TF of interest. Finally, variations in gene expression are equally observed for genes either directly or indirectly controled by a given TF. In the following, B, S and E will respectively stand for binding, sequence and expression data.

8.5.1 Integrating Multiple Genomic Data Sources with Multiple Gene Networks

To cross evidences from multiple types of genomic data, two categories of methods have been investigated. In regression approaches, where a data type is regressed against another, a large number of observations is required. This is a severe limitation in the case of gene expression microarray data. In mixture model[4] methods, the probabilistic framework allows inference based on the posterior probability of being a target conditional on the multiple data evidences. In the mixture model developed in [39], integration includes only two data types—(S, E) or (S, B) -. This model was further adapted in [30], to jointly handle the three data types B, S and E. So far, the mixture models used assumed conditional independence: conditional on a gene being a target or not, the different data types are independent. Nevertheless, for the pair (B, S), such an hypothesis is not consistent with experimental results: the higher the similarity with the canonical site (S), the higher the binding strength (B).

This section describes the model of Wei and Pan [41]. Therein, the multiple sources of genomic data are modeled through a multivariate normal mixture model, and integration of multiple gene networks with these genomic data types relies on a Markov random field (MRF). Besides relaxing the constraint on conditional independence of genomic data types, another major contribution of Wei and Pan lies in incorporating biological prior knowledge stating that neighboring genes tend to be co-regulated by a TF. Thus, not only does Wei and Pan's approach integrate several genomic data types; it allows to automatically incorporate knowledge from multiple gene networks (see Fig. 8.3). More and more gene

[4] A mixture model is a probabilistic model that represents a population of k groups, with random proportions $\pi_1, \ldots, \pi_k$.

networks are made available, such as protein–protein interaction networks . On the other hand, novel networks may be inferred, such as co-expression gene networks and networks derived from gene ontologies.

Without loss of generality, the presentation here restrains to the three genomic data types *B*, *S* and *E*. It is important to note that regions which associate with TFs according to ChIP-chip and ChIP-seq assays are not determined with single-nucleotide resolution. Wei and Pan computed the binding data (*B*) from ChIP-chip assay data: given two replicates in presence of the antibody appropriate for the TF of interest and given two control replicates, four log2 intensity ratios (LIR) were measured for the four combinations Immuno-Precipitation/control. The binding score B_i of a given gene *i* was computed as the average of the four LIR peaks on the coding region. If there were probes in the intergenic region, B_i was then calculated as the maximum of the average over the coding region and the average over the intergenic region. The sequence data (*S*) used by Wei and Pan was obtained as follows: first, a consensus sequence was produced from 10 known binding sites of the TF of interest; then the genome was scanned with respect to this consensus. Fixing a very low threshold allowed the detection of at least one match per gene. For gene *i*, S_i was calculated as the maximum of all its matching scores.

8.5.2 The Unified MRF-Based Mixture Joint Model

For a specified TF, and given a set of *n* genes, the aim is to estimate whether gene *i* is a target for a factor transcription of interest: $T_i = 1$ denotes a target, otherwise $T_i = 0$. The gene *i* is described by (B_i, S_i, E_i), summarizing observations for B, S and E data. In this approach, conditional normal distributions are described for the observed genomic data (*B*, *S*, *E*):

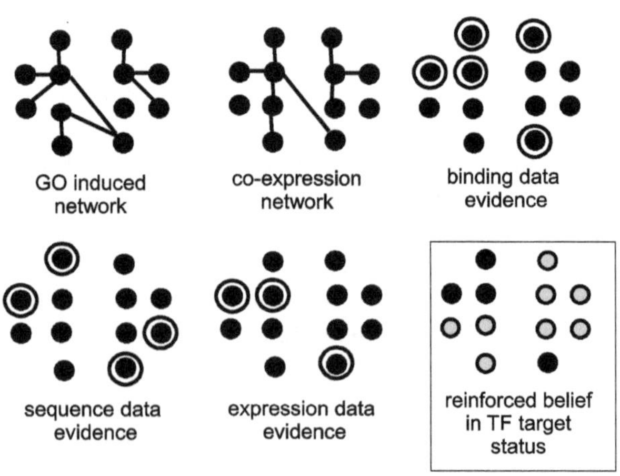

Fig. 8.3 Integration scheme for two networks and three types of data evidence in the approach of Wei and Pan [41]. A *circled node* indicates that strong evidence is observed for the corresponding node (gene). *Bottom right section* reinforced belief in transcription factor (*TF*) target status is indicated in *black*.

$$\mathbb{P}((B_i, S_i, E_i) \mid T_i = j) = \phi((B_i, S_i, E_i); \mu_j, \Sigma_j) = \phi_j(B_i, S_i, E_i),$$

where $j = 0, 1$ and ϕ is a trivariate normal density function of mean μ_j and covariance matrix Σ_j. A mixture model is then depicted as:

$$\mathbb{P}((B_i, S_i, E_i) \mid T_i) = (1 - \pi_1)\ \phi_0(B_i, S_i, E_i) + \pi_1\ \phi_1(B_i, S_i, E_i), \qquad (8.6)$$

where $\pi_1 = \mathbb{P}(T_i = 1)$ is the prior probability of gene i being a target (and, symmetrically, $(1 - \pi_1)$ is the prior probability of gene i not being a target). The model in Eq. 8.6 has to be understood as the "superimposition" of two normal densities, one under the assumption that gene i is a target (ϕ_1), and one under the assumption that gene i is not a target (ϕ_0).

The knowledge from the N_{net} gene networks is incorporated through a Markov random field that rules the states $T_1, \ldots T_n$ of the n genes according to their N_{net} neighborhoods. Wei and Pan formalized an MRF-based mixture joint model (MRF-MJM), which writes as the following logistic regression model:

$$logit\left(\mathbb{P}(T_i = 1 \mid T_{\bigcup_{k=1}^{N_{net}} neigh(i,k)}, \theta)\right) = \gamma + \sum_{k=1}^{N_{net}} \beta_k\ (n_1(i,k) - n_0(i,k))/m(i,k), \qquad (8.7)$$

where $neigh(i, k)$ designates the neighborhood of gene i in network k, parameter θ stands for $(\gamma, \beta_1, \ldots, \beta_{N_{net}})$, $n_j(i, k)$ is the number of genes in $neigh(i, k)$ that have state T_j ($j = 0, 1$) and $m(i, k) = n_0(i, k) + n_1(i, k)$. The contribution of each network k is weighted by the non negative regression coefficient β_k, which therefore measures how informative network k is. In Eq. 8.7, conditioning by $T_{\bigcup_{k=1}^{N_{net}} neigh(i,k)}$ indicates that the TF target status of gene i depends on the statuses of all its neighbor genes, considered over all the N_{net} networks.

In this case, estimating the likelihood is intractable. In this framework, a tractable approximation to the joint distribution, the pseudolikelihood [1], is used instead. Tractability is ensured by the conditional independence assumption which leads to the following factorization:

$$\begin{aligned}\mathbb{P}(T) \simeq L_{pseudo}(T, \theta) &= \prod_{i=1}^{n} \mathbb{P}(T_i \mid T_{\bigcup_{k=1}^{N_{net}} neigh(i,k)}, \theta) \\ &= \prod_{i=1}^{n} \frac{exp(\gamma + \sum_{k=1}^{N_{net}} \beta_k\ (n_1(i,k) - n_0(i,k))/m(i,k))}{1 + exp(\gamma + \sum_{k=1}^{N_{net}} \beta_k\ (n_1(i,k) - n_0(i,k))/m(i,k))}.\end{aligned} \qquad (8.8)$$

Besides the factorization, the transition from Eq. 8.7 to 8.8 uses the conversion $y = logit(x) = log\ \left(\frac{x}{1-x}\right) \Rightarrow x = \frac{e^y}{e^y+1}$.

The Bayes theorem states that $\mathbb{P}(T \mid (B,S,E)) \propto \mathbb{P}((B,S,E) \mid T)\ \mathbb{P}(T)$. The two ingredients on the right hand side are available from Eqs. 8.6 and 8.8, respectively. An MCMC is used to estimate the posterior probability of genes being targets of a specified TF.

8.5.3 Performances

A tremendous variety of alternative computational approaches are available. Some pointers to general surveys are provided in [29, p. 584]. In particular, Elnitski et al. wrote a summary on the synergism between *in silico*, *in vitro* and *in vivo* identification of TF binding sites [8]. On the other hand, the influential role of data integration is stressed in the surveys provided in [17, 28].

The MRF-MJM approach was evaluated with the LexA transcription factor of *Escherichia coli*. It was first noticed that allowing conditional dependence by assuming a general conditional variance structure in the MRF-MJM model does not increase the predictive power over assuming conditional independence. However, as binding data and sequence data are highly correlated for target genes, this result appears to go against intuition. It might be explained by moderate predictive power of sequence data and a simpler model in the assumption of conditional independence. All subsequent analyses were then run incorrectly assuming conditional independence.

Wei and Pan tested six different integration schemes. Six instances of the MRF-MJM approach, including simplified ones, were run: $(E;N_{CoE})$, $(E;N_{GO})$, $(E;N_{CoE}+N_{GO})$, where N_{CoE} and N_{GO} are gene networks respectively derived from gene co-expression and a gene ontology (GO), and the three previous instances with the full set of genomic data (B,S,E) instead of E. Besides, instances of the standard mixture model (SMM), which considers a single genomic data type, were also run for comparison: $SMM(B)$, $SMM(S)$, $SMM(E)$. The genes were ranked according to their posterior probabilities. The variation in the ranking across these instances was studied for the genes supported by experimental evidence or annotated with "strong evidence" in the RegulonDB database [11].

It was confirmed using ROC curves that mixed integration of both networks and various genomic data types greatly improves over considering a single genomic data type alone. Besides, in a mixed scheme, the improvement is less drastic when increasing the number of genomic data types or when increasing the number of networks. The GO-derived network constantly showed a β coefficient lower than the co-expression network's: it is explained by a higher connectivity of the GO network, which entails that a target and a non target genes are more likely to be neighbors in this network.

8.6 Inference of Causal Relationships Among Phenotypes via Integration of QTL Genotypes

A quantitative phenotype (or trait) is defined as any physical, physiological or biochemical quantitative feature that may be observed for organisms. Quantitative trait loci (QTL) mapping aims at identifying the genomic regions, or QTLs, where genotype variation is correlated with phenotype variation. Deciphering the causal relationships among *expression* traits involved in the same biological pathways—and therefore correlated—is a current research topic. To this aim, the identification of the eQTLs (expression QTLs) causal to each phenotype is of prime importance. In the following, we will denote by genetic architecture (GA) of a given phenotype the locations and effects of its (directly) causal QTLs. Conversely, GA inference has to benefit from the information borne by the network that links the phenotypes. Though, standard QTL mapping merely addresses one single trait at a time, not considering a possible causal network structure among traits. Thus, QTLs that exert a direct effect on the trait under study cannot be distinguished from QTLs with an indirect effect (see Fig. 8.4a). To reconstruct a causal phenotype network (CPN), several approaches in the literature include QTLs in a probabilistic framework. However, the common feature of these approaches lies in that GA inference and CPN reconstruction are conducted separately [3, 34, 47]. In general, the GA is first inferred, to further help the determination of the CPN. In the QTLnet approach, Chaibub Neto and co-authors pioneered the principle of joint inference of CPN and GA [4].

8.6.1 Joint Inference of Causal Phenotype Network and Genetic Architecture

Chaibub Neto et al. showed that performing the mapping analysis of a phenotype conditional on its parents in the CPN is the way to avoid detecting QTLs with indirect effects on this phenotype as directly causal QTLs. Namely, whereas standard mapping analysis would test the dependence between phenotype φ_1 and QTL candidate Q_1 ($\varphi_1 \perp\!\!\!\perp Q_1$), *conditional mapping* assesses or invalidates the dependence relation $\varphi_1 \perp\!\!\!\perp Q_1 | \mathbf{Pa}(\varphi_1)$ where $Pa(\varphi_1)$ is the set of parents of φ_1 in the CPN. As the CPN is itself unknown, the QTLNet approach jointly infers the CPN and the GA: the procedure iterates a process where updating the CPN alternates with updating the GA. Thus, GA inference will benefit from information on the CPN. The core idea is to learn a Bayesian network whose structure coincides with the candidate CPN, using the current information available about causal QTLs. It has to be noted that the central dogma of biology constrains unidirectionality for causality, from QTL to phenotype: arcs $\varphi \rightarrow Q$ are not allowed.

Adding information about causal QTLs is crucial to distinguish between candidate phenotype networks, when learning a phenotype network. The network

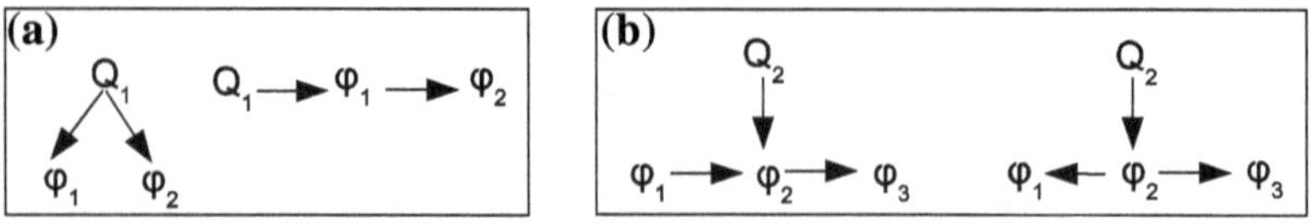

Fig. 8.4 Disambiguisation of causal relationships. **a** *Left* direct effect of QTL Q_1 on phenotype φ_2; *right* indirect effet of Q_1 on φ_2. In both cases, $\varphi_2 \not\perp\!\!\!\perp Q_1$. **b** The two models have the same joint probability $\mathbb{P}(\varphi_1, \varphi_2, \varphi_3)$ but have different conditional probabilities $\mathbb{P}(\varphi_1, \varphi_2, \varphi_3 \mid Q_2)$ given the QTL data

which best fits the data is that which maximizes some given criterion. In a probabilistic framework without QTL data integration, this criterion would rely on the joint probability $\mathbb{P}(\varphi_1, \ldots, \varphi_n)$. With QTL integration, the probability to be taken into account is $\mathbb{P}(\varphi_1, \ldots, \varphi_n \mid GA)$ where GA stands for the r QTLs available: $Q_1, \ldots, Q_r$. Let us consider a toy-example where the networks (1) $\varphi_1 \rightarrow \varphi_2 \rightarrow \varphi_3$ and (2) $\varphi_1 \leftarrow \varphi_2 \rightarrow \varphi_3$ cannot be distinguished without QTL data integration since they have the same joint probability $\mathbb{P}(\varphi_1, \varphi_2, \varphi_3)$.[5] We now incorporate QTL knowledge as Q_2 affecting φ_2 but neither φ_1 nor φ_3 directly and obtain two mixed models (see Fig. 8.4b). Then, the conditional probabilities of the two networks are: $\mathbb{P}_{(1)}(\varphi_1, \varphi_2, \varphi_3 \mid Q_2) = \mathbb{P}(Q_2)\ \mathbb{P}(\varphi_1)\ \ \mathbb{P}(\varphi_2 \mid Q_2, \varphi_1)\ \mathbb{P}(\varphi_3 \mid \varphi_2)$ and $\mathbb{P}_{(2)}(\varphi_1, \varphi_2, \varphi_3 \mid Q_2) = \mathbb{P}(Q_2)\ \mathbb{P}(\varphi_2 \mid Q_2)\ \mathbb{P}(\varphi_1 \mid \varphi_2)\ \mathbb{P}(\varphi_3 \mid \varphi_2)$. In the general case, the previous conditional probabilities are not equal.

8.6.2 The Mixed Model

To model continuous phenotypes that are involved in a causal phenotype network while also being under the dependence of discrete QTLs, a conditional Gaussian regression model is used: conditional on the genotypes and, possibly, covariates, the phenotypes follow a multivariate normal distribution.

Given n individuals, t phenotypes, let $\varphi = (\varphi_1, \ldots, \varphi_n)^T$ represent all phenotype values, with $\varphi_i = (\varphi_{1i}, \ldots, \varphi_{ti})^T$ representing the t phenotype values for individual i. Let $\epsilon_i = (\epsilon_{1i}, \ldots, \epsilon_{ti})^T$ be independent normal error terms. The regression model for the phenotype p of individual i writes:

$$\varphi_{pi} = \mu^*_{pi} + \sum_{v \in Pa(\varphi_p)} \beta_{pv}\ \varphi_{vi} + \epsilon_{pi}, \epsilon_{pi} \sim \mathcal{N}(0, \sigma_p^2). \tag{8.9}$$

The genetic contribution describes the effects of QTLs and possibly covariates: $\mu^*_{pi} = \mu_p + X_{pi}\ \theta_p$, where μ_p is the overall mean for phenotype p, X_{pi} represents the

[5] $\mathbb{P}_{(1)}(\varphi_1, \varphi_2, \varphi_3) = \mathbb{P}(\varphi_1)\ \mathbb{P}(\varphi_2 \mid \varphi_1)\ \mathbb{P}(\varphi_3 \mid \varphi_2)$ and $\mathbb{P}_{(2)}(\varphi_1, \varphi_2, \varphi_3) = \mathbb{P}(\varphi_2)\ \mathbb{P}(\varphi_1 \mid \varphi_2)\ \mathbb{P}(\varphi_3 \mid \varphi_2)$. Equality is assessed from the Bayes theorem.

row vector of genetic effect predictors derived from the QTL genotypes along with any covariates, and θ_p is a column vector of all genetic effects defining the genetic architecture of phenotype p augmented with any covariates. In the phenotypic contribution (second term of Eq. 8.9), $Pa(\varphi_p)$ designates the set of parents of phenotype p in the phenotype network and β_{pv} models the effect of parent phenotype v on phenotype p.

8.6.3 Causal Phenotype Network Reconstruction

Since the graph space grows super-exponentially with the number of phenotypes, reconstructing a CPN requires a heuristic. An MCMC is implemented, that combines sampling over CPN structures and QTL mapping. However, conceptually, a mixed structure $G = G_\varphi \cup GA$, is considered, which is the CPN G_φ augmented with the genetic architecture GA connecting QTLs to phenotypes (see Fig. 8.5). The posterior probability of a candidate G_φ is estimated as explained below.

From Eq. 8.9, we know that $\mathbb{P}(\varphi_{pi} \mid G_\varphi, GA, \gamma)$ is $\mathcal{N}(\mu^*_{pi} + \sum_{v \in Pa(\varphi_p)} \beta_{pv}\varphi_{vi}, \sigma_p^2)$,[6] where γ stands for the set of parameters of the mixed model (i.e. the coefficients β). Under the assumption of independence between the n individuals, the likelihood of the candidate CPN factorizes as:

$$\mathbb{P}(\varphi \mid G_\varphi, GA, \gamma) = \prod_{i=1}^{n} \prod_{p=1}^{t} \mathbb{P}(\varphi_{pi} \mid G_\varphi, GA, \gamma).$$

In this case, it is straightforward to compute the marginal likelihood by integrating the previous expression with respect to γ:

$$\mathbb{P}(\varphi \mid G_\varphi, GA) = \int_\gamma \mathbb{P}(\varphi \mid G_\varphi, GA, \gamma)\ \mathbb{P}(\gamma \mid G_\varphi, GA)\ d\gamma.$$

Finally, the posterior probability of structure G_φ conditional on the data may be computed from:

$$\mathbb{P}(G_\varphi \mid \varphi, GA) \propto \mathbb{P}(\varphi \mid G_\varphi, GA)\ \mathbb{P}(G_\varphi),$$

where $P(G_\varphi)$ is a prior on the CPNs.

Thus, integrating knowledge about QTLs actually modifies the likelihood landscape for the search space of G_φ structures.

To navigate in this search space, three moves are implemented in the MCMC scheme of Chaibub Neto et al.: addition of a directed edge, removal or direction

---

[6] If $X = y + E$, with $E \sim \mathcal{N}(0, \sigma^2)$, then $X \sim \mathcal{N}(y, \sigma^2)$.

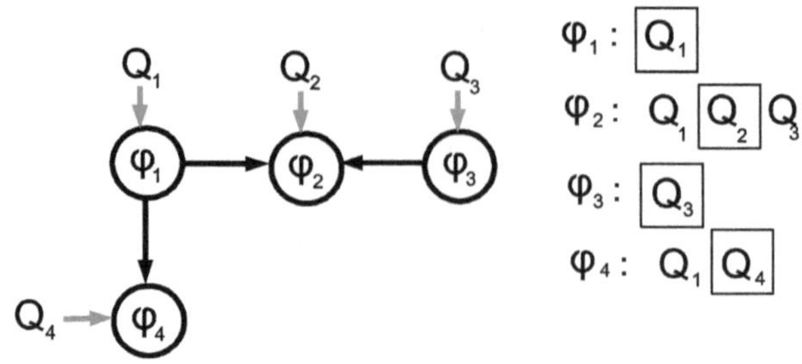

Fig. 8.5 The mixed model in the approach of Chaibub Neto et al. [4]. **a** The genetic architecture defines the QTL mapping (*arrows* in *light grey*); the causal network defines the dependences between the phenotypes (transcripts) (*arrows* in *dark*). **b** Comparison of the genetic architectures inferred without and with conditioning (in the latter case, the QTLs are framed) (see Sect. 8.6.1)

reversal. Subsequent to a move, (conditional) QTL mapping is replayed for those phenotypes whose set of parent nodes was modified by the move. Finally, a posterior probability for the causal relationship $\varphi_i \rightarrow \varphi_j$ $(1 \leq i, j \leq t)$ involving each pair of phenotypes is assessed through *Bayesian model averaging*: for each directed edge $\varphi_i \rightarrow \varphi_j$, the posterior probability is estimated as the frequence of occurrence observed over all the models sampled by the MCMC process.

8.6.4 Performances

First, 1,000 tests were performed based on simulated data generated under two conditions: respectively weak and strong dependences between the phenotypes and their eQTLs. The genetic architectures produced were compared with those obtained through standard QTL mapping. Conditional mapping (see first paragraph in Sect. 8.6.1) revealed the true architecture in both conditions. To estimate the quality of the phenotype network inference, the authors measured the frequency that the posterior probability of the true network was the highest, second highest, *etc*. Under the strong dependence condition, the true network is identified as the best one in 84 % of the cases. The results are more subdued under the weak dependence condition.

The QTLnet method was then used on real data (132 mice of a F2 intercross, 3,421 transcripts, 1,065 markers), to derive the causal phenotype network relative to 14 highly correlated transcripts. A consensus network was constructed through Bayesian model averaging. Interestingly, this consensus network suggests a key role of one of the transcripts in the regulation of the other transcripts in the phenotype network.

8.7 Prediction of Protein Function Through GO-Enriched Networks of Multiple Related Species

In Sect. 8.5, a gene ontology (GO) was used to derive a functional coupling gene network, to enhance the identification of transcription factor targets. Therein, a preprocessing step merely derived a gene network, based on some similarity measure in the ontology. In the present section, we outline an approach which benefits from GO knowledge on the fly. As the most developed biological ontology is the Gene Ontology [36], it is not surprising that this approach addresses the prediction of protein function.

Improving the coverage and accuracy for functional annotation of proteins is an active field in post-genomics research. On the one hand, only labor intensive small-scale experiments are able to provide direct evidence about the functions of proteins such as energy and RNA metabolism, signal transduction, translation initiation, enzymatic catalysis and immune response. In contrast, though numerous high-throughput technologies allow large-scale experimental investigations, the various types of molecular data but only yield indirect clues about protein function. To reach the objectives of coverage and accuracy, much is expected from computational methods.

Established prediction methods use sequence or structure similarity to transfer functional annotation from protein to protein [22]. However, it is well known that sequence similarity does not obligatorily entail functional identity. More reliable evidence is derived from indirect information provided by the biological context of the protein. Such contextual information includes physical protein–protein interactions (PPI), genetic interactions and co-expression of the genes coding for the proteins. These contextual data are commonly represented as networks. Thus, a wide category of methods predicts the function of a protein from the known functions of its neighbors in the network [2, 14, 45]. Besides, incorporation of heterogeneous data has been proven useful to increase the power of automated predictive systems [26].

Probabilistic graphical models offer an appealing framework to propagate functional annotations through neighborhoods; this explains that approaches based on these models are not new to protein function inference (e.g. [6, 19, 26]). However, severe limitations hamper these approaches in the (frequent) case of proteins that are isolated in the network or whose neighborhood is poorly annotated. Refined GO-based strategies have been proposed to overcome these issues. Amongst them, the probabilistic approach of Mitrofanova and collaborators combines random Markov models and Bayesian networks into a single model [23].

In classical approaches, probabilistic inference relies on partial knowledge of functional annotations to discover the missing functions by passing on and handling uncertain information over a large network. For instance, this network may be derived from knowledge on physical interactions (PPIs). One of the original concepts of Mitrofanova and co-workers' model lies in connecting the networks of two (or more) related species into a single computational model. The rationale behind this approach exploits the fact that in most cases, proteins of different related species that share high similarity—orthologs—exerted the same established function before the speciation event. The second original concept of the approach described is the direct integration of an ontology or rather, of a sub-ontology (GOS), into the graphical model. This integration allows the simultaneous prediction for the multiple functional categories—or terms—described by the GOS. In the combined model, each protein is represented by its own GOS. As a consequence, during function inference, not only is the information passed between protein neighbors within a species, information also percolates within the GOS. Moreover, due to inter-species connections between orthologs, such information is diffused in an enlarged network.

8.7.1 The GO-Enriched Intra-Species Model

For the sake of a progressive exposition, we first present a model deprived of inter-species relationships. In the model, each protein is represented by a Bayesian network whose structure is a replicate of the GO sub-ontology (GOS) of interest (see Fig. 8.6a). Each protein has its own annotation (positive, negative, unknown) for each of the GOS terms. A positive annotation means that the protein has the function represented by the GOS term. The final objective of the probabilistic inference is to assign an annotation (positive/negative) to each term (GOS node) labeled unknown in the combined model. The GOS is a directed acyclic graph where the relationship between child c and parent p may be "IS A" or "IS PART OF". The GOS information is naturally modeled as a Bayesian network (BN). The so-called *true-path rule* for gene ontologies requires that if a protein i is positively annotated at a child node t (denoted by $x_i^t = +$), then it must also be at all the ancestor nodes of this child. Consequently, positive annotations may be expanded up within a GOS whereas negative annotations are expanded down if all the parent terms of a child term are annotated negative. It follows that conditional probabilities $\mathbb{P}(x_i^t = + \mid pa_{it})$ and $\mathbb{P}(x_i^t = - \mid pa_{it})$ need be estimated only if one parent at least is annotated positive within a possible assignment pa_{it} of the parents (for instance, $pa_{it} = (+, +, -)$ in the case when node t has three parents in the GOS).

On the other hand, a pairwise Markov random field (MRF) is used to encode connections between the proteins, based on some similarity measure between the proteins. Such measures may be derived from PPIs or orthology (i.e. sequence similarity). In the model resulting from GOS and MRF combination, a potential function, ψ^{intra}, is defined; this potential expresses the probability of joint annotation of two proteins i and j at a GOS term t, conditional on their being similar. In the case of a PPI-based measure, similar proteins are defined as interacting proteins: then, the probabilities $\psi^{intra}(x_i^t,\ x_j^t) = \mathbb{P}(x_i^t, x_j^t \mid interaction)$, with $x_i^t, x_j^t \in \{+, -\}$ are estimated from a training set. In the case of a sequence similarity-based measure, a potential is derived from a pairwise normalized BLAST score s_B:

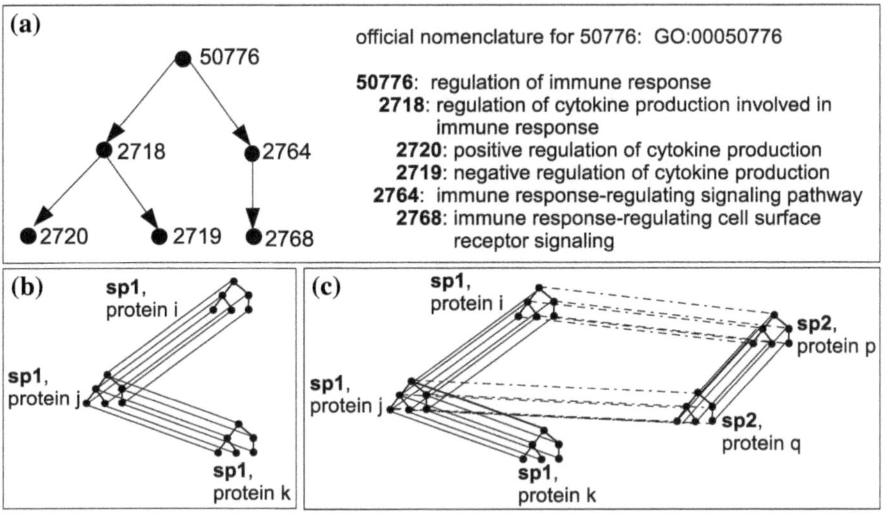

Fig. 8.6 The combined model of Mitrofanova et al. [23] for protein function prediction. **a** The Gene Ontology (GO) substructure. **b** The GO-enriched intra-species model. **c** The combined model obtained through inter-species homology (Two species are considered: sp1 and sp2)

$\psi^{intra}(+,+) = \psi^{intra}(-,-) = s_B(i,j);\quad \psi^{intra}(+,-) = \psi^{intra}(-,+) = 1 - s_B(i,j)$. If both similarity measures are available for a given pair of proteins, thus defining two potentials, the resulting potential is defined as the product of the two former.

The knowledge about the annotation information of protein i, at GOS term t is modeled through function ϕ: $\phi(+) = 1; \phi(-) = 0$ for a positive annotation; $\phi(-) = 1; \phi(+) = 0$ for a negative annotation; equiprobability for an unkown annotation ($\phi(?) = 0.5$).

The MRF and the GO-based BNs are combined into a single hybrid model [18]—(see Fig. 8.6b). Based on the material above defined, the joint distribution of the functional term annotations (X_i^t) over the set of proteins $\mathscr{P}$ is defined as a pairwise MRF distribution (see Eq. 8.2), whose statement is simplified as follows for the sake of conciseness:

$$\mathbb{P}(\{x_i^t\}_{t\in\mathscr{S},i\in\mathscr{P}}) = \frac{1}{Z}\prod\nolimits_{t\in\mathscr{S}} \prod_{i\in\mathscr{P}} \phi(x_i^t) \prod_{i,j\in edges(MRF(\mathscr{P}))} \psi^{intra}(x_i^t, x_j^t) \prod_{i\in\mathscr{P}} (x_i^t \mid pa_{it}). \tag{8.10}$$

$\mathscr{S}$ is the sub-ontology of interest and Z is the so-called normalizing constant (see Definition 2, Sect. 8.2). In the above joint distribution, it is easy to identify the contribution of the Markov random field defined by the similarity relation between proteins, and the contribution of the Bayesian networks. The flow of information about annotation is propagated through the hybrid model using a message-passing mechanism tailored for such hybrid models.

8.7.2 The GO-Enriched Inter-Species Model

The extension to the inter-species case is straightforward when two related species are considered. This time, when sequence similarity is ascertained for protein i in the first species and protein j in the second species, a corresponding potential function ψ^{inter} is defined. Similarly to the scheme described by Eq. 8.10, the combined model is merely augmented with undirected edges connecting the identical GOS term nodes of i and j proteins (see Fig. 8.6c). The joint distribution is readily extended as follows: a second intra-species factor is added, $\prod_{i\in\mathscr{P}'} \phi(x_i^t) \prod_{i,j\in edges(MRF(\mathscr{P}'))} \psi^{intra}(x_i^t, x_j^t)$ where $\mathscr{P}'$ is the set of proteins for the second species; an inter-species factor is also added, that accounts for the (valid) edges connecting some protein in $\mathscr{P}$ to some other similar protein in $\mathscr{P}'$: $\prod_{i,j\in edges(MRF(\mathscr{P}\cup\mathscr{P}'))} \psi^{inter}(x_i^t, x_j^t)$.

8.7.3 Performances

Mitrofanova and collaborators performed tests on Yeast and Fly. Respectively 6,008 and 12,199 proteins were considered for Yeast and Fly species. Various tests were performed based on (1) executions (S) of the approach without integration of the gene ontology, that is single-term prediction, and (2) runs (GO) of the approach with the integration of the gene ontology. Besides, annotation transfer by similarity was considered either within a single species (1), or within two species (2). We will denote S1, S2, GO1 and GO2 these four kinds of tests. Figure 8.7 recapitulates the experimental protocole.

The comparison S1 versus GO1 is meant to evaluate the impact of using a gene ontology. In this baseline test, predictions were then compared for a single term. The improvement in the prediction is outstanding in all cases: a gain of 26 % (from

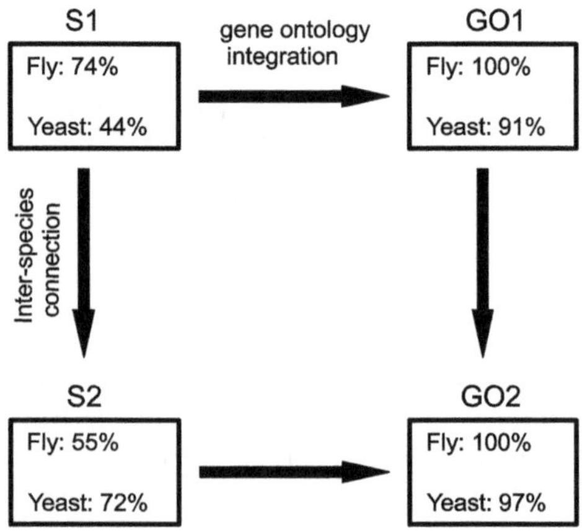

Fig. 8.7 Evaluation of the impact of gene ontology integration and inter-species connection on accuracy, in Mitrofanova et al.'s method. S1 denotes a run with intra-species connection only, whereas S2 indicates intra- and inter-species connection. GO_1 indicates the integration of gene ontology knowledge to the basic scheme S_1 (and symmetrically for GO_2 and S_2)

74 to 100 %) is observed for the Fly species; an increase of 47 % (from 44 to 91 %) is observed for the Yeast species. Thus, in the case of the Fly species, the integration of GO knowledge suffices to produce the accuracy of 100 %.

The comparison S1 versus S2 aims at evaluating the influence of annotation transfer between genomes, through inter-species connection. Mitrofanova and collaborators performed 5-cross validation for the Fly, Yeast and combined Yeast-Fly networks. The results are contrasted: an under-performance is obtained in the case of the Fly species, for which the accuracy decreases by 19 % (from 74 to 55 %); a gain of 28 % (from 44 to 72 %) is observed for the Yeast species. Thus, inter-species connection alone may be counter-productive (Fly). If a gain is observed through inter-species connection, it is more subdued than the gain obtained through integration of a gene ontology (Yeast).

The aim of comparing S2 against GO2 is to measure the impact of the integration of GO knowledge in presence of inter-species connection. This time, in the case of the Fly species, inter-species connection does not interfere to lower the performance, which confirms the prominent role of the GO integration (55 % to 100 %). A gain of 25 % (from 72 to 97 %) is observed for the Yeast species (to be compared to the increase from 44 to 91 % without inter-species connection).

A significant gain of 8 % (from 91 to 97 %) is thus observed for the Yeast species in the GO1 versus GO2 test.

The main conclusion is that the GO integration exerts the most influential role. Inter-species connection may perform worse than merely considering a single genome. However, it is always beneficial to integrate both GO knowledge and inter-species connection. Yeast species shows more substantial improvements compared to Fly species: this may be explained by the higher quality of Fly data and hence better neighborhoods for the Fly proteins. Annotation transfer is enhanced through two independent principles: simultaneous consideration of multiple but related functional GO categories, higher connectivity due to orthology or PPI knowledge. Expanded protein coverage is another observed advantage.

In the spirit of the comparison S2 versus GO2, Mitrofanova and collaborators also compared their full approach (GO2: GO integration and inter-species connection) to the method of Naria et al. [26] which can be seen as a variant of S2. The method of Naria et al. relies on a probabilistic Bayesian framework that integrates networks (e.g. PPI and/or expression networks) with categorical features (i.e. presence of protein domains, knockout phenotype (e.g. "starvation sensitivity") and cellular location categories). The case of lack of information about categorical features is taken into account in [26], which thus allows the comparison. Besides, for comparability, both PPI and sequence similarity were used to build the networks input to the two methods. The method of Mitrofanova et al. improves over that of Naria et al.: for the Fly species, the accuracies observed are respectively 100 and 45 %; for the Yeast species, the accuracies are 97 and 50 %. Again, GO integration is shown to play a more prominent role than inter-species connection. This improved performance can be attributed to the increased connectivity endowered in the GO structure. However, it has to be noted that the S2 executions of Mitrofanova and co-workers' method already outperformed the (S2) runs of Naria et al.'s approach: 55 versus 45 % for the Fly species, and 72 versus 50 % for the Yeast species. It is difficult to speculate on the reasons why annotation information percolates more efficiently in the probabilistic model of [23] (without GO integration) than in that of [26]. Unfortunately, no common types of results are available (such as accuracy, false positive rate, or number of true positives) that could allow the comparison of the methods both at full integration level (GO2 for Mitrofanova et al.'s method, and integration of categorical features for Naria et al.'s approach.).

Finally, with a Gene Ontology subtree of size 8, the running times observed for each five-cross validation round on Yeast, Fly and Yeast-Fly models were 35, 59 and 28 mn on average on a standard personal computer. The third low execution time is explained by faster convergence in the combined network, probably due to denser sources of evidence.

8.8 Discussion and Future Directions

In this chapter, we have presented different approaches based on probabilistic graphical models, to illustrate the use of this class of models as an integrative framework for systems biology. In particular, various forms of Markov random fields were described, that were used to model the propensity to share a common state for neighbor nodes in a single network or in multiple networks. For instance, in the illustration devoted to genetic association study (GAS), the MRF models a single network—a biological pathway—and the state accounts for association with the disease.

One of the simplest Bayesian networks that can be imagined, the naive Bayes classifier, also represents one of the most flexible tools to integrate multiple data types. In this line, the method presented here to detect protein–protein interactions (PPIs) assigns equal weights to the genomic data types. However, a limitation lies in that the positive and negative sets of examples and counter-examples requested by this simple method do not necessarily benefit from equal covers across the data types.

Enhancement through mixed integration of genomic data types and gene networks is shown for the identification of the target genes of transcription factors (TF). It was emphasized that the key to improvement is much more mixing data sources than multiplying either the number of genomic data types or the number of networks. In the probabilistic model, a single Markov random field integrates and weights the contributions of the gene networks. Neighbor genes therein are expected to share a common state (target or non target). In the global model, genomic data types are integrated through a prior distribution. In the GAS application, the prior distribution accounted for the integration of pathway knowledge.

The gene networks mentioned above provide *qualitative* knowledge to rely upon. This time, for causal phenotype network (CPN) reconstruction, a conditional Gaussian regression model was used to integrate *quantitative* characteristics (continuous phenotypes) and *qualitative* assumptions (latent relationships between the phenotypes). In contrast with the preceding approaches, prior knowledge—consisting in the genetic architecture (GA)—is not fixed from the start but is instead refined throughout the CPN inference procedure: feedback from the most recent incumbent CPN offers opportunity to update the GA and *vice versa*.

The second (PPIs) and third (TF target genes) models presented both rely on shared functional annotation. Raw data is used in the second model whereas the third one may incorporate a gene network induced from a gene ontology. In contrast, accounting for ontological knowledge thoroughly impacts the statistical inference scheme in the last approach presented, that addresses protein function prediction. This approach combines ontology replication with intra- and -inter-species homology knowledge. Again, as for the GAS illustration above, a Markov random field (MRF) is built from a known structure, here a network connecting similar proteins. Similarity is assessed from PPI knowledge as well as intra- and inter-species homology. Unlike the GAS approach, neighbors in the network tend to share a common hierarchy of function annotations instead of a single variable. The originality of the mixed model arises from the expansion of the protein nodes of the MRF into Bayesian networks (BNs), each replicating the gene ontology substructure. The completion by links between identical term nodes in similar protein meta-nodes provides a highly connected network. Thus boosted information propagation is expected.

Among the five integrative methods reviewed, the one addressing PPI prediction and the one predicting TF gene targets are perhaps the most exemplary in that they take advantage of various genomic data and/or networks. In the case of the TF gene target application, integrating genomic data and networks outstandingly improves the results but then, increasing the number of genomic data types or networks does not provide much improvement. On the other hand, the illustration on the prediction of protein functions reveals the prominent role of gene ontology (GO) knowledge. GO integration exerts the most influential role. However, in this context, it is always beneficial to integrate both GO knowledge and inter-species connection.

The previous paragraph raises in particular the question on the possible dependence of the various data sources and on how this dependence is ignored or modeled. In the illustration of the PPI detection, the naives Bayes classifier requires independence of the data types conditional on the state variable $(PPI/\neg PPI)$. Robustness to deviation from this rule was not evaluated in this framework. However, in the case of another model and for another application (identification of TF target genes), the conclusion was that the simplifying assumption of conditional independence does not decrease performance. The PPI detection illustrates here a case where multiple data sets may be examined within a common data type. Retaining the empirical maximum likelihood computed over all data sets of the same data type avoids the dependence bias for this type. Again, an open question remains the significance of a high likelihood obtained for some data type if there are cover biases between data types, in terms of positive and negative sets.

Further progress in the field will mainly depend on improving implementations and allowing actual flexibility. For instance, MCMC implementations rely on hyperparameters whose tuning can hardly be delegated to the end-user. Besides, it is worth examining how to incorporate additional biological knowledge in priors, as in the case of causal phenotype network inference. The reported advantages of probabilistic graphical networks in promoting highly integrative approaches combining various heterogeneous data sources may be sometimes offset by the computational burden. From the theoretical viewpoint, for all models presented here, generalization to multiple data types is straightforward. Mitrofanova et al.'s method readily generalizes to more than two species but scalability might be an issue. The method designed to predict protein functions was shown tractable for gene ontology substructures of size below 20, which might appear insufficient to some end-users and therefore requires further work. The next-generation sequencing era is also that of grid and cloud computing. For example, three of the models presented here use an MCMC scheme. MCMCs are amenable to distributed implementations. As more data and more data types will become available, adding a novel data type should be automatically handled by the models' implementations. Therefore, the dissemination in the biological community of integrated PGM-based approaches also implies that service-oriented integration accompanies theoretical developments.

Acknowledgments The author wishes to thank the anonymous reviewer for constructive comments on her manuscript, and feedback most helpful to produce the final version.

References

1. Besag J (1986) On the statistical analysis of dirty pictures. J Roy Statist Soc Ser B 48:259–302
2. Carroll S, Pavlovic V (2006) Protein classification using probabilistic chain graphs and the Gene Ontology structure. Bioinformatics 22(15):1871–1878
3. Chaibub Neto E, Ferrara CT, Attie AD, Yandell BS (2008) Inferring causal phenotype networks from segregating populations. Genetics 179(2):1089–1100. doi:10.1534/genetics.107.085167
4. Chaibub Neto E, Keller MP, Attie AD, Yandell BS (2010) Causal graphical models in systems genetics: a unified framework for joint inference of causal network and genetic architecture for correlated phenotypes. Ann Appl Stat 4(1):320–339
5. Chen M, Cho J, Zhao H (2011) Incorporating biological pathways via a Markov random field model in genome-wide association studies. PLOS Genet 7(4):e1001353. doi:10.1371/journal.pgen.1001353
6. Deng M, Chen T, Sun F (2003) An integrated probabilistic model for functional prediction of proteins. In: Proceedings of the seventh annual international conference on research in computational molecular biology (RECOMb), pp 95–103
7. Eisen MB, Spellman PT, Brown PO, Botstein D (1998) Cluster analysis and display of genome-wide expression patterns. Proc Natl Acad Sci USA 95(25):14863–14868
8. Elnitski L, Jin VX, Farnham PJ, Jones SJ (2006) Locating mammalian transcription factor binding sites: a survey of computational and experimental techniques. Genome Res 16(12):1455–1464
9. Enright AJ, Iliopoulos I, Kyripides NC, Ouzounis CA (1999) Protein interaction maps for complete genomes based on gene fusion events. Nature 402:86–90

10. Franke L, van Bakel H, Fokkens L, de Jong ED, Egmont-Petersen M, Wijmenga C (2006) Reconstruction of a functional human gene network, with an application for prioritizing positional candidate genes. Am J Hum Gen 78(6):1011–1025
11. Gama-Castro S, Jimánez-Jacinto V, Peralta-Gil M et al (2008) RegulonDB (version 6.0): Gene regulation model of Escherichia coli K-12 beyond transcription, active (experimental) annotated promoters and Textpresso navigation. Nucleic Acids Res 36:D120–D124. doi:10.1093/nar/gkm994
12. Gavin AC, Bosche M, Krause R, Grandi P, Marzioch M, Bauer A, Schultz J, Rick JM, Michon AM, Cruciat CM et al (2002) Functional organization of the yeast proteome by systematic analysis of protein complexes. Nature 415(6868):141–147
13. Hutz JE, Kraja AT, McLeod HL, Province MA (2008) CANDID: a flexible method for prioritizing candidate genes for complex human traits. Genet Epidemiol 32(8):779–790
14. Karaoz U, Murali T, Letovsky S, Zheng Y, Ding C, Cantor CR, Kasif S (2004) Whole-genome annotation by using evidence integration in functional-linkage networks. Proc Natl Acad Sci USA 101:2888–2893
15. Kindermann R, Snell JL (1980) Markov random fields and their applications. American Mathematical Society
16. Köhler S, Bauer S, Horn D, Robinson PN (2008) Walking the interactome for prioritization of candidate disease genes. Am J Hum Genet 82:949–958
17. Ladunga I (2010) An overview of the computational analyses and discovery of transcription factor binding sites. Methods Mol Biol 674:1–22
18. Lauritzen SL (1996) Graphical models. Oxford University Press, New York
19. Letovsky S, Kasif S (2003) Predicting protein function from protein/protein interaction data: a probabilistic approach. Bioinformatics 19:i197–i204
20. Li H, Wei Z, Maris J (2010) A hidden Markov random field model for genome-wide association studies. Biostatistics 11:139–150
21. Marcotte EM (2000) Computational genetics: finding protein function by nonhomology methods. Curr Opin Struct Biol 10(3):359–365
22. Mering CV, Jensen LJ, Snel B et al (2005) String: known and predicted protein-protein associations, integrated and transferred across organisms. Nucleic Acids Res 33:433–437
23. Mitrofanova A, Pavlovic V, Mishra B (2011) Prediction of protein functions with Gene Ontology and interspecies protein homology data. EEE/ACM Trans Comput Biol Bioinf 8(3):775–784
24. Mourad R, Sinoquet C, Leray P (2011) A hierarchical Bayesian network approach for linkage disequilibrium modeling and data-dimensionality reduction prior to genome-wide association studies. BMC Bioinform 12:16+
25. Mourad R, Sinoquet C, Dina C, Leray P (2011) Visualization of pairwise and multilocus linkage disequilibrium structure using latent forests. PLOS ONE 6(12):e27320
26. Nariai N, Kolaczyk ED, Kasif S (2007) Probabilistic protein function prediction from heterogeneous genome-wide data. PLOS ONE 2(3):e337
27. Ng SK, Zhang Z, Tan SH, Lin K (2003) InterDom: a database of putative interacting protein domains for validating predicted protein interactions and complexes. Nucleic Acids Res 31(1):251–254
28. Nguyen TT, Androulakis IP (2009) Recent advances in the computational discovery of transcription factor binding sites. Algorithms 2(1):582–605. doi:10.3390/a2010582
29. Oshchepkov DY, Levitsky VG (2011) In silico prediction of transcriptional factor-binding sites. In: Series. Methods in molecular biology, vol 760, pp 251–267. doi:10.1007/978-1-61779-176-5_16
30. Pan W, Wei P, Khodursky A (2008) A parametric joint model of DNA-protein binding, gene expression and DNA sequence data to detect target genes of a transcription factor. Pacific Symp Biocomput 13:465–476
31. Peng G, Luo L, Siu H, Zhu Y et al (2010) Gene and pathway-based second-wave analysis of genome-wide association studies. Eur J Hum Genet 18:111–117

32. Peri S, Navarro JD, Amanchy R et al (2003) Development of human protein reference database as an initial platform for approaching systems biology in humans. Genome Res 13(10):2363–2371
33. Rhodes DR, Tomlins SA, Varambally S et al (2005) Probabilistic model of the human protein-protein interaction network. Nature Biotechnol 23:951–959. doi:10.1038/nbt1103
34. Schadt EE, Lamb J, Yang X et al (2005) An integrative genomics approach to infer causal associations between gene expression and disease. Nat Genet 37(7):710–717
35. Sinoquet C, Mourad R, Leray P (2012) Forests of latent tree models for the detection of genetic associations. In: International conference on bioinformatics models, methods and algorithms (Bioinformatics), 5–14
36. The Gene Ontology Consortium, Ashburner M, Ball CA, Blake JA et al (2000) Gene Ontology: tool for the unification of biology. Nat Genet 25(1):25–29. doi:10.1038/75556
37. Verzilli CJ, Stallard N, Whittaker JC (2006) Bayesian graphical models for genome-wide association studies. Am J Hum Genet 79:100–112
38. von Mering C, Krause R, Snel B, Cornell M, Oliver SG, Fields S, Bork P (2002) Comparative assessment of large-scale data sets of protein-protein interactions. Nature 417(6887):399–403
39. Wang W, Cherry JM, Nochomovitz Y, Jolly E, Botstein D, Li H (2005) Inference of combinatorial regulation in Yeast transcriptional networks: a case study of sporulation. Proc Natl Acad Sci USA 102:1998–2003
40. Wang K, Li M, Bucan M (2007) Pathway-based approaches for analysis of genomewide associations studies. Am J Hum Genet 81:1278–1283
41. Wei P, Pan W (2012) Bayesian joint modeling of multiple gene networks and diverse genomic data to identify target genes of a transcription factor. Ann Appl Stat 6(1):334–355
42. Wu X, Jiang R, Zhang MQ, Li S (2008) Network-based global inference of human disease genes. Mol Syst Biol 4:189
43. Xia K, Dong D, Han J-DJ (2006) IntNetDB v1.0: an integrated protein-protein interaction network database generated by a probabilistic model. BMC Bioinform 7:508. doi:10.1186/1471-2105-7-508
44. Xia JF, Wang S-L, Lei Y-K (2010) Computational methods for the prediction of protein-protein interactions. Protein Pept Lett 17(9):1069–1078
45. Yosef N, Sharan R, Stafford Noble W (2008) Improved network-based identification of protein orthologs. Bioinformatics 24(16):i200–i206
46. Yu H, Luscombe NM, Lu HX, Zhu X, Xia Y, Han JD, Bertin N, Chung S, Vidal M, Gerstein M (2004) Annotation transfer between genomes: protein-protein interologs and protein-DNA regulogs. Genome Res 14(6):1107–1118
47. Zhu J, Zhang B, Smith EN, Drees B, Brem RB, Kruglyak L, Bumgarner RE, Schadt EE (2008) Integrating large-scale functional genomic data to dissect the complexity of Yeast regulatory networks. Nat Genet 40(7):854–861. doi:10.1038/ng.167

Chapter 9
Innovations of the Rule-Based Modeling Approach

Lily A. Chylek, Edward C. Stites, Richard G. Posner and William S. Hlavacek

Abstract New modeling approaches are needed to tackle the complexity of cell signaling systems. An emerging approach is rule-based modeling, in which protein-protein interactions are represented at the level of functional components. By using rules to represent interactions, a modeler can avoid enumerating the reachable chemical species in a system, which is a necessity in traditional modeling approaches. A set of rules can be used to generate a reaction network, or to perform simulations with or without network generation. Although the rule-based approach is a relatively recent development in biology, it is based on concepts that have proven useful in other fields. In this chapter, we discuss innovations of the rule-based modeling approach, relative to traditional approaches for modeling chemical kinetics. These innovations include the use of rules to concisely capture the dynamics of molecular interactions, the view of models as programs, and agent-based computational approaches that can be applied to simulate the

L. A. Chylek
Department of Chemistry and Chemical Biology, Cornell University,
Ithaca, NY 14853, USA
e-mail: lily.chylek@gmail.com

E. C. Stites · R. G. Posner
Clinical Translational Research Division, Translational Genomics Research Institute,
Phoenix, AZ 85004, USA
e-mail: edstites@gmail.com

R. G. Posner
Department of Biological Sciences, Northern Arizona University, Flagstaff,
AZ 86011, USA
e-mail: rposner@tgen.org

W. S. Hlavacek (✉)
Department of Biology, University of New Mexico, Albuquerque, NM 87131, USA
e-mail: wish@lanl.gov

W. S. Hlavacek
Theoretical Biology and Biophysics Group, Theoretical Division, Los Alamos National
Laboratory, Los Alamos, NM 87545, USA

A. Prokop and B. Csukás (eds.), *Systems Biology*,
DOI: 10.1007/978-94-007-6803-1_9,

chemical kinetics of a system characterized by a large traditional model. These innovations should enable the development of models that can relate the molecular state of a cell to its phenotype, even though vast and complex networks bridge perturbations at the molecular level to fates and activities at the cellular level. In the future, we expect that validated rule-based models will be useful for model-guided studies of cell signaling mechanisms, interpretation of temporal phosphoproteomic data, and cell engineering applications.

Keywords Computational modeling · Combinatorial complexity · Protein-protein interactions · Cell signaling · Rule-based modeling · Formal languages · Simulation algorithms · Chemical kinetics

Acronyms

BEM	Bond electron matrix
BNGL	BioNetGen Language
ODE	Ordinary differential equation
SBGN	Systems Biology Graphical Notation
SBML	Systems Biology Markup Language

9.1 Introduction

An important aim of systems biology is to understand phenomena that arise from the interactions of the component parts of cellular regulatory systems [1], such as genes, proteins, and metabolites. Key components of many regulatory systems have been studied extensively in isolation, which remains a common approach for investigating cellular regulation. Synthesis of the knowledge gained from reductionist studies, and accompanying development of systems-level understanding, necessitates the use of computational models that can account for the complexity of cellular regulatory networks [2–6]. Models are useful because they can make testable predictions and elucidate the logical consequences of the assumptions upon which a model is based. Models can advance understanding in other ways [7], for example, by consolidating available knowledge, visualizing this knowledge to make it more accessible, and revealing knowledge gaps. For a model to be useful, it need not capture all known mechanistic details, but the level of detail included in a model should be suitable for the system of interest and the questions that a modeler intends to ask.

Here, we focus on cell signaling systems. These systems consist of interacting molecules that coordinate responses to changes in the environment (signals). Aspects of these responses may not always be possible to predict using intuition alone. Indeed, molecularly targeted therapies, such as RAF inhibitors for cancer treatment [8], may lead to unexpected and even harmful outcomes due to complex repercussions emanating from perturbed molecular states. To better understand how cell signaling systems process information and respond to stimuli, we need mathematical/computational models that capture the chemical kinetics of molecular interactions in these systems. These physical interactions have been found to be dynamic [9, 10], regulated (viz., protein-protein interactions that are affected by post-translational modifications [11]), and mediated by modular components (e.g., domains and linear motifs [12]). Thus, it seems worthwhile to develop models that can account for these mechanistic details.

However, mechanistic details of protein-protein interactions in cell signaling systems give rise to at least two significant challenges for modelers. The first challenge is size: a signaling system typically contains numerous proteins [13]. The second challenge is combinatorial complexity [14, 15]: a protein may participate in multiple interactions and undergo post-translational modifications at multiple sites. As a result, a large number of chemical species can potentially be populated. Traditional modeling approaches, such as those indicated in Fig. 9.1, are poorly suited to cope with combinatorial complexity because they require enumeration of every reachable species. An alternative approach more suited for modeling of cell signaling systems, and other types of biochemical systems, is that of rule-based modeling, which is distinguished from traditional modeling approaches in several

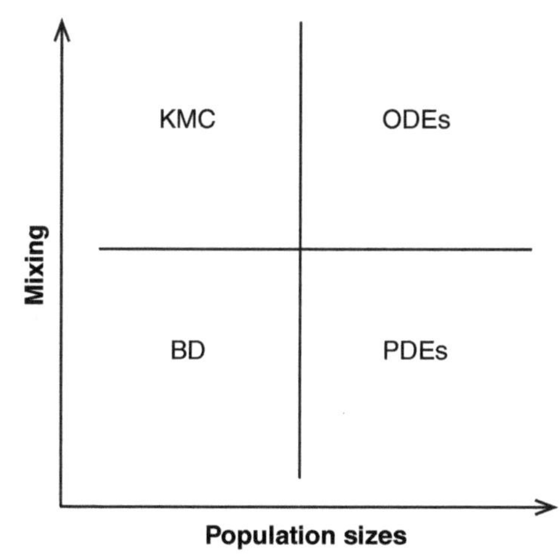

Fig. 9.1 Traditional modeling approaches. For well-mixed systems with large population sizes (i.e., populations that are large enough for concentrations to be continuous), ordinary differential equations (ODEs) can be used. For well mixed-systems with small population sizes, kinetic Monte Carlo (KMC) methods (e.g., Gillespie's method) may be more appropriate. If the rate of mixing in a system is slower than the rate of chemical reactions, spatial effects can be expected to be important. In these cases, partial differential equations (PDEs) and Brownian dynamics (BD) can be used for systems with large and small population sizes, respectively. For each of these traditional modeling approaches, there is a corresponding rule-based approach. BioNetGen [16] and Smoldyn [17] can perform ODE-based and KMC-based simulations. Other simulators, including DYNSTOC [18], KaSim [19], NFsim [20], and RuleMonkey [21] can also perform KMC-based simulations. BNG@VCell [22] and Simmune [23, 24] can perform PDE-based simulations. Smoldyn [17] can also use BD to model diffusion of molecules

ways. Here, we review key innovative features of the rule-based modeling approach. It is a method of systems biology that is likely to grow in importance in the future, in part because of the number of sophisticated software tools now available to support it. For example, see [16–24]. There is also a large body of knowledge available about a number of cell signaling systems and a need to formalize this knowledge.

To illustrate the size and combinatorial complexity of a well-studied cell signaling system, let us consider a subset of the proteins involved in signaling via the epidermal growth factor receptor (EGFR). Specifically, let us focus on 21 proteins included in the model of Chen et al. [25]; it is worth noting that the NetPath database lists over 400 proteins involved in EGFR signaling [13]. Based on information available in public databases, on average each of the 21 proteins contains 8.2 sites of phosphorylation [26], 2.6 domains [27], and 0.6 motifs [28], and has 6.5 interaction partners among the other 20 proteins [29]. These statistics are summarized in Fig. 9.2. Enumeration of every possible species of interest that could arise in this subsystem would be impractical, if not impossible, without the use of simplifying assumptions to reduce combinatorial complexity. For example, consider Gab1, Raf-1, and EGFR. According to Phospho.ELM, these proteins have 14, 21, and 35 sites of phosphorylation, respectively [26]. As a result, Gab1 has $2^{14} = 16{,}384$ possible phosphorylation states, Raf-1 has $2^{21} \approx 2 \times 10^6$ possible phosphorylation states, and EGFR has $2^{35} \approx 3.4 \times 10^{10}$ possible phosphorylation states.

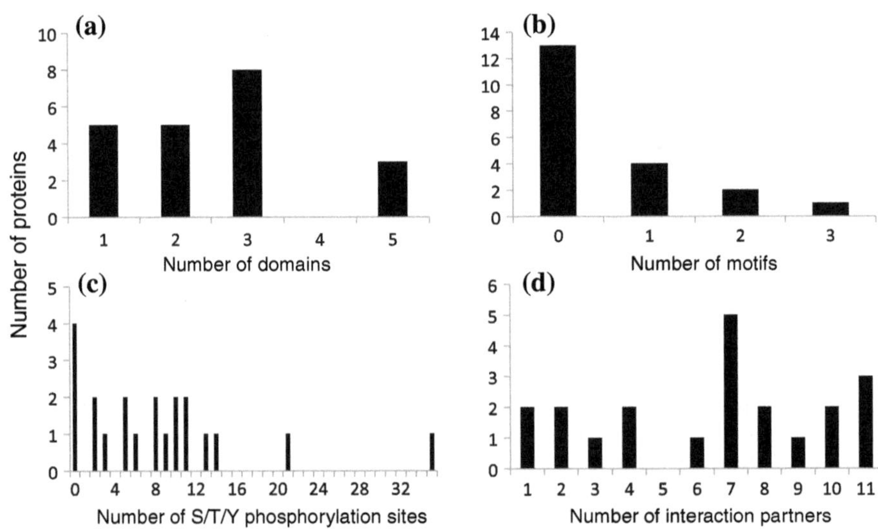

Fig. 9.2 Sites, modifications and interactions of proteins involved in EGFR signaling. Gene names of the proteins considered here are EGF, NRG1, EGFR, ERBB2, ERBB3, ERBB4, SHC1, GRB2, SOS1, GAB1, PIK3RI, PIK3CA, PDPK1, AKT1, KRAS, RASA1, RAF1, MAP2K1, MAPK1, PTEN, and PTPN11. **a** Domains considered are those documented in the Pfam database [27]. **b** Similarly, motifs were obtained from ELM [28]. **c** Phosphorylation sites were obtained from Phospho.ELM [26]. **d** Interaction partners were obtained from HPRD [29]

The challenge of combinatorial complexity can be addressed using the rule-based modeling approach [30–32]. In this approach, proteins are represented as structured objects whose components can interact independently of one another unless otherwise specified. Contextual constraints on protein-protein interactions can be captured in rules, which include necessary and sufficient conditions for firing of reaction events. One can view reactants as satisfying conditions required at specific sites, as specified in rules. Fewer simplifying assumptions are typically required and a more comprehensive picture of a signaling system can be developed that is more aligned with mechanistic understanding.

In this chapter, we discuss innovations of the rule-based modeling approach. The first innovation that we discuss is the use of rules, which builds on concepts that have proven useful in other fields. A second innovation is the use of formal languages to specify models, allowing models to be viewed as programs. A third innovation is network-free algorithms for stochastic simulation of agent-based models consistent with the law of mass action. These algorithms are needed for mechanistic modeling of cell signaling on a large scale.

9.2 Use of Rules to Represent Molecular Interactions in Cellular and Molecular Biology

The network motifs (e.g., the writer, reader, eraser motif, which consists of tyrosine phosphorylation, SH2 domain binding, and dephosphorylation [33]) and subsystems that constitute a signaling system may each involve only a few different proteins. However, interactions among these proteins may give rise to far larger numbers of distinct chemical species through combinations of the different possible interactions and modifications [14, 15, 34, 35]. To capture these effects, a number of tools and modeling frameworks have been developed that use rules to represent molecular interactions at the level of molecular components, or sites.

Among the first software tools developed for rule-based modeling of biological systems were OLIGO [37] and StochSim [38, 39]. OLIGO is capable of generating reaction networks for assembly of oligomeric complexes, but does not capture regulation of interactions through post-translational modifications. This capability is provided in StochSim, where proteins are represented as multi-state entities. A protein is encoded as a set of "flags" that represent binding or modification states. During a simulation, molecules are selected randomly and a list of rules is used to determine whether a reaction can occur between them (i.e., whether states can change). Although StochSim can be used to effectively capture changes in state, it is poorly suited for explicitly tracking the connectivity of molecular complexes.

Another early approach, developed by Regev et al. [40], uses π-calculus to model a cell signaling system as a concurrent computational system. In this approach, molecules and sites are treated as parallel processes that can behave independently of one another, in accordance with a set of rules. Stochastic

π-calculus [41] and tools implementing this method, such as BioSPI [42], BlenX [43], and SPiM [44, 45], enable simulation of biochemical kinetics. However, the use of π-calculus introduces artifacts from the study of concurrency, such as directionality of communication.

An early example of a non-trivial rule-based model is that of Goldstein et al. [46] and Faeder et al. [47]. This model is equivalent to 354 ODEs with 3,680 distinct right-hand-side terms, making it tedious to specify using traditional approaches. The model was used to investigate early events in signaling via the high-affinity receptor for IgE. The rule-based approach has since been applied extensively to study immunoreceptor signaling [48–55]. However, these and other applications are not the main subject of this chapter; instead, we focus on methodology.

To demonstrate the use of rules, let us consider a system in which a scaffold, *S*, may bind ligand *A* with a forward rate constant of k_{+A} and a reverse rate constant of k_{-A}. The scaffold *S* may also independently bind ligand *B* with a forward rate constant of k_{+B} and a reverse rate constant of k_{-B}. Thus, the system has six species: *S, A, B*, a complex of *S* and *A*, a complex of *S* and *B*, and a ternary complex of *S, A*, and *B*. Figure 9.3 illustrates traditional formulations of a model of this system. Panels A and B show the reactions of the model as a list and as a reaction scheme, respectively. Panel C is a visualization of the model using the conventions of Systems Biology Graphical Notation (SBGN) [36]. Panel D shows the six ODEs of the model that follow from mass-action kinetics. The ODEs characterize the

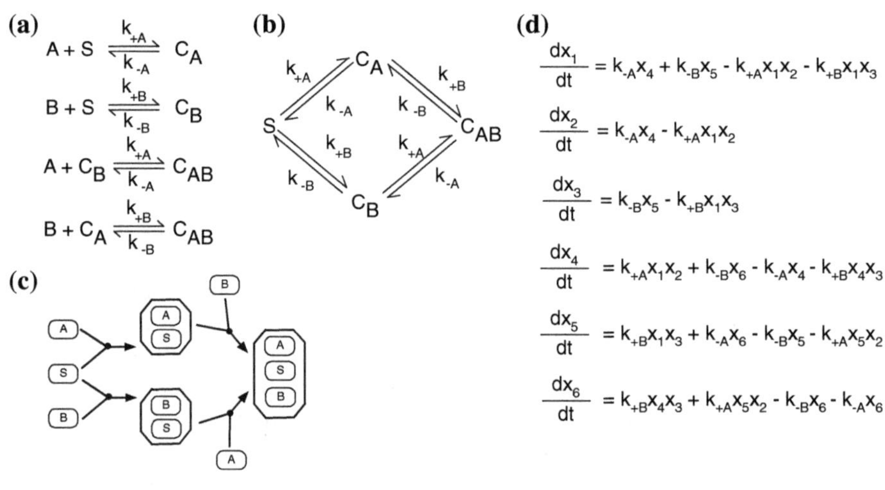

Fig. 9.3 A model of scaffold-ligand interactions traditionally formulated. Ligands *A* and *B* bind non-cooperatively to scaffold *S*. **a** A list of reactions with associated rate constants for forward and reverse reactions. *A*, *B*, and *S* represent unbound proteins. *CA* and *CB* represent *S* bound to *A* and *B*, respectively. *CAB* represents the ternary complex of *S, A*, and *B*. **b** A reaction scheme, i.e., an organized layout of the reactions of Panel A. **c** An SBGN (Process Diagram) [36] representation of the model of Panels *A* and *B*. **d** The model in the form of a system of ODEs: the variables $x_1, \ldots, x_6$ represent the concentrations of *S, A, B, CA, CB,* and *CAB*, respectively

change with time of each of the six concentrations for a well-mixed reaction compartment and continuous population levels (i.e., large numbers of molecules).

A rule-based formulation of the same model is illustrated graphically in Fig. 9.4. This model can be encoded in a number of rule-based modeling languages. As we will discuss, the most commonly used languages for rule-based modeling are BioNetGen Language (BNGL) [16, 59] and Kappa [31, 60, 61], and we will use shared conventions of these languages in our description of the model. The scaffold is represented as a structured object, `S`, with two components, `a` and `b`. These components are binding sites that recognize ligands `A` and `B`, respectively. Ligand `A` contains a component `s` that binds `a` in `S`. Similarly, ligand `B` contains a component `s` that binds `b` in `S`. Figure 9.4a illustrates two rules that capture the interactions among these molecules. The first rule specifies the conditions necessary for `S` to bind `A`: `S` must have an unbound component `a` and `A` must have an unbound component `s`. We assume that the state of site `b` does not affect the interaction between `S` and `A`, so it is omitted from the rule. If `b` could affect the interaction between `S` and `A` (e.g., through an allosteric mechanism), it would be possible to express such an effect by appropriate modification of the rules that comprise the model. The second rule specifies the conditions necessary for `S` to bind `B`, which are similar to those of the first rule. These two rules represent the same set of interactions as the eight unidirectional reactions and the six ODEs shown in Fig. 9.3. Figure 9.4b shows the model visualized as a contact map, which in general provides an illustration of all molecules, components, modifications (none in this model) and interactions that are included in a model. Figure 9.4c is an alternative rendering of the contact map.

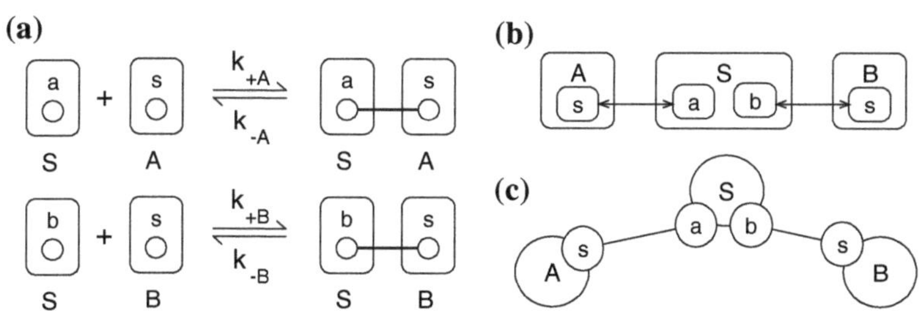

Fig. 9.4 A rule-based model of scaffold-ligand interactions. **a** Two rules visualized using the graphical conventions of Faeder et al. [56]. Components not affecting an interaction are omitted from a rule. Proteins are represented as simple colored graphs. The "color" of a graph is the name of the protein that the graph represents. By convention, *boxes* enclose vertices of the *same color*. Bonds are represented as edges, which connect vertices that represent cognate binding sites. A BioNetGen Language (BNGL) encoding of the first rule is `S(a) + A(s) <-> S(a!1).A(s!1) kpa,kma`. A BNGL encoding of the second rule is `S(b) + B(s) <-> S(b!1).B(s!1) kpb, kmb`. **b** The model visualized as an extended contact map [57]. Boxes represent proteins and components. A double-headed arrow represents a noncovalent bond. Contact maps can be generated using RuleBender [58]. **c**. An alternative rendering of a contact map, consistent with conventions of Danos et al. [31]

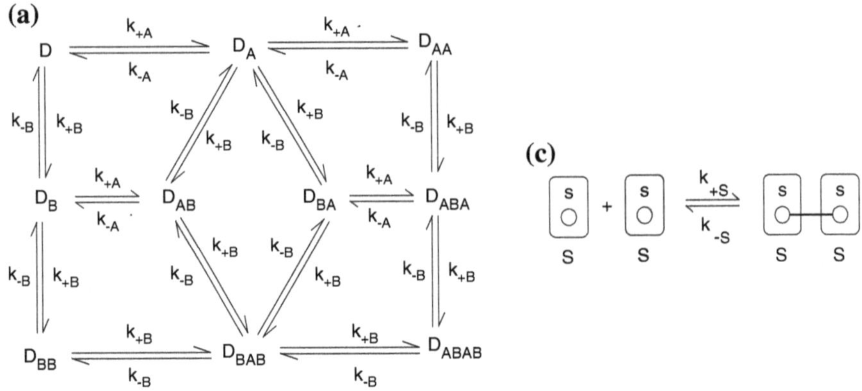

(b)

$$\frac{dx_1}{dt} = k_{-A}x_4 + k_{-B}x_5 - k_{+A}x_1x_2 - k_{+B}x_1x_3 \boxed{- k_{+S}x_1x_1 - k_{+S}x_1x_4 - k_{+S}x_1x_5 + 2k_{-S}x_6 - k_{+S}x_1x_7 + k_{-S}x_8 + k_{-S}x_8 + k_{-S}x_{13}}$$

Fig. 9.5 Illustration of an extension of the scaffold model of Figs. 9.3 and 9.4 to allow the scaffold to dimerize. **a** A partial reaction scheme showing the ten additional species that arise when scaffold dimerization is allowed. **b** An example of an equation from the scaffold model of Fig. 9.3d that must be modified if scaffold dimerization is allowed. The terms that must be added to account for scaffold dimerization are highlighted. **c**. The rule that is added to the rule-based model of Fig. 9.4a to account for scaffold dimerization. A BNGL encoding of this rule is `S(s) + S(s) <-> S(s!1).S(s!1) kps,kms`

From the simple example given above, the benefits of the rule-based approach may not be evident. However, consider addition of one more interaction to the model: dimerization of the scaffold protein. This added interaction gives rise to ten additional species (Fig 9.5a). Thus, ten additional ODEs must be added to capture this one additional interaction. Furthermore, the original six equations must be modified to account for additional reactions that each of the original six species can now undergo. Figure 9.5b shows a modified equation from the original model; the highlighted terms are added to account for interactions that become possible if the scaffold dimerizes.

In contrast, scaffold dimerization can be incorporated into a rule-based model through single addition of the rule of Fig. 9.5c without modification of the rules of Fig. 9.4a. Thus, rule-based modeling is more extensible than traditional modeling for chemical kinetics. It is worth noting that the rule-based approach has been used to study scaffold effects in cell signaling [62, 63]. A complete specification of the rule-based model in which scaffold molecules may dimerize is provided in Fig. 9.6, wherein a BNGL [16, 59] encoding (Listing 1) and an equivalent Kappa [31, 60, 61] encoding (Listing 2) are given. Listing 1 can be used to generate a MATLAB (The MathWorks, Natick, MA) definition of a system of ODEs (i.e., a traditional model specification) by adding the command "`writeMfile();`" to the end of the listing (i.e., in the actions block of the code, which is not shown). Listing 1 can be processed by BioNetGen [16, 59] to generate an M-file consisting

Listing 1

```
begin molecule types
S(s,a,b)
A(s)
B(s)
end molecule types
begin seed species
S(s,a,b) S_init
A(s) A_init
B(s) B_init
end seed species
begin reaction rules
S(a) + A(s) <-> S(a!1).A(s!1) kpa,kma
S(b) + B(s) <-> S(b!1).B(s!1) kpb,kmb
S(s) + S(s) <-> S(s!1).S(s!1) kps,kms
end reaction rule
```

Listing 2

```
%agent: S(s,a,b)
%agent: A(s)
%agent: B(s)
%init: 1e5 * S(s,a,b)
%init: 1e5 * A(s)
%init: 1e5 * B(s)
S(a),A(s) ->S(a!1),A(s!1) @0.1
S(a!1),A(s!1) ->S(a),A(s) @0.1
S(b),B(s) ->S(B!1),B(s!1) @0.1
S(B!1),B(s!1) ->S(b),B(s) @0.1
S(S),S(s) ->S(s!1),S(s!1) @0.1
S(s!1),S(s!1) ->S(S),S(s) @0.1
```

Fig. 9.6 The model composed of the rules illustrated in Figs. 9.4a and 9.5c encoded in two formal languages, the BioNetGen Language (BNGL) and Kappa. **Listing 1**. BNGL-encoded model specification. This excerpt from a BioNetGen input file illustrates the definition of molecule types, seed species (initial conditions for a simulation), and rules. Lines of code for setting parameter values and actions are not shown. The model specification can be simulated using different methods available in BioNetGen [16, 59] or other BNGL-compliant software tools [18, 20–22]. Note that the *first rule* corresponds to the top rule of Fig. 9.4a, the second rule corresponds to the *bottom rule* of Fig. 9.4a, and the *third rule* corresponds to the rule of Fig. 9.5c. **Listing 2**. Kappa-encoded model specification. This model specification, which is equivalent to Listing 1, can be processed by KaSim [61]

of 113 lines of code. The M-file defines a system of 16 ODEs with 60 different right-hand-side terms. In contrast, the model specification of Listing 1 in Fig. 9.6 consists of only three rules and three molecule type definitions (molecule type definitions are not illustrated in earlier figures), as well as specifications of parameter values and initial conditions.

Although the rule-based modeling approach is a relatively recent development in biology, similar concepts have long been used in other fields. Below, we briefly discuss related approaches that have been developed for a variety of problems in physics, chemistry, and computer science. The success of this approach in other fields suggests that it will also be useful for studying the systems of cellular and molecular biology.

9.2.1 Precedents in Physics

The Ising model was originally developed to study ferromagnetism: the emergence of a magnetic moment through alignment of atomic spin states. The model, which has a number of other applications, involves a lattice of sites, each of which has one of two states, e.g., spin-up or spin-down. The state or spin of a site can be reversed. The probability of a site's spin reversing depends on the spin states of its neighbors. The Ising model can be simulated using a number of methods. In the classic Metropolis method [64], a site (if spin is flipped) or pair of sites (if spins are exchanged) is first selected at random. For the purposes of this discussion, we assume that a single site is chosen. The probability of spin reversal is then computed based on temperature and the configuration of a site's neighbors. This probability is then compared to a random number. If the random number is less than or equal to the probability of flipping, the spin of the site is reversed and time is incremented. If the random number is greater than the probability of flipping, spin is reversed with a probability equal to the ratio of the probabilities of the initial and final states. If spin is not reversed, a null event occurs and time is incremented. A drawback of this method is that a high frequency of null events causes simulation to slow significantly.

An alternative approach is the n-fold way [65], a kinetic Monte Carlo (KMC) method [66], in which null events are avoided. In this algorithm, a site is classified based on its spin state and the spin states of its nearest neighbors. A classification scheme for a square lattice is shown in Fig. 9.7a. Use of this scheme is illustrated in Fig. 9.7b; white squares represent spin-up sites, gray squares represent spin-down sites, and the number of a square indicates its class. Rather than selecting a site randomly, the probability of a site being selected is related to the probability of its spin flipping. Once a site is selected, its spin is flipped immediately. Thus, null events do not arise. The n-fold way for the example of Fig. 9.7 consists of the following steps:

1. Assign each site to one of ten possible classes.
2. Choose a class $r \in [1, 2, \ldots, 10]$. A class is chosen by first calculating cumulative rates $Q_1, Q_2, \ldots, Q_{10}$, where

Class	Spin	Number of spin-up nearest neighbors
1	Up	4
2	Up	3
3	Up	2
4	Up	1
5	Up	0
6	Down	4
7	Down	3
8	Down	2
9	Down	1
10	Down	0

(a)

5	7	3	9	8
6	3	3	8	4
3	2	8	8	2
7	4	7	4	3
9	8	4	8	9

(b)

Fig. 9.7 Classes used in KMC simulation of the Ising model. **a**. Scheme used for classification of lattice sites. A site is classified based on its spin and the number of its nearest neighbors that are spin-up. **b**. In this example, white squares are used for spin-up sites and dark squares are used for spin-down sites. Class numbers are shown on squares in accordance with the scheme of Panel A. The lattice is assumed to have periodic boundary conditions, i.e., the lower boundary is replicated above the upper boundary and the left boundary is replicated after the right boundary

$$Q_r = \sum_{j=1}^{r} n_j P_j$$

In the above expression, n_j is the number of sites in class j and P_j is the probability of spin reversal for a site in class j. Then, a random number R_1, uniformly distributed between 0 and Q_{10}, is chosen, and a class r is chosen such that $Q_{r-1} \leq R_1 < Q_r$.

3. Randomly choose a site i within r.
4. Flip the spin of site i.
5. Update classes and the rates $Q_1, Q_2, \ldots, Q_{10}$ based on the new configuration of the lattice.
6. Increment time. The time step is calculated as $\Delta t = \frac{\tau \log_2 R_2}{Q_{10}}$ where R_2 is a random number and τ is the expectation value (i.e., the average time per spin flip). Recall that Q_{10} is the overall rate of spin flipping.

This procedure applied to stochastic simulation of chemical reaction systems is known as Gillespie's method [67–69], which is discussed below. The similarity between the n-fold way and the rule-based modeling approach lies in the use of classes. In the n-fold way, a class defines a set of lattice sites that have a particular spin state and configuration of neighbors. Sites within a class all have the same probability of undergoing a transition. Similarly, a rule defines a class of reactions whose reactants share certain local component properties and reactions that are defined by a rule are taken to have the same kinetic parameters.

9.2.2 Precedents in Chemistry

In modeling chemical reactions, matrices and matrix operations can serve as useful abstractions for representing molecular structures and functional group transformations. Ugi and co-workers developed a formalism in which a bond electron matrix (BEM) is used to represent the atoms present in a molecule (or set of molecules) and the sharing of electrons between them. In this formalism, a chemical reaction is viewed as converting a BEM into an isomeric BEM by redistributing valence electrons. A BEM for n atoms contains n rows and columns. The ith row and column correspond to the ith atom of the molecule or set of molecules. The matrix entry b_{ij} is the number of bonds between atoms i and j and the diagonal matrix entry b_{ii} is the number of free valence electrons of atom i [71]. (When applied to reactions on surfaces, b_{ii} can also represent the number of electrons backdonated to the absorbate). Electrons are redistributed (i.e., a chemical reaction is executed) by addition of a reaction matrix R to a reactant matrix B. An entry in a reaction matrix corresponds to the number of bonds formed (positive numbers) or broken (negative numbers) between atoms as a result of a reaction. The matrix $E \equiv B + R$ represents the product molecule(s) of a reaction.

The BEM formalism can be used to generate reaction networks and elucidate possible synthetic routes between reactant and target molecules [72–75]. This method has also been used for time-scale analysis of rule-based models in which reactions within the same class have different kinetic parameters [76].

BEMs have been applied by Broadbelt and co-workers to investigate reaction mechanisms for heterogeneous catalytic chemistry [77] and novel metabolites and pathways in metabolic networks [78, 79]. The assumption underlying the latter is that the large number of reactions found in a metabolic network can be represented by a smaller number of rules for common functional group transformations in metabolism [78, 80]. Functional groups can be encoded as BEMs and associated with reaction matrices. An input molecule can also be encoded as a BEM and compared to the BEM of a functional group to determine whether the molecule contains the functional group necessary to undergo a reaction. If so, a reaction matrix is added to the appropriate part of the reactant matrix to yield a matrix for a product molecule or set of molecules. If the product is a chemical species that has not yet been generated, it is evaluated to determine whether it can undergo further reactions. A maximum number of generations can be specified as a stopping criterion. In this way, a set of rules can be identified that can generate potentially novel reaction paths from reactants to products. This approach has been complemented by thermodynamic studies to evaluate the feasibility of possible reaction paths [79].

An example of the use of BEMs to model a chemical reaction is shown in Fig. 9.8. Panel A shows a rule that specifies the functional groups involved in an esterification reaction. Panel B shows the same functional groups in the form of BEMs, with atoms numbered to correspond to Panel A. Panels C and D show two instances of the rule acting on specific molecules.

9.2.3 Precedents in Computer Science

A concurrent computational system is one in which multiple processes are executed in parallel and can potentially influence each other. Interaction among processes can lead to many possible outcomes. The complexity of concurrent systems necessitates a language that can be used to analyze and reason about a system's behavior. This need is addressed by process algebras [81, 82]. Here, we focus on π-calculus, a process algebra that has been intensely studied in computer science and that has also been applied to model biological systems, as noted earlier. A notable feature of π-calculus is that it allows explicit representation of communication channels and allows system components to be modeled independently.

An example of the use of π-calculus is shown in Fig. 9.9. In π-calculus, "+" designates choice, "." designates sequence, and "|" designates processes executed in parallel. The symbol 0 designates an inert process (i.e., a process that does nothing further). A process can contain one or more channels, which can be used to communicate with other processes. Channels that can communicate with one another are referred to as complementary channels. Complementary channels

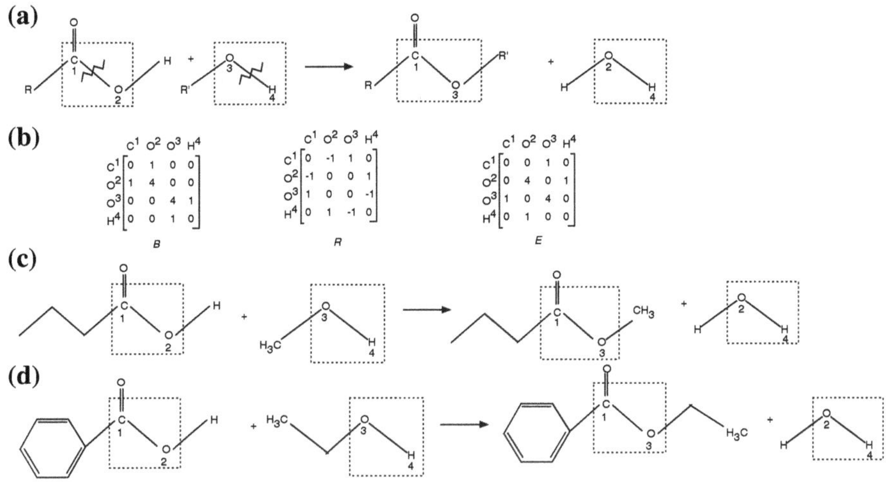

Fig. 9.8 Bond electron matrices and matrix operations can be used to model functional group transformations in organic chemistry. **a** A general Fischer esterification reaction between a carboxylic acid and an alcohol. A *box* is placed around functional groups that participate in the reaction, and a jagged line is used to mark the bonds that are broken. **b** *B* is a bond electron matrix for the reactants. Rows and columns are labeled to correspond to labeling of atoms in Panel A. *R* is a reaction matrix showing the bonds that are broken and formed as a result of esterification. *E* is a bond electron matrix for the products. **c, d** Two instances of rule application. Functional groups are enclosed within a *box* and atoms are numbered to correspond to panels A and B

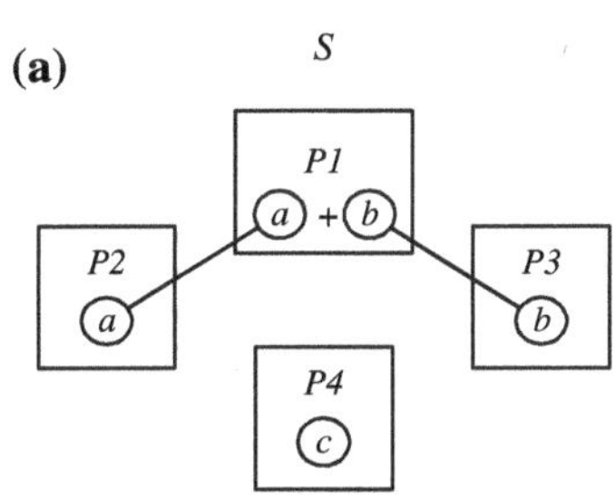

(b)
$$P1 ::= \overline{a}\langle x\rangle.R + \overline{b}\langle y\rangle.T$$
$$P2 ::= a(u).0$$
$$P3 ::= b(v).0$$
$$P4 ::= c(w).0$$
$$S ::= P1 \mid P2 \mid P3 \mid P4$$

(c) $S' ::= R \mid 0 \mid P3 \mid P4$

(d) $S' ::= T \mid P2 \mid 0 \mid P4$

Fig. 9.9 The process algebra π-calculus is used to model concurrent computational systems. **a** System *S* contains processes *P1, P2, P3*, and *P4*. *P1* can communicate with *P2* using channel *a*. *P1* can communicate with *P3* using channel *b*. *P4* can receive information along channel *c*; however, there is no complementary channel in *S*. **b** Representation of the system in π-calculus. **c** The system that *S* reduces to if *P1* sends a message on channel *a*. **d** The system that *S* reduces to if *P1* sends a message on channel *b*

share the same name, and prefixing conventions are used to distinguish an input channel from an output channel. For example, *a* <*x*> is an output channel named *a* that sends a piece of information named *x*. A complementary input channel can be designated *a(u)*. When *a(u)* receives information, the name received (e.g., *x*) becomes bound to *u*.

In the example of Fig. 9.9, *P1* is a process that can send *x* on channel *a*. *P1* then behaves as *R*. Alternatively, *P1* can send *y* on channel *b* and then behave as *T*. In the first scenario, *P1* becomes *R*, and *P2* uses channel *a* to receive *x* from *P1*. The message *x* becomes bound to *u*. Then, *P2* becomes 0. The processes *P3* and *P4* are unaffected by this communication event. As a result, *P1|P2|P3|P4* becomes *R|0|P3|P4*. Similarly, if *P1* chooses the second option, the system becomes *T|P2|0|P4*.

The similarity between π-calculus in computer science and rule-based modeling approaches in systems biology lies in modularity. In a rule-based model, one may specify an interaction using only the sites that participate in the interaction. In π-calculus, one may likewise specify the effect of communication using expressions that only include the relevant (sub)processes and channels. For example, in the system of Fig. 9.9, communication between *P1* and *P2* or *P3* is expressed without the inclusion of *P4*, which does not communicate with the other processes. Rule-based modeling approaches and process algebras share context-free properties, meaning that context can be omitted from a rule. Omitted contexts have no affect on the transformation specified in a rule, so that the rule can be applied in multiple contexts that need not be specified by the modeler. However, in some cases it is necessary for rule application to be restricted by context (e.g., when a reaction can only occur intramolecularly). In these cases, features of rule-based modeling languages, such as the dot-plus notation of BNGL (see Sect. 9.3 below), can be used to impose contextual constraints. Different but functionally equivalent notation is available in Kappa.

9.3 Models as Programs

A model can be formalized using mathematical expressions. A different approach is to formalize a model as an executable program, which can potentially facilitate analysis [61], extensibility [83], and high-level abstractions [84–86]. A number of languages for modeling biological systems have been developed, including languages designed for specification of rule-based models. As we will discuss below, BNGL [16, 59] and Kappa [60, 61] are the most widely used rule-based modeling languages. A BNGL-encoded model and an equivalent Kappa-encoded model can be found in Listings 1 and 2 of Fig. 9.6, respectively.

A model, once specified using rules, may be simulated in a number of different ways without modification of the model specification *per se*. The model of Listing 1 can be simulated deterministically with the command `simulate_ode`, or stochastically with `simulate_ssa`. (For a complete description of BNGL syntax, see Faeder et al. [16]). Thus, model specification is separated from simulation. For an example of a rule-based model simulated in multiple ways, see Lipniacki et al. [87].

Methods for simulating rule-based models include generate-first, on-the-fly, and network-free methods. In generate-first methods, rules are iteratively applied to a set of species to generate new reactions and new species. This process continues until the full network is generated (i.e., no new species are generated) or until a stopping criterion is satisfied [88]. The dynamics of the network can then be simulated through numerical integration of a corresponding system of ODEs or a stochastic simulation algorithm. In on-the-fly methods, a network is generated as simulation progresses rather than before simulation begins [88, 89]. When a species is first populated, rules are applied to it, and new reactions and species may be generated. This approach can be useful if a set of rules implies a large number of possible species, some of which might never become populated. However, this method still relies on a computer's memory to store the network, which can be a limitation. The step of network generation is avoided entirely with network-free methods, which are discussed in more detail in the next section. In short, in a network-free method, all components of a system are tracked individually and rules are used directly to advance the state of a system by modifying states of components. Thus, network-free methods are particle- or agent-based.

A number of software tools compatible with BNGL and/or Kappa implement the simulation methods described above, in addition to providing other capabilities. These tools are listed in Table 9.1. Other languages that may be used to specify rule-based models include cBNGL, a form of BNGL that allows for explicit representation of compartments [90]; ML-rules, designed for multi-level rule-based modeling [91]; and SBML-multi, which is in development. See the Systems Biology Markup Language (SBML) website (http://sbml.org).

BNGL and Kappa are closely related but differ in several details. One difference is the treatment of indistinguishable sites. In BNGL, a molecule is allowed to have two or more sites that have the same name. Such sites are taken to be indistinguishable. This capability is useful for molecules such as an IgG or IgE antibody, which contains two antigen-combining sites that are essentially identical. A bivalent antibody can be captured in BNGL with a molecule type definition such as `IgE(Ag,Ag)`. In contrast, Kappa requires that every site have a unique name. Thus, the same molecule would necessarily require a definition of the form `IgE(Ag1,Ag2)`.

Reaction rules in BNGL constrain the molecularity of reactions using "dot-plus" notation. This notation does not exist in Kappa; however, equivalent distinctions can now be made through other conventions [107].

The dot-plus notation is used to distinguish molecules that are part of the same chemical species (i.e., molecules that are directly or indirectly connected) from molecules that are part of separate species (i.e., not connected). For example, the following rule states that a bond forms between molecules `L` and `R`.

$$\texttt{R(l)} + \texttt{L(r)} - > \texttt{R(l!1).L(r!1)}$$

The "+" sign specifies that the two reactant sites must be part of separate species for the rule to be applied. Thus, the rule defines only bimolecular association reactions. In general, the molecularity of a reaction is $1 + p$, where p is the number of "+" signs on the left-hand-side of a rule. In contrast, the following rule

Table 9.1 Software tools that use BNGL and/or Kappa

Tool	Language	Reference
BioNetGen	BNGL	[16, 59, 88]
BNG@VCell	BNGL	[22]
little b	little b	[92]
Smoldyn/libMoleculizer	BNGL/Kappa-like	[17, 89]
SSC	SSC	[93]
DYNSTOC	BNGL	[18]
NFsim	BNGL	[20, 94]
RuleMonkey	BNGL	[21]
KaSim	Kappa	[19, 61]
SRSim	BNGL	[95]
RuleBender	BNGL	[58]
RuleStudio	Kappa	[19]
RuleBase	BNGL and Kappa	[96]
GetBonNie	BNGL	[97]
BioLab	BNGL	[98]
complx	Kappa	[19]
PySB	BNGL and Kappa	[86]

Capabilities: BioNetGen is capable of network generation, ODE-based simulation, and generate-first and on-the-fly stochastic simulation. BNG@VCell has the additional capability of PDE-based simulation. The little b environment uses BioNetGen to perform network generation. Smoldyn/libMoleculizer and SSC can perform particle-based reaction diffusion calculations. BioNetGen can convert BNGL-specified rules into SSC format. DYNSTOC, NFsim, and RuleMonkey perform network-free simulation. KaSim performs network-free simulations. SRSim combines rule-based modeling with atomistic modeling (i.e., molecular dynamics simulation). RuleBender and RuleStudio are modeling interfaces, and RuleBender provides visualization capabilities. RuleBase and GetBonnie are model databases. BioLab is a model-checking tool, complx is a tool for static analysis, and PySB is tool for model building and analysis. Other software tools for rule-based modeling that do not use BNGL or Kappa include ALC [99], ANC [100], BIOCHAM [101], BioSPI [40], BlenX4Bio [102], CplexA [103], Meredys [104], ML-Rules [91], Moleculizer [89], Pathway Logic Assistant [105], PottersWheel [106], Simmune [23, 24], and StochSim [39]

states that a bond forms between molecules `L` and `R` only when they are part of the same species.

```
R(l).L(r) - > R(l!1).L(r!1)
```

The absence of a "+" sign is an application condition of the rule, which indicates that the rule generates only unimolecular reactions. This rule defines reactions that form intramolecular bonds.

The dot-plus notation of BNGL allows a modeler to not only impose molecularity constraints but also specify that a pair of molecules are connected without explicitly specifying connectivity. For example, to obtain the number of complexes that contain two receptors, one may specify an observable `R().R()`, which encompasses all complexes that contain at least two receptors.

9.4 Agent-Based Modeling Consistent with the Law of Mass Action

Traditional models are usually simulated via population-based methods, which require explicit tracking of all potentially populated chemical species. A rule-based model can also be simulated with population-based methods; however, combinatorial complexity can give rise to a large number of species, which makes the approach impractical or, in some cases, impossible. An alternative method is network-free simulation. Algorithms for network-free simulation are agent-based simulation protocols consistent with the law of mass action. Agent-based models are used in a variety of fields [108], and most algorithms for agent-based simulation are not guided by physicochemical principles. Thus, the innovation of network-free methods is that agents behave according to rules that recapitulate chemical kinetics.

To illustrate agent-based simulation of a rule-based model, let us consider a model of a bivalent ligand and a bivalent cell-surface receptor (Fig. 9.10). The ligand contains two identical, independent sites that can bind receptors. The receptor contains two identical, independent sites that can bind ligands. Interactions between ligands and receptors can give rise to chains (i.e., linear aggregates) and rings (i.e., cyclic aggregates). The two molecule type definitions and three rules that form this model are shown in Figs. 9.10a and b, respectively.

The rate for a free ligand binding a free receptor site (Fig. 9.10b, Rule 1) is given by the following Equation [109]:

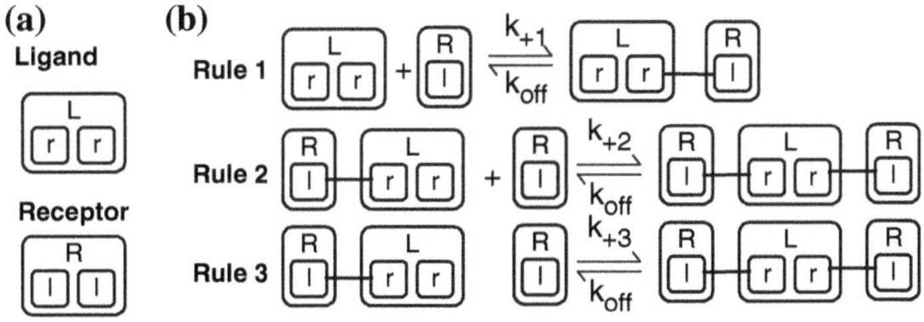

Fig. 9.10 A rule-based model for multivalent ligand-receptor binding. **a** Molecule type definitions for a bivalent ligand and bivalent receptor. **b** Rules for interactions between a bivalent ligand and a bivalent receptor. Note that the receptor R contains two identical sites (Panel A), but only one site is shown in Rules 1–3 (in accordance with the conventions of Faeder et al. [56]) because we assume that the bound state of the second site does not affect interactions represented by these rules. Rule 1 characterizes binding of a free ligand to a receptor. Rule 2 characterizes binding of a tethered ligand to a second receptor, thereby cross-linking a pair of receptors. Rule 3 characterizes ring closure and opening. The reactant sites of Rule 3 are required to be (indirectly) connected (i.e., they must exist within the same complex). Rules 2 and 3 differ on their left-hand sides. The plus sign in Rule 2 indicates that the rule defines reactions with molecularity of 2, whereas the absence of a plus sign in Rule 3 indicates that the rule defines reactions with molecularity of 1

$$\eta_1 = 4k_{+1}F_L(N_R - N_B) \quad (1)$$

where k_{+1} is the forward rate constant associated with the rule, F_L is the number of free ligands, N_R is the number of receptors, and N_B is the number of bonds. The statistical factor of four arises from the two identical binding sites per receptor and two identical binding sites per ligand.

The rate of dissociation of ligand from receptor (Fig. 9.10b, Rules 1, 2, and 3), including breaking of a cyclic aggregate, is proportional to the number of ligand-receptor bonds [109, 110]:

$$\eta_{1r} = k_{\text{off}}N_B \quad (2)$$

We assume that a single dissociation rate constant, k_{off}, applies for all dissociation reactions.

The rate for a tethered ligand binding a receptor site that is not part of the same complex as the ligand (Fig. 9.10b, Rule 2) is given by the following equation [109]:

$$\eta_2 = k_{+2}\sum_{i=1}^{N_A} l_i(2N_R - N_B - r_i) \quad (3)$$

where N_A is the number of aggregates, l_i is the number of free ligand sites in the ith aggregate, and r_i is the number of free receptor sites in the ith aggregate.

The rate for ring closure (Fig. 9.10b, Rule 3) is given by the following equation:

$$\eta_3 = k_{+3}\sum_{i=1}^{N_A} \frac{l_i r_i}{L_i} \quad (4)$$

The rate constant for ring closure can be taken to be inversely proportional to the length of a chain [110]. Here, we assume that rings of size one (i.e., containing one ligand and one receptor) are prohibited, and that k_{+3} is the rate constant for closure of a chain that yields a ring of size two (i.e., containing two ligands and two receptors). L_i is proportional to the length of a chain. For a ring of size two, we take $L_i = 1$.

Information about rates is used by network-free simulation algorithms to select rules to apply. Sequential rule application produces a system trajectory. Figure 9.11 shows an example of a trajectory in a network-free simulation of the model of Fig. 9.10. It is worth noting that Rule 1 is executed twice, in Panel B and Panel F. The two instances represent different reactions, but both reactions are captured by the same rule. A rule can be viewed as a generalized reaction, and algorithms for network-free simulation can be viewed as generalizations of Gillespie's method, which we briefly present below before reviewing different network-free algorithms reported in the literature.

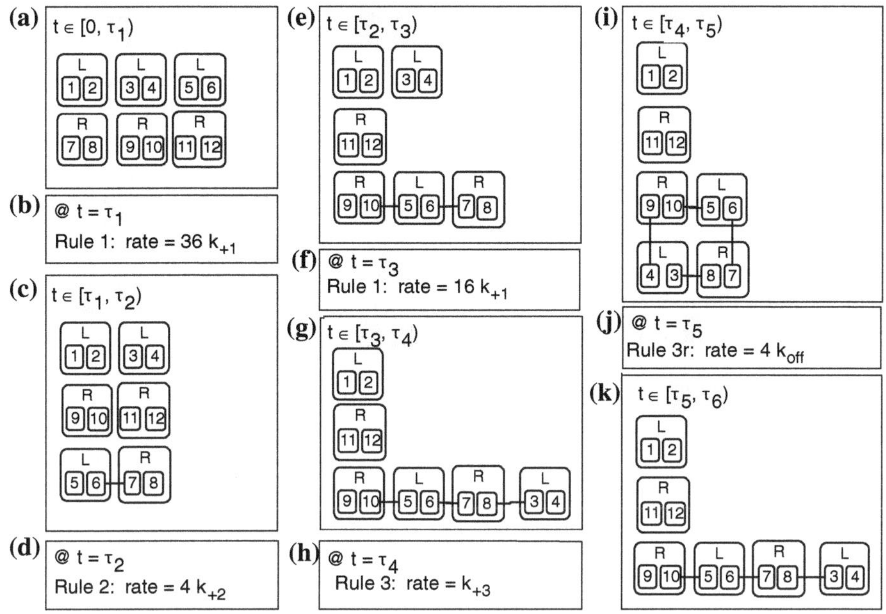

Fig. 9.11 An example trajectory in agent-based simulation of the model of Fig. 9.10. Each site is labeled with a unique number because sites are tracked individually. Reaction rates are calculated using Eqs. 1–4. **a** The initial state of the system contains three free ligands and three free receptors. **b** At time τ_1, Rule 1 is fired and a ligand binds a receptor. **c** Between times τ_1 and τ_2, the system contains two free receptors, two free ligands, and a ligand-receptor complex. **d** At time τ_2, Rule 2 is fired and a pair of receptors are cross-linked. **e** Between times τ_2 and τ_3, the system contains two free ligands, one free receptor, and a complex of two receptors and one ligand. **f** At time τ_3, Rule 1 is fired and a ligand binds a receptor. This reaction differs from the instance of Rule 1 in Panel B because the receptor that undergoes reaction is already part of a complex. **g** Between times τ_3 and τ_4, the system contains one free ligand, one free receptor, and a chain of two ligands and two receptors. **h** At time τ_4, Rule 3 is fired and a ring or cyclic aggregate forms. **i** Between times τ_4 and τ_5, the system contains one free ligand, one free receptor, and a cyclic aggregate of two ligands and two receptors. **j** At time τ_5, the reverse of Rule 3 is fired and the cyclic aggregate is transformed into a chain or linear aggregate. **k** Between times τ_5 and τ_6, the system contains one free ligand, one free receptor, and a chain of two ligands and two receptors. This state is identical to the state of Panel G

9.4.1 Gillespie's Method

Gillespie's method [67–69], a method for stochastic simulation of chemical reaction systems, is useful because it takes into account two facts that a deterministic method is not designed to capture: a system contains a whole number of molecules, and reactions among molecules are subject to randomness. These qualities are likely to be important in systems where population sizes are small.

Gillespie's method consists of essentially the same steps as the n-fold way, described above. Both methods belong to the class of kinetic Monte Carlo methods

[66]. An implicit assumption of Gillespie's method is the assumption that an explicit list of the reactions that can occur in a system is available. A simulation proceeds as follows. First, initial population sizes and reaction rates are calculated. Reaction rates are calculated based on rate constants and numbers of reactant species. Rates are used to select the next event time and the next reaction. A reaction is then fired. Populations and rates are updated for the new state of the system, and simulation continues until a stopping criterion is satisfied. Variations of this method have been developed to increase its speed. For example, efficiency of simulation can be improved through use of a reaction classification scheme, as demonstrated in the method of Blue et al. [111] or Gibson and Bruck [112]. More recently, the method of Slepoy et al. [113] groups together reactions that share similar rates.

Reaction classification is an inherent feature of the rule-based approach: as a coarse-graining assumption, reactions implied by the same rule are assigned the same rate law. Thus, Gillespie's method is well-suited for simulation of rule-based models, if rates of all reactions implied by a rule can be calculated without explicitly deriving the reactions. These calculations are performed in network-free methods, of which there are multiple variants.

9.4.2 Algorithms for Network-Free Simulation

Gillespie's method has been generalized for simulation of rule-based models. These simulation methods are termed "network-free" because rules are used directly to advance the state of a system, thereby avoiding network generation. Currently, four related algorithms have been described for network-free simulation. These algorithms are summarized in Fig. 9.12. A main point of difference between them lies in the handling of non-local site properties. An example of a non-local site property is connectivity. The non-local environment of two sites must be examined if they are connected indirectly. Determining if two sites are indirectly connected is important for enforcing rule application conditions that place constraints on molecularity of rule-defined reactions. In general, non-local properties are more difficult to evaluate than local properties (e.g., whether a site is bound or free).

In the method of Danos et al. [31], rates are assumed to depend on local properties only. A waiting time is determined, a rule is selected, and sites are selected for rule application. The system is updated, rates and populations are recalculated, time is incremented, and simulation continues. The method of Yang et al. [94] performs the same calculation of rates as the method of Danos et al. [31]. However, after sites are selected based on local properties, non-local properties are checked. If a site is found to lack permissive non-local properties, it is rejected and a null event occurs. The method of Colvin et al. [21] avoids the rejection step by calculating rates exactly (i.e., with consideration of both local and non-local properties) before selecting rules and sites. Lastly, the method of

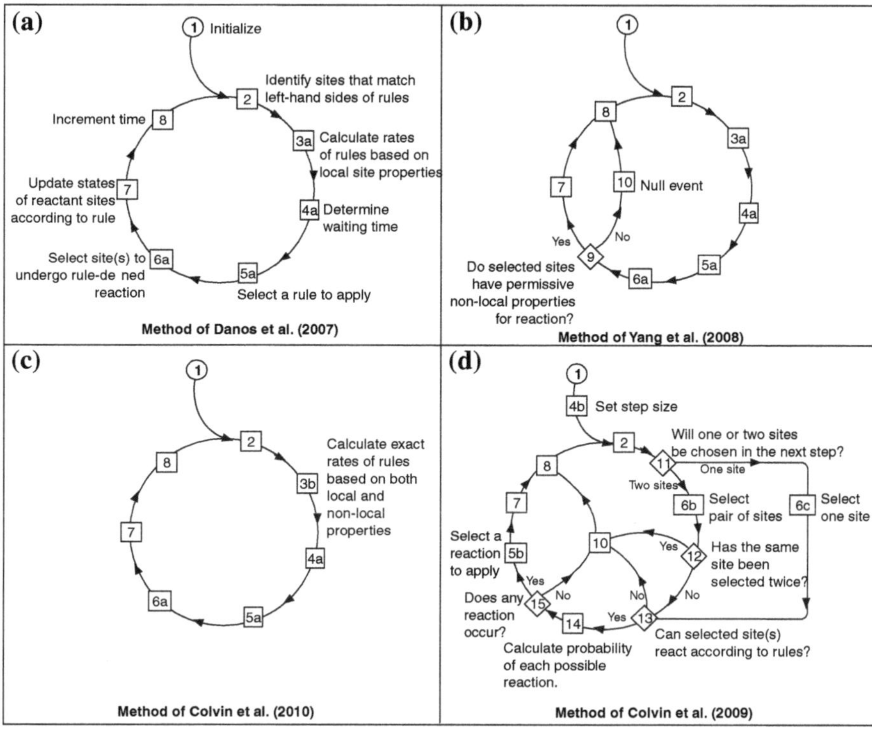

Fig. 9.12 A comparison of algorithms for network-free simulation. **a** The method of Danos et al. [114] can be applied if rule application depends on local context only. **b** The method of Yang et al. [94] introduces a rejection step to account for non-local site properties. **c** The method of Colvin et al. [21] calculates exact rule rates considering both local and non-local site properties. Thus, it has no rejection step. **d** The method of Colvin et al. [18] is a generalization of the StochSim method [38, 115], which has a number of distinguishing features, including a fixed time step and reversal of the site and rule selection steps. However, the method yields results consistent with other methods, as long as step size is below a threshold

Colvin et al. [18] differs from the other algorithms in a number of ways. The time step is fixed, and sites are selected before rules are selected. This method yields results consistent with the other methods as long as the time step is below a certain threshold, which is checked during simulation. The performance of tools implementing these methods has been compared, to a limited extent [20, 21, 109].

9.5 Outlook and Closing Remarks

Our accumulated knowledge about the mechanisms of cell signaling motivates the development of models that can capture these details. Current experimental capabilities that allow us to characterize the functional roles of specific protein sites and to monitor the dynamics of protein-protein interactions [116–122] makes the development of complementary modeling methods especially timely. A method that addresses this need is the rule-based modeling approach. By using rules to represent interactions, a modeler can avoid enumerating the reachable chemical species in a system, which is required in traditional approaches and which is a severe limitation given the typical size and combinatorial complexity of cell signaling (sub)systems. The rule-based approach allows models to be specified compactly, with simplifying assumptions that are more transparent and arguably less *ad hoc* than traditional modeling approaches [123]. With recent advances in simulation methodology, simulation of models that imply large reaction networks has become feasible. As a result, it is now possible to develop models that capture site-specific details of a large number of protein-protein interactions.

These capabilities are relevant for the study and, potentially, manipulation of cell signaling mechanisms. For example, different residues in the same protein may have different kinetics of phosphorylation, and each phosphorylated residue may regulate a distinct set of interactions (for example, see Houtman et al. [124]). As a result, perturbations that affect phosphorylation kinetics of specific sites (e.g., therapeutic kinase inhibitors, such as imatinib [125]) may be difficult to analyze without a model in which individual sites of phosphorylation are distinguished. However, traditional modeling approaches often necessitate a "virtual phosphorylation site" assumption [126], meaning that multiple sites are lumped together as a single, virtual site of phosphorylation. Roles for individual sites are not distinguished. This assumption can be lifted in a rule-based model more easily than in a traditional model.

Rule-based models can be specified using formal domain-specific languages (i.e., programming languages specialized for modeling). In contrast, traditional models for chemical kinetics formulated in terms of equations are more suitable for analysis (e.g., integration or differentiation) than for computation. Traditional modeling forms are used by many software tools, including tools that bridge equations to numerical methods of analysis (e.g., numerical integration), such as MATLAB (The MathWorks, Natick, MA). However, departure from traditional forms can be advantageous [83] and for mechanistic modeling of cell signaling systems, it is necessary. This need arises from the size and combinatorial complexity of signaling systems, which can be better captured if a model is viewed as a program. The reason is that a programming language can be tailored for the problem at hand. A model specified as a program has a number of other advantages over a set of equations. One advantage that has perhaps not yet been fully appreciated is greater extensibility and a potential for clearer annotation. As demonstrated by Thomson et al. [126], the formal elements of a rule-based model can be specified incrementally (i.e., one at a time), annotated independently, and then later assembled to address specific questions about system properties, which can also be formalized [97, 127, 128]. Guidelines for annotating rule-based models have been proposed [57], which if adopted, could make models more understandable and reusable. Rule-based modeling provides a general paradigm for modeling interactions of structured objects, with proven applications in physics, chemistry, and computer science. The approach is being used increasingly often in systems biology. In the future, we expect it to be a foundational method of the field because its extensibility addresses large network size, and the use of rules addresses combinatorial complexity, which are two inherent features of cell signaling systems.

Acknowledgments We thank Michael L. Blinov, James R. Faeder, David J. Klinke II, Jean Krivine, and Carlos F. Lopez for helpful discussions.

References

1. Kitano H (2002) Systems biology: a brief overview. Science 295:1662–1664
2. Kitano H (2002) Computational systems biology. Nature 420:206–210
3. Lazebnik Y (2002) Can a biologist fix a radio?–or, what I learned while studying apoptosis. Cancer Cell 2:179–182
4. Kreeger PK, Lauffenburger DA (2010) Cancer systems biology: a network modeling perspective. Carcinogenesis 31:2–8
5. Chakraborty AK, Das J (2010) Pairing computation with experimentation: a powerful coupling for under-standing T cell signalling. Nat Rev Immunol 10:59–71
6. Germain RN, Meier-Schellersheim M, Nita-Lazar A, Fraser IDC (2011) Systems biology in immunology–a computational modeling merspective. Annu Rev Immunol 29:527–585
7. Lander AD (2010) The edges of understanding. BMC Biol 8:40
8. Downward J (2011) Targeting RAF: trials and tribulations. Nat Med 17:286–288
9. Kholodenko BN (2006) Cell-signalling dynamics in time and space. Nat Rev Mol Cell Biol 7:165–176
10. Kholodenko BN, Hancock JF, Kolch W (2010) Signalling ballet in space and time. Nat Rev Mol Cell Biol 11:414–426
11. Hunter T (2000) Signaling–2000 and beyond. Cell 100:113–127
12. Pawson T, Nash P (2003) Assembly of cell regulatory systems through protein interaction domains. Science 300:445–452
13. Kandasamy K, Mohan SS, Raju R, Keerthikumar S, Kumar G, Venugopal A, Telikicherla D, Navarro JD, Mathivanan S, Pecquet C, Gollapudi S, Tattikota S, Mohan S, Padhukasahasram H, Subbannayya Y, Goel R, Jacob H, Zhong J, Sekhar R, Nanjappa V, Balakrishnan L, Subbaiah R, Ramachandra Y, Rahiman BA, Prasad TK, Lin JX, Houtman J, Desiderio S, Renauld JC, Constantinescu S (2010) NetPath: a public resource of curated signal transduction pathways. Genome Biol 11:R3
14. Hlavacek WS, Faeder JR, Blinov ML, Perelson AS, Goldstein B (2003) The complexity of complexes in signal transduction. Biotechnol Bioeng 84:783–794
15. Mayer BJ, Blinov ML, Loew LM (2009) Molecular machines or pleiomorphic ensembles: signaling complexes revisited. J Biol 8:81
16. Faeder JR, Blinov ML, Hlavacek WS (2009) Rule-based modeling of biochemical systems with BioNetGen. Methods Mol Biol 500:113–167
17. Andrews SS, Addy NJ, Brent R, Arkin AP (2010) Detailed simulations of cell biology with smoldyn 2.1. PLoS Comput Biol 6:e1000705
18. Colvin J, Monine MI, Faeder JR, Hlavacek WS, Von Hoff DD, Posner RG (2009) Simulation of large-scale rule-based models. Bioinformatics 25:910–917
19. Website about Kappa and Kappa-based software tools [http://kappalanguage.org/]
20. Sneddon MW, Faeder JR, Emonet T (2011) Efficient modeling, simulation, and coarse-graining of biological complexity with NFsim. Nat Methods 8:177–183

21. Colvin J, Monine MI, Gutenkunst R, Hlavacek WS, Von Hoff DD, Posner RG (2010) RuleMonkey: software for stochastic simulation of rule-based models. BMC Bioinf 11:404
22. Moraru II, Schaff JC, Slepchenko BM, Blinov ML, Morgan F, Lakshminarayana A, Gao F, Li Y, Loew LM (2008) Virtual Cell modelling and simulation software environment. IET Syst Biol 2:352–362
23. Meier-Schellersheim M, Xu X, Angermann B, Kunkel E, Jin T, Germain RN (2006) Key role of local regulation chemosensing revealed by a new molecular interaction-based modeling method. PLoS Comput Biol 2:e82
24. Angermann BR, Klauschen F, Garcia AD, Prustel T, Zhang F, Germain RN, Meier-Schellersheim M (2012) Computational modeling of cellular signaling processes embedded into dynamic spatial contexts. Nat Methods
25. Chen WW, Schoeberl B, Jasper PJ, Niepel M, Nielse UB, Lauffenburger D, Sorger PK (2009) Input–output behavior of ErbB signaling pathways as revealed by a mass action model trained against dynamic data. Mol Syst Biol 5:239
26. Dinkel H, Chica C, Via A, Gould CM, Jensen LJ, Gibson TJ, Diella F (2011) Phospho.ELM: a database of phosphorylation sites—update 2011. Nucleic Acid Res 39:D261
27. Bateman A, Coin L, Durbin R, Finn RD, Hollich V, GriffithsJones S, Khanna A, Marshall M, Moxon S, Sonnhammer ELL, Studholme DJ, Yeats C, Eddy SR (2004) The Pfam protein families database. Nucleic Acids Res 32:D138–D141
28. Dinkel H, Michael S, Weatheritt RJ, Davey NE, Van Roey K, Altenberg B, Toedt G, Uyar B, Seiler M, Budd A, Jdicke L, Dammert MA, Schroeter C, Hammer M, Schmidt T, Jehl P, McGuigan C, Dymecka M, Chica C, Luck K, Via A, Chatr-aryamontri A, Haslam N, Grebnev G, Edwards RJ, Steinmetz MO, Meiselbach H, Diella F, Gibson TJ (2012) ELM—the database of eukaryotic linear motifs. Nucleic Acids Res 40:D242–D251
29. Keshava Prasad TS, Goel R, Kandasamy K, Keerthikumar S, Kumar S, Mathivanan S, Telikicherla D, Raju R, Shafreen B, Venugopal A, Balakrishnan L, Marimuthu A, Banerjee S, Somanathan DS, Sebastian A, Rani S, Ray S, Harrys Kishore CJ, Kanth S, Ahmed M, Kashyap MK, Mohmood R, Ramachandra YL, Krishna V, Rahiman BA, Mohan S, Ranganathan P, Ramabadran S, Chaerkady R, Pandey A (2009) Human protein reference database—2009 update. Nucleic Acids Res 37:D767–D772
30. Hlavacek WS, Faeder JR, Blinov ML, Posner RG, Hucka M, Fontana W (2006) Rules for modeling signal transduction systems. Sci STKE, 2006:re6
31. Danos V, Feret J, Fontana W, Harmer R, Krivine J (2007) Rule-based modelling of cellular signalling. Lect Notes Comput Sci 4703:17–41
32. Fisher J, Harel D, Henzinger TA (2011) Biology as reactivity. Commun ACM 54:72–82
33. Lim WA, Pawson T (2010) Phosphotyrosine signaling: evolving a new cellular communication system. Cell 142:661–667
34. Endy D, Brent R (2001) Modeling cellular behavior. Nature 409:391–395
35. Bray D (2003) Molecular prodigality. Science 299:1189–1190
36. Le Novere N, Hucka M, Mi H, Moodie S, Schreiber F,Sorokin A, Demir E, Wegner K, Aladjem MI, Wimalaratne SM, Bergman FT, Gauges R, Ghazal P, Kawaji H, Li L, Matsuoka Y, Villeger A, Boyd SE, Calzone L, Courtot M, Dogrusoz U, Freeman TC, Funahashi A, Ghosh S, Jouraku A, Kim S, Kolpakov F, Luna A, Sahle S, Schmidt E, Watterson S, Wu G, Goryanin I, Kell DB, Sander C, Sauro H, Snoep JL, Kohn K, Kitano H (2009) The systems biology graphical notation. Nat Biotechnol 27:735–741
37. Bray D, Lay S (1997) Computer-based analysis of the binding steps in protein complex formation. Proc Natl Acad Sci 94:13493–13498
38. Morton-Firth CJ, Bray D (1998) Predicting temporal fluctuations in an intracellular signalling pathway. J Theor Biol 192:117–128
39. Le Novere N, Shimizu TS (2001) StochSim: modelling of stochastic biomolecular processes. Bioinformatics 17:575–576
40. Regev A, Silverman W, Shapiro E (2001) Representation and simulation of biochemical processes using the π-calculus process algebra. Pac Symp Biocomput 2001:459–470

41. Priami C, Regev A, Shapiro E, Silverman W (2001) Application of a stochastic name-passing calculus to representation and simulation of molecular processes. Inf Process Lett 80:25–31
42. The BioSPI project homepage [http://www.wisdom.weizmann.ac.il/~biospi/]
43. Dematte L, Priami C, Romanel A (2008) The BlenX language: a tutorial. Lect Notes Comput Sci 5016:313–365
44. Kahramanogullari O, Cardelli L, Caron E: An Intuitive Automated Modelling Interface for Systems Biology. In DCM 2009:73–86
45. Phillips A, Cardelli L (2007) Efficient, correct simulation of biological processes in the stochastic pi-calculus. Lect Notes Comput Sci 4695:184–199
46. Goldstein B, Faeder JR, Hlavacek WS, Blinov ML, Redondo A, Wofsy C (2002) Modeling the early signaling events mediated by FcεRI. Mol Immunol 38:1213–1219
47. Faeder JR, Hlavacek WS, Reischl I, Blinov ML, Metzger H, Redondo A, Wofsy C, Goldstein B (2003) Investigation of early events in FcεRI-mediated signaling using a detailed mathematical model. J Immunol 170:3769–3781
48. Nag A, Monine MI, Faeder JR, Goldstein B (2009) Aggregation of membrane proteins by cytosolic cross-linkers: theory and simulation of the LAT-Grb2-SOS1 system. Biophys J 96:2604–2623
49. Nag A, Monine MI, Blinov ML, Goldstein B (2010) A detailed mathematical model predicts that serial engagement of IgE-FcεRI complexes can enhance Syk activation in mast cells. J Immunol 185:3268–3276
50. Nag A, Blinov ML, Goldstein B (2010) Shaping the response: the role of FcεRI and Syk expression levels in mast cell signaling. IET Syst Biol 4:334–347
51. Monine MI, Posner RG, Savage PB, Faeder JR, Hlavacek WS (2010) Modeling multivalent ligand-receptor interactions with steric constraints on configurations of cell-surface receptor aggregates. Biophys J 98:48–56
52. Lee KH, Dinner AR, Tu C, Campi G, Raychaudhuri S, Varma R, Sims TN, Burack WR, Wu H, Wang J, Kanagawa O, Markiewicz M, Allen PM, Dustin ML, Chakraborty AK, Shaw AS (2003) The immunological synapse balances T cell receptor signaling and degradation. Science 302:1218–1222
53. Li QJ, Dinner AR, Qi S, Irvine DJ, Huppa JB, Davis MM, Chakraborty AK (2004) CD4 enhances T cell sensitivity to antigen by coordinating Lck accumulation at the immunological synapse. Nat Immunol 5:791–799
54. Altan-Bonnet G, Germain RN (2005) Modeling T cell antigen discrimination based on feedback control of digital ERK responses. PLoS Biol 3:e356
55. Nag A, Monine M, Perelson AS, Goldstein B (2012) Modeling and simulation of aggregation of membrane protein LAT with molecular variability in the number of binding sites for cytosolic Grb2-SOS1- Grb2. PLoS ONE 7:e28758
56. Faeder JR, Blinov ML, Hlavacek WS (2005) Graphical rule-based representation of signal transduction net- works. In: Liebrock L (ed.) Proceedings 2005 ACM Symposium on Applied Computing, ACM Press, New York, pp 133–140
57. Chylek LA, Hu B, Blinov ML, Emonet T, Faeder JR, Goldstein B, Gutenkunst RN, Haugh JM, Lipniacki T, Posner RG, Yang J, Hlavacek WS (2011) Guidelines for visualizing and annotating rule-based models. Mol BioSyst 7:2779–2795
58. Xu W, Smith AM, Faeder JR, Marai GE (2011) RuleBender: a visual interface for rule-based modeling. Bioinformatics 27:1721–1722
59. Blinov ML, Faeder JR, Goldstein B, Hlavacek WS (2004) BioNetGen: software for rule-based modeling of signal transduction based on the interactions of molecular domains. Bioinformatics 20:3289–3291
60. Danos V, Laneve C (2004) Formal molecular biology. Theoret Comput Sci 325:69–110
61. Feret J, Danos V, Krivine J, Harmer R, Fontana W (2009) Internal coarse-graining of molecular systems. Proc Natl Acad Sci USA 106:6453–6458
62. Barua D, Faeder JR, Haugh JM (2009) A bipolar clamp mechanism for activation of Jak-family protein tyrosine kinases. PLoS Comput Biol 5:e1000364

63. Dushek O, Das R, Coombs D (2009) A role for rebinding in rapid and reliable T cell responses to antigen. PLoS Comput Biol 5:e1000578
64. Metropolis N, Rosenbluth AW, Rosenbluth MN, Teller AH, Teller E (1953) Equation of state calculations by fast computing machines. J Chem Phys 21:1087–1092
65. Bortz AB, Kalos M, Lebowitz J (1975) A new algorithm for Monte Carlo simulations of Ising spin systems. J Comput Phys 17:10–18
66. Voter AF (2007) Introduction to the kinetic Monte Carlo method. In: Sickafus KE, Sickafus KE (eds) Radiation Effects in Solids. Springer, Kotomin, pp 1–21
67. Gillespie DT (1976) A general method for numerically simulating the stochastic time evolution of coupled chemical reactions. J Comput Phys 22:403–434
68. Gillespie DT (1977) Exact stochastic simulation of coupled chemical reactions. J Phys Chem 81:2340–2361
69. Gillespie DT (2007) Stochastic simulation of chemical kinetics. Annu Rev Phys Chem
70. Dugundji J, Ugi I (1973) An algebraic model of constitutional chemistry as a basis for chemical computer programs. Top Curr Chem 39:19–64
71. Ugi I, Bauer J, Bley K, Dengler A, Dietz A, Fontain E, Gruber B, Herges R, Knauer M, Reitsman K, Stein N (1993) Computer-assisted solution of chemical problems—the historic development and the present state of the art of a new discipline of chemistry. Agnew Chem Int Ed Engl 32:201–227
72. Green WH Jr (2007) Predictive kinetics: a new approach for the 21st century. Adv Chem Eng 32:1–50
73. Faulon JL, Carbonell P (2010) Reaction network generation. In: Faulon JL, Bender A (eds.) Handbook of Chemoinformatics Algorithms, Chapman & Hall/CRC Press, Boca Raton, pp 317–341
74. Rangarajan S, Bahn A, Daoutidis P (2010) Rule-based generation of thermochemical routes to biomass conversion. Ind Eng Chem Res 49:10459–10470
75. Klinke DJ II, Finley SD (2012) Timescale analysis of rule-based biochemical reaction networks. Biotechnol Progr
76. Klinke DJ II, Broadbelt LJ (1999) Construction of a mechanistic model of Fischer-Tropsch synthesis on Ni(1 1 1) and Co (0 0 0 1) surfaces. Chem Eng Sci 54:3379–3389
77. Broadbelt LJ, Pfaendtner J (2005) Lexicography of kinetic modeling of complex reaction networks. AIChE J 51:2112–2121
78. Hatzimanikatis V, Li C, Ionita JA, Henry CS, Jankowski MD, Broadbelt LJ (2005) Exploring the diversity of complex metabolic networks. Bioinformatics 21:1603–1609
79. Mu F, Unkefer CJ, Unkefer PJ, Hlavacek WS (2011) Prediction of metabolic reactions based on atomic and molecular properties of small-molecule compounds. Bioinformatics 27:1537–1545
80. Milner R, Parrow J, Walker D (1992) A calculus of mobile processes, I. Inform Comput 100:1–40
81. Fokkink W (2000) Introduction to process algebra. Springer, Berlin
82. Fisher J, Henzinger TA (2007) Executable cell biology. Nat Biotechnol 25:1239–1249
83. Harmer R (2009) Rule-based modeling and tunable resolution. EPTCS 9:65–72
84. Harmer R, Danos V, Feret J, Krivine J, Fontana W (2010) Intrinsic information carriers in combinatorial dynamical systems. Chaos 20:037108
85. PySB Python framework for Systems Biology modeling [http://pysb.org/]
86. Lopez CF, Muhlich JL, Bachman JA, Sorger PK (2013) Programming biological models in Python with PySB. Mol Syst Biol 9:646
87. Faeder JR, Blinov ML, Goldstein B, Hlavacek WS (2005) Rule-based modeling of biochemical networks. Complexity 10:22–41
88. Lok L, Brent R (2005) Automatic generation of cellular reaction networks with Moleculizer 1.0. Nat Biotechnol 23:131–136
89. Harris LA, Hogg JS, Faeder JR (2009) Compartmental rule-based modeling of biochemical systems. In: Rossetti M, Hill R, Johansson B, Dunkin A, Ingallls R (eds.) Proceedings of the 2009 Winter Simulation Conference. pp 908–919

90. Maus C, Rybacki S, Uhrmacher AM (2011) Rule-based multi-level modeling of cell biological systems. BMC Syst Biol 5:166
91. Mallavarapu A, Thomson M, Ullian B, Gunawardena J (2009) Programming with models: modularity and abstraction provide powerful capabilities for systems biology. J R Soc Interf 6:257
92. Lis M, Artyomov MN, Devadas S, Chakraborty AK (2009) Efficient stochastic simulation of reaction-diffusion processes via direct compilation. Bioinformatics 25:2289–2291
93. Yang J, Monine MI, Faeder JR, Hlavacek WS (2008) Kinetic Monte Carlo method for rule-based modeling of biochemical networks. Phys Rev E 78:031910
94. Gruenert G, Ibrahim B, Lenser T, Lohel M, Hinze T, Dittrich P (2010) Rule-based spatial modeling with diffusing, geometrically constrained molecules. BMC Bioinf 11:307
95. RuleBase [http://rulebase.org/]
96. Hu B, Fricke GM, Faeder JR, Posner RG, Hlavacek WS (2009) GetBonNie for building, analyzing and sharing rule-based models. Bioinformatics 25:1457–1460
97. Clarke EM, Faeder JR, Harris LA, Langmead CJ, Legay A, Jha SK (2008) Statistical model checking in BioLab: applications to the automated analysis of T-cell receptor signaling pathway. Lect Notes Comput Sci 5307:231–250
98. Koschorreck M, Gilles E (2008) ALC: automated reduction of rule-based models. BMC Syst Biol 2:91
99. Ollivier JF, Shahrezaei V, Swain P (2010) Scalable rule-based modeling of allosteric proteins and biochemical networks. PLoS Comput Biol 6:e1000975
100. Fages F, Soliman S, Chabrier-Rivier N (2004) Modelling and querying interaction networks in the biochemical abstract machine BIOCHAM. J Biol Phys Chem 4:64–73
101. Priami C, Ballarini P, Quaglia P (2009) BlenX4Bio–BlenX for Biologists. In: Computational Methods in Systems Biology, Springer, pp 26–51
102. Vilar JMG, Saiz L (2010) CplexA: a Mathematica package to study macromolecular-assembly control of gene expression. Bioinformatics 26:2060–2061
103. Tolle D, Nov'ere L (2010) Meredys, a multi-compartment reaction-diffusion simulator using multistate realistic molecular complexes. BMC Syst Biol 4:24
104. Eker S, Knapp M, Laderoute K, Lincoln P, Talcott C (2004) Pathway Logic: Executable models of biological networks. Electron Notes Theor Comput Sci. 71:125–142
105. Maiwald T, Timmer J (2008) Dynamical modeling and multi-experiment fitting with PottersWheel. Bioinformatics 24:2037–2043
106. The KaSim user manual [http://cloud.github.com/downloads/jkrivine/KaSim/KaSim_manual.pdf]
107. Bonabeau E (2002) Agent-based modeling: methods and techniques for simulating human systems. Proc Natl Acad Sci USA, 99:7280–7287
108. Yang J, Hlavacek WS (2011) Efficiency of reactant site sampling in network-free simulation of rule-based models for biochemical systems. Phys Biol 8:055009
109. Goldstein B (1988) Desensitization, histamine release and the aggregation of IgE on human basophils. In: Perelson AS (ed.) Theoretical immunology, part one, SFI studies in the sciences of complexity. Addison-Wesley, Reading, MA, pp 3–40
110. Blue JL, Beichl I, Sullivan F (1995) Faster Monte Carlo simulations. Phys Rev E 51:R867–R868
111. Gibson MA, Bruck J (2000) Efficient exact stochastic simulation of chemical systems with many species and many channels. J Phys Chem A 104:1876–1889
112. Slepoy A, Thompson AP, Plimpton SJ (2008) A constant-time kinetic Monte Carlo algorithm for simulation of large biochemical reaction networks. J Chem Phys 128:205101
113. Danos V, Feret J, Fontana W, Krivine J (2007) Scalable simulation of cellular signalling networks. Lect Notes Comput Sci 4807:139–157
114. Shimizu TS, Bray D (2001) Computational cell biology—the stochastic approach. In: Kitano H (ed.) Foundations of systems biology. MIT Press
115. Houtman JCD, Barda-Saad M, Samelson LE (2005) Examining multiprotein signaling complexes from all angles. FEBS J 500:5426–5435

116. Schulze WX, Deng L, Mann M (2005) Phosphotyrosine interactome of the ErbB-receptor kinase family. Mol Syst Biol 2005(1):0008
117. Zhang Y, Wolf-Yadlin A, Ross PL, Pappin DJ, Rush J, Lauffenburger DA (2005) Phosphotyrosine interactome of the ErbB-receptor kinase family. Mol Cell Proteomics 4:1240–1250
118. Jones RB, Gordus A, Krall JA, MacBeath G (2006) A quantitative protein interaction network for the ErbB receptors using protein microarrays. Nature 439:168–174
119. VanMeter AJ, Rodriguez AS, Bowman ED, Jen J, Harris CC, Deng J, Calvert VS, Silvestri A, Fredolini C, Chandhoke V, Petricoin EF, Liotta LA, Espina V (2008) Laser capture microdissection and protein microar- ray analysis of human non-small cell lung cancer: differential epidermal growth factor receptor (EGPR) phosphorylation events associated with mutated EGFR compared with wild type. Mol Cell Proteomics 7:1902–1924
120. Ciaccio MF, Wagner JP, Chuu CP, Lauffenburger DA, Jones RB (2010) Systems analysis of EGF receptor signaling dynamics with microwestern arrays. Nat Methods 7:148–155
121. Cox J, Mann M (2011) Quantitative, high-resolution proteomics for data-driven systems biology. Annu Rev Biochem 80:273–299
122. Blinov ML, Faeder JR, Golstein B, Hlavacek WS (2006) A network model of early events in epidermal growth factor receptor signaling that accounts for combinatorial complexity. Biosystems 83:136–151
123. Houtman JCD, Houghtling RA, Barda-Saad M, Toda Y, Samelson LE (2005) Early phosphorylation kinetics of proteins involved in proximal TCR-mediated signaling pathways. J Immunol 175:2449
124. Sawyers C (2004) Targeted cancer therapy. Nature 432:294–297
125. Birtwistle MR, Hatakeyama M, Yumoto N, Ogunnaike BA, Hoek JB, Kholodenko BN (2007) Ligand-dependent responses of the ErbB signaling network: experimental and modeling analyses. Mol Syst Biol 3:144
126. Thomson T, Benjamin KR, Bush A, Love T, Pincus D, Resnekov O, Yu RC, Gordon A, Colman-Lerner A, Endy D, Brent R (2011) Scaffold number in yeast signaling system sets tradeoff between system output and dynamic range. Proc Natl Acad Sci USA 13:20265–20270
127. Calzone L, Fages F, Soliman S (2006) BIOCHAM: an environment for modeling biological systems and formalizing experimental knowledge. Bioinformatics 22:1805–1807
128. Gong H, Zuliani P, Komuravelli A, Faeder JR, Clarke EM (2010) Analysis and verification of the HMGB1 signaling pathway. BMC Bioinf 11:S10

Chapter 10
Reproducibility of Model-Based Results in Systems Biology

Dagmar Waltemath, Ron Henkel, Felix Winter and Olaf Wolkenhauer

Abstract Science requires that results are reproducible. This is naturally expected for wet-lab experiments and it is equally important for model-based results published in the literature. Reproducibility, in general, requires standards that provide the information necessary and tools that enable others to re-use this information. In computational biology, reproducibility requires not only a coded form of the model but also a coded form of the experimental setup to reproduce the analysis of the model. Well-established databases and repositories store and provide mathematical models. Recently, these databases started to distribute simulation setups together with the model code. These developments facilitate the reproduction of results. In this chapter, we outline the necessary steps towards reproducing model-based results in computational biology. We exemplify the workflow using a prominent example model of the Cell Cycle and state-of-the-art tools and standards.

Keywords Reproducibility · Simulation experiments · Standards · SED-ML · SBML · CellML · Model management

Acronyms

SBML	Systems Biology Markup Language
ChEBI	Chemical Entities of Biological Interest
RDF	Resource Description Framework
GO	Gene Ontology
TEDDY	TErminology for the Description of DYnamics

D. Waltemath (✉) · R. Henkel · F. Winter · O. Wolkenhauer
Department of Systems Biology and Bioinformatics, Rostock University, 18051 Rostock, Germany
e-mail: dagmar.waltemath@uni-rostock.de, ron.henkel@uni-rostock.de, felix.winter@uni-rostock.de

O. Wolkenhauer
Stellenbosch Institute for Advanced Study (STIAS), Wallenberg Research Centre at Stellenbosch University, Marais Street, Stellenbosch 7600, South Africa
e-mail: olaf.wolkenhauer@uni-rostock.de

A. Prokop and B. Csukás (eds.), *Systems Biology*,
DOI: 10.1007/978-94-007-6803-1_10,

COMBINE	The COmputational Modeling in BIology NEtwork
XML	Extensible Markup Language
OWL	Web Ontology Language
MIRIAM	Minimum Information Required in the Annotation of Models
URN	Uniform Resource Name
PMR2	Physiome Model Repository
COPASI	Complex Pathway SImulator
SED-ML	Simulation Experiment Description Markup Language
MIASE	Minimum Information About a Simulation Experiment
IR	Information Retrieval
DDMoRe	Drug Disease Model Resources

10.1 Introduction

Computer science technologies and methods support research in various areas. In systems biology, they provide means of modeling and simulation to increase the understanding of biological systems. This support accelerates the scientific process, and allows for sophisticated analyses of complex biological systems. Furthermore, testing hypotheses computationally reduces time and costs for experimental biologists. Similar to wet-lab experiments, results obtained from computational models must be reproducible.

One example for the successful application of modeling and simulation on a biological question is the analysis of the cell cycle, which investigates the processes a cell undergoes to divide and replicate itself. The cell is a so-called autopoietic system, i.e. it is capable of self-creating the processes that reproduce itself; a key process in cell replication is the division of the cell. One of the earliest attempts to simulate the molecular details of the cell cycle is the well-known computational model published by John Tyson in 1991 [1]. The model explains the complex cell cycle mechanism in an abstract way by using only six interacting proteins and nine reactions. It shows that the oscillating concentration of the Maturation Promotion Factor (MPF) is dependent on the Cyclin concentration. Over the past years, Tyson's lab developed a number of enhanced computational models based on their first representations of the cell cycle, e.g. [2–4]. We will here use a computational representation of Tyson's original model of the cell cycle to exemplify the usefulness of concepts and tools for the reproduction of model-based results.

A *computational model of a biological system* is an abstract representation of the living system, simplified by a number of assumptions, and instantiated with a certain set of parameter values. Computational models of biological systems can be diverse in scale and complexity, ranging from 'omics'-scale (i.e. modeling whole genomes and proteomes) to modeling small sub-circuits of a network [5].

During development, these models are often written in software tools such as Matlab, or are directly programmed in languages such as R, Python, Java or C. One way of studying the models and thereby obtaining results is through *simulation*. A simulation mimics the behavior of the system, for example by calculating the changes in concentration of a particular entity over time. It is important to understand that simulations are not part of a model, but can be applied on a model to study its behavior. To reproduce a model-based result, one does therefore not only need the model itself, but one must also have available information about the simulation that was run on the model to show a desired effect, for example oscillating behavior. When models and associated simulations need to be exchanged (e.g. together with a publication, or in large-scale collaborative projects) model databases help distributing code. Here, specifically designed, computer-readable *standard representation formats* allow for model exchange across different software tools and projects. Exporting model code in a standard format is a common practice in many modeling communities, for example when developing biochemical models, or in physiology.

One feature of many cell cycle models is the reproduction of the oscillatory limit cycle[1] which is observable in wet-lab experiments. To reproduce this feature on a computer, an executable version of the model must be run. In many cases it is possible to manually transform the original model equations (as given in the respective publication) into an executable file. However, it is much more efficient to reuse an existing implementation of a model in a standard format. Computational encodings of the Tyson models are already available from open databases. For example, one encoding of the 1991 Tyson model [1] can be obtained from BioModels Database [6], an open repository of computational biology models. The encoding[2] of that model (in SBML format) contains the model structure and initial parameterization. To understand the intention behind the Tyson model one may read the reference publication, study related models, explore the information available from the BioModels Database site, or investigate simulation experiments performed on the model. In addition, if the model is sufficiently annotated with information from *biology ontologies* one can infer semi-automatically the model's scope and level of detail together with the implied assumptions. Annotations link to entries in openly available biology ontologies such as the Gene Ontology [7] or ChEBI [8]. These links are stored using semantic technologies such as the Resource Description Framework (RDF) [9]. For example, a modeler can semantically enrich the cell cycle models with entries from the Gene Ontology. She could link the above-mentioned Tyson model to the biological term "Opisthokonta", the broad group of eukaryotes to which the modeled yeast belongs to. Thereby, she supports the interpretation of the model's biological scope. A model with a detailed explanation of *all* involved species can be faster understood; a

[1] A limit cycle is "a closed orbit which is isolated, i.e. neighboring orbits are not closed". (Definition: TEDDY ontology, http://www.biomodels.net/teddy)

[2] http://www.ebi.ac.uk/biomodels-main/BIOMD0000000005

model with associated simulation experiments and simulation results can be better interpreted, and modifications on the simulation settings can be easier reconstructed.

The provision of annotated models in standard format is particularly important as the modeling process more often than not demands to incorporate previously obtained findings from collaborators or literature, as has been exemplified with the yeast metabolic network model [10]. Standards for model code ease the exchange of models across work groups, as they make model code usable in different software environments. The exchange of computational models enables the reproduction of other researchers' works, the study of results that have already been obtained for a given biological question, and even the adjustment or extension of models to test own hypotheses. In summary, to reproduce model-based results in computational biology,

1. Models should be encoded in standard formats;
2. Meta-information should be provided to help understanding the models' intention; and
3. Associated simulation experiments should be encoded in standard formats.

All information must be made available through open repositories to foster reuse and consequently allow for result reproducibility. In this chapter we describe existing standard formats for model encoding, and how models encoded in these formats are made available through model repositories. We introduce a standard format for simulation experiment descriptions and we exemplify how the existing infrastructure of model databases and exchange formats allows modelers to reproduce other researcher's results.

10.2 Standard Formats for Computational Models

People can't share knowledge if they don't speak a common language
Tom Davenport, Lawrence Prusak [11]

Standardization plays a central role in facilitating the exchange and interpretation of the outcomes of scientific research, and in particular of computational modeling [12]. Already in 1969, David Garfinkel reported on the development of standard formats to describe "what a model should be like, how it should be described and documented [...] to facilitate communication of information about models" [13]. Standardization is crucial to data exchange, information exchange and knowledge exchange.[3] Standardized, machine-readable formats facilitate model exchange between users, databases and different simulation tools, and they enable the unified description of models. Nicolas Le Novère, who contributed to

[3] Please refer to Bellinger et al. [14] for an overview about the distinction between data, information, and knowledge

many standardization efforts, even goes as far as saying that "quantitative models will be only as useful as their access and reuse is easy for all scientists" [15]. Recently launched, cross-standard community meetings such as the annual COMBINE[4] meeting allow close interaction between researchers from the different fields dealing with computational biology. These meetings foster the development of standard formats for models and simulation descriptions. Model reuse is already practiced, but it is limited. Examples for phylogenies of models include models on MAPK, Glycolysis, or the aforementioned cell cycle. In these examples, existing models evolved into more complex representations of a system.

Exchangeable models are supplied in standard representation formats and through open repositories. *Representation formats* are typically defined in the Extensible Markup Language (XML) [16]; sometimes the Web Ontology Language (OWL) [17] is used. XML is a markup-language that permits model encodings to be stored in a structured way, using pre-defined XML elements, or tags. Owing to the diversity of modeling frameworks used in computational biology, several standard formats exist to cover the various aspects of biology [18]. We will here introduce two of them, namely the *Systems Biology Markup Language* (SBML) [19] and *CellML* [20]. Further formats are being established, e.g. for the encoding of neurophysiological models in *Neuro ML* [21], or for discrete event based modeling frameworks [22].

SBML is a community effort encompassing researchers and software developers from diverse backgrounds in academia and industry. It is a standard representation format for the description of biochemical reaction networks, including cell signaling pathways, but also metabolic pathways, gene regulation networks etc. To date SBML has been adopted by more than 200 software systems from simulators to modeling tools and databases. As such it can be considered the most successful standard in the field so far [6]. SBMLs major revisions are called levels. A level represents substantial changes to the composition and structure of the language. The current language specification is called SBML Level 3, Version 1 core. With Level 3, SBML has switched to a modular, *plug-in* type of language specification: In addition to the core language, so-called "packages" are developed for specific modeling needs, e.g. a spatial package. A core SBML model incorporates the following main parts [19]:

- *Function definition*: A named mathematical function available for use in the model,
- *Unit definition*: A named new unit of measurement,
- *Compartment*: A container for species location, either representing physical structures or not. It is assumed to be a well-stirred one,
- *Species*: A pool of any kind of entities of the same kind,
- *Parameter*: A quantity with a symbolic name (constants or variable, global or local),

[4] http://co.mbine.org/

- *Initial assignment*: A mathematical expression defining the initial condition of the model,
- *Rule*: A mathematical expression defining how to calculate the value or the rate of change of a variable,
- *Constraint*: A mathematical expression computing a true/false value from model variables, parameters and constants,
- *Reaction*: A statement with an associated rate expression and describing a process that might change the amount of one or more species,
- *Event*: A statement describing instantaneous, discontinuous change in one or more variables when a condition is triggered.

An extract of the SBML representation of the "Tyson model", which is available from BioModels Database, is shown in Fig. 10.1. The example encodes Cyclin-dependent kinase cdc2 as an SBML species. Species define pools of biological entities within an SBML model. Each species has an id (id = "C2"), a location (compartment = "cell"), and an initial amount or concentration (initialAmount = "0"). The optional name (cdc2 k) helps when displaying the model code to the users. SBML reuses the MathML standard to encode the mathematical expressions contained in the model, including the kinetic rates, equations, or function definitions (not shown in the code snippet). The SBML species is additionally annotated with meta-information in RDF format.

CellML is a model description language for specifying and exchanging biological processes. The current version is CellML 1.1. It has a modular structure and focuses on the mathematical description of biological processes. The explicit representation of modularity and the extensibility of the language both permit the

```
<species metaid="_000004" id="C2" name="cdc2k" compartment="cell"
initialAmount="0" xmlns="http://www.sbml.org/sbml/level2/version4">
  <annotation>
    <rdf:RDF
    xmlns:rdf="http://www.w3.org/1999/02/22-rdf-syntax-ns#"
xmlns:dc="http://purl.org/dc/elements/1.1/"
xmlns:dcterms="http://purl.org/dc/terms/"
xmlns:vCard="http://www.w3.org/2001/vcard-rdf/3.0#"
xmlns:bqbiol="http://biomodels.net/biology-qualifiers/"
xmlns:bqmodel="http://biomodels.net/model-qualifiers/">
      <rdf:Description rdf:about="#_000004">
        <bqbiol:isVersionOf>
          <rdf:Bag>
            <rdf:li rdf:resource="urn:miriam:uniprot:P04551" />
          </rdf:Bag>
        </bqbiol:isVersionOf>
      </rdf:Description>
    </rdf:RDF>
</annotation>
</species>
```

Fig. 10.1 SBML code snippet from the Tyson Cell Cycle model, showing the encoding of a biological entity, the Cyclin-dependent kinase cdc2 (SBML species), and its annotation description of a range of cellular and sub-cellular systems, including biochemistry, electrophysiology, system physiology and the mechanics of the intracellular environment [23]. CellML, in contrast to SBML, provides generic components which may be abstract groupings, or be representing biological entities and physical compartments. A CellML model is built as a network of connections between self-contained elements [24]:

- *Model*: The CellML root element, contains all the following elements,
- *Component*: Smallest functional unit of a model, contains the variables and mathematics to describe the behavior of the subsystem,
- *Connection*: Connects components to each other, and maps variables in one component to variables in another,
- *Import*: Allows for import of further valid CellML models,
- *Unit*: Allows for the definition of units, apart from standard units already provided; every variable and number has to have a unit assigned,
- *Group*: Allows for the definition of logical (encapsulation) and physical (containment) relationships between components to form hierarchical structures (i.e. a tree of components linked by parent–child relationships of the same type).

The Tyson model is not only available in SBML format, but also in CellML. The code snippet in Fig. 10.2 represents the afore-mentioned Cdc2 kinase, which is here modeled as a CellML component. The component is additionally annotated with meta-information in RDF format. Similarly to SBML, an initial value is assigned to the component, and its interaction within the network is specified using variable declarations and MathML (not shown in the code snippet).

The MIRIAM guidelines. The *Minimum Information Requested in the Annotation of Models* (MIRIAM) [25] describes the minimum of information necessary to be provided together with a computational model in order for it to be useful to others. Both the SBML and CellML format follow these guidelines. Annotations provide links to information that is not covered in the XML

```
<component cmeta:id="C2" name="C2"
xmlns:cmeta=http://www.cellml.org/metadata/1.0#
xmlns="http://www.cellml.org/cellml/1.0#">
  <rdf:RDF xmlns:rdf="http://www.w3.org/1999/02/22-rdf-syntax-ns#">
    <rdf:Description rdf:about="#C2">
      <dc:title xmlns:dc="http://purl.org/dc/elements/1.1/">C2</dc:title>
      <dcterms:alternative
xmlns:dcterms="http://purl.org/dc/terms/">cdc2</dcterms:alternative>
    </rdf:Description>
  </rdf:RDF>
  <variable cmeta:id="C2_C2" initial_value="0.001" name="C2"
   public_interface="out" units="millimolar" />
  <variable name="k6" public_interface="in" units="first_order_rate_constant" />
  <variable name="k8" public_interface="in" units="second_order_rate_constant" />
  <variable name="k9" public_interface="in" units="first_order_rate_constant" />
  <variable name="M" public_interface="in" units="millimolar" />
  <variable name="P" public_interface="in" units="millimolar" />
  <variable name="CP" public_interface="in" units="millimolar" />
  <variable name="time" public_interface="in" units="minute" />
</component>
```

Fig. 10.2 CellML component representing Cyclin-dependent kinase cdc2 in the Tyson model representation, but must be made available. These links point to entries in external biology ontologies, databases or classifications. A human-understandable interpretation of the SBML code shown in Fig. 10.1 can be generated by resolving the model annotation, which is also shown in the figure: The species cdc2k is linked to a MIRIAM URN which can be resolved (urn:miriam:uniprot:P04551) using the MIRIAM query services.[5] The URN points to the entry with ID P04551 in the UniProt ontology.[6] P04551 represents "Cyclin-dependent kinase". A detailed description of the biological entity is available online from the term entry in Uniprot.

The MIRIAM guidelines are a textual description—they are not computer-readable. To get a better understanding of the information encoded in standard formats and annotations, we recommend reading the publication [25], or at least browsing through the listed "rules".

10.3 Models in Public Repositories

Many published models are openly available from model repositories, and most journals today demand the submission of model code together with the manuscript. We will here briefly describe the aforementioned *BioModels Database* [6], the *CellML Physiome Model Repository* [26], and the *JWS Online Model repository* [27]. These databases have developed into valuable knowledge resources over the past years to guarantee modelers easy access to published model code. All three repositories use standard representation formats for code export. As a result, support is restricted to models that can be encoded using these standards. However, the exchangeability of models is ensured.

Bio Models Database [6, 28] is a repository of freely accessible model code.[7] It is an open-source project for commercial and academic use. BioModels Database accepts models submitted by modelers (e.g. for reference in a publication), but it also imports models from collaborative model repositories such as the CellML Model Repository. While the main focus is on SBML-encoded models, the import of other formats is possible using specifically developed converters. The number of models available from open model repositories grows quickly. BioModels Database very well reflects this tendency [29]: The number of stored, curated models has increased from 22 in the year 2005 when BioModels Database was launched to 854 models in May 2012. In the 20th release 142050 more pathway models were added to the repository. However, BioModels Database does not only grow in terms of the amount of models, but also with regard to the variety of modeled entities and the size of encoded reaction networks. Cell cycle models in BioModels

[5] http://www.ebi.ac.uk/miriam/main

[6] http://www.uniprot.org/

[7] www.ebi.ac.uk/biomodels-main/

Database, for example, have over the years increased in complexity and size: An XML-encoded version of the first "Tyson model" represents the published model in SBML format using nine physical entities. One of its successors, the 1993 Novak/Tyson model [30] already encodes 14 species and 23 reactions; both models are single-compartmental. Since then, more models on the cell cycle have been published; one example is the 2004 cell cycle model by Chen et al. [31]; its computational encoding is built of 74 species and 94 reactions; it is still a single-compartmental model. The most complex model in BioModels Database in terms of number of encoded species and reactions as well as size of the model file is a model for the yeast molecular interaction network. It contains 36,263 species and 30,965 reactions in one compartment; the XML file has a size of 39.7 MB.

BioModels Database provides several services, including model curation and annotation. Models can easily be accessed through the web interface, and even be simulated using one of the embedded simulation tools [6]. The SBML file history is tracked using a version control system. The so-called model metadata is stored separately in a MySQL database. Metadata that is available from the SBML file includes information on the submission and modification dates of a model file, authors' information, references and annotations encoded in a MIRIAM-compliant manner. Stored models can either be browsed from a list of available models (sorted by BioModels Database ID, model name, publication ID, or modification date) or in a tree-structured classification that is based on Gene Ontology terms.

The **Physiome Model Repository** (PMR2) [26] is the standard repository for CellML models at different stages of curation. PMR2 implements a model management system based on the content management system *Plone*. It can be used either as an online application[8] or as a standalone application. Available models cover different biological processes, including signal transduction pathways, metabolic pathways, electrophysiology, immunology, cell cycle, muscle contraction and mechanical models [23]. It is the intention of the maintainers of the repository to bring forward the model curation and annotation process so that ideally all models replicate the results in the published paper. The CellML model repository contains about 500 model exposures encoded in CellML format (as of April 2012). A CellML exposure consists of a model and its associated documentation and meta-information. Models in PMR2 are browsable by different (physiological) categories, including cell cycle, signal transduction, or metabolism. A system-wide full-text search is offered that permits simple free text search. Additionally, models of different states of curation can be searched.

JWS Online Model Database [27] is part of the JWS Online Simulator [32], a web-based simulator for biochemical kinetic models.[9] The model repository serves as the maintainer for kinetic models that can be interactively run online. It supports the search for SBML models by a limited number of characteristics, including the author, publication title and journal, organism or model type. A web-based tool

[8] http://www.cellml.org/models

[9] http://jjj.biochem.sun.ac.za/database

offers a searchable categorization of models in the repository, distinguishing, for example, between "cell cycle" models and "metabolism". JWS Online is used as a model repository within collaborative European projects. For example, it is integrated with the SEEK platform [33] which is used by the SysMO consortium.

Encodings of the 1991 cell cycle model are available in all three databases described above. BioModels Database and the JWS Online Model Database both provide SBML versions of the model, while PMR2 offers a CellML encoding. Based on the users' preferences, the model could either be run in an SBML-aware simulation tool, e.g. COPASI [34], or one that reads CellML, e.g. OpenCOR [35]. The three repositories, apart from the format they naturally support, have different foci: BioModels Database offers a rich set of model-related information together with the model, including lists of encoded biological entities and links to the ontology entries associated to them, visualization of the model structure, and information about the original publication. All models available from the curated branch have been tested to be reproducing the results mentioned in the publication. A detailed search system allows searching for models by key words and model content. In contrast, JWS Online Model Database provides a lighter interface to its models, but links more tightly to the JWS Online simulator which allows users to run the model directly from the web page. The CellML Model repository then considers the model as part of a bundle of files (so-called exposures) in a version-controlled repository. In PMR2, the basic information about a model is presented on the web site, all files associated with the model and its different versions are provided through the content management system.

The list of model repositories named above is by far not complete. ModelDB[10], for example, is a valuable resource of open model code in computational neuroscience. Models stored in ModelDB can be searched for based on the meta-information which had been provided at model upload. ModelDB, in contrast to the databases mentioned above, does not put restrictions on the format of submitted model code, and it does not require standard formats. Consequently, ModelDB can handle different types of model code very flexibly. However, the free choice of model formats hampers reuse. Whenever downloading a model from ModelDB, the user must check if she has available the facilities to run the model code on her machine, or if she needs to install additional simulation tools capable of running the code. For example, ModelDB offers a large number of models dealing with synaptic plasticity. Some of them require Matlab to execute m-files, others demand XPP or C++. Consequently, users may have to set up a whole computational environment in order to reuse the model code. The NeuroML standard is the equivalent in computational neuroscience of what SBML is for computational biology. The NeuroML community aims at making models in NeuroML format available from ModelDB. Similar developments in other realms of biology and medicine, such as ecology or pharmacometrics, will benefit from the experiences already gained in the SBML and CellML communities.

---

[10] http://senselab.med.yale.edu/modeldb/

In order to reuse a model, it is essential to find the model in the first place. Hence, sophisticated model retrieval techniques are a prerequisite to model reuse. The model retrieval problem relates to the well-known problem of information retrieval (IR) which deals with techniques to efficiently find relevant data [36]. Together with techniques and algorithms to rank the results of a search, relevant models can be sorted according to the user's demands, thereby improving the model retrieval process [29]. The approach has been implemented as a two-step semantic search in BioModels Database.[11]

10.4 Simulation Experimental Setups

> An unplanned, hit-or-miss course of experimentation with a simulation model can often be frustrating, inefficient, and ultimately unhelpful. David Kelton [37]

Model representation formats are widely accepted and used to describe model structure, but they do not cover the description of simulations or analyses performed with the models. However, once a model has been successfully retrieved from a model repository in standard format, the next step is to simulate that model to obtain a desired output. The above-mentioned MIRIAM guidelines more formally state that (rule 6):

> The model, when instantiated within a suitable simulation environment, must be able to reproduce all relevant results given in the reference description that can readily be simulated.

This statement corresponds with Pawlikowski's earlier demand that "any scientific activity should be based on controlled and independently repeatable experiments" [38]. In summary, whenever performing experiments one must ensure that for many repetitions of a simulation the same final results can be obtained, with acceptably small statistical errors for stochastic simulations.

One prerequisite for result reproducibility is the existence of a valid, useful and documented computational encoding of the experiment. To address this need, the *Simulation Experiment Description Markup Language* (SED-ML, [39]) has been designed. It is an XML-based format for the encoding of simulation experiments that can be applied on a set of computational models. SED-ML follows the *Minimum Information About a Simulation Experiment* (MIASE) guidelines [40] in the same way that SBML obeys to the MIRIAM guidelines. MIASE defines, in a textual description, the information that is to provide together with a model in order to ensure the reproducibility of the experiments run on that model. SED-ML is the computational format to store and exchange that information.

SED-ML. To enable result reproducibility in the life sciences, model authors should ideally provide SED-ML files together with their model code, describing

[11] http://www.ebi.ac.uk/biomodels-demo/search

how to reproduce the presented simulation results. End-users can thereupon run the simulation experiments in simulation software of their choice. They might furthermore share their own simulation experiment descriptions by exporting SED-ML descriptions from their simulation tool. SED-ML is agnostic about the underlying model representation formats and the software tool that ran the experiment. The model variables that a SED-ML model needs to be aware of are addressed directly by XPath; the only limitation is for the model code to be XML.

The SED-ML format is built of five main elements: (1) the models used in the experiment; (2) the simulation algorithms and their configurations; (3) the combination of algorithm and model into a numerical experiment; (4) post-processing of results; (5) and output of results. Each SED-ML file holds a list of references containing all models used in the simulation experiment. Ideally, the reference is an unambiguous link to a model in an open model repository, making code retrieval easier. For example, a SED-ML file could contain a link to the aforementioned encoding of the Tyson model in BioModels Database. Alternatively, the link to the model in PMR2 could be used. If adjustments of model parameters are necessary before simulation (pre-processing), they can be encoded in SED-ML as well. The format furthermore allows storing changes in the model's XML structure. All changes are defined by linking to a particular model entity with XPath [41], which is the standard technology to address nodes within an XML file. A SED-ML file furthermore contains all information on the type of simulation (e.g. time course), the solver that has been used (e.g. CVODE) and the additional settings of that solver. The *Kinetic Simulation Algorithm Ontology* (KiSAO) [42] has been specifically designed to structure the knowledge about simulation algorithms used in computational biology, and about their characteristics. Post-processing steps necessary in between simulation and output of the experimental results are encoded in a specific XML structure within the SED-ML file. In the case of the Tyson model, for example, the reproduction of one plot in the manuscript required the calculation of the ratio of two modeled entities (Cdc13/Cdc2). Finally, the type of output can be stored, including information about the assignment of plotted entities to the axes of the plots.

To date, several communities establish SED-ML as a standard to exchange simulation setups; the main supporters so far are the SBML and CellML communities. Software support for SED-ML has been implemented in simulation tools, libraries, validation tools, and a SED-ML visual editor [43]. All software is openly available. An up-to-date list is available at the SED-ML website.[12] More details about the SED-ML language are given in in the reference publication [39].

SED-ML support and availability of SED-ML files. One way of contributing to improved result reproducibility is through providing SED-ML files together with model files when publishing the model code in an open repository. BioModels Database, for example, welcomes the submission of supplementary material with each published model, and SED-ML could become one such type of supplement.

[12] http://sed-ml.org

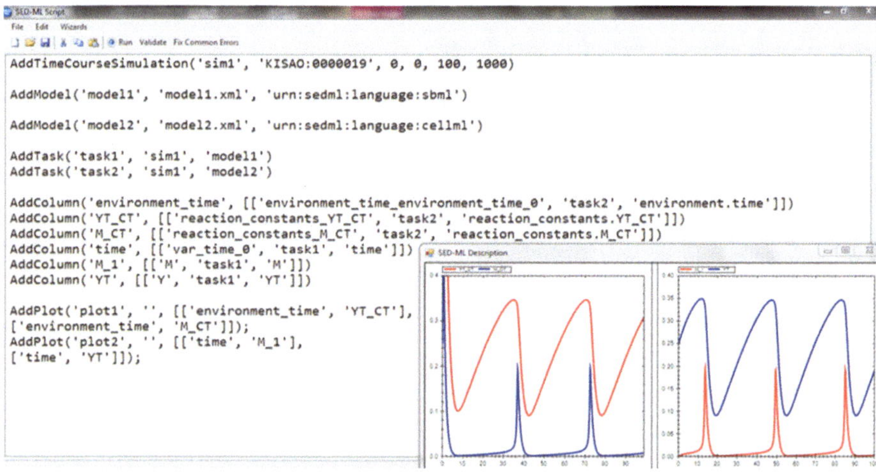

Fig. 10.3 Simulation of the CellML (*left*) and SBML (*right*) encoding of the Tyson model using the script language and online simulator provided by SED-ML Web Tools

SED-ML support is still under development in many simulation tools. SED-ML documents can meanwhile be created by the help of the SED-ML script editor.[13] It allows modelers to design SED-ML files without writing the XML code by hand, but using a more comprehensive, script-like language. An example is shown in Fig. 10.3: The script takes both the SBML and the CellML versions of the Tyson model (AddModel) and applies the same simulation experimental setup to them (AddTimeCourseSimulation). The output then shows the result curve for the two models (AddPlot). The CellML and SBML model instantiate the simulation with different initial values for the concentration of YT and M, which leads to a shift in the CellML model by 20 time units. A repository of simulation experiments applicable to a biological question supports modelers in retrieving existing experiments for a model. The SEMS project currently develops methods for integrated management of model code and associated simulation files [44].

Users may download complete SED-ML descriptions for the simulation of the Tyson model. The results shown in Fig. 3a of the publication describing the Tyson model [1] can be reproduced with the SED-ML file that is available from the curation tab in BioModels Database (for simulating the SBML model) or from the PMR2 workspace (for simulating the CellML model). Figure 10.4 in this chapter shows an extract from the SED-ML file that is available from BioModels Database; the expected simulation result is shown in Fig. 10.6.

---

[13] http://libsedml.sourceforge.net/libSedML/SedMLScript.html

```
<?xml version="1.0" encoding="utf-8"?>
<!-- Written by libSedML v1.1.4521.18154 see http://libsedml.sf.net -->
<sedML level="1" version="1" xmlns="http://sed-ml.org/">
  <listOfSimulations> [..] </listOfSimulations>
  <listOfModels>
    <model id="model1" language="urn:sedml:language:sbml" source="urn:miriam:biomodels.db:BIOMD0000000005" />
  </listOfModels>
  <listOfTasks> [..] </listOfTasks>
  <listOfDataGenerators>
    <dataGenerator id="total_cyclin" name="total_cyclin">
      <listOfVariables>
        <variable id="c1" name="YT" taskReference="task1"
        target="/sbml:sbml/sbml:model/sbml:listOfSpecies/sbml:species[@id='YT']" />
      </listOfVariables>
      <math xmlns="http://www.w3.org/1998/Math/MathML">
        <ci> c1 </ci>
      </math>
    </dataGenerator>
    <dataGenerator id="time" name="time"> [..]
    </dataGenerator>
    <dataGenerator id="pcyclin_cdc2" name="pcyclin_cdc2"> [..]
    </dataGenerator>
  </listOfDataGenerators>
  <listOfOutputs>
    <plot2D id="plot1">
      <listOfCurves>
        <curve id="curve_0" logX="false" logY="false" xDataReference="time" yDataReference="total_cyclin" />
        <curve id="curve_1" logX="false" logY="false" xDataReference="time" yDataReference="pcyclin_cdc2" />
      </listOfCurves>
    </plot2D>
  </listOfOutputs>
</sedML>
```

Fig. 10.4 Extract from the SED-ML file defining a simulation experiment on the Tyson model: The result is a plot with *two curves*, one displaying the total amount of cyclin and the other one displaying the amount of active MPF, relative to total cdc2. The simulation result is shown in Fig. 10.6 in this manuscript. The SED-ML file reproduces Fig. 3a in [1] of the original publication.

10.5 Use Cases

In the following, two use cases for standardized model and simulation encoding are given. The first example shows the retrieval of computational code for a published model and the exploration of possible modifications. The second example describes the necessary steps to reproduce asserted findings using SED-ML.

10.5.1 Example 1: Exploring Existing Models

To get a better understanding of the biological mechanisms behind the cell cycle, one may explore Tyson's cell cycle model [1]. The capabilities of the model and their different experimental setups may best be studied in a familiar simulation tool.

1. **Choose a model repository and search for a computational encoding of the Tyson model**. Here we choose BioModels Database. The search field on the main page is used to search for "cell cycle Tyson". This search results in a number of curated models, identifiable by their model IDs. Clicking on the model link for BIOMD0000000005 displays detailed information on the model, including a PubMed link, model authors and model code submitters,

submission date, a brief description of the model, and a summary of all model entities, their interconnection (through reactions) and parameters as encoded in the model. All this information is valuable for a first understanding of the model's complexity and what aspects it covers.
2. **Download the available model code in the SBML standard format**. A model can best be explored in familiar software tools. Therefore, the model code should be retrieved from the repository in a format that can be read by the software. BioModels Database offers downloads in all SBML levels; here we choose a model file in SBML Level 2 Version 3 format, which is the current standard level, and store it locally on our computer.
3. **Explore the model**. To explore the model, it should be loaded in an SBML-aware simulation software. Effects of different parameterizations and changes in the network structure may be studied (extending it or replacing parts of it to model an alternative assumption). These experiments can finally be stored in SED-ML format, if export is supported by the simulation tool. SED-ML files can be uploaded to the Web, and thereby be made available to colleagues for further elaboration of the performed experiments.

10.5.2 Example 2: Reproducing Existing Simulations

Another application of SED-ML is the reproduction of published results. The following steps, also depicted in Fig. 10.5, enable the reproduction of experiments already published in a paper, given that the model and the simulation setup file had been made publicly available:

1. **Obtain the SED-ML code for the model of interest**. To reproduce the findings described in a publication one needs to re-run the experiments. In BioModels Database, some results are already shown as screenshots in the "curation" tab. These screenshots show the plots obtained by the curator when simulating the model to verify the model's correctness at the time of publication in BioModels Database. In the perfect case, the simulation description is linked to the model code, e.g. by referring to a database of virtual simulation experiments, or by directly providing the simulation description files in the model database. For the Tyson model, the simulation description can be downloaded directly from the BioModels Database site (from the "curation" tab).

2. **Load the simulation description into the simulation tool**. The SED-ML file can be loaded with a simulation tool that supports the SED-ML format. For example, a SED-ML file can be opened in the online simulation tool *SED-ML Web Tools*.[14] The software then executes the SED-ML files, fetches the necessary models from BioModels Database and performs the simulations.

[14] http://sysbioapps.dyndns.org/SED-ML_Web_Tools/

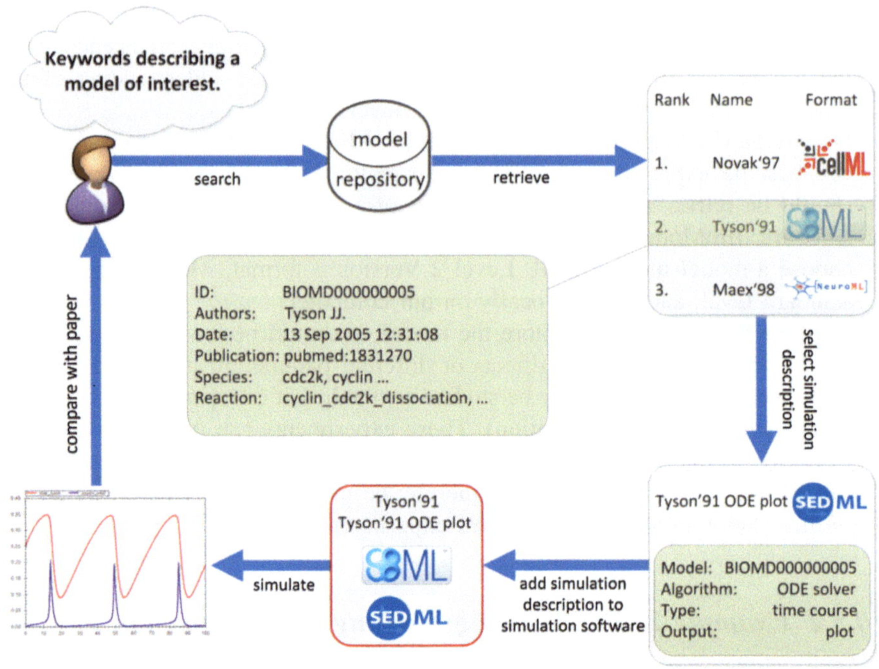

Fig. 10.5 Steps towards reproducible results in computational biology

3. **Verify the model behaves as expected**. A SED-ML file typically contains a set of experiments to be run on the model. When successfully having run all experiments, one can be sure that the model behaves as expected. The SED-ML file available for the Tyson model reproduces the results shown in Fig. 10.4a of the publication. When running the simulation description in SED-ML Web Tools, or another simulation tool running SED-ML files, the result should look identical to the one provided in Fig. 10.6. Once the correct behavior of a model is ensured, one can continue working with the model, e.g. trying different simulation setups, or extending the model.

10.6 Summary

> What is the difference between a live cat and a dead one? One scientific answer is 'systems biology'. A dead cat is a collection of its component parts. A live cat is the emergent behavior of the system incorporating those parts. Nature Editors [45]

Developing models is a time-consuming process; sometimes it can even be life-long. A model on the function of a single protein can result in a PhD thesis. Some researchers dedicate their career to "just one" model. The American physiologist

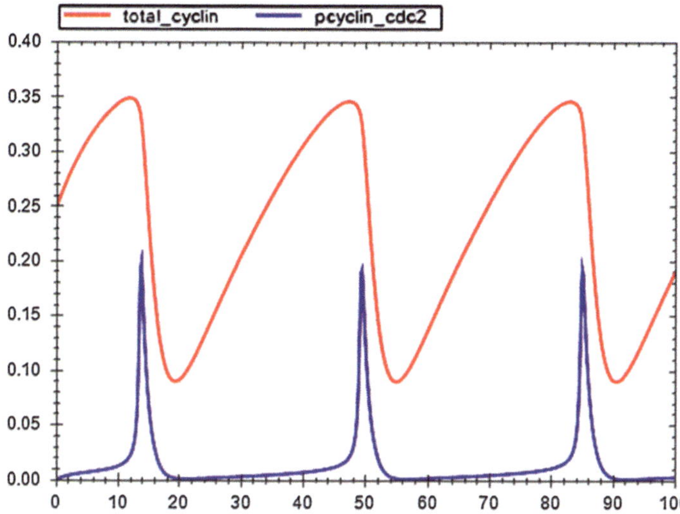

Fig. 10.6 Simulation output for the SBML Tyson model on Cyclin-based kinase, loading the SED-ML provided through BioModels Database and simulated in SED-ML Web Tools

Arthur C. Guyton, for example, started developing a model on the "Relative importance of venous and arterial resistance in controlling venous return and cardiac output" in the 1950s [46]; he continued working on that model for 30 years, until his retirement in the 1980s [47]. The great effort put into model development is one reason to reuse a model in other biological contexts. To test a new hypothesis in silica, it is desirable to take an existing model as a starting point. Considering previously developed models and components timely and qualitatively accelerates the model creation process. Since 2006, model reuse has been repeatedly discussed as one major issue in computational biology [15, 48, 49]. Model repositories support finding and reusing models. However, providing relevant information about a model and reusing it are two different issues. In order to bring systems biology to life, the models' emergent behavior needs to be described in standard format to ensure result reproducibility. The information how to simulate a model is also necessary to reuse a model.

The major concepts on which model reuse and result reproducibility can be built are models encoded in standardized exchange formats, meta-information to find models of interest from public repositories, and simulation experiments encoded in standardized exchange formats. We would like to encourage modelers to make their model code and their simulation setup available to colleagues in standard formats and through publication in open repositories. To ensure reproducibility of model-based results we recommend the following "best practice":

1. When preparing to publish a model, first try to reproduce the results from the information intended to publish.

2. Ask a colleague to try and reproduce the results before publication. Every piece of information necessary for that person to reproduce the findings should be included in the supplementary information.
3. Publish the model code in a standard format, e.g. SBML. Publication should not be limited to the supplementary material, but the model should also be put in a model database, e.g. BioModels Database.
4. If unsure how to annotate the model, consult the MIRIAM guidelines.
5. Provide simulation experiment descriptions for every simulation result mentioned in the publication. Use a standard format, e.g. SED-ML.
6. If unsure what to include, consult the MIASE guidelines.

References

1. Tyson J (1991) Modeling the cell division cycle: cdc2 and cyclin interactions. Proc Natl Acad Sci 88(16):7328
2. Novak B, Tyson J (1997) Modeling the control of DNA replication in fission yeast. Proc Natl Acad Sci 94(17):9147
3. Gardner TS, Dolnik M, Collins JJ (1998) A theory for controlling cell cycle dynamics using a reversibly binding inhibitor. Proc Natl Acad Sci 95(24):14190–14195
4. Csikasz-Nagy A, Battogtokh D, Chen KC et al (2006) Analysis of a generic model of eukaryotic cell-cycle regulation. Biophys J 90(12):4361–4379
5. Ferrell J Jr (2009) Q&A: systems biology. J Biol 8:2
6. Li C, Donizelli M, Rodriguez N et al (2010) BioModels database: an enhanced, curated and annotated resource for published quantitative kinetic models. BMC Syst Biol 4(1):92
7. Ashburner M, Ball C, Blake J et al (2000) Gene ontology: tool for the unification of biology. Nat Genet 25(1):25
8. de Matos P, Alcántara R, Dekker A et al (2010). Chemical entities of biological interest: an update. Nucleic acids research, 38(suppl 1):D249-D254.
9. Lassila O, Swick R et al (1998) Resource description framework (RDF) model and syntax specification. http://www.w3.org/TR/REC-rdf-syntax/. Accessed 19 Mar 2013
10. Herrgård MJ, Swainston N, Dobson P et al (2008) A consensus yeast metabolic network reconstruction obtained from a community approach to systems biology. Nat Biotechnol 26:1155–1160
11. Davenport TH, Prusak L (2000) Working knowledge: How organizations manage what they know. Harvard Business Press, Boston
12. Klipp E, Liebermeister W, Helbig A et al (2007) Systems biology standards—the community speaks. Nat Biotechnol 25(4):390–391
13. Garfinkel D (1969) Construction of biochemical computer models. FEBS Lett 2:S9–S13
14. Bellinger G, Castro D, Mills A (2004) Data, information, knowledge, and wisdom. http://www.systems-thinking.org/dikw/dikw.htm. Accessed 13 Mar 2013
15. Le Novère N (2006) Model storage, exchange and integration. BMC Neurosci 7:S11
16. Bray T, Paoli J, Sperberg-McQueen CM et al (2010) Extensible markup language (XML) 1.0. http://www.w3.org/TR/REC-xml/. Accessed 13 Mar 2013
17. McGuinness D, van Harmelen F et al (2004) OWL web ontology language overview. http://www.w3.org/TR/owl2-overview/. Accessed 13 Mar 2013
18. Strömbäck L, Hall D, Lambrix P (2007) A review of standards for data exchange within systems biology. Proteomics 7(6):857–867

19. Hucka M, Bergmann F, Keating S et al (2010) The systems biology markup language (SBML): language specification for level 3 Version 1. Nature proceedings
20. Cuellar A, Lloyd C, Nielsen P et al (2003) An overview of CellML 1.1, a biological model description language. Simulation 79(12):740–747
21. Gleeson P, Crook S, Cannon R et al (2010) NeuroML: a language for describing data driven models of neurons and networks with a high degree of biological detail. PLoS Comput Biol 6(6):e1000815
22. Chen M, Freier A, Köhler J et al (2002) The biology Petri net markup language. In: J Desel, M Weske (ed) Proceedings of the Promise 2002, Potsdam, 2002
23. Lloyd CM, Halstead MDB, Nielsen PF (2004) CellML: its future, present and past. Prog Biophys Mol Biol 85:433–450
24. Cuellar A, Nielsen P, Halstead M et al (2006) Cellml 1.1 specification. http://www.cellml.org/specifications/cellml_1.1. Accessed 13 Mar 2013
25. Le Novère N, Finney A, Hucka M et al (2005) Minimum information requested in the annotation of biochemical models (MIRIAM). Nat Biotechnol 23(12):1509–1515
26. Yu T, Lloyd C, Nickerson D et al (2011) The Physiome model repository 2. Bioinformatics 27(5):743–744
27. Snoep J, Olivier B (2003) JWS online cellular systems modelling and microbiology. Microbiology 149(11):3045–3047
28. Le Novère N, Bornstein B, Broicher A et al (2006) BioModels database: a free, centralized database of curated, published, quantitative kinetic models of biochemical and cellular systems. Nucleic Acids Res 34(suppl 1):D689–D691
29. Henkel R, Endler L, Peters A, Le Novère N et al (2010) Ranked retrieval of computational biology models. BMC Bioinformatics 11(1):423
30. Novak B, Tyson J (1993) Numerical analysis of a comprehensive model of M-phase control in Xenopus oocyte extracts and intact embryos. J Cell Sci 106(4):1153–1168
31. Chen K, Calzone L, Csikasz-Nagy A et al (2004) Integrative analysis of cell cycle control in budding yeast. Mol Biol Cell 15(8):3841–3862
32. Olivier B, Snoep J (2004) Web-based kinetic modelling using JWS online. Bioinformatics 20(13):2143–2144
33. Wolstencroft K, Owen S, du Preez F et al (2011) The SEEK: a platform for sharing data and models in systems biology. Methods Enzymol 500:629–655
34. Mendes P, Hoops S, Sahle S et al (2006) Computational modeling of biochemical networks using COPASI. Methods Mol Biol 500(2):17–59
35. Miller AK, Marsh J, Reeve A et al (2010) An overview of the CellML API and its implementation. BMC Bioinformatics 11:178
36. Baeza-Yates R, Ribeiro-Neto B (1999) Modern information retrieval. Addison-Wesley, New York
37. Kelton W, Barton R (2003) Experimental design for simulation: experimental design for simulation. In: Proceedings of the 35th conference on Winter Simulation: driving innovation, New Orleans, 7–10 December 2003
38. Pawlikowski K, Jeong H, Lee J (2002) On credibility of simulation studies of telecommunication networks. Communications Magazine, IEEE 40(1):132–139
39. Waltemath D, Adams R, Bergmann F et al (2011) Reproducible computational biology experiments with SED-ML—The simulation experiment description markup language. BMC Syst Biol 5(1):198
40. Waltemath D, Adams R, Beard D et al (2011) Minimum information about a simulation experiment (MIASE). PLoS Comput Biol 7(4):e1001122
41. Clark J, DeRose S (1999) XML path language (XPath) version 1.0. http://www.w3.org/TR/1999/REC-xpath-19991116. Accessed 13 Mar 2013
42. Courtot M, Juty N, Knüpfer C et al (2011) Controlled vocabularies and semantics in systems biology. Mol Syst Biol 7:543
43. Adams R (2012) SED-ED, a workflow editor for computational biology experiments written in SED-ML. Oxf Bioinform 28(8):1180–1181

44. Henkel R, Le Novère N, Wolkenhauer O et al (2012) Considerations of graph-based concepts to manage computational biology models and associated simulations. In: Goltz U (ed) Proceedings of the INFORMATIK2012. Brunswig, Germany, pp 16–21
45. Nature editors (2005) In pursuit of systems. Nature 435(7038):1
46. Guyton A, Abernathy B, Langston J et al (1959) Relative importance of venous and arterial resistances in controlling venous return and cardiac output. Am J Physiol-Legacy Content 196(5):1008–1014
47. Montani JP, Mizelle HL, Adair TH et al (1989) Regulation of cardiac output during aldosterone-induced hypertension. J Hypertens 7(6):S206
48. Bergmann F, Sauro H (2006) SBW-a modular framework for systems biology. In: Proceedings of the 38th conference on Winter Simulation, Monterey, 3–6 Dec 2006
49. Nordlie E, Gewaltig M, Plesser H (2009) Towards reproducible descriptions of neuronal network models. PLoS Comput Biol 5(8):e1000456

Chapter 11
Parameter Identifiability and Redundancy, with Applications to a General Class of Stochastic Carcinogenesis Models

Mark P. Little, Wolfgang F. Heidenreich and Guangquan Li

Abstract Models for complex biological systems may involve a large number of parameters. It may well be that some of these parameters (in particular any value of those parameters) cannot be derived from observed data via regression techniques. Such parameters are said to be unidentifiable, the remaining parameters being identifiable. Closely related to this idea is that of redundancy, that a set of parameters can be expressed in terms of some smaller set. Before data is analysed it is critical to determine which model parameters are identifiable or redundant to avoid ill-defined and poorly-convergent regression. This problem has been considered from a number of points of view in the literature. One distinct recent application has been to biologically-based cancer models. Heidenreich et al. (*Risk Anal* 1997 **17** 391–399) considered parameter identifiability in the context of the two-mutation cancer model and demonstrated that combinations of all but two of the model parameters are identifiable. Here we outline general considerations on parameter identifiability, and introduce the notion of weak local identifiability and gradient weak local identifiability. These are based on local properties of the likelihood, in particular the rank of the Hessian matrix. We relate these to the notions of parameter identifiability and redundancy previously introduced by Rothenberg (*Econometrica* 1971 **39** 577–591) and Catchpole and Morgan (*Biometrika* 1997 **84** 187–196). Within the widely-used exponential family,

M. P. Little (✉)
Radiation Epidemiology Branch, National Cancer Institute, 9609 Medical Center Drive, MSC 9778, Bethesda, MD 20892–9778, USA
e-mail: mark.little@nih.gov

W. F. Heidenreich
Institut für Strahlenbiologie, Helmholtz Zentrum München, Ingolstädter Landstrasse 1, 85758 Neuherberg, Germany
e-mail: heidenreich@helmholtz-muenchen.de

G. Li
Department of Epidemiology and Biostatistics, School of Public Health, Imperial College Faculty of Medicine, Norfolk Place, London, W2 1PG, UK
e-mail: guang.li04@imperial.ac.uk

A. Prokop and B. Csukás (eds.), *Systems Biology*,
DOI: 10.1007/978-94-007-6803-1_11,

parameter irredundancy, local identifiability, gradient weak local identifiability and weak local identifiability are shown to be largely equivalent. We consider applications to a recently developed class of cancer models of Little and Wright (*Math Biosciences* 2003 **183** 111–134) and Little et al. (*J Theoret Biol* 2008 **254** 229–238) that generalize a large number of other recently used quasi-biological cancer models, in particular the two-mutation model of Heidenreich et al. (*Risk Anal* 1997 **17** 391–399). We show that in the simpler model proposed by Little and Wright (*Math Biosciences* 2003 **183** 111–134) the number of identifiable combinations of parameters is at most two less than the number of biological parameters, thereby generalizing previous results of Heidenreich et al. (*Risk Anal* 1997 **17** 391–399) for the two-mutation model. For the more general model of Little et al. (*J Theoret Biol* 2008 **254** 229–238) the number of identifiable combinations of parameters is at most $r + 1$ less than the number of biological parameters, where r is the number of destabilization types (types of genomic instability), thereby also generalizing all these results. Numerical evaluations suggest that these bounds are sharp. We also identify particular combinations of identifiable parameters.

Keywords Cancer · Stochastic model · Quasi-biological model · Exponential family · Parameter redundancy · Parameter identifiability · Local identifiability · Weak local identifiability · Gradient weak local identifiability · Genomic instability

Acronyms

GLM Generalized linear model
GSD Geometric standard deviation

11.1 Introduction

Models for complex biological systems typically involve many parameters, some of which cannot be derived from observed data via regression techniques. Such parameters are said to be unidentifiable or non-identifiable, the remaining parameters being identifiable. A closely related idea is that of redundancy, that a set of parameters can be expressed in terms of some smaller set. To avoid ill-defined and poorly convergent regression one must determine which model parameters are identifiable or redundant.

There is a substantial literature on identifiability in stochastic models in various contexts [1–3]. Catchpole and Morgan [3] defined a set of model parameters in an exponential family model to be *redundant* if the likelihood can be written using a strictly smaller parameter vector; otherwise they are *irredundant*. Rothenberg [1], Jacquez and Perry [4] and Catchpole and Morgan [3] also defined a notion of *local identifiability*, to mean that within a neighbourhood of each set of parameter values the likelihood differs for at least some data points. This notion has been extended by Little et al. [5]—they defined a set of parameters to be *weakly locally identifiable* if the maxima of the likelihood are isolated; they defined parameters to be *gradient weakly locally identifiable* if the turning points (those for which the likelihood derivative with respect to the parameters is zero) are isolated. The results obtained by Little et al. [5], show that, subject to some regulatory conditions, the number of locally identifiable or (gradient) weakly locally identifiable parameter combinations is equal to the rank of the Hessian matrix, or equivalently the rank of the Fisher information matrix. The notions of identifiability in stochastic models [1–3, 5], within which framework this paper is set, should be contrasted with the consideration of identifiability in non-stochastic settings considered by some [4, 6, 7].

In this chapter we outline some general considerations on parameter identifiability. We shall demonstrate that the concepts of parameter local identifiability and redundancy are closely related to apparently weaker properties of weak local identifiability and gradient weak local identifiability, as shown elsewhere [5]. Within the widely-used exponential family we demonstrate that these concepts (local identifiability, redundancy, weak local identifiability, gradient weak local identifiability) largely coincide [5].

We go on to consider applications of all these ideas to a recently developed general class of carcinogenesis models of Little and Wright [8] and Little et al. [9]. These models generalize a large number of other quasi-biological cancer models, in particular those of Armitage and Doll [10] and other multistage models [11–13]. Most of the carcinogenesis models developed in the last thirty years are special cases of the class of models considered here. We shall show that via a specific reparameterization, in the model of Little and Wright [8] in principle combinations of all but two of the model parameters are identifiable, thereby generalizing previous results of Heidenreich et al. [14, 15] for a simple special case. For the more general model of Little et al. [9] combinations of all but $r+1$ of the model parameters are identifiable, where r is the number of destabilization types (the number of types of genomic instability), thereby also generalizing all these results. We also identify particular forms of identifiable parameters. These are outlined in the later parts of the Analysis and the Discussion, also in a related paper [16].

11.2 Methods

11.2.1 General Considerations on Parameter Identifiability

Jacquez and Perry [4] defined a notion of local identifiability, which is that in a local region of the parameter space, there is a unique θ_0 that fits some specified body of data, $(x_i, y_i)_{i=1}^n$, i.e., for which the model predicted mean $h(x|\theta)$ is such that the residual sum of squares:

$$S = \sum_{l=1}^{n} [y_l - h(x_l|\theta)]^2 \tag{11.1}$$

has a unique minimum. We present here a straightforward generalization of this to other error structures. If the model prediction $h(x) = h(x|\theta)$ for the observed data y is a function of some vector parameters $\theta = (\theta_j)_{j=1}^p$ then under the equivalence of likelihood maximization and iteratively reweighted least squares for generalized linear models [17] (Chap. 2), parameter minimization via maximum likelihood implies that one is trying to minimize:

$$S = \sum_{l=1}^{n} \frac{1}{v_l} \left[y_l - h(x_l|\theta_0) - \sum_{j=1}^{p} \frac{\partial h(x_l|\theta)}{\partial \theta_j}\bigg|_{\theta=\theta_0} \cdot \Delta\theta_j \right]^2 \tag{11.2}$$

where y_l $(1 \le l \le n)$ $(n \ge p)$ is the observed measurement (e.g., the numbers of observed cases in the case of binomial or Poisson models) at point l and the v_l $(1 \le l \le n)$ are the current estimates of variance at each point. Heuristically, this has a unique minimum in the perturbing $\Delta\theta = (\Delta\theta_j)_{j=1}^p$ $(\theta = \theta_0 + \Delta\theta)$ given by $H^T D H \Delta\theta = H^T D \delta$, where $(\delta_l)_{l=1}^n = (y_l - h(x_l|\theta_0))_{l=1}^n$, $(H_{lj})_{l=1,j=1}^{n,p} = \left(\frac{\partial h(x_l|\theta)}{\partial \theta_j}\Big|_{\theta=\theta_0} \right)_{l=1,j=1}^{n,p}$, $D = diag[1/v_1, 1/v_2, \ldots, 1/v_n]$, whenever $H^T D H$ has full rank (=p).

More formally:

Definitions 1 Suppose that the likelihood associated with observation x_l is $l(x_l|\theta)$ and let $L(x_l|\theta) = \ln[l(x_l|\theta)]$ for $\theta \in \Omega \subset R^p$. A set of parameters $(\theta_i)_{i=1}^p$ is identifiable if for any $\theta \in \Omega$ there are no $\delta \in \Omega \backslash \{\theta\}$ for which $L(x|\delta) = L(x|\theta)$ (x almost everywhere ($a.e.$)). A set of parameters $(\theta_i)_{i=1}^p$ is locally identifiable at that point if there exists a neighborhood $N \in \aleph_\theta$ such that for no $\delta \in N \backslash \{\theta\}$ is $L(x|\delta) = L(x|\theta)$ (x $a.e.$). A set of parameters $(\theta_i)_{i=1}^p$ is weakly locally identifiable at that point if there exists a neighborhood $N \in \aleph_\theta$ and data $x = (x_1, \ldots, x_n) \in \Sigma^n$ such that the log-likelihood $L = L(x|\theta) = \sum_{l=1}^{n} L(x_l|\theta)$ is maximized by at most one set of $\widehat{\theta} \in N$. If $L = L(x|\theta)$ is C^1 as a function of $\theta \in \Omega$ a set of parameters $(\theta_i)_{i=1}^p \in \mathrm{int}(\Omega)$ is gradient weakly locally identifiable at that point if there exists a neighborhood $N \in \aleph_\theta$ and data $x = (x_1, \ldots, x_n) \in \Sigma^n$ such that $\left(\frac{\partial L(x|\widehat{\theta})}{\partial \widehat{\theta}_i} \right)_{i=1}^p = 0$ (i.e., $\widehat{\theta}$ is a turning point of $L(x|\theta)$) for at most one set of $\widehat{\theta} \in N$.

Our definitions of identifiability and local identifiability coincide with those of Rothenberg and others [1–3]. Rothenberg [1] proved that if the Fisher information matrix, $I = I(\theta)$, in a neighborhood of $\theta \in \mathrm{int}(\Omega)$ is of constant rank and satisfies various other more minor regularity conditions, then $\theta \in \mathrm{int}(\Omega)$ is locally identifiable if and only if $I(\theta)$ is non-singular. Clearly identifiability implies local identifiability. By the Mean Value Theorem [18] (p.107) gradient weak local identifiability implies weak local identifiability. We have the following key result.

Theorem 1 *Suppose that the log-likelihood $L(x|\theta)$ is C^2 as a function of the parameter vector $\theta \in \Omega \subset R^p$, for all $x = (x_1, \ldots, x_n) \in \Sigma^n$.*

1. Suppose that for some x and $\theta \in \text{int}(\Omega)$ it is the case that $rk\left[\left(\frac{\partial^2 L(x|\theta)}{\partial\theta_i \partial\theta_j}\right)_{i,j=1}^p\right] = p$. Then turning points of the likelihood in the neighborhood of θ are isolated, i.e., there is an open neighborhood $N \in \aleph_\theta \subset \Omega$ for which there is at most one $\widehat{\theta} \in N$ that satisfies $\left(\frac{\partial L(x|\theta)}{\partial\theta_i}\right)_{i=1}^p\Big|_{\theta=\widehat{\theta}} = 0$.
2. Suppose that for some x and $\theta \in \text{int}(\Omega)$ it is the case that $rk\left[\left(\frac{\partial^2 L(x|\theta)}{\partial\theta_i \partial\theta_j}\right)_{i,j=1}^p\right] = p$ then local maxima of the likelihood in the neighborhood of θ are isolated, i.e., there is an open neighborhood $N \in \aleph_\theta \subset \Omega$ for which there is at most one $\widehat{\theta} \in N$ that is a local maximum of $L(x|\theta)$.
3. Suppose that for some x and all $\theta \in \text{int}(\Omega)$ it is the case that $rk\left[\left(\frac{\partial^2 L(x|\theta)}{\partial\theta_i \partial\theta_j}\right)_{i,j=1}^p\right] = r < p$ then all local maxima of the likelihood in $\text{int}(\Omega)$ are not isolated, as indeed are all $\theta \in \text{int}(\Omega)$ for which $\left(\frac{\partial L(x|\theta)}{\partial\theta_i}\right)_{i=1}^p = 0$.

We prove this result in Appendix A. As an immediate consequence we have the following result.

Corollary 1 *For a given $x = (x_1, \ldots, x_n) \in \Sigma^n$, a sufficient condition for the likelihood $L(x|\theta) = \sum_{l=1}^n L(x_l|\theta)$ to have at most one maximum and one turning point in the neighborhood of a given $\theta = (\theta_1, \ldots, \theta_p) \in \text{int}(\Omega)$ is that $rk\left[\left(\frac{\partial^2 L(x|\theta)}{\partial\theta_i \partial\theta_j}\right)_{i,j=1}^p\right] = p$. In particular, if this condition is satisfied θ is gradient weakly locally identifiable (and therefore weakly locally identifiable) ($\Omega \subset R^p$ is the parameter space).*

That this condition is not necessary is seen by consideration of the likelihood $l(x|\theta) = C \cdot \exp\left(-\sum_{i=1}^p [x_i - \theta_i]^4\right)$, where C is chosen so that this has unit mass. Then $\frac{\partial^2 L(x|\theta)}{\partial\theta_i \partial\theta_j} = -12 \cdot [x_i - \theta_i]^2 \cdot \delta_{ij}$ which has rank 0 at $\theta = x$ and a unique maximum there. In particular, this shows that the result claimed by Viallefont et al. [19] (proposition 2) is incorrect.

Definitions 2 A subset of parameters $(\theta_{\pi(i)})_{i=1}^k$ (for some permutation $\pi : \{1, 2, \ldots, p\} \to \{1, 2, \ldots, p\}$) is <u>weakly maximal</u> (respectively <u>weakly gradient maximal</u>) if for any <u>permissible</u> fixed $(\theta_{\pi(i)})_{i=k+1}^p$ (such that $\Omega_{k,\pi}^{(\theta_{\pi(i)})_{i=k+1}^p} = \{(\theta_{\pi(i)})_{i=1}^k : (\theta_1, \ldots, \theta_k, \theta_{k+1}, \ldots, \theta_p) \in \Omega\} \neq \emptyset$) $(\theta_{\pi(i)})_{i=1}^k$ is weakly locally identifiable (respectively gradient weakly locally identifiable) at that point (in relation to the restricted likelihood $L_{(\theta_{\pi(i)})_{i=k+1}^p}(x|(\theta_{\pi(i)})_{i=1}^k) = L(x|(\theta_i)_{i=1}^p)$ but that this is not the case for any larger number of parameters. A subset of parameters $(\theta_{\pi(i)})_{i=1}^{k}$ is strongly maximal (respectively strongly gradient maximal) if for any permissible fixed $(\theta_{\pi(i)})_{i=k+1}^{p}$ and any open $U \subset \Omega_{k,\pi}^{(\theta_{\pi(i)})_{i=k+1}^{p}}$, $(\theta_{\pi(i)})_{i=1}^{k}$ restricted to the set U is weakly maximal (respectively weakly gradient maximal), i.e., all $(\theta'_{\pi(i)})_{i=1}^{k} \in U$ are weakly maximal (respectively weakly gradient maximal).

From this it easily follows that a strongly (gradient) maximal set of parameters $(\theta_{\pi(i)})_{i=1}^{k}$ is *a fortiori* weakly (gradient) maximal at all points $(\theta'_{\pi(i)})_{i=1}^{k} \in \Omega_{k,\pi}^{(\theta_{\pi(i)})_{i=k+1}^{p}}$ for any permissible $(\theta_{\pi(i)})_{i=k+1}^{p}$. Assume now that k of the p θ_i are a weakly maximal set of parameters. So for some permutation $\pi : \{1,2,\ldots,p\} \to \{1,2,\ldots,p\}$ and for any permissible fixed $(\theta_{\pi(i)})_{i=k+1}^{p}$ and any $(\theta_{\pi(i)})_{i=1}^{k} \in \Omega_{k,\pi}^{(\theta_{\pi(i)})_{i=k+1}^{p}} \subset R^k$ there is an open neighborhood $N \in \aleph_{(\theta_{\pi(i)})_{i=1}^{k}} \subset \Omega_{k,\pi}^{(\theta_{\pi(i)})_{i=k+1}^{p}}$ and some data $x = (x_1,\ldots,x_n) \in \Sigma^n$ for which $L_{(\theta_{\pi(i)})_{i=k+1}^{p}}(x|(\theta_{\pi(i)})_{i=1}^{k})$ is maximized by at most one set of $(\widehat{\theta}_{\pi(i)})_{i=1}^{k} \in N$, but that this is not the case for any larger number of parameters. Assume that $r = \max\left\{ rk\left[\left(\frac{\partial^2 L_{(\theta_{\pi(i)})_{i=k+1}^{p}}(x|(\theta_{\pi(i)})_{i=1}^{k})}{\partial\theta_{\pi(i)}\partial\theta_{\pi(j)}}\right)_{i,j=1}^{k}\right] : (\theta_{\pi(i)})_{i=1}^{k} \in N \right\} < k$. If L is C^2 as a function of θ then it follows easily that $\Omega_{k,r} = \left\{ (\theta_{\pi(i)})_{i=1}^{k} \in N : rk\left[\left(\frac{\partial^2 L_{(\theta_{\pi(i)})_{i=k+1}^{p}}(x|(\theta_{\pi(i)})_{i=1}^{k})}{\partial\theta_{\pi(i)}\partial\theta_{\pi(j)}}\right)_{i,j=1}^{k}\right] = r \right\}$ must be an open non-empty subset of N. By Theorem 1 (3) any $\widehat{\theta} \in \Omega_{k,r}$ which maximizes $L_{(\theta_{\pi(i)})_{i=k+1}^{p}}$ in $\Omega_{k,r}$ cannot be isolated, a contradiction (unless there are no maximizing $\widehat{\theta} \in \Omega_{k,r}$). Therefore, either there are no maximizing $\widehat{\theta} \in \Omega_{k,r}$ or for at least one $\widehat{\theta} \in N$ $rk\left[\left(\frac{\partial^2 L_{(\theta_{\pi(i)})_{i=k+1}^{p}}(x|(\theta_{\pi(i)})_{i=1}^{k})}{\partial\theta_{\pi(i)}\partial\theta_{\pi(j)}}\right)_{i,j=1}^{k}\Bigg|_{(\theta_{\pi(i)})_{i=1}^{k}=\widehat{\theta}}\right] = k$. This implies that $rk\left[\left(\frac{\partial^2 L(x|\theta)}{\partial\theta_i\partial\theta_j}\right)_{i,j=1}^{p}\Bigg|_{\theta=\widehat{\theta}'}\right] \geq k$, where $\widehat{\theta}' = (\widehat{\theta}) \cup (\theta_{\pi(i)})_{i=k+1}^{p}$ in the obvious sense.

Assume now that the $(\theta_{\pi(i)})_{i=1}^{k}$ are strongly maximal. Suppose that for some $\theta_1 = (\theta_{1i})_{i=1}^{p} \in \Omega$ and some $x = (x_1,\ldots,x_n) \in \Sigma^n$ it is the case that $rk\left[\left(\frac{\partial^2 L(x|\theta)}{\partial\theta_i\partial\theta_j}\right)_{i,j=1}^{p}\Bigg|_{\theta=\theta_1}\right] > k$. Because $\left(\frac{\partial^2 L(x|\theta)}{\partial\theta_i\partial\theta_j}\right)_{i,j=1}^{p}\Bigg|_{\theta=\theta_1}$ is symmetric, there is a permutation $\pi' : \{1,\ldots,p\} \to \{1,\ldots,p\}$ for which $rk\left[\left(\frac{\partial^2 L(x|\theta)}{\partial\theta_{\pi'(i)}\partial\theta_{\pi'(j)}}\right)_{i,j=1}^{k+1}\Bigg|_{\theta=\theta_1}\right] = k+1$ [20] (p.79). If L is C^2 as a function of θ this will be the case in some open neighborhood $N' \in \aleph_{(\theta_{1\pi'(i)})_{i=1}^{k+1}} \subset R^{k+1}$. By Theorem 1 (2) this implies that the parameters $\left(\theta_{\pi'(i)}\right)_{i=1}^{k+1}$ have at most one maximum in N', so that $\left(\theta_{\pi(i)}\right)_{i=1}^{k}$ is not a strongly maximal set of parameters in N'. With small changes everything above also goes through with "weakly gradient maximal" substituted for "weakly maximal" and "strongly gradient maximal" substituted for "strongly maximal". Therefore we have proved the following result.

Theorem 2 *Let $L(x|\theta)$ be C^2 as a function of $\theta \in \Omega \subset R^p$ for all $x \in \sum^n$.*

1. If there is a weakly maximal (respectively weakly gradient maximal) subset of k parameters, $(\theta_{\pi(1)}, \theta_{\pi(2)}, \ldots, \theta_{\pi(k)})$ for some permutation $\pi : \{1, 2, \ldots, p\} \to \{1, 2, \ldots, p\})$, and for fixed $(\theta_{\pi(i)})_{i=k+1}^{p}$ and some $x = (x_1, \ldots, x_n) \in \Sigma^n$ $L_{(\theta_{\pi(i)})_{i=k+1}^{p}}(x|(\theta_{\pi(i)})_{i=1}^{k})$ has a maximum (respectively turning point) on the set of θ where $rk\left[\left(\frac{\partial^2 L_{(\theta_{\pi(i)})_{i=k+1}^{p}}(x|(\theta_{\pi(i)})_{i=1}^{k})}{\partial\theta_{\pi(i)}\partial\theta_{\pi(j)}}\right)_{i,j=1}^{k}\right]$ is maximal then $\max\left\{rk\left[\left(\frac{\partial^2 L_{(\theta_{\pi(i)})_{i=k+1}^{p}}(x|(\theta_{\pi(i)})_{i=1}^{k})}{\partial\theta_{\pi(i)}\partial\theta_{\pi(j)}}\right)_{i,j=1}^{k}\right] : (\theta_{\pi(i)})_{i=1}^{k} \in \Omega_{k,\pi}^{(\theta_{\pi(i)})_{i=k+1}^{p}}\right\} = k$ and $\max\left\{rk\left[\left(\frac{\partial^2 L(x|\theta)}{\partial\theta_i\partial\theta_j}\right)_{i,j=1}^{p}\right] : \theta \in \Omega\right\} \geq k.$
2. If there is a strongly maximal (respectively strongly gradient maximal) subset of k parameters, $(\theta_{\pi(1)}, \theta_{\pi(2)}, \ldots, \theta_{\pi(k)})$ (for some permutation $\pi : \{1, 2, \ldots, p\} \to \{1, 2, \ldots, p\})$ then $rk\left[\left(\frac{\partial^2 L(x|\theta)}{\partial\theta_i\partial\theta_j}\right)_{i,j=1}^{p}\right] \leq k \quad \forall\theta \in \Omega$.

All further results in this Section assume that the model is a member of the exponential family, so that if the observed data $x = (x_l)_{l=1}^{n} \in \sum^n$ then the log-likelihood is given by $L(x|\theta) = \sum_{l=1}^{n}\left[\frac{x_l\varsigma_l - b(\varsigma_l)}{a(\phi)} + c(x_l, \phi)\right]$ for some functions $a(\phi), b(\varsigma), c(x, \phi)$. We assume that the natural parameters $\varsigma_l = \varsigma_l[(\theta_i)_{i=1}^{p}, z_l]$ are functions of the model parameters $(\theta_i)_{i=1}^{p}$ and some auxiliary data z_l, but that the scaling parameter ϕ is not. Let $\mu_l = b'(\varsigma_l) = E[x_l]$, so that $\mu_l = b'(\varsigma_l[(\theta_i)_{i=1}^{p}, z_l])$. In all that follows we shall assume that the function $b(\varsigma)$ is C^2. The following definition was introduced by Catchpole and Morgan [3].

Definition 3 With the above notation, a set of parameters $(\theta_i)_{i=1}^{p} \in \Omega$ is parameter redundant for an exponential family model if $\mu_l = b'(\varsigma_l[(\rho_i)_{i=1}^{q}, z_l])$ can be expressed in terms of some strictly smaller parameter vector $(\rho_i)_{i=1}^{q}$ $(q < p)$. Otherwise, the set of parameters $(\theta_i)_{i=1}^{p}$ is parameter irredundant or full rank.

Catchpole and Morgan [3] proved (their Theorem 1) that a set of parameters is parameter redundant if and only if $rk\left[\left(\frac{\partial\mu_l}{\partial\theta_i}\right)_{l=1,i=1}^{np}\right] < p$. They defined full rank models to be essentially full rank if $rk\left[\left(\frac{\partial\mu_l}{\partial\theta_i}\right)_{l=1,i=1}^{np}\right] = p$ for every $(\theta_i)_{i=1}^{p} \in \Omega$; if $rk\left[\left(\frac{\partial\mu_l}{\partial\theta_i}\right)_{l=1,i=1}^{np}\right] = p$ only for some $(\theta_i)_{i=1}^{p} \in \Omega$ then the parameter set is conditionally full rank. They also showed (their Theorem 3) that if $I = I(\theta)$ is the Fisher information matrix then $rk\left[\left(\frac{\partial\mu_l}{\partial\theta_i}\right)_{l=1,i=1}^{np}\right] = rk[I(\theta)]$, and that parameter redundancy implies lack of local identifiability; indeed their proof of Theorems 2 and 4 showed that there is also lack of weak local identifiability (respectively gradient weak local identifiability) for all $(\theta_i')_{i=1}^{p} \in \Omega$ which for some $x = (x_l)_{l=1}^{n} \in \sum^n$ are local maxima (respectively turning points) of the likelihood.

Assume that $\theta = (\theta_i)_{i=1}^{p}$ are an essentially full rank set of parameters for the model. From the above result for every $\theta = (\theta_i)_{i=1}^{p} \in \Omega$ $rk\left[\left(\frac{\partial\mu_l}{\partial\theta_i}\right)_{l=1,i=1}^{np}\right] = rk(I(\theta)) = p$. Therefore, since $E\left[\frac{\partial^2 L(x|\theta)}{\partial\theta_i\partial\theta_j}\right] = -E\left[\frac{\partial L(x|\theta)}{\partial\theta_i}\frac{\partial L(x|\theta)}{\partial\theta_j}\right] = -I(\theta)$ is of full rank and so negative definite, so by the strong law of large numbers we can choose $x = (x_l)_{l=1}^{n} \in \sum^n$ so that the same is true of $\frac{\partial^2 L(x|\theta)}{\partial\theta_i\partial\theta_j} = \sum_{l=1}^{n}\left\{\left[\frac{x_l - b'(\varsigma_l)}{a(\phi)}\right]\frac{\partial^2\varsigma_l}{\partial\theta_i\partial\theta_j} - \frac{b''(\varsigma_l)}{a(\phi)}\frac{\partial\varsigma_l}{\partial\theta_i}\frac{\partial\varsigma_l}{\partial\theta_j}\right\}$. This implies that on some $N \in \aleph_\theta \subset R^p$ $\frac{\partial^2 L(x|\theta)}{\partial\theta_i\partial\theta_j} = \sum_{l=1}^{n}\left\{\left[\frac{x_l - b'(\varsigma_l)}{a(\phi)}\right]\frac{\partial^2\varsigma_l}{\partial\theta_i\partial\theta_j} - \frac{b''(\varsigma_l)}{a(\phi)}\frac{\partial\varsigma_l}{\partial\theta_i}\frac{\partial\varsigma_l}{\partial\theta_j}\right\}$ is of full rank, and therefore by Corollary 1 $\theta = (\theta_i)_{i=1}^{p}$ is (gradient) weakly locally identifiable. Furthermore, the above argument shows that if $\theta = (\theta_i)_{i=1}^{p}$ are a conditionally full rank set of parameters then on the (open) set $\Omega_p = \left\{\theta = (\theta_i)_{i=1}^{p} \in \Omega : rk\left[\left(\frac{\partial\mu_l}{\partial\theta_i}\right)_{l=1,i=1}^{np}\right] = p\right\}$, $\theta = (\theta_i)_{i=1}^{p}$ is gradient weakly locally identifiable. We have therefore proved:

Theorem 3 *Let $L(x|\theta)$ belong to the exponential family and be C^2 as a function of $\theta \in \Omega \subset R^p$ for all $x \in \sum^n$.*

1. If the parameter set $\theta = (\theta_i)_{i=1}^{p}$ is parameter redundant then it is not locally identifiable, and is not weakly locally identifiable (respectively gradient weakly locally identifiable) for all $(\theta_i')_{i=1}^{p} \in \Omega$ which for some $x = (x_l)_{l=1}^{n} \in \sum^n$ are local maxima (respectively turning points) of the likelihood.
2. If the parameter set $\theta = (\theta_i)_{i=1}^{p}$ is of essentially full rank then for some $x = (x_l)_{l=1}^{n} \in \sum^n$ $\frac{\partial^2 L(x|\theta)}{\partial\theta_i\partial\theta_j}$ is of full rank and therefore $\theta = (\theta_i)_{i=1}^{p}$ is gradient weakly locally identifiable (and so weakly locally identifiable) for all $\theta = (\theta_i)_{i=1}^{p} \in \Omega$.
3. If the parameter set $\theta = (\theta_i)_{i=1}^{p}$ is of conditionally full rank then it is gradient weakly locally identifiable on the open set $\Omega_p = \left\{\theta = (\theta_i)_{i=1}^{p} \in \Omega : rk\left[\left(\frac{\partial\mu_l}{\partial\theta_i}\right)_{l=1,i=1}^{np}\right] = p\right\}$.

Remarks It should be noted that part (1) of this generalizes part (i) of Theorem 4 of Catchpole and Morgan [3]. However, some components of part (2) (that being essentially full rank implies gradient weak local identifiability) is weaker than the other result, proved in part (ii) of Theorem 4 of Catchpole and Morgan [3], namely that if a model is of essentially full rank it is locally identifiable. As noted by Catchpole and Morgan [3] there are exponential-family models that are conditionally full rank, but not locally identifiable, so part (3) is about as strong a result as can be hoped for.

From Theorem 3 we deduce the following.

Corollary 2 *Let $L(x|\theta)$ belong to the exponential family and be C^2 as a function of $\theta \in \Omega \subset R^p$ for all $x \in \sum^n$. Then*

1. If for some subset of parameters $(\theta_{\pi(i)})_{i=1}^k$ and some $x = (x_1, \ldots, x_n) \in \Sigma^n$ it is the case that $rk\left[\left(\frac{\partial^2 L(x|\theta)}{\partial\theta_{\pi(i)}\partial\theta_{\pi(j)}}\right)_{i,j=1}^k\right] = k$ then this subset is gradient weakly locally identifiable at this point.
2. If a subset of parameters $(\theta_{\pi(i)})_{i=1}^k$ is weakly locally identifiable and for some $x \in \sum^n$ this point is a local maximum of the likelihood then it is parameter irredundant, i.e., of full rank, so $rk[I(\theta)] = k$, so that for some $x' \in \sum^{n'}$ $rk\left[\left(\frac{\partial^2 L(x'|\theta)}{\partial\theta_{\pi(i)}\partial\theta_{\pi(j)}}\right)_{i,j=1}^k\right] = k$. In particular, if this holds for all $\theta \in \Omega$ then parameter irredundancy, local identifiability, gradient weak local identifiability and weak local identifiability are all equivalent.

Proof This is an immediate consequence of the remarks after Definition 1, Corollary 1, Theorem 3 (1) and Theorems 1 and 3 of Catchpole and Morgan [3]. **QED.**

Remarks

1. By the remarks preceding Theorem 3 the conditions of part (1) (that for some $x = (x_1, \ldots, x_n) \in \Sigma^n$ it is the case that $rk\left[\left(\frac{\partial^2 L(x|\theta)}{\partial\theta_i\partial\theta_j}\right)_{i,j=1}^k\right] = k$) are automatically satisfied if $\theta = (\theta_i)_{i=1}^k$ are an essentially full rank set of parameters for the model.
2. Assume the model is constructed from a stochastic cancer model embedded in the exponential family, in the sense outlined in Appendix B, so that the natural parameters $\varsigma_l = \varsigma_l[(\theta_i)_{i=1}^p, z_l]$ are functions of the model parameters $(\theta_i)_{i=1}^p$ and some auxiliary data $(z_l)_{l=1}^n$, and the means are given by $\mu_l = b'(\varsigma_l[(\theta_i)_{i=1}^p, z_l]) = z_l \cdot h[(\theta_i)_{i=1}^p, y_l]$, where $h[(\theta_i)_{i=1}^p, y_l]$ is the cancer hazard function. In this case, as shown in Appendix B, $\frac{\partial^2 L(x|\theta)}{\partial\theta_i\partial\theta_j} = \sum_{l=1}^n \left[\begin{array}{l} \frac{[x_l - b'(\varsigma_l)]z_l}{a(\phi)b''(\varsigma_l)} \frac{\partial^2 h(\theta, y_l)}{\partial\theta_i\partial\theta_j} \\ -\frac{z_l^2}{a(\phi)} \frac{\partial h(\theta, y_l)}{\partial\theta_i} \frac{\partial h(\theta, y_l)}{\partial\theta_j} \left\{\frac{[b''(\varsigma_l)]^2 + b'''(\varsigma_l)[x_l - b'(\varsigma_l)]}{[b''(\varsigma_l)]^3}\right\} \end{array}\right]$. The second term

inside the summation $\left(-\frac{z_l^2}{a(\phi)}\frac{\partial h(\theta,y_l)}{\partial\theta_i}\frac{\partial h(\theta,y_l)}{\partial\theta_j}\left\{\frac{[b''(\varsigma_l)]^2+b'''(\varsigma_l)[x_l-b'(\varsigma_l)]}{[b''(\varsigma_l)]^3}\right\}\right)_{i,j=1}^{p}$ is a rank 1 matrix and can be made small in relation to the first term, e.g., by making z_l small. Therefore finding data $(x,y,z)=(x_1,\ldots,x_n,y_1,\ldots,y_n,z_1,\ldots,z_n)\in\Sigma^n$ for which $rk\left[\left(\frac{\partial^2 L(x|\theta)}{\partial\theta_{\pi(i)}\partial\theta_{\pi(j)}}\right)_{i,j=1}^{k}\right]=k$ is equivalent to finding data for which $rk\left[\left(\frac{\partial^2 h(\theta,y_l)}{\partial\theta_{\pi(i)}\partial\theta_{\pi(j)}}\right)_{i,j=1}^{k}\right]=k$, or by the result of Dickson [20] (p.79) for which $rk\left[\left(\frac{\partial^2 h(\theta,y_l)}{\partial\theta_i\partial\theta_j}\right)_{i,j=1}^{p}\right]=k$.

11.2.2 Hessian vs Fisher Information Matrix as a Method of Determining Redundancy and Identifiability in Generalised Linear Models

We, as with Catchpole and Morgan [3], emphasise use of the Hessian of the likelihood rather than the Fisher information matrix considered by Rothenberg [1]. In the context of generalized linear models (GLM), we have $L(x|\theta)=\sum_{l=1}^{n}\left[\frac{x_l\varsigma_l-b(\varsigma_l)}{a(\phi)}+c(x_l,\phi)\right]$ and $g(\mu_i)=g(b'(\varsigma_i))=\sum_{j=1}^{p}A_{ij}\theta_j$ for some link function $g()$ and fixed matrix A. We define $D_{ij}=\frac{\partial\mu_j}{\partial\theta_i}=\frac{1}{g'(\mu_j)}A_{ji}=(A^TG^{-1})_{ij}$ where $G=diag(g'(\mu_1),g'(\mu_2),\ldots,g'(\mu_n))$. Theorem 1 of Catchpole and Morgan [3] states that a model is parameter irredundant if and only if $rk[D]=p$. The score vector is given by $U_i=\frac{\partial L(x|\theta)}{\partial\theta_i}=\sum_{l=1}^{n}\frac{[x_l-\mu_l]}{a(\phi)}\frac{\partial\varsigma_l}{\partial\theta_i}=\sum_{l=1}^{n}\frac{[x_l-\mu_l]}{b''(\varsigma_l)a(\phi)}\frac{\partial\mu_l}{\partial\theta_i}=\frac{1}{a(\phi)}(D\Delta(x-\mu))_i$ where $\Delta=diag\left(\frac{1}{b''(\varsigma_1)},\frac{1}{b''(\varsigma_2)},\ldots,\frac{1}{b''(\varsigma_n)}\right)$. The Fisher information is therefore given by $I(\theta)=E[UU^T]=\frac{1}{a(\phi)^2}D\Delta V\Delta D^T$ where $V=\left(E\left[[x_i-\mu_i][x_j-\mu_j]\right]\right)_{i,j}$ is the data variance. Theorem 1 of Rothenberg [1] states that a model is locally identifiable if and only if $rk[I(\theta)]=p$. As above [Corollary 2 (2)], heuristically parameter irredundancy, local identifiability, gradient weak local identifiability and weak local identifiability are all equivalent and occur whenever $rk(D\Delta V\Delta D^T)=rk(D)=p$. Clearly evaluating the rank of D is generally much easier than that of $D\Delta V\Delta D^T$.

However, for certain applications, both the Fisher information and the Hessian must be employed, as we now outline. Assume that the model is constructed from a stochastic cancer model embedded in an exponential family model in the sense outlined in Appendix B. The key to showing that such an embedded model has no more than N irredundant parameters is to construct some scalar functions

$G_1(.), G_2(.), \ldots, G_N(.)$ such that the cancer hazard function $h(\theta)$ can be written as $h(G_1(\theta), G_2(\theta), \ldots, G_N(\theta))$. Since the cancer model is embedded in a member of the exponential family (in the sense outlined in Appendix B) the same will be true of the total log-likelihood $L(x|\theta) = L(x|G_1(\theta), G_2(\theta), \ldots, G_N(\theta))$. By means of the Chain Rule [18] (p.215) we obtain $\frac{\partial^2 L(x|\theta)}{\partial\theta_i \partial\theta_j} = \sum_{l,k=1}^{N} \frac{\partial^2 L(x|G_1,\ldots,G_N)}{\partial G_l \partial G_k} \frac{\partial G_l}{\partial\theta_i} \frac{\partial G_k}{\partial\theta_j} + \sum_{l=1}^{N} \frac{\partial L(x|G_1,\ldots,G_N)}{\partial G_l} \frac{\partial^2 G_l}{\partial\theta_i \partial\theta_j}$, so that the Fisher information matrix is given by:

$$\begin{aligned} I(\theta) &= -E_\theta\left[\frac{\partial^2 L(x|\theta)}{\partial\theta_i\partial\theta_j}\right] = -E\left[\sum_{l,k=1}^{N} \frac{\partial^2 L(x|G_1,\ldots,G_N)}{\partial G_l \partial G_k} \frac{\partial G_l}{\partial\theta_i} \frac{\partial G_k}{\partial\theta_j}\right] \\ &= -\sum_{l,k=1}^{N} \frac{\partial G_l}{\partial\theta_i} E\left[\frac{\partial^2 L(x|G_1,\ldots,G_N)}{\partial G_l \partial G_k}\right] \frac{\partial G_k}{\partial\theta_j} \end{aligned} \tag{11.3}$$

which therefore has rank at most N. Therefore by Corollary 2 there can be at most N irredundant parameters, or indeed (gradient) weak locally identifiable parameters. [A similar argument shows that if one were to reparameterise (via some invertible C^2 mapping $\theta = f(\omega)$) then the embedded log-likelihood $L(x|f^{-1}(\theta)) = L(x|\omega)$ associated with $h(f^{-1}(\theta)) = h(\omega)$ must also have Fisher information matrix of rank at most N.] By remark (2) after Corollary 2, to show that a subset of cardinality N of the parameters $(\theta_i)_{i=1}^p$ is (gradient) weak locally identifiable parameters, requires that one show that $\left[\frac{\partial^2 h(\theta, y_l)}{\partial\theta_i \partial\theta_j}\right]_{i,j=1}^{p}$ has rank at least N for some (θ, y_l). This is the approach adopted in the paper of Little et al. [16], which we now outline.

11.3 Parameter Identifiability in the Context of a Stochastic Cancer Model with Genomic Instability

We consider the problem of parameter identifiability in a particular class of stochastic cancer models [8, 9], generalizing previously developed ideas [15]. Throughout this section we shall assume that this model is embedded in a member of the exponential family so that the log-likelihood is given by $L(x|\theta) = \sum_{l=1}^{n} \left[\frac{x_l \varsigma_l - b(\varsigma_l)}{a(\phi)} + c(x_l, \phi)\right]$ where the natural parameters $\varsigma_l = \varsigma_l[(\theta_i)_{i=1}^p, z_l]$ are functions of the model parameters $(\theta_i)_{i=1}^p$ and some auxiliary data $(z_l)_{l=1}^n$, but that the scaling parameter ϕ is not. We shall assume that the $\mu_l = b'(\varsigma_l[(\theta_i)_{i=1}^p, z_l]) = z_l \cdot h[(\theta_i)_{i=1}^p, y_l]$, where $h[(\theta_i)_{i=1}^p, y_l]$ is the cancer hazard function, and that the $(z_l)_{l=1}^n$ are all non-zero. This is generally the case, in particular when cohort data are analysed using Poisson regression models [8, 21]. By the remarks following Corollary 2 above, proving weak local identifiability of a subset of cardinality k of the biological parameters $(\theta_i)_{i=1}^{p}$ is equivalent to showing that for this subset of parameters $rk\left[\left(\frac{\partial^2 h}{\partial\theta_i\partial\theta_j}\right)_{i,j=1}^{p}\right] = k$.

The model of Little et al. [9], generalizing various others [8, 10–13, 22], assumes that cells can acquire up to k successive cancer-stage mutations, and any of r (mutually exclusive) types of destabilization mutation(s). Cells become malignant when k cancer-stage mutations have occurred, no matter how many destabilizing mutations there have been. Once a cell has acquired a destabilizing mutation of type d ($1 \le d \le r$), it and its daughter cells can acquire up to $m_d - 1$ further destabilizing mutations of the same type. We define r to be the multiplicity of destabilization mutation types (types of genomic instability). It is to be expected that the more destabilizing mutations cells acquire of each type, the higher the cancer stage mutation rate is, but this is not intrinsic to the model. We write $(m_1 - m_2 - \cdots - m_r)$ as the *signature of the destabilizing mutation types*. We habitually describe this model as of type $k - r - (m_1 - m_2 - \cdots - m_r)$ for short. The model is illustrated schematically in Figs. 11.1 and 11.2. Table 11.1 lists the biological parameters that are used in the model, and their multiplicity.

Cells at different stages of the process are labelled by $I_{(\alpha,\beta,d)}$, where the first subscript, α, represents the number of cancer stage mutations that the cell has accumulated, the second subscript, β, represents the number of destabilizing mutations acquired, their type being given by the third subscript, d. At all stages other than $I_{(0,0,0)}$, cells are allowed to divide symmetrically or differentiate/

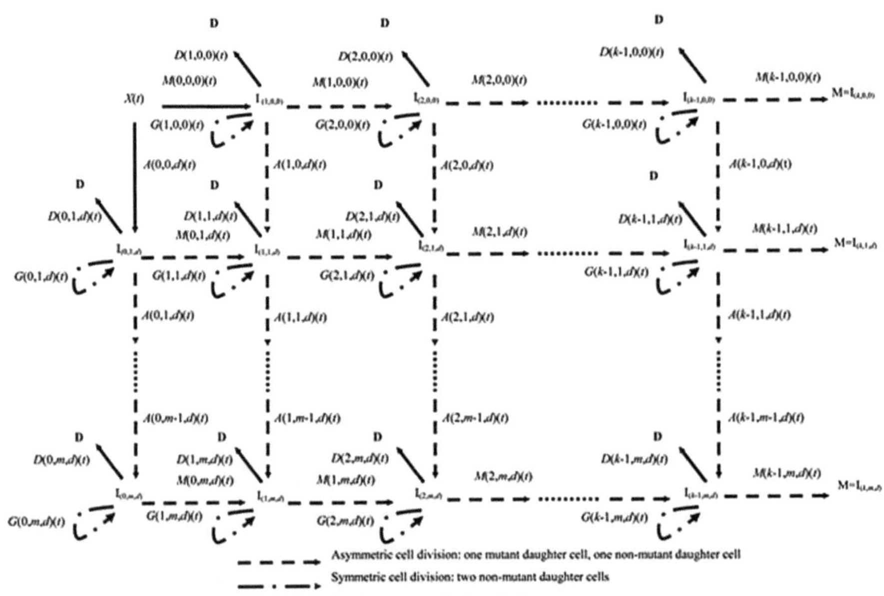

Fig. 11.1 Diagram of cancer model with k cancer-stage mutations and m destabilizing mutations, as in [9]

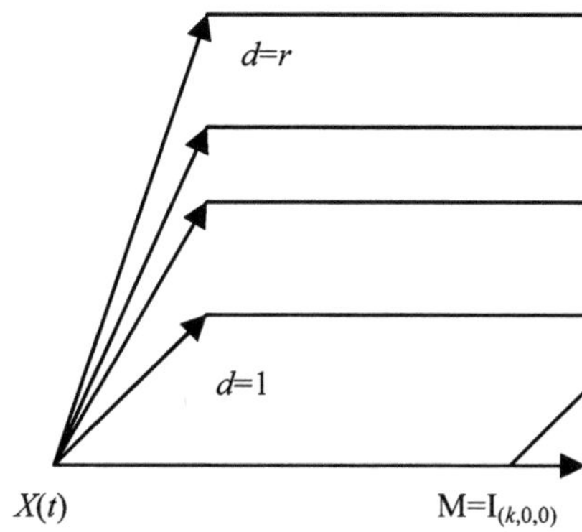

Fig. 11.2 Destabilizing-mutation planes in model, each plane with structure of Fig. 11.1, as in [9]

Table 11.1 The number of biological parameters in a model with k cancer stages, r types of GI and m_d $((d = 1, \ldots, r))$ levels of destabilizations

Model parameter descriptions	Model parameters	Number of such parameters in the model
Stem cell population number	$X(t)$	1
Growth rate	$G(\alpha, \beta, d)(t)$	$k - 1 + k \cdot \sum_{d=1}^{r} m_d$
Death/differentiation rate	$D(\alpha, \beta, d)(t)$	$k - 1 + k \cdot \sum_{d=1}^{r} m_d$
Cancer-stage mutation rate	$M(\alpha, \beta, d)(t)$	$k + k \cdot \sum_{d=1}^{r} m_d$
Destabilizing mutation rate	$A(\alpha, \beta, d)(t)$	$k \cdot \sum_{d=1}^{r} m_d$
Total		$3 \cdot k - 1 + 4 \cdot k \cdot \sum_{d=1}^{r} m_d$

apoptose at rates $G(\alpha, \beta, d)$ and $D(\alpha, \beta, d)$, respectively. Each cell can divide into an equivalent daughter cell and another cell with an extra cancer stage mutation at rate $M(\alpha, \beta, d)$. Likewise, cells can also divide into an equivalent daughter cell and another cell with an additional destabilizing mutation of type d at rate $A(\alpha, \beta, d)$. The model assumes that there are $X(t)$ susceptible stem cells at age t. Further details on derivation of the hazard function are given elsewhere [9].

11.3.1 *Assessment of Parameter Identifiability for a Stochastic Cancer Model with Genomic Instability*

In Appendix C we derive the hazard function and show that it can be written in terms of certain combinations of the biological parameters given in Table 11.1. From equations (11.C12–11.C16) in Appendix C it is seen that the characteristics and ψ are governed by certain parameter combinations. Table 11.2 summarizes the maximum number of identifiable parameter combinations and their forms associated with each cell compartment. The maximum number of identifiable parameters associated with each destabilization zone, $I_{\alpha,\beta,d}$, are 4 when $\alpha < k-1$ and $0 < \beta < m_d$; 4 when $\alpha = k-1$ and $0 < \beta < m_d$; 3 when $\alpha < k-1$ and $\beta = m_d$ and 2 when $\alpha = k-1$ and $\beta = m_d$. The function ψ is governed by at most $r+1$ parameter combinations. Therefore, we have shown that the hazard function $h(\theta)$ can be written as $h(G_1(\theta), G_2(\theta), \ldots, G_N(\theta))$ for some scalar functions $G_1(.), G_2(.), \ldots, G_N(.)$, where $N = (k-2)\cdot(3+r)\ +(3+r)\cdot 1\ +(1+r)\cdot 1\ +4\cdot(k-1)\cdot\sum_{d=1}^{r}(m_d-1)\ +4\cdot\sum_{d=1}^{r}(m_d-1)\ +3\cdot(k-1)\cdot r\ \ +2\cdot r\ = 3k-2-r+4k\cdot\sum_{d=1}^{r} m_d$ (Table 11.2). Assuming that the cancer model is embedded in a member of the exponential family (in the sense outlined in Appendix B) the same will be true of the total log-likelihood $L(x|\theta) = L(x|G_1(\theta), G_2(\theta), \ldots, G_N(\theta))$. By means of the Chain Rule [18] (p.215) we obtain $\frac{\partial^2 L(x|\theta)}{\partial\theta_i\partial\theta_j} = \sum_{l,k=1}^{N} \frac{\partial^2 L(x|G_1,\ldots,G_N)}{\partial G_l \partial G_k}\frac{\partial G_l}{\partial\theta_i}\frac{\partial G_k}{\partial\theta_j} + \sum_{l=1}^{N}\frac{\partial L(x|G_1,\ldots,G_N)}{\partial G_l}\frac{\partial^2 G_l}{\partial\theta_i\partial\theta_j}$, so that the Fisher information matrix is given by

$$
\begin{aligned}
I(\theta) = -E_\theta\left[\frac{\partial^2 L(x|\theta)}{\partial\theta_i\partial\theta_j}\right] &= -E\left[\sum_{l,k=1}^{N}\frac{\partial^2 L(x|G_1,\ldots,G_N)}{\partial G_l\partial G_k}\frac{\partial G_l}{\partial\theta_i}\frac{\partial G_k}{\partial\theta_j}\right] \\
&= -\sum_{l,k=1}^{N}\frac{\partial G_l}{\partial\theta_i}E\left[\frac{\partial^2 L(x|G_1,\ldots,G_N)}{\partial G_l\partial G_k}\right]\frac{\partial G_k}{\partial\theta_j}
\end{aligned}
\tag{11.4}
$$

which therefore has rank at most N. A similar argument shows that if one were to reparameterise [via some invertible C^2 mapping $\theta = f(\omega)$] then the embedded log-likelihood $L(x|f^{-1}(\theta)) = L(x|\omega)$ associated with $h(f^{-1}(\theta)) = h(\omega)$ must also have Fisher information matrix of rank at most N. By Theorems 1 and 3 of Catchpole and Morgan [3], for this embedded exponential family model therefore there can be at most N irredundant parameters. Therefore, of the theoretically available $1+2\cdot[k-1+k\cdot\sum_{d=1}^{r} m_d]+k+2\cdot k\cdot\sum_{d=1}^{r} m_d = 3k-1+4k\cdot\sum_{d=1}^{r} m_d$ biological parameters (Table 11.1), at most $N = 3k-2-r+4k\cdot\sum_{d=1}^{r} m_d$ parameter combinations are identifiable, indicating a minimum of $(r+1)$ parameter redundancies in the model. Also, from Corollary 2 (2) and the subsequent Remark (2) above,

Table 11.2 Parameter combinations associated with each cell compartment

Compartment $I_{\alpha,\beta,d}$	Number of such compartments	Forms of identifiable parameter combinations	Maximum number of identifiable parameter combinations	Total maximum number of identifiable parameter combinations
Principal axis (non-destabilization) $I_{\alpha,0,0}$ ($\alpha = 0,\ldots,k-1,\ \beta = d = 0$)				
$0<\alpha<k-1$	$(k-2)$	$G(\alpha,0,0), D(\alpha,0,0)-G(\alpha,0,0), \frac{M(\alpha,0,0)}{G(\alpha+1,0,0)}, \left(\frac{A(\alpha,0,d')}{G(\alpha,1,d')}\right)_{d'=1}^{r}$	$3+r$	$(k-2)\cdot(3+r)$
$\alpha=k-1$	1	$G(\alpha,0,0), D(\alpha,0,0)-G(\alpha,0,0)+M(\alpha,0,0), M(\alpha,0,0), \left(\frac{A(\alpha,0,d')}{G(\alpha,1,d')}\right)_{d'=1}^{r}$	$3+r$	$3+r$
$\psi(\alpha=\beta=d=0)$	1	$\frac{X\cdot M(0,0,0)}{G(1,0,0)}, \left(\frac{X\cdot A(0,0,d')}{G(0,1,d')}\right)_{d'=1}^{r}$	$1+r$	$1+r$
r destabilization zones ($0\le\alpha\le k-1,\ 1\le\beta\le m_d,\ 1\le d\le r$)				
$\alpha<k-1$, $1\le\beta<m_d$	$(k-1)\cdot\sum_{d=1}^{r}(m_d-1)$	$G(\alpha,\beta,d),\ D(\alpha,\beta,d)-G(\alpha,\beta,d),\ \frac{A(\alpha,\beta,d)}{G(\alpha,\beta+1,d)}, \frac{M(\alpha,\beta,d)}{G(\alpha,\beta+1,d)}$	4	$4\cdot(k-1)\cdot\sum_{d=1}^{r}(m_d-1)$
$\alpha=k-1$, $1\le\beta<m_d$	$\sum_{d=1}^{r}(m_d-1)$	$G(\alpha,\beta,d),\ M(\alpha,\beta,d),\ D(\alpha,\beta,d)-G(\alpha,\beta,d)+M(\alpha,\beta,d),\ \frac{A(\alpha,\beta,d)}{G(\alpha,\beta+1,d)}$	4	$4\cdot\sum_{d=1}^{r}(m_d-1)$
$\alpha<k-1,\ \beta=m_d$	$(k-1)\cdot r$	$\frac{M(\alpha,\beta,d)}{G(\alpha+1,\beta,d)},\ D(\alpha,\beta,d)-G(\alpha,\beta,d),\ G(\alpha,\beta,d)$	3	$3\cdot(k-1)\cdot r$
$\alpha=k-1,\ \beta=m_d$	r	$D(\alpha,m_d,d)-G(\alpha,m_d,d)+M(\alpha,m_d,d),\ G(\alpha,m_d,d)\cdot M(\alpha,m_d,d)$	2	$2\cdot r$
Total				$3k-2-r+4k\cdot\sum_{d=1}^{r}m_d$

The forms of these combinations are extracted from Eqs. (11.C12–11.C16) in Appendix C subject to some regulatory conditions, the number of locally identifiable or (gradient) weakly locally identifiable parameter combinations is equal to the rank of the Fisher information matrix, so $\leq N$. For example, in the case of the two-mutation model [11], with $k = 2, r = 1, d = 0$ and $m_d = 0$ (and so $m = \sum_{d=1}^{r} m_d = 0$), there are $k \cdot (m+1) - 1 = 2 \cdot 1 - 1 = 1$ G's (namely G(1,0,0)), $k \cdot (m+1) - 1 = 2 \cdot 1 - 1 = 1$ D's (namely $D(1,0,0)$), $k \cdot m = 2 \cdot 0 = 0$ A's, $k \cdot (m+1) = 2 \cdot 1 = 2$ M's (namely $M(0,0,0), M(1,0,0)$), and a single X, giving a total of five biological parameters. It is known from the results of Heidenreich et al. [14, 15] that for the two-mutation model only three combinations of these are estimable, i.e., that there are two redundancies, precisely in agreement with the result given here for $r = 1$. This result therefore precisely generalizes the results and approach of Heidenreich et al. [14, 15]. Unfortunately, analytical methods for proving that precisely this number of parameters are estimable, including some recently outlined [23], cannot be used for the model considered here. Nevertheless, we conjecture that in fact precisely this number of parameters are estimable, so that the upper bound on the number of estimable parameter combinations that we have proved above is in fact sharp. This is supported by numerical evaluation of the Hessian in a couple of example cases, which we now outline.

11.3.2 Numerical Evaluation of Hessian and Determination of its Rank

That there are likely to be exactly this number of estimable parameters is supported by numerical evaluation of the Hessian matrix of the hazard function. We make use of the solution of the system of ordinary differential equations defining the Hessian, outlined in Appendix E. We will show in two cases that the Hessian has rank two less than the number of biological parameters, w. By the above results (also of Catchpole and Morgan [3]) this suggests that precisely $w - 2$ parameters are (gradient) weakly locally identifiable. In order to show that the Hessians are of rank two less than the number of biological parameters, w, we evaluate the eigenvalues of the Hessian matrix, and establish that the smallest eigenvalue among the $w - 2$ largest eigenvalues in absolute value exceeds the likely magnitude of the error by at least an order of magnitude. We know the likely size of the error in numerical evaluations of each element, h_{ij}, of the Hessian from the Boerlisch-Stoer integrator that is employed, namely $\max(10^{-10}, 10^{-10} \cdot |h_{ij}| : 1 \leq i,j \leq w)$ (**bsstep** routine, Press et al. [24]). It is known that if two symmetric matrices H and $\tilde{H}$ have eigenvalues $\lambda_1 \leq \lambda_2 \leq \ldots \lambda_{w-1} \leq \lambda_w$ and $\tilde{\lambda}_1 \leq \tilde{\lambda}_2 \leq \ldots \tilde{\lambda}_{w-1} \leq \tilde{\lambda}_w$ then $|\lambda_i - \tilde{\lambda}_i| \leq ||H - \tilde{H}||_2$, $1 \leq i \leq w$, where $||H||_2 = \sup[||Hx||_2/||x||_2 : x \neq 0]$ [25]. Since the approximate Hessian that we calculate, $\tilde{H}$, differs from the true Hessian, H, by an amount $||H - \tilde{H}||_2 \leq \sqrt{w} \cdot \max[|h_{ij} - \tilde{h}_{ij}| : 1 \leq i,j \leq w]$, we know that:

$$|\lambda_i - \tilde{\lambda}_i| \leq \sqrt{w} \cdot \max[|h_{ij} - \tilde{h}_{ij}| : 1 \leq i,j \leq w] \leq \sqrt{w} \cdot \max[10^{-10}, 10^{-10} \cdot |\tilde{h}_{ij}| : 1 \leq i,j \leq w] \quad (11.5)$$

There is also the issue of numerical roundoff error in the QR algorithm (Numerical Algorithms Group (NAG) routine **F02FAF** [26]) used to compute eigenvalues. If we write now $\tilde{\lambda}_i, \widehat{\tilde{\lambda}}_i$ for the true and approximate eigenvalues associated with the approximate Hessian, $\tilde{H}$, this is bounded by:

$$|\tilde{\lambda}_i - \widehat{\tilde{\lambda}}_i| \leq c(w) \cdot \varepsilon \cdot ||\tilde{H}||_2 \leq c(w) \cdot \varepsilon \cdot \sqrt{w} \cdot \max[|\tilde{h}_{ij}| : 1 \leq i,j \leq w], \quad 1 \leq i \leq w \quad (11.6)$$

where $c(w)$ is a modestly increasing function of the dimension, w, of the approximate Hessian $\tilde{H}$ and ε is the machine precision [26]. Since the machine precision (in double precision) is of the order 10^{-15}, expression (11.6) will be dominated by the error associated with the approximation to the Hessian, given by expression (11.5).

We evaluated the Hessian matrix for a model with three cancer-stage mutations and one destabilizing mutation, and a model with two cancer-stage mutations and one destabilizing mutation; log-normal perturbations of all parameters were performed, assuming a geometric standard deviation (GSD) of 4, centred on models with cancer-stage mutation rates of 4.0×10^{-3} year^{-1}, destabilizing mutation rates of 3.0×10^{-3} year^{-1}, intermediate cell proliferation rates of 1.0×10^{-1} year^{-1}, and intermediate cell death rates of 5.0×10^{-1} year^{-1}. For each of 1,000 random sets of parameters we evaluated the Hessian by numerical integration, as outlined in Appendix E. We calculated the eigenvalues of the Hessian using the QR algorithm, specifically the NAG FORTRAN subroutine **F02FAF** [26]. For each model we selected the set of random parameters for which the ratio of minimum to maximum among the $w - 2$ largest eigenvalues (w being the number of biological parameters) in absolute value was greatest. These are given in Tables 11.3 and 11.4, for the three-stage and two-stage models, respectively. The associated eigenvalues are given in Table 11.5. The absolute value of the $w - 2$th smallest eigenvalue associated with each set exceeds the error bound (11.5) by at least an order of magnitude in each case. This strongly suggests that the Hessians calculated for these two examples really are of rank $w - 2$ for each model.

11.4 Discussion

We have introduced various novel notions of identifiability, related to ideas previously introduced by Rothenberg and others [1, 3]. In particular we have shown that within the exponential family models parameter irredundancy, local

Table 11.3 Example coefficients of model with three cancer stage mutations and one destabilizing mutation

Coefficient	Value
$G(1,0,0)$	$8.64714335947694 \times 10^{-2}$
G(2,0,0)	$1.06188950764276 \times 10^{-3}$
$D(1,0,0)$	$4.25556779736062 \times 10^{-2}$
$D(2,0,0)$	$2.68975909218019 \times 10^{-1}$
$M(0,0,0)$	$1.33167380928588 \times 10^{-2}$
$M(1,0,0)$	$1.08841503240502 \times 10^{0}$
$M(2,0,0)$	$9.79093689335407 \times 10^{-2}$
$A(0,0,1)$	$1.33537580655960 \times 10^{-1}$
$A(1,0,1)$	$7.65789029061483 \times 10^{-2}$
$A(2,0,1)$	$3.73742902997137 \times 10^{-2}$
G(0,1,1)	$5.31044255713088 \times 10^{-1}$
$G(1,1,1)$	$1.32418227810710 \times 10^{1}$
$G(2,1,1)$	$6.88863709884594 \times 10^{-2}$
$D(0,1,1)$	$1.14118194976730 \times 10^{-2}$
$D(1,1,1)$	$2.99644035332771 \times 10^{-1}$
$D(2,1,1)$	$8.92155178101449 \times 10^{-1}$
$M(0,1,1)$	$7.55711980917015 \times 10^{0}$
$M(1,1,1)$	$6.58304546585478 \times 10^{0}$
$M(2,1,1)$	$4.33636256393215 \times 10^{-3}$
X	$4.06993305645860 \times 10^{0}$

Table 11.4 Example coefficients of model with two cancer stage mutations and one destabilizing mutation

Coefficient	Value
$G(1,0,0)$	$2.22095885699822 \times 10^{-3}$
$D(1,0,0)$	$1.31378739613141 \times 10^{-6}$
$M(0,0,0)$	$8.12022029775447 \times 10^{-4}$
$M(1,0,0)$	$1.40674010365097 \times 10^{-5}$
$A(0,0,1)$	$2.06668108660923 \times 10^{-1}$
$A(1,0,1)$	$4.57214970326658 \times 10^{-3}$
G(0,1,1)	$1.56644835664010 \times 10^{-2}$
$G(1,1,1)$	$3.16379145991048 \times 10^{-4}$
$D(0,1,1)$	$1.29917705679554 \times 10^{0}$
$D(1,1,1)$	$1.92969737536413 \times 10^{-1}$
$M(0,1,1)$	$9.58173133172697 \times 10^{0}$
$M(1,1,1)$	$2.26339224702545 \times 10^{-1}$
X	$2.78141105650539 \times 10^{-1}$

identifiability, gradient weak local identifiability and weak local identifiability are largely equivalent.

The slight novelty of our approach, as with that of Catchpole and Morgan [3], is that for reasons of greater analytic tractability the notions of identifiability that we introduce are related more to the Hessian of the likelihood rather than the Fisher information matrix that was considered elsewhere [1]. The use of this approach is motivated by the application, namely to determine identifiable parameter combinations in the very general class of stochastic cancer models [8–13, 22] outlined

Table 11.5 Eigenvalues in ascending order of Hessian matrix associated with a model with three cancer stage mutations and one destabilizing mutation (as in Table 11.3), and with a model with two cancer stage mutations and one destabilizing mutation (as in Table 11.4)

Number	Eigenvalues (Table 11.3)	Eigenvalues (Table 11.4)
1	$-1.20726415206490 \times 10^{1}$	$-1.45810346778189 \times 10^{0}$
2	$-4.92487558715060 \times 10^{0}$	$-7.77741441881355 \times 10^{-1}$
3	$-1.11648980088601 \times 10^{0}$	$-2.77127189259301 \times 10^{-1}$
4	$-2.44711976272777 \times 10^{-1}$	$-6.66243518532325 \times 10^{-3}$
5	$-9.84288250086772 \times 10^{-2}$	$-3.53209777682867 \times 10^{-4}$
6	$-1.23814589706358 \times 10^{-2}$	$-2.86471102388267 \times 10^{-4}$
7	$-2.95522329598474 \times 10^{-3}$	$\mathbf{-9.25930409562877 \times 10^{-6}}$
8	$-1.53669876331947 \times 10^{-3}$	$\mathbf{-1.78637642487767 \times 10^{-11}}$
9	$-9.80139032107413 \times 10^{-5}$	$2.74342908757636 \times 10^{-4}$
10	$-3.36238129341872 \times 10^{-5}$	$4.98697524563660 \times 10^{-4}$
11	$-2.14105771381677 \times 10^{-6}$	$1.11215731049368 \times 10^{-2}$
12	$\mathbf{-1.86967299054058 \times 10^{-7}}$	$8.18426507233826 \times 10^{-1}$
13	$\mathbf{5.01559183858810 \times 10^{-12}}$	$1.45195703291853 \times 10^{0}$
14	$9.44044820094881 \times 10^{-7}$	–
15	$4.05661818962605 \times 10^{-4}$	–
16	$1.92220119614334 \times 10^{-3}$	–
17	$1.11042617352459 \times 10^{-2}$	–
18	$1.03277102432191 \times 10^{-1}$	–
19	$1.12667702944003 \times 10^{0}$	–
20	$1.08248991510735 \times 10^{1}$	–

Non-significant eigenvalues are underlined in bold

above. In certain applications the Fisher information may be best for estimating the upper bound to the number of irredundant parameters, but the Hessian may be best for estimating the lower bound of this quantity.

The model of Little et al. [9] that we consider generalizes many other well known cancer models [10–13, 22], indeed most of the widely-used cancer models developed in the last 30 years; these and other cancer models are generally embedded in an exponential family model in the sense outlined in Appendix B, in particular when cohort data are analysed using Poisson regression models, e.g., as in Little et al. [8, 9, 21]. As we show at the end of the Analysis Section, proving (gradient) weak local identifiability of a subset of cardinality k of the parameters $(\theta_i)_{i=1}^{p}$ can be done by showing that for this subset of parameters $rk\left[\left(\frac{\partial^2 h(\theta,y)}{\partial\theta_i\partial\theta_j}\right)_{i,j=1}^{p}\right] = k$ where h is the cancer hazard function. We have demonstrated (by exhibiting a particular parameterization) that there is redundancy in the parameterization for this model: the number of theoretically estimable parameters in the models of Little and Wright [8] and Little et al. [9] is at most two less than the number that are theoretically available, demonstrating (by Corollary 2) that there can be no more than this number of irredundant parameters. Two numerical examples suggest that this bound is sharp—we show that the rank of the Hessian,

$rk\left[\left(\frac{\partial^2 h(\theta,y)}{\partial\theta_i\partial\theta_j}\right)^p_{i,j=1}\right]$, is two less than the row dimension of this matrix. This result generalizes previously derived results of Heidenreich et al. [14, 15] and Hanin et al. [27, 28] for the two-mutation model, a special case of this model. For the more general genomic-instability cancer model of Little et al. [9] the number of identifiable combinations of parameters is at most $r+1$ less than the number of biological parameters, where r is the number of destabilization types, thereby also generalizing all these results. Numerical evaluations in two special cases (with $r=1$) suggest that this bound is tight: a combination of parameters with cardinality two less than the number of biological parameters is of full rank, and so is not redundant.

A weakness of the paper is that one cannot be absolutely sure (because of the uncertainty implicit in any numerical evaluation) that the bound demonstrated by the mathematics of Sect. 11.3.1 and Appendix C is sharp. Nevertheless, we have clearly established a maximum number of identifiable parameter combinations. We have also specified particular combinations of identifiable parameters, and these should be used in model fitting to avoid obvious numerical problems, of lack of convergence and absence of a unique set of parameters maximizing the likelihood.

Our results imply that for the very general class of cancer models considered here, only certain specific parameter combinations should be estimated in principle, and this is the case whatever the size of the dataset being considered. Whether for complex models for even this theoretically available number of parameters there is useful information is of course uncertain, and may well depend on the particular dataset and on the likely size of the parameters to be estimated. However, fits to a large population-based registry of colon cancer, as recently analysed by Little and Li [21], suggests that, for example, the model with two cancer-stage and one destabilizing mutations can be fitted to the dataset and yields stable parameter estimates for certain combinations of 11 parameters, in accordance with the results of this chapter.

Acknowledgments This work was supported by the Intramural Research Program of the National Institutes of Health, the National Cancer Institute, Division of Cancer Epidemiology and Genetics.

11.5 Appendix A

Proof of Theorem 1 In this Section we outline a proof of Theorem 1 in the main text. To prove this result we need the following lemma of Rudin [18] (p.229).

Lemma A1 Suppose m, n, r are non-negative integers such that (s.t.) $m \geq r$, $n \geq r$ and F is a C^1 function $E \subset R^n \to R^m$ where E is an open set. Suppose that $rk(F'(x)) = r\ \forall x \in E$. Fix $a \in E$ and put $A = F'(a)$, and let $Y_1 = A(R^n)$ and let

$P : R^m \to R^m$ be a linear projection operator ($P^2 = P$) s.t. $Y_1 = P(R^m)$ and $Y_2 = null(P)$. Then $\exists U, V \subset R^n$, open sets and a bijective C^1 function $H : V \to U$ whose inverse is also C^1 and s.t. $F(H(x)) = Ax + \varphi(Ax)$, $\forall x \in V$ where $\varphi : AV \subset Y_1 \to Y_2$ is a C^1 function.

We now restate Theorem 1 here.

Theorem A2 *Suppose that the log-likelihood $L(x|\theta)$ is C^2 as a function of the parameter vector $\theta \in \Omega \subset R^p$, and for all $x = (x_1, \ldots, x_n) \in \Sigma^n$.*

1. Suppose that for some x and $\theta \in \mathrm{int}(\Omega)$ it is the case that $rk\left[\left(\frac{\partial^2 L(x|\theta)}{\partial\theta_i \partial\theta_j}\right)_{i,j=1}^p\right] = p$. Then turning points of the likelihood in the neighborhood of θ are isolated, i.e., there is an open neighborhood $N \in \aleph_\theta \subset \Omega$ for which there is at most one $\widehat{\theta} \in N$ that satisfies $\left(\frac{\partial L(x|\theta)}{\partial\theta_i}\right)_{i=1}^p\Big|_{\theta=\widehat{\theta}} = 0$.
2. Suppose that for some x and $\theta \in \mathrm{int}(\Omega)$ it is the case that $rk\left[\left(\frac{\partial^2 L(x|\theta)}{\partial\theta_i \partial\theta_j}\right)_{i,j=1}^p\right] = p$ then local maxima of the likelihood in the neighborhood of θ are isolated, i.e., there is an open neighborhood $N \in \aleph_\theta \subset \Omega$ for which there is at most one $\widehat{\theta} \in N$ that is a local maximum of $L(x|\theta)$.
3. Suppose that for some x and all $\theta \in \mathrm{int}(\Omega)$ it is the case that $rk\left[\left(\frac{\partial^2 L(x|\theta)}{\partial\theta_i \partial\theta_j}\right)_{i,j=1}^p\right] = r < p$ then all local maxima of the likelihood in $\mathrm{int}(\Omega)$ are not isolated, as indeed are all $\theta \in \mathrm{int}(\Omega)$ for which $\left(\frac{\partial L(x|\theta)}{\partial\theta_i}\right)_{i=1}^p = 0$.

Proof

1. Let $F : \Omega \subset R^p \to R^p$ be defined by $F(\theta_1, \theta_2, \ldots, \theta_p) = \left(\frac{\partial L(x|\theta)}{\partial\theta_1}, \frac{\partial L(x|\theta)}{\partial\theta_2}, \ldots, \frac{\partial L(x|\theta)}{\partial\theta_p}\right)$. Since L is C^2, F is C^1 on $\mathrm{int}(\Omega) \subset R^p$. By assumption $\frac{\partial F(\theta)_i}{\partial\theta_j} = \left(\frac{\partial^2 L(x|\theta)}{\partial\theta_j \partial\theta_i}\right)_{i,j=1}^p$ is of full rank at θ. By the inverse function theorem [18] (p.221–223) there are open $N, M \subset R^p$ such that $\theta \in N$ and a C^1 bijective function $G : M \to N$ such that $G(F(\widehat{\theta})) = \widehat{\theta}$ for all $\widehat{\theta} \in N$. In particular there can be at most a single $\widehat{\theta} \in N$ for which $F(\widehat{\theta}) = 0$. **QED.**
2. By (1) there is an open neighborhood $N \in \aleph_\theta \subset \Omega$ for which if $\widehat{\theta} \in N$ is such that $\left(\frac{\partial L(x|\theta)}{\partial\theta_i}\right)_{i=1}^p\Big|_{\theta=\widehat{\theta}} = 0$ then for $\theta' \neq \widehat{\theta} \in N$ $\left(\frac{\partial L(x|\theta)}{\partial\theta_i}\right)_{i=1}^p\Big|_{\theta=\theta'} \neq 0$. Suppose now that $\widehat{\theta} \in N$ is a local maximum of $L(x|\theta)$. Any member of this neighborhood other than $\widehat{\theta}$ cannot be a turning point, and so by the Mean Value Theorem [18] (p.107) cannot be a local maximum. **QED.**

3. Let $F: \Omega \subset R^p \to R^p$ be defined by $F(\theta_1, \theta_2, \ldots, \theta_p) = \left(\frac{\partial L(x|\theta)}{\partial \theta_1}, \frac{\partial L(x|\theta)}{\partial \theta_2}, \ldots, \frac{\partial L(x|\theta)}{\partial \theta_p}\right)$. Since L is C^2, F is C^1 on $\mathrm{int}(\Omega) \subset R^p$. By assumption $rk(F) = rk\left[\left(\frac{\partial^2 L(x|\theta)}{\partial \theta_i \partial \theta_j}\right)_{i,j=1}^p\right] = r$ for all $\theta \in \mathrm{int}(\Omega) \subset R^p$. Suppose that $\theta_0 \in \mathrm{int}(\Omega)$ is a local maximum of L. Let $A = \frac{\partial F}{\partial \theta}\Big|_{\theta=\theta_0}: R^p \to R^p$ $(A \in L(R^p, R^p))$, and choose some arbitrary projection $P \in L(R^p, R^p)$ s.t. $P(R^p) = Y_1 = A(R^p)$, and let $Y_2 = null(P)$. By Lemma A1 there are open sets $U, V \subset R^p$ with $\theta_0 \in U \subset \mathrm{int}(\Omega)$ and a bijective C^1 mapping with C^1 inverse $H: V \to U$ s.t. $F(y) = AH^{-1}y + \varphi(AH^{-1}y)$, $\forall y \in V$ where $\varphi: AV \subset Y_1 \to Y_2$ is a C^1 function.

Since $\theta_0 \in \mathrm{int}(\Omega)$ is a local maximum of $L(x;\theta)$, by the Mean Value Theorem [18] (p.107) $F(\theta_0) = 0$. Now choose some non-trivial vector $k \in null(A)$ and define a function, as we can, on some interval $\delta: (-\varepsilon, \varepsilon) \to R^p$ by $\delta(t) = H(H^{-1}(\theta_0) + tk)$. Because $H: V \to U$ is bijective and k is non-trivial $\delta(t) = \delta(t') \Leftrightarrow t = t'$. Also, it is the case that:

$$F(\delta(t)) = AH^{-1}(H[H^{-1}(\theta_0) + tk]) + \varphi(AH^{-1}(H[H^{-1}(\theta_0) + tk])) = \\ A[H^{-1}(\theta_0) + tk] + \varphi(A[H^{-1}(\theta_0) + tk]) = A[H^{-1}(\theta_0)] + \varphi(A[H^{-1}(\theta_0)]) = F(\theta_0) = 0 \tag{11.A1}$$

Define $G: (-\varepsilon, \varepsilon) \to R$ by $G(t) = L(\delta(t)) = L((\delta_1(t), \delta_2(t), \ldots, \delta_n(t)))$. By the Chain Rule [18] (p.215) $\frac{dG}{dt} = \sum_{i=1}^{p} \frac{\partial L(x|\theta)}{\partial \theta_i} \frac{d\delta_i}{dt} = 0$ $\forall t \in (-\varepsilon, \varepsilon)$. Finally, by the Mean Value Theorem [18] (p.107) G must be constant; in particular $L(x|\delta(t)) = L(x|\delta(0)) = L(x|\theta_0)$ $\forall t \in (-\varepsilon, \varepsilon)$ and so all points $\delta(t)$ must also be local maxima of $L(x|\theta)$. Therefore θ_0 is not an isolated local maximum. Since all we used about $\theta_0 \in \mathrm{int}(\Omega)$ was that $F(\theta_0) = 0$, $F((\theta_{0i})_{i=1}^p) = \left(\frac{\partial L(x|\theta_0)}{\partial \theta_{01}}, \frac{\partial L(x|\theta_0)}{\partial \theta_{02}}, \ldots, \frac{\partial L(x|\theta_0)}{\partial \theta_{0p}}\right) = 0$, the above argument also shows that turning points cannot be isolated: $F(\delta(t)) = 0$. **QED.**

Appendix B

Specification of Embedded Exponential Family Model

In this Section we outline the specification of an embedding of a stochastic cancer model in a general class of statistical models, the so-called exponential family [17]. This is often done in fitting cancer models to epidemiological and biological data (e.g., see references [8, 9, 16, 21]). Recall that a model is a member of the exponential family if the observed data $x = (x_l)_{l=1}^n \in \sum^n$ is such that the log-likelihood is given by $L(x|\theta) = \sum_{l=1}^{n} \left[\frac{x_l \varsigma_l - b(\varsigma_l)}{a(\phi)} + c(x_l, \phi)\right]$ for some functions $a(\phi), b(\varsigma), c(x, \phi)$. We assume that the natural parameters $\varsigma_l = \varsigma_l[(\theta_i)_{i=1}^{p}, z_l]$ are functions of the model parameters $(\theta_i)_{i=1}^{p}$ and some auxiliary data $(z_l)_{l=1}^{n}$, and that $\mu_l = b'(\varsigma_l[(\theta_i)_{i=1}^{p}, z_l]) = z_l \cdot h[(\theta_i)_{i=1}^{p}, y_l]$. Here $h[(\theta_i)_{i=1}^{p}, y_l]$ is the cancer hazard function (for example, that of Little et al. [9], as also specified in Appendix C), $(y_l)_{l=1}^{n}$ are some further auxiliary data, and we assume that the $(z_l)_{l=1}^{n}$ are all non-zero. [Note: this is not necessarily a GLM.] In this case it is seen that

$$\frac{\partial^2 L(x|\theta)}{\partial\theta_i \partial\theta_j} = \sum_{l=1}^{n} \left[\begin{array}{l} \frac{[x_l - b'(\varsigma_l)] z_l}{a(\phi) b''(\varsigma_l)} \frac{\partial^2 h(\theta, y_l)}{\partial\theta_i \partial\theta_j} \\ - \frac{z_l^2}{a(\phi)} \frac{\partial h(\theta, y_l)}{\partial\theta_i} \frac{\partial h(\theta, y_l)}{\partial\theta_j} \left\{ \frac{[b''(\varsigma_l)]^2 + b'''(\varsigma_l)[x_l - b'(\varsigma_l)]}{[b''(\varsigma_l)]^3} \right\} \end{array} \right] \tag{11.B1}$$

so that the Fisher information matrix is given by

$$I(\theta)_{ij} = -E_\theta \left[\frac{\partial^2 L(x|\theta)}{\partial\theta_i \partial\theta_j} \right] = \frac{1}{a(\phi)} \sum_{l=1}^{n} \frac{z_l^2}{b''(\varsigma_l)} \frac{\partial h(\theta, y_l)}{\partial\theta_i} \frac{\partial h(\theta, y_l)}{\partial\theta_j} \tag{11.B2}$$

Appendix C

Derivation of Hazard Function in Terms of Specific Parameter Combinations for the Cancer Model of Little et al. [9]

In this Appendix we derive the hazard function for the cancer model of Little et al. [9] and show that it can be written in terms of certain combinations of parameters, given in Table 11.2. The hazard function is defined as:

$$h(t) = -\frac{d}{dt} \ln \psi(1, 1, \ldots, 1, 0; t, 0) \tag{11.C1}$$

where

$$\begin{aligned} &\psi(y_{1,0,0}, y_{2,0,0}, \ldots, y_{k-1,0,0}, y_{0,1,1}, \ldots, y_{k-1,1,1}, y_{0,2,1}, y_{1,2,1}, \ldots, y_{k-1,m_r,r}, y_k; t, 0) \\ &\equiv \psi(t) \\ &= \sum_n y_{1,0,0}^{n_{1,0,0}} \cdot \ldots \cdot y_{k-1,0,0}^{n_{k-1,0,0}} \cdot y_{0,1,1}^{n_{0,1,1}} \cdot \ldots \cdot y_{k-1,1,1}^{n_{k-1,1,1}} \cdot \ldots \cdot y_{k-1,m_r,r}^{n_{k-1,m_r,r}} \cdot y_k^{n_k} \times \\ &\quad P\big(Y_{1,0,0}(t) = n_{1,0,0}, \ldots, Y_k(t) = n_k \big| N(0) = X(0), Y_{1,0,0}(0) = \ldots = Y_k(0) = 0\big) \end{aligned} \tag{11.C2}$$

is the full probability generating function (PGF) starting with $X(0)$ cell(s) in the normal compartment at time 0. The number of biological parameters in this specific model is summarized in Table 11.1.

By straightforward generalizations of material in Little and Wright [8] (given in Appendix D) it is seen that ψ satisfies a Kolmogorov forward equation:

$$\frac{d\psi}{dt} = \psi \cdot [y_{1,0,0} - 1] \cdot X(t) \cdot M(0,0,0)(t) + \psi \cdot \sum_{d=1}^{r} [y_{0,1,d} - 1] \cdot X(t) \cdot A(0,0,d)(t) + \sum_{1 \le \alpha \le k-1} \begin{bmatrix} -y_{\alpha,0,0} \cdot [D(\alpha,0,0)(t) + G(\alpha,0,0)(t) + M(\alpha,0,0)(t) \\ + \sum_{d'=1}^{r} A(\alpha,0,d')(t)] + y_{\alpha,0,0}^2 \cdot G(\alpha,0,0)(t) + D(\alpha,0,0)(t) \\ + y_{\alpha,0,0} \cdot y_{\alpha+1,0,0} \cdot M(\alpha,0,0)(t) + \sum_{d'=1}^{r} y_{\alpha,0,0} \cdot y_{\alpha,1,d'} \cdot A(\alpha,0,d')(t) \end{bmatrix} \cdot \frac{\partial\psi}{\partial y_{\alpha,0,0}} + \sum_{\substack{0 \le \alpha \le k-1 \\ 1 \le \beta \le m_d \\ 1 \le d \le r}} \begin{bmatrix} -y_{\alpha,\beta,d} \cdot [D(\alpha,\beta,d)(t) + G(\alpha,\beta,d)(t) + M(\alpha,\beta,d)(t) \\ + A(\alpha,\beta,d)(t)] + y_{\alpha,\beta,d}^2 \cdot G(\alpha,\beta,d)(t) + D(\alpha,\beta,d)(t) \\ + y_{\alpha,\beta,d} \cdot y_{\alpha+1,\beta,d} \cdot M(\alpha,\beta,d)(t) + y_{\alpha,\beta,d} \cdot y_{\alpha,\beta+1,d} \cdot A(\alpha,\beta,d)(t) \end{bmatrix} \cdot \frac{\partial\psi}{\partial} \tag{11.C3}$$

with the conventions that $y_{\alpha,\beta,d} \equiv 0$ for $\beta > m_d$, $A(\alpha,\beta,d) \equiv 0$ for $\beta \ge m_d$ and $y_{k,\beta,d} \equiv y_k$ for all defined β and d. We solve the equation by means of Cauchy's method of characteristics. Suppose $y_{\alpha,\beta,d} \equiv y_{\alpha,\beta,d}(u)$ and $t \equiv t(u)$, then $\psi \equiv \psi(y_{\alpha,\beta,d}(u), t(u))$. This implies that:

$$\begin{aligned} \frac{\partial\psi}{\partial u} &= \frac{\partial\psi}{\partial t} \cdot \frac{dt}{du} + \frac{\partial\psi}{\partial y_k} \cdot \frac{dy_k}{du} + \sum_{1 \le \alpha \le k-1} \frac{\partial\psi}{\partial y_{\alpha,0,0}} \cdot \frac{dy_{\alpha,0,0}}{du} + \sum_{\substack{0 \le \alpha \le k-1 \\ 1 \le \beta \le m_d \\ 1 \le d \le r}} \frac{\partial\psi}{\partial y_{\alpha,\beta,d}} \cdot \frac{dy_{\alpha,\beta,d}}{du} \\ &= \frac{dt}{du} \cdot \begin{bmatrix} \psi \cdot [y_{1,0,0} - 1] \cdot X(t) \cdot M(0,0,0)(t) + \psi \cdot \sum_{d=1}^{r} [y_{0,1,d} - 1] \cdot X(t) \cdot A(0,0,d)(t) + \\ \sum_{1 \le \alpha \le k-1} \begin{bmatrix} -y_{\alpha,0,0} \cdot [D(\alpha,0,0)(t) + G(\alpha,0,0)(t) + M(\alpha,0,0)(t) \\ + \sum_{d'=1}^{r} A(\alpha,0,d')(t)] + y_{\alpha,0,0}^2 \cdot G(\alpha,0,0)(t) + D(\alpha,0,0)(t) \\ + y_{\alpha,0,0} \cdot y_{\alpha+1,0,0} \cdot M(\alpha,0,0)(t) + \sum_{d'=1}^{r} y_{\alpha,0,0} \cdot y_{\alpha,1,d'} \cdot A(\alpha,0,d')(t) \end{bmatrix} \cdot \frac{\partial\psi}{\partial y_{\alpha,0,0}} + \\ \sum_{\substack{0 \le \alpha \le k-1 \\ 1 \le \beta \le m_d \\ 1 \le d \le r}} \begin{bmatrix} -y_{\alpha,\beta,d} \cdot [D(\alpha,\beta,d)(t) + G(\alpha,\beta,d)(t) + M(\alpha,\beta,d)(t) \\ + A(\alpha,\beta,d)(t)] + y_{\alpha,\beta,d}^2 \cdot G(\alpha,\beta,d)(t) + D(\alpha,\beta,d)(t) \\ + y_{\alpha,\beta,d} \cdot y_{\alpha+1,\beta,d} \cdot M(\alpha,\beta,d)(t) + y_{\alpha,\beta,d} \cdot y_{\alpha,\beta+1,d} \cdot A(\alpha,\beta,d)(t) \end{bmatrix} \cdot \frac{\partial\psi}{\partial y_{\alpha,\beta,d}} \end{bmatrix} \\ &+ \frac{\partial\psi}{\partial y_k} \cdot \frac{dy_k}{du} + \sum_{1 \le \alpha \le k-1} \frac{\partial\psi}{\partial y_{\alpha,0,0}} \cdot \frac{dy_{\alpha,0,0}}{du} + \sum_{\substack{0 \le \alpha \le k-1 \\ 1 \le \beta \le m_d \\ 1 \le d \le r}} \frac{\partial\psi}{\partial y_{\alpha,\beta,d}} \cdot \frac{dy_{\alpha,\beta,d}}{du} \end{aligned} \tag{11.C4}$$

A solution is therefore given by:

$$\begin{aligned} \frac{\partial\psi}{\partial u} &= \psi \cdot [y_{1,0,0}(u) - 1] \cdot X(u) \cdot M(0,0,0)(u) \\ &+ \psi \cdot \sum_{d=1}^{r} [y_{0,1,d}(u) - 1] \cdot X(u) \cdot A(0,0,d)(u) \end{aligned} \tag{11.C5}$$

$$\frac{\partial t}{\partial u} = 1 \tag{11.C6}$$

$$\frac{\partial y_k}{\partial u} = 0 \tag{11.C7}$$

and for $d = 0$:

$$\begin{aligned}\frac{dy_{\alpha,0,0}}{du} &= y_{\alpha,0,0} \cdot [D(\alpha,0,0)(t) + G(\alpha,0,0)(t) + M(\alpha,0,0)(t) \\ &\quad + \sum_{d'=1}^{r} A(\alpha,0,d')(t)] - y^2_{\alpha,0,0} \cdot G(\alpha,0,0)(t) - D(\alpha,0,0)(t) \\ &\quad - y_{\alpha,0,0} \cdot y_{\alpha+1,0,0} \cdot M(\alpha,0,0)(t) - \sum_{d'=1}^{r} y_{\alpha,0,0} \cdot y_{\alpha,1,d'} \cdot A(\alpha,0,d')(t)\end{aligned} \tag{11.C8}$$

while for $d \neq 0$:

$$\begin{aligned}\frac{dy_{\alpha,\beta,d}}{du} &= y_{\alpha,\beta,d} \cdot [D(\alpha,\beta,d)(t) + G(\alpha,\beta,d)(t) + M(\alpha,\beta,d)(t) \\ &\quad + A(\alpha,\beta,d)(t)] - y^2_{\alpha,\beta,d} \cdot G(\alpha,\beta,d)(t) - D(\alpha,\beta,d)(t) \\ &\quad - y_{\alpha,\beta,d} \cdot y_{\alpha+1,\beta,d} \cdot M(\alpha,\beta,d)(t) - y_{\alpha,\beta,d} \cdot y_{\alpha,\beta+1,d} \cdot A(\alpha,\beta,d)(t)\end{aligned} \tag{11.C9}$$

For the hazard, a solution is required for $\psi(1,1,1,\ldots,1,0;t,0)$, i.e., $\psi(y_{1,0,0} = 1, y_{2,0,0} = 1, y_{3,0,0} = 1, \ldots, y_{k-1,m_r,r} = 1, y_k = 0; t, s = 0)$, so that a particular characteristic must have the boundary value $y_{\alpha,\beta,d}(t) = 1$ and $y_k(t) = 0$ [implying by (11.C7) $y_k(u) \equiv 0$], so that $y_{\alpha,\beta,d}(u)$ is a function of both u and t, i.e., $y_{\alpha,\beta,d}(u) \equiv y_{\alpha,\beta,d}(u,t)$. Integrating (11.C5) over $u \in [0,t]$ yields

$$\psi(t) = \exp\left\{\int_0^t \left[\begin{array}{l}\{y_{1,0,0}(u,t) - 1\} \cdot X(u) \cdot M(0,0,0)(u) \\ + \sum_{d=1}^{r} \{y_{0,1,d}(u,t) - 1\} \cdot X(u) \cdot A(0,0,d)(u)\end{array}\right] du\right\} \tag{11.C10}$$

Assume now that the model parameters $G(\alpha,\beta,d)(t)$, $D(\alpha,\beta,d)(t)$, $M(\alpha,\beta,d)(t)$, $A(\alpha,\beta,d)(t)$ and $X(t)$ are constant over time. By substituting $z_{\alpha,\beta,d}(u,t) = [y_{\alpha,\beta,d}(u,t) - 1] \cdot G(\alpha,\beta,d)$ into (11.C8) and (11.C9), the following can be obtained

$$\begin{cases}\dfrac{dz_{\alpha,\beta,d}}{ds} = -z^2_{\alpha,\beta,d} + z_{\alpha,\beta,d} \cdot N\left[\alpha,\beta,d,z_{\alpha,\beta+1,d},z_{\alpha+1,\beta,d}\right] & \textit{when } d \neq 0 \\ \qquad + P\left[\alpha,\beta,d,z_{\alpha,\beta+1,d},z_{\alpha+1,\beta,d}\right] & \\ \dfrac{dz_{\alpha,0,0}}{ds} = -z^2_{\alpha,0,0} + z_{\alpha,0,0} \cdot N'\left[\alpha,z_{\alpha,1,1},\cdots,z_{\alpha,1,r},z_{\alpha+1,0,0}\right] & \textit{when } d = 0 \\ \qquad + P'\left[\alpha,z_{\alpha,1,1},\cdots,z_{\alpha,1,r},z_{\alpha+1,0,0}\right] & \end{cases} \tag{11.C11}$$

where

$$N[\alpha,\beta,d,v,w]=\begin{cases} D(\alpha,\beta,d)-G(\alpha,\beta,d)-w\cdot\frac{M(\alpha,\beta,d)}{G(\alpha+1,\beta,d)} & \\ -v\cdot\frac{A(\alpha,\beta,d)}{G(\alpha,\beta+1,d)} & (\alpha<k-1,\beta<m_d) \\ D(\alpha,\beta,d)-G(\alpha,\beta,d)+M(\alpha,\beta,d) & \\ -v\cdot\frac{A(\alpha,\beta,d)}{G(\alpha,\beta+1,d)} & (\alpha=k-1,\beta<m_d) \\ D(\alpha,\beta,d)-G(\alpha,\beta,d)-w\cdot\frac{M(\alpha,\beta,d)}{G(\alpha+1,\beta,d)} & (\alpha<k-1,\beta=m_d) \\ D(\alpha,\beta,d)-G(\alpha,\beta,d)+M(\alpha,\beta,d) & (\alpha=k-1,\beta=m_d) \end{cases} \tag{11.C12}$$

$$P[\alpha,\beta,d,v,w]=\begin{cases} -G(\alpha,\beta,d)\cdot\left[w\cdot\frac{M(\alpha,\beta,d)}{G(\alpha+1,\beta,d)}+v\cdot\frac{A(\alpha,\beta,d)}{G(\alpha,\beta+1,d)}\right] & (\alpha<k-1,\beta<m_d) \\ G(\alpha,\beta,d)\cdot\left[M(\alpha,\beta,d)-v\cdot\frac{A(\alpha,\beta,d)}{G(\alpha,\beta+1,d)}\right] & (\alpha=k-1,\beta<m_d) \\ -w\cdot\frac{G(\alpha,\beta,d)\cdot M(\alpha,\beta,d)}{G(\alpha+1,\beta,d)} & (\alpha<k-1,\beta=m_d) \\ G(\alpha,\beta,d)\cdot M(\alpha,\beta,d) & (\alpha=k-1,\beta=m_d) \end{cases} \tag{11.C13}$$

$$N'[\alpha,v_1,\cdots,v_r,w]=\begin{cases} D(\alpha,0,0)-G(\alpha,0,0)-w\cdot\frac{M(\alpha,0,0)}{G(\alpha+1,0,0)} & \\ -\sum_{d=1}^{r}v_d\cdot\frac{A(\alpha,0,d)}{G(\alpha,1,d)} & (0<\alpha<k-1) \\ D(\alpha,0,0)-G(\alpha,0,0)+M(\alpha,0,0) & \\ -\sum_{d=1}^{r}v_d\cdot\frac{A(\alpha,0,d)}{G(\alpha,1,d)} & (\alpha=k-1) \end{cases} \tag{11.C14}$$

and

$$P'[\alpha,v_1,\cdots,v_r,w]=\begin{cases} -G(\alpha,0,0)\cdot\left[w\cdot\frac{M(\alpha,0,0)}{G(\alpha+1,0,0)}+\sum_{d=1}^{r}v_d\cdot\frac{A(\alpha,0,d)}{G(\alpha,1,d)}\right] & (0<\alpha<k-1) \\ G(\alpha,0,0)\cdot\left[M(\alpha,0,0)-\sum_{d=1}^{r}v_d\cdot\frac{A(\alpha,0,d)}{G(\alpha,1,d)}\right] & (\alpha=k-1) \end{cases} \tag{11.C15}$$

Likewise, with the same trick, (11.C10) can be rewritten as

$$\psi(t)=\exp\left\{\int_0^t z_{1,0,0}(s,t)\cdot\frac{X\cdot M(0,0,0)}{G(1,0,0)}+\sum_{d=1}^{r}z_{0,1,d}(s,t)\cdot\frac{X\cdot A(0,0,d)}{G(0,1,d)}\,ds\right\} \tag{11.C16}$$

Appendix D

Derivation of the Kolmogorov Forward Differential Equation for the Cancer Model of Little et al. [9]

In this Appendix we derive the Kolmogorov forward differential equation defining the generating function of the cancer model of Little et al. [9]. The generating function ψ is given by

$$\begin{aligned}
&\psi(y_{1,0,0}, y_{2,0,0}, \ldots, y_{k-1,0,0}, y_{0,1,1}, \ldots, y_{k-1,1,1}, y_{0,2,1}, y_{1,2,1}, \ldots, y_{k-1,m_r,r}, y_k; t, s) \\
&\equiv \psi(t,s) \\
&= \sum_n y_{1,0,0}^{n_{1,0,0}} \cdot \ldots \cdot y_{k-1,0,0}^{n_{k-1,0,0}} \cdot y_{0,1,1}^{n_{0,1,1}} \cdot \ldots \cdot y_{k-1,1,1}^{n_{k-1,1,1}} \cdot \ldots \cdot y_{k-1,m_r,r}^{n_{k-1,m_r,r}} \cdot y_k^{n_k} \times \\
&\quad P\left[Y_{1,0,0}(t) = n_{1,0,0}, \ldots, Y_k(t) = n_k \middle| N(s) = X(0), Y_{1,0,0}(s) = \ldots = Y_k(s) = 0\right]
\end{aligned} \tag{11.D1}$$

By differentiating term by term (justified by the absolute convergence of the derivative power series) ψ satisfies:

$$
\begin{aligned}
&\frac{\partial\psi}{\partial t}[t,s] = \sum_{n_{1,0,0},n_{2,0,0},\ldots,n_k} y_{1,0,0}^{n_{1,0,0}}\cdot\ldots\cdot y_{k-1,m_r,r}^{n_{k-1,m_r,r}}\cdot y_k^{n_k}\cdot\frac{dP}{dt}[Y_{1,0,0}(t)=n_{1,0,0},\ldots,Y_{k-1,m_r,r}(t)=n_{k-1,m_r,r},Y_k(t)=n_k]\\
&= \sum_{n_{1,0,0},n_{2,0,0},\ldots,n_k} y_{1,0,0}^{n_{1,0,0}}\cdot\ldots\cdot y_{k-1,m_r,r}^{n_{k-1,m_r,r}}\cdot y_k^{n_k}\cdot X(t)\cdot M(0,0,0)(t)\cdot P[Y_{1,0,0}(t)=n_{1,0,0}-1,\ldots,Y_{k-1,m}(t)=n_{k-1,m_r,r},Y_k(t)=n_k]+\\
&\sum_{d=1}^{r}\sum_{n_{1,0,0},n_{2,0,0},\ldots,n_k} y_{1,0,0}^{n_{1,0,0}}\cdot\ldots\cdot y_{k-1,m_r,r}^{n_{k-1,m_r,r}}\cdot y_k^{n_k}\cdot X(t)\cdot A(0,0,d)(t)\cdot P[Y_{1,0,0}(t)=n_{1,0,0},\ldots,Y_{0,1,d}(t)=n_{0,1,d}-1,\ldots,Y_{k-1,m_r,r}(t)=n_{k-1,m_r,r},Y_k(t)=n_k]-\\
&\sum_{n_{1,0,0},n_{2,0,0},\ldots,n_k} y_{1,0,0}^{n_{1,0,0}}\cdot\ldots\cdot y_{k-1,m_r,r}^{n_{k-1,m_r,r}}\cdot y_k^{n_k}\cdot X(t)\cdot M(0,0,0)(t)\cdot P[Y_{1,0,0}(t)=n_{1,0,0},\ldots,Y_{k-1,m_r,r}(t)=n_{k-1,m_r,r},Y_k(t)=n_k]-\\
&\sum_{d=1}^{r}\sum_{n_{1,0,0},n_{2,0,0},\ldots,n_k} y_{1,0,0}^{n_{1,0,0}}\cdot\ldots\cdot y_{k-1,m_r,r}^{n_{k-1,m_r,r}}\cdot y_k^{n_k}\cdot X(t)\cdot A(0,0,d)(t)\cdot P[Y_{1,0,0}(t)=n_{1,0,0},\ldots,Y_{k-1,m_r,r}(t)=n_{k-1,m_r,r},Y_k(t)=n_k]+\\
&\sum_{1\le\alpha\le k-1}\sum_{n_{1,0,0},n_{2,0,0},\ldots,n_k} y_{1,0,0}^{n_{1,0,0}}\cdot\ldots\cdot y_{k-1,m_r,r}^{n_{k-1,m_r,r}}\cdot y_k^{n_k}\cdot
\left[\begin{array}{l}
G(\alpha,0,0)(t)\cdot[n_{\alpha,0,0}-1]\cdot P[Y_{1,0,0}(t)=n_{1,0,0},\ldots,Y_{\alpha,0,0}(t)=n_{\alpha,0,0}-1,\ldots,Y_k(t)=n_k]+\\
D(\alpha,0,0)(t)\cdot[n_{\alpha,0,0}+1]\cdot P[Y_{1,0,0}(t)=n_{1,0,0},\ldots,Y_{\alpha,0,0}(t)=n_{\alpha,0,0}+1,\ldots,Y_k(t)=n_k]+\\
M(\alpha,0,0)(t)_{\alpha,0,0}\cdot P[Y_{1,0,0}(t)=n_{1,0,0},\ldots,Y_{\alpha,0,0}(t)=n_{\alpha,0,0},Y_{\alpha+1,0,0}(t)=n_{\alpha+1,0,0}-1,\ldots,Y_k(t)=n_k]+\\
\sum_{d'=1}^{r}A(\alpha,0,d')(t)_{\alpha,0,0}\cdot P[Y_{1,0,0}(t)=n_{1,0,0},\ldots,Y_{\alpha,0,0}(t)=n_{\alpha,0,0},\ldots,Y_{\alpha,1,d'}(t)=n_{\alpha,1,d'}-1,\ldots,Y_k(t)=n_k]-\\
n_{\alpha,0,0}\cdot P[Y_{1,0,0}(t)=n_{1,0,0},\ldots,Y_k(t)=n_k]\cdot\left[G(\alpha,0,0)(t)+D(\alpha,0,0)(t)+M(\alpha,0,0)(t)+\sum_{d'=1}^{r}A(\alpha,0,d')(t)\right]
\end{array}\right]+\\
&\sum_{\substack{0\le\alpha\le k-1\\1\le\beta\le m_d\\1\le d\le r}}\sum_{n_{1,0,0},n_{2,0,0},\ldots,n_k} y_{1,0,0}^{n_{1,0,0}}\cdot\ldots\cdot y_{k-1,m_r,r}^{n_{k-1,m_r,r}}\cdot y_k^{n_k}\cdot
\left[\begin{array}{l}
G(\alpha,\beta,d)(t)\cdot[n_{\alpha,\beta,d}-1]\cdot P[Y_{1,0,0}(t)=n_{1,0,0},\ldots,Y_{\alpha,\beta,d}(t)=n_{\alpha,\beta,d}-1,\ldots,Y_k(t)=n_k]+\\
D(\alpha,\beta,d)(t)\cdot[n_{\alpha,\beta,d}+1]\cdot P[Y_{1,0,0}(t)=n_{1,0,0},\ldots,Y_{\alpha,\beta,d}(t)=n_{\alpha,\beta,d}+1,\ldots,Y_k(t)=n_k]+\\
M(\alpha,\beta,d)(t)\cdot n_{\alpha,\beta,d}\cdot P[Y_{1,0,0}(t)=n_{1,0,0},\ldots,Y_{\alpha,\beta,d}(t)=n_{\alpha,\beta,d},Y_{\alpha+1,\beta,d}(t)=n_{\alpha+1,\beta,d}-1,\ldots,Y_k(t)=n_k]+\\
A(\alpha,\beta,d)(t)\cdot n_{\alpha,0,0}\cdot P[Y_{1,0,0}(t)=n_{1,0,0},\ldots,Y_{\alpha,\beta,d}(t)=n_{\alpha,\beta,d},\ldots,Y_{\alpha,\beta+1,d}(t)=n_{\alpha,\beta+1,d}-1,\ldots,Y_k(t)=n_k]-\\
n_{\alpha,\beta,d}\cdot P[Y_{1,0,0}(t)=n_{1,0,0},\ldots,Y_k(t)=n_k]\cdot[G(\alpha,\beta,d)(t)+D(\alpha,\beta,d)(t)+M(\alpha,\beta,d)(t)+A(\alpha,\beta,d)(t)]
\end{array}\right]\\
&=\psi\cdot[y_{1,0,0}-1]\cdot X(t)\cdot M(0,0,0)(t)+\psi\cdot\sum_{d=1}^{r}[y_{0,1,d}-1]\cdot X(t)\cdot A(0,0,d)(t)+\\
&\sum_{1\le\alpha\le k-1}\left[\begin{array}{l}
-y_{\alpha,0,0}\cdot[D(\alpha,0,0)(t)+G(\alpha,0,0)(t)+M(\alpha,0,0)(t)\\
+\sum_{d'=1}^{r}A(\alpha,0,d')(t)]+y_{\alpha,0,0}^2\cdot G(\alpha,0,0)(t)+D(\alpha,0,0)(t)\\
+y_{\alpha,0,0}\cdot y_{\alpha+1,0,0}\cdot M(\alpha,0,0)(t)+\sum_{d'=1}^{r}y_{\alpha,0,0}\cdot y_{\alpha,1,d'}\cdot A(\alpha,0,d')(t)
\end{array}\right]\cdot\frac{\partial\psi}{\partial y_{\alpha,0,0}}+\\
&\sum_{\substack{0\le\alpha\le k-1\\1\le\beta\le m_d\\1\le d}}\left[\begin{array}{l}
-y_{\alpha,\beta,d}\cdot[D(\alpha,\beta,d)(t)+G(\alpha,\beta,d)(t)+M(\alpha,\beta,d)(t)\\
+A(\alpha,\beta,d)(t)]+y_{\alpha,\beta,d}^2\cdot G(\alpha,\beta,d)(t)+D(\alpha,\beta,d)(t)\\
+y_{\alpha,\beta,d}\cdot y_{\alpha+1,\beta,d}\cdot M(\alpha,\beta,d)(t)+y_{\alpha,\beta,d}\cdot y_{\alpha,\beta+1,d}(\alpha,\beta,d)(t)
\end{array}\right]\cdot\frac{\partial\psi}{\partial y_{\alpha,\beta,d}}
\end{aligned}
\tag{11.D2}
$$

Appendix E

Derivation of System of Differential Equations Defining the Hessian of the Hazard Function for the Cancer Model of Little and Wright [8] and Little et al. [9]

In this Section we derive the set of differential equations defining the Hessian (with respect to the model parameters) for the cancer model of Little and Wright [8] and Little et al. [9] in the case when all model parameters are constant. For simplicity we present only the derivation for the simpler model of Little and Wright [8]; the derivation for the more complex model of Little et al. [9] is straightforward but lengthy. This allows us to drop the final identifying label in each of $G(\alpha,\beta,d), D(\alpha,\beta,d), M(\alpha,\beta,d), A(\alpha,\beta,d)$, which we will henceforth write as $G(\alpha,\beta), D(\alpha,\beta), M(\alpha,\beta), A(\alpha,\beta)$, respectively. The hazard function of the cancer model with k cancer-stage mutations and m destabilizing mutations developed by Little and Wright [8] may be written as:

$$h(t) = -\int_0^t \left\{ \frac{\partial\phi_{1,0}[t,s]}{\partial t} \cdot M(0,0) + \frac{\partial\phi_{0,1}[t,s]}{\partial t} \cdot A(0,0) \right\} \cdot X ds \tag{11.E1}$$

where the PGFs $\phi_{i,j}$ also satisfy the following Kolmogorov backward equations (for $0 \le i \le k-1$, $0 \le j \le m$, $(i,j) \ne (0,0)$):

$$\begin{aligned} \frac{\partial\phi_{i,j}}{\partial s}[t,s] = {} & [D(i,j) + G(i,j) + M(i,j) + A(i,j)] \cdot \phi_{i,j}[t,s] \\ & - G(i,j) \cdot \phi_{i,j}[t,s]^2 - M(i,j) \cdot \phi_{i,j}[t,s] \cdot \phi_{i+1,j}[t,s] \\ & - A(i,j) \cdot \phi_{i,j}[t,s] \cdot \phi_{i,j+1}[t,s] - D(i,j) \end{aligned} \tag{11.E2}$$

Differentiating (11.E1) gives:

$$\begin{aligned} \frac{\partial h(t)}{\partial X} &= -\int_0^t \left\{ \frac{\partial\phi_{1,0}[t,s]}{\partial t} \cdot M(0,0) + \frac{\partial\phi_{0,1}[t,s]}{\partial t} \cdot A(0,0) \right\} ds \\ &= [1 - \phi_{1,0}[t,0]] \cdot M(0,0) + [1 - \phi_{0,1}[t,0]] \cdot A(0,0) \end{aligned} \tag{11.E3}$$

$$\frac{\partial h(t)}{\partial M(0,0)} = -\int_0^t \frac{\partial\phi_{1,0}[t,s]}{\partial t} \cdot X ds = [1 - \phi_{1,0}[t,0]] \cdot X \tag{11.E4}$$

$$\frac{\partial h(t)}{\partial A(0,0)} = -\int_0^t \frac{\partial\phi_{0,1}[t,s]}{\partial t} \cdot X ds = [1 - \phi_{0,1}[t,0]] \cdot X \tag{11.E5}$$

and for all other model parameters, β_k:

$$\begin{aligned}\frac{\partial h(t)}{\partial \beta_k} &= -\int_0^t \left\{ \frac{\partial^2 \phi_{1,0}[t,s]}{\partial \beta_k \partial t} \cdot M(0,0) + \frac{\partial^2 \phi_{0,1}[t,s]}{\partial \beta_k} \cdot A(0,0) \right\} X ds \\ &= -\frac{\partial \phi_{1,0}[t,0]}{\partial \beta_k}(0,0) - \frac{\partial \phi_{0,1}[t,0]}{\partial \beta_k} \cdot A(0,0)\end{aligned} \tag{11.E6}$$

Likewise, we can evaluate the second derivatives by differentiating (11.E3–11.E5) further:

$$\frac{\partial^2 h(t)}{\partial X \partial M(0,0)} = 1 - \phi_{1,0}[t,0] \tag{11.E7}$$

$$\frac{\partial^2 h(t)}{\partial X \partial A(0,0)} = 1 - \phi_{0,1}[t,0] \tag{11.E8}$$

$$\frac{\partial^2 h(t)}{\partial^2 M(0,0)} = \frac{\partial^2 h(t)}{\partial^2 A(0,0)} = \frac{\partial^2 h(t)}{\partial A(0,0) \partial M(0,0)} = 0 \tag{11.E9}$$

and for all model parameters, $\beta_k, \beta_l \notin \{X, M(0,0), A(0,0)\}$:

$$\frac{\partial^2 h(t)}{\partial \beta_k \partial X} = -\frac{\partial \phi_{1,0}[t,0]}{\partial \beta_k} \cdot M(0,0) - \frac{\partial \phi_{0,1}[t,0]}{\partial \beta_k} \cdot A(0,0) \tag{11.E10}$$

$$\frac{\partial^2 h(t)}{\partial \beta_k \partial M(0,0)} = -\frac{\partial \phi_{1,0}[t,0]}{\partial \beta_k} \cdot X \tag{11.E11}$$

$$\frac{\partial^2 h(t)}{\partial \beta_k \partial A(0,0)} = -\frac{\partial \phi_{0,1}[t,0]}{\partial \beta_k} \cdot X \tag{11.E12}$$

$$\frac{\partial^2 h(t)}{\partial \beta_k \partial \beta_l} = -\frac{\partial^2 \phi_{1,0}[t,0]}{\partial \beta_k \partial \beta_l} \cdot M(0,0) - \frac{\partial^2 \phi_{0,1}[t,0]}{\partial \beta_k \partial \beta_l} \cdot A(0,0) \tag{11.E13}$$

We can evaluate $\frac{\partial \phi_{i,j}[t,s]}{\partial \beta_k}$ by differentiating (11.E2), for $\beta_k \notin \{D(i,j), G(i,j), M(i,j), A(i,j)\}$:

$$\begin{aligned}\frac{\partial^2 \phi_{i,j}}{\partial \beta_k \partial s}[t,s] &= [D(i,j) + G(i,j) + M(i,j) + A(i,j)] \cdot \frac{\partial \phi_{i,j}}{\partial \beta_k}[t,s] \\ &\quad -2 \cdot G(i,j) \cdot \phi_{i,j}[t,s] \cdot \frac{\partial \phi_{i,j}}{\partial \beta_k}[t,s] - M(i,j) \cdot \left[\frac{\partial \phi_{i,j}}{\partial \beta_k}[t,s] \cdot \phi_{i+1,j}[t,s] + \frac{\partial \phi_{i+1,j}}{\partial \beta_k}[t,s] \cdot \phi_{i,j}[t,s]\right] \\ &\quad -A(i,j) \cdot \left[\frac{\partial \phi_{i,j}}{\partial \beta_k}[t,s] \cdot \phi_{i,j+1}[t,s] + \frac{\partial \phi_{i,j+1}}{\partial \beta_k}[t,s] \cdot \phi_{i,j}[t,s]\right] \\ &= \Omega_{i,j}[\beta_k, t, s]\end{aligned} \tag{11.E14}$$

with appropriate initial conditions (discussed later). For $\beta_k \in \{D(i,j), G(i,j), M(i,j), A(i,j)\}$ we have:

$$\frac{\partial^2 \phi_{i,j}}{\partial \beta_k \partial s}[t,s] = \begin{cases} \Omega_{i,j}[\beta_k,t,s] + \phi_{i,j}[t,s] - 1 & \beta_k = D(i,j) \\ \Omega_{i,j}[\beta_k,t,s] + \phi_{i,j}[t,s] - \phi_{i,j}[t,s]^2 & \beta_k = G(i,j) \\ \Omega_{i,j}[\beta_k,t,s] + \phi_{i,j}[t,s] - \phi_{i,j}[t,s]\phi_{i+1,j}[t,s] & \beta_k = M(i,j) \\ \Omega_{i,j}[\beta_k,t,s] + \phi_{i,j}[t,s] - \phi_{i,j}[t,s]\phi_{i,j+1}[t,s] & \beta_k = A(i,j) \end{cases} \tag{11.E15}$$

Likewise, we can evaluate $\frac{\partial^2 h(t)}{\partial \beta_k \partial \beta_l}$ by differentiating (11.E14), for $\beta_k, \beta_l \notin \{D(i,j), G(i,j), M(i,j), A(i,j)\}$:

$$\begin{aligned} \frac{\partial^3 \phi_{i,j}}{\partial \beta_k \partial \beta_l \partial s}[t,s] &= [D(i,j) + G(i,j) + M(i,j) + A(i,j)] \cdot \frac{\partial^2 \phi_{i,j}}{\partial \beta_k \partial \beta_l}[t,s] \\ &\quad -2 \cdot G(i,j) \cdot \left[\frac{\partial \phi_{i,j}}{\partial \beta_k}[t,s] \cdot \frac{\partial \phi_{i,j}}{\partial \beta_l}[t,s] + \phi_{i,j}[t,s] \cdot \frac{\partial^2 \phi_{i,j}}{\partial \beta_k \partial \beta_l}[t,s]\right] \\ &\quad -M(i,j) \cdot \left[\begin{array}{l} \frac{\partial^2 \phi_{i,j}}{\partial \beta_k \partial \beta_l}[t,s] \cdot \phi_{i+1,j}[t,s] + \frac{\partial \phi_{i,j}}{\partial \beta_k}[t,s] \cdot \frac{\partial \phi_{i+1,j}}{\partial \beta_l}[t,s] + \frac{\partial \phi_{i,j}}{\partial \beta_l}[t,s] \cdot \frac{\partial \phi_{i+1,j}}{\partial \beta_k}[t,s] + \\ \frac{\partial^2 \phi_{i+1,j}}{\partial \beta_k \partial \beta_l}[t,s] \cdot \phi_{i,j}[t,s] \end{array}\right] \\ &\quad -A(i,j) \cdot \left[\begin{array}{l} \frac{\partial^2 \phi_{i,j}}{\partial \beta_k \partial \beta_l}[t,s] \cdot \phi_{i,j+1}[t,s] + \frac{\partial \phi_{i,j}}{\partial \beta_k}[t,s] \cdot \frac{\partial \phi_{i,j+1}}{\partial \beta_l}[t,s] + \frac{\partial \phi_{i,j}}{\partial \beta_l}[t,s] \cdot \frac{\partial \phi_{i,j+1}}{\partial \beta_k}[t,s] + \\ \frac{\partial^2 \phi_{i,j+1}}{\partial \beta_k \partial \beta_l}[t,s] \cdot \phi_{i,j}[t,s] \end{array}\right] \\ &= \Psi_{i,j}[\beta_k, \beta_l, t, s] \end{aligned} \tag{11.E16}$$

For $\beta_k \notin \{D(i,j), G(i,j), M(i,j), A(i,j)\}$ we have:

$$\frac{\partial^3 \phi_{i,j}}{\partial \beta_k \partial \beta_l \partial s}[t,s] = \begin{cases} \Psi_{i,j}[\beta_k,\beta_l,t,s] + \frac{\partial \phi_{i,j}[t,s]}{\partial \beta_k} & \beta_l = D(i,j) \\ \Psi_{i,j}[\beta_k,\beta_l,t,s] + \frac{\partial \phi_{i,j}[t,s]}{\partial \beta_k} & \beta_l = G(i,j) \\ \quad -2 \cdot \frac{\partial \phi_{i,j}[t,s]}{\partial \beta_k} \cdot \phi_{i,j}[t,s] & \\ \Psi_{i,j}[\beta_k,\beta_l,t,s] + \frac{\partial \phi_{i,j}[t,s]}{\partial \beta_k} - \frac{\partial \phi_{i,j}[t,s]}{\partial \beta_k} \cdot \phi_{i+1,j}[t,s] & \beta_l = M(i,j) \\ \quad - \frac{\partial \phi_{i+1,j}[t,s]}{\partial \beta_k} \cdot \phi_{i,j}[t,s] & \\ \Psi_{i,j}[\beta_k,\beta_l,t,s] + \frac{\partial \phi_{i,j}[t,s]}{\partial \beta_k} - \frac{\partial \phi_{i,j}[t,s]}{\partial \beta_k} \cdot \phi_{i,j+1}[t,s] & \beta_l = A(i,j) \\ \quad - \frac{\partial \phi_{i,j+1}[t,s]}{\partial \beta_k} \cdot \phi_{i,j}[t,s] & \end{cases} \tag{11.E17}$$

Finally, we have that:

$$\frac{\partial^3 \phi_{i,j}}{\partial D(i,j)^2 \partial s}[t,s] = \Psi_{i,j}[D(i,j), D(i,j), t, s] + 2 \cdot \frac{\partial \phi_{i,j}[t,s]}{\partial D(i,j)} \tag{11.E18}$$

$$\begin{aligned} \frac{\partial^3 \phi_{i,j}}{\partial D(i,j) \partial G(i,j) \partial s}[t,s] &= \Psi_{i,j}[D(i,j), G(i,j), t, s] + \frac{\partial \phi_{i,j}[t,s]}{\partial D(i,j)} + \frac{\partial \phi_{i,j}[t,s]}{\partial G(i,j)} \\ &\quad - 2 \cdot \frac{\partial \phi_{i,j}[t,s]}{\partial D(i,j)} \cdot \phi_{i,j}[t,s] \end{aligned} \tag{11.E19}$$

$$\frac{\partial^3 \phi_{i,j}}{\partial D(i,j)\partial M(i,j)\partial s}[t,s] = \Psi_{i,j}[D(i,j),M(i,j),t,s] + \frac{\partial \phi_{i,j}[t,s]}{\partial D(i,j)} + \frac{\partial \phi_{i,j}[t,s]}{\partial M(i,j)} - \frac{\partial \phi_{i,j}[t,s]}{\partial D(i,j)} \cdot \phi_{i+1,j}[t,s] \tag{11.E20}$$

$$\frac{\partial^3 \phi_{i,j}}{\partial D(i,j)\partial A(i,j)\partial s}[t,s] = \Psi_{i,j}[D(i,j),A(i,j),t,s] + \frac{\partial \phi_{i,j}[t,s]}{\partial D(i,j)} + \frac{\partial \phi_{i,j}[t,s]}{\partial A(i,j)} - \frac{\partial \phi_{i,j}[t,s]}{\partial D(i,j)} \cdot \phi_{i,j+1}[t,s] \tag{11.E21}$$

$$\frac{\partial^3 \phi_{i,j}}{\partial G(i,j)^2\partial s}[t,s] = \Psi_{i,j}[G(i,j),G(i,j),t,s] + 2 \cdot \frac{\partial \phi_{i,j}[t,s]}{\partial G(i,j)} - 4 \cdot \frac{\partial \phi_{i,j}[t,s]}{\partial G(i,j)} \cdot \phi_{i,j}[t,s] \tag{11.E22}$$

$$\frac{\partial^3 \phi_{i,j}}{\partial G(i,j)\partial M(i,j)\partial s}[t,s] = \Psi_{i,j}[G(i,j),M(i,j),t,s] + \frac{\partial \phi_{i,j}[t,s]}{\partial G(i,j)} + \frac{\partial \phi_{i,j}[t,s]}{\partial M(i,j)} - 2 \cdot \frac{\partial \phi_{i,j}[t,s]}{\partial M(i,j)} \cdot \phi_{i,j}[t,s] - \frac{\partial \phi_{i,j}[t,s]}{\partial G(i,j)} \cdot \phi_{i+1,j}[t,s] \tag{11.E23}$$

$$\frac{\partial^3 \phi_{i,j}}{\partial G(i,j)\partial A(i,j)\partial s}[t,s] = \Psi_{i,j}[G(i,j),A(i,j),t,s] + \frac{\partial \phi_{i,j}[t,s]}{\partial G(i,j)} + \frac{\partial \phi_{i,j}[t,s]}{\partial A(i,j)} - 2 \cdot \frac{\partial \phi_{i,j}[t,s]}{\partial A(i,j)} \cdot \phi_{i,j}[t,s] - \frac{\partial \phi_{i,j}[t,s]}{\partial G(i,j)} \cdot \phi_{i,j+1}[t,s] \tag{11.E24}$$

$$\frac{\partial^3 \phi_{i,j}}{\partial M(i,j)^2\partial s}[t,s] = \Psi_{i,j}[M(i,j),M(i,j),t,s] + 2 \cdot \frac{\partial \phi_{i,j}[t,s]}{\partial M(i,j)} - 2 \cdot \frac{\partial \phi_{i,j}[t,s]}{\partial M(i,j)} \cdot \phi_{i+1,j}[t,s] \tag{11.E25}$$

$$\frac{\partial^3 \phi_{i,j}}{\partial M(i,j)\partial A(i,j)\partial s}[t,s] = \Psi_{i,j}[M(i,j),A(i,j),t,s] + \frac{\partial \phi_{i,j}[t,s]}{\partial M(i,j)} + \frac{\partial \phi_{i,j}[t,s]}{\partial A(i,j)} - \frac{\partial \phi_{i,j}[t,s]}{\partial M(i,j)} \cdot \phi_{i,j+1}[t,s] - \frac{\partial \phi_{i,j}[t,s]}{\partial A(i,j)} \cdot \phi_{i+1,j}[t,s] \tag{11.E26}$$

$$\frac{\partial^3 \phi_{i,j}}{\partial A(i,j)^2 \partial s}[t,s] = \Psi_{i,j}[A(i,j),A(i,j),t,s] + 2 \cdot \frac{\partial \phi_{i,j}[t,s]}{\partial A(i,j)} - 2 \cdot \frac{\partial \phi_{i,j}[t,s]}{\partial A(i,j)} \cdot \phi_{i,j+1}[t,s] \tag{11.E27}$$

As in Little and Wright [8], the following boundary conditions must be satisfied, for all i,j, β_k, β_l:

$$\phi_{i,j}[t,t] = 1 \quad 0 \le i \le k-1 \tag{11.E28}$$

$$\frac{\partial \phi_{i,j}[t,t]}{\partial \beta_k} = \frac{\partial^2 \phi_{i,j}[t,t]}{\partial \beta_k \partial \beta_l} = 0 \tag{11.E29}$$

This system of ordinary differential equations (in the variable s) for $\phi_{i,j}[t,s]$, $\frac{\partial \phi_{i,j}[t,s]}{\partial \beta_k}$, $\frac{\partial^2 \phi_{i,j}[t,s]}{\partial \beta_k \partial \beta_l}$ were integrated using the Boerlisch-Stoer algorithm with adaptive stepsize control [24]. Very similar results were obtained using a Runge-Kutta integrator with adaptive stepsize control [24].

References

1. Rothenberg TJ (1971) Identification in parametric models. Econometrica 39(3):577–591
2. Silvey SD (1975) Statistical inference. London, Chapman and Hall, pp 1–191
3. Catchpole EA, Morgan BJT (1997) Detecting parameter redundancy. Biometrika 84(1):187–196
4. Jacquez JA, Perry T (1990) Parameter estimation: local identifiability of parameters. Am J Physiol 258(4 Pt 1):E727–E736
5. Little MP, Heidenreich WF, Li G (2010) Parameter identifiability and redundancy: theoretical considerations. PLoS One 5(1):e8915
6. Chappell MJ, Gunn RN (1998) A procedure for generating locally identifiable reparameterisations of unidentifiable non-linear systems by the similarity transformation approach. Math Biosci 148(1):21–41
7. Evans ND, Chappell MJ (2000) Extensions to a procedure for generating locally identifiable reparameterisations of unidentifiable systems. Math Biosci 168(2):137–159
8. Little MP, Wright EG (2003) A stochastic carcinogenesis model incorporating genomic instability fitted to colon cancer data. Math Biosci 183(2):111–134
9. Little MP, Vineis P, Li G (2008) A stochastic carcinogenesis model incorporating multiple types of genomic instability fitted to colon cancer data. J Theor Biol 254(2):229–238, Sept 21; Nov 21; 255(2):268
10. Armitage P, Doll R (1954) The age distribution of cancer and a multi-stage theory of carcinogenesis. Br J Cancer 8(1):1–12
11. Moolgavkar SH, Venzon DJ (1979) Two-event models for carcinogenesis: incidence curves for childhood and adult tumors. Math Biosci 47(1–2):55–77
12. Little MP (1995) Are two mutations sufficient to cause cancer? Some generalizations of the two-mutation model of carcinogenesis of Moolgavkar, Venzon, and Knudson, and of the multistage model of Armitage and Doll. Biometrics 51(4):1278–1291
13. Nowak MA, Komarova NL, Sengupta A, Jallepalli PV, Shih IM, Vogelstein B et al (2002) The role of chromosomal instability in tumor initiation. Proc Natl Acad Sci USA 99(25):16226–16231 Dec 10

14. Heidenreich WF (1996) On the parameters of the clonal expansion model. Radiat Environ Biophys 35(2):127–129
15. Heidenreich WF, Luebeck EG, Moolgavkar SH (1997) Some properties of the hazard function of the two-mutation clonal expansion model. Risk Anal 17(3):391–399
16. Little MP, Heidenreich WF, Li G (2009) Parameter identifiability and redundancy in a general class of stochastic carcinogenesis models. PLoS One 4(12):e8520
17. McCullagh P, Nelder JA (1989) Generalized linear models. Monographs on statistics and applied probability, vol 37, 2nd edn. Chapman and Hall/CRC, Boca Raton, pp 1–526
18. Rudin W (1976) Principles of mathematical analysis, 3rd edn. McGraw Hill, Auckland, pp 1–352
19. Viallefont A, Lebreton J-D, Reboulet A-M, Gory G (1998) Parameter identifiability and model selection in capture-recapture models: a numerical approach. Biometrical J 40:313–325
20. Dickson LE (1926) Modern algebraic theories. B.J.H. Sanborn, Chicago, pp 1–273
21. Little MP, Li G (2007) Stochastic modelling of colon cancer: is there a role for genomic instability? Carcinogenesis 28(2):479–87
22. Armitage P, Doll R (1957) A two-stage theory of carcinogenesis in relation to the age distribution of human cancer. Br J Cancer 11(2):161–169
23. Cole D, Morgan BJT, Titterington DM (2010) Determining the parametric structure of models. Math Biosci 228(1):16–30
24. Press WH, Teukolsky SA, Vetterling WT, Flannery BP (1992) Numerical recipes in Fortran 77. The art of scientific computing, 2nd edn. Cambridge University Press, Cambridge, pp 1–934
25. Golub GH, van Loan CF (1996) Matrix computations, 3rd edn. The Johns Hopkins University Press, Baltimore, pp 1–728
26. Numerical Algorithms Group (2009) NAG FORTRAN Library Mark 22, Oxford
27. Hanin LG, Yakovlev AY (1996) A nonidentifiability aspect of the two-stage model of carcinogenesis. Risk Anal 16(5):711–715
28. Hanin LG (2002) Identification problem for stochastic models with application to carcinogenesis, cancer detection and radiation biology. Discrete Dyn Nat Soc 7(3):177–189

Chapter 12
Semantic Systems Biology: Formal Knowledge Representation in Systems Biology for Model Construction, Retrieval, Validation and Discovery

Michel Dumontier, Leonid L. Chepelev and Robert Hoehndorf

Abstract With the publication of the human genome, scientists worldwide opened champagne and let out a collective cheer for progress in biology. After all, the untold number of interactions of tens of thousands of genes, a greater number of their products and product derivatives, and tens of thousands of chemicals came much closer to complete characterization. Paradoxically however, while individual efforts produced important biological results, an integrated view of biology from systems perspective seemed ever more distant due to the complexity of data integration from multiple knowledge representation forms, formalisms, modeling paradigms, and conflicting scientific statements. To address this, semantic technologies have risen over the past decade with the promise of truly unifying biological knowledge and allowing cross-domain queries and model integration. In this chapter, we shall examine Semantic Web technologies and their applications to build, publish, query, discover, compare, validate, reason about, and evaluate models and knowledge in Systems Biology. We shall specifically address biological ontologies, open data repositories, modeling and annotation tools, and selected promising applications of Semantic Systems Biology. We firmly believe that it shall soon be possible to completely close the gap between facts, models, and results, and to fully apply the accrued models and facts to evaluate biological hypotheses on a system level, discovering meaning within the vast collection of biological knowledge and taking Systems Biology research to a new, unprecedented level.

Keywords Systems biology · Semantic web · Bioinformatics · Computational biology · Ontology · Semantic annotation · Integration · Query · Validation · Discovery · Models · Simulation · Sustainability · Protein · Small molecule · Pathway · SBML · RDF · SPARQL · OWL · Type · Relation · Identifier ·

M. Dumontier (✉) · L. L. Chepelev · R. Hoehndorf
Carleton University, Ottawa, Canada
e-mail: michel.dumontier@gmail.com, leonid.chepelev@gmail.com, leechuck@leechuck.de

A. Prokop and B. Csukás (eds.), *Systems Biology*,
DOI: 10.1007/978-94-007-6803-1_12,

Description · Formal knowledge representation · Automated reasoning · Biology · Biochemistry · Artificial intelligence · World Wide Web

Acronyms

BFO	Basic formal ontology
ChEBI	Chemical entities of biological interest
DL	Description logic
EBI	European bioinformatics institute
FMA	Foundation model of anatomy
GO	Gene ontology
HCLS	Health care and life sciences
IAO	Information artifact ontology
IRI	International resource identifier
IUBMB	The international union of biochemistry and molecular biology
KiSAO	Kinetic simulation algorithm ontology
MIRIAM	Minimum information required in the annotation of models
MUO	Measurement unit ontology
NCBI	National center for biotechnology information
OBI	Ontology for biomedical investigations
OPB	Ontology for physics and biolog
OWL	Web ontology language
PATO	Phenotype and trait ontology
RDF	Resource description framework
RDFS	Resource description framework schema
RNAO	RNA ontology
SBML	Systems biology markup language
SBO	Systems biology ontology
SIO	Semanticscience integrated ontology
SPARQL	SPARQL query language
TEDDY	Terminology for the description of dynamics
XML	Extensible markup language
W3C	World Wide Web consortium

12.1 Introduction

In the quest to improve our current understanding of biological systems, life scientists devise, develop and test a variety of computational models based on the latest knowledge to explain and predict biological phenomena. In their simplest formulation, models that aim to describe biological phenomena do so in terms of the participant physical entities and their functions at desired granularity of detail that is best understood or is relevant for a given purpose. In the past 60 years, thousands of computational models have been constructed to examine everything from action potentials [1], to synthetic oscillatory network of transcriptional regulators [2], and complete yeast metabolic networks [3]. The construction of effective models involves rational selection of components, retrieval of as many experimental values as possible and the nearly unavoidable estimation or fitting of unmeasured parameters [4]. Clearly, building reusable or "sustainable" models [5] requires careful attention to accurate description of model components so that it becomes possible to recognize and test assumptions against observed biological behaviours [6]. Indeed, an increased interest in sustainable computational modelling has led to a search for more effective methods to construct, publish, share, compare, validate and evaluate models [7].

Here we focus on semantic systems biology as an interdisciplinary approach that combines computational biology with formal semantics in order to accurately build, annotate, publish, share and validate facts and computational models concerning biological entities and phenomena. As a first step, the semantic annotation of models makes it possible to find models with *particular* model components or simulating certain biological processes. As a second and more sophisticated approach, formal knowledge representation and automated reasoning make it possible to not only retrieve models containing certain *kinds* of information, but also can be used to check consistency of knowledge, determine the implicit relationships between data and can also be used to classify data. We will examine current approaches to organize systems biology knowledge and describe applications related to searching, querying, comparing model similarity, integration of simulation results, and validation of model annotations. Finally, we will speculate as to how the simulation of semantically-annotated models could be directly used as evidence in the evaluation of formalized biological hypotheses.

12.2 The Semantic Web as a Platform for Large Scale Data Access and Integration

Having access to high quality data is of paramount importance in formulating well supported hypotheses, and in developing computational models to examine these hypotheses. Unfortunately, the current state of things makes it rather difficult to find the right data or software in a timely fashion, largely owing to the lack of coordination around publishing these on the internet [8]. Even the absence of coordination might be manageable if it were not for the continuous and exponential growth in scientific literature, the thousands of biological databases where raw data and curated facts are being stored and served from, and the daily myriads of new bioinformatics applications and web services to process data and generate new structural/functional annotations.

With the aim to address the outstanding bioinformatic challenge of large scale integration of heterogeneous data, life science researchers are examining the possibilities afforded by the standards and technologies emerging from the World

Wide Web Consortium's Semantic Web effort to produce a web of data [9, 10]. The Semantic Web's stated aim is to make it easier to use the internet to publish, share, discover, link to and query data. At the core of this effort lie the Resource Description Framework (RDF), RDF schema (RDFS), SPARQL query language, and the Web Ontology Language (OWL).

The Resource Description Framework (RDF) offers a simple mechanism to (1) identify and (2) describe entities in terms of their types, attributes and relations to other entities. Entities are unambiguously identified by Internationalized Resource Identifiers (IRIs) which allows for de-referenceable web-based identifiers (HTTP URIs), that is to say, when users paste the identifier in their browsers, they *will* get back information about it. The information returned takes the form of a document containing one or more statements that describes the entity. Simple statements are specified as "triples" comprised of an IRI-identified subject, an IRI-identified relation, and an IRI-specified object or an XML-Schema compliant literal (e.g. integer, decimal, string, etc.). For instance, to state that a nucleus is part of a cell, we formulate a triple composed of the subject "nucleus", the predicate "is part of" and the object "cell". Identifiers for such entities and relations can be drawn from specific terminologies such as the Gene Ontology [11]. The Gene Ontology's identifier for "nucleus" is GO:0005634 and the corresponding IRI as specified by the OBO Foundry initiative is http://purl.obolibrary.org/obo/GO_0005634. Identifiers.org [21] is a recent effort from the European Bioinformatics Institute (EBI) to provide stable HTTP identifiers for life science datasets such that their resolution gives the set of services that provide information for that identifier—http://identifiers.org/obo.go/GO:0005634.

RDF Schema (RDFS) builds on RDF and provides additional vocabulary for naming resources (rdfs:label) and specifying simple type and relation hierarchies (rdfs:subClassOf, rdfs:subPropertyOf). Type hierarchies specify that all the members of one type are also members of another type. For instance, we know that both the Golgi Apparatus (GO:0005794) and the Endoplasmic Reticulum (GO:0005793) are kinds of intracellular membrane-bounded organelles (GO:0043231). We can assert this more formally by indicating that 'Golgi Apparatus [GO:0005794]' is a rdfs:subClassOf 'intracellular membrane-bounded organelle [GO:0043231]'.

RDF-based knowledge can be queried using the SPARQL query language. SPARQL queries may contain triple patterns that can be conjunctively (AND) or disjunctively (OR) combined with mandatory or OPTIONAL triple query patterns. A key feature of SPARQL is that it uses the standard web protocol (HTTP) to query any SPARQL-compliant database on the internet. A recent feature delivered in SPARQL 1.1 makes it possible to formulate a query that can be executed over multiple SPARQL-compliant databases, thus supporting a decentralized (non-warehouse) solution for access to up-to-date resources.

A number of efforts now provide access to RDF data of interest to systems biology researchers including Bio2RDF, Chem2Bio2RDF, BioGateway and the W3C HCLS. Of these Bio2RDF [13] is the oldest and largest linked open data provider that makes available over 4 billion RDF statements for over 40 datasets including UniProt, PDB, Entrez Gene, iRefIndex, OMIM, PubMed, Genbank, and RefSeq among others. A key aspect of Bio2RDF is that it helps build an interlinked network of resources by following a simple naming convention:http://bio2rdf.org/namespace:identifier where *namespace* refers to short name provided to a given dataset and the *identifier* is that which is provided in the dataset. By using HTTP URIs, Bio2RDF identifiers resolve to documents that provide information about named entities, whether it comes from statements obtained in the original dataset, or by virtue of the fact that there are resources that link to it. Bio2RDF's architecture supports select mirroring of resources (one or more), and mirrors currently exist in Canada, Australia and Ireland. Bio2RDF has been used in a systems biology investigation to identify a protein interaction network that arises from significantly expressed genes during the first hours of an HIV infection of primary human macrophages as monitored by a time-course microarray experiment [17].

Among the other RDF data providers, Chem2Bio2RDF focuses on chemical interactions and has been used to support investigations of polypharmacology, potential multiple pathway inhibitors, and the association of pathways with adverse drug reactions [14]. BioGateway [15, 16] has developed a lightweight ontology to facilitate linked life science data queries. The W3C Semantic Web for Health Care and Life Sciences Interest Group (HCLS) has developed several prototype databases that integrate key health and life science datasets and ontologies to demonstrate, including scientific articles and patient records for translational medicine [12–14].

12.3 Bio-Ontologies

Ontologies have long been regarded as a panacea to problems relating to accurate definitions, semantic annotation, data integration, search and retrieval, and (autonomous) agent-based discovery. In its simplest formulation, an ontology provides a description of entities in terms of their attributes or relationships they hold with other entities [15]. Formal ontologies are those ontologies that have been constructed using a formal language that has a clearly defined syntax matched with unambiguous (e.g. mathematically defined) machine-understandable semantics. By using a formally defined ontology language to describe entities of interest, one makes an ontological commitment as to the meaning of the phrases constructed using terminology from the domain. Making an ontological commitment enables automated reasoners to correctly generate inferences.

There are well over 300 bio-ontologies available from the National Center for Biomedical Ontology (NCBO)'s BioPortal resource [16]. These cover (1) material entities such as molecules (ChEBI for small molecules, PRO for proteins, LiPro for lipids, RNAO for (RNA) nucleic acids), macromolecular structures and organelles (GO cellular component), cells and tissues (cell type ontology), organs and organisms (FMA for human anatomy, NCBI taxonomy of species),

(2) non-material entities such as qualities (PATO), functions (GO molecular function) and roles (ChEBI molecular roles), (3) natural processes (GO biological process, Ontology of physics for biology), and (4) informational entities and ontologies that combine or provide a scaffold for all of these (SIO—Semantic-science Integrated Ontology). However, significant overlap exists between ontologies, as a search yielding 20 different terms for "protein" will attest. In an effort to develop a set of orthogonal ontologies, the OBO Foundry [17] hopes to coordinate ontology development so as to minimize overlap and build on a common set of upper level ontologies for types (BFO) and relations (RO). However, proponents of this approach face the overwhelming task of adapting complex philosophical concerns to support the needs of the bio-ontologies community [18]. Regardless, this social engineering effort has not prevented progress in the development of ontologies of interest to the Systems Biology community, including the Systems Biology Ontology (SBO) for describing model components including rate equations and parameters, the Kinetic Simulation Algorithm Ontology (KiSAO) to specify simulation algorithms and Terminology for the Description of Dynamics (TEDDY) to characterize systems behavior from simulation results [7].

12.4 Semantic Annotation of Biological Models

The Systems Biology Markup Language (SBML) is a world-wide standard for describing computational models [19, 20]. As an XML-based markup language, SBML has the advantage of being specified in a standard machine-accessible format and may be combined with MathML for the mathematical description of rate equations and their parameters. SBML can be used to specify compartments, species, reactions, events, functions, parameters and units. In what is perhaps the greatest validation of the success of SBML is the large number (>700) and range of models published in the Biomodels database [21, 22] or JWS Online [23]. However, just having the models in a common syntax or having natural language descriptions are not sufficient to identify shared components across models or to classify models in terms of what they represent. For instance, some models fail to include sufficiently accurate names for species (po1 in BIOMD0000000060) or reactions (re1 to re76 in BIOMD0000000227) [24]. Efforts have since been directed to semantically annotate models using common resources (databases, ontologies) [22].

The semantic annotation of biological models largely depends on selecting the closest and most specific term that matches the element to be annotated. The Minimum Information Required in the Annotation of Models (MIRIAM) specifies what metadata is required for systems biology models [25]. Annotations are made using an identifier from external resources such as ontology or database entries. There are over 45 different resources that have been used to annotate EBI's biomodels. Biomodel curators typically pick terms from the NCBI taxonomy for denoting the species, Chemical Entities of Biological Interest (ChEBI) for small molecules, UniProt for proteins/enzymes and the Gene Ontology for cellular components. Reactions may be annotated with IUBMB Enzyme Classification codes, KEGG reaction codes, or interaction identifiers from IntAct or BIND. Annotators pick among a set of basic relations to indicate the relevance of the annotation. For instance, models that represent specific species use the “qualifier” relation “is”, but models that represent more general types of organism that are built using organism-specific data would be related with “hasVersion”, while models that represent specific organisms but are built using data from other organisms would be specified with “isVersionOf”. A number of other relations are also provided in order to specify parts (hasPart) or wholes (isPartOf), homologs (isHomologTo), and gene-protein relationships (encodes, isEncodedBy). Of the 249 annotated biomodels (May 2010), ~69 % of compartments, species and reactions were annotated. Semantic annotations for SBML bio-models are represented as RDF statements in an “annotation” sub-element. In Fig. 12.1, we see the RDF annotation for a SBML species named “GLCi” in a compartment “cyto”, which presumably represents the internal pool of glucose in the cytoplasm. The element is annotated with two resources—ChEBI and KEGG Compound.

Although Biomodel curators have exclusive access to an internal system to semantically annotate the models, there are a number of other software that can also do this including SBML editor [26], COPASI [27], CellDesigner [28], semanticSBML [29] and the SEEK [30].

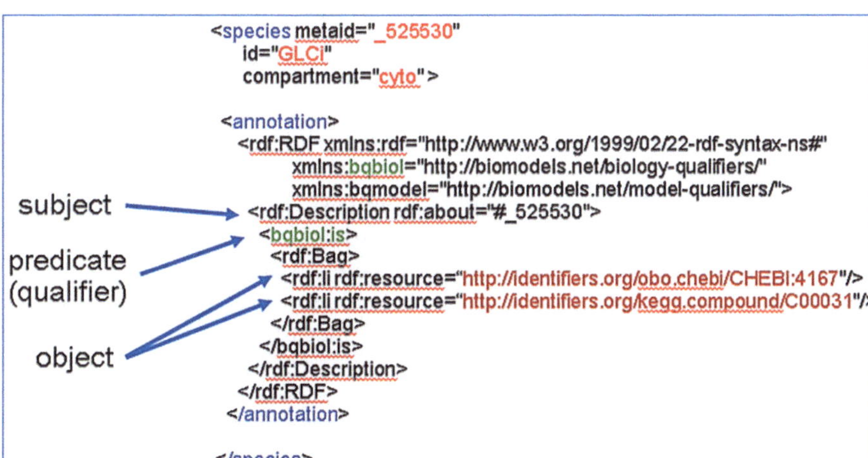

Fig. 12.1 Embedded RDF annotation of a species element in an SBML model. The annotation indicates that the species is of type glucose as specified by ChEBI:4167 and KEGG compound C00031

12.5 Search, Comparison and Retrieval of Models

The semantic annotation of SBML models makes it possible to more accurately search, compare and retrieve models. Ranked retrieval of models has been incorporated into the biomodels database and was first demonstrated in [24] using (1) query by value in which a subset of keywords are used to initiate the search, or (2) query by example where a model forms the basis to search for similar models. Ranking occurs by weighing the component terms with respect to their frequency in the database. A more recent approach [31] investigated the use of a feature vector-based model similarity which performs well in semantic search and model alignment. Semantic search results yield the significance of matches in terms of p-values calculated from a set of randomly annotated models. Model alignment using semantic annotations (as opposed to purely structural elements), makes it possible to align models describing identical or similar phenomena, but possibly at different levels of detail. For instance, while both Biomodel 9 and 84 detail MAP kinase cascades, Biomodel 9 captures the phosphorylation of MAP kinase enzymes. Interestingly, through iterative model alignments, the authors report that 8 kinetic models cover $\sim$15 % of the yeast consensus metabolic network, demonstrating that there is much more work to be done for a whole cell kinetic simulation.

12.6 Validating Model Annotations

Given human curated annotations on a bio-model, how do we automatically check whether these are consistent among each other? One approach, termed the SBML Harvester [32], involves converting semantic annotations into formal representations of knowledge that can be automatically reasoned about. A key aspect is the use of the Web Ontology Language (OWL), a language for building ontologies on the Semantic Web.

OWL is a formal knowledge representation language that offers an enhanced vocabulary to more accurately express knowledge relating to types, relations, individuals and data values. Among many features, it provides vocabulary to (1) quantify the number and type of relations that hold from one type of individual to another (existential, cardinality and universal quantifiers), (2) to discriminate between things that are the same or different (negation), (3) to add specific implications to relations (transitive, functional, inverse functional, symmetric, antisymmetric, reflexive, irreflexive), and (4) to formulate simple property chains (DL-safe rules). All of these features become exceedingly important when trying to get at the heart of what is meant by certain statements. For instance, consider the RDF triple: 'nucleus' 'part of' 'cell'.

On first read you might not acknowledge the ambiguity of the statement. Is it that every nucleus is part of exactly one cell? Can a nucleus be part of more than one cell? Is it that when a nucleus is a part, it can only be part of a cell? Can the nucleus completely overlap with a cell (e.g. can they be the same or are they necessarily different?) Interpreted in the other direction, this may be a statement about cells: Is it that every cell has a nucleus as a part? Do only some cells have a nucleus as a part? Do cells contain one nucleus or more? Is a nucleus the only part that cells can have? The list of possible questions is indeed quite long. In using OWL, we make an ontological commitment to precisely state what we mean by constructing phrases that can be consistently interpreted by both humans and machines carrying out automated reasoning.

Indeed, it is not just the ability to more accurately express knowledge that makes OWL interesting; it is that OWL-based reasoners offer a number of salient services including:

- Consistency checking: to determine whether the ontology contains contradictions.
- Satisfiability: to determine whether classes can have instances.
- Computation of subsumption: to determine whether one class is an implicit subclass of another.
- Classification: to discover all implicit subclass links by the repetitive application of subsumption.
- Realization: to find the most specific class that an individual belongs to.

Thus, by converting model annotations into formal representations of knowledge, we anticipate the following benefits:

- Accurate capture of the nature of models and the biological systems they represent
- The ability to leverage knowledge explicit in externally linked ontologies
- The ability to validate the consistency of the annotations
- The ability to discover biological implications inherent in the models.

The SBML Harvester approach converts biomodel annotations into OWL ontologies through the use of different OWL axiom patterns, depending on the kind of entity the annotation denotes. The first pattern distinguishes SBML components from the physical systems and processes that they represent.

E subClassOf represents some Rep(E)

If an element annotation C is a material entity, then we write an axiom to indicate that the species represents some material entity.

E subClassOf represents some C

If an element annotation F is a function, then we write an axiom that indicates the species represents a material entity that has the function F.

E subClassOf represents some (MaterialEntity that has-function some F)

If an element annotation P is a process, then we write an axiom that indicates the species represents a material entity which has a function that is realized only in instances of type P.

E subClassOf represents some (MaterialEntity that has-function some (Function and realized-by only P))

Thus, for a model that is annotated with the process heterotrimeric G-protein complex cycle (GO:0031684), we write an axiom that indicates that the model represents an object that has the capability F which is realized in processes of the type heterotrimeric G-protein complex cycle.

M subClassOf represents some (MaterialEntity that has-function some (realized-by only GO:0031684))

SBML Harvester uses libSBML to access model structure and extract RDF annotations, Jena RDF API to parse RDF annotations and the OWLAPI to create OWL axioms and reason with a top-level ontology that made models, model components, material entities, functions, processes and qualities different from one another. Application to BioModels repository yields an OWL ontology with more than 300,000 classes, 800,000 axioms and includes all referenced ontologies: GO (functions, compartments, processes), ChEBI (molecules), Celltype Ontology (cell types), FMA (anatomy) and PATO (qualities). Reasoning over the integrated ontology resulted in 27 inconsistent models, for which most could be attribute to errors in the annotation. In two models, BIOMODELS 176 and 177, we uncovered the incorrect species annotation for an ATPase reaction [32]. The Gene Ontology defined ATPase activity maintained that the input to the reaction was an ATP and water and the output to the reaction was ADP and inorganic phosphate, to which we added that these were the only inputs and outputs. The chemical classes were taken from the ChEBI ontology, to which we added that all chemical classes were different from one another (disjointness). Automated reasoning over the ontology uncovered that the species annotation of 'alpha-D-glucose-phosphate' is not only different from ATP, but also is not an allowed chemical in an ATPase reaction.

The generation of a large-scale, formalized knowledge base also makes it possible to answer questions about models and what they represent across multiple levels of spatial granularity. For example, one can create a sophisticated query for model entities (models or model components) that represent something capable of catalyzing the conversion of a sugar in the pancreas:

```
represents some (
    has-function some (
     realized-by only (
            realizes some 'catalytic activity'
            and has-participant some (
               sugar and contained-in some (part-of some 'endocrine pancreas')))))
```

These and other queries are reported in [32].

12.7 Integration of Simulation Parameters and Results

While most of the discussion thus far has been centered on the semantic annotations of biological models, we must not forget that the principal reason to create models is to use them in simulations, thereby providing insight with respect to their accuracy or to make predictions about unmeasured phenomena.

Quantitative biochemical simulations generate quantities concerning the number or concentration of species, along with values related to changes with time such as flux (Fig. 12.2). In order to link this data to the model and physical system descriptions described in the past section, it is necessary to convert tabular or xml files into semantically annotated RDF. For any (generated/measured) quantity, it is important to keep track of (1) the value of the quantity, (2) the unit of measure (if any), (3) the kind of quantity, (4) the entity (entities) that it is an attribute of, (5) the time at which the quantity was generated/measured and (6) the process that the quantity resulted from. Although there exists a number of ontologies to describe quantities including the Ontology for Biomedical Investigations (OBI) and the Information Artifact Ontology (IAO), the Measurement Unit Ontology (MUO), Ontology for Physics and Biology (OPB), and the Semanticscience Integrated Ontology (SIO). Each of these ontologies has their particular advantages and disadvantages. In OBI and the IAO, it is necessary that the measurement values refer to actual measured entities, as opposed to the postulated entities that may be specified in a model. MUO captures the relation to the value, type and unit. OPB provides a well thought out view of physical dependencies between rate variables and flow variables. SIO, however, considers all of these and provides the vocabulary to capture time-course values as RDF linked data. As illustrated in Fig. 12.3, quantities in SIO can be ascribe a specific value or fall within a range of values (uncertainty) with an optional unit of measure, obtained at a particular time, be associated with specific information content entity (model/model component), object or process, and be linked as an output of a process such as a simulation.

Since SIO is fully compatible with the formalization made by the SBML Harvester, the RDFized simulation results can be queried inline with the models

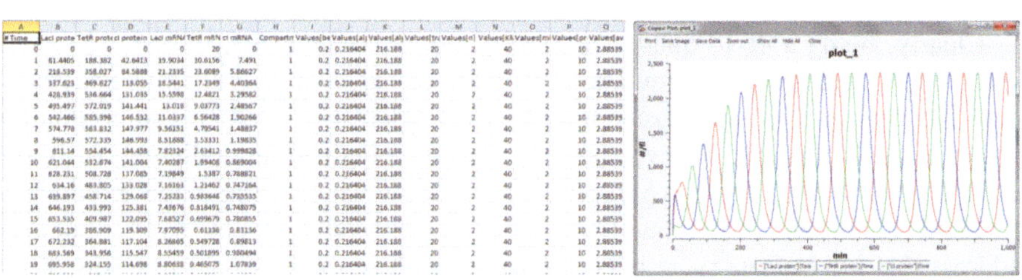

Fig. 12.2 **a** tabular data, and **b** 2d plot time course data of the repressilator, an oscillating gene regulatory system

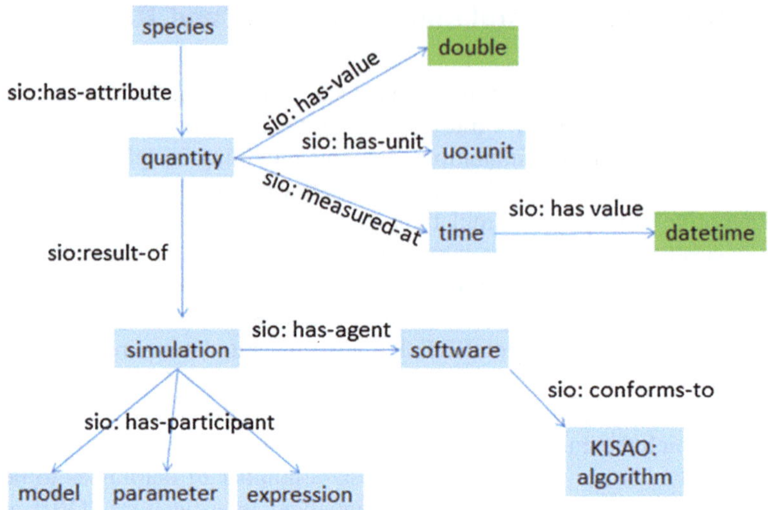

Fig. 12.3 Entity-relation diagram to specify time-indexed quantities resulting from parametric simulations

and what they represent. Consider the following DL query which obtains concentrations of species containing ribonucleotide residues during the 20 and 40 s interval of time.

```
'concentration'
and ('measured at' some double[>20.0, <40.0])
and 'is attribute of' some (
   'species'
   and 'represents' some ('has part' some 'ribonucleotide residue')
)
```

Clearly, being able to query the model, model components, biological system and the raw data form the simulation results in one system is of obvious benefit for data retrieval or even data exchange [33]. Yet, more work remains to be done in order to fully understand whether the results obtained are aligned with our expectations. For instance, we might want to know whether the model produces a periodic oscillation, or in general, the behaviour generated by any given model. To address this kind of query, it is necessary to analyze the raw data in terms of the curve and trend features. Vocabulary for the semantic annotation of dynamic features is provided by an OWL ontology called Terminology for the Description of Dynamics (TEDDY) [7]. Thus, through simple multi-curve analysis, it becomes possible to identify curves that are monotonic (TEDDY_0000144) such as strictly increasing (TEDDY_0000008) or strictly decreasing (TEDDY_0000009), or non-monotonic (TEDDY_0000005) such as periodically oscillating (TEDDY_000 0066) or damped oscillation (TEDDY_0000063). After analysis, we can contribute new information to the knowledge base regarding the dynamical behaviour of the systems in terms of the curves and line segments and the time-indexed points they are composed of, and create new queries that integrate these with the model entities and their meaning. As shown below, we can query for periodic oscillations stemming from concentration data of species that have DNA binding function.

```
'periodic oscillation'
and 'has part' some (
  'concentration'
 and 'is attribute of' some (
   'species'
   and 'has function' some 'dna binding')
)
```

12.8 Towards Model-Based Hypothesis Testing

If the goal of building models and running simulations is ultimately to better understand biology, then there must be a place for models to provide evidence for scientific hypotheses. HyQue [34] is a project to facilitate the formulation and evaluation of scientific hypotheses using linked open data and Semantic Web technologies. Through the use of customizable SPARQL-based rules [35], HyQue retrieves relevant data that can be used to support or dispute claims made by the hypothesis. HyQue tries to find supporting evidence depending on the kind of molecular event specified: transport, binding, modulation of function, and positive and negative regulation of biological processes. For instance, if a claim of the hypothesis involves the expression of a gene by a protein, HyQue will attempt to determine whether this fact is known, and also determine whether the agent is indeed a protein, the target is indeed a gene, whether the protein has been annotated as having the capability to bind DNA or exhibits transcription factor activity, and whether the gene is known to be regulated. Finally, HyQue scores a hypothesis based on the total level of support garnered and produces an RDF-based representation that links hypothesis to its scores, along with the rules and data used to generate those scores. This data can then be republished on the semantic web for others to discover. Thus, with some measure of excitement, we can imagine that semantically annotated systems biology models and the analysis of the results of model simulation could potentially be used as evidence for HyQue-based biological hypotheses.

Table 12.1 Summary of selected formats, software, and resources of use in semantic systems biology efforts

Resource		Description	Further information
Formats and languages	RDF	A language offering a simple mechanism to identify and describe entities in terms of their types, attributes and relations to other entities	http://www.w3.org/TR/rdf-primer/
	RDF schema	A language that extends RDF, provides additional vocabulary for naming resources (rdfs:label) and specifying simple type and relation hierarchies	http://www.w3.org/TR/rdf-schema/
	SPARQL	A language designed to query RDF datasets	http://www.w3.org/TR/rdf-sparql-query/
	OWL	A formal knowledge representation language that offers an enhanced vocabulary to more accurately express knowledge relating to types, relations, individuals and data values	http://www.w3.org/TR/owl2-overview/
	SBML	An XML-based markup language that is a world-wide standard for describing computational models	http://sbml.org
Linked open data providers	Bio2RDF	The largest linked open data provider with over 4 billion RDF statements for over 40 datasets including UniProt, PDB, Entrez Gene, iRefIndex, OMIM, PubMed, Genbank, and RefSeq among others	http://bio2rdf.org/
	Chem2Bio2RDF	A chemical open data collection focusing on chemical interactions geared for use primarily in drug development	http://cheminfov.informatics.indiana.edu:8080/
	BioGateway	Facilitates facilitate linked life science data queries with the use of a lightweight ontology	http://www.semantic-systems-biology.org/biogateway
	W3C HCLS	A consortium effort from which multiple linked open data repositories were produced	http://www.w3.org/blog/hcls/

(continued)

Table 12.1 (continued)

Resource		Description	Further information
Selected ontologies	GO	A broad ontology describing genes, gene products, and a vast array of their features	http://www.geneontology.org/
	SIO	The semanticscience integrated ontology aims to provide a description of informational entities to provide a scaffold for other ontologies	http://semanticscience.org
	BFO	Aims to provide an overarching ontology framework for other ontologies to reuse	http://ontology.buffalo.edu/bfo/
	ChEBI	Describes small molecules and their classes, as well as numerous chemical relationships between these classes, as well as chemical features of chemicals	http://www.ebi.ac.uk/chebi/
	PRO	The protein ontology describes proteins and related entities as well as their relationships and features	http://pir.georgetown.edu/pro/pro.shtml
	LiPro	An ontology describing lipid classes and their relations	http://www.lipidprofiles.com/LipidOntology
	RNAO	An ontology describing ribonucleic acids, their relations, and structures	http://code.google.com/p/rnao/
	PATO	The phenotypic quality ontology provides a description for features of biological organisms	http://obofoundry.org/wiki/index.php/PATO:Main_Page
	KiSAO	Specifies kinetic simulation algorithms and corresponding parameters with the aim of disambiguation and model integration	http://biomodels.net/kisao/
	TEDDY	Characterizes system behavior from simulation results	http://www.ebi.ac.uk/compneur-srv/teddy/
	SBO	Describes model components including rate equations and parameters	http://www.ebi.ac.uk/sbo/main/

(continued)

Table 12.1 (continued)

Resource		Description	Further information
Data repositories	Biomodels DB	Database of semantically annotated SBML encoded systems biology models	http://www.ebi.ac.uk/biomodels-main/
	Bioportal	A comprehensive directory of ontologies of relevance for life sciences research	http://bioportal.bioontology.org/
	JWS online	Database of semantically annotated SBML encoded systems biology models	http://jjj.mib.ac.uk/
Annotation tools	SBML editor	Provides libSBML for use in other projects, and is the primary tool for SBML model annotation and editing	http://www.ebi.ac.uk/compneur-srv/SBMLeditor.html
	COPASI	Allows model creation, simulation, and analysis using numerous methods and approaches. Permits detailed model component annotation. Compatible with annotated SBML models	http://www.copasi.org
	CellDesigner	Allows the creation, annotation, and simulation of biochemical models	http://www.celldesigner.org/
	SemanticSBML	Provides software for model building, annotation, and checking	http://semanticsbml.org/semanticSBML/simple/index
	SEEK	An online tool for model annotation, part of a broader model integration effort	http://www.sysmo-db.org/seek
Practical applications of semantic technologies	SBML harvester	Allows the conversion of semantic annotations into formal representations of knowledge that can be automatically reasoned about to check consistency or integrate disparate models	http://code.google.com/p/sbmlharvester/
	HyQue	Allows the evaluation of scientific hypotheses based on the available linked data. An example based on the galactose gene network of *S. cerevisiae* is provided	http://semanticscience.org/projects/hyque/

12.9 Conclusion

In this chapter we have attempted to paint the landscape of semantic systems biology from the perspective of how formalized knowledge constructed from Semantic Web technologies can be used to build, publish, query, discover, compare, validate, and evaluate models and knowledge in systems biology. At the core of the Semantic Web effort is the use of Web technology as a means to disseminate and discover information, and this is complimented with formal knowledge representation for consistency checking and more sophisticated question answering through automated reasoning. A key benefit for systems biologists is that the gap between facts, models and results is quickly closing due to the inevitable integration arising from a common framework for representing heterogeneous knowledge. As we look to the future, we envision models being more routinely incorporated in the evaluation of biological hypotheses, and for which easier discovery may lead to a renaissance in model reuse and focused extension Table 12.1.

References

1. Hodgkin AL, Huxley AF (1952) A quantitative description of membrane current and its application to conduction and excitation in nerve. J physiol 117(4):500–544
2. Elowitz MB, Leibler S (2000) A synthetic oscillatory network of transcriptional regulators. Nature 403(6767):335–338
3. Herrgard MJ, Swainston N, Dobson P, Dunn WB, Arga KY, Arvas M, Bluthgen N, Borger S, Costenoble R, Heinemann M, Hucka M, Le Novere N, Li P, Liebermeister W, Mo ML, Oliveira AP, Petranovic D, Pettifer S, Simeonidis E, Smallbone K, Spasic I, Weichart D, Brent R, Broomhead DS, Westerhoff HV, Kirdar B, Penttila M, Klipp E, Palsson BO, Sauer U, Oliver SG, Mendes P, Nielsen J, Kell DB (2008) A consensus yeast metabolic network reconstruction obtained from a community approach to systems biology. Nat Biotechnol 26(10):1155–1160
4. Endler L, Rodriguez N, Juty N, Chelliah V, Laibe C, Li C, Le Novere N (2009) Designing and encoding models for synthetic biology. J R Soc Interface/R Soc 6(4):S405–S417
5. Krause F, Schulz M, Swainston N, Liebermeister W (2011) Sustainable model building the role of standards and biological semantics. Methods Enzymol 500:371–395
6. May RM (2004) Uses and abuses of mathematics in biology. Science 303(5659):790–793
7. Courtot M, Juty N, Knupfer C, Waltemath D, Zhukova A, Drager A, Dumontier M, Finney A, Golebiewski M, Hastings J, Hoops S, Keating S, Kell DB, Kerrien S, Lawson J, Lister A, Lu J, Machne R, Mendes P, Pocock M, Rodriguez N, Villeger A, Wilkinson DJ, Wimalaratne S, Laibe C, Hucka M, Le Novere N (2011) Controlled vocabularies and semantics in systems biology. Mol Syst Biol 7:543
8. Goble C, Stevens R (2008) State of the nation in data integration for bioinformatics. J Biomed Inform 41(5):687–693
9. Berners-Lee T, Hendler J, Lassila O (2001) The semantic web. Sci Am 284:28–37
10. Shadbolt N, Hall W, Berners-Lee T (2006) The semantic web revisited. Intell Syst IEEE 21(3):96–101
11. The Gene Ontology project in (2008) Nucleic acids res 36(Database issue):D440–444

12. Ruttenberg A, Rees JA, Samwald M, Marshall MS (2009) Life sciences on the semantic web: the neurocommons and beyond. Briefings Bioinform 10(2):193–204
13. Ruttenberg A, Clark T, Bug W, Samwald M, Bodenreider O, Chen H, Doherty D, Forsberg K, Gao Y, Kashyap V, Kinoshita J, Luciano J, Marshall MS, Ogbuji C, Rees J, Stephens S, Wong GT, Wu E, Zaccagnini D, Hongsermeier T, Neumann E, Herman I, Cheung KH (2007) Advancing translational research with the semantic web. BMC Bioinform 8(3):S2
14. Luciano JS, Andersson B, Batchelor C, Bodenreider O, Clark T, Denney CK, Domarew C, Gambet T, Harland L, Jentzsch A, Kashyap V, Kos P, Kozlovsky J, Lebo T, Marshall SM, McCusker JP, McGuinness DL, Ogbuji C, Pichler E, Powers RL, Prud'hommeaux E, Samwald M, Schriml L, Tonellato PJ, Whetzel PL, Zhao J, Stephens S, Dumontier M (2011) The translational medicine ontology and knowledge base: driving personalized medicine by bridging the gap between bench and bedside. J Biomed Seman 2(2):S1
15. Rubin DL, Shah NH, Noy NF (2008) Biomedical ontologies: a functional perspective. Briefings Bioinform 9(1):75–90
16. Whetzel PL, Noy NF, Shah NH, Alexander PR, Nyulas C, Tudorache T, Musen MA (2011) BioPortal: enhanced functionality via new web services from the national center for biomedical ontology to access and use ontologies in software applications. Nucleic Acids Res 39(Web Server issue):W541–545
17. Smith B, Ashburner M, Rosse C, Bard J, Bug W, Ceusters W, Goldberg LJ, Eilbeck K, Ireland A, Mungall CJ, Leontis N, Rocca-Serra P, Ruttenberg A, Sansone SA, Scheuermann RH, Shah N, Whetzel PL, Lewis S (2007) The OBO foundry: coordinated evolution of ontologies to support biomedical data integration. Nat Biotechnol 25(11):1251–1255
18. Lord P, Stevens R (2010) Adding a little reality to building ontologies for biology. PLoS One 5(9):e12258
19. Hucka M, Finney A, Sauro HM, Bolouri H, Doyle JC, Kitano H, Arkin AP, Bornstein BJ, Bray D, Cornish-Bowden A, Cuellar AA, Dronov S, Gilles ED, Ginkel M, Gor V, Goryanin II, Hedley WJ, Hodgman TC, Hofmeyr JH, Hunter PJ, Juty NS, Kasberger JL, Kremling A, Kummer U, Le Novere N, Loew LM, Lucio D, Mendes P, Minch E, Mjolsness ED, Nakayama Y, Nelson MR, Nielsen PF, Sakurada T, Schaff JC, Shapiro BE, Shimizu TS, Spence HD, Stelling J, Takahashi K, Tomita M, Wagner J, Wang J (2003) The systems biology markup language (SBML): a medium for representation and exchange of biochemical network models. Bioinformatics 19(4):524–531
20. Hucka M, Finney A, Bornstein BJ, Keating SM, Shapiro BE, Matthews J, Kovitz BL, Schilstra MJ, Funahashi A, Doyle JC, Kitano H (2004) Evolving a lingua franca and associated software infrastructure for computational systems biology: the systems biology markup language (SBML) project. Syst Biol 1(1):41–53
21. Le Novere N (2006) Model storage, exchange and integration. BMC Neurosci 7(1):S11
22. Le Novere N, Bornstein B, Broicher A, Courtot M, Donizelli M, Dharuri H, Li L, Sauro H, Schilstra M, Shapiro B, Snoep JL, Hucka M (2006) BioModels database: a free, centralized database of curated, published, quantitative kinetic models of biochemical and cellular systems. Nucleic Acids Res 34(Database issue):D689–691
23. Olivier BG, Snoep JL (2004) Web-based kinetic modelling using JWS online. Bioinformatics 20(13):2143–2144
24. Henkel R, Endler L, Peters A, Le Novere N, Waltemath D (2010) Ranked retrieval of computational biology models. BMC Bioinform 11:423
25. Le Novere N, Finney A, Hucka M, Bhalla US, Campagne F, Collado-Vides J, Crampin EJ, Halstead M, Klipp E, Mendes P, Nielsen P, Sauro H, Shapiro B, Snoep JL, Spence HD, Wanner BL (2005) Minimum information requested in the annotation of biochemical models (MIRIAM). Nat Biotechnol 23(12):1509–1515
26. Rodriguez N, Donizelli M, Le Novere N (2007) SBMLeditor: effective creation of models in the systems biology markup language (SBML). BMC Bioinform 8:79
27. Hoops S, Sahle S, Gauges R, Lee C, Pahle J, Simus N, Singhal M, Xu L, Mendes P, Kummer U (2006) COPASI–a complex pathway simulator. Bioinformatics 22(24):3067–3074

28. Florez LA, Lammers CR, Michna R, Stulke J (2010) Cell publisher: a web platform for the intuitive visualization and sharing of metabolic, signalling and regulatory pathways. Bioinformatics 26(23):2997–2999
29. Krause F, Uhlendorf J, Lubitz T, Schulz M, Klipp E, Liebermeister W (2010) Annotation and merging of SBML models with semanticSBML. Bioinformatics 26(3):421–422
30. Wolstencroft K, Owen S, du Preez F, Krebs O, Mueller W, Goble C, Snoep JL (2011) The SEEK: a platform for sharing data and models in systems biology. Methods Enzymol 500:629–655
31. Schulz M, Krause F, Le Novere N, Klipp E, Liebermeister W (2011) Retrieval, alignment, and clustering of computational models based on semantic annotations. Mol Syst Biol 7:512
32. Hoehndorf R, Dumontier M, Gennari JH, Wimalaratne S, de Bono B, Cook DL, Gkoutos GV (2011) Integrating systems biology models and biomedical ontologies. BMC Syst Biol 5:124
33. Dada JO, Spasic I, Paton NW, Mendes P (2010) SBRML: a markup language for associating systems biology data with models. Bioinformatics 26(7):932–938
34. Callahan A, Dumontier M, Shah NH (2011) HyQue: evaluating hypotheses using semantic web technologies. J Biomedl Seman 2(2):S3
35. Callahan A, Dumontier M (2012) Evaluating scientific hypotheses using the SPARQL inferencing notation. In: 9th extended semantic web conference (ESWC2012)

Part III
Critical Analysis of Multi-Scale Computational Methods and Tools, Computational Tools for Crossing Levels and Applications

Chapter 13
Computational Infrastructures for Data and Knowledge Management in Systems Biology

Fotis Georgatos, Stéphane Ballereau, Johann Pellet, Moustafa Ghanem, Nathan Price, Leroy Hood, Yi-Ke Guo, Dominique Boutigny, Charles Auffray, Rudi Balling and Reinhard Schneider

Abstract The volume, complexity and heterogeneity of data originating from high throughput functional genomics technologies have created challenges and opportunities for Information technology (IT) departments. These increased demands have also led to increasing costs for IT infrastructure such as necessary computing power and storage devices, as well as further costs for manpower effort, required for maintenance. This chapter describes some of the challenges for computational analysis infrastructure, including bottlenecks and most pressing needs that have to be addressed to effectively support the development of systems biology and its application in medicine.

Keywords Bioinformatics · Infrastructures · Information technology · Knowledge management · Systems biology · Translational research · Data centers · Storage · Computational science · Computational infrastructures · Computing · Scientific computing · e-science · Data management · High performance computing · Cluster · Grid · Desktop grid · Cloud

---

The authors Fotis Georgatos and Stéphane Ballereau contributed equally to this chapter.

F. Georgatos · R. Balling · R. Schneider (✉)
Luxembourg Centre for Systems Biomedicine, University of Luxembourg, 7 Avenue des Hauts-Fourneaux, 4362 Esch-sur-Alzette, Luxembourg
e-mail: reinhard.schneider@uni.lu

F. Georgatos
e-mail: fotis.georgatos@uni.lu

R. Balling
e-mail: rudi.balling@uni.lu

S. Ballereau · J. Pellet · C. Auffray
European Institute for Systems Biology and Medicine—CNRS-UCBL-ENS, Université de Lyon, 50 avenue Tony Garnier, 69007 Lyon, France
e-mail: sballereau@eisbm.org

A. Prokop and B. Csukás (eds.), *Systems Biology*,
DOI: 10.1007/978-94-007-6803-1_13,

Acronyms

3D	Three dimensional
4D	Four dimensional typically 3D plus time dimension
BOINC	Berkeley Open Infrastructure for Network Computing
CERN	European Council for Nuclear Research
CPU	Central Processing Unit
EBI	European Bioinformatics Institute
EGEE	Enabling Grids for E-sciencE
EGI	European Grid infrastructure
ELIXIR	European Life Sciences Infrastructure for Biological Information
EMBL	European Molecular Biology Laboratory
Flops	FLoating-point Operations Per Second
GPU	Graphical Processing Unit
HPC	High Performance Computing
HTC	High Throughput Computing
IaaS	Infrastructure as a Service
IT	Information Technology
I/O	Input/Output—typically used in the context of software processing of data
LHC	Large Hadron Collider
MPI	Message Passing Interface
Omics	A collective term to refer to -omics keywords like metabolomics, genomics, proteomics etc.
PaaS	Platform as a Service

J. Pellet
e-mail: jpellet@eisbm.org

C. Auffray
e-mail: cauffray@eisbm.org

M. Ghanem · Y.-K. Guo
Department of Computing, Imperial College London, London SW7 2AZ, UK
e-mail: mmg@doc.ic.ac.uk

Y.-K. Guo
e-mail: y.guo@imperial.ac.uk

N. Price · L. Hood
Institute for Systems Biology, 401 Terry Avenue North, Seattle, WA 98109-5234, USA
e-mail: Nathan.price@systemsbiology.org

L. Hood
e-mail: lhood@systemsbiology.org

D. Boutigny
Centre de Calcul de l'IN2P3, USR6402 CNRS/IN2P3, 43 Bd du 11 Novembre 1918,
69622 Villeurbanne Cedex, France
e-mail: boutigny@in2p3.fr

PRACE	Partnership for Advanced Computing in Europe
ROI	Return On Investment
SaaS	Software as a Service
SBML	Systems Biology Markup Language
WLCG	Worldwide LHC Computing Grid

13.1 Introduction

Technological advances in biomedical research have provided powerful means for investigating complex molecular processes and phenotypes. Each single high-throughput functional genomics technique leads to the generation of a substantial amount of data. Combinations of such 'omics' approaches [1] are becoming common practice in multidisciplinary systems biology and systems medicine projects [2, 3].

The size, complexity and heterogeneity of the data thus raise a number of challenges requiring diverse computational support:

1. What type of hardware (computational, storage and network) infrastructure is needed? How should it be acquired and managed?
2. What types of software services (data management, analytics, etc.) are needed and how to provide them?
3. How to provide seamless support to biological scientists in using the hardware and software infrastructure?

Such increase in size, heterogeneity and complexity of data clearly expands the need for large-scale data and knowledge management, including communication via networks, solutions for data confidentiality, open standards and open source software and harmonised analysis workflows. Last but not least, such efforts also rely on collaboration and training.

13.2 Infrastructure

Popular infrastructures for Scientific Computing typically involve permutations and combinations of the following list of types:

- HPC systems
- Computing Grids
- Dedicated Clusters
- Cloud Computing
- Desktop Grids.

The exact characteristics of the different resource types are provided in detail in Sect. 13.11, seen later in this chapter, with their advantages and limitations analysed.

Costs for large data and compute facilities have become a hurdle for many academic institutions and even national centers, typically accumulating investments that fall in the millions of euros range [4]. Several collaborative attempts therefore aim to join forces and to manage the data and computing infrastructures on a broader basis [5]. As regards the European landscape, the following projects have had to deal with scientific computing for life sciences: EGI/EGEE [6] with the Biomed VO [7], HealthGrid [8], BioInfoGrid [9], ItalianGrid [10], ELIXIR [11]; to be sure, there are more efforts on-going, both within and outside of Europe.

The most prominent multinational effort in Life Science domains currently is the pan-European ELIXIR project [11]. ELIXIR's goal is to operate a sustainable distributed infrastructure for managing biological data to support research and its translation into the medical and environment sciences as well as the biological industries and society. In order to achieve this mission, ELIXIR is constructing and enhancing a distributed research infrastructure in accordance with the requirements of the scientific community. The ELIXIR Hub will be connected to ELIXIR Nodes to provide infrastructure for data, computing tools, standards and training as well as support for the European Strategy Forum on Research Infrastructures (ESFRI) in the biological and medical sciences.

The Innovative Medicines Initiative (IMI), a joint initiative of the European Union and the European Federation of Pharmaceutical Industry Associations, has recognized the need of a common infrastructure to support its portfolio of public–private partnership projects dedicated to overcome bottlenecks in the development of biomarkers of drug efficacy and toxicity for the treatment of human diseases. As a result, IMI has launched eTRIKS [12], a project dedicated to the delivery of European Translational Research Information and Knowledge management Services, first for all IMI projects, then for other European projects in translational research with similar needs.

13.3 Network Management

Computing systems available on a network may be produced by different vendors, provide different interfaces and/or support different network protocols. Indeed, a programmer wishing to exploit a collection of networked computers must usually contend with different types of heterogeneity: architecture, data format, computational speed, and even dynamic aspects such as machine load, and network load. Security, Performance and Reliability are all objectives that have to be met, yet, with varying weight ratio in the mix, since such a ratio is often specific to the particular application at hand.

Unlike other situations in which systems and networks are managed in isolation, heavy-duty scientific workflows call for tight integration between these two resources. For example, while 10 Gb/s network links are now common and 100 Gb/s equipment is readily available, the optimal use of 10 Gb/s links over a long distance, implies big packets equivalent to jumbo frames, and requires special network tuning on both the endpoint systems and the intermediate equipment. More often than not, this proves to be a delicate exercise in that system and network teams cannot realistically work apart from each other effectively [13].

Furthermore, the end-to-end nature of data connections across multiple sites and the need to replicate datasets across organizations imply that too strict network policies may hinder data management. Such is the case for sites linked with no true IPv4 end-to-end connectivity [14]. When IPv4 traffic restrictions are in place, standard Internet protocols and tools such as ftp, scp, rsync and even http may therefore not readily be supported, in both directions concurrently. The issue may be more pronounced with the more recent and better scalable grid technologies, e.g. Globus and gridftp, which depend on open networks using multiple ports concurrently in both traffic flow directions and dynamically spawning client–server pairs.

Organisations of different sizes such as laboratories, institutes, multi-center companies and collaborating centers have requirements that can be met by varying solutions, which is the reason why there is much divergence in how many critical network environments are being defined; for end-users this typically implies extra effort in order to adjust in the particularities of a given institutional environment.

Massive data-producing scientific domains, such as High Energy Physics [15] (HEP), have chosen a distributed computing model relying on a sophisticated worldwide network infrastructure. This capability enables high bandwidth virtual private networks to move large chunks of data from multiple remote locations. The bioinformatics community has started to benefit from this technology through the European multi-science grid Enabling Grids for E-science (EGEE) and, currently, its follow-up effort [15], the European Grid Infrastructure [16] (EGI).

13.4 Data Management and Integration

Systems biology and systems medicine studies incorporate multiple levels of information including genomics, epigenomics, transcriptomics, proteomics, metabolomics, lipidomics, various phenotypic assays (including imaging), as well as pre-clinical and clinical studies. Over the years, a large number of specialized online databases have emerged providing public access to a variety of such data. Data on different public sources tends to focus on different organisms, species, diseases, etc. Such data is typically heterogeneous in its format, e.g. tabular, network structure, imaging, textual, etc. Moreover, in many cases the data on different sites uses different terminologies to describe the same concepts. The problems and historical approaches concerning syntactic and semantic data integration have been well described [17, 18]. Data management systems aims are to collect, integrate and deliver data, and most rely on guidelines and standards defined by the Minimum Information for Biological and Biomedical Investigations (MIBBI) project [37], other initiatives [19] and on numerous ontologies [20].

In a typical study in systems biology data accessed from remote sites needs to be integrated with in-house experimental data. Furthermore, analyses of the different data types rely on specific analytical methods and software. An effective data infrastructure must thus render such integration seamless. Current efforts include Garuda Alliance, tranSMART [21] and proprietary systems from Ingenuity (IPA), IDBS (ClinicalSense), Oracle (Translational research Solution) and BioMax Informatics (BioXM) [22–24].

The translational medicine data warehouse tranSMART [21] has been designed and implemented as an open-source system that addresses some of the above challenges. It allows users to curate public and internal data from multiple modalities to align them using standard ontologies and store them in a central data warehouse. It also provides a variety of user interfaces that are implemented with open source components to enable flexible data query, analysis and mining. It was built initially by the pharmaceutical company Johnson & Johnson using the Galaxy portfolio of open-source tools developed by the i2b2 consortium [25]. Its source code has now been placed in the public domain through the tranSMART consortium [21]. After its successful deployment in the IMI U-BIOPRED project [3] tranSMART is the core component for the construction of the IMI eTRIKS platform.

13.5 Use of Open Source Software and Open Standards

In a typical Systems biology computational environment, a plethora of open source as well as commercial software packages co-exist. Both types can and will undergo a constant and sometimes fast update pattern. Given that the lifecycle of software usually does not exceed a decade, being able to replace it as soon as the need arises may be more important than owning the source code itself. It is therefore essential that software- commercial or not-follows open standards to allow interoperability of systems and its components [26]. Such standards are either inherent or easy to implement with open source software, but have to be explicitly requested with commercial codes. When the latter requirement is met, the software would thus enable the essential management and standardisation of data across sites and projects.

13.6 Scientific Workflow Reproducibility

Bioinformatics as a field has one unusual record, versus other disciplines based on computer science: the proliferation of small custom-made tools each tailored for a particular task. Indeed, bioinformatics has a wealth of code trees and forks, unlike other sciences where a relatively small set of software codes can be parameterized to simulate given (e.g. physical) systems. This is, at least in part, a consequence of the fact that new types of data drive the development of analytical tools and each bioinformatician has often a tendency to develop his own tools, independently of others. What is needed are integrated efforts and platforms to better unify these tools so that more collective value can be gained by the community from these activities. In the US, large efforts such as the Department of Energy's Knowledgebase (K-base) are seeking to provide a common platform for such bioinformatics and computational biology models. Additionally, tools such as the Systems Biology Markup Language (SBML) [27] are used to make models from individual groups more amenable to sharing and more precise replication.

The combination of these code trees with multiple compilers, optimization options, parallelization frameworks [28] [e.g. Message Passing Interface (MPI) libraries] launchers and execution environments make the repeatability of these workflows partly an experiment of luck. This aspect can be controlled though, by meticulously documenting the exact versions of the software used, and how it was delivered in production so that the result can be predicted rather than hoped for.

The 'modules' framework and EasyBuild are elegant modern tools to support the 'Scientific Workflow reproducibility' argument, against high-performance computing (HPC), Grid, Cloud and other such computational environments. They allow users to manage the configuration of any given system in a structured way, even across multiple versions of software, varying MPI stacks, compiler tool chains etc. This is extremely important for resources that are transient in nature and provided within shared systems, whereby more than one particular version of software may be required. Such environments can nowadays count beyond 400 different instances of software versions, which need to be available in a given setup [29].

Finally, workflow engines such as Taverna [30], Galaxy [31], Tavaxy [32] or the GenePattern system [33] for documentation and use of complex job dependencies are becoming essential tools and will undoubtedly enable reproduction of future complex executions across different infrastructures. Another key endeavour is the development of Synapse [34] by Sage Bionetworks that tracks data provenance while housing data and models together in an open source and community catalysing way. In this way, it can create joint data and model packets that can be transferred between researchers with straightforward reproducibility.

A critical, yet often overlooked, issue is the need to create software workflows for the assessment and validation of large data sets. Such tools aim at countering "lab-specific" effects and facilitating the identification and dismissal of poor quality data. In addition, it is important to note that there are two fundamental types of noise in all large-scale biological data—technical noise and biological noise. Technical noise arises from the process of generating the data—and can generally be handled by mathematical approaches. Biological noise is due to stochasticity, diversity and heterogeneity of biological processes leading to a phenotype (e.g. the transcriptome of a given brain area). How one subtracts away the various non-relevant biological variations to identify the genes or proteins that control the biological process, function or dysfunction of interest is a very challenging problem. Indeed, whether any of the various machine learning techniques can manage to separate the various types of biological processes involved is not known. Such endeavour requires a systems approach to biology.

13.7 Data Confidentiality

The need for data confidentiality in medical applications has far reaching consequences, which are hard to grasp from the outset. For example, protection of information derived from a given individual at the highest possible level implies that participating IT infrastructure must undergo full inventory management, including tracing equipment lifecycle, including end-of-life components and verifying third-party services. An old backup or broken disk can therefore not just be disposed of, and personnel passwords must be protected very carefully while third-party systems and maintenance personnel have to be vetted for compliance in accordance to the desired confidentiality level, while this should not happen at the expense of scientific results.

Techniques such as anonymisation or pseudo-anonymisation [35] might help in certain cases. For example, full anomymisation prevents anybody from tracking back to the original data while with pseudo-anonymisation the data-owner can trace back to the individual via an ID-number. Early definition of constraints and appropriate and/or available solutions is crucial, as with genetic variation profiles, which are themselves unique identifiers.

These issues concern not only data centers in the commercial or health sectors but also IT in academic institutions involved in research projects. Indeed, current large research consortia typically include partners in the pharmaceutical/biotech sector and clinical research departments in particular in translational research. Last but not least, the physical environment of systems and their configuration also have to be carefully controlled.

13.8 Collaboration

A most important and difficult issue in operating the IT infrastructure of a systems biology center is that there are no defined standard guidelines. Very often minimum prerequisites to serve such a center, e.g. network features, and system services are not well defined. Similarly, needs for storage and computing change rapidly and substantially without much time for preparation. Centers sometimes act without consulting technical teams, but when explicit projects' needs force collaboration. This often leads to successive ad-hoc activities and often results in unwanted dependencies.

In an ideal case, operators of such IT infrastructures should be able to exchange information in electronic (emails, mailing lists), scholarly (annual conferences, journals) and physical channels (books, hard media) just like in other fields, notably High Energy Physics.

A potential collaboration opportunity exists in pooling the experience of current state-of-the-art centers and documenting: (1) Services provided internally and externally; (2) Systems on which they are based; (3) Type and amount of human effort to support such services and systems; (4) High performance computing (HPC) environments and computational techniques across different centers. This would ideally become a "Best Practice" document.

It is worthy to mention that European Union countries have already built a long history of collaborative projects pooling resources e.g. ELIXIR [11], EGEE [6] Biomed VO [7], HealthGrid [8], BioInfoGrid [9], ItalianGrid [10]. Current trends will most likely encourage continuation of this tradition.

13.9 Training

To use the high-end hardware and input/output (I/O) systems in an efficient if optimal way, a substantial insight into the technical details is required. This may be an extra hurdle for bioinformaticians who have to follow two or more fast moving fields like biology and IT especially since many have moderate experience with HPC/high-throughput computing (HPC/HTC) systems. In addition, because tasks are very specialized and knowledge is fragmented, training is required not only for end-users but also for service providers such as systems engineers. Technical training and roles assignment are therefore becoming increasingly important to ensure efficient use of available resources [36].

Building on the experience of past and existing schools [37] in providing trainings in both physical and electronic formats [38], future efforts should focus on skills matching current and upcoming infrastructures e.g. "HPC-", or "Grid-", or "Cloud-based bioinformatics". Such training would help improve the design of these environments and, in the end, benefiting the end-users.

Last but not least, creation of cross-disciplinary environments with readily available experts in diverse types of HPC/HTC is essential. Indeed each bioinformatician must not only have appropriate understanding of the IT requirements but most importantly also must have access to individuals that collectively have all of the required information. These cross-disciplinary environments will be essential for biology of the future.

The leading project in European High performance computing, PRACE, has identified the training priorities in the life sciences [41]:

- Method development
- Memory management
- Integration of optimised libraries in scripts
- Data storage
- Data integration and analysis

- Code parallelisation
- Benchmarking support
- Computational methods training.

13.10 A Diversity of Scientific Workflows

Many parameters can be fine-tuned for a given workflow on a chosen infrastructure (HPC, grid, cloud). In order to assist in a reliable and justified choice among and/or to adapt to available infrastructures, a given workflow should ideally describe a minimum set of nine requirements:

1. Number of *cores* needed;
2. Amount of *memory* and *memory bandwidth* needed;
3. Amount of temporary *scratch data storage* needed, *at run-time*;
4. Amount of permanent *work data storage* needed, *in-between runs*;
5. Local *disk I/O throughput* level required;
6. *Backbone bandwidth* level required;
7. *Maximum time span* expected per single job;
8. *Data Confidentiality* level required;
9. Temporal *characteristics of each usage pattern* (e.g. batch, interactive/service oriented, calendar-based, dedicated etc.).

13.11 Categories of Computational Analysis Infrastructures: Characteristics and Limitations

Here we briefly describe the five main computational infrastructures: HPC, computing grids, dedicated clusters, clouds and desktop grids.

13.11.1 HPC Systems

HPC is synonymous with Supercomputing and originated in the late 1970s to spread dramatically in the current decade. Today HPC is used in a wide variety of applications outside the classic fields such as oil exploration, structural analysis, computational fluid dynamics, atmospheric sciences and defence applications. New applications include: medicine, computational chemistry, virtual reality and many more [39–42].

13.11.2 Computing Grids

The idea to bring together multiple, possibly heterogeneous, computing resources in order to build a global architecture usable as a single super computing platform has been brought by Ian Foster and Karl Kesselman in 1998 [43], notably encapsulated in the Globus grid stack. The Grid concept [44] has also been considerably advanced through various international projects [DataGrid [45]—Enabling Grids for EsciencE (EGEE) [6]—EGI [46]—Open Science Grid (OSG) [47]—NORDUGRID [48]—etc.] and several scientific projects have been able to successfully make use of it. One of the most remarkable success is the Grid deployed in the framework of the Large Hadron Collider (LHC) [15] which spans all over the world and allows the quasi-online treatment of ~15 PBytes of raw data each year; this infrastructure is often cited as *WLCG*.

Grids are especially well adapted for trivially parallel computing tasks accessing small chunks of data, or even larger ones if they have been copied locally in advance. For example, Grids are very successful in the area of molecular docking where a multitude of configurations may be tested in each workflow execution.

On the other hand, Grids have drawbacks and adapting a computational code to run on Grid architecture is complex and sometimes not possible for small scientific communities or for users lacking a strong computing background and not having access to a portal service, automating the procedure. While substantial effort has been successfully invested to provide high-level security in Grids, they typically cannot guarantee a level of confidentiality matching medical or industrial standards. These limitations are not inherent to the Grid paradigm but have simply not been addressed by the early developers and further organizational effort is required before they become daily practice.

13.11.3 Dedicated Clusters

The resource category of dedicated clusters is a common solution preferred by both academic and commercial environments, especially for teams in early growth stages; they involve either resources that are self-owned by the resource user or, owned by a third party but delivered in the form of a dedicated service. This type of resource can be a prime technology to build expertise on but its maintenance can prove to be a burden for the smaller teams. It can even be a distraction for bigger teams, whose objective is not HPC and, would rather opt to outsource the task to more specialized parties.

13.11.4 Cloud Computing

While the processing power of single servers was increasing following Moore's law, virtualization techniques started to appear. Virtualization allows to instantiate on demand several virtual machines on a single server. It is then possible to fully de-correlate the physical architecture from the virtual one that is exposed to the users. A complete virtual infrastructure (CPU, storage, network, applications, etc.) can be deployed on the fly with all the characteristics matching the users' needs. The use of these on-demand virtual computers is known as cloud computing [49].

With the past experience with Grids and taking advantages of virtualization techniques, the Cloud paradigm [4] emerged in 2006 with strong support of a few Internet companies which were able to build an economical model based on Cloud services. Within the Cloud model, various classes of services are offered to the users:

- Infrastructure as a Service (IaaS) provides access to a large number of on-demand virtualized resources.
- Platform as a Service (PaaS) provides a toolkit for application development, deployment and management on a virtual infrastructure (database management system for instance).
- Software as a Service (SaaS) where the user's application itself can be executed on the virtual infrastructure through a web interface.

The Cloud model may have certain advantages over the Grid paradigm: (1) It better hides the complexity to the end user; (2) It owns a built-in mechanism to adapt the amount of resources to the needs (elasticity); (3) Virtualization allows creation of a virtual infrastructure strictly reserved to a given set of users, enabling predictable implementation of security and confidentiality; (4) It provides the ability to prepare virtual machines embedding users' software and all necessary external software and libraries that can be deployed and run without any pre-requisite on the physical layer.

13.11.5 Desktop Grids

Desktop grids are in effect the reincarnation of an older technology that has been deployed for many years in the form of distributed computing projects, BOINC [50] being the most notable one among them. Such technologies recently got more attention due to increased reachability thanks to increased network bandwidth, new types of platforms such as GPUs, plus advances in organizing both the collection of resources as well as the scheduling aspects of them, that are practically volatile. Because, by definition, Desktop Grids imply that at least some physical access on the computing systems exists for 3rd parties and perhaps by unsuspecting computer users, this has implications for security and reliability and at least some level of results checking is necessary. In computer science terms the literature refers to these techniques as *sabotage tolerance* and *fault-tolerance*. One notable comment as regards Desktop Grids is that they can overlap but may not be the same as volunteer computing:

- A Desktop grid may be a committed set by the, otherwise idle, resources of a university campus, intending to serve internal needs of researchers. That would be hardly voluntary, though calling it part-time (non-dedicated) resource is appropriate.
- A volunteer may donate the GPUs and CPUs inside servers of a computer room, to a third party bigger network; that would certainly be accounted as volunteer computing, yet, it is not Desktop Grid (Table 13.1).

Taking into account special characteristics of different IT platforms, one can deduce patterns of use-cases in which each one can excel and when it may be better to seek for alternatives:

- HPC systems have the best bulk capabilities but their cost is prohibitive for the simplest tasks, esp. pre- or post-production work.
- Grids can cope well with volume, e.g. drug screening, but once resolution has to be raised, HPC systems are performing better.
- Dedicated clusters have very predictable performance aspects but they are unsuitable for non-flat loads, whereby demand is not fixed.
- Clouds may suit one-off experiments but can become expensive when data volume (storage, network throughput) increases.
- Desktop grids can be cost-effective in data mining, assuming the datasets are fixed and do not pose a confidentiality risk.

In short, matching an infrastructure and its specific characteristics to a scientific project with specific requirements can only be done on a case-by-case basis and no single individual best solution exists for all possible problems.

Concluding, a bioinformatics infrastructure therefore also ought to be diverse and flexible to accommodate the wide variety of applications and their specific requirement.

13.11.6 Conclusions and Future Work

Computational infrastructures are enabling technology for biomedical domain applications with much certainty as regards the timeframe up to year 2020, whereby Exascale infrastructures are expected to be in place and in daily exploitation.

Exascale class systems are often understood and planned as Exaflop infrastructures; frankly put, for biomedical sciences that's just not sufficient enough: The nature of applications in this family of scientific domains may not allow to

Table 13.1 1: Advantages (+) and drawbacks (–) of scientific computing infrastructures

High performance computing (HPC) systems
+ Scalability possible even for "tight" computations, e.g. high resolution molecular docking @ O(1PFlops) and beyond
+ Thread-to-thread data transfers at fast backbone interconnect speeds, e.g. @ O(40Gbps)
+ Typically connected to fast local I/O filesystems, e.g. O(1PBs) @ O(1GBytes per second per thread)
+ Homogeneous and tightly controlled environment
+ Can provide guarantees for timely execution, upon agreement
+ Can provide guarantees for data confidentiality aspects, upon agreement
– Unfavourable from techno-economic point, for big workflows of low-resolution and high-throughput
– Complex computer architectures may imply that idea-to-implementation can take a longer time frame
Computing grids
+ Suitable from techno-economic point, for big workflows of low-resolution and high-throughput
+ High throughput is possible at extreme ranges, e.g. 100,000 jobs per day for molecular docking
+ Distributed storage of replicated datasets is typically well handled in a grid environments
+ Scalability is moderate but throughput is high, e.g. Molecular docking @ O(10TFlops * N), where N is number of sites
+ Thread-to-thread data transfers at -varying- backbone interconnect speeds, typically @ O(1Gbps)
+ May be connected to fast local I/O file systems, e.g. O(100TBs) @ O(100MBytes per second)
+ Composite workflows running at multiple sites a theoretical possibility; not yet widely adopted
– Heterogeneous infrastructure, by its definition—it has multiple owners
– A typical grid infrastructure cannot provide guarantees for timely execution; exception: special agreement with a virtual organization owner (e.g. BIOMED)
– A typical grid infrastructure cannot provide guarantees for data confidentiality; notable exception: secure storage over insecure grid is possible by deploying n-out-of-m KeyStores as per Shamir's work
Dedicated clusters
+ Scalability usually at bare minimum, e.g. molecular docking @ O(10TFlops)
+ Thread-to-thread data transfers at backbone interconnect speeds, typically @ O(1Gbps)
+ May be connected to fast local I/O file systems, e.g. O(10TBs) @ O(100MBytes per second)
+ Homogeneous and tightly controlled environment
+ Can provide guarantees for timely execution; very predictable performance aspects
+ Can provide guarantees for data confidentiality aspects
– Unsuitable for high-resolution workflows or very high-throughput applications
– Unsuitable for scientific workflows which exhibit peak demand; or else, there is often underutilization
– Domain experts may need to build or purchase local expertise in clusters administration
Cloud computing
+ On demand allocation, cost fully linear to activity—as regards computational aspects
+ Scalability usually at bare minimum, e.g. Molecular docking @ O(1Tflops), yet allows throughput

(continued)

Table 13.1 (continued)

+	Thread-to-thread data transfers at backbone interconnect speeds, typically @ O(1Gbps)
+	May be connected to local I/O file systems, e.g. O(10TBs) @ O(100Mbytes per second)
+	Homogeneous and tightly controlled environment
+	Can provide guarantees for timely execution, upon agreement and availability
+	Can provide guarantees for data confidentiality aspects, upon agreement
–	Data transfers and storage tend to be prohibitively expensive, if scaling is a requirement
–	Unsuitable for high-resolution workflows or very high-throughput applications with significant bidirectional data management
–	Requires negotiation for legal liability brokerage as regards data with confidentiality requirements
–	Systems reliability is truly optimized for the needs of a 3rd party, those of the resource provider
Desktop grids and resources	
+	Suitable from techno-economic point of view for workflows of low-resolution & low-throughput; e.g. data mining
+	Can allow for notable cost savings by collating resources which idle outside of working hours
+	Large pool of such candidate resources is available, and growing
–	Scalability is moderate to low—but may allow for interesting interactive workflows
–	Thread-to-thread data transfers at lower network speeds, typically @ O(100Mbps) or less
–	Heterogeneous- and significantly so- infrastructure, by its definition—it may have multiple owners
–	A typical desktop grid infrastructure may not provide guarantees for timely execution
–	A typical desktop grid infrastructure may not provide guarantees for data confidentiality

Disclaimer: *numbers in brackets are indicative as of December 2012*; performance can improve significantly within a few months period. Table has been republished with permission by own author (F. Georgatos)

oversimplify the algorithmic aspects of the problems at hand, nor readjust code optimally against, for instance, a specific ratio of CPU horsepower (Tflops) per memory throughput (GBs per second). Aspects related to data management are going to be as important as pure processing capacity. Furthermore, mapping of physical processes (e.g. protein molecules) or, bioinformatics algorithmic (e.g. comparative genomics) to supercomputing architectures is not always straight-forward, if possible at all with a given codebase.

A certain subset of such applications are foreseen to influence the design of Exascale class systems; the reason for this is their practical applicability combined with special needs, as regards the balancing of computational resources (CPU, memory, I/O). Here are the very ones, which are expected to drive the HPC landscape in the next decade:

- Genomics (next-gen sequencing, comparative genomics)
- Systems biology (epigenomics, proteomics, metabolomics, regulatory networks, protein pathways)
- Molecular Simulations (drug design, structural biology, nanotechnology
- Biomedical Simulations (drug impact, ADME/Tox modeling).

Furthermore, we hereby describe the grand challenges, which lie ahead and areas in which innovations are expected to occur over next decade.

13.11.7 Big Data, Ever-Growing and, with Fragmented Datasets

Biomedical sciences are producing nowadays data, in a pace, which increases faster than so far known IT technological curves. This development implies that there is a trend towards inflating the IT costs and this impacts the relative cost fractions. IT costs are expected to become heavier -or may do so in the future- in the total balance sheet.

This in effect may also imply the need to filter, re-process or throw away data, as operational needs dictate.

Fragmented datasets of many small files, put serious pressure on modern Operating Systems and Networks, as regards latency aspects and optimization techniques; many concepts for speedup that are trivial in other scientific domains (such as read-ahead algorithms to address slower speeds in the lower levels of the memory hierarchy) will just not be able to address the needs of some biomedical applications. One potential solution is to "packetize" information in bigger chunks, but not all applications are amenable to this practice.

13.11.8 Long-Term Sustainable Funding, Coupled with Justifiable ROI

Many institutes cannot fund the necessary infrastructure in isolation; this kind of challenge is already know and addressed in other fields; notably, High Energy Physics institutes had to pool resources together to cope with the needs for the computational aspects of LHC, Large Hadron Collider, what is often called the biggest physics experiment ever. The European ELIXIR project can be seen as a first step in a similar direction. One current re-established fact is that institutes in isolation have limited financial power to work on the grand challenges of the current scientific frontier problems.

Scientific funding does not occur in isolation of the economic environment at large; this implies that as countries' resources get constrained, there is increasing pressure in justifying the research investments in terms of potential Return On Investment (ROI) and provide in advance measurable parameters to evaluate the outcomes of scientific endeavour. This is easy to put in a sentence but hard to achieve in practice, because of the complexities of running scientific projects, the need for long-term vision and the uncertainties that accompany them.

13.11.9 Interpretation of Big Data

The challenge with Big Data will be a challenge for at least the next decade. Already well established technologies like next generation sequencing will produce a significantly higher output of data, but even more data will come from the other omics data and 3D and 4D images and time series. The challenge will not only be related to the practical issues of storage management but will also require new ways to analyse and more important to visualize and distribute the findings.

Here we can foresee the need to setup collaboration tools and visualization systems which will allow the domain specialists spread over continents to work on the same datasets and to discuss and interpret the data at the same time. This will require much better networks and real-time behaviour than what we experience today.

The interpretation of Big Data is a multi-faceted matter and touches on subjects such as:

- image analysis/visualization
- data analysis & pattern recognition
- adoption of new visualization technologies
- distribution of results
- fusion with domain expertise, e.g. from clinician practitioners.

The grand challenge is to be able to turn computational efforts into scientific insights and progress, which corresponds to pending societal needs. It can be argued that this is what science is about but the extra hurdle, that relates to data volumes should not be overlooked too easily.

13.11.10 Publishing Scientific Workflows and, Reproducibility Thereof

The data accessibility challenge was briefly touched in previous chapters. Another challenge and where we don't have a clear picture yet how to tackle it, is the reproducibility of research. In a recent paper by Mesirov the issue was discussed and a first attempt was made how a scientist can make his/her results not only accessible but also have it repeated by other scientists.

As per Mesirov [51] "Scientific publications have at least two goals: (1) to announce a result and (2) to convince readers that the result is correct". It is true and it has to be acknowledged early on, that confining the results of certain biological experiments can be an elusive target; this though is not a convincing excuse for not confining the results of the computational aspects of the scientific workflows, which relate to such data. For certain researchers the in silico investigations may be should not be admissible unless results can be verified by another party once given the same input datasets and access to the same infrastructure.

This notion is becoming increasingly recognized in the life sciences community and individual groups have already started setting up related standards to ease and enhance collaboration; one such case could be APBioNet, Asia–Pacific Bioinformatics Network [52], whereby the BioDB100 initiative, involving 100 Bio-Databases, is highlighting the validation of scientific claims and standards compliance as key.

In fact, certain researchers go as far as fully automating the complete workflow of generating scientific publications, automatically confining the information within papers and other publications. According to Mesirov [51], Reproducible Research System (RRS), consists of two components: RRE & RRP. Reproducible Research Environment (RRE) is for the computational work and provides the computational tools addressing provenance of data, analyses, and results; Reproducible Research Publisher (RRP) is a document preparation system, including standard word-processing software that functions in collaboration with the RRE.

Regardless of how far someone is willing to go in automatically producing the scientific products, including publications, every effort in this direction has to be agreed as commendable and of great value to the scientific effort. The true challenge in this area is to ensure that the overheads of the reproducible methods correspond to benefits for the community and convince peers that such approaches have high merit.

In such scenario the impact for the IT management is not only to ensure data management for maybe years but also make sure that the execution of the analysis pipeline will be archived and able to be re-activated if and when necessary. This will require employment of modern facilities like data and file system snapshots, versioning and archiving capabilities, which clearly are beyond what a typical data center now provides.

13.11.11 Collection of Data and, Confidentiality Thereof

There is common agreement among biomedical researchers that data is indispensable for science progress and not all objectives can be achieved if the data, including data that refers to specific individuals, cannot be pooled together. Especially, the individual should always have the right to choose to receive information that can improve his/her own health. For instance, if a common gene- or gene combination- is found that corresponds to a disease -or risk factor- and that is eventually connected with some medical treatment, the donor of information should be able to receive back the benefits of scientific progress. This runs against the currently favoured concept of complete anonymisation, which defeats this basic principle.

The issue is a challenge because there are many choices possible, in a spectrum defined between two extremes:

– One line of thinking is over full anonymization; this counts on the complete anonymisation of data, e.g. patient data gets isolated from the information on the person and then "lives a life of its own". On one hand, this gives good freedom for data distribution, yet, at the same time the potential benefit to the original data donor is suppressed, which may act as counter-incentive for building a big pool of datasets.

– Another line of thinking is that people should to be willing, after depersonalization, to make their data open and available at large, for the data mining that will transform medicine of the future. The concern is that far too much lip service is given to confidentiality and legalistic formalisms like Institutional Review Boards (IRBs), which indeed imply overheads for routine scientific activities. Going to such direction would require a fundamental change in mentality and concessions from individuals that not everyone is willing to take, at short notice.

Perhaps a middle way to go is that in order to make the best with information, it must be collected and protected with appropriate measures, so that the needed capabilities for scientific processing become possible, including identifying and informing the individual, if that has been requested. There are technical and organizational challenges for such issues since they come at the cost of complexity overheads: that would be the consequence of trying to improve on both flexibility and security; and doing so does imply certain compromises.

Acknowledgements This work was supported by the CNRS, and in part by the EU grants to CA in the context of the MeDALL consortium (*Mechanisms of the Development of Allergy,* Grant Agreement FP7 N°264357), the U-BIOPRED consortium (*Unbiased Biomarkers for the PRE-Diction of respiratory disease outcomes*, Grant Agreement IMI 115010), and the eTRIKS consortium to CA, DB, MG, YG, RS, RB (*European Translational research Information & Knowledge management Services*, Grant Agreement n°115446). The formation of the European Institute for Systems biology & Medicine hosted at Claude Bernard University is supported by the Lyonbiopole competitive cluster and its academic, industrial and local authority partners, including Grand Lyon, Région Rhône-Alpes, Direction de la Recherche et de la Technologie, and the Finovi Foundation (CA). We would like to acknowledge the support of the Luxembourg Centre for Systems Biomedicine and the University of Luxembourg (LH and RB), the NIH General Medical Sciences Center for Systems Biology GM076547 (LH) and a Department of Defense contract on Liver Toxicity W911SR-09-C-0062 (LH).

References

1. Chen C, McGarvey PB, Huang H, Wu CH (2010) Protein bioinformatics infrastructure for the integration and analysis of multiple high-throughput 'omics' data. Adv Bioinform, 19p
2. Bousquet J et al (2011) MeDALL (mechanisms of the development of ALLergy): an integrated approach from phenotypes to systems medicine. Allergy 66:596–604
3. Bel EH et al (2011) Diagnosis and definition of severe refractory asthma: an international consensus statement from the innovative medicine initiative (IMI). Thorax 66:910–917
4. Rosenthal A et al (2010) Cloud computing: a new business paradigm for biomedical information sharing. J Biomed Inform 43:342–353
5. Ruusalepp R (2008) Infrastructure planning and data curation: acomparative study of international approaches to enabling the sharing of research data. At http://www.jisc.ac.uk/media/documents/programmes/preservation/national_data_sharing_report_final.pdf

6. Twiki—a web-based collaboration for EGEE project. At https://twiki.cern.ch/twiki/bin/view/EGEE/LifeSciences
7. Biomedical applications description. At http://proton.polytech.unice.fr/biomed/egee2-applications.html#medimg
8. HealthGrid Portal—A Human Grid Initiative. At http://healthgrid.org/
9. The BioinfoGRID Project. At http://www.bioinfogrid.eu/
10. IGI—Italian Grid Infrastructure. List of scientific application for VO biomed at http://www.italiangrid.it/appdb/listbyvo/6
11. Crosswell LC, Thornton JM (2012) ELIXIR: a distributed infrastructure for European biological data. Trends Biotechnol 30:241–242
12. eTRIKS European Transnational Information and Knowledge Management Services. At http://www.etriks.org/
13. Wu Y, Kumar S, Park S-J (2010) Measurement and performance issues of transport protocols over 10 Gbps high-speed optical networks. Comput Netw 54:475–488
14. Saltzer JH, Reed DP, Clark DD (1984) End-to-end arguments in system design. ACM Trans Comput Syst 2:277–288
15. Welcome to the Worldwide LHC Computing Grid. At http://wlcg.web.cern.ch/
16. Newhouse S. D2.3 EGI-InSPIRE Paper, European Grid Infrastructure. At http://go.egi.eu/pdnon
17. Sujansky W (2001) Heterogeneous database integration in biomedicine. J Biomed Inform 34:285–298
18. Alonso-Calvo R et al (2007) An agent- and ontology-based system for integrating public gene, protein, and disease databases. J Biomed Inform 40:17–29
19. Brazma A, Krestyaninova M, Sarkans U (2006) Standards for systems biology. Nat Rev Genet 7:593–605
20. Courtot M et al (2011) Controlled vocabularies and semantics in systems biology. Mol Syst Biol 7:543
21. Szalma S, Koka V, Khasanova T, Perakslis ED (2010) Effective knowledge management in translational medicine. J Transl Med 8:68
22. Stein LD (2008) Towards a cyberinfrastructure for the biological sciences: progress, visions and challenges. Nat Rev Genet 9:678–688
23. Ghosh S, Matsuoka Y, Asai Y, Hsin K-Y, Kitano H (2011) Software for systems biology: from tools to integrated platforms. Nat Rev Genet 12:821–832
24. Wruck W, Peuker M, Regenbrecht CRA (2012) Data management strategies for multinational large-scale systems biology projects. Brief Bioinform. doi:10.1093/bib/bbs064
25. Blankenberg D et al (2010) Galaxy: a web-based genome analysis tool for experimentalists. Curr Protoc Mol Biol Chapter 19, Unit 19.10.1–21
26. Chervitz SA et al (2011) Data standards for omics data: the basis of data sharing and reuse. Methods Mol Biol 719:31–69
27. Hucka M et al (2003) The systems biology markup language (SBML): a medium for representation and exchange of biochemical network models. Bioinformatics 19:524–531
28. Foster I, Kohr DR, Krishnaiyer R, Choudhary A (1997) A library-based approach to task parallelism in a data-parallel language. J Parallel Distrib Comput 45:148–158
29. VitalIT tools—High Performance Computing Center. At http://www.vital-it.ch/software/tools.php
30. Hull D et al (2006) Taverna: a tool for building and running workflows of services. Nucleic Acids Res 34:729–732
31. Hillman-Jackson J et al (2012) Using galaxy to perform large-scale interactive data analyses. Curr Protoc Bioinform Chapter 10, Unit10.5
32. Abouelhoda M, Issa SA, Ghanem M (2012) Tavaxy: integrating taverna and galaxy workflows with cloud computing support. BMC Bioinform 13:77
33. Reich M et al (2006) GenePattern 2.0. Nat Genet 38:500–501
34. Sage Synapse: Contribute to the Cure. At https://synapse.sagebase.org

35. Kushida CA et al (2012) Strategies for De-identification and anonymization of electronic health record data for use in multicenter research studies. Med Care 50:S82–S101
36. Lyon L (2007) Dealing with data: roles, rights, responsibilities and relationships. Consultancy report, UKOLN, University of Bath, UK
37. Biosapiens network—A European Virtual Institute for Genome Annotation. At http://www.biosapiens.info
38. Training at EMBL-EBI. At http://www.ebi.ac.uk/training/
39. Laxminarayan S, Michelson L (1988) Perspectives in biomedical supercomputing. IEEE Eng Med Biol Mag 7:12–15
40. Böhm K (1997) Supercomputing in cancer research. Stud Health Technol Inform 43 Pt A:104–108
41. Maizel JR (1988) Supercomputing in molecular biology: applications to sequence analysis. IEEE Eng Med Biol Mag 7:27–30
42. Orphanoudakis SC (1988) Supercomputing in medical imaging. IEEE Eng Med Biol Mag 7:16–20
43. Kesselman C, Foster I (1998) The grid: blueprint for a new computing infrastructure. Morgan Kaufmann Publishers, Burlington. At http://www.amazon.ca/exec/obidos/redirect?tag=citeulike09-20&path=ASIN/1558604758
44. Szolovits P (2007) What is a grid? J Am Med Inform Assoc 14:386
45. Breton V, Medina R, Montagnat J (2003) DataGrid, prototype of a biomedical grid. Methods Inf Med 42:143–147
46. European Grid Infrastructure. For further information, kindly refer to the EGI-InSPIRE paper. *EGI* at http://go.egi.eu/pdnon
47. The Open Science Grid Homepage. At http://www.opensciencegrid.org
48. The NorduGrid Collaboration, Web site. http://www.nordugrid.org
49. Armbrust M et al (2009) Above the clouds: a berkeley view of cloud computing. EECS Department, University of California, Berkeley. At http://www.eecs.berkeley.edu/Pubs/TechRpts/2009/EECS-2009-28.html
50. Anderson DP (2004) Boinc: a system for public-resource computing and storage. In: 5th IEEE/ACM international workshop on grid computing 4–10
51. Mesirov J (2010) Computer science: accessible reproducible research. Science 327(5964):415–416. doi:10.1126/science.1179653. 22 Jan 2010
52. Tan TW et al (2010) Advancing standards for bioinformatics activities: persistence, reproducibility, disambiguation and minimum information about a bioinformatics investigation (MIABi). BMC Genomics 11(4):S27. doi:10.1186/1471-2164-11-S4-S27. http://www.ncbi.nlm.nih.gov/pubmed/21143811. 2 Dec 2010
53. Kenneth H et al (2012) EasyBuild: building software with ease, PyHPC 2012, Supercomputing 2012, Salt Lake City

Chapter 14
Computational Tools and Resources for Integrative Modeling in Systems Biology

Christoph Wierling and Hendrik Hache

Abstract Mathematical modeling is key for systems level understanding of cellular processes. The development of mathematical models demands advanced computational tools that keep track of heterogeneous data of molecules and their interactions. Especially the integration of experimental data and pre-existing knowledge into computational models of biological systems is of considerable importance. *In silico* simulations of model behavior under similar conditions as in the experiment give the possibility for model validation regarding specific experimental data. Such an integrative approach leads eventually to a more accurate and consistent description of the observed biological system. We review several resources and computational tools which support the investigation of biological networks and describe several resources and methods for integrative modeling.

Keywords Omics data · Mathematical modeling · Software tools · Network analysis · Reverse engineering

Acronyms/Abbreviations

ARACNE	Algorithm for the Reconstruction of Accurate Cellular Networks
ATP	Adenosine TriPhosphate
BioPAX	Biological Pathway Exchange
BRENDA	BRaunschweig ENzyme Database
CCLE	Cancer Cell Line Excyclopedia
CellML	Cell Markup Language
ChEBI	Chemical Entities of Biological Interest
ChiP	Chromatin immunoprecipitation
COPASI	COmplex PAthway Simulator
DNA	Deoxyribonucleic Acid

C. Wierling (✉) · H. Hache
Max Planck Institute for Molecular Genetics, Ihnestr. 63-73, 14195 Berlin, Germany
e-mail: wierling@molgen.mpg.de, hache@molgen.mpg.de

A. Prokop and B. Csukás (eds.), *Systems Biology*,
DOI: 10.1007/978-94-007-6803-1_14,

DNase	Deoxyribonuclease
DREAM	Dialogue on Reverse Engineering Assessment and Methods
FBA	Flux-Balance Analysis
GEDAS	Gene Expression Data Analysis Suite
GENT	Gene Expression database of Normal and Tumor tissues
GEO	Gene Expression Omnibus
GO	Gene Ontology
GXD	Gene eXpression Database
HGNC	HUGO Gene Nomenclature Committee
HMDB	Human Metabolite DataBase
HUGO	Human Genome Organisation
KEGG	Kyoto Encyclopedia of Genes and Genomes
MCA	Metaboloc Control Analysis
MeDIP	Methylated DNA immunoprecipitation
MMMDB	Mouse Multiple tissue Metabolome DataBase
MOPED	Model Organism Protein Expression Database
mRNA	messenger RNA
MS	Mass Spectrometry
NEST	Neighborhood-based Entity SeT
NMR	Nuclear Magentic Resonace
ODE	Ordinary Differential Equation
PaxDB	Protein Abundance across organisms DataBase
PCA	Principal Component Analysis
RNA	Ribonucleic Acid
SABIO-RK	System for the Analysis of Biochemical Pathways - Reaction Kinetics)
SBGN	Systems Biology Graphical Notation
SBML	Systems Biology Markup Language
SMD	Stanford Microarray Database
SVM	Support Vector Machine
TCGA	The Cancer Genome Atlas
TRED	Transcriptional Regulatory Element Database
VANTED	Visualization and Analysis of Networks containing Experimental Data
XML	eXtensible Markup Language

14.1 Introduction

Mathematical modeling has a long history in science in particular in physics and engineering. In recent years, due to the huge amount of qualitative and quantitative data it became also very popular in molecular biology leading to the establishment of the new research subject systems biology. On the heart of systems biology is always a mathematical model. The development of such a molecular, cellular, or physiological model often requires a lot of data about the structure of the interaction network defining the network topology, the rules or concepts defining the types of interactions, such as kinetic laws and respective kinetic parameters, as well as values defining the state of the system. Moreover, usually additional data is required for model adaptation and model refinement, e.g., experimental data of time course analysis or specific genetic perturbations or drug treatments.

Since biological systems are usually composed of a vast number of often very heterogeneous components, the size of such models often becomes quite large, with a lot of different constraints and parameters depending on the used modeling approach. Hence, the development of such large models requires advanced computational tools for evaluation and integration of large amounts of data, for model analysis and simulation, and for data visualization.

Within the last decades many comprehensive databases were developed for the integration of any kind of experimental and molecular data. This comprises databases for many kinds of sequence data, such as genomic DNA, mRNA, microRNA, or protein sequences, interaction data, such as protein–protein interactions or biochemical reactions, quantitative data of kinetic parameters, gene expression or metabolite and protein amounts, and last but not least scientific literature. In parallel, also a lot of different analysis tools were developed and integrated, such as tools for any kind of sequence analysis and statistical evaluation of, e.g., high-throughput data.

Besides the integration and analysis of primary data, also the need for advanced algorithms and software for the analysis and integration of interaction data becomes more and more relevant. Two important approaches in this respect are forward modeling and reverse engineering. While the concept of forward modeling first of all tries to integrate existing knowledge with new data to derive a revised model that can be used for the generation of new hypothesis by means of simulation, reverse engineering is less biased by pre-existing knowledge and tries to deduce the internal structure of a system based on perturbation or time course data.

Finally, the visual representation of large amounts of data and results either coming from wet lab experiments or *in silico* analysis is of significant importance in this context. For instance, the visualization of clustering results of expression data by heatmaps and hierarchical trees is a very useful tool for the understanding of such large-scale data sets. Furthermore, visualization of interaction networks is crucial for modeling of molecular or physiological systems and 2D- or 3D-images of, e.g., molecules, cells and cellular interaction networks as well as tissues or organs are highly demanded.

Developing mathematical models of biological systems is always an iterative approach (Fig. 14.1). Pre-existing data, such as data about the model structure and model topology as well as data on reaction kinetics and the model state, can be used for the development of a model prototype. Subsequently, by *in silico* analysis such as simulation studies new hypothesis can be generated and validated in follow up experiments used for model refinement in an iterative sequence. This underlines

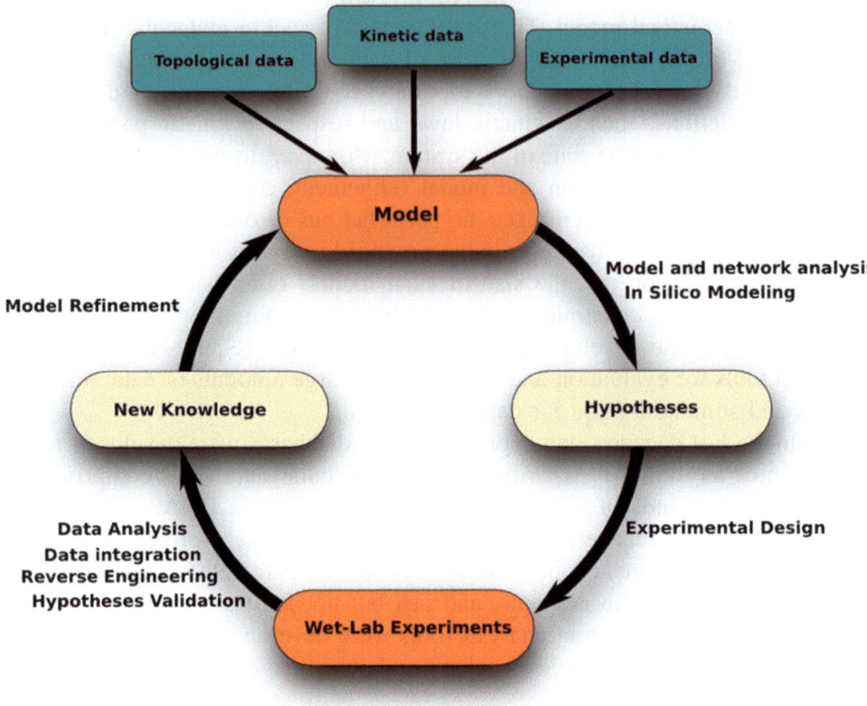

Fig. 14.1 Modeling biological systems is an iterative approach. A model prototype based on pre-existing knowledge is used for the formulation of hypothesis that are going to be validated by further wet lab experiments whose results can subsequently be used for model refinement

the importance of a strong interaction between wet lab experiments, data analysis, and *in silico* modeling.

In the following we give an overview of several resources for experimental data and pre-existing know-how that is relevant for model development. Moreover, we discuss several computational tools for subsequent data analysis, data integration, and modeling in systems biology.

14.2 Data Resources

The integration of experimental data is crucial for the development of meaningful biological computer models. Advances in high-throughput technologies in molecular biology yield comprehensive data, e.g., about the tempo-spatial behavior of the system under study. Such diverse data can be integrated into mathematical models in order to perform several operations such as validating the model's topology, fitting its kinetic parameters, initializing its components' concentrations or abundances for *in silico* simulations, or testing its predictions.

A biological system can be investigated by the accurate measurement of the system constituents at various levels, such as the genome, transcriptome, proteome, metabolome, epigenome, or interactome. Various experimental techniques, featuring high precision and coverage, have been developed in recent years to provide measurements at many of these levels. In contrast to single component analysis, the routine practical use of global measurements has dramatically accelerated the progress in systems biology. For this integrative approach quantitative data is of particular interest. Table 14.1 summarizes several data resources for systems biology that are discussed in the following in more detail.

Table 14.1 Data and model resources for systems biology

Name	Web-link	Content or short description
Data resources		
COSMIC	http://www.sanger.ac.uk/cosmic/	Somatic mutations information
ArrayExpress	http://www.ebi.ac.uk/arrayexpress/	Gene expression data
GEO	http://www.ncbi.nlm.nih.gov/geo/	Gene expression data
SMD	http://smd.stanford.edu/	Gene expression data
GENT	http://medical-genome.kribb.re.kr/GENT/	Gene expression data
GXD	http://www.informatics.jax.org/expression.shtml	Gene expression data
CCLE	http://www.broadinstitute.org/ccle/home/	Diverse data of cancer cell lines
TCGA	http://cancergenome.nih.gov/	Diverse data of cancer patient's samples
Human Protein Atlas	http//www.proteinatlas.org/	Gene and protein expression data
PeptideAtlas	http://www.peptideatlas.org/	Peptides data
MOPED	http://moped.proteinspire.org/	Protein expression data
PaxDB	http://pax-db.org/	Protein expression data
HMDB	http://www.hmdb.ca/	Metabolites in human
MMMDB	http://mmmdb.iab.keio.ac.jp/	Metabolites in mice
Model resources		
PathGuide	http://www.pathguide.org/	Pathways
KEGG	http://www.genome.jp/kegg	Pathways
Reactome	http://www.reactome.org/	Pathways
ConsensusPathDB	http://consensuspathdb.org/	Pathways and interactions
Pathway Commons	http://www.pathwaycommons.org/	Pathways and interactions
BRENDA	http://www.brenda-enzymes.org/	Enzyme information
SABIO-RK	http://sabio.villa-bosch.de/	Biochemical reaction kinetics
JWS	http://jjj.biochem.sun.ac.za/	Biochemical models
BioModels	http://www.ebi.ac.uk/biomodels/	Biochemical models
CellML	http://www.cellml.org/models/	Biochemical models
Transfac	http://www.gene-regulation.com/pub/databases.html	Gene regulation
TRED	http://rulai.cshl.edu/TRED	Gene regulation

14.2.1 Genomics Data

The advent of next-generation sequencing technologies in 2008 has revolutionized genomics and transcriptomics within only a few years. Their applications are widespread, including the sequencing of complete genomes [1, 14, 64, 163, 169], the detection, identification, and quantification of rare transcripts and variants with RNA-seq [97, 166], and the genome-wide profiling of epigenetic marks and chromatin structures with ChIP-seq [162], MeDIP-seq [96], and DNase-seq [18]. Despite the impact of next-generation technologies on genomics, the usage and integration of large amounts of genomics data into systems biology is still ambitious. Systems biology requires information about the functional impact of genomic aberrations, such as mutations, methylations, and changes in copy number in order to translate this information into mathematical models of the cellular system. For instance, over 30,000 mutations have been identified in a human malignant melanoma [127]. Many of them and even more somatic mutations in human cancers together with related details are collected in the catalogue of somatic mutations in cancer (COSMIC; [49]). However, functions and phenotypes of most of the mutations are unknown.

14.2.2 Transcriptomics Data

Since its origin in the mid-1990s, high-throughput gene expression measurement technologies have been applied on large-scale to determine the activity of genes in parallel, for instance in different tissues, disease states, after perturbations, or in different organisms. DNA array techniques have been successfully applied in many studies [45, 72, 141, 154, 161, 170] and nowadays are taken over by deep RNA sequencing technologies [30, 65, 113, 158, 173]. The majority of experimental data is stored and provided in web-accessible repositories, such as the three major databases ArrayExpress with currently more than 30,000 publicly accessible experiments in about 1,600 organisms, Gene Expression Omnibus (GEO) with almost 31,000 public experiments in around 2,000 organisms, and Stanford Microarray Database (SMD) with more than 23,000 public experiments in 280 organisms [43, 70, 124]. Besides these general databases, there are species-, tissue-, or disease-specific databases, which collect and compile data from public resources and experiments. For instance, GENT is a web-accessible database that provides more than 40,000 gene expression samples across diverse human cancer and normal tissues [149] and GXD is a resource for gene expression data particularly focused on endogenous gene expression during mouse development [47]. Cancer Cell Line Encyclopedia (CCLE) stores not only gene expression but also genomic data for almost 1,000 human cancer cell lines [11]. Pharmacological profiles were taken from around 500 cell lines after treatment with 24 anticancer drugs. Additionally, mutation and copy number variation data for all cell lines were obtained via massively parallel sequencing and proteomic as well as metabolic data will be added in the near future. Recently, the cancer genome atlas (TCGA) has massively generated data of cancer patient samples by genomic DNA copy numbers arrays, DNA methylation, exome sequencing, mRNA arrays, microRNA sequencing, and reverse-phase protein arrays. The data of around 30 cancer types is provided on the data portal to search, download, and analyze the data sets. Many recent high impact studies using this data have significantly extended our knowledge about diverse cancer types [22, 59, 89].

14.2.3 Proteomics Data

A more detailed view on cellular processes is provided by quantitative global proteome measurements. Large-scale high-accuracy proteomics is considered as powerful as other established omics technologies such as genomics and transcriptomics [33]. Nevertheless, it faces several limitations, such as the uncertainty about the precise number of different human proteins. Due to the vast amount of possibilities of post-transcriptional and post-translational modifications, the number of different proteins is thought to range from 20,000 to several millions [33]. However, the number of expressed proteins in a single cell is much smaller. A recent large-scale proteome study identified around 10,000 different expressed proteins in HeLa cell lines [114]. A further difficulty of quantitative proteomics technologies is the wide range of protein abundances from a few per cell for signaling proteins to millions for structural proteins [117, 126, 164].

Several technologies such as expression profiling based on antibodies, and mass spectrometry for quantitative proteomics have been developed in the last decades to enable the identification, quantification, and analysis of proteins in cells, tissues, and biological fluids.

A technology for systematic analysis of cellular distributions and subcellular localizations of proteins is based on the large-scale generation of specific protein antibodies. Currently, over 15,500 antibodies, targeting proteins from more than 12,200 human genes (around 61% of the human protein-coding genes) are stored in the public database Human Protein Atlas [160]. These antibodies were used on tissue microarrays to examine the spatial distribution and the relative expression values of proteins (categorized into four levels) in different cell populations of various human tissues and cancer types [128]. Results are accessible over the web-interface of the Human Protein Atlas.

High-resolution mass spectrometry-based proteomics technologies have been rapidly developed in recent years such that the vast majority of proteomics studies to date have used mass spectrometric techniques to identify and quantify proteins in mixtures [117]. Three main mass spectrometry-based proteomic strategies, shotgun or discovery, directed, and targeted strategies, have emerged [40]. These can be distinguished by the detailed work-flow of a proteomics experiment and their application.

There are various resources and repositories for quantitative protein expression data coming from high-throughput proteomics measurements. All of them have to deal with the wide spectrum of technologies, applications, and experimental protocols, which hampers the establishment of easy-to-use databases. A proteomics database example is PeptideAtlas, which is a publicly accessible database of peptides identified in many tandem mass spectrometry proteomics studies [38]. Raw data from mass spectrometry experiments are collected from the community, processed through a consistent analysis pipeline with several established algorithms, and made available together with additional information on PeptideAtlas. However, not all data is publicly accessible and the expression values of the peptides corresponding to a particular protein over various experiments are given only as images.

Model Organism Protein Expression Database (MOPED) is another proteomics data resource of publicly available studies in human and model organisms [90]. Over the web-interface users are able to access protein level expression data together with meta-analysis results with standardized methods. MOPED also supports the comparison of proteomics data and visualization of patterns of expression values within and across sample sets.

A meta resource for absolute protein abundance levels is PaxDB [164]. Publicly available experimental data from various tissues and organisms are reprocessed and averaged over various samples, conditions, and cell types. Contributing data sets are ranked based on a calculated score of consistency against externally provided protein network information. The protein abundance values are expressed in parts per million, i.e., each protein entity is enumerated relative to all other protein molecules in the same sample. All data, including expression data and functional information linked to other resources is accessible via the website.

14.2.4 Metabolomics and Metabonomics Data

Metabolites are small molecules acting as intermediates and products of the cellular metabolism. Metabolite's concentrations can directly be influenced by changes of transcriptome, proteome, and interactome, e.g., due to responses to a stimulus. Therefore, the metabolome provides an instantaneous snapshot of the cellular state of the system under study. Metabolomics and metabonomics studies seek to analyze and characterize the metabolome. Whereas metabolomics is a comprehensive and quantitative analysis of all metabolites in a biological sample, metabonomics focus on quantitative measurements of the dynamic metabolic response of living systems to pathophysiological stimuli or genetic modification. However, these terms are often used interchangeably [116].

Two main techniques, nuclear magnetic resonance (NMR) spectroscopy and mass spectrometry (MS), exist to measure, analyze, and quantify the metabolome of biological fluids, tissues, and cell extracts [39]. Both methods have been applied to many multivariate metabolic profilings (e.g., [15, 26, 66, 145]). NMR spectroscopy has the advantage that the biological sample does not require any physical or chemical treatment prior to the analysis, therefore, it does not destruct the nature of the metabolites in the fluid. In contrast, in mass spectrometry studies, the metabolites have to be separated from the biological fluid, typically by using liquid or gas chromatography, in order to be analyzed [116]. However, mass spectrometry technology is considered to be more sensitive than NMR.

Although a number of genomics, transcriptomics, and proteomics databases are available until now, metabolomics databases being appropriate for systems biology applications are still rare and limited. Most databases provide structural information of metabolites but no quantitative data. However, the Human Metabolite Database (HMDB) contains detailed information of about 8,000 metabolites including chemical, clinical, and enzymatic data (linked to external databases, e.g., BioCyc, ChEBI, and KEGG, among others) as well as concentration data in various human biofluids, such as urine, blood, or sweat [174]. A large collection of metabolites and their quantification in multiple tissues in mice is covered by Mouse Multiple Tissue Metabolome Database (MMMDB; [157]). Non-targeted analyzes were performed and over 200 metabolites were successfully identified. Users can upload normalized data to visualize them in parallel to other data in the MMMDB.

14.2.5 Pathway Data

Besides experimental data, the development of models of biological systems builds also on the basis of existing knowledge about molecular entities like genes, enzymes, metabolites etc. as well as their interaction networks. This is usually documented in the primary scientific literature, which is difficult to be used directly for the development of models on a large scale and which is also very costly and error-prone. In recent years also a lot of this information became integrated in interaction and pathway databases. PathGuide is providing a comprehensive list of pathway databases [5]. It covers, e.g., lists of databases for protein–protein interactions, metabolic, signaling, and gene regulatory networks as well as other information on molecular interactions. This kind of data is an ideal resource for the development of the topological structure of a model.

One of the earliest and most popular pathway database is KEGG, the Kyoto Encyclopedia of Genes and Genomes [82, 83]. A detailed description of the usage of the KEGG pathway database is given in Tanabe and Kanehisa [159]. KEGG provides a broad spectrum of genomic data from multiple species and organisms and it integrates comprehensive data on metabolic, signal transduction, and gene regulatory processes and pathways. Pathways in KEGG are displayed as reference maps that can be overlaid with species specific information of enzymes and genes. Another comprehensive and well curated pathway database is Reactome. Reactome is an online database of predominantly human biological processes [34, 106]. It provides a comprehensive description of mostly metabolic and signal transduction pathways and it is well annotated by reference literature. Besides these databases on metabolic and signaling data, there are also databases providing information on gene regulation such as Transfac and TRED [75, 108].

Furthermore, meta-databases can integrate different types of functional interactions from heterogeneous interaction data resources. One of such meta-databases is the ConsensusPathDB database that has currently integrated data of 31 different pathway and interaction databases covering information about protein interactions, signaling and metabolic reactions, gene regulations, drug-target interactions, and pathway annotations [80, 81]. ConsensusPathDB provides methods to search for components and pathways and to visualize individual sub-networks. Moreover, ConsensusPathDB has methods for gene set and metabolite set over-representation and enrichment analysis to identify, e.g., pathways, network neighborhood-based entity sets (NESTs) or gene ontology (GO) categories based on a list of, e.g., differentially expressed genes or gene expression data. For instance, this is useful to identify pathways or network modules that are relevant in a certain study hence subject to modeling. Another collection of publicly available pathway data from multiple organisms is Pathway Commons [25].

14.2.6 Quantitative Data for Kinetic Modeling

Important for the set up and development of quantitative models is also information about appropriate abundances of molecules and enzyme properties, such as enzyme-specific kinetic parameters or dissociation constants. One of the main collections of enzyme function and property data is the BRENDA database integrating manually extracted data from the primary literature [140]. Another resource providing comprehensive information about biochemical reactions and their kinetic properties is the SABIO-RK database [175]. The database integrates kinetic parameters in relation to biochemical reactions and their biological sources.

14.3 Modeling Biological Systems

Modeling starts usually with a description of the relevant components and interactions of the system under study. Different mathematical and algorithmic approaches have been proposed for the description, modeling, and simulation of molecular and cellular processes. As a first approximation, it can be described by a Boolean network where the state of a component can be "on" (1) or "off" (0) indicating whether the component is present or absent or active or inactive. Interactions can be represented by Boolean functions calculating the state of the components from the activity state of other components. The Boolean approach has frequently been used for the description of gene regulatory and signaling networks (e.g., [27, 46]). Latter and in particular metabolic networks are often modeled by ordinary differential equation (ODE) systems. Therefore, for a set of reactions the concentration changes of a single component S_i over time t is given by the sum of in- and out-fluxes as follows:

$$\frac{d[S_i]}{dt} = \sum_{j=1}^{r} n_{ij} v_j \quad \text{with } i = 1, \ldots, m,$$

where m is the number of components, n_{ij} the stoichiometric coefficient of the ith component in the jth reaction, and v_j the reaction rate of the jth reaction. Reaction rates can be described, e.g., by mass action kinetics $v = k \prod [S_l]$ with $l = 1, \ldots, p$, where $[S_l]$ is the concentration of the lth substrate and k is a specific kinetic constant, or by more complex kinetic laws such as the Michaelis–Menten equation.

The Boolean and the ODE approach are deterministic descriptions in which the exact system dynamics can be predicted from the knowledge of the initial state. In contrast to this a stochastic description gives a probability distribution for the succeeding states. The stochastic approach is frequently used to model gene regulatory processes.

Petri nets are another widely used approach for the analysis of biochemical systems. A Petri net is a graphical and mathematical modeling tool for concurrent systems, in which several processes can occur at the same time. Basic elements of a Petri net are places, shown as circles and representing objects (e.g., small molecules or proteins), transitions, shown as bars and representing events (e.g., a molecular reaction), and arcs connecting places with transitions and vice versa. Places can hold tokens indicating the number of objects at a certain state and transitions can fire depending on the number of tokens of the places of incoming arcs leading to a state transition of the system. Many tools are available to explore Petri nets. The Petri Nets World website (http://www.informatik.uni-hamburg.de/TGI/PetriNets) is a good starting point for this. For a further introduction to different modeling techniques we refer to other literature, such as [77, 88].

14.3.1 Standards for Systems Biology

The description of large and heterogeneous data types and subject areas, such as experimental data of expression analysis or a graphical or mathematical model requires standards with controlled vocabularies and semantics. Standards in life sciences can be divided in three different categories, namely content standards defining what should be stored for a particular data type or subject area, syntax standards defining structures for formatting the information (e.g., the extensible markup language, XML, that is frequently used), and semantics standards providing a unified common definition or vocabulary of the data type or subject area [32]. Standards are also very important for the computational work, since software tools usually require data in a well defined machine-readable format.

Different standards have been developed for the definition and implementation of mathematical models of molecular interaction networks in systems biology. Most popular are the Systems Biology Markup Language (SBML) and the Cell Markup Language (CellML) [71, 98]. Moreover, a standard for the exchange of pathway data has been defined by BioPAX [37].

Furthermore, standards for the graphical representation of models has been proposed. Traditionally molecular models of biological systems are described by block-and-arrow diagrams. While useful for the description of hypotheses, such diagrams often lack comprehensive definitions for the description of complex interaction networks with types of interactions and interaction partners. In recent years the systems biology graphical notation (SBGN) was developed [118]. It is a standard graphical notation to foster the reuse of information of cellular interaction networks and it defines comprehensive sets of symbols syntactic rules regarding the construction of interaction maps. SBGN has been accepted as a general notation standard and is used by several computational tools.

14.3.2 Model Databases

In recent years several databases collecting mathematical models of biochemical and molecular processes have been developed. JWS Online is one of the first repositories of kinetic models describing biochemical systems [121]. Besides acting as a model resource, JWS also provides online methods for model simulation and analysis via its website. Other comprehensive model collections are BioModels and the CellML model repository providing models either as SBML or CellML [95, 99, 119].

14.4 Tools for Data Analysis

The thorough analysis of omics data including data collection, processing, normalization, exploration, selection, classification, and interpretation is essential for the integration of such data into cellular network models. Many software tools, ranging from large software packages to small and specialized tools, have been developed to accomplish theses bioinformatic tasks. Table 14.2 summarizes several tools for data analysis that are discussed in the following in more detail.

14.4.1 Topological Analysis

The representation of cellular networks, such as gene regulatory networks, signal transduction networks, protein–protein interaction networks, or metabolic

Table 14.2 General purpose and bioinformatics tools for data analysis and modeling in systems biology

Name	Web-link	Short description
Analysis tools		
Bioconductor	http://www.bioconductor.org/	R packages for bioinformatics
Biopython	http://biopython.org/	Python package for biological computation
SciPy	http://www.scipy.org/	Python package for scientific computing
Matplotlib	http://matplotlib.sourceforge.net/	Python graph library
MATLAB	http://www.mathworks.com/products/matlab/	Programming environment
SystemsBiologyToolbox	http://www.sbtoolbox.org/	Matlab package for systems biology
COBRA	http://opencobra.sourceforge.net/	Matlab package for constraint-based modeling
ManLab	http://manlab.lma.cnrs-mrs.fr/	Matlab package for bifurcation analysis
SensSB	http://www.iim.csic.es/~gingproc/SensSB.html	Matlab package for sensitivity analysis
CellNetAnalyzer	http://www.mpi-magdeburg.mpg.de/projects/cna/cna.html	Matlab package for structural and functional analysis
DBSolve	http://insysbio.ru/en/soft/dbsolveoptimum.html	Model analysis
WebCell	http://www.webcell.org/	Model analysis
BioMet	http://www.sysbio.se/BioMet/	Model analysis
J-Express	http://jexpress.bioinfo.no/site/	Gene expression data analysis
GEDAS	http://sourceforge.net/projects/gedas/	Microarry data analysis
GenePattern	https://genepattern.genome.duke.edu/	Analysis platform
Expander	https://genepattern.genome.duke.edu/	Gene expression data analysis
BioMart	http://www.biomart.org/	Data service
Cytoscape	http://www.cytoscape.org/	Data visualization
Simulation tools		
CellDesigner	http://www.celldesigner.org/	Modeling and simulation
PyBioS	http://pybios.molgen.mpg.de/	Modeling and simulation
Copasi	http://www.copasi.org	Modeling and simulation
Petri nets world	http://www.informatik.uni-hamburg.de/TGI/PetriNets/	Modeling and simulation overview
SBML software guide	http://sbml.org/SBML_Software_Guide	Modeling and simulation overview

networks as graphs makes it possible to investigate the topology and function of these networks using graph-theoretical topological measures. These measures are used to characterize and classify biological networks and to identify certain universal laws governed by networks or subnetworks with similar cellular functions [2, 68, 139, 177, 178, 182].

The exact graphical representation of a cellular network depends on the type of network, e.g., protein–protein interaction, signaling, gene regulatory, or metabolic network [100]. Protein–protein interaction networks are modeled as undirected graphs, where the nodes are proteins and two nodes are connected by an undirected edge if the corresponding proteins physically interact. In contrast, signaling networks are often represented as directed graphs with nodes corresponding to molecules and directed links representing activation or inhibition of another molecule's activity. Similar to signaling networks, gene regulatory networks are commonly represented as directed graphs. A node within such a graph represents a gene and its coded protein simultaneously. A directed link between two nodes indicates a regulatory interaction of the source node protein on the transcription of the target node gene. Furthermore, bipartite graphs are used to represent metabolic networks, whose two types of nodes are metabolites/enzymes and reactions, and two types of edges are representing mass flow and catalytic regulations, respectively. Mass flow edges connect substrates to reactions and reactions to products, whereas regulatory edges correspond to catalytic regulation of the connected reaction by an enzyme node.

Given a network topology, several local and global structural properties can be calculated to characterize the network [10, 101]. Local properties include in- and out-degree, closeness centrality [136], betweenness centrality [50], and clustering coefficient [167]. Global topological characteristics of networks include degree distributions, clustering coefficient distribution, characteristic path length, averaged clustering coefficient, and network diameter (Fig. 14.2). These topological measures can be used to capture cellular network organization, providing new insights into function, stability, dynamic behavior, and evolution. For instance, compared to random networks most biological networks are identified as scale-free (power law degree distribution), small-world (small characteristic length), and hierarchical (high average clustering coefficient and scale-free clustering distribution). Typically, for real world networks the degree exponent of the power law degree distribution ranges between two and three, the characteristic length is proportional to the logarithm of the network size, and the degree exponent of the power law clustering coefficient distribution is around one [10]. Networks with such topology, e.g., protein–protein interaction and gene regulatory networks, capture specific features, such as complexity, stability, robustness against perturbations, and modularity, i.e., occurrence of subnetworks with specific cellular functions [10]. However, considering only the topology has limitations in explaining the entire functional or dynamical behavior of biological networks [7, 74, 179].

Several global analyses of transcriptional regulatory networks in various organisms revealed that small network motifs, i.e., subnetworks with defined characteristics, occur more often in gene regulatory networks than expected by chance [148, 111, 94, 103, 120, 17]. Some motifs show certain dynamical characteristics, e.g., single-input motifs control temporal order of expression [180], feedback loops directly affect robustness [85, 86] and fragility [92], and feed-

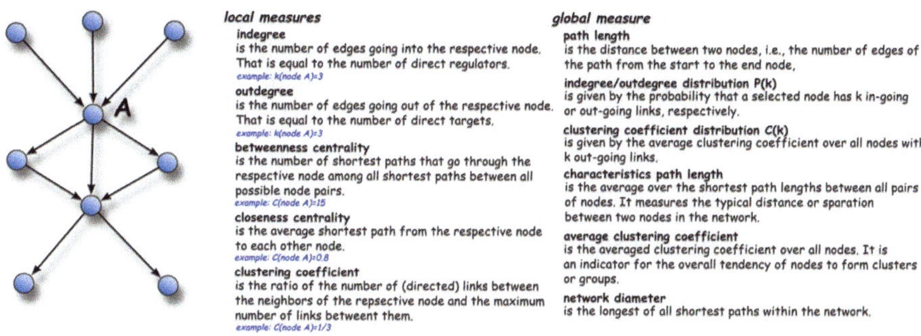

Fig. 14.2 Local and global topological measures. Local measure values are given for the example network from the left panel. See Albert and Barabasi [3, 10, 102] for further details

forward loops provide temporal control, dynamical stability, and signal amplification [103].

Several tools have been developed to calculate various topological measures for a given network structure. For instance, Cytoscape [146, 152] enables the visualization and analyses of complex networks. Its features are extended by several plugins, for instance, NetworkAnalyzer which computes and visualizes several topological parameters including some of the measures mentioned above [4]. Many stand-alone and web-based tools support visualization even of large networks (see [52] for a comprehensive overview of tools). Some of them integrate advanced network analysis functions, such as VisANT which is a web-based application [69] and GraphCrunch 2 which compares properties with random networks [91]. Programming languages can be extended by specialized libraries, such as igraph for Python and R [35], graph-tool for Python, and Boost Graph Library for C++ [151] which support the development of simple as well as complex network analysis algorithms.

14.4.2 Flux-Balance Analysis

An analysis approach for metabolic systems is flux-balance analysis (FBA). It is a mathematical approach to optimize the fluxes of the metabolic system under a number of constraints, such as steady state or homeostasis assumption, thermodynamic consistency, or limitations of certain enzyme reactions. Together with one or more of such constraints, an objective function, e.g., ATP production, nutrient uptake, or total biomass, will be maximized with respect to the flux distribution. FBA does not require metabolite concentrations or enzyme kinetics of the systems for optimization, but it assumes a steady state or homeostasis and it makes use of the stoichiometic matrix of the system (see [84] for a brief history and [93, 123, 133] for more details about FBA). FBA has been applied in many metabolic studies, for instance in order to accurately predict *in silico* the growth rates of *Escherichia coli* under different culture conditions [44], to characterize the optimal flux distributions for maximal ATP production in the mitochondrion [132], or to predict the metabolic steady state and overall lethality after gene knockouts [150].

14.4.3 Sensitivity Analysis

Sensitivity analysis of a biochemical network reveals the behavior of the system, e.g., the steady state or frequency of an oscillation, against various parameter changes [137]. For instance, robustness and fragility of a system, i.e., stability against various parameter perturbations, can be investigated given the model as an ODE system. Usually, the equations are linearly approximated around a fixed point in the parameter space (local sensitivity analysis). Therefore, only small changes around that point are considered. In contrast to local sensitivity analysis, global sensitivity analysis addresses the system's behavior over a wide range of parameter values using statistical models [138]. To capture the dynamical behavior after parameter perturbations a time-varying sensitivity analysis has to be carried out [125, 176].

14.4.4 Metabolic Control Analysis

A classical approach for sensitivity analysis is metabolic control analysis (MCA) [62, 79]. It is a powerful quantitative and qualitative framework for understanding the relationship between steady state properties of a biochemical network and component's concentrations, reaction's fluxes, or model parameters. In contrast to FBA, MCA utilizes not only the stoichiometic matrix but also the detailed kinetic description of the reactions. In order to quantify its behavior, the system is linearized at a fixed point and only small changes are considered. MCA is widely applied in the regulation of metabolic systems, e.g., see [112, 172].

MCA distinguishes several coefficients, namely elasticity, control, and response coefficients, to reflect and quantify local and global effects of changes and perturbations on network properties, e.g., steady state concentrations or reaction fluxes. The elasticity coefficients are local coefficients to quantify the local sensitivity of a reaction rate to changes of a component's concentration or a model parameter, such as enzyme concentration. Other parameters and concentrations in the network are kept fixed. Elasticity coefficients can be calculated in any given state of the system, even in a non-steady state. In contrast, control coefficients and response coefficients are global measures and require a steady state of the entire system. Flux and concentration control coefficients measure the changes of the system's fluxes and concentrations, respectively, in response to a perturbation of a parameter and, hence, a small change of an individual reaction rate. The response coefficient describes the direct dependence of steady state variables on model parameters. Please consult textbooks, such as Klipp et al. [88] for more details.

14.4.5 Bifurcation Analysis

Dynamical systems can show different types of qualitative behavior, such as oscillation or equilibrium, under different parameter values. Transitions between such states, i.e., bifurcations, are studied within bifurcation analysis given the ordinary differential equation system of the model. Critical parameter values where bifurcations occur are identified and help to predict systems behavior under parameter perturbations [16, 28, 63].

Various software tools support model analysis by providing implementations of one or more of the above mentioned methods. For instance MATLAB is widely used in the community due to its rich analysis functionalities and expandability by specialized toolboxes, such as Systems Biology Toolbox (multitude of functions/ methods for analysis and simulation of biological and biochemical systems) [142], COBRA (scripts for constraint-based modeling, including FBA), ManLab (stability and bifurcation analysis), SensSB (local and global sensitivity analysis methods) [135], and CellNetAnalyzer (structural and functional analysis of biochemical networks) [87]. Moreover, several stand-alone or web-based tools are available for model analysis, for instance DBSolve (development and analysis of kinetic models as well as parameter estimation with experimental data) [55], WebCell (web-based and SBML compliant simulation environment which provides analysis techniques, such as MCA), and BioMet Toolbox (web-based resource for stoichiometric analysis and data integration) [36]. Software tools for the modeling and simulation of cellular systems, such as PyBioS [88, 171] and Copasi [67], also often provide powerful collections of tools and algorithms for structural and functional model analysis.

14.4.6 General Purpose Software Packages for Data Analysis

The statistical software package R provides a wide variety of statistical (linear and nonlinear modeling, classical statistical tests, time-series analysis, classification, clustering, and many more) and graphical techniques [131]. R is also a programming language and is extensible through self written functions or a multitude of user-submitted software packages for specific tasks. There are currently almost 4,000 packages available in the R package repository. More R packages appropriate for biological data analyses are provided for instance by the Bioconductor project [53]. It is an open source software for bioinformatics and provides a variety of R packages for the analysis and comprehension of high-throughput genomic data.

Another large software suite is MATLAB. It is a programming environment for algorithm development, data analysis, visualization, and numerical computation. It is used, e.g., in systems biology for mathematical analysis and modeling. MATLAB can be extended by toolboxes, such as Bioinformatics or SimBiology Toolbox to provide algorithms and visualization techniques for next generation sequencing, microarray, and mass spectrometry data analysis and visualization.

Programming languages, such as Python, Perl, and C++, are often used for the development of data analysis functions and tools. For instance, Python together with packages such as Biopython [31], SciPy [76], Pandas, and Matplotlib [73] is commonly used in bioinformatics for instance for data management, statistical data analysis, and graphical visualization.

Besides these large software suites, many specialized tools are widely used in the context of biological data analysis. Some examples are J-Express (a java application with a free of charge license to explore gene expression data equipped with a user-friendly interface) [42] and GEDAS (a gene expression data analysis suite which integrates standardized tools and techniques, such as hierarchical clustering, k-means, SVM, and PCA in one system) [129]. GenePattern is a powerful genomic analysis platform that provides via a web-based interface more than 150 tools for a wide range of studies, including gene expression analysis, pathway analysis, RNA-seq analysis, and data processing tasks [134]. Data tables can be uploaded or retrieved from several data resources and multi-step data analyses are performed on GenePattern's computer cluster. Expander is also a software tool for analysis and visualization of gene expression data [147]. Network-based analyses of such data has been integrated.

Not only comparative genomics but also systems biology requires a consistent handling and annotation of data from different databases and experiments [6]. An accurate mapping between different experimental platforms is crucial for a comprehensive biological interpretation and integration of such data [8]. For instance, BioMart has been developed to make data available over a single interface [58]. It links currently more than 40 databases; among them are Ensembl [48], HGNC [144], Reactome [107], and COSMIC [49]. BioMart enables users to perform advanced querying of biological data sources through biomaRt [41], an R package, or a web-interface.

14.4.7 Data Visualization

Another important aspect in systems biology is data visualization. Besides the generation of standard diagrams as provided by spreadsheet or general purpose software, also the visualization of omics data on the top of network graphs is important for data interpretation and significantly important in systems biology

[52]. Several specialized software tools for this purpose have been developed, such as Cytoscape [146, 152], VANTED [78], VisANT [68], and NAViGaTOR [19].

Cytoscape is a stand-alone application that allows the visualization of diverse types of attribute data on the nodes and edges of a biological network. This mapping of data to visual attributes allows to study multiple types of data in a network context. Moreover, a large number of different plugins allows the user to extend Cytoscape with additional functionalities such as network query and download services, additional network layout algorithms, Gene Ontology (GO) enrichment analysis, or network motif and functional module detection. Cline et al. [29] describe how to obtain gene and protein networks, displaying networks using different network layout algorithms, integrating gene expression or other functional attributes, identifying putative complexes of functional modules or enriched GO annotations within a network.

14.4.8 Reverse Engineering

Data-driven approaches that elucidate network structures from experimental (temporal) observations are classified as reverse engineering approaches. They aim to reveal interactions between cellular entities from high throughput data all at once rather than one interaction at time. To tackle this problem regarding gene regulatory networks, a battery of reverse engineering methods has been developed based on different mathematical approaches [9, 20, 56, 60, 61, 77, 115, 153, 168]. These methods are adapted to different problem domains, such as perturbation and time series analyzes, and have to cope with several difficulties, such as small size, noisy, and incomplete data. The underlying models of each approach are of varying level of detail. They are static or dynamic, continuous or discrete, linear or nonlinear, deterministic or stochastic. Furthermore, they can differ in the information they provide and, thus, have to be interpreted differently. Some methods result in correlation measures (e.g., relevance networks), some calculate conditional independencies (e.g., Bayesian networks), and others infer regulation strengths (e.g., ODE models).

Reverse engineering approaches usually use data from high-throughput gene expression measurements to infer network structures of transcriptional regulation. However, several recent studies have integrated heterogeneous data, such as gene and protein expression data, protein–protein interaction data, and DNA–protein binding data, to identify regulatory interactions. For instance, Belcastro et al. [13] collected heterogeneous data from human and mouse samples to identify both functional and physical interactions among genes; Zhao et al. [181] determined posterior probabilities for all gene-pair interactions based on chromatin immunoprecipitation data and utilized this additional information in combination with gene expression data in gene network reconstruction; Wang and Chen [165] integrated different kinds of omics data and reconstructed a cellular network of *S. cerevisiae* based on coupling dynamic models; and Bauer et al. [12] employed support vector machines trained on known interactions from public databases in order to predict (unknown) regulatory interactions.

Almost every reverse engineering application is adapted and specialized in a certain problem domain and often is not suitable for other applications. Furthermore, a well-performing method might be a result of over-fitting to the test data. Application parameters have to be chosen accurately in order to obtain best learning results. Therefore, a significant amount of time is required not only for data collection and processing but also for adjustment of application parameters. Hence, simple-to-use software tools are rare. A few of them are ARACNE [106], GeneNet [122], and qp-graph [24], among others.

Besides the algorithmic developments, the actual assessment of methods performance remains a challenge, primarily due to the lack of experimental benchmark data (gold-standards), with only a few exceptions, e.g., Cantone et al. [23] built a five genes yeast synthetic network and measured time series and steady state expression data after multiple perturbations. Therefore, simulated data is often used to perform systematic validations showing strength and weaknesses of the individual methods [21, 57].

Several reverse engineering comparison studies have revealed that there is no definitive best-performing method or approach over various problem domains [104, 115]. To foster a concerted effort to address a critical and independent performance assessment of reverse engineering methods the Dialogue on Reverse Engineering Assessment and Methods (DREAM) project has been initiated [130, 155, 156]. Within this project the performances of multitude of reverse engineering methods developed by participating teams are rigorously assessed in an annual community challenge. This challenge, based on anonymized *in silico* and real gene expression data, evaluates the submitted network predictions by several statistical measures. The results revealed that the majority of predictions were equivalent to random and that even the best performing methods show weaknesses [105]. However, it was shown that combining the predictions of all methods improves in some cases predictive power beyond that of any single method. Even if not best-performing the community prediction still ranked under the best methods. Moreover, the community prediction is also robust to inclusion of methods with low performance. It seems that the methods complement each other and compensate their weaknesses [104].

14.5 Tools for Modeling and Simulation

Several tools for the design, simulation, and analysis of mathematical models of biological systems have been developed. CellDesigner is one of the most popular specialized stand-alone tool for systems biology [51]. It is a free and easy-to-use application implemented in Java and hence running on all major operating systems. As a native model format it uses SBML and supports SBGN for model visualization. The graphical user interface of CellDesigner has a comprehensive collection of predefined node icons such as nodes for genes, proteins, receptors etc. and edge types for the definition of reactions that can be used for the design of a model in the main window. It provides also several icons for special reaction types such as transport, catalysis, inhibition, and activation. For reaction definition CellDesigner allows entering arbitrary kinetic equations. Simulations in CellDesigner can be performed using its integrated simulation engine. CellDesigner is a very user-friendly software with a good set-up for model development and simulation.

Another platform-independent and user-friendly software for modeling is COPASI [67] the successor of Gepasi [109, 110]. COPASI offers rich functionalities for model simulation and analysis and it provides methods for stochastic and deterministic simulation. It supports the computation of steady states, the analysis of the stoichiometry of a model, e.g., for the computation of elementary modes [143], metabolic flux analysis and methods for parameter estimation and optimization.

Besides stand-alone tools, like CellDesigner and Copasi, there exist also several web-based tools for modeling and simulation. Web-based tools are easily

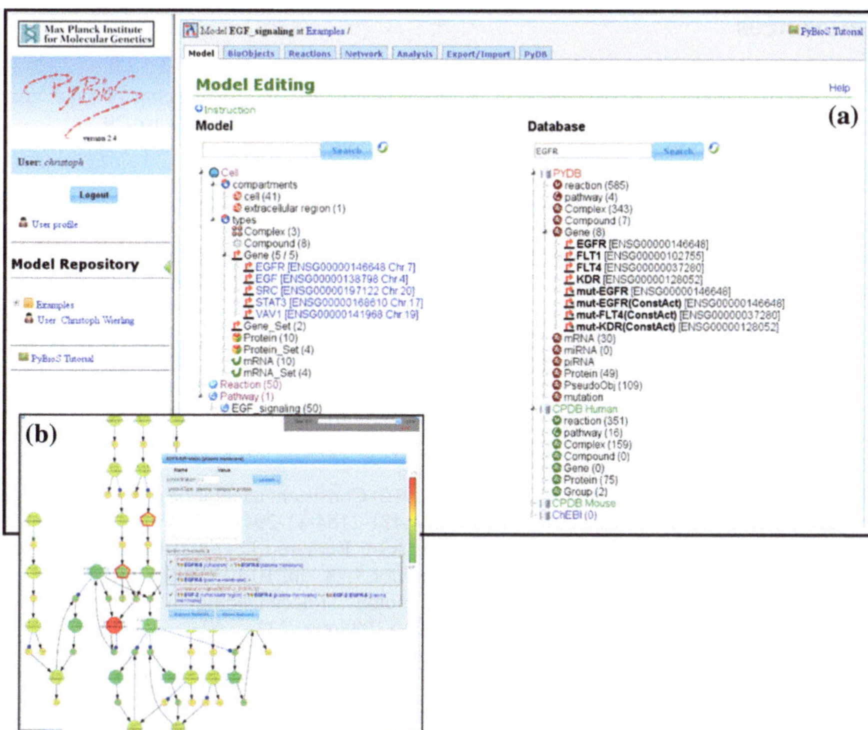

Fig. 14.3 (**a**) PyBioS (http://pybios.molgen.mpg.de) acts as a repository for mathematical models of biological systems. Different tabs of the web interface provide functions for model design, simulation, analysis, and visualization (**b**)

accessible via the web browser without any previous installation. One web-based tool tailored for modeling and simulation is PyBioS (Fig. 14.3). PyBioS is designed for modeling and simulation of cellular and biochemical systems [88, 171]. It offers a high level of automation in model design integrating a number of public interaction resources via ConsensusPathDB [81], an integrated resource of human and mouse interaction information. Models can be simulated as deterministic ordinary differential equation systems (ODEs) and as Petri nets. Simulation results can be visualized and explored in the context of the network interaction graph. Moreover, PyBioS provides functions for network and sensitivity analysis. PyBioS is an SBML-compliant application, supports direct import of models from BioModels and keeps track of model annotation using Ensembl, UniProt and ChEBI as component references.

Ghosh et al. [54] give a comprehensive overview of multiple software tools and integrated platforms for systems biology ranging from tools for data management and analysis and data-driven network-inference to tools for physiological modeling.

References

1. Akao T, Yashiro I, Hosoyama A et al (2011) Whole-genome sequencing of sake Yeast Saccharomyces cerevisiae Kyokai no. 7. DNA Res. doi:10.1093/dnares/dsr029
2. Albert R (2005) Scale-free networks in cell biology. J Cell Sci 118:4947–4957. doi:10.1242/jcs.02714
3. Albert R, Barabási A-L (2002) Statistical mechanics of complex networks. Rev Mod Phys 74:47–97. doi:10.1103/RevModPhys.74.47
4. Assenov Y, Ramírez F, Schelhorn S-E et al (2008) Computing topological parameters of biological networks. Bioinformatics 24:282–284. doi:10.1093/bioinformatics/btm554
5. Bader GD, Cary MP, Sander C (2006) Pathguide: a pathway resource list. Nucleic Acids Res 34:D504–D506. doi:10.1093/nar/gkj126
6. Baker M (2012) Quantitative data: learning to share. Nat Methods 9:39–41. doi:10.1038/nmeth.1815
7. Balaji S, Iyer LM, Aravind L, Babu MM (2006) Uncovering a hidden distributed architecture behind scale-free transcriptional regulatory networks. J Mol Biol 360:204–212. doi:10.1016/j.jmb.2006.04.026
8. Ballester B, Johnson N, Proctor G, Flicek P (2010) Consistent annotation of gene expression arrays. BMC Genomics 11:294. doi:10.1186/1471-2164-11-294
9. Bansal M, Belcastro V, Ambesi-Impiombato A, di Bernardo D (2007) How to infer gene networks from expression profiles. Mol Syst Biol 3:78. doi:10.1038/msb4100120
10. Barabási A-L, Oltvai ZN (2004) Network biology: understanding the cell's functional organization. Nat Rev Genet 5:101–113. doi:10.1038/nrg1272
11. Barretina J, Caponigro G, Stransky N et al (2012) The cancer cell line encyclopedia enables predictive modelling of anticancer drug sensitivity. Nature 483:603–607. doi:10.1038/nature11003
12. Bauer T, Eils R, König R (2011) RIP: the regulatory interaction predictor—a machine learning-based approach for predicting target genes of transcription factors. Bioinformatics 27:2239–2247. doi:10.1093/bioinformatics/btr366

13. Belcastro V, Siciliano V, Gregoretti F et al (2011) Transcriptional gene network inference from a massive dataset elucidates transcriptome organization and gene function. Nucl Acids Res 39:8677–8688. doi:10.1093/nar/gkr593
14. Bentley DR, Balasubramanian S, Swerdlow HP et al (2008) Accurate whole human genome sequencing using reversible terminator chemistry. Nature 456:53–59. doi:10.1038/nature07517
15. Bollard ME, Contel NR, Ebbels TMD et al (2009) NMR-based metabolic profiling identifies biomarkers of liver regeneration following partial hepatectomy in the rat. J Proteome Res 9:59–69. doi:10.1021/pr900200v
16. Borisuk Tyson (1998) Bifurcation analysis of a model of mitotic control in frog eggs. J Theor Biol 195:69–85. doi:10.1006/jtbi.1998.0781
17. Boyer LA, Lee TI, Cole MF, Johnstone SE, Levine SS, Zucker JP, Guenther MG, Kumar RM, Murray HL, Jenner RG, Gifford DK, Melton DA, Jaenisch R & Young RA (2005) Core transcriptional regulatory circuitry in human embryonic stem cells. Cell 122:947–956
18. Boyle AP, Davis S, Shulha HP et al (2008) High-resolution mapping and characterization of open chromatin across the genome. Cell 132:311–322. doi:10.1016/j.cell.2007.12.014
19. Brown KR, Otasek D, Ali M et al (2009) NAViGaTOR: network analysis, visualization and graphing Toronto. Bioinformatics 25:3327–3329. doi:10.1093/bioinformatics/btp595
20. Camacho D, Licona PV, Mendes P, Laubenbacher R (2007) Comparison of reverse-engineering methods using an in silico network. Ann N Y Acad Sci 1115:73–89. doi:10.1196/annals.1407.006
21. Di Camillo B, Toffolo G, Cobelli C (2009) A gene network simulator to assess reverse engineering algorithms. Ann N Y Acad Sci 1158:125–142. doi:10.1111/j.1749-6632.2008.03756.x
22. Cancer Genome Atlas Network (2012) Comprehensive molecular characterization of human colon and rectal cancer. Nature 487:330–337. doi:10.1038/nature11252
23. Cantone I, Marucci L, Iorio F et al (2009) A yeast synthetic network for in vivo assessment of reverse-engineering and modeling approaches. Cell 137:172–181. doi:10.1016/j.cell.2009.01.055
24. Castelo R, Roverato A (2009) Reverse engineering molecular regulatory networks from microarray data with qp-graphs. J Comput Biol 16:213–227. doi:10.1089/cmb.2008.08TT
25. Cerami EG, Gross BE, Demir E et al (2011) Pathway commons, a web resource for biological pathway data. Nucleic Acids Res 39:D685–D690. doi:10.1093/nar/gkq1039
26. Chan ECY, Koh PK, Mal M et al (2008) Metabolic profiling of human colorectal cancer using high-resolution magic angle spinning nuclear magnetic resonance (HR-MAS NMR) spectroscopy and gas chromatography mass spectrometry (GC/MS). J Proteome Res 8:352–361. doi:10.1021/pr8006232
27. Chaouiya C, Naldi A, Thieffry D (2012) Logical modelling of gene regulatory networks with GINsim. Methods Mol Biol 804:463–479. doi:10.1007/978-1-61779-361-5_23
28. Chen KC, Csikasz-Nagy A, Gyorffy B et al (2000) Kinetic analysis of a molecular model of the budding yeast cell cycle. Mol Biol Cell 11:369–391
29. Cline MS, Smoot M, Cerami E et al (2007) Integration of biological networks and gene expression data using Cytoscape. Nat Protoc 2:2366–2382. doi:10.1038/nprot.2007.324
30. Cloonan N, Forrest ARR, Kolle G et al (2008) Stem cell transcriptome profiling via massive-scale mRNA sequencing. Nat Methods 5:613–619. doi:10.1038/nmeth.1223
31. Cock PJA, Antao T, Chang JT et al (2009) Biopython: freely available Python tools for computational molecular biology and bioinformatics. Bioinformatics 25:1422–1423. doi:10.1093/bioinformatics/btp163
32. Courtot M, Juty N, Knüpfer C et al (2011) Controlled vocabularies and semantics in systems biology. Mol Syst Biol 7:543. doi:10.1038/msb.2011.77
33. Cox J, Mann M (2011) Quantitative, high-resolution proteomics for data-driven systems biology. Annu Rev Biochem 80:273–299. doi:10.1146/annurev-biochem-061308-093216
34. Croft D, O'Kelly G, Wu G et al (2011) Reactome: a database of reactions, pathways and biological processes. Nucleic Acids Res 39:D691–D697. doi:10.1093/nar/gkq1018

35. Csardi G, Nepusz T (2006) The igraph software package for complex network research. Int J Complex Syst 1695
36. Cvijovic M, Olivares-Hernandez R, Agren R et al (2010) BioMet Toolbox: genome-wide analysis of metabolism. Nucleic Acids Res 38:W144–W149. doi:10.1093/nar/gkq404
37. Demir E, Cary MP, Paley S et al (2010) The BioPAX community standard for pathway data sharing. Nat Biotechnol 28:935–942. doi:10.1038/nbt.1666
38. Deutsch EW, Lam H, Aebersold R (2008) PeptideAtlas: a resource for target selection for emerging targeted proteomics workflows. EMBO Rep 9:429–434. doi:10.1038/embor.2008.56
39. Dieterle F, Riefke B, Schlotterbeck G, et al. (2011) NMR and MS methods for metabonomics. In: Gautier J-C, Walker JM (eds) Drug safety evaluation. Humana Press, New York, pp 385–415
40. Domon B, Aebersold R (2010) Options and considerations when selecting a quantitative proteomics strategy. Nat Biotechnol 28:710–721. doi:10.1038/nbt.1661
41. Durinck S, Moreau Y, Kasprzyk A et al (2005) BioMart and bioconductor: a powerful link between biological databases and microarray data analysis. Bioinformatics 21:3439–3440. doi:10.1093/bioinformatics/bti525
42. Dysvik B, Jonassen I (2001) J-Express: exploring gene expression data using Java. Bioinformatics 17:369–370. doi:10.1093/bioinformatics/17.4.369
43. Edgar R, Domrachev M, Lash AE (2002) Gene Expression Omnibus: NCBI gene expression and hybridization array data repository. Nucl Acids Res 30:207–210. doi:10.1093/nar/30.1.207
44. Edwards JS, Ibarra RU, Palsson BO (2001) In silico predictions of Escherichia coli metabolic capabilities are consistent with experimental data. Nat Biotechnol 19:125. doi:10.1038/84379
45. Eisen MB, Spellman PT, Brown PO, Botstein D (1998) Cluster analysis and display of genome-wide expression patterns. Proc Natl Acad Sci U S A 95:14863–14868
46. Fauré A, Naldi A, Chaouiya C, Thieffry D (2006) Dynamical analysis of a generic Boolean model for the control of the mammalian cell cycle. Bioinformatics 22:e124–e131. doi:10.1093/bioinformatics/btl210
47. Finger JH, Smith CM, Hayamizu TF et al (2010) The mouse Gene Expression Database (GXD): 2011 update. Nucl Acids Res. doi:10.1093/nar/gkq1132
48. Flicek P, Amode MR, Barrell D et al (2011) Ensembl 2012. Nucleic Acids Res 40:D84–D90. doi:10.1093/nar/gkr991
49. Forbes SA, Bindal N, Bamford S et al (2010) COSMIC: mining complete cancer genomes in the catalogue of somatic mutations in cancer. Nucleic Acids Res 39:D945–D950. doi:10.1093/nar/gkq929
50. Freeman L (1977) A set of measures of centrality based on betweenness. Sociometry 40:35–41
51. Funahashi A, Matsuoka Y, Jouraku A et al (2008) Cell designer 3.5: a versatile modeling tool for biochemical networks. Proc IEEE 96:1254–1265. doi:10.1109/JPROC.2008.925458
52. Gehlenborg N, O'Donoghue SI, Baliga NS et al (2010) Visualization of omics data for systems biology. Nat Methods 7:S56–S68. doi:10.1038/nmeth.1436
53. Gentleman RC, Carey VJ, Bates DM et al (2004) Bioconductor: open software development for computational biology and bioinformatics. Genome Biol 5:R80. doi:10.1186/gb-2004-5-10-r80
54. Ghosh S, Matsuoka Y, Asai Y et al (2011) Software for systems biology: from tools to integrated platforms. Nat Rev Genet 12:821–832. doi:10.1038/nrg3096
55. Gizzatkulov NM, Goryanin II, Metelkin EA et al (2010) DBSolve Optimum: a software package for kinetic modeling which allows dynamic visualization of simulation results. BMC Syst Biol 4:109. doi:10.1186/1752-0509-4-109
56. Hache H, Lehrach H, Herwig R (2009) Reverse engineering of gene regulatory networks: a comparative study. EURASIP J Bioinf Syst Biol 2009:617281. doi:10.1155/2009/617281

57. Hache H, Wierling C, Lehrach H, Herwig R (2009) GeNGe: systematic generation of gene regulatory networks. Bioinformatics 25:1205–1207. doi:10.1093/bioinformatics/btp115
58. Haider S, Ballester B, Smedley D et al (2009) BioMart Central Portal—unified access to biological data. Nucleic Acids Res 37:W23–W27
59. Hammerman PS, Hayes DN, Wilkerson MD et al (2012) Comprehensive genomic characterization of squamous cell lung cancers. Nature 489:519–525. doi:10.1038/nature11404
60. He F, Balling R, Zeng A-P (2009) Reverse engineering and verification of gene networks: principles, assumptions, and limitations of present methods and future perspectives. J Biotechnol 144:190–203. doi:10.1016/j.jbiotec.2009.07.013
61. Hecker M, Lambeck S, Toepfer S et al (2009) Gene regulatory network inference: data integration in dynamic models-a review. BioSystems 96:86–103. doi:10.1016/j.biosystems.2008.12.004
62. Heinrich R, Rapoport TA (1974) A linear steady-state treatment of enzymatic chains. Eur J Biochem 42:89–95. doi:10.1111/j.1432-1033.1974.tb03318.x
63. Higham CF (2009) Bifurcation analysis informs Bayesian inference in the Hes1 feedback loop. BMC Syst Biol 3:12. doi:10.1186/1752-0509-3-12
64. Hillier LW, Marth GT, Quinlan AR et al (2008) Whole-genome sequencing and variant discovery in C. elegans. Nat Methods 5:183–188. doi:10.1038/nmeth.1179
65. Ho DWY, Yang ZF, Yi K et al (2012) Gene expression profiling of liver cancer stem cells by RNA-sequencing. PLoS ONE 7:e37159. doi:10.1371/journal.pone.0037159
66. Holmes E, Loo RL, Stamler J et al (2008) Human metabolic phenotype diversity and its association with diet and blood pressure. Nature 453:396–400. doi:10.1038/nature06882
67. Hoops S, Sahle S, Gauges R et al (2006) COPASI–a COmplex PAthway SImulator. Bioinformatics 22:3067–3074. doi:10.1093/bioinformatics/btl485
68. Hu Z, Hung J-H, Wang Y et al (2009) VisANT 3.5: multi-scale network visualization, analysis and inference based on the gene ontology. Nucleic Acids Res 37:W115–W121. doi:10.1093/nar/gkp406
69. Hu Z, Mellor J, Wu J et al (2005) VisANT: data-integrating visual framework for biological networks and modules. Nucleic Acids Res 33:W352–W357. doi:10.1093/nar/gki431
70. Hubble J, Demeter J, Jin H et al (2009) Implementation of genepattern within the Stanford microarray database. Nucleic Acids Res 37:D898–D901. doi:10.1093/nar/gkn786
71. Hucka M, Finney A, Sauro HM et al (2003) The systems biology markup language (SBML): a medium for representation and exchange of biochemical network models. Bioinformatics 19:524–531
72. Hughes TR, Marton MJ, Jones AR et al (2000) Functional discovery via a compendium of expression profiles. Cell 102:109–126
73. Hunter J (2007) Matplotlib: a 2D Graphics Environment. Comput Sci Eng 9:90–95
74. Ingram PJ, Stumpf MP, Stark J (2006) Network motifs: structure does not determine function. BMC Genomics 7:108. doi:10.1186/1471-2164-7-108
75. Jiang C, Xuan Z, Zhao F, Zhang MQ (2007) TRED: a transcriptional regulatory element database, new entries and other development. Nucleic Acids Res 35:D137–D140. doi:10.1093/nar/gkl1041
76. Jones E, Oliphant T, Peterson P (2001) SciPy: open source scientific tools for Python. In: http://www.scipy.org/. http://www.scipy.org/Citing_SciPy. Accessed 6 Aug 2012
77. de Jong H (2002) Modeling and simulation of genetic regulatory systems: a literature review. J Comput Biol 9:67–103. doi:10.1089/10665270252833208
78. Junker BH, Klukas C, Schreiber F (2006) VANTED: a system for advanced data analysis and visualization in the context of biological networks. BMC Bioinformatics 7:109. doi:10.1186/1471-2105-7-109
79. Kacser H, Burns JA (1973) The control of flux. Symp Soc Exp Biol 27:65–104
80. Kamburov A, Pentchev K, Galicka H et al (2011) ConsensusPathDB: toward a more complete picture of cell biology. Nucleic Acids Res 39:D712–D717. doi:10.1093/nar/gkq1156

81. Kamburov A, Wierling C, Lehrach H, Herwig R (2009) ConsensusPathDB–a database for integrating human functional interaction networks. Nucleic Acids Res 37:D623–D628. doi:10.1093/nar/gkn698
82. Kanehisa M, Goto S (2000) KEGG: Kyoto Encyclopedia of Genes and Genomes. Nucl Acids Res 28:27–30. doi:10.1093/nar/28.1.27
83. Kanehisa M, Goto S, Sato Y et al (2012) KEGG for integration and interpretation of large-scale molecular data sets. Nucleic Acids Res 40:D109–D114. doi:10.1093/nar/gkr988
84. Kauffman KJ, Prakash P, Edwards JS (2003) Advances in flux balance analysis. Curr Opin Biotechnol 14:491–496. doi:10.1016/j.copbio.2003.08.001
85. Kitano H (2004) Biological robustness. Nat Rev Genet 5:826–837. doi:10.1038/nrg1471
86. Kitano H (2004) Cancer as a robust system: implications for anticancer therapy. Nat Rev Cancer 4:227–235. doi:10.1038/nrc1300
87. Klamt S, Saez-Rodriguez J, Gilles ED (2007) Structural and functional analysis of cellular networks with Cell NetAnalyzer. BMC Syst Biol 1:2. doi:10.1186/1752-0509-1-2
88. Klipp E, Liebermeister W, Wierling C et al (2009) Systems biology: a textbook. Wiley-VCH, Weinheim
89. Koboldt DC, Fulton RS, McLellan MD et al (2012) Comprehensive molecular portraits of human breast tumours. Nature. doi:10.1038/nature11412
90. Kolker E, Higdon R, Haynes W et al (2011) MOPED: model organism protein expression database. Nucleic Acids Res 40:D1093–D1099. doi:10.1093/nar/gkr1177
91. Kuchaiev O, Stevanović A, Hayes W, Pržulj N (2011) GraphCrunch 2: software tool for network modeling, alignment and clustering. BMC Bioinform 12:24. doi:10.1186/1471-2105-12-24
92. Kwon Y-K, Cho K-H (2008) Quantitative analysis of robustness and fragility in biological networks based on feedback dynamics. Bioinformatics 24:987–994. doi:10.1093/bioinformatics/btn060
93. Lee JM, Gianchandani EP, Papin JA (2006) Flux balance analysis in the era of metabolomics. Brief Bioinform 7:140–150. doi:10.1093/bib/bbl007
94. Lee TI, Rinaldi NJ, Robert F, Odom DT, Bar-Joseph Z, Gerber GK, Hannett NM, Harbison CT, Thompson CM, Simon I, Zeitlinger J, Jennings EG, Murray HL, Gordon DB, Ren B, Wyrick JJ, Tagne J-B, Volkert TL, Fraenkel E, Gifford DK, et al (2002) Transcriptional regulatory networks in Saccharomyces cerevisiae. Science 298:799–804
95. Li C, Donizelli M, Rodriguez N et al (2010) BioModels database: an enhanced, curated and annotated resource for published quantitative kinetic models. BMC Syst Biol 4:92. doi:10.1186/1752-0509-4-92
96. Li N, Ye M, Li Y et al (2010) Whole genome DNA methylation analysis based on high throughput sequencing technology. Methods 52:203–212. doi:10.1016/j.ymeth.2010.04.009
97. Lipson D, Raz T, Kieu A et al (2009) Quantification of the yeast transcriptome by single-molecule sequencing. Nat Biotechnol 27:652–658. doi:10.1038/nbt.1551
98. Lloyd CM, Halstead MDB, Nielsen PF (2004) CellML: its future, present and past. Prog Biophys Mol Biol 85:433–450. doi:10.1016/j.pbiomolbio.2004.01.004
99. Lloyd CM, Lawson JR, Hunter PJ, Nielsen PF (2008) The CellML model repository. Bioinformatics 24:2122–2123. doi:10.1093/bioinformatics/btn390
100. Ma'ayan A (2008) Network integration and graph analysis in mammalian molecular systems biology. Systems Biology, IET 2:206–221. doi:10.1049/iet-syb:20070075
101. Ma'ayan A (2011) Introduction to Network Analysis in Systems Biology. Sci Signal 4:tr5. doi:10.1126/scisignal.2001965
102. Ma'ayan A (2009) Insights into the organization of biochemical regulatory networks using graph theory analyses. J Biol Chem 284:5451–5455. doi:10.1074/jbc.R800056200
103. Mangan S, Alon U (2003) Structure and function of the feed-forward loop network motif. Proc Natl Acad Sci U S A 100:11980–11985. doi:10.1073/pnas.2133841100
104. Marbach D, Costello JC, Küffner R et al (2012) Wisdom of crowds for robust gene network inference. Nat Methods 9:796–804. doi:10.1038/nmeth.2016

105. Marbach D, Prill RJ, Schaffter T et al (2010) Revealing strengths and weaknesses of methods for gene network inference. Proc Natl Acad Sci U S A 107:6286–6291. doi:10.1073/pnas.0913357107
106. Margolin AA, Wang K, Lim WK et al (2006) Reverse engineering cellular networks. Nat Protoc 1:662–671. doi:10.1038/nprot.2006.106
107. Matthews L, Gopinath G, Gillespie M et al (2009) Reactome knowledgebase of human biological pathways and processes. Nucleic Acids Res 37:D619–D622. doi:10.1093/nar/gkn863
108. Matys V, Kel-Margoulis OV, Fricke E et al (2006) TRANSFAC and its module TRANSCompel: transcriptional gene regulation in eukaryotes. Nucleic Acids Res 34:D108–D110. doi:10.1093/nar/gkj143
109. Mendes P (1993) GEPASI: a software package for modelling the dynamics, steady states and control of biochemical and other systems. Comput Appl Biosci 9:563–571
110. Mendes P (1997) Biochemistry by numbers: simulation of biochemical pathways with Gepasi 3. Trends Biochem Sci 22:361–363
111. Milo R, Shen-Orr S, Itzkovitz S, Kashtan N, Chklovskii D & Alon U (2002) Network motifs: simple building blocks of complex networks. Science 298:824–827
112. Moreno-Sánchez R, Saavedra E, Rodríguez-Enríquez S, Olín-Sandoval V (2008) Metabolic control analysis: a tool for designing strategies to manipulate metabolic pathways. J Biomed Biotechnol 2008:1–31. doi:10.1155/2008/597913
113. Nagalakshmi U, Wang Z, Waern K et al (2008) The transcriptional landscape of the yeast genome defined by RNA sequencing. Science 320:1344–1349. doi:10.1126/science.1158441
114. Nagaraj N, Wisniewski JR, Geiger T et al (2011) Deep proteome and transcriptome mapping of a human cancer cell line. Molecular Systems Biology. doi:10.1038/msb.2011.81
115. Narendra V, Lytkin NI, Aliferis CF, Statnikov A (2011) A comprehensive assessment of methods for de-novo reverse-engineering of genome-scale regulatory networks. Genomics 97:7–18. doi:10.1016/j.ygeno.2010.10.003
116. Nicholson JK, Lindon JC (2008) Systems biology: metabonomics. Nature 455:1054–1056. doi:10.1038/4551054a
117. Nilsson T, Mann M, Aebersold R et al (2010) Mass spectrometry in high-throughput proteomics: ready for the big time. Nat Methods 7:681–685. doi:10.1038/nmeth0910-681
118. Le Novère N, Hucka M, Mi H et al (2009) The systems biology graphical notation. Nat Biotechnol 27:735–741. doi:10.1038/nbt.1558
119. Le Novère N, Bornstein B, Broicher A et al (2006) BioModels Database: a free, centralized database of curated, published, quantitative kinetic models of biochemical and cellular systems. Nucleic Acids Res 34:D689–D691. doi:10.1093/nar/gkj092
120. Odom DT, Zizlsperger N, Gordon DB, Bell GW, Rinaldi NJ, Murray HL, Volkert TL, Schreiber J, Rolfe PA, Gifford DK, Fraenkel E, Bell GI & Young RA (2004) Control of pancreas and liver gene expression by HNF transcription factors. Science 303:1378–1381
121. Olivier BG, Snoep JL (2004) Web-based kinetic modelling using JWS Online. Bioinformatics 20:2143–2144. doi:10.1093/bioinformatics/bth200
122. Opgen-Rhein R, Strimmer K (2007) From correlation to causation networks: a simple approximate learning algorithm and its application to high-dimensional plant gene expression data. BMC Syst Biol 1:37. doi:10.1186/1752-0509-1-37
123. Orth JD, Thiele I, Palsson BØ (2010) What is flux balance analysis? Nat Biotechnol 28:245. doi:10.1038/nbt.1614
124. Parkinson H, Sarkans U, Kolesnikov N et al (2010) ArrayExpress update–an archive of microarray and high-throughput sequencing-based functional genomics experiments. Nucleic Acids Res 39:D1002–D1004. doi:10.1093/nar/gkq1040
125. Perumal TM, Gunawan R (2011) Understanding dynamics using sensitivity analysis: caveat and solution. BMC Syst Biol 5:41. doi:10.1186/1752-0509-5-41
126. Picotti P, Bodenmiller B, Mueller LN et al (2009) Full dynamic range proteome analysis of S. cerevisiae by targeted proteomics. Cell 138:795–806. doi:10.1016/j.cell.2009.05.051

127. Pleasance ED, Cheetham RK, Stephens PJ et al (2009) A comprehensive catalogue of somatic mutations from a human cancer genome. Nature 463:191–196. doi:10.1038/nature08658
128. Pontén F, Gry M, Fagerberg L et al (2009) A global view of protein expression in human cells, tissues, and organs. Mol Syst Biol. doi:10.1038/msb.2009.93
129. Prasad TV, Babu RP, Ahson SI (2006) GEDAS—gene expression data analysis suite. Bioinformation 1:83–85
130. Prill RJ, Marbach D, Saez-Rodriguez J et al (2010) Towards a rigorous assessment of systems biology models: the DREAM3 challenges. PLoS ONE 5:e9202. doi:10.1371/journal.pone.0009202
131. R Development Core Team (2012) R: a language and environment for statistical computing
132. Ramakrishna R, Edwards JS, McCulloch A, Palsson BO (2001) Flux-balance analysis of mitochondrial energy metabolism: consequences of systemic stoichiometric constraints. Am J Physiol Regul Integr Comp Physiol 280:R695–R704
133. Raman K, Chandra N (2009) Flux balance analysis of biological systems: applications and challenges. Brief Bioinform 10:435–449. doi:10.1093/bib/bbp011
134. Reich M, Liefeld T, Gould J et al (2006) GenePattern 2.0. Nat Genet 38:500–501. doi:10.1038/ng0506-500
135. Rodriguez-Fernandez M, Banga JR (2010) SensSB: a software toolbox for the development and sensitivity analysis of systems biology models. Bioinformatics 26:1675–1676. doi:10.1093/bioinformatics/btq242
136. Sabidussi G (1966) The centrality index of a graph. Psychometrika 31:581–603. doi:10.1007/BF02289527
137. Saltelli A, Chan K, Scott E (2000) Sensitivity analysis. Wiley, Chichester
138. Saltelli A, Ratto M, Andres T et al (2008) Global sensitivity analysis: the primer. Wiley, Chichester
139. Saraç ÖS, Pancaldi V, Bähler J, Beyer A (2012) Topology of functional networks predicts physical binding of proteins. Bioinformatics. doi:10.1093/bioinformatics/bts351
140. Scheer M, Grote A, Chang A et al (2011) BRENDA, the enzyme information system in 2011. Nucleic Acids Res 39:D670–D676. doi:10.1093/nar/gkq1089
141. Schena M, Shalon D, Davis RW, Brown PO (1995) Quantitative monitoring of gene expression patterns with a complementary DNA microarray. Science 270:467–740
142. Schmidt H, Jirstrand M (2006) Systems Biology Toolbox for MATLAB: a computational platform for research in systems biology. Bioinformatics 22:514–515. doi:10.1093/bioinformatics/bti799
143. Schuster S, Dandekar T, Fell DA (1999) Detection of elementary flux modes in biochemical networks: a promising tool for pathway analysis and metabolic engineering. Trends Biotechnol 17:53–60
144. Seal RL, Gordon SM, Lush MJ et al (2011) genenames.org: the HGNC resources in 2011. Nucleic Acids Res 39:D514–D519. doi:10.1093/nar/gkq892
145. Shaham O, Wei R, Wang TJ et al (2008) Metabolic profiling of the human response to a glucose challenge reveals distinct axes of insulin sensitivity. Molecular Systems Biology. doi:10.1038/msb.2008.50
146. Shannon P, Markiel A, Ozier O et al (2003) Cytoscape: a software environment for integrated models of biomolecular interaction networks. Genome Res 13:2498–2504. doi:10.1101/gr.1239303
147. Sharan R, Maron-Katz A, Shamir R (2003) CLICK and EXPANDER: a system for clustering and visualizing gene expression data. Bioinformatics 19:1787–1799. doi:10.1093/bioinformatics/btg232
148. Shen-Orr SS, Milo R, Mangan S & Alon U (2002) Network motifs in the transcriptional regulation network of Escherichia coli. Nat Genet 31:64–68
149. Shin G, Kang T-W, Yang S et al (2011) GENT: gene expression database of normal and tumor tissues. Cancer Inform 10:149–157. doi:10.4137/CIN.S7226

150. Shlomi T, Berkman O, Ruppin E (2005) Regulatory on/off minimization of metabolic flux changes after genetic perturbations. PNAS 102:7695–7700. doi:10.1073/pnas.0406346102
151. Siek J, Lee L-Q, Lumsdaine A (2001) The boost graph library: user guide and reference manual (C++ In-Depth Series). Addison-Wesley Professional
152. Smoot ME, Ono K, Ruscheinski J et al (2011) Cytoscape 2.8: new features for data integration and network visualization. Bioinformatics 27:431–432. doi:10.1093/bioinformatics/btq675
153. Soranzo N, Bianconi G, Altafini C (2007) Comparing association network algorithms for reverse engineering of large-scale gene regulatory networks: synthetic versus real data. Bioinformatics 23:1640–1647. doi:10.1093/bioinformatics/btm163
154. Spellman PT, Sherlock G, Zhang MQ et al (1998) Comprehensive identification of cell cycle-regulated genes of the yeast Saccharomyces cerevisiae by microarray hybridization. Mol Biol Cell 9:3273–3297
155. Stolovitzky G, Monroe D, Califano A (2007) Dialogue on reverse-engineering assessment and methods: the DREAM of high-throughput pathway inference. Ann N Y Acad Sci 1115:1–22. doi:10.1196/annals.1407.021
156. Stolovitzky G, Prill RJ, Califano A (2009) Lessons from the DREAM2 Challenges. Ann N Y Acad Sci 1158:159–195. doi:10.1111/j.1749-6632.2009.04497.x
157. Sugimoto M, Ikeda S, Niigata K et al (2011) MMMDB: mouse multiple tissue metabolome database. Nucleic Acids Res 40:D809–D814. doi:10.1093/nar/gkr1170
158. Sultan M, Schulz MH, Richard H et al (2008) A global view of gene activity and alternative splicing by deep sequencing of the human transcriptome. Science 321:956–960. doi:10.1126/science.1160342
159. Tanabe M, Kanehisa M (2012) Using the KEGG Database Resource. Curr Protoc Bioinformatics Chapter 1: Unit1.12. doi:10.1002/0471250953.bi0112s38
160. Uhlen M, Oksvold P, Fagerberg L et al (2010) Towards a knowledge-based Human Protein Atlas. Nat Biotechnol 28:1248–1250. doi:10.1038/nbt1210-1248
161. Visco C, Li Y, Xu-Monette ZY et al (2012) Comprehensive gene expression profiling and immunohistochemical studies support application of immunophenotypic algorithm for molecular subtype classification in diffuse large B-cell lymphoma: a report from the International DLBCL Rituximab-CHOP Consortium Program Study. Leukemia. doi:10.1038/leu.2012.83
162. Visel A, Blow MJ, Li Z et al (2009) ChIP-seq accurately predicts tissue-specific activity of enhancers. Nature 457:854–858. doi:10.1038/nature07730
163. Wang J, Wang W, Li R et al (2008) The diploid genome sequence of an Asian individual. Nature 456:60–65. doi:10.1038/nature07484
164. Wang M, Weiss M, Simonovic M et al (2012) PaxDb, a database of protein abundance averages across all three domains of life. Mol Cell Proteomics. doi:10.1074/mcp.O111.014704
165. Wang Y-C, Chen B-S (2010) Integrated cellular network of transcription regulations and protein–protein interactions. BMC Syst Biol 4:20. doi:10.1186/1752-0509-4-20
166. Wang Z, Gerstein M, Snyder M (2009) RNA-Seq: a revolutionary tool for transcriptomics. Nat Rev Genet 10:57–63. doi:10.1038/nrg2484
167. Watts DJ, Strogatz SH (1998) Collective dynamics of "small-world" networks. Nature 393:440–442. doi:10.1038/30918
168. Werhli AV, Grzegorczyk M, Husmeier D (2006) Comparative evaluation of reverse engineering gene regulatory networks with relevance networks, graphical Gaussian models and Bayesian networks. Bioinformatics 22:2523–2531. doi:10.1093/bioinformatics/btl391
169. Wheeler DA, Srinivasan M, Egholm M et al (2008) The complete genome of an individual by massively parallel DNA sequencing. Nature 452:872–876. doi:10.1038/nature06884
170. Whitfield ML, Sherlock G, Saldanha AJ et al (2002) Identification of genes periodically expressed in the human cell cycle and their expression in tumors. Mol Biol Cell 13:1977–2000. doi:10.1091/mbc.02-02-0030

171. Wierling C, Herwig R, Lehrach H (2007) Resources, standards and tools for systems biology. Brief Funct Genomic Proteomic 6:240–251. doi:10.1093/bfgp/elm027
167. Wildermuth MC (2000) Metabolic control analysis: biological applications and insights. Genome Biol 1: reviews1031. doi:10.1186/gb-2000-1-6-reviews1031
173. Wilhelm BT, Marguerat S, Watt S et al (2008) Dynamic repertoire of a eukaryotic transcriptome surveyed at single-nucleotide resolution. Nature 453:1239–1243. doi:10.1038/nature07002
174. Wishart DS, Knox C, Guo AC et al (2009) HMDB: a knowledgebase for the human metabolome. Nucleic Acids Res 37:D603–D610. doi:10.1093/nar/gkn810
175. Wittig U, Kania R, Golebiewski M et al (2012) SABIO-RK–database for biochemical reaction kinetics. Nucleic Acids Res 40:D790–D796. doi:10.1093/nar/gkr1046
176. Wu WH, Wang FS, Chang MS (2008) Dynamic sensitivity analysis of biological systems. BMC Bioinformatics 9:S17. doi:10.1186/1471-2105-9-S12-S17
177. Wunderlich Z, Mirny LA (2006) Using the topology of metabolic networks to predict viability of mutant strains. Biophys J 91:2304–2311. doi:10.1529/biophysj.105.080572
178. Xu J, Li Y (2006) Discovering disease-genes by topological features in human protein–protein interaction network. Bioinformatics 22:2800–2805. doi:10.1093/bioinformatics/btl467
179. Yu H, Greenbaum D, Xin LuH et al (2004) Genomic analysis of essentiality within protein networks. Trends Genet 20:227–231. doi:10.1016/j.tig.2004.04.008
180. Zaslaver A, Mayo AE, Rosenberg R et al (2004) Just-in-time transcription program in metabolic pathways. Nat Genet 36:486–491. doi:10.1038/ng1348
181. Zhao W, Serpedin E, Dougherty ER (2008) Recovering Genetic Regulatory Networks from Chromatin Immunoprecipitation and Steady-State Microarray Data. EURASIP J Bioinf Syst Biol 2008:248747. doi:10.1155/2008/248747
182. Zhu M, Gao L, Li X et al (2009) The analysis of the drug-targets based on the topological properties in the human protein–protein interaction network. J Drug Target 17:524–532. doi:10.1080/10611860903046610

Chapter 15
Agent-Based Modeling Approaches to Multi-Scale Systems Biology: An Example Agent-Based Model of Acute Pulmonary Inflammation

Gary An, Michael Wandling and Scott Christley

Abstract Implicit in systems biology is the concept that the whole is greater than the sum of its parts. Agent-based modeling, an object-oriented, discrete event, population-based computational modeling method, is well suited to meeting this goal. By viewing systems as aggregates of populations of interacting components, agent-based models (ABMs) map well to biological conceptual models and present an intuitive means by which biomedical researchers can represent their knowledge in a dynamic computational form. ABMs are particularly suited for representing the behaviour of populations of cells (i.e. "cell-as-agents"), but ABMs have also been used to model molecular interactions, particularly when spatial and structural properties are involved. Presented herein are a series of ABMs of biomedical systems that cross multiple scales of biological organization, as well as a detailed description of an example ABM of acute pulmonary inflammation. Because of these characteristics agent-based modeling is a useful addition to the suite of equation-based mathematical modeling methods found in systems biology, and can serve as an integrating framework for dynamic knowledge representation of biological systems.

Keywords Inflammation · Agent-based modeling · Translational systems biology · Complex systems analysis

G. An (✉) · S. Christley
Department of Surgery, University of Chicago, Chicago, USA
e-mail: docgca@gmail.com

S. Christley
e-mail: schristley@uchicago.edu

M. Wandling
Department of Surgery, Northwestern University, Chicago, USA
e-mail: m-wandling@md.northwestern.edu

A. Prokop and B. Csukás (eds.), *Systems Biology*,
DOI: 10.1007/978-94-007-6803-1_15,

Abbreviations

ABM	Agent-Based Modeling
ABMF	Agent-Based Modeling Format
AI	Artificial Intelligence
ALI	Acute Lung Injury
APIABM	Acute Pulmonary Injury Agent-Based Model
ARDS	Acute Respiratory Distress Syndrome
CMA	Computational Modeling Assistant
DAMP	Damage-Associated Molecular Products
EINISI	Enteric Immunity Simulator
I-κB	I-kappa-B
NCBO	National Center for Biomedical Ontology
NEC	Necrotizing enterocolitis
NF-κB	Nuclear Factor kappa-B
ODD	Overview, Design and Detail Protocol
ODE	Ordinary differential equation
PMN	Polymorphonuclear neutrophils
TGF-β1	Transforming growth factor-β1
TNF-α	Tumor necrosis factor-α
VILI	Ventilator Induced Lung Injury

15.1 The Translational Dilemma in Biomedical Research

The greatest challenge facing the biomedical research community is the ability to translate the successes at obtaining basic mechanistic knowledge about biological processes into clinically effective therapeutics. There is a growing gap between the capability to acquire and analyze data and the ability to effectively and efficiently evaluate the hypotheses generated from that data. This is the Translational Dilemma. Recognition of this gap was made evident in 2004 when the United States Food and Drug Administration released a white paper titled: "Innovation or Stagnation: Pathways for the Future of Biomedical Research" [1]. This report noted that while there has been steady increase in funding for basic biomedical research there has been a concurrent steady decrease in the number of new clinically effective therapeutics brought to the bedside. These divergent trends are not sustainable. A recent review analyzed the roots of the Translational Dilemma and defined it as: *the inability to efficiently translate data into viable mechanistic hypotheses across levels of biological organization, and limitations in the ability to test those hypotheses in a meaningful and efficient way* (see Fig. 15.1) [2]. Currently, biomedical research faces two fundamental limits to its goal of being able to develop new interventions that can beneficially affect human health:

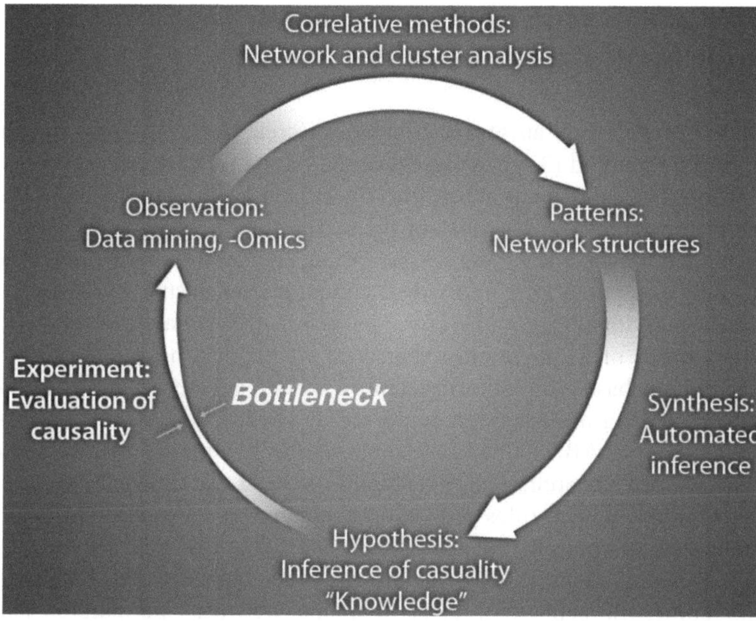

Fig. 15.1 The current imbalance in the scientific cycle. Technological advances in the past few decades have greatly increased the ability to generate, collect and correlate data, but a process bottleneck has developed at the point of being able to evaluate hypotheses via experiment. This bottleneck restricts the ability of the biomedical research community to efficiently and systematically reduce the space of possible hypotheses to those that are plausible. Augmenting this iterative cycle will identify those hypotheses that will be targeted for further investigation and refinement, and serve as potential points of therapeutic control. Reprinted with permission from Ref. [2]

(1) achieving the breadth of hypothesis testing necessary to deal with the multiplicity of possible explanations of high-resolution data (throughput problem) and (2) adequately representing the complexity of integrative hypotheses (multi-scale problem). Both of these issues are directly related to the need to greatly increase the ability to evaluate the plausibility of mechanistic hypotheses, and will almost certainly involve using computational modeling and simulation for dynamic knowledge representation and hypothesis instantiation. The ability to execute in silico experiments offers the potential to substantially accelerate and enhance the Scientific Cycle by providing a plausibility filter for putative hypotheses to help direct traditional experimental design to separate sets of plausible hypotheses and provide a wider search capability for plausible solutions. This chapter will discuss the use of agent-based modeling (also known as individual based modeling), for dynamic knowledge representation, and provide specific examples in the area of acute inflammation.

15.2 Dynamic Knowledge Representation with Agent-Based Modeling

Agent-based modeling is an object-oriented, discrete-event, rule-based computational modeling method [3–7]. An agent-based model (ABM) represents a system as populations of components where the simulation agent level of the ABM corresponds to the primary component of the system being studied. An ABM *agent class* is defined by specific properties governing its identity and behavior, and an ABM creates a population of individual computational instances of each agent-class. Each individual agent therefore possesses the behavioral rule sets and defined properties of its agent class, but once created can have diverging behavioral trajectories based on differing inputs within a heterogeneous simulation environment. ABM rules are often expressed as conditional statements ("if-then" statements), making ABM suited to expressing the hypotheses that are generated from basic science research, though it should be noted that the general conditional nature of simulation agent rules does not preclude the encapsulation of other types of mathematical or computational models (i.e. differential equation, stochastic or network) as rule systems [8–10]. Regardless of the specific ABM rules, ABMs offer the ability to achieve a close mapping between the natural language expression of hypotheses present in publications (the current means by which this knowledge is communicated within the community), and the structure of ABM [11, 12]. This property facilitates the use of agent-based modeling as a means of dynamic knowledge representation, particularly for non-mathematicians/computational scientists. ABMs are also intrinsically multi-scale, utilizing behavioral rules (Scale #1) to determine individual agent behavior (Scale #2) and then aggregating individuals into population dynamics of the global system (Scale #3). These levels can theoretically be nested, to provide a comprehensive depiction of a multi-scale biological system (see Fig. 15.2), making ABMs well suited for creating modular models [7, 8, 13–15].

15.2.1 Properties of Agent-Based Models

ABMs are related to other spatially discrete modeling methods, most notably cellular automata, though the mobile capability ABM agents and ability to represent a wider range of model topologies could lead to consideration of cellular automata as a special type of ABM. Similarly, neural nets can be considered ABMs, with the nodes representing instances of an agent class, and the network structure being the model topology. However, in practice, many ABMs have several characteristics of agent-based modeling that set it apart from other object-oriented, rule-based modeling systems (such as Petri nets, Boolean or Bayesian Networks), even though at its purest definition, they could all be potentially viewed as ABMs:

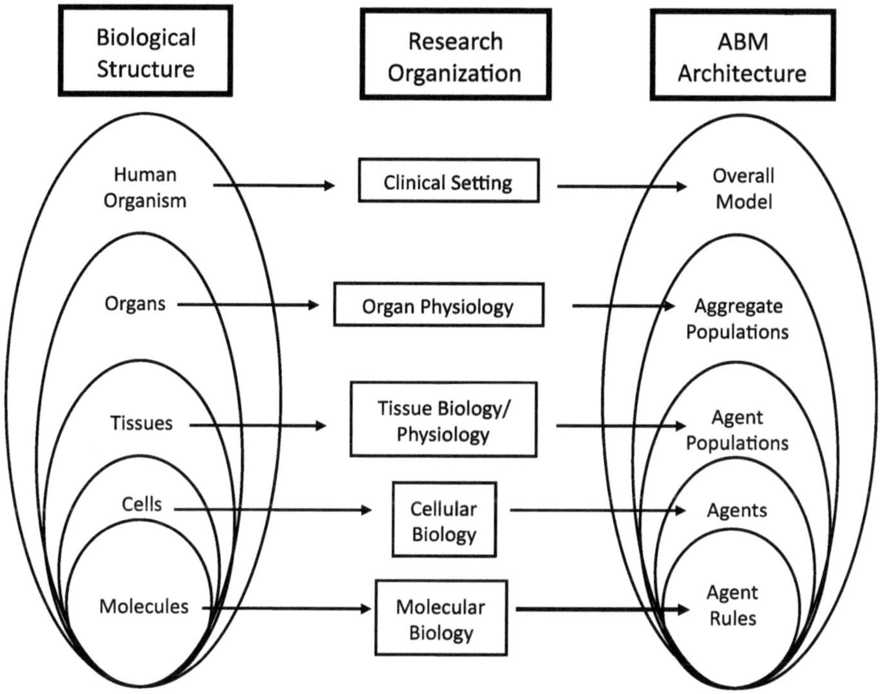

Fig. 15.2 The mapping between scales of biological organization, research community structure and agent-based models. This diagram maps the similar structure of organizational scales present in biological systems, the research communities studying them and the architecture of an ABM. Note that scales of organization are nested in the biological system and the ABM, reflecting the trans-scale coupling seen in both systems. Alternatively, the research community structure is disparate and compartmentalized, arising from both social and pragmatic logistical factors. Reprinted with permission from Ref. [12]

1. Agent-based models (ABMs) readily incorporate *space*. In an ABM agent behavior is driven by interactions determined by agent neighborhoods defining the communication and interaction network for each agent. An agent neighborhood can be represented as a two-dimensional square grid (very common), a 3-dimensional cubic space [8, 13], 2- or 3-dimensional hexagon cal space [12, 16] or as a network topology, as a neighborhood does not necessarily mean physical proximity but rather the configuration of some set of other agents with whom an agent can interact. This definition of an agent neighborhood is consistent with the bounded nature of the sense-and-respond and message passing capabilities of biological objects.
2. ABMs utilize *parallelism*. In general, each ABM agent class has multiple computational instantiations that form a population of agents, each capable of having different behavioral trajectories. These heterogeneous behaviors produce population dynamics that are the observable, system-level output of the ABM. A classic example of this phenomenon is the behavior of flocks of birds,

in which simulations utilizing relatively simple interaction rules among birds can lead to sophisticated flocking patterns without an overall controller [17].

3. ABMs incorporate *stochasticity*. Many biological systems have behaviors that appear to be random [18, 19]. Probabilities of a particular behavior can be determined for the population as a whole, and used to generate a probability function for the behavior of a single agent that is then incorporated into the agent's rules. As a population of agents executes their rules during the course of a simulation, each agent follows a particular behavioral trajectory as its behavior rules' probabilities are resolved as the simulation progresses. A set of behavioral outputs is thusly generated from a single ABM, producing system behavioral state spaces representing the set of population-level biological observations.
4. ABMs are *modular*. Agents represent a modular level into which new information can be added either through the introduction of new agent-types or by the modification of existing agent rules without having to re-engineer the entire simulation. Agent classes representing generic cell types can be subdivided and expanded to include a finer degree of detail with respect to sub-categories of cells while the remainder of the ABM remains essentially intact. New mediators can be similarly added by creating new cellular-state or environmental variables and rules. Multiple ABMs can be aggregated, providing that their points of contact and interaction are consistent across the incorporated ABMs [12, 13].
5. ABMs produce *emergent properties*. A central hallmark of ABM is that they generate system-level behaviors that could not have been reasonably inferred from, and often may be counter-intuitive to, examination of the rules of the agents alone. This is our definition of *emergent* behavior. ABMs are able to generate this type of behavior due to the locally constrained and stochastic nature of agent rules, and the population effects of their aggregated interactions. For example, in the bird flock an initial observation would suggest an overall leader, thereby requiring a means of determining rules for flock-wide command and control communication. This, however, is not the actual case; birds function on a series of locally-constrained, neighborhood-defined interaction rules, and the flocking behavior emerges from the aggregate of these interactions [17]. The capacity to generate non-intuitive behavior is a vital advantage of using ABM for conceptual model verification, as often the translation of generative mechanisms to system-level behavior produces paradoxical and un-anticipated results that break a conceptual model.
6. ABMs can be readily constructed using incomplete and abstracted knowledge. When constructing an ABM it is advantageous at the outset to keep the rules as simple and verifiable as possible, even at the expense of some detail. As such, meta-analyses of existing basic research often guide the development of an ABM [20]. ABMs constructed with admittedly incomplete and uncertain mechanisms representing statements of hypotheses can provide qualitative verification of those hypotheses [21]. As with all computational models, the greater fidelity of mapping between the ABM and its biological counterparts

enhances the correlation between simulation results and the real-world behaviors. An iterative process of refinement of an ABM will lead to increased detail, possibly a stronger correlation to real-world data and a greater confidence in the ability of the ABM to describe observable phenomena.

Agent-based modeling is an integrative modeling framework that can readily be used for communicable dynamic knowledge representation [11–13, 22] (see Fig. 15.2). Agent-based modeling, because of its emphasis on "things doing things", is generally more intuitive for non-mathematicians/computer scientists than more formal mathematical modeling methods such as ordinary differential equations, partial differential equations, and their stochastic variants. Agent-based modeling presents a lower threshold barrier for researchers to "bring to life" their conceptual models and integrate in silico methods with traditional in vitro and in vivo experiments [22].

Since ABMs are knowledge-based models, constructed by instantiating bottom-up mechanisms (as opposed to inductive models, where mechanisms are inferred with the goal of explaining data), agent-based modeling addresses different modeling questions than equation-based inductive models. For instance, ABMs are not readily developed directly from a mass of raw data; they require that the modeler have a mechanistic hypothesis that, when instantiated in an ABM, can be used to generate simulated data, which can then be compared to the real-world data set. One can envision an iterative process by which inductive models are applied to large data sets, wet lab experiments are carried out to investigate the mechanisms inferred from the inductive model, and the experimentally confirmed mechanisms are used as a basis of an ABM which would close the discovery loop by recapitulating the original data set.

Agent-based modeling was pioneered in the areas of ecology, social science and economics, but in the last decade they have been used to in the biomedical arena to study sepsis [12, 13, 23, 24] cancer [8, 16, 25–27] cellular trafficking [28–32] wound healing [33–35] and intracellular processes and signaling [9, 36–42]. Many biomedical ABMs focus on cells as the primary simulation agent level (with the notable exceptions of modeling intracellular processes from Refs. [9, 36–42] above). From a knowledge translation standpoint, cells form an easily identifiable level of "encapsulated complexity" that is both highly studied as a unit (i.e. cellular biology) and can be addressed with relatively straightforward input-output rules [7]. As noted above, while ABM agent rules are often logical or algebraic statements, rules can be a mathematical model in itself. There are multiple examples of embedding complex mathematical models within a cell-level ABM agent [7–10, 15, 35, 43]. These examples emphasize the potential unifying role of agent-based modeling as a means of "wrapping" different simulation methodologies. This suggests that the meta-structure of an ABM can be used as a template into which structured biomedical knowledge can be integrated to facilitate the instantiation of multiple mechanistic hypotheses [44].

15.2.2 Tools for Agent-Based Modeling

Agent-based modeling environments require addressing certain software issues beyond the basic capabilities of more traditional object-oriented programming tools. These issues include emulating parallel processing to represent the actions of multiple agents within populations, dealing with associated execution concurrency issues within those populations, establishing means of defining model topology (i.e. agent interaction neighborhood), and the development of task schedulers to account for the multiple iterations that constitute an ABM run. As a result of these issues, along with the case that many researchers who utilize ABMs are not trained computer scientists or programmers, many biomedical ABMs are created using existing ABM development software packages. These agent-based modeling environments attempt to strike a balance between representational capacity, computational efficiency, and user-friendliness. A non-comprehensive list of such ABM toolkits includes Swarm (http://www.swarm.org/index.php/Swarm_main_page), Mason (http://cs.gmu.edu/∼eclab/projects/mason/), RePast (http://repast.sourceforge.net/), NetLogo (http://ccl.northwestern.edu/netlogo/), StarLogo (http://education.mit.edu/starlogo/) and SPARK (Simple Platform for Agent-based Representation of Knowledge www.pitt.edu/∼cirm/spark [45]). All these platforms represent some trade-off among the triad of goals mentioned above. For an excellent review and comparison of many of these agent-based modeling toolkits, see Ref. [46].

15.2.3 Agent-Based Modeling of Inflammation

The use of agent-based modeling has dramatically increased since the year 2000, and is now a generally accepted means of performing computational biology. As is the case when discussing any specific modeling method, it should be reemphasized that agent-based modeling is only one of an array of methods that can be used to represent and investigate biological systems (such as those covered in other chapters in this book). Each of these modeling techniques has its strengths and weaknesses, and potential modelers need to recognize that the modeling method chosen should be tailored to the question(s) being asked of the model [47]. One of the most effective ways of communicating the capabilities (and limitations) of a particular modeling method is through the use of examples. Towards this end, the following sections list a series of ABMs of different aspects of inflammation, followed by a more detailed description of the development and use of an ABM directed at a specific issue, that of acute pulmonary inflammation.

We focus on ABMs of the inflammatory response because inflammation is one of the most basic and ubiquitous processes in biology: in addition to growth, metabolism and replication, the response to injury leading into repair is a core function of all organisms. It is highly evolutionarily conserved, and in multicellular organisms is a well-coordinated network consisting of specialized cell types and molecular mediators [47, 48]. The inflammatory response can be described simply as: (1) Sensing of damage or threat, (2) Containment and clearance of the threat, and (3) Repair of the damaged tissue. Intrinsic to all of these steps are counter-regulatory controls intended to limit and modulate the response. Evolution has operated on the components of the inflammatory response to produce systems that are robust over a wide range of heterogeneous insults, with trade offs on the efficacy of the pro-inflammatory response versus the negative consequences of an overly sensitive and exuberant response. While this balance generally operates well it is subject to disordered behaviour with significant consequences on the development of disease. Diseases such as sepsis, trauma, inflammatory bowel diseases, chronic wounds, autoimmune diseases and asthma are the direct result of disordered and inappropriate inflammation, while many other diseases, such as cancer, diabetes, atherosclerosis, Alzheimer's, and obesity are associated with inflammatory processes as either a generative mechanism or a means of perpetuating the disease. This is because inflammation can damage normal healthy tissues, which in turn leads to the production of molecules that re-stimulate inflammation. Acceleration of this forward feedback loop can lead to disordered inflammation that promotes organ dysfunction and death [47–51], Inflammation may also manifest in slower degenerative processes that share many common mediators with acute pro-inflammatory insults [52]. However, experience has shown that caution must be exercised in targeting inflammation with pharmacological agents. Because of its ubiquitous role in homeostasis, modulation of inflammation is fraught with unintended systemic consequences, such as gastro-intestinal toxicity of cyclooxygenase-2 inhibitors [53, 54] or increased susceptibility to infection in persons taking TNF-α inhibitors [55, 56] or the general failure of anti-cytokine therapies for sepsis [24]. The difficulty in engineering safe and effective therapeutic agents directed at inflammation is a primary example of the Translational Dilemma in biomedical research. Because of these characteristics inflammation represents perhaps the ideal target for systems biology and computational modelling with agent-based modelling, and the following sections list a series of ABMs of different aspects of inflammation across a range of organizational scales. This brief survey of inflammation-related ABMs is followed by an example describing in more detail the development and use of an ABM directed at a specific issue, namely that of acute pulmonary inflammation.

15.2.3.1 ABMs of Inflammation-Related Intracellular Processes

The characterization of intracellular pathways is the traditional focus of systems biology, with a long history of work and achievement in the development of mathematical models of cellular signaling and metabolic control. These models are generally biochemical kinetic models, utilizing differential equations and stochastic methods based on the Gillespie Algorithm. However, the use of discrete-event, particle based modeling, exemplified by agent-based modeling, is growing in this arena. With increasing awareness of the influence of the complex, compartmentalized environment of the intracellular milieu on intracellular dynamics, there is a need to account for issues of molecular crowding and spatial heterogeneity of the reaction milieu and how they affect enzymatic reactions within the intracellular environment. Additionally, the presence of sub-cellular structures, cytoskeletal elements, organelles, and compartments call for the increasing incorporation of spatial properties and detail. Ridgway et al. [40] used an ABM of intracellular signaling to demonstrate that the reaction dimension determining biochemical kinetics within a prokaryotic cytoplasm was reduced from the expected three dimensions to nearly two, with significant consequences for the dynamic modeling of control loops in which subtle changes in feedback determine the direction of a molecular switch. Pogson et al. [39] developed an ABM of control pathways affecting the transcription factor Nuclear Factor kappa B (NF-kB). These studies demonstrating the importance of the spatial distribution in terms of nuclear translocation of the constitutive inhibitor of NF-kB, I-kappa-B (IkB), and the binding of IkB to actin, a cytoskeletal protein, a mechanism subsequently identified in their laboratory [38]. We developed an agent-based architecture called Spatially Configured Stochastic Reaction Chambers to demonstrate that even an abstract representation of enzyme kinetics could, if sufficient pathway component detail was included, reproduce canonical behavior at the cellular level, as in the effect of preconditioning on the behavior of the Toll-like Receptor 4 (TLR-4) signaling pathway [36]. Similarly, an ABM of NF-kB response to endotoxin utilized molecular level agents nested within "mega-agents" representing different inflammatory cell types to reproduce recognizable dynamics of endotoxin response, including priming and tolerance at both the transcription factor and cellular activation level [42].

15.2.3.2 Cell-Level ABMs of Systemic Inflammation and Simulated Trials for Sepsis

The cell-as-agent level of component representation provides perhaps the most intuitive link between the laboratory-derived basic mechanistic knowledge and the structure of an ABM. Some of the earliest examples of biomedical ABMs were focused at this level [23, 24, 26], leading to the realization that even abstract agent-rules could produce very recognizable dynamics that could provide deep insights into the essential characterization of a disease process. For example, an early ABM of systemic inflammation and sepsis viewed the inflammatory process as being governed by interactions at the endothelial blood interface [23]. This ABM generated four clusters of distinct trajectories of model-system behavior purely by altering the degree of initial perturbation, trajectories that matched the four primary clinical scenarios associated with systemic inflammatory response. This ABM also demonstrated that the mechanistic basis of inflammation was the same whether the initiating insult was infectious, as in classical sepsis, or tissue damage, as in severe trauma.

The endothelial-surface systemic inflammation ABM was further extended to perform in silico clinical trials based on published and hypothetical inflammatory-mediator-based interventions [24]. Published pharmacologic properties of a series of mediator-targeting compounds were inputted into the ABM simulating a sepsis population. The efficacies of the interventions were then evaluated against a simulated control population. None of the mediator-directed interventions led to a statistically significant improvement in simulated patient outcome, including a set of immune augmenting interventions (e.g. addition of Granulocyte Colony Stimulating Factor) and combination anti-cytokine therapy (intended to overcome possible pathway redundancy). While these results were not totally unexpected, the exercise demonstrated that the ABM could be used as a means of assessing the veracity of the proposed intervention: i.e. what are the global consequences of intervening in a particular pathway, and is it actually a good idea to intervene at this point? The confirmation that what appeared to be intuitively plausible points of mechanistic intervention did *not* in fact behave as expected when placed in a systemic context demonstrated the potential usefulness of agent-based modeling and dynamic knowledge representation for hypothesis verification. We suggest that one of the primary roles of dynamic knowledge representation is exactly this type of hypothesis evaluation and verification, intended to reduce the set of plausible hypotheses and thereby help direct future investigation by eliminating therapeutic dead-ends.

15.2.3.3 Cell-Level ABMs of Wound Healing of Skin and Soft Tissue

As a system of response and repair, inflammation is intimately tied to healing. Many cellular and molecular mediators are shared between acute inflammation and healing; for instance the anti-inflammatory mediators that limit and contain the propagation of the pro-inflammatory response, such as Interleukin-10 and transforming growth factor-β1(TGF-β1) are themselves growth factors. Wound healing is also an intrinsically spatial process, as damaged tissue is removed and replaced by surrounding "normal" tissue. Therefore, ABMs of wound healing represented a natural direction of development arising from the early inflammatory ABMs. Wound healing ABMs have been used to shed basic insights on the spatial nature of skin wounds and their healing [34, 57], to represent the mechanistic pathophysiology of diabetic wounds and to posit potential mechanistic targets for therapeutics development [33], and offer the potential for personalized medicine by modeling individual responses to injury and therapy in vocal chord trauma [58, 59]. The diabetic wound ABM [33] was used to determine the phenotypic effects of under-activation of latent TGF-β1 and over-production of tumor necrosis factor-α (TNF-α), both associated with diabetes, and generated a host of emergent features characteristic of diabetic ulcers. Moreover, this ABM was used to test in silico the effects of both current therapies for diabetic ulcers (namely wound debridement and treatment with platelet-derived growth factor) as well as novel interventions (e.g. inhibition of TNF-α or addition of TGF-β1) [33]. The ABM of vocal fold inflammation and healing attempted to create personalized sets of models by calibrating parameters using data on cytokine levels in laryngeal secretions of individual human volunteers subjected to experimental phonotrauma. Patient-specific computational simulations were created based on baseline levels of cytokines as well as at 1 and 4 h after phonotrauma. These simulations generally predicted the levels of cytokines at much later time points (24 h), and were used as the basis for simulated therapy [58, 59].

15.2.3.4 ABMs of Organ-Level Inflammation

A critical point of the translational dilemma is the transfer of cellular and molecular mechanisms, which are measured and characterized in the laboratory environment, to the level of organ level physiology and phenotype, which is the primary means by which disease is defined and diagnosed. It is here that the population-oriented capabilities of agent-based modeling can serve an important translational role. As a result there has been a great deal of interest in producing ABMs that represent organ-level manifestations of inflammation.

Intestinal Inflammation

The intestinal tract is subject to a variety of inflammatory conditions, both acute, such as in systemic shock, gut-derived sepsis and necrotizing enterocolitis, as well as in more chronic diseases, such as inflammatory bowel disease. The nature of the inflammatory processes in the gut is particularly notable due the persistent presence of huge numbers of microbes that can initiate and propagate inflammation. While the study of the gut ecology has been traditionally divided into those who study the host (epithelial biology and immunology) and those who study the microbes (microbiology), there is an increasing recognition that these two fields need to be merged into a comprehensive characterization of the host-microbe environment [60]. The integrative capabilities of agent-based modeling may play a particularly important role in this arena, and there has already been some preliminary work in this direction. A group at the Virginia Bioinformatics Institute has developed the Enteric Immunity Simulator (EINISI), an ABM environment to investigate the pathogenesis of enteric diseases related to the immune response to pathogen and reproduced the dynamics of bacterial dysentery [61]. Our group at the University of Chicago has developed an ABM of gut host-pathogen interactions specifically related to virulence activation of *Pseudomonas aeruginosa*, an important nosocomial pathogen, and the development of gut-derived sepsis [62]. This ABM contains a detailed representation of *P. aeruginosa* virulence activation pathways integrated with an abstracted gut epithelial surface. The ABM's output is mapped to in vitro and in vivo experimental platforms of gut-derived sepsis, used to simulated a more clinically relevant manifestation of intestinal ischemia resulting from systemic shock than currently possible using in vivo techniques (i.e. non-lethal systemic shock), and has been used to identify gaps in the low-phosphate-sensing model of *P. aeruginosa* virulence activation in circumstances of major abdominal surgical stress. Additional laboratory experiments are in the process of being performed to more comprehensively characterize the factors involved in low-phosphate-related *P. aeruginosa* virulence activation. Finally, we have also developed an ABM that represents a unifying hypothesis underlying the pathogenesis of necrotizing enterocolitis (NEC), the leading cause of gastrointestinal morbidity and mortality in the premature infant population [63]. NEC is a complex, multi-factorial disease that involves prematurity, enteral feeding and a bacterial component resulting in bowel inflammation and necrosis. The research community has found it extremely challenging to create laboratory models that can comprehensively reproduce the range of pathogenic components associated with NEC, mainly related to the extreme degree of experimental perturbations required to generate the NEC phenotype in vivo. We have formulated a minimally sufficient unifying hypothesis of NEC that posits that the fundamental deficit in infants susceptible to NEC is immaturity of the ability of the neonatal gut epithelial cells to manage reactive oxygen species, including those produced as a byproduct of cellular respiration. When this basic feature was instantiated in the NEC ABM, and then overlaid with the other recognized contributing factors, a recognizable pattern of cascading systems failure was demonstrated to be necessary for the generation of the NEC phenotype. Specifically, immature neonatal gut epithelial cells had increased fragility to inflammation propagating challenges, such as metabolic stress (from feeding), decreased mucus barrier integrity and bacterial contacts. It is hoped that this ABM can be used to integrate the multiple theories and mechanisms currently studied concerning the pathogenesis of NEC.

Pulmonary Inflammation

The lung is an organ that is commonly subjected to inflammatory insults and responses, either through direct infection, inhalation of particulate matter, or in a "bystander" role associated with systemic inflammation. One type of pulmonary infection that has been the subject of extensive agent-based modeling is tuberculosis. ABMs have been used to study inflammatory cell control mechanisms associated with the generation of pulmonary granulomas [64], and the pathogenesis of pulmonary tuberculosis has been modeled using a multi-scale architecture where ODEs representing the molecular dynamics of TNF-α signaling were embedded within inflammatory cell agents [10]. Another ABM examined the pulmonary inflammatory response to inhaled particulate matter and the subsequent transition from acute inflammation to fibrosis [65]. While relatively simple in terms of cellular agent rules and types of mediators represented, this ABM was able to reproduce histological patterns of pulmonary inflammation and fibrosis seen in a clinically relevant murine model of particulate inhalation. Finally, in Sect. 15.2.4 we present a detailed description of an ABM of acute pulmonary inflammation designed to examine the dynamics of acute lung injury from trauma, pneumonia, and systemic sepsis.

15.2.3.5 Multi-Organ Inflammation and Failure

The structural/anatomic approach to multi-scale modeling can be taken one step further by using the modular property of agent-based modeling to link individual organ ABMs in a multi-scale architecture. The approach was introduced in an ABM of the gut-lung axis of systemic acute inflammation and multiple organ failure [13]. This ABM incorporates multiple structural and anatomic spaces, e.g. endothelial and epithelial surfaces as aggregated by cell-type into organ-specific tissues and finally to organ-to-organ interconnections and cross-talk. This architecture also *translates* knowledge across domain specialties (molecular biology to clinical critical care), representing molecular and cellular mechanisms and behaviors derived from in vitro studies, extrapolated to ex vivo tissue experiments and observations, leading to patterns of organ-specific physiology, and finally simulating clinically relevant, interconnected, multi-organ physiology including the response to ventilator support of acute respiratory failure. This ABM also posited certain characteristics of the gut-derived pro-inflammatory compound that is circulated in the mesenteric lymph and induces pulmonary inflammation. Examining the time course of pulmonary inflammation and comparing that to generated factors following intestinal ischemia suggested that the mesenteric lymph inflammatory compound was not an initial inflammatory cytokine, nor a translocating luminal compound manifesting decreased intestinal permeability, but rather a substance reflecting cellular damage of gut tissue with properties consistent with damage-associated molecular patterns (DAMPs). This last hypotheses remains to be completely confirmed by the sepsis research community, but at this time appears to be consistent with ongoing research in this area [66].

15.2.4 An Example ABM of Acute Pulmonary Injury

Herein we present a description of the development of an ABM focused on representing existent knowledge concerning acute pulmonary inflammation and the dynamics of various types of acute lung injury. We term this ABM the Acute Pulmonary Injury ABM (APIABM). The primary goal of this example is to demonstrate some of the steps and modeling issues related to the development and use of an ABM. While the APIABM is a relatively simple model and its output is qualitative in nature, these characteristics actually emphasize one of the greatest advantages of agent-based modeling, namely the ability to relatively quickly and with limited computational overhead instantiate mechanistic biological knowledge into a computational model that can produce recognizable behaviors. There is a significant role for qualitative modeling within the greater context of the discovery phase of science [2, 36] and in particular the ease with which agent-based modeling maps to biological knowledge and can be performed with a "low threshold, high ceiling" strategy [22] that allows for future modular expansion of the ABMs. As all modeling can be considered as selective abstraction, we have abstracted out a fair amount of molecular detail in the APIABM as to not distract from the cellular functions of interest. Additionally, rather than presenting detailed simulation experiments we instead emphasize the calibration/validation steps, and then discuss future directions that can be taken with this model. For interested readers, the entire APIABM can be downloaded from http://bionetgen.org/SCAI-wiki.

The process of ABM construction and use is described in the general context of the Overview, Design Concepts and Details (ODD protocol), an attempt to help standardize the description of ABMs and their uses [67]. The ODD protocol was originally developed for ABMs studying ecological and social systems, and though it does not present an exact fit with the use of agent-based modeling as a means of biomedical dynamic knowledge representation (notable discrepancies include: format-driven redundancies; potential disruption of explanatory flow, particularly in terms of describing the mapping between the biology and the ABM; non-applicability of certain categories, such as learning and adaptation; the inherent imprecision of the term "emergence"; and lack of section concerning calibration) it does provide a useful framework in which the rationale and process behind the design of an ABM can be communicated. We utilize a modified version of the ODD protocol as the organizational framework for the description of the APIABM.

15.2.4.1 Purpose

The modelling purpose of this ABM is to dynamically represent the molecular, cellular and organ-level dynamics of acute pulmonary inflammation and provide a unifying basis for the response to multiple types of acute lung injury, namely direct trauma (pulmonary contusion), bacterial infection (primary pneumonia), and systemic inflammation (acute lung injury/acute respiratory distress syndrome or ALI/ARDS). These disease processes represent a major source of morbidity and mortality in the acutely and critically ill patient, and present significant diagnostic and therapeutic challenges to medical practitioners. The complexity of the inflammatory response means that effective modulating therapies need to be the "right drug for the right condition at the right time," a criteria that requires disease characterization at a level of resolution not currently achieved (and this includes –omic characterization, which just provides for a series of high-dimensional snapshots). By integrating existing mechanistic knowledge, down to the scale of putative molecularly targeted interventions, to produce a recognizable organ-level phenotype in the form of edema patterns, the APIABM can serve as a dynamic bridge to fill in the gaps in existing knowledge and data.

15.2.4.2 Entities, State Variables, and Scales

The Entities in the ODD refer to the objects from the reference system represented in the ABM. Entities can refer to the active components of the system (i.e. the agents/individuals), subcomponents sensed/used/manipulated by the agents (i.e. variables in the agent or the environment) or aggregated collections of agents or sections of space that make up the ABM's environment. Each entity has a set of state variables that defines its current state. For agents these state variables would correspond to molecular components such as receptors, enzymes and genes, for the spatial environment, these state variables would represent levels of secreted mediators or extracellular structures. The set of state variables is consistent for a particular entity type, but individual values of the state variables distinguish one entity from other entity of the same type and are used to track how a particular entity changes over time [67]. In the APIABM entities range from cellular mediators to alveolar space, and are discussed in more detail in the sections below. In order to distinguish computational components in the APIABM from their biological referents, we will use a different font to denote `APIABM components`.

We have elected to abstract the large number of specific molecular species into functional groups, which are then assigned to aggregated descriptive variables. For instance, the plethora of pro-inflammatory cytokines involved in pulmonary inflammation is represented by a single variable called `pro-inflammatory cytokine`. We justify this modelling decision based on the fact that we are not interested in high-resolution examination of molecular interactions, but rather what the overall consequences of these types of interactions have on the behaviour of cellular populations. This is one example of how the "encapsulated complexity" offered by agents allows investigation of higher-level system properties even given incomplete knowledge, as is often the case, of lower-level detail.

Agents/Individuals

In developing an ABM one of the first modelling decisions to be made involves selection of the agent level. As noted above, the agent-level should represent a level of "encapsulated complexity" that exists in sufficient numbers such that a population of agents can be modelled, but not too many numbers such that the population size abuts computational limitations. The cell types represented include alveolar epithelial cells, monocytes, macrophages, neutrophils, and bacteria. The behaviours exhibited by each cell type reproduce those that are known to exist in situ and vary in response to changes in the inflammatory milieu of the tissue in which they are located. A description of the agent classes and their state variables with their process flow can be seen in Table 15.1. A more detailed explanation of the rules for each agent class is found in the Process Overview and Scheduling Sect. 15.2.4.3.

Table 15.1 Agent types and their state variables

Cell type/agent class	State variables
Monocytes: Circulating precursors to macrophages	• Age • Chemotaxis-Threshold: Threshold value of damage-signal and pro-inflammatory-cytokine to stop movement and transform into a macrophage
Monocyte-makers: Simulate bone marrow generation of monocytes	• Generation-Rate
Macrophages: Inflammatory cells with controller functions, primary source of pro- and anti-inflammatory mediators	• Age • MP-activation: Activation state of the macrophage • Pro-inflam-receptor: Activated by the presence of pro-inflammatory-cytokine • Pro-inflam-signal-kinase: Activated by the activation of pro-inflam-receptor, activates pro-inflam-gene • Pro-inflam-gene: Produces intracellular-pro-inflam-stimulus • Intracellular-pro-inflam-stimulus: Produced by pro-inflam-gene, produces and secretes pro-inflammatory-cytokine • Anti-inflam-receptor: Activated by the presence of anti-inflammatory-cytokine • Anti-inflam-signal-kinase-p: Activated by activation of anti-inflam-receptor, activates anti-inflam-gene-p • Anti-inflam-signal-kinase-a: Activated by activation of anti-inflam-receptor, activates anti-inflam-gene-a • Anti-inflam-gene-p: Produces intracellular-anti-inflam-stimulus-p • Anti-inflam-gene-a: Produces intracellular-anti-inflam-stimulus-p • Intracellular-anti-inflam-stimulus-p: Produced by anti-inflam-gene-p, inhibits amount of pro-inflammatory-cytokine produced and secreted • Intracellular-anti-inflam-stimulus-a produced by anti-inflam-gene-a, produces and secretes anti-inflammatory-cytokine

(continued)

Table 15.1 (continued)

Cell type/agent class	State variables
Neutrophils (PMNs): Inflammatory cells that are the initial responders to inflammatory insult and primary actors on the lung tissue	• Age • PMN-apoptosis • Pro-inflam-receptor: Activated by the presence of pro-inflammatory-cytokine • Pro-inflam-signal-kinase: Activated by the activation of pro-inflam-receptor, activates pro-inflam-gene • Pro-inflam-gene: Produces intracellular-pro-inflam-stimulus • Intracellular-pro-inflam-stimulus: Produced by pro-inflam-gene, produces and secretes pro-inflammatory-cytokine • Anti-inflam-receptor: Activated by the presence of anti-inflammatory-cytokine • Anti-inflam-signal-kinase-p: Activated by activation of anti-inflam-receptor, activates anti-inflam-gene-p • Anti-inflam-signal-kinase-a: Activated by activation of anti-inflam-receptor, activates anti-inflam-gene-a • Anti-inflam-gene-p: Produces intracellular-anti-inflam-stimulus-p • Anti-inflam-gene-a: Produces intracellular-anti-inflam-stimulus-p • Intracellular-anti-inflam-stimulus-p: Produced by anti-inflam-gene-p, inhibits amount of pro-inflammatory-cytokine produced and secreted • Intracellular-anti-inflam-stimulus-a produced by anti-inflam-gene-a, produces and secretes anti-inflammatory-cytokine
PMN-makers: Simulate bone marrow generation of Neutrophils (PMNs)	• Generation-Rate
Alveolar epithelial cells: Epithelial cells lining the alveolar airspace, they provide barrier function between the airspace and the lung interstitium. In the APIABM these agents also represent the barrier between the circulation and the interstitial tissue	• Damage
Bacteria: Generic infectious agents that are introduced to simulation pneumonia	• Bacteria-Age • Bacteria-Energy

Spatial Units and Environment

The topology of the APIABM is a 2-dimensional square grid with edges that wrap forming a torus. This world structure was selected in great part due to the constraints placed by the software system, NetLogo [68], in which the APIABM was implemented. We were willing to accept these limitations given the ease with which models of biomedical systems can be rapidly implemented and prototyped in NetLogo [22]. The two-dimensional grid spaces ("patches" in NetLogo terminology) are the fundamental spatial units of the APIABM, each possessing state variables representing extracellular mediators and structures that make up the microenvironment experienced by the cellular agents occupying them. A screenshot of the APIABM can be seen in Fig. 15.3. The patches represent an abstract cross-sectional depiction of the lung parenchyma, with specific focus on representing the alveolar air spaces and their interposing interstitial tissue. Patches at x- and y-coordinates that are multiples of 5 are given the state variable "alveolar interstitium", while the rest are given the state variable "alveolar space". For a complete list of the state variables of the spatial units see Table 15.2. A detailed explanation of these can be found below in the Process Overview and Scheduling section.

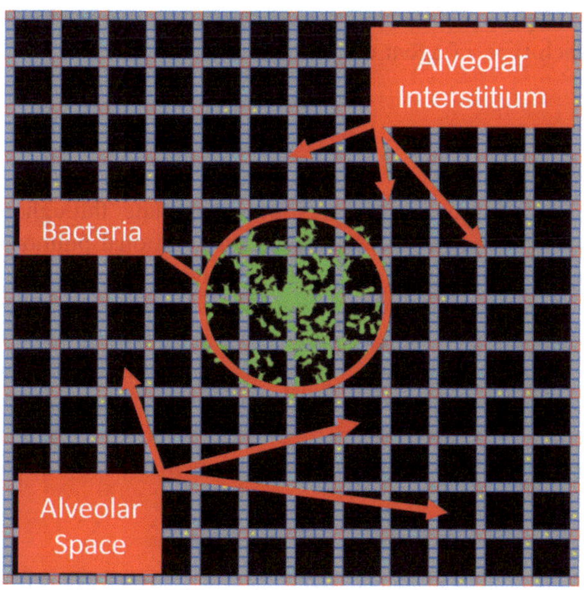

Fig. 15.3 Screenshot of acute pulmonary injury agent-based model (*APIABM*). This screenshot displays the overall architecture of the APIABM, which includes a regular lattice of alveolar interstitium, on which move the inflammatory cells, with interposed areas corresponding to alveolar space. This screenshot also displays an initial localized inoculum of bacteria prior to the execution of the model. Pulmonary edema is seen as *bluish-white patches* within the alveolar spaces (see Figs. 15.4, 15.5, and 15.6), with brighter areas corresponding to higher levels of edema fluid

Scale

Since ABMs iteratively execute a set of rules and commands, the time scale of the ABM is often tied to the length of time it takes for the reference system to perform the actions reflected in the ABM rules. For the ALIABM, each iteration of the program (or "tick") represents $\sim$7 min in reference system time. The disease processes being simulated have general time courses of $\sim$72 h for development, with recovery (should it happen) taking $\sim$14 days. Based on these timeframes the simulations were run for 14 days of simulated time.

15.2.4.3 Process Overview and Scheduling

The dynamics of the pulmonary inflammation arise from the actions and interactions of the cellular agents in response to the conditions of the patch on which they are located. Cellular agents are also able to sense certain variables on the patches immediately adjacent to them (such as for allowing the simulation of chemotaxis). As noted above the cells of interest are alveolar epithelial cells, monocytes, macrophages, neutrophils and generic bacteria; the rule sets for each of these agent-classes constitute a submodel of the APIABM. An overview of these cell submodels is presented in this section. For a comprehensive list of the state variables for each type of agent class, refer to Table 15.2.

Monocytes

Under baseline conditions, monocytes move/circulate throughout the alveolar interstitium and represent a potential source of additional pulmonary tissue macrophages with a differentiation rate corresponding to the lifespan of the macrophages (see Macrophages). After an insult is applied, `pro-inflammatory cytokines` and damage signals are released secondary to inflammation and when they are present above a set chemotaxis threshold, the transformation rate is accelerated as `monocytes` migrate to the focus of inflammation and subsequently differentiate into `macrophages`. `Monocytes` are repleted by an "off-screen" `monocyte-maker` that represents the hematopoietic activity of the bone marrow.

Macrophages

Under baseline conditions, `macrophages` move randomly through the alveolar interstitium. When they encounter pro-inflammatory stimuli they migrate towards the focus of inflammation. Additionally, in response to inflammatory mediators, `macrophages` release pro-inflammatory, anti-inflammatory, and tissue repair cytokines.

Table 15.2 Spatial units and patch variables

Spatial unit	State variables
Alveolar space: Represents air-fill spaces of the lung parenchyma. Volume and surface area represent the gas-exchange surface of the lung	• Capillary-Leak: Represents the rate at which edema fluid is produced and transferred into the alveolar space. Determined by the presence of damaged alveolar epithelial cells
Alveolar interstitium: Represents the tissue of the lung, forms the walls of the alveolar space	• Fluid: Represents edema fluid that has leaked from the interstitium into the airspace
General patch variables: These are extracellular variables, generally representing secreted/produced mediators that are sensed by and responded to by the different cell types	• Pro-inflammatory-cytokine: Produced and sensed by macrophages, monocytes and neutrophils • Anti-inflammatory-cytokine: Produced and sensed by macrophages and neutrophils • Damage-signal: Produced by alveolar-epithelial-cells and sensed by macrophages, monocytes and neutrophils • Cytotoxic-compound: Produced by neutrophils and results in damage to alveolar-epithelial-cells and kills bacteria • Nutrients: Produced by bacteria damaging alveolar-epithelial-cells and consumed by bacteria to increase their energy

As seen in Table 15.1, the macrophages in the APIABM have numerous state variables. Macrophages have an age, which is initially set at 3,000 ticks, which corresponds approximately to a 14 days lifespan [69]. This value decreases by 1 with every tick until it reaches 0, at which time the macrophage dies and is removed from the simulation. We note that given the time frame of the current set of simulations (14 days) we could have excluded age as a macrophage state variable; however, our goal is to not produce "one-off" models, but rather incorporate selected aspects from the reference system with an eye towards additional simulation experiments in the future. For instance, a natural next set of simulation experiments using the APIABM would examine the immunocompromised phase of sepsis, which extends the simulated time frame out to 28 days or beyond. Additionally, the inclusion of the age variable eases the possible inclusion of mechanisms that may either speed or attenuate programmed cell death (apoptosis).

Macrophages include representation of both pro-inflammatory and anti-inflammatory state signalling pathways. The state variables that make up these pathways represent the various components of the molecular signalling cascades that drive the response to and the release of cytokines during the inflammatory response. These include representations of receptors, signalling kinases, genes and transcription/translational events. Patch variables representing pro-inflammatory stimuli are sensed by macrophages, which respond by activating pro-inflammatory receptor variables, which in turn leads to activation of pro-inflammatory-kinases, activation of pro-inflammatory-genes, with subsequent production and release pro-inflammatory cytokines. Similarly, a macrophage's anti-inflammatory receptors can be activated, leading to activation of anti-inflammatory-kinases of two types leading to two sets of genes; those associated with inhibiting pro-inflammatory cytokine production, and those associated with the production of anti-inflammatory cytokines. These two pathways represent positive and negative feedback control systems, respectively, on macrophage function.

Neutrophils

Under baseline conditions, neutrophils move randomly throughout the alveolar interstitium. Neutrophils respond to pro-inflammatory stimuli by turning on their activation state variable. Activated neutrophils migrate towards the pro-inflammatory signals, which triggers the activation of signal cascades, pro-inflammatory-kinases, that result in the release of further pro-inflammatory cytokines as well as cytotoxic-compounds representing reactive oxygen species (ROS). Neutrophils also have an age, which is set to 1,000 ticks, approximating a life span of 5 days, and are repleted by an "off-screen" neutrophil-maker representing the hematopoietic activity of the bone marrow.

(Generic) Bacteria

Bacteria represent the introduced pathogens that cause primary pneumonia. Bacteria induce tissue damage, leading to the release of tissue damage compounds that stimulate the activation of the host inflammatory response. The primary bacteria state variable is energy. Bacteria acquire energy through their tissue damage induction, and when they reach a set energy threshold they will replicate. If they are prevented from inducing tissue damage, their energy degrades at a rate of 1 per tick until it reaches 0, at which time the bacteria die. Bacteria are also killed by activated neutrophils and the presence of cytotoxic-compound.

Alveolar Epithelial Cells

Alveolar epithelial cells are stationary cells representing the cellular components of the alveolar interstitium comprising the lung parenchymal tissue. Alveolar epithelial cells sense and respond to inflammatory stimuli in their local microenvironment, and also form the barrier between the fluid in the interstitial space and the gas-exchange spaces of the alveoli.

The primary state variable determining the function of the `alveolar epithelial cells` is `damage`. When high levels of `pro-inflammatory cytokines` or `cytotoxic-compounds` are present, the `alveolar epithelial cells` become damaged, releasing their own pro-inflammatory damage signal molecules, which in turn leads to further propagation of inflammation. Also, damaged `alveolar epithelial cells` release fluid into the surrounding alveolar space, simulating the formation of alveolar edema. It is the spatial distribution of the alveolar edema pattern that forms the qualitative metric used for validation of the APIABM.

Pulmonary Compartment Spatial Units

The APIABM abstractly depicts the gas-exchange structure of the lung, and is divided into patches that are either `alveolar interstitium` or `alveolar space`. Under normal conditions, mobile cellular agents have their movement confined to the patches possessing the `alveolar interstitium` state variable and therefore do not enter patches possessing the `alveolar space` state variable. The patches comprising the `alveolar interstitium` further possess a `capillary leak` state variable. In response to a set level of pro-inflammatory mediators at a given patch, the `capillary leak` state variable activates, allowing inflammatory mediators to leave the interstitium and enter the alveolar space, as occurs in situ. Additionally, the `alveolar space` patches have a `fluid level` state variable, which represents the degree of fluid leaking from the `alveolar interstitium` into the `alveolar space` through the damaged `alveolar epithelial cells`. The distribution and degree of alveolar edema represents the qualitative metric used for validation of the APIABM. The spatial unit categories and their respective state variables can be seen in Table 15.2.

15.2.4.4 Design Concepts and Initialization

In initiating a modeling project, it is of the utmost importance to define the experimental frame, thereby establishing what can and cannot be examined by the particular model. The experimental frame is defined by the scientific questions at hand, and provides direction as to the degree of abstraction used in the development of the model [70]. The APIABM is a highly abstracted representation of acute inflammation of the pulmonary parenchyma. The parenchymal focus of the APIABM is directed by the scientific goal of understanding and mechanistically unifying diseases such as pulmonary contusion (i.e. direct lung trauma), bacterial pneumonia and acute lung injury/acute respiratory distress syndrome (ALI/ARDS). There are many details of the real lung that are left out. The APIABM does not incorporate the mechanical forces associated with ventilation, spontaneous or mechanical, and therefore cannot be used to examine the effects of ventilator associated lung injury (VILI). Our modeling focus does not require addressing the bronchial airways, and therefore specifically excludes the consequences of inflammation in the airways as is seen in asthma. The focus on acute inflammation also excludes the ability of the APIABM to represent more chronic processes such as pulmonary fibrosis, or the development of chronic obstructive pulmonary disease. While some may consider such restrictions as highly limiting the potential utility of the APIABM, the fact is that one should strive to develop the simplest model that can address a defined scientific focus and provide a recognized use for the researcher. In this case, our interest is in the acute processes that might affect the lung in a acutely ill patient, and given the role of the inflammatory response in this setting, we make the modeling decision to focus on the consequences of inflammation on the gas-exchanging parenchymal aspect of the lung, specifically manifest in the patterns of production of alveolar edema.

15.2.4.5 Initialization

One critical point to remember when using ABMs for biomedical processes is that the baseline state is one of dynamic equilibrium, i.e. health. This means that the state of the system prior to any perturbation that would lead to disease is dynamically stable. The corollary to this fact is that biomedical ABMs are not models of disease, but rather models of health that can be subsequently perturbed to generate system trajectories that correspond to disease. As such, part of the initialization process involves making sure that the APIABM produces stable behaviour absent an invoked perturbation, including stability of those cellular populations that have their life-cycle represented (namely `monocytes`, `macrophages` and `neutrophils`).

15.2.4.6 Simulations

The simulations carried out here using APIABM are geared towards demonstrating calibration and validation. Calibration involves the adjustment of parameters of the ABM to attempt to fit some set of defined descriptors of the reference system, be they a quantitative data set or some more qualitative pattern/phenotype. This latter approach, called Pattern Oriented Modeling [21], is very commonly used as a means of calibrating and validating ABMs. Initial validation of an ABM is accomplished when calibration results in satisfactory matching between the ABM and its referent with parameter values that are not clearly implausible, a level of validation is termed *face validity* [71]. Despite being the lowest level of validation possible for a simulation, establishing face validity is of extreme importance in the use of computational models for dynamic knowledge representation of biomedical systems. This is because biomedical research is primarily a discovery-oriented endeavour, where the primary procedural challenge is being able to separate plausible hypotheses from those that are not [2, 44]. Conversely, the inability to identify a set of parameters that can achieve plausible behaviour represents a failure of face validity; in these cases the underlying rules of the ABM need to be re-evaluated. Unfortunately, there are not clear guidelines about how to identify the transition point between inadequate sampling of parameter space and determination of model insufficiency, and the fact remains that this is a heuristic process that is enhanced by modelling experience.

We utilize pattern oriented analysis in the evaluation of the APIABM, focusing on two primary system patterns: (1) matching between the time courses of the modelled processes and the known disease pathophysiology, and (2) matching between the spatial patterns of alveolar edema generated by the APIABM and those recognized in the clinical setting. Each of the simulated disease processes below will include a brief description of the nature of the perturbation, confirmation of the expected time course and APIABM screenshots demonstrating the resulting patterns of pulmonary edema. Of note, other than the code changes to implement the specific type of perturbation, there were no differences or alteration in the code of the APIABM between the different disease state simulations.

Simulation of Pulmonary Contusion

A pulmonary contusion arises from direct trauma to the chest wall with force transmitted to the pulmonary parenchyma. It is, literally, a bruising of the lung. The traumatic force leads to locally distributed tissue damage, with subsequent activation of inflammation. Pulmonary contusion was simulated in the APIABM by applying a roughly circular injury pattern centered on the Cartesian coordinates of the APIABM with increasing radius of the applied injury pattern representing progressively increasing trauma. The dynamics of the inflammatory response followed the expected trajectory, peaking at approximately 3 days for those runs able to recover. A sequence of APIABM pulmonary contusion screenshots can be seen in Fig. 15.4.

Simulation of Pneumonia

Pneumonia arises from the introduction of pathogenic bacteria into the lung, with subsequent bacterial growth, tissue damage and inflammatory response. Pneumonia was simulated in the APIABM by applying a roughly circular distribution of `bacteria` agents, where increasing number of `bacteria` and corresponding size of the inoculated area represent progressively increasing inoculum. The dynamics of the inflammatory response followed the expected trajectory, with development of a significant "infiltrate" by 3 days in those levels of initial inoculum not spontaneously cleared. A sequence of APIABM pneumonia screenshots can be seen in Fig. 15.5.

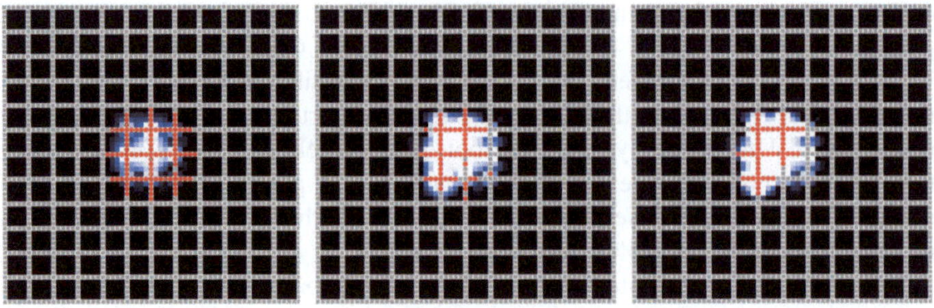

Fig. 15.4 Screenshots of 3-day course of pulmonary contusion simulated in the APIABM. This series of screenshots demonstrate the progression of alveolar edema resulting from a localized injury (sterile) corresponding to blunt pulmonary trauma. This is consistent with the time course seen both clinically and radiographically

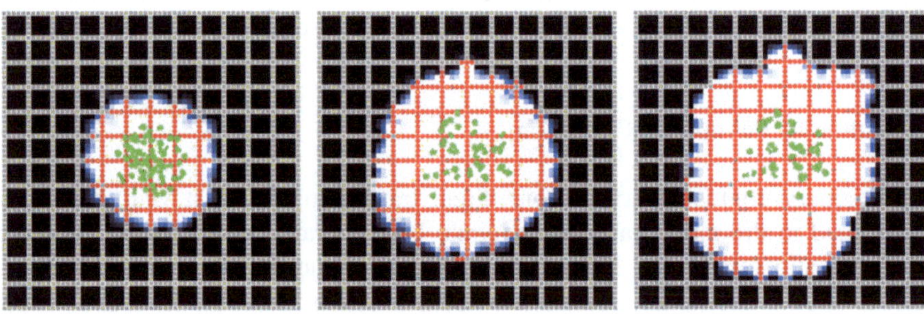

Fig. 15.5 Screenshots of 3-day course of bacterial pneumonia simulated in the APIABM. This series of screenshots demonstrate the progression of pneumonia resulting from a localized inoculation of bacteria. The pattern of alveolar edema corresponds to the evolution of a pneumonia-induced infiltrate seen radiographically

Simulation of Acute Lung Injury/Acute Respiratory Distress Syndrome

Acute lung injury/acute respiratory distress syndrome (ALI/ARDS) arises from activation of pulmonary inflammation by circulating inflammatory products generated by non-pulmonary systemic inflammation, such as sepsis. The multi-scale gut-lung ABM mentioned in Sect. 15.2.3.5 [13] examines the role of the mesenteric lymph in activating pulmonary inflammation, and we use the putative mechanism described by that ABM to simulate the effects of remote systemic inflammation on the lung. Systemic inflammation and subsequent production of inflammatory mesenteric lymph were abstractly represented by introducing a probability of spontaneous activation of neutrophils; this reflects both the priming of neutrophils and activation of pulmonary endothelium by inflammatory

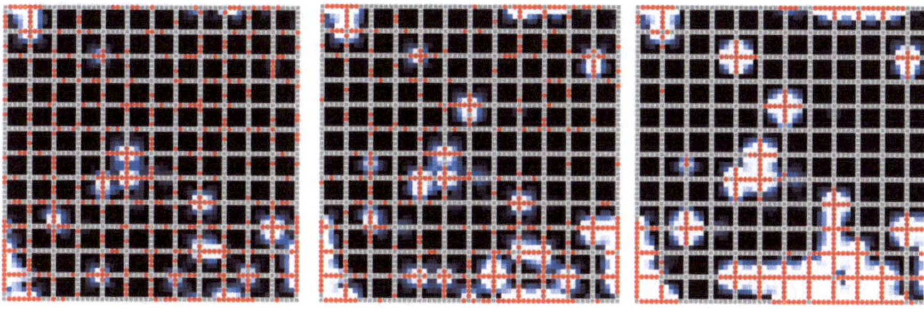

Fig. 15.6 Screenshots of 3-day course of acute lung injury/acute respiratory distress syndrome (*ALI/ARDS*) simulated in the APIABM. This series of screenshots demonstrate the progression of diffusely heterogeneously distributed alveolar edema arising for diffuse pro-inflammatory activation of the alveolar epithelium. This perturbation is consistent with the pulmonary effects of acute systemic inflammation as would be seen in sepsis or severe trauma. The time course and qualitative pattern of edema formation are consistent with the development of ARDS in the clinical setting

mesenteric lymph. The dynamics of the inflammatory response followed the expected trajectory, with the development of extensive patchy infiltrates by Day 3. A sequence of APIABM ALI/ARDS screenshots can be seen in Fig. 15.6.

15.2.4.7 Possible Extensions of the APIABM

The APIABM is a very abstract model, but due to the modular nature of ABMs it is readily extensible along a series of future development paths. Certainly more molecular detail can be included into the representation of the pro- and anti-inflammatory pathways; this could be driven by a researcher's particular interest area and desire to examine/confirm higher order behaviour related to that particular pathway. While the APIABM currently represented the alveolar airspaces involved in gas exchange, there is no functional consequence of the alveolar edema; it would be relatively straightforward to tie the edema state of each represented airspace to a gas exchange function, thereby being able to tie the inflammatory biology to a functional output of the lung. Pharmacological interventions can also be simulated: standard therapies such as antibiotics could be represented by a culling function applied to the bacterial populations, while anti-mediator interventions could be simulated as has been previously shown in in silico clinical trials [24]. More detail concerning the functions and characteristics of bacteria can be added where the specific virulence properties can be embedded into the bacterial agents to more closely approximate the complex of host-pathogen interactions in the face of inflammation [62]. Finally, the APIABM can be linked to other organ-level ABMs in a modular fashion [13], in order to capture the broader, systemically oriented genesis and consequences of pulmonary inflammation. Interested readers are encouraged to download the APIABM from http://bionetgen.org/SCAI-wiki and explore the possibilities available from agent-based modelling.

15.3 Discussion

15.3.1 Challenges to the use of Agent-Based Modeling

As with all modeling methods, agent-based modeling is not without its limitations. One common issue shared with all computational and mathematical modeling methods is that the quality and reliability of the models are directly related to the reliability of the underlying assumptions of the model and the quality of their implementation during construction of the model. This issue can be addressed by emphasizing transparency of both underlying assumptions and implementation details with respect to the construction of an ABM. The ODD protocol, while not developed specifically with biomedical ABMs in mind, provides a useful reference point with respect to documenting the structure and goals associated with an agent-based modeling project [67].

One shortcoming of agent-based modeling is the difficulty in applying formal analysis to the relationship between the agent-rules and the behavior of the system. Due to the combined stochastic behaviour of agents and the difficulty in assigning scalar metrics to account for the spatial aspects of an ABM's output it can be very challenging to evaluate the effect of parameter values and model structure on an ABM's behaviour. Alternatively, equation-based models have well-established procedures for analytical tasks such as parameter sensitivity analysis, bifurcation analysis, and behaviour-state-space determination. Work on developing mathematical descriptions of ABMs offer the prospect that formal analysis may be available in the future [72]. In the meantime, ABM researchers use a variety of strategies, such as heuristics [6, 24], literature-based constraints [28, 31] and Latin Hypercubes [10, 64] for parameter estimation and sensitivity analysis.

Some of the apprehension associated with the analysis of ABMs can be addressed by viewing ABMs as objects more akin to wet lab experimental platforms rather than more traditional, equation-based mathematical models. Pattern-oriented analysis, in which corresponding patterns of dynamic behaviour are used to relate the computational ABM to its real-world referent, allows ABMs to be evaluated much in the same way as wet lab systems or model organisms [21]. From this regard, the stochastic and emergent properties of ABMs reinforce their ability to capture the robustness of dynamic behaviour seen in complex systems, thereby allowing more insight into their core organizational structure.

ABMs are, in general, more computationally intensive than equation based models. The increased computational requirements place constraints on both the size of ABMs in terms of number of agents as well as the complexity of their internal rule systems. The natural solution to this bottleneck is to implement very large scale ABMs on current high performance computing platforms. However, there are intrinsic properties of ABMs, primarily related to the high degree of dynamics in the agent-to-agent interaction and communication network, that challenge the ability to implement ABM on highly distributed memory systems. Certain types of model architectures, mostly incorporating limited or relatively static interaction neighbourhoods with a high ratio of intra-agent computation (i.e. very complex mathematical rules) to inter-agent communication, are more suited to implementation on these massively parallel computer architectures. These types of models are also suited to implementation using Graphical Processing Units (GPUs), which offers the possibility of "supercomputer on a desk" computational power for selected types of ABMs [73–75]. It should be noted that there are also nontrivial modeling issues associated with parallel implementation of ABMs, aside from the computer science challenges just noted above. The selection of the scale of process to be distributed across multiple processors may have consequences with respect to concurrency and event scheduling and to the mapping of the simulation behaviour back to the biological referent; for instance attempting to distribute a single agent's rules over a series of processors. Thus far parallel ABM implementations have not explored the distribution of a single agent's execution across multiple processors, and have opted for a more organizationally defined distribution strategy that expands the overall size of the ABM (i.e. more agents) and keeps the implementation of agent-scale behaviour at the processor and sub-processor level.

15.3.2 Conclusion

The Translational Dilemma is the greatest challenge facing the biomedical research community today. Future operational procedures for biomedical science should involve technological augmentation of all the steps of the scientific cycle and allow the knowledge generated from such research to manifest in multiple areas. These include the development of highly predictive, personalized simulations to streamline the development and design of therapies, simulating the clinical application of these therapies in population studies (in silico clinical trials), and predicting the effects of drugs on individuals. We suggest that the agent-based paradigm, incorporating knowledge encapsulation, modularity and parallelism, can play an important role in the development of this meta-engineering process. Agent-based modeling can provide an integrative architecture for the computational representation of biological systems. Expanding the tools for AI-augmentation of computational dynamic knowledge representation and ties to biomedical ontologies [44, 76] can significantly reduce the threshold for the general researcher to utilize computational modelling and allow investigators to "see" the consequences of a particular hypothesis-structure/conceptual model, such that the mechanistic consequences of each component of the hypothesis can be probed and evaluated. Dynamic knowledge representation enables the instantiation of "thought experiments:" the exploration of possible alternative solutions and identifying those that are plausible, i.e. consistent with the observed data. These models can aid in the scientific process by providing a transparent framework for this type of speculation, which can then be used as jumping off points for the planning and design of further laboratory experiments and measurements. It is hoped that the increasing use of this type of knowledge representation and communication will foster the further development of "virtual laboratories" and in silico investigations.

References

1. Innovation or stagnation: challenge and opportunity on the critical path to new medical products (2004) [cited 1 May 2008]. Available from: http://www.fda.gov/oc/initiatives/criticalpath/whitepaper.html
2. An G (2010) Closing the scientific loop: bridging correlation and causality in the petaflop age. Sci Transl Med 2(41): 41ps34
3. An G et al (2009) Agent-based models in translational systems biology. Wiley Interdisc Rev Syst Biol Med. doi:10:1002/wsbm.45
4. Bankes SC (2002) Agent-based modeling: a revolution? Proc Natl Acad Sci U S A 99(3): 7199–7200
5. Bonabeau E (2002) Agent-based modeling: methods and techniques for simulating human systems. Proc Natl Acad Sci U S A 99(3):7280–7287
6. Hunt CA et al (2009) At the biological modeling and simulation frontier. Pharm Res
7. Walker DC, Southgate J (2009) The virtual cell: a candidate co-ordinator for 'middle-out' modeling of biological systems. Brief Bioinform 10(4):450–461
8. Zhang L, Athale CA, Deisboeck TS (2007) Development of a three-dimensional multiscale agent-based tumor model: simulating gene-protein interaction profiles, cell phenotypes and multicellular patterns in brain cancer. J Theor Biol 244(1):96–107
9. Santoni D, Pedicini M, Castiglione F (2008) Implementation of a regulatory gene network to simulate the TH1/2 differentiation in an agent-based model of hypersensitivity reactions. Bioinformatics 24(11):1374–1380
10. Fallahi-Sichani M et al (2011) Multiscale computational modeling reveals a critical role for TNF-alpha receptor 1 dynamics in tuberculosis granuloma formation. J Immunol 186(6):3472–3483
11. An G (2009) Dynamic knowledge representation using agent-based modeling: ontology instantiation and verification of conceptual models. Methods Mol Biol 500:445–468
12. Hunt CA et al (2006) Physiologically based synthetic models of hepatic disposition. J Pharmacokinet Pharmacodyn 33(6):737–772
13. An G (2008) Introduction of an agent-based multi-scale modular architecture for dynamic knowledge representation of acute inflammation. Theor Biol Med Model 5(1):11
14. Kirschner DE et al (2007) Toward a multiscale model of antigen presentation in immunity. Immunol Rev 216:93–118
15. Christley S, Alber MS, Newman SA (2007) Patterns of mesenchymal condensation in a multiscale, discrete stochastic model. PLoS Comput Biol 3(4):e76
16. Engelberg JA, Ropella GE, Hunt CA (2008) Essential operating principles for tumor spheroid growth. BMC Syst Biol 2(1):110
17. Reynolds CW (1987) Flocks, herds, and schools: a distributed behavioral model. SIGGRAPH 87 Comput Graph

18. Lipniacki T et al (2006) Stochastic regulation in early immune response. Biophys J 90(3):725–742
19. Lipniacki T et al (2006) Transcriptional stochasticity in gene expression. J Theor Biol 238(2):348–367
20. Vodovotz Y et al (2007) Evidence-based modeling of critical illness: an initial consensus from the society for complexity in acute illness. J Crit Care 22(1):77–84
21. Grimm V et al (2005) Pattern-oriented modeling of agent-based complex systems: lessons from ecology. Science 310:987–991
22. An G, Wilensky U (2009) From artificial life to in silico medicine: netlogo as a means of translational knowledge representation in biomedical research. In: Adamatsky A, Komosinski M (eds) Artificial life in software, vol 2. Springer, London, pp 183–209
23. An G (2001) Agent-based computer simulation and sirs: building a bridge between basic science and clinical trials. Shock 16(4):266–273
24. An G (2004) In silico experiments of existing and hypothetical cytokine-directed clinical trials using agent-based modeling. Crit Care Med 32(10):2050–2060
25. Mansury Y, Diggory M, Deisboeck TS (2006) Evolutionary game theory in an agent-based brain tumor model: exploring the 'genotype-phenotype' link. J Theor Biol 238(1):146–156
26. Deisboeck TS et al (2001) Pattern of self-organization in tumour systems: complex growth dynamics in a novel brain tumour spheroid model. Cell Prolif 34(2):115–134
27. Chen S, Ganguli S, Hunt CA (2004) An agent-based computational approach for representing aspects of in vitro multi-cellular tumor spheroid growth. Conf Proc IEEE Eng Med Biol Soc 1:691–694
28. Thorne BC et al (2006) Modeling blood vessel growth and leukocyte extravasation in ischemic injury: an integrated agent-based and finite element analysis approach. J Crit Care 21(4):346
29. Tang J, Ley KF, Hunt CA (2007) Dynamics of in silico leukocyte rolling, activation, and adhesion. BMC Syst Biol 1:14
30. Tang J et al (2004) Simulating leukocyte-venule interactions: a novel agent-oriented approach. Conf Proc IEEE Eng Med Biol Soc 7:4978–4981
31. Bailey AM, Thorne BC, Peirce SM (2007) Multi-cell agent-based simulation of the microvasculature to study the dynamics of circulating inflammatory cell trafficking. Ann Biomed Eng 35(6):916–936
32. Bailey AM et al (2009) Agent-based model of therapeutic adipose-derived stromal cell trafficking during ischemia predicts ability to roll on P-selectin. PLoS Comput Biol 5(2):e1000294
33. Mi Q et al (2007) Agent-based model of inflammation and wound healing: insights into diabetic foot ulcer pathology and the role of transforming growth factor-beta1. Wound Repair Regen 15(5):671–682
34. Walker DC et al (2004) Agent-based computational modeling of wounded epithelial cell monolayers. IEEE Trans Nanobioscience 3(3):153–163
35. Adra S et al (2010) Development of a three dimensional multi scale computational model of the human epidermis. PLoS ONE 5(1):e8511
36. An G (2009) A model of TLR4 signaling and tolerance using a qualitative, particle-event-based method: introduction of spatially configured stochastic reaction chambers (SCSRC). Math Biosci 217(1):43–52
37. Broderick G et al (2005) A life-like virtual cell membrane using discrete automata. In Silico Biol 5(2):163–178
38. Pogson M et al (2008) Introducing spatial information into predictive NF-kappaB modelling: an agent-based approach. PLoS ONE 3(6):e2367
39. Pogson M et al (2006) Formal agent-based modelling of intracellular chemical interactions. Biosystems 85(1):37–45
40. Ridgway D et al (2008) Coarse-grained molecular simulation of diffusion and reaction kinetics in a crowded virtual cytoplasm. Biophys J 94(10):3748–3759

41. Troisi A, Wong V, Ratner MA (2005) An agent-based approach for modeling molecular self-organization. Proc Natl Acad Sci U S A 102(2):255–260
42. Dong X et al (2010) Agent-based modeling of endotoxin-induced acute inflammatory response in human blood leukocytes. PLoS ONE 5(2):e9249
43. Hoehme S, Drasdo D (2010) A cell-based simulation software for multi-cellular systems. Bioinformatics 26(20):2641–2642
44. An G, Christley S (2011) Agent-based modeling and biomedical ontologies: a roadmap. Wiley Interdisc Rev Comput Stat 3(4):343–356
45. Solovyev A et al (2011) SPARK: a framework for multi-scale agent-based biomedical modeling. Int J Agent Technol Syst 2(3):18–31
46. Railsback SF, Lytinen SL, Jackson SK (2006) Agent-based simulation platforms: review and development recommendations. Simulation 82(9):609–623
47. Vodovotz Y et al (2009) Mechanistic simulations of inflammation: current state and future prospects. Math Biosci 217(1):1–10
48. Nathan C (2002) Points of control in inflammation. Nature 420(6917):846–852
49. Schlag G, Redl H (1996) Mediators of injury and inflammation. World J Surg 20(4):406–410
50. Matzinger P (2002) The danger model: a renewed sense of self. Science 296(5566):301–305
51. Santos CC et al (2005) Bench-to-bedside review: biotrauma and modulation of the innate immune response. Crit Care 9(3):280–286
52. Medzhitov R (2008) Origin and physiological roles of inflammation. Nature 454(7203): 428–435
53. Oviedo JA, Wolfe MM (2001) Clinical potential of cyclo-oxygenase-2 inhibitors. BioDrugs 15(9):563–572
54. Borer JS, Simon LS (2005) Cardiovascular and gastrointestinal effects of COX-2 inhibitors and NSAIDs: achieving a balance. Arthritis Res Ther 7(4):S14–S22
55. Rychly DJ, DiPiro JT (2005) Infections associated with tumor necrosis factor-alpha antagonists. Pharmacotherapy 25(9):1181–1192
56. Calabrese L (2006) The yin and yang of tumor necrosis factor inhibitors. Cleve Clin J Med 73(3):251–256
57. An GC (2010) Translational systems biology using an agent-based approach for dynamic knowledge representation: an evolutionary paradigm for biomedical research. Wound Repair Regen 18(1):8–12
58. Li NY et al (2008) A patient-specific in silico model of inflammation and healing tested in acute vocal fold injury. PLoS ONE 3(7):e2789
59. Li NY et al (2010) Bio simulation of inflammation and healing in surgically injured vocal folds. Ann Otol Rhinol Laryngol 119(6):412–423
60. Seal JB et al (2010) The molecular Koch's postulates and surgical infection: a view forward. Surgery 147(6):757–765
61. Wendelsdorf K et al (2011) Enteric immunity simulator: a tool for in silico study of gut immunopathologies. Virginia Bioinformatics Institute, Blacksburg, p 1–27
62. Seal JB et al (2011) Agent-based dynamic knowledge representation of pseudomonas aeruginosa virulence activation in the stressed gut: towards characterizing host-pathogen interactions in gut-derived sepsis. Theor Biol Med Model 8:33
63. Kim M et al (2012) Immature oxidative stress management as a unifying principle in the pathogenesis of necrotizing enterocolitis: insights from an agent-based model. Surg Infect (Larchmt) 13(1):18–32
64. Segovia-Juarez JL, Ganguli S, Kirschner D (2004) Identifying control mechanisms of granuloma formation during M. tuberculosis infection using an agent-based model. J Theor Biol 231(3):357–376
65. Brown BN et al (2011) An agent-based model of inflammation and fibrosis following particulate exposure in the lung. Math Biosci 231(2):186–196
66. Deitch EA (2010) Gut lymph and lymphatics: a source of factors leading to organ injury and dysfunction. Ann N Y Acad Sci 1207(Suppl 1):E103–E111

67. Grimm V et al (2010) The ODD protocol: a review and first update. Ecol Model 221:2760–2768
68. Wilensky U, Rand W (2009) An introduction to agent-based modeling: modeling natural, social and engineered complex systems with NetLogo. MIT Press, Cambridge
69. van oud Alblas AB, van Furth R (1979) Origin, kinetics, and characteristics of pulmonary macrophages in the normal steady state. J Exp Med 149(6):1504–1518
70. Zeigler B, Praehofer H, Kim TG (2000) Theory of modeling and simulation: integrating discrete event and continuous complex dynamic systems, vol 2. Elsevier, Sas Diego, p 510
71. Balci O (2001) A methodology for certification of modeling and simulation applications. ACM Trans Model Comput Simul 11(4):352–377
72. Hinkelmann F et al (2011) A mathematical framework for agent based models of complex biological networks. Bull Math Biol 73(7):1583–1602
73. Richards RS et al (2008) Data-parallel techniques for agent-based tissue modeling on graphical processing units. In: Design engineering technical conference and computers and information in engineering conference. New York
74. Richmond P et al (2010) High performance cellular level agent-based simulation with FLAME for the GPU. Briefings Bioinform 11(3):334–347
75. Christley S et al (2010) Integrative multicellular biological modeling: a case study of 3D epidermal development using GPU algorithms. BMC Syst Biol 4:107
76. Christley S, An G (2011) A proposal for augmenting biological model construction with a semi-intelligent computational modeling assistant. Comput Math Organ Theor 17(4):1–24

Chapter 16
Reconstruction and Comparison of Cellular Signaling Pathway Resources for the Systems-Level Analysis of Cross-Talks

Máté Pálfy, László Földvári-Nagy, Dezső Módos, Katalin Lenti and Tamás Korcsmáros

Abstract Signaling pathways control a large variety of cellular processes and their defects are often linked with diseases. Reliable analyses of these pathways need uniform pathway definitions and curation rules applied to all pathways. Here, we compare KEGG, Reactome, Netpath and SignaLink pathway databases and examine their usefulness in systems-level analysis. Further on, we show that the integration of various bioinformatics databases allows a comprehensive understanding of the regulatory processes that control signaling pathways. We also discuss the drug target relevance of cross-talking (i.e., multi-pathway) proteins and signal transduction regulators (e.g., phophatases and miRNAs). Accordingly, modern integrated databases are not only essential for studying signaling processes at the systems level, but will also serve as invaluable tools for pharmacology and network-based medicine.

Keywords Signaling · Cross-talk · Regulation · Drug discovery · Network · Pathway · miRNA · Drug targeting · Pathway database

Acronyms

HTP	High-throughput
PPI	Protein-protein interaction
TF	Transcription factor
TFBS	Transcription factor binding site
miRNA	microRNA

M. Pálfy · L. Földvári-Nagy · D. Módos · T. Korcsmáros (✉)
Department of Genetics, Eötvös Loránd University, Budapest, Hungary
e-mail: korcsmaros@netbiol.elte.hu, palfy.mate@gmail.com, foldvari-nagi@netbiol.elte.hu, dezso.modos@netbiol.elte.hu

D. Módos · K. Lenti
Department of Morphology and Physiology, Faculty of Health Sciences, Semmelweis University, Budapest, Hungary
e-mail: dr.lenti.kata@gmail.com

A. Prokop and B. Csukás (eds.), *Systems Biology*,
DOI: 10.1007/978-94-007-6803-1_16,

16.1 Signaling Pathways and Cross-Talks

Intracellular signaling, from the simplest cascades to the highly intertwined networks of kinases, contributes to the diversity of developmental programs and adaptation responses in metazoans [1]. In humans, defects in intracellular signaling can cause various diseases, e.g., cancer, neurodegeneration, or diabetes. Thus, understanding the structure, function, and evolution of signal transduction is an important task for both basic research and medicine.

Signaling pathways, the functional building blocks of intracellular signaling, transmit extracellular information from ligands through receptors and mediators to transcription factors, which induce specific gene expression changes. In contrast to the wide variety of signaling functions and the macroscopic and microscopic diversity of living forms, the number of signaling pathway types are relatively low (a few dozen) [2]. The basic mechanisms of each pathway are conservative, characteristic to large taxon groups, and present ubiquitously in different tissues [1, 2]. Interestingly, most of the pathways have a maximum of 10–20 protein components [1]. These numbers apparently contradict to the number of cell types that signaling pathways can create and maintain. The major sources to generate diverse and complex signaling flow with such few pathways are specific co-factors and positive/negative feedback loops [3, 4]. Over the past decade, it has been realized that signaling pathways are highly structured and rich in cross-talks (where cross-talk is defined here as a directed physical interaction between pathways) ([84], [5]). Cross-talks can form and change more frequently than the interactions within pathways [6, 7]. As the number and combination of transducable signals are limited, new cross-talks between pathways can create novel input/output combinations, which increase the possible ways of signaling flow and thus contribute to diverse phenotypes.

However, to ensure that an appropriate response is elicited, the signaling system has to maintain the pathways' output specificity (inputs preferentially activate their own output) and input fidelity (outputs preferentially respond to their own input) [8]. Thus, new interactions between pathways need to be precisely regulated. Regulation of cross-talks to prevent 'leaking' or 'spillover' can be achieved with different insulating mechanisms [8]. Signaling cross-talks are controlled mainly by scaffold proteins, cross-pathway inhibitions, kinetic insulation, and the spatial and temporal expression patterns of proteins [4, 9–11]. One can find all these mechanisms in the concept of critical nodes, defined by Kahn and co-workers, and demonstrated for the insulin pathway [12]. Critical nodes are defined as protein groups, where the members are (1) essential in the signal transduction of a given pathway, (2) related to each other (isoforms), (3) regulated and function in a partially different way, and where (4) at least one of the members participates in a cross-talk with another pathway [12]. The relative concentration of the critical node members and their differential regulation determine the way of the signaling flow [12]. The default way of signaling flow is from a pathway-specific ligand via a critical node to a pathway-specific transcription factor. But when a critical node contains multiple protein isoforms, which include a member that can cross-talk, the signal can be switched to another pathway, i.e., generating another output. Consequently, two pathways can specifically cross-talk with a shared protein group, where the partially regulated protein isoforms serve as a source of divergence [12].

16.2 Challenges to Study Cross-Talk at the Systems-Level

Despite the general prevalence of network approaches, the definition of pathways has seen little change. As structural, functional, tissue- and disease-specific aspects come into consideration while identifying individual pathways, these differing aspects also have to be taken into account when studying cross-talks. Different studies address the role of cross-talks in the context of distinct cell fates, cell types, single or multiple pathways. In Table 16.1 we list some examples for these different approaches.

Nowadays, systems-level and network-based methods have started to dominate the study of signaling pathways, accordingly the systems level analyses of cross-talks has become a major task. First of all, this requires a precise definition of pathways and pathway borders. By reviewing the major issues of studying cross-talks at the systems level, Gerstein and colleges point out that pathways compiled from different systems and constructed for distinct purposes are not suitable for examining cross-talks [13]. Bauer-Mehren et al. came to the same conclusion while testing the cPath integrated database [14] and argue the need of new databases that make the study of cross-talks possible at the systems level [15]. These require a compilation based on general principles and importantly, the use of standardized methods. Among these, high-throughput (HTP) methods provide the greatest number of protein–protein interactions (PPIs) and are therefore commonly used in network biology research. However, for methodological reasons, these HTP screens are unable to reveal interactions of extracellular, membrane-bound and nuclear proteins—all of them important players in signal transduction. A further problem of PPIs from HTP data is that they are mostly undirected, while most of the reactions in the signaling network are directed.

Due to these limitations, manually curated databases have emerged as indispensable tools for systems-level research of signaling pathways. Although usually containing less information, they are more detailed and reliable. However, most of these curated signaling databases both lack a precise definition of the pathways and a standardized curation protocol. Consequently, it is difficult to compare the distinct pathways even within the same database, or to analyze interactions between pathways. For extensive cross-talk analysis, a signaling database is required, that: (1) has a structure fulfilling the modern requirements of systems biology; (2) is objective and contains uniformly defined pathways; (3) contains sufficient and reliable network information. Additionally, if the above criteria apply to multiple species, this further allows prediction of new proteins, protein functions, and PPIs based on orthology.

Table 16.1 Examples of different approaches for the analysis of cross-talks ordered in growing complexity

Cross-talk studies	Cross-talking signaling pathways	References	Type and details of the reference(s)
Modeling cross-talks in a single pathway	Hyperosmolar and pheromone MAPK pathway	[70]	Research article on mathematical modeling and experimental validation
Cross-talks of a single pathway in healthy cell types	Cross-talks of TGF-β/BMP with MAPK, PI3K/Akt, WNT, Hh, notch, IL/TNF-β/IFN-γ pathways; Cross-talks of notch with Hh, JAK-STAT,TGF-β, RTK, WNT pathways	[71, 72]	Review articles
Cross-talks of a single pathway in stem-cells	Cross-talk of WNT pathway with FGF, notch pathways	[73]	Review article
Cross-talks of two pathways	EGF and Insulin pathways; PI3K and ERK(MAPK) pathways	[74, 75]	Research articles on computational modeling and experimental validation
Cross-talks of specific pathways in a specific tissue	(many)	[76]	Research article
Interaction of multiple pathways in stem cells	Notch, WNT, TGF-β, BMP pathways	[77]	Review article
Coordination of multiple pathways during organ development	Hh, WNT, FGF, WNT, IGF; EGF, notch, WNT	[86], [78, 79]	Review article
Cross-talk of multiple pathways in an organ	Notch and WNT pathways	[80]	Research article on experimental data
Cross-talk of multiple pathways in the development of tumors	WNT, BMP, FGF, notch and Hh pathways	[81]	Review article
Interaction of multiple pathways in normal and stem cell differentiation	JAK-STAT, notch, MAPK, PI3 K/AKT, NF-κB, WNT, TGF-β pathways	[85]	Review article
Cross-talk of multiple (9) pathways in a general protein network	MAPK, TGF-β, notch, WNT, Hh, mTOR, TLR, JAK-STAT, VEGF pathways	[13]	Review article
Extensive cross-talk (580) of multiple pathways in a general protein network	(many)	[82]	Research article on bioinformatic data
Cross-talks in intercellular communication of two pathways in an organ	FGF and BMP pathways	[83]	Research article on experimental data

16.3 Benchmarking Signaling Resources to Study Cross-Talks

We examined 3 widely used, freely available general signaling pathway databases, KEGG, Reactome and Netpath [16–19], and compared it with SignaLink, a recently developed signaling pathway database intended for the analysis of signaling cross-talks [6]. All four databases were constructed by utilizing different sources and applying distinct methods, hence they greatly vary in a number of aspects. KEGG contains pathway information from a large number of species, whereas SignaLink deals only with data from the model organisms *Caenorhabditis elegans, Drosophila melanogaster* and from human. In contrast, the data collected in the Reactome and Netpath databases are restricted to human signaling pathways. In case of KEGG there is no clear pathway definition, thus, what is considered as an individual pathway is decided by the curator. In Netpath 10 immune and 10 cancer signaling pathways were curated based on PPI data from the HPRD resource [18, 19]. In contrast, the Reactome and SignaLink databases feature a unified and available protocol for data collection. The pathways in SignaLink are biochemically and evolutionarily defined and are identical with the pathway grouping of [1]. It is important to note, that solely in virtue of the number of pathways, these databases are not comparable. For example, in the SignaLink database, the EGF/MAPK pathway contains the proteins and interactions between the EGF ligand and the terminal MAPK proteins. While the grouping of these interactions and proteins into a single pathway is biochemically and evolutionarily reasonable, many databases scatter this pathway across many (sub)pathways (e.g., EGFR, RAS, p38, JNK, ERK, ASK). Although a relevant and objective pathway definition decreases the overall number of pathways in the database, it avoids artificial and biased pathway grouping.

An important aspect in manually curated databases is the assignment of proteins to signaling pathways. In the Reactome and Netpath databases, this is entirely dependent on individual experts who construct the pathways, but no references are provided for the users. Similarly, in the KEGG and SignaLink pathways, to which pathway a protein is annotated is decided by curators, but importantly, their decision is based on published review papers from experts of the given pathway. While KEGG collects the information from only a few (usually 5–10) reviews, SignaLink uses 20–25 reviews per pathway and also adds additional PPI information based on orthology. The reliability and utility of databases greatly depends on the availability of published references, which underlie every single protein–protein interaction. This is accessible for every interaction in the Reactome, NetPath and SignaLink databases, however, KEGG only refers to review papers.

By comparing all 4 databases, SignaLink showed the largest overlap with the other databases and contained the most references from the literature. Therefore, we set SignaLink against the other databases comparing 7 human signal transduction pathways in SignaLink (EGF/MAPK, IGF, Hedgehog, JAK/STAT, Notch, TGF-β, WNT) with 7 human signaling pathways from KEGG (MAPK, Insulin,

Hedgehog, JAK/STAT, Notch, TGF-β, WNT), 5 pathways from Reactome (EGFR, Insulin receptor, Notch, TGF, WNT) and 5 pathways from NetPath (EGFR1, Hedgehog, Notch, TGF, WNT). Regarding the number of proteins found in the pathways, the 7 pathways in KEGG have 17 % less, the 5 pathways in Reactome have 84 % less, while the 5 pathways in Netpath contain approximately the same number of proteins as in the corresponding pathways in SignaLink. In the case of so-called multi-pathway proteins [20], which participate in multiple pathways and function in cross-talks, KEGG contains about the same number as SignaLink, whereas only half of these types of proteins can be found in Reactome and Netpath.

In comparing the number of PPIs, KEGG contains 52 % more interactions than SignaLink, but notably, most of these interactions are artificial, as they were obtained indirectly using a matrix method [21]. Interestingly, the number of cross-talks linking distinct signaling pathways is about the same in SignaLink and KEGG, therefore, the relative amount of cross-talks in SignaLink is probably higher. In Reactome, when including all interactions within protein complexes, this database has up to two times as many PPIsas SignaLink in overall, however, without the protein complexes, the number of interactions is roughly equal. Regarding the number of cross-talks, SignaLink contains almost three times as many as Reactome, and this is not influenced by the presence of interactions within protein complexes. In comparison to Netpath, SignaLink contains about three times as many cross-talks, 1.5 times as many PPIs, while the number of proteins is approximately the same, albeit with only about 30 % overlap between the two databases [6].

Based on this comparison we can conclude, that the major advantage of SignaLink over the other three databases is that it features precisely defined signaling pathways, has detailed criteria for assigning proteins to pathways and uses a unified curation method which makes a systems-wide analysis of signaling pathways possible. Furthermore, within the signaling pathways shared by all four databases, SignaLink contains the most proteins, interactions and references. This makes SignaLink an excellent resource for taking on the new challenges of signal transduction research and for the efficient study of cross-talks.

16.4 Extending Signaling Pathways with Regulatory Processes

Signaling networks can be divided into upstream and downstream subnetworks. The upstream subnetwork contains the intertwined network of signaling pathways, presented earlier, while the downstream, gene regulatory subnetwork (GRN) contains transcription factor binding sites, transcription factors and microRNAs, ultimately controlling global gene expression and the dynamics of protein output in a living cell [22] (Fig. 16.1). The GRN can further be divided into

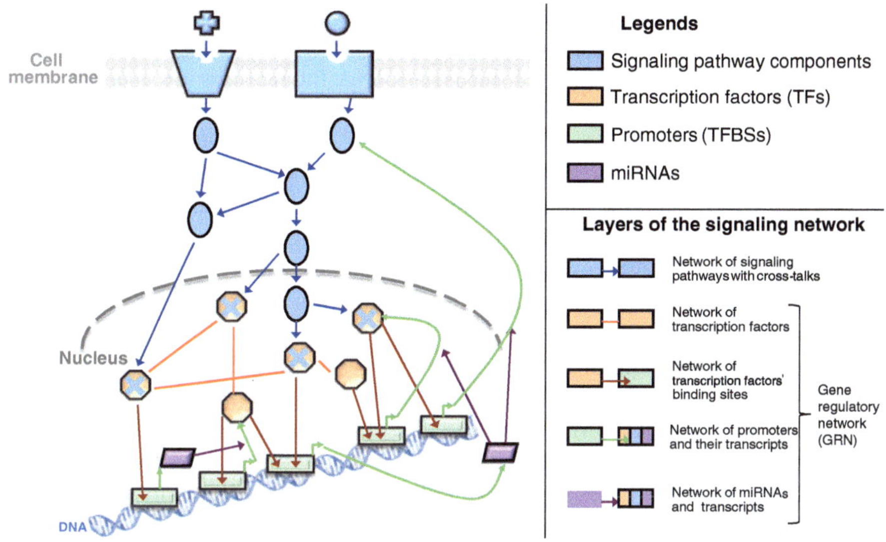

Fig. 16.1 Layers of the signaling network. In the upstream part the network of signaling pathways contains the incoming signals (ligands) that activate receptors and mediator proteins to reach the transcription factors in the nucleus. In the nucleus, the downstream, gene regulatory network (*GRN*) contains four layers (networks): the network of transcription factors (*TFs*), the network of TFs and their binding site in the promoter region of certain genes, the network between these regions and their transcripts, and the network of microRNAs (*miRNAs*) and their target mRNAs

transcriptional and post-transcriptional subnetworks. At the transcriptional level, transcription factors (TFs) bind specific regions of DNA sequences (called transcription factor binding sites (TFBS) or response elements) and regulate the mRNA expression of transcription factor target genes. Post-transcriptionally, microRNAs (miRNAs) regulate gene expression by binding to complementary sequences (i.e., miRNA binding-sites) on target mRNAs. The specific binding of a miRNA to its target mRNA can suspend or permanently repress the translation of a given transcript, thereby specifically inhibiting protein production [23, 24]. Despite the difficulties of identifying miRNA targets, it is predicted that nearly all human genes can be controlled by at least one miRNA [25] and mutations in many miRNA coding genes have pathological consequences [26]. The importance of miRNAs in the regulation of protein–protein networks was highlighted by a positive correlation between the number of repressing miRNAs and the protein partners (i.e., degree) of a given protein [27]. Thus, proteins having many interactors (i.e., protein hubs) are more tightly regulated than proteins with less interactors [27]. In addition, a comprehensive analysis suggested that specific biological processes are regulated by miRNAs through targeting the hub and bottleneck proteins of the protein interaction network [28].

Recently, many databases comprising the downstream regulatory subnetwork components of signaling pathways have been created. A compendium of human

TFs have been collected and analyzed in [29], while their regulatory interactions can be acquired from the resources JASPAR, MPromDB, PAZAR and OregAnno [30–33]. Experimentally validated miRNA-mRNA interactions are available from TarBase [34], while predicted interactions can be accessed at TargetScan and PicTar [35, 36]. TransMir and PutMir contain TF-miRNA regulatory information to examine how miRNAs are regulated [37, 38]. In addition, miRecords and miRGen provide an integrated resource from where different miRNA-related resources can be accessed [39, 40]. To examine the signaling network in a unified fashion, integrated resources including IntegromeDB and TranscriptomeBrowser 3.0 have been developed, which allow the examination of all layers from signaling pathways to miRNAs through TFs [41, 42].

As an update for SignaLink, we have recently developed an integrated database on the regulation of signaling, containing information from *C. elegans*, *D. melanogaster*, and humans (BMC Syst Biol. 7:1752-0509-7-7). Signaling pathway information from SignaLink was integrated with major processes that regulate signaling. First, on the bases of manual curation of primary literature and reviews, we linked scaffold proteins, specific ubiquitin-ligases, and proteins involved in endocytosis to pathway proteins. Next, we extended the network with the first neighbors of the proteins based on directed protein–protein interactions (PPI). The PPI data was retrieved from BioGRID, DroID, and WI8. The direction and the confidence for each interaction was calculated based on domain–domain and domain-motif interactions. In the next step, we included the underlying regulatory network: (1) downstream transcription factors and their subnetworks, based on manual curation of primary literature; (2) interactions between transcription factors and transcription factor binding sites of genes, using OregAnno, JASPAR, and MPromDB; (3) mRNA transcripts (from ENSEMBL), miRNA transcripts (from miRBase, miRGen and PutmiR), and their interactions (from miRecords and Tarbase). The database can be freely downloaded for academic purposes in various network file formats (BioPAX, SBML, CSV, etc.) via a BioMART-like download page, where users can filter the datasets.

16.5 Pharmacological Relevance of Signaling Networks

Understanding the structure and mechanism of normal signaling networks can reveal important targets for drug discovery. In many cases, these targets have no direct relation to a particular disease but their stimulation or inhibition can have beneficial systems-level effects on the cellular network, and lead to the survival of the organism. Pharmacological modulation of key proteins of the signaling network can influence the robustness of the cells for therapeutic purposes, e.g., increasing robustness in healthy cells and decreasing robustness in cancerous cells during chemotherapy [43, 44]. Three members of the insulin signaling pathway (PI3 kinase, AKT and IRS families) have already been identified as 'critical nodes' having distinctive roles in the junctions of signaling pathways and effecting the behavior of the cell during diabetes [12]. Hwang et al. developed the network parameter bridging centrality to identify key proteins in signal flow-modulation as promising drug targets [45]. The major strength of bridging centrality is that it effectively combines local and global network properties. Proteins with high bridging centrality (i.e., bridging nodes) are located in the critical sites of the signaling network and connect different parts (regions or modules) to one another [45]. They also found that many bridging nodes (e.g., SHC, JAK2, cortisol) had a track record as effective drug targets [45].

On the other hand, gene expression and sequencing studies on pathologically altered signaling networks can uncover possible drug targets whose malfunction directly cause disease. For example, during tumorigenesis when cells acquire continuous cell division and often increased mutation rate [46] most of the (driver) mutations affect a limited number of central pathways [47]. Drug targeting of these specific pathways could potentially prevent tumor growth. However, the development of aggressive and malignant tumor cells cause a systems-level change in the signaling network [48], thus their therapeutic treatment poses a major challenge. The pathological rewiring of the signaling network allows the appearance of cancer hallmarks, including sustained angiogenesis and metastatic tissue invasion capabilities [49], as well as the deregulation of cellular metabolism and avoidance from immune destructions [50]. The effect of signaling rewiring on cancer hallmarks was shown in prostate cancer [51]. Several works demonstrated that changes of cross-talk (i.e., multi-pathway) proteins are important for the rewiring of the signaling network [48, 52, 53]. Mutation even in one multi-pathway protein can have a systems-level effect as it can significantly alter the signaling flow, for example, transducing a 'death' signal to a 'survival' transcription factor [49, 54]. Similarly, we found a significant change in the expression level of multi-pathway proteins in hepatocellular carcinoma [6]. Accordingly, multi-pathway proteins are often altered in systems diseases such as cancer, thus, they are among the most promising drug targets [20]. Pharmacological modification of these proteins can re-transform the rewired cancerous signaling network.

Kinases are traditionally among the most targeted proteins of the cellular signaling network [55] although their selective targeting is a challenge for drug development. Kinase domains and their target motifs (i.e., specific amino acid sequences in the substrate proteins) are well-known and comprehensively compiled in resources such as Phosphosite [56], NetworKIN and NetPhorest [57, 58]. Regulatory domains of these kinases and scaffold proteins are also important to maintain kinase-substrate or scaffold-substrate specificity [59] but our systems-level knowledge on these (undirected) protein–protein interactions are less limited than the directed phosphorylation data.

It is important to highlight that less attention has been taken on the other players of the phosphorylation system: the protein phoshatases. As reviewed by Kholodenko and colleagues [60], protein phosphatases can play a dominant role in determining the spatio-temporal behavior of protein phosphorylation systems in the cell as both immediate and delayed negative regulators. Thus, pharmacological targeting of phosphatases can modify the signaling network at the systems-level. Despite their promising effect, only a few protein tyrosine phosphatases are currently used as therapeutic targets [61]. The development of drugs specifically targeting phosphatases is much more complicated than the development of anti-kinase drugs for the following reasons: (1) high-level of homology between phosphatase domains limits the development of selective compounds; (2) contrary to kinases, phosphatase substrate specificity is achieved through docking of the phosphatase complex at a site distant from the dephosphorylated amino acid [62, 63]; (3) the targeted sequences are highly charged, and many of the developed drug compounds cannot cross the membrane [64]. Resolved phosphatase-complex structures and detailed knowledge of their enzymatic activity will allow effective drug development and their utilization as systems-level drug targets.

Recently, miRNAs have been recognized as highly promising, non-protein intervention points in the signaling network. Therapeutic targeting of regulatory components is a challenging task because of specificity and pharmacological availability (i.e., therapeutic agents often have off-target effects and hardly enter the nucleus). Pharmacological modulation of protein and miRNA expression with an antisense strategy appears to be more specific than targeting TFs, TFBSs and miRNA promoters [65]. Specificity comes from the fact that antisense strategy affects single miRNAs and miRNA families that are specific for a given mRNA (or mRNA cluster), while TFs and promoters have less specific effects on the whole transcriptome [65].

Besides specificity, miRNAs can be important therapeutic targets, as their down- or up-regulation is implicated in more than 270 diseases according to the the Human MicroRNA Disease Database [66]. The diseases where altered expression of miRNAs have been reported include cardiovascular, neurodegenerative diseases, viral infections like HIV and various types of cancer [67]. The development of therapeutic strategies involving miRNAs requires the exploration of the signaling network. Therapeutic miRNAs can only be selected if their mRNA-interactions have been confidently identified and experimentally validated. These interactions can be accessed in specific and integrated resources listed in the previous section. In addition, evaluation of the cellular processes that are affected by the given miRNA is also necessary to avoid side-effects and unwanted drug effects. Web-services, such as Pathway Linker (http://PathwayLinker.org; [68]) have been developed for this purpose. As miRNAs often have multiple targets analysis of the network of the affected proteins (encoded by target mRNAs) can facilitate pharmacological development: identification of proteins whose knock-down has limited side-effects and toxicity profiles can be promising agents for miRNA-based therapeutics [65]. Such side-effects can be analyzed by databases such as SIDER [69].

16.6 Conclusion

The study of cross-talks has emerged as an important field in signal transduction research. To identify cross-talks and understand their roles in development and disease, one needs to analyze signaling networks at the systems level. Decades of research on signaling pathways and modern high-throughput methods have provided large data sets on the signaling components. Still, only a small number of databases fulfill the requirements of analyzing cross-talks at the systems level. By comparing 4 signaling databases (KEGG, Reactome, Netpath and SignaLink) in terms of pathway definition, curation methods, protein number, PPI number and the number of cross-talks, we point out that the SignaLink database is a valuable resource for cross-talk research. Signaling pathways are strictly regulated by downstream components, including transcription factors and miRNAs. Information on this gene regulatory subnetwork has been compiled into various databases which serve specific needs. For a comprehensive analysis of signaling from ligand binding to alterations in gene expression, integrated databases containing a great number of regulatory components (including both posttranscriptional and post-translational modifications) of signaling proteins are needed. These will contribute to the understanding of systems biology diseases such as cancer, and help predict more efficient drug targets for fighting against these diseases.

Acknowledgments The authors thank the members of the NetBiol Group (http://netbiol.elte.hu) for their useful comments and Peter Csermely for discussions. Work of TK was supported by a János Bolyai Scholarship of the Hungarian Academy of Sciences.

References

1. Pires-daSilva A, Sommer RJ (2003) The evolution of signalling pathways in animal development. Nat Rev Genet 4:39–49
2. Gerhart J (1999) Warkany lecture: signaling pathways in development. Teratology 60:226–239
3. Bray SJ (2006) Notch signalling: a simple pathway becomes complex. Nat Rev Mol Cell Biol 7:678–689
4. Freeman M (2000) Feedback control of intercellular signalling in development. Nature 408:313–319
5. Papin JA, Hunter T, Palsson BO, Subramaniam S (2005) Reconstruction of cellular signalling networks and analysis of their properties. Nat Rev Mol Cell Biol 6:99–111
6. Korcsmaros T, Farkas IJ, Szalay MS, Rovo P, Fazekas D, Spiro Z, Bode C, Lenti K, Vellai T, Csermely P (2010) Uniformly curated signaling pathways reveal tissue-specific cross-talks and support drug target discovery. Bioinformatics 26:2042–2050
7. Ulitsky I, Shamir R (2007) Pathway redundancy and protein essentiality revealed in the Saccharomyces cerevisiae interaction networks. Mol Syst Biol 3:104
8. Haney S, Bardwell L, Nie Q (2010) Ultrasensitive responses and specificity in cell signaling. BMC Syst Biol 4:119

9. Behar M, Dohlman HG, Elston TC (2007) Kinetic insulation as an effective mechanism for achieving pathway specificity in intracellular signaling networks. Proc Natl Acad Sci U S A 104:16146–16151
10. Bhattacharyya RP, Remenyi A, Yeh BJ, Lim WA (2006) Domains, motifs, and scaffolds: the role of modular interactions in the evolution and wiring of cell signaling circuits. Annu Rev Biochem 75:655–680
11. Kholodenko BN (2006) Cell-signalling dynamics in time and space. Nat Rev Mol Cell Biol 7:165–176
12. Taniguchi CM, Emanuelli B, Kahn CR (2006) Critical nodes in signalling pathways: insights into insulin action. Nat Rev Mol Cell Biol 7:85–96
13. Lu LJ, Sboner A, Huang YJ, Lu HX, Gianoulis TA, Yip KY, Kim PM, Montelione GT, Gerstein MB (2007) Comparing classical pathways and modern networks: towards the development of an edge ontology. Trends Biochem Sci 32:320–331
14. Cerami EG, Bader GD, Gross BE, Sander C (2006) cPath: open source software for collecting, storing, and querying biological pathways. BMC Bioinf 7:497
15. Bauer-Mehren A, Furlong LI, Sanz F (2009) Pathway databases and tools for their exploitation: benefits, current limitations and challenges. Mol Syst Biol 5:290
16. Ogata H, Goto S, Sato K, Fujibuchi W, Bono H, Kanehisa M (1999) KEGG: kyoto encyclopedia of genes and genomes. Nucleic Acids Res 27:29–34
17. Joshi-Tope G, Gillespie M, Vastrik I, D'Eustachio P, Schmidt E, de Bono B, Jassal B, Gopinath GR, Wu GR, Matthews L, Lewis S, Birney E, Stein L (2005) Reactome: a knowledgebase of biological pathways. Nucleic Acids Res 33:D428–D432
18. Kandasamy K, Mohan S, Raju R, Keerthikumar S, Kumar GS, Venugopal AK, Telikicherla D, Navarro DJ, Mathivanan S, Pecquet C, Gollapudi SK, Tattikota SG, Mohan S, Padhukasahasram H, Subbannayya Y, Goel R, Jacob HK, Zhong J, Sekhar R, Nanjappa V et al (2010) NetPath: a public resource of curated signal transduction pathways. Genome Biol 11:R3
19. Keshava Prasad TS, Goel R, Kandasamy K, Keerthikumar S, Kumar S, Mathivanan S, Telikicherla D, Raju R, Shafreen B, Venugopal A, Balakrishnan L, Marimuthu A, Banerjee S, Somanathan DS, Sebastian A, Rani S, Ray S, Harrys Kishore CJ, Kanth S, Ahmed M et al (2009) Human protein reference database–2009 update. Nucleic Acids Res 37:D767–D772
20. Farkas IJ, Korcsmaros T, Kovacs IA, Mihalik A, Palotai R, Simko GI, Szalay KZ, Szalay-Beko M, Vellai T, Wang S, Csermely P (2011) Network-based tools for the identification of novel drug targets. Sci Signal 4:3
21. Orchard S, Salwinski L, Kerrien S, Montecchi-Palazzi L, Oesterheld M, Stumpflen V, Ceol A, Chatr-aryamontri A, Armstrong J, Woollard P, Salama JJ, Moore S, Wojcik J, Bader GD, Vidal M, Cusick ME, Gerstein M, Gavin AC, Superti-Furga G, Greenblatt J et al (2007) The minimum information required for reporting a molecular interaction experiment (MIMIx). Nat Biotechnol 25:894–898
22. Lin CC, Chen YJ, Chen CY, Oyang YJ, Juan HF, Huang HC (2012) Crosstalk between transcription factors and microRNAs in human protein interaction network. BMC Syst Biol 6:18
23. Doench JG, Sharp PA (2004) Specificity of microRNA target selection in translational repression. Genes Dev 18:504–511
24. Guo H, Ingolia NT, Weissman JS, Bartel DP (2010) Mammalian microRNAs predominantly act to decrease target mRNA levels. Nature 466:835–840
25. Lewis BP, Shih IH, Jones-Rhoades MW, Bartel DP, Burge CB (2003) Prediction of mammalian microRNA targets. Cell 115:787–798
26. Calin GA, Croce CM (2006) MicroRNA signatures in human cancers. Nat Rev Cancer 6:857–866
27. Liang H, Li WH (2007) MicroRNA regulation of human protein protein interaction network. RNA 13:1402–1408
28. Hsu CW, Juan HF, Huang HC (2008) Characterization of microRNA-regulated protein–protein interaction network. Proteomics 8:1975–1979

29. Vaquerizas JM, Kummerfeld SK, Teichmann SA, Luscombe NM (2009) A census of human transcription factors: function, expression and evolution. Nat Rev Genet 10:252–263
30. Griffith OL, Montgomery SB, Bernier B, Chu B, Kasaian K, Aerts S, Mahony S, Sleumer MC, Bilenky M, Haeussler M, Griffith M, Gallo SM, Giardine B, Hooghe B, Van Loo P, Blanco E, Ticoll A, Lithwick S, Portales-Casamar E, Donaldson IJ et al (2008) ORegAnno: an open-access community-driven resource for regulatory annotation. Nucleic Acids Res 36:D107–D113
31. Gupta R, Bhattacharyya A, Agosto-Perez FJ, Wickramasinghe P, Davuluri RV (2011) MPromDb update 2010: an integrated resource for annotation and visualization of mammalian gene promoters and ChIP-seq experimental data. Nucleic Acids Res 39:D92–D97
32. Portales-Casamar E, Kirov S, Lim J, Lithwick S, Swanson MI, Ticoll A, Snoddy J, Wasserman WW (2007) PAZAR: a framework for collection and dissemination of cis-regulatory sequence annotation. Genome Biol 8:R207
33. Portales-Casamar E, Thongjuea S, Kwon AT, Arenillas D, Zhao X, Valen E, Yusuf D, Lenhard B, Wasserman WW, Sandelin A (2010) JASPAR 2010: the greatly expanded open-access database of transcription factor binding profiles. Nucleic Acids Res 38:D105–D110
34. Vergoulis T, Vlachos IS, Alexiou P, Georgakilas G, Maragkakis M, Reczko M, Gerangelos S, Koziris N, Dalamagas T, Hatzigeorgiou AG (2012) TarBase 6.0: capturing the exponential growth of miRNA targets with experimental support. Nucleic Acids Res 40:D222–D229
35. Krek A, Grun D, Poy MN, Wolf R, Rosenberg L, Epstein EJ, MacMenamin P, da PI, Gunsalus KC, Stoffel M, Rajewsky N (2005) Combinatorial microRNA target predictions. Nat Genet 37:495–500
36. Lewis BP, Burge CB, Bartel DP (2005) Conserved seed pairing, often flanked by adenosines, indicates that thousands of human genes are microRNA targets. Cell 120:15–20
37. Bandyopadhyay S, Bhattacharyya M (2010) PuTmiR: a database for extracting neighboring transcription factors of human microRNAs. BMC Bioinformatics 11:190
38. Wang J, Lu M, Qiu C, Cui Q (2010) TransmiR: a transcription factor-microRNA regulation database. Nucleic Acids Res 38:D119–D122
39. Alexiou P, Vergoulis T, Gleditzsch M, Prekas G, Dalamagas T, Megraw M, Grosse I, Sellis T, Hatzigeorgiou AG (2010) miRGen 2.0: a database of microRNA genomic information and regulation. Nucleic Acids Res 38:D137–D141
40. Xiao F, Zuo Z, Cai G, Kang S, Gao X, Li T (2009) miRecords: an integrated resource for microRNA-target interactions. Nucleic Acids Res 37:D105–D110
41. Baitaluk M, Kozhenkov S, Dubinina Y, Ponomarenko J (2012) IntegromeDB: an integrated system and biological search engine. BMC Gen 13:35
42. Lepoivre C, Bergon A, Lopez F, Perumal NB, Nguyen C, Imbert J, Puthier D (2012) TranscriptomeBrowser 3.0: introducing a new compendium of molecular interactions and a new visualization tool for the study of gene regulatory networks. BMC Bioinformatics 13:19
43. Kitano H (2007) A robustness-based approach to systems-oriented drug design. Nat Rev Drug Discov 6:202–210
44. Korcsmaros T, Szalay MS, Bode C, Kovacs IA, Csermely P (2007) How to design multi-target drugs: target-search options in cellular networks. Exp Op Drug Discovery 2:799–808
45. Hwang WC, Zhang A, Ramanathan M (2008) Identification of information flow-modulating drug targets: a novel bridging paradigm for drug discovery. Clin Pharmacol Ther 84:563–572
46. Tomlinson IP, Novelli MR, Bodmer WF (1996) The mutation rate and cancer. Proc Natl Acad Sci U S A 93:14800–14803
47. Ali MA, Sjoblom T (2009) Molecular pathways in tumor progression: from discovery to functional understanding. Mol BioSyst 5:902–908
48. Hornberg JJ, Bruggeman FJ, Westerhoff HV, Lankelma J (2006) Cancer: a systems biology disease. Biosystems 83:81–90
49. Hanahan D, Weinberg RA (2000) The hallmarks of cancer. Cell 100:57–70
50. Hanahan D, Weinberg RA (2011) Hallmarks of cancer: the next generation. Cell 144:646–674

51. Papatsoris AG, Karamouzis MV, Papavassiliou AG (2007) The power and promise of rewiring the mitogen-activated protein kinase network in prostate cancer therapeutics. Mol Cancer Ther 6:811–819
52. Kim D, Rath O, Kolch W, Cho KH (2007) A hidden oncogenic positive feedback loop caused by crosstalk between Wnt and ERK pathways. Oncogene 26:4571–4579
53. Torkamani A, Schork NJ (2009) Identification of rare cancer driver mutations by network reconstruction. Genome Res 19:1570–1578
54. Mimeault M, Batra SK (2010) Frequent deregulations in the hedgehog signaling network and cross-talks with the epidermal growth factor receptor pathway involved in cancer progression and targeted therapies. Pharmacol Rev 62:497–524
55. Pawson T, Linding R (2008) Network medicine. FEBS Lett 582:1266–1270
56. Hornbeck PV, Chabra I, Kornhauser JM, Skrzypek E, Zhang B (2004) PhosphoSite: a bioinformatics resource dedicated to physiological protein phosphorylation. Proteomics 4:1551–1561
57. Linding R, Jensen LJ, Ostheimer GJ, van Vugt MA, Jorgensen C, Miron IM, Diella F, Colwill K, Taylor L, Elder K, Metalnikov P, Nguyen V, Pasculescu A, Jin J, Park JG, Samson LD, Woodgett JR, Russell RB, Bork P, Yaffe MB et al (2007) Systematic discovery of in vivo phosphorylation networks. Cell 129:1415–1426
58. Miller ML, Jensen LJ, Diella F, Jorgensen C, Tinti M, Li L, Hsiung M, Parker SA, Bordeaux J, Sicheritz-Ponten T, Olhovsky M, Pasculescu A, Alexander J, Knapp S, Blom N, Bork P, Li S, Cesareni G, Pawson T, Turk BE et al (2008) Linear motif atlas for phosphorylation-dependent signaling. Sci Signal 1:ra2
59. Remenyi A, Good MC, Lim WA (2006) Docking interactions in protein kinase and phosphatase networks. Curr Opin Struct Biol 16:676–685
60. Nguyen LK, Matallanas D, Croucher DR, von Kriegsheim A, Kholodenko BN (2012) Signalling by protein phosphatases and drug development: a systems-centred view. FEBS J. 280:751–765
61. Alonso A, Sasin J, Bottini N, Friedberg I, Friedberg I, Osterman A, Godzik A, Hunter T, Dixon J, Mustelin T (2004) Protein tyrosine phosphatases in the human genome. Cell 117:699–711
62. Roy J, Cyert MS (2009) Cracking the phosphatase code: docking interactions determine substrate specificity. Sci Signal 2:re9
63. Shi Y (2009) Serine/threonine phosphatases: mechanism through structure. Cell 139:468–484
64. Barr AJ (2010) Protein tyrosine phosphatases as drug targets: strategies and challenges of inhibitor development. Future Med Chem 2:1563–1576
65. Gambari R, Fabbri E, Borgatti M, Lampronti I, Finotti A, Brognara E, Bianchi N, Manicardi A, Marchelli R, Corradini R (2011) Targeting microRNAs involved in human diseases: a novel approach for modification of gene expression and drug development. Biochem Pharmacol 82:1416–1429
66. Lu M, Zhang Q, Deng M, Miao J, Guo Y, Gao W, Cui Q (2008) An analysis of human microRNA and disease associations. PLoS ONE 3:e3420
67. McDermott AM, Heneghan HM, Miller N, Kerin MJ (2011) The therapeutic potential of microRNAs: disease modulators and drug targets. Pharm Res 28:3016–3029
68. Farkas IJ, Szanto-Varnagy A, Korcsmaros T (2012) Linking proteins to signaling pathways for experiment design and evaluation. PLoS ONE 7:e36202
69. Kuhn M, Campillos M, Letunic I, Jensen LJ, Bork P (2010) A side effect resource to capture phenotypic effects of drugs. Mol Syst Biol 6:343
70. McClean MN, Mody A, Broach JR, Ramanathan S (2007) Cross-talk and decision making in MAP kinase pathways. Nat Genet 39:409–414
71. Guo X, Wang XF (2009) Signaling cross-talk between TGF-beta/BMP and other pathways. Cell Res 19:71–88
72. Hurlbut GD, Kankel MW, Lake RJ, Artavanis-Tsakonas S (2007) Crossing paths with Notch in the hyper-network. Curr Opin Cell Biol 19:166–175

73. Katoh M, Katoh M (2007) WNT signaling pathway and stem cell signaling network. Clin Cancer Res 13:4042–4045
74. Borisov N, Aksamitiene E, Kiyatkin A, Legewie S, Berkhout J, Maiwald T, Kaimachnikov NP, Timmer J, Hoek JB, Kholodenko BN (2009) Systems-level interactions between insulin-EGF networks amplify mitogenic signaling. Mol Syst Biol 5:256
75. Wang CC, Cirit M, Haugh JM (2009) PI3 K-dependent cross-talk interactions converge with Ras as quantifiable inputs integrated by Erk. Mol Syst Biol 5:246
76. Natarajan M, Lin KM, Hsueh RC, Sternweis PC, Ranganathan R (2006) A global analysis of cross-talk in a mammalian cellular signalling network. Nat Cell Biol 8:571–580
77. Blank U, Karlsson G, Karlsson S (2008) Signaling pathways governing stem-cell fate. Blood 111:492–503
78. Robinson GW (2007) Cooperation of signalling pathways in embryonic mammary gland development. Nat Rev Genet 8:963–972
79. Sternberg PW (2005) Vulval development. WormBook 1–28
80. Yan SJ, Gu Y, Li WX, Fleming RJ (2004) Multiple signaling pathways and a selector protein sequentially regulate Drosophila wing development. Development 131:285–298
81. Katoh M (2007) Networking of WNT, FGF, Notch, BMP, and Hedgehog signaling pathways during carcinogenesis. Stem Cell Rev 3:30–38
82. Li Y, Agarwal P, Rajagopalan D (2008) A global pathway crosstalk network. Bioinformatics 24:1442–1447
83. Boswell BA, Lein PJ, Musil LS (2008) Cross-talk between fibroblast growth factor and bone morphogenetic proteins regulates gap junction-mediated intercellular communication in lens cells. Mol Biol Cell 19:2631–2641
84. Fraser ID, Germain RN (2009) Navigating the network: signaling cross-talk in hematopoietic cells. Nat Immunol 10:327-331
85. Dreesen O, Brivanlou AH (2007) Signaling pathways in cancer and embryonic stem cells. Stem Cell Rev 3:7–17
86. Fisher J, Piterman N, Hajnal A, Henzinger TA (2007) Predictive modeling of signaling crosstalk during C. elegans vulval development. PLOS Comput Biol 3:e92

Chapter 17
A Survey of Current Integrative Network Algorithms for Systems Biology

Andrew K. Rider, Nitesh V. Chawla and Scott J. Emrich

Abstract The goal of systems biology is to gain a more complete understanding of biological systems by viewing all of their components and the interactions between them simultaneously. Until recently, the most complete global view of a biological system was through the use of gene expression or protein-protein interaction data. With the increasing number of high-throughput technologies for measuring genomic, proteomic, and metabolomic data, scientists now have the opportunity to create complex network-based models for drug discovery, protein function annotation, and many other problems. Each technology used to measure a biological system inherently presents a limited view of the system. However, the combination of multiple technologies can provide a more complete picture. Much recent work has studied integrating these heterogeneous data types into single networks. Here we provide a survey of integrative network-based approaches to problems in systems biology. We focus on describing the variety of algorithms used in integrative network inference. Ultimately, the survey of current approaches leads us to the conclusion that there is an urgent need for a standard set of evaluation metrics and data sets in this field.

Keywords Network inference · Integrative networks · Systems biology

Acronyms

PPI	Protein-protein interaction
GO	Gene ontology
TF	Transcription factor
TFBS	Transcription factor binding site
eQTL	Expression quantitative trait locus

A. K. Rider · N. V. Chawla (✉) · S. J. Emrich
Department of Computer Science and Engineering, University of Notre Dame, Notre Dame, IN, USA
e-mail: nvchawla@nd.edu

A. Prokop and B. Csukás (eds.), *Systems Biology*,
DOI: 10.1007/978-94-007-6803-1_17,

17.1 Introduction

The history of genetics has been a process of uncovering increasing amounts of complexity and depth in biological systems. In the past, we knew that DNA was transcribed into RNA and then translated to proteins. Our growing knowledge of alternative splicing and other post-transcriptional regulation complicated this view. We knew that transcription factors were the primary regulators of gene expression. This view became complicated by our increasing knowledge of the regulating effect of phosphorylation on transcription factors. Given the complexity of biological systems and the certain knowledge that we do not fully understand fundamental aspects of biology, it is important to carefully consider how prior knowledge and diverse data types are incorporated into computational models.

As we learn more about genetics, it is becoming increasingly clear that the traits and behaviors of organisms are emergent: they are the product of complex interactions between numerous biological components. In systems biology, networks are used to capture this complexity by modeling an entire biological system. This approach gives scientists a global view of a biological system that can enable further understanding of the nature of human disease as well as new tools to understand the processes driving life [1].

Networks are versatile tools that have been used to model interactions between numerous different biological concepts. Nodes can be used to represent genes, proteins, metabolites, or any other discrete biological component or concept. Edges in the network may represent the relationship between a gene and a protein, similarity of function between genes, or any other pair of biological concepts. Edges may represent multiple types of relationships simultaneously. Each type of relationship reveals unique information about an organism. For example, protein-protein interaction (PPI) data reveals which proteins can physically interact, but alone it does not impart knowledge about how an organism will react to stimuli. Similarly, gene expression data can reveal how an organism responds to stimuli in terms of the amount of RNA produced but it does not impart any knowledge about the physical mechanisms that cause change in the organisms behavior. Therefore, the key to furthering our understanding of biological systems the integration of diverse data types.

Differences in the underlying architecture of networks can affect their utility. Directed networks such as Bayesian networks or networks that use asymmetric edge weighting metrics implicitly contain some indication of causality [2, 3]. These methods are well suited for making specific inferences about how the effects of a perturbation to one or more genes will propagate through the network. Undirected networks make fewer assumptions about how nodes are connected and are often less computationally demanding to construct but may yield less specific information.

17.1.1 Contributions

There are a number of review articles that cover network inference. De Smet and Marchal [4] reviews network inference and integrative methods in the context of how they approach the problem of underdetermination. Sharan et al. [5] reviews several integrative network methods in the context of clustering. Hecker et al. covers network models for time course behavior of gene expression data and integration of heterogeneous data sources. They discuss a wide range of network inference algorithms both within and outside of the context of integrative approaches. They cover the inclusion of previous biological knowledge such as expected network topology. In terms of the integration heterogeneous data types, they primarily cover Bayesian networks [6]. Gitter et al. [7] survey a number of approaches to integrate time series data with various heterogeneous data types gathered from single time points to create dynamic regulatory networks. Califano et al. [8] review a number of integrative networks approaches in terms of the combinations of data used. They describe how different approaches use different combinations of data types to uncover specific relationships in the data. They also address the need for more focus on awareness of context specific regulation in network models. Bebek et al. [9] focus on integrative approaches specifically used for the identification of biomarkers and the betterment of clinical science. In this work, we focus on presenting a wide range of integrative model types and exclusively on the integration of heterogeneous data. Our purpose is to provide a familiarity with the variety of algorithms used for integration in network models.

In Sect. 17.2 we discuss some of the most commonly used data types in integrative network models. Section 17.3 covers the problem of network inference in the abstract. In Sects. 17.3.1 through 17.3.4 we discuss various approaches to network inference and cover examples of each in some detail. Sections 17.3.1 and 17.3.2 cover Bayesian and other probabilistic networks. Section 17.3.3 discusses integration methods based on machine learning techniques. In Sect. 17.3.4 we cover techniques that rely on identifying modules in networks and context specific regulatory patterns. Finally, in Sect. 17.4 we discuss some of the patterns that emerge from examining the variety of methods discussed in the previous sections. We conclude that there is an urgent need for consensus about how to evaluate and compare models.

17.2 Data

Each data type is measured in a unique way. Additionally, two data sets describing the same type of data may not be comparable due to differences in scale, noisy data, or measurement errors. Therefore, normalization and the use of well curated data are essential for meaningful comparisons between data sets and the integration of diverse data types. For example, microarray results have been shown to vary based on the location of probes on the chip, complicating comparisons between results gathered with different chips [10]. Even assuming identical chips, different approaches to normalization can have significant impact on the meaning of the data and on the validity of comparisons between data sets [11]. Another complication is that there are multiple methods to measure the same data type. There are many distinct methods to measure PPI, each with different strengths and weaknesses [12]. For example, Affinity Capture-MS protein interactions are determined by using a "bait" protein that is "captured" by a polyclonal antibody or an epitope tag. The associated partner is then identified by mass spectrometry. An alternative approach, co-immunoprecipitation, isolates a protein with antibodies. Interacting partner proteins are then detected with western blotting. Different methodologies in data collection add noise or bias in different ways that must be accounted for in the analysis.

Precursors to integrative networks used microarray expression data alone to infer regulatory and other types of relationships between genes. Microarrays enable high-throughput measurement of the expression level of genes. Expression levels measure the relative amount of RNA produced from the transcription of genes. RNA levels give some indication about the amount of protein that is expected to be produced. Since proteins are the primary causes of change in a cell, expression data can give indirect evidence towards answering many different questions in systems biology. Many studies have relied on clustering and network models to identify functionally similar genes or infer regulatory networks based on expression data [13–17].

Protein-protein interactions provide more direct information in the form of which proteins physically interact. Like expression data, PPI data is commonly used to cluster genes or proteins or to infer networks in order to identify novel interactions or determine function [5, 18].

Some data types are themselves integrative. The ChIP-chip technique combines microarrays with chromatin immunoprecipitation to allow the identification of protein binding sites on DNA [19]. This is particularly useful for the study of transcription factors (TFs) which are proteins that transcribe DNA into RNA and are thought to play a major role in the regulation of gene expression. Motifs or identifiable strings of DNA can also be located computationally from sequence data to identify potential transcription factor binding sites (TFBS). Expression quantitative trait loci (eQTLs) use genetic variation between individuals in combination with gene expression data to measure the association between expression levels and genotypes. An expression trait refers to the amount of RNA produced by a gene. Each eQTL represents a strong association between a position or locus in the genome and the expression level of a gene. eQTLs describe the relationship between genotype and phenotype and enable inferences about the regulatory interactions between genes [20].

Annotation data can come from many sources and can describe experimentally or computationally derived knowledge such as functions associated with biological components or pathways that components are a part of. The Gene Ontology (GO) keeps curated functional annotations for genes [21]. Annotations exist in a hierarchy such that a gene my have a number of general and specific functions. Pathway information describes the chain of biological components involved in causing some event or fulfilling some function in a cell. Many databases exist to curate pathway and other data types, often for specific organisms [12, 22, 23].

One consideration that affects many data types is the experimental conditions under which measurements are taken. For example, the expression level of genes can change drastically based on environmental and genetic conditions [20, 24]. Common genetic conditions include gene knockout experiments, in which a gene is made inoperative, and chemical or environmental treatments. Measurements may also involve an element of time under a condition or after a treatment.

17.3 Network Inference

The basic problem of network inference is to create a network that has a meaningful topology. Ultimately this means creating a sparse network in which only important edges are present. This is accomplished in various ways by different algorithms. In the abstract, there are two general types of networks: distance-based networks and probabilistic networks.

Network inference algorithms universally depend on some measure of dependence or distance between biological components. The approach used to calculate edge weight can have a significant effect on what is contained in the resulting network [25, 26]. Mason et al. [27] compared co-expression networks based on Pearson's correlation to co-expression networks based on the absolute value of Pearson's correlation and showed that modules in the signed network are more biologically coherent. Probabilistic network inference faces a similar problem in that conditional probabilities can be calculated in a number of different ways.

The fundamental assumption in relevance networks and other distance-based networks is that relationships between biological components can be accurately ranked in some meaningful way. Once the relationships between all components have been quantified, edges are removed from the network. This results in a sparse network with some meaningful topology that is determined in part by the edge weighting method and in part by the pruning criterion. Each approach makes different underlying assumptions that can impact the information contained in the network. Relevance networks make inherent assumptions in the choice of weighting method and the pruning approach. The underlying assumption is that the weighting method correctly ranks edges in terms of importance. Zhou et al. [28] use Pearson's correlation to infer a co-expression network for yeast. They use the shortest paths between all nodes in the network to identify functionally related genes. This approach assumes that transitive relationships that are represented in the network may be as important to understand relationships between genes as direct relationships. ARACNE makes the opposite assumption and explicitly disallows triangles in the network, assuming that all triangles contain an indirect relationship that should not be explicitly represented in the network [16].

Additional approaches in this category use LASSO (Least Absolute Shrinkage and Selection Operator) or related linear methods to explicitly penalize and eliminate weak relationships [29]. LASSO and related approaches optimize the parameter vector of linear equations such that their ℓ_p norm is less than or equal to a given value. LASSO's constraint is based on the ℓ_1 norm whereas other approaches may use different norms or a combination of norms as is the case with elastic nets [30]. Such approaches are used to discover a sparse topology and replace an arbitrary threshold with a more principled one [31–33]. One recent related integrative approach uses Multi-Block Partial Least Squares (sMBPLS) to find sets of input variables from multiple data types including copy number variation, DNA methylation, and microRNA expression that together explain the gene expression in cancer data [34]. Li et al. [35] use a tensor-based approach to identify sets of recurring subgraphs from large sets of heterogeneous biological networks. This approach is similar to LASSO and similar approaches in that the sparseness of the resulting networks are controlled primarily through the choice of ℓ_p norm in the objective function.

Other approaches that can be described by the category of distance-based networks focus on machine learning techniques such as feature selection and decision trees. MRNET uses a maximum-relevance minimum-redundancy feature selection method to identify important neighbors for every node. After the pairwise mutual information between expression levels of all genes is calculated, edges are effectively pruned by the feature selection algorithm. For each node, the algorithm selects the neighbor with the highest mutual information that has the lowest redundancy with the neighbors already selected. Neighbor selection stops when the score of the next best neighbor is below a threshold [17].

Probabilistic or graphical models represent the dependence between random variables as nodes in a network. Edge weights represent conditional probabilities. This approach naturally captures the noise and stochastic nature of biological data.

17.3.1 Bayesian Networks

Bayesian networks are one of the most commonly used methods of integrating diverse biological data types. Using this approach, measurements of a gene's expression levels may be interpreted as samples from a random variable. Relationships in Bayesian networks are directed, reflecting the conditional dependence between variables. As such, they are often interpreted as causal. This interpretation allows Bayesian networks to represent pathways and to be used to predict the effect of perturbations to the system. Bayesian networks can be discrete, continuous, or a mixture of both.

Discrete Bayesian networks model the probability of discrete states. For example, an edge between nodes A and B can indicate the probability that gene B is highly expressed given the state of gene A. Discrete Bayesian networks may require that each node have a prior distribution to represent the possible prior states of the variable. A model relying only on the frequency of observed values may be unable to assign a probability to new observations if they do not fall within the observed range. Discrete Bayesian networks can model relationships in the data relatively concisely with a conditional probability table for each node that lists the probability of each state given the inputs. One drawback is that discretization of the data may lead to information loss. Bayesian networks that use continuous variables rely on conditional probability densities instead of conditional probability tables. Continuous variables may also be modeled using linear conditional densities, in which the conditional density of a node X is dependent on its parents as shown in Eq. 17.1. The equation shows that the conditional density of X given its parents p is linearly dependent on the values of the parents. It is common to use a normal distribution in this approach. Continuous Bayesian networks do not lose information due to discretization but it is more computationally complex to infer the continuous model than the discrete model.

$$P(X|p_1,\ldots,p_n) = N(\beta_0 + \sum_i^n \beta_i * p_i, \sigma^2) \tag{17.1}$$

There are three major steps in Bayesian network inference. First, a structure must be proposed. Second, the parameters or probabilities associated with edges and nodes must be set. Third, networks must be evaluated to determine how well they model the data. These steps are commonly used iteratively to propose a structure and parameters, then evaluate the model against further structural changes. This process allows a search through potential Bayesian network models.

Identifying edges in the network is a critical step in Bayesian network inference, as the direction of edges can greatly affect the interpretation of the model. The presence or lack of edges between nodes can also have a large effect as it determines the conditional relationships between variables. The most straightforward method to infer network structure would be to exhaustively compare every possible network. This approach is prohibitively expensive, as the number of possible networks grows super exponentially with the number of nodes [36]. Practical methods rely on sampling or heuristics to reduce the search space dramatically.

The sparse candidate algorithm relies on simple local statistics such as correlation to identify potential parents for each gene [2]. It greatly reduces the search space by evaluating edges only between a node and its candidate set. The algorithm can then use hill-climbing or a divide and conquer approach to determine edges. Choices made early in the assignment of edges can result in a restricted search space. Therefore, the algorithm iteratively creates a network then updates the candidate parent sets for each node by replacing nodes in node X's candidate set with a transitive relationship with nodes that had a weaker dependency with X.

Sampling methods such as the Metropolis-Hastings algorithm can be used to reduce computational cost of structure learning at the expense of an accurate description of the data. Sampling and other inexact techniques are often used repeatedly and then averaged to form a single network. Alternatively, one model or a few 'good' models can be selected as representative of all possible models. This process is called model selection when one network is chosen or selective model averaging if multiple representative networks are averaged [37].

Model parameters in Bayesian networks are conditional probability distributions or tables. A continuous node may assume that the observed data come from a normal distribution. However, the parameters of the distribution, the mean and standard deviation, may be incorrect. If the assumed distribution or prior is incorrect then the calculated probability of an observed instance and the fit of the network to the data will be incorrect. Parameter fitting is the process of calculating the priors and conditional probabilities in the network.

In Eq. 17.2, D is the data, E is background knowledge, and θ is the model. $p(\theta|E)$ and $p(\theta|D,E)$ are the prior and posterior probability distributions for the model θ, respectively. The prior describes the agreement between the prior knowledge and the network. The posterior describes how well the model fits the observed data. We direct the reader to Heckerman [38] and Needham et al. [36] for a more thorough treatment of parameter fitting and the selection of priors.

$$p(\theta|D,E) = \frac{p(\theta|E)p(D|\theta,E)}{p(D|E)} \tag{17.2}$$

There are two primary ways to include prior knowledge in Bayesian networks. The first is to constrain the edges in the structure learning step. This is a commonly used approach to integrate heterogeneous biological knowledge [39, 40]. The second is to update the priors in an iterative process. Often, a Bayesian network will be inferred and the parameters fitted to one type of biological knowledge, then priors are updated to take into account additional sources of data iteratively [41, 42].

Zhu et al. use a mixture of constrained and prior-updated techniques to integrate data types into a Bayesian network. They use the sparse candidate algorithm to infer structure in Bayesian networks based on only expression data, based on eQTL data, and based on expression data, eQTL data, TFBS, and PPI data [39]. For each network type, they learned 1,000 networks and determined a consensus network that consisted of edges that were present in at least 30 % of the networks. Loops were resolved by removing the weakest edge. Prior knowledge gained from eQTL data was incorporated by constraining edge direction such that genes with cis-acting eQTLs (as defined in Doss et al. [43]) are considered as potential parent nodes for genes with trans-acting eQTLs in the same region of the genome. Representative genes were used to incorporate TFBS and PPI data. They used a set of genes that were determined to be the most strongly associated with a transcription factor to represent each transcription factor in the network. The prior probability that the gene associated with a transcription factor is the parent of other genes that carry the TFBS was proportional to the number of expression traits correlated with the transcription factor's expression levels. The inferred networks were evaluated in terms of predicting functional categories from the Gene Ontology, predicting genes regulated by various transcription factors, and predicting the response of gene expression to gene knockout experiments.

17.3.2 Other Probabilistic Networks

While Bayesian networks are a popular approach to integrating diverse data types, there are many other network models that rely on a probabilistic interpretation of the data. As is the case with Bayesian networks, learning the structure of probabilistic models in general can be computationally prohibitive. Structure learning for probabilistic graphical models has been the subject of much recent research. Wainwright et al. [44] use ℓ_1 regularized logistic regression to learn the structure of each node's neighborhood in a Markov network. Other approaches make the structure learning problem tractable by restricting the model's structure. Choi et al. [45] propose algorithms to learn tree-structured probabilistic models. Srebro [46] controls the tree-width (maximum clique size) of Markov networks in order to limit the computational cost of inferring network structure while providing a provable performance bound. This is by no means an exhaustive list of approaches to infer probabilistic networks. Many approaches fit a prior distribution to the data in order to measure explanatory power. Friedman and Nachman [47] use Gaussian processes to learn the structure of Bayesian networks. Gaussian processes model the relationship between a set of variables and an output variable by defining a mean function and a covariance function over the random input variables. In this approach, response variables are modeled as mixtures of related Gaussians. The structure of a candidate network can be evaluated by computing the marginal likelihood of the data given the structure.

Tu et al. [48] use a stochastic network to integrate PPI, TFBS, phosphorylation, eQTL, and expression data in order to identify causal genes and regulatory pathways. Their model works under the assumption that causal or regulating genes in the network regulate their targets through either direct or indirect affects on the activity of transcription factors. They take into account the possibility that transcription factors can be regulated at the protein level. They also make the common assumption that gene activity correlates with gene expression. Protein-protein interactions are represented in the network as undirected edges, protein phosphorylation and TFBS are represented as directed edges. Each node has a set of transcription factors that bind to it and a set of genes with eQTLs that are candidate regulators. For each node in the network they estimate the likelihood that every neighboring gene is the cause for its expression by calculating Pearson's correlation between the expression level of the two genes. The algorithm determines the causal regulator of gene G by taking random walks without cycles along the edges in the network until it reaches a candidate eQTL gene. They used this algorithm on subsets of expression data from specific treatments as well as with bootstrapped samples to observe variation in transcription factor activity and account for variation in expression levels. The method was evaluated by comparing predicted relationships against a compendium of gene knock-out expression data.

Lee et al. propose a method to represent functional associations between biological components. They use a Bayesian statistics approach to determine the likelihood that genes are functionally linked based on evidence from heterogeneous data sources [49]. They use microarray data, philogenetic profiles, PPI, functional linkages from text mining, as well as four other data types. Their log-likelihood score compares the frequency of linkages in each data type between genes that share a pathway to the frequency of linkages between genes that do not share a pathway. In Eq. 17.3, $P(L|E)$ is the frequency of linkages (L) in a data type (E) between genes in the same pathway, $\sim$P(L|E) is the frequency of linkages between genes in different pathways for the data type. P(L) and $\sim$P(L) are the total frequency across data types of all linkages between genes sharing a pathway and not sharing a pathway, respectively.

$$LLS = \frac{P(L|E)/\sim P(L|E)}{P(L)/\sim P(L)} \tag{17.3}$$

This method relies on the use of the KEGG (Kyoto Encyclopedia of Genes and Genomes) pathway and sub-cellular location data as ground truth data for the calculation of LLS [22]. The use of a common ground truth allows scores for different types of data to be meaningfully compared. The resulting integrative network showed improved accuracy in terms of linking genes that share pathways in the KEGG database over other methods.

Other methods integrate diverse data types and model the stochastic nature of biological systems use hidden Markov models, Markov networks, and naïve Bayes models [50–52].

17.3.3 Statistical and Machine Learning Approaches

Machine learning and statistical approaches are distance based as many provide some confidence or probability that a prediction is correct. They tend to be different from other distance based methods in that the distances are often determined in a supervised manner.

SEREND is a semi-supervised network construction method that integrates TFBS, DNA sequence binding motifs, and gene expression data to predict transcription factor-gene interactions [53]. It uses a logistic regression classifier for expression data and sequence motif data, then combines the two in a hierarchical classification scheme by training a third logistic regression classifier on the output of the other two classifiers. In experimentation, features for the classification of expression data were from 455 expression experiments from a compendium of treatment experiments. Each instance corresponded to a gene. Class labels were activated by a transcription factor, repressed by a transcription factor, or not regulated by a transcription factor. The motif classifier used only a single feature to classify genes as regulated by the transcription factor or not regulated by the transcription factor. If the meta-classifier found that there was enough evidence that a non-regulated gene was regulated by a transcription factor, then the algorithm would switch the label from not regulated to regulated and update the weights for all classifiers. This process allows SEREND to iteratively expand its predictions about transcription factor-gene relationships until they converge. SEREND was evaluated in terms of how well it recovered gene targets that were verified in a ChIP-chip data set.

Hwang et al. [54] use a few statistical methods to combine p-values from different data sets. They use an ensemble of Fisher's weighted F, Mudholkar-George's weighted T, and Liptak-Stouffer's weighted Z where the weight is a measure of the relative statistical power for each data set. They determine a combined weight by comparing a hypothetical weight distribution to an observed distribution. The resulting integrative network has a p-value for each node and edge that indicates the confidence that the node or edge belongs in the network. Multiple approaches were tested on simulated data sets, which allowed a comparison on the basis of ground truth data.

The modENCODE Consortium is a group that collects a great deal of diverse data about model organisms [55]. They use correlated activity patterns from over 700 data sets to define a functional regulatory network. They use logistic regression to classify promoters as active or inactive based on chromatin modification, TFBS, and nucleosome physical properties. The resulting probabilities are used to weight the confidence of each regulatory edge in the network. They evaluated inferred networks based on the enrichment in the network compared to randomized networks of GO terms, correlation of gene expression across time, frequency of protein-protein interactions in the network, and other metrics.

The STRING (Search Tool for the Retrieval of Interacting Genes) database is a collection of data for the understanding of functional interactions among proteins [56]. Interactions in the database come from many curated data sets from multiple organisms as well as from text mining the literature, predicted interactions from gene co-expression and cross-genome homology. Each interaction in the database has a confidence score assigned to it based on benchmarks against a trusted PPI data source, the KEGG database. Each data source is individually benchmarked and then combined in a naïve Bayesian approach by simply multiplying the normalized scores together. Interactions with more support from multiple sources of data will naturally have a higher combined score. STRING is properly a search tool rather than an integrative network inference method. As such, it does not attempt to evaluate the resulting network but provides the ability to alter the data types included, as well as access the raw data.

An alternative approach to modeling heterogeneous data in a single network is to use multiple edge types in what is called a multi-relational network. Davis and Chawla [57] use this approach to make predictions about disease occurrence in patients and study the relationship between diseases and genes. They combine a network of disease co-morbidity data with a network of genes related to each other by their relationship to the same disease. They then use a link prediction method that uses a triad census (counting the occurrences of sets of three nodes with each possible combination of edges) as the basis to predict unknown genetic links.

Predicted links were benchmarked against a number of canonical link prediction methods and performance was measured in terms of area under the ROC curve, and the precision-recall curve.

17.3.4 Modular Networks and Condition Specific Regulators

One of the fundamental problems in creating a network model for regulatory interactions in the genome is that the regulatory program of a cell appears to change under different conditions [58]. Network modules can be viewed as discrete groups composed of many types of molecules whose function is separable from other modules. The aggregate expression of these modules may have condition specific regulators. Integrative network approaches to modeling condition specific regulatory networks rely on compendiums of expression data from different experimental conditions and commonly use TFBS, ChIP-chip, or other protein-DNA interaction data [59–61].

SAMBA integrates heterogeneous data from gene expression, PPI, phenotypic sensitivity, and TFBS sources into a probabilistic bipartite network in order to identify genes with common behavior across experiments [62]. The nodes on one side of the network are genes and the other side are properties of genes or proteins. Weighted edges in the network between node N and property P are interpreted as the probability that node N has property P. Property nodes can indicate anything from interaction with a specific protein to different levels of discretized gene expression. Subgraphs are scored based on the log ratio of the observed topology under two statistical models, a model for the dependency expected in modules and a model for the background dependency. Biclustering is used to identify gene sets that share sets of properties. Modules are evaluated in terms of functional enrichment based on the Gene Ontology. It finds complete bipartite subgraphs with high density by using a hashing technique to find 'seed' nodes and then using a local search to identify other nodes in the module.

DISTILLER is an integrative framework to identify condition-dependent modularity and regulatory relationships [63]. It uses an efficient item set mining algorithm to identify modules. It starts with "seed" modules, consisting of a small number of genes that are co-expressed in a sufficiently large number of conditions and share motifs for the same regulators. Seed modules are expanded to nodes that do not violate the module properties. A drawback of the item set mining approach is that it can be difficult to identify the most interesting modules from the large amount of potentially redundant output. DISTILLER ranks modules by a measure that takes into account how much they help to cover the entire condition space and their redundancy with already ranked modules. DISTILLER was evaluated in terms of precision and recall on a ChIP-chip gold standard data set.

17.4 Discussion

While there are many benefits to integrating diverse data types, integration of prior knowledge may reinforce bias in network models to the detriment of new discoveries. For example, a number of networks papers have observed that many biological networks appear to have scale-free topology [64, 65]. In response, methods to infer or evaluate networks based on their topology have been developed. Networks inferred using this criterion will systematically overlook possible networks with alternative architectures [66]. There is evidence that this may be happening as many of the observed scale-free topologies in biological networks may not truly be scale free. Clauset et al. [67] showed that the methods used to measure scale-free topology in many preceding studies of biological networks were unable to distinguish between power-law distributions (such as scale-free) and a number of other distinct distributions. Bias may also enter into models through other prior knowledge. For example, Zhu et al. [39] and Tu et al. [48] both constrain their models to use trans-acting eQTLs to constrain edges but the definition of trans acting is different.

Network inference methods that are constrained to include edges from PPI, TFBS, eQTL, or other data may reinforce bias in the models as they do not allow room for error in the data. Less constrained approaches avoid this problem but may add a more subtle bias to the model. Many integrative network approaches construct a single network by integrating data based on a single algorithm [39, 40, 53]. As is the case with different types of data, different algorithms contain different biases. Bayesian approaches that create an ensemble or consensus model with Monte Carlo techniques may suffer less from this type of bias but may reduce bias further by use of fundamentally different algorithms.

The problem of evaluation is made extraordinarily difficult in systems biology by the scarcity of ground truth data. Even curated data sets such as PPI data from KEGG that are used to benchmark novel methods are based on uncertain data. The problem of network evaluation has been noted before in the single data type network inference problem [68]. Marbach et al. propose a unifying approach to the evaluation of network models that includes common evaluation metrics and simulated data. While these are excellent suggestions, the problem is made much more complicated by the diversity of data involved in integrative methods.

Any single type of data presents a one-dimensional view of a biological system. Therefore, evaluation based on a single data type may not be a baseline for the performance of an integrative method. Furthermore, different approaches tend to use different amounts and types of data, making the actual methods themselves very difficult to compare. There are, of course, high-confidence experimentally derived interactions, but it can be difficult to locate and identify them. Databases such as STRING, KEGG, and modENCODE will be critical for the future progress of integrative network models because they provide this service. The creation of a common body of data for evaluation and a standard for evaluation methods for integrative network approaches would allow integrative network algorithms to be truly compared. This in turn could help us to better understand the complex interplay of diverse data types.

References

1. Schadt EE, Friend SH, Shaywitz DA (2009) A network view of disease and compound screening. Nat Rev Drug Discov 8:286–295
2. Friedman N, Linial M, Nachman I, Pe'er D (2000) Using Bayesian networks to analyze expression data. J Comput Biol 7:601–620
3. Rao A, Hero AO, States DJ, Engel JD (2007) Using directed information to build biologically relevant influence networks. Comput Syst Bioinform/Life Sci Soc Comput Syst Bioinform Conf 6:145–156
4. De Smet R, Marchal K (2010) Advantages and limitations of current network inference methods. Nat Rev Micro 8:717–729
5. Sharan R, Ulitsky I, Shamir R (2007) Network-based prediction of protein function. Mol Syst Biol 3
6. Hecker M, Lambeck S, Toepfer S, Van Someren E, Guthke R (2009) Gene regulatory network inference: data integration in dynamic models: a review. Biosystems 96:86–103
7. Gitter A, Siegfried Z, Klutstein M, Fornes O, Oliva B et al (2009) Backup in gene regulatory networks explains differences between binding and knockout results. Mol Syst Biol 5
8. Califano A, Butte A, Friend S, Ideker T, Schadt EE (2011) Integrative network-based association studies: leveraging cell regulatory models in the post-GWAS era. Nat Precedings 10
9. Bebek G, Koyutürk M, Price ND, Chance MR (2012) Network biology methods integrating biological data for translational science. Briefings Bioinform
10. Canales R, Luo Y, Willey J, Austermiller B, Barbacioru C et al (2006) Evaluation of dna microarray results with quantitative gene expression platforms. Nat Biotechnol 24:1115–1122
11. Quackenbush J (2002) Microarray data normalization and transformation. Nat Genet 32:496
12. Christie KR, Hong EL, Cherry JM (2009) Functional annotations for the Saccharomyces cerevisiae genome: the knowns and the known unknowns. Trends Microbiol 17:286–294
13. Hanisch D, Zien A, Zimmer R, Lengauer T (2002) Co-clustering of biological networks and gene expression data. Bioinformatics 18:S145–S154
14. Datta S, Datta S (2003) Comparisons and validation of statistical clustering techniques for microarray gene expression data. Bioinformatics 19:459–466
15. Eisen MB, Spellman PT, Brown PO, Botstein D (1998) Cluster analysis and display of genome-wide expression patterns. Proc Natl Acad Sci 95:14863–14868
16. Margolin A, Nemenman I, Basso K, Wiggins C, Stolovitzky G et al (2006) ARACNE: an algorithm for the reconstruction of gene regulatory networks in a mammalian cellular context. BMC Bioinform 7:S7
17. Meyer P, Lafitte F, Bontempi G (2008) Minet: AR/Bioconductor package for inferring large transcriptional networks using mutual information. BMC Bioinform 9:461
18. Sen T, Kloczkowski A, Jernigan R (2006) Functional clustering of yeast proteins from the protein-protein interaction network. BMC Bioinform 7:355
19. Aparicio O, Geisberg JV, Sekinger E, Yang A, Moqtaderi Z et al (2005) Chromatin immunoprecipitation for determining the association of proteins with specific genomic sequences in vivo. In: Ausubel FM et al Current protocols in molecular biology. Chapter 21
20. Jansen R (2001) Genetical genomics: the added value from segregation. Trends Genet 17:388–391

21. Ashburner M, Ball CA, Blake JA, Botstein D, Butler H et al (2000) Gene Ontology: tool for the unification of biology. Nat Genet 25:25–29
22. Kanehisa M, Goto S, Sato Y, Furumichi M, Tanabe M (2012) KEGG for integration and interpretation of large-scale molecular data sets. Nucleic Acid Res 40:D109–D114
23. Keseler IM, Collado-Vides J, Gama-Castro S, Ingraham J, Paley S et al (2005) EcoCyc: a comprehensive database resource for Escherichia coli. Nucleic Acids Res 33
24. Hughes TR, Marton MJ, Jones AR, Roberts CJ, Stoughton R et al (2000) Functional discovery via a compendium of expression profiles. Cell 102:109–126
25. Steuer R, Kurths J, Daub CO, Weise J, Selbig J (2002) The mutual information: detecting and evaluating dependencies between variables. Bioinformatics 18:S231–S240
26. de Matos Simoes R, Emmert-Streib F (2011) Influence of statistical estimators of mutual information and data heterogeneity on the inference of gene regulatory networks. PLoS ONE 6:e29279
27. Mason M, Fan G, Plath K, Zhou Q, Horvath S (2009) Signed weighted gene co-expression network analysis of transcriptional regulation in murine embryonic stem cells. BMC Genomics 10:327
28. Zhou X, Kao MCC, Hung W (2002) Transitive functional annotation by shortest-path analysis of gene expression data. Proc Natl Acad Sci U S A 99:12783–12788
29. Tibshirani R (1996) Regression shrinkage and selection via the lasso. J R Stat Soc Series B (Methodol):267–288
30. Li C, Li H (2008) Network-constrained regularization and variable selection for analysis of genomic data. Bioinformatics 24:1175–1182
31. Friedman J, Hastie T, Tibshirani R (2008) Sparse inverse covariance estimation with the graphical lasso. Biostatistics 9:432–441
32. Shimamura T, Imoto S, Yamaguchi R, Miyano S (2007) Weighted lasso in graphical gaussian modeling for large gene network estimation based on microarray data. Genome Inform 19:142–153
33. Gustafsson M, Hornquist M, Lombardi A (2005) Constructing and analyzing a large-scale gene-to-gene regulatory network lasso-constrained inference and biological validation. IEEE/ACM Trans Comput Biol Bioinform 2:254–261
34. Li W, Zhang S, Liu C, Zhou X (2012) Identifying multi-layer gene regulatory modules from multi-dimensional genomic data. Bioinformatics on line
35. Li S, Hsu L, Peng J, Wang P (2011) Bootstrap inference for network construction. Arxiv, preprint arXiv:11115028
36. Needham CJ, Bradford JR, Bulpitt AJ, Westhead DR (2007) A primer on learning in Bayesian networks for computational biology. PLoS Comput Biol 3:e129
37. Maxwell Chickering D, Heckerman D (1997) Efficient approximations for the marginal likelihood of Bayesian networks with hidden variables. Mach Learn 29:181–212
38. Heckerman D (2008) A tutorial on learning with Bayesian networks. Innovations in Bayesian networks, pp 33–82
39. Zhu J, Zhang B, Smith EN, Drees B, Brem RB et al (2008) Integrating large-scale functional genomic data to dissect the complexity of yeast regulatory networks. Nat Genet 40:854–861
40. Hartemink AJ, Gifford DK, Jaakkola TS, Young RA (2002) Combining location and expression data for principled discovery of genetic regulatory network models. Pacific Symp Biocomput:437–449
41. Tamada Y, Kim S, Bannai H, Imoto S, Tashiro K et al (2003) Estimating gene networks from gene expression data by combining Bayesian network model with promoter element detection. Bioinformatics 19:2
42. Imoto S, Higuchi T, Goto T, Tashiro K, Kuhara S et al (2003) Combining microarrays and biological knowledge for estimating gene networks via Bayesian networks. Proc IEEE Comput Soc Bioinform Conf 2:104–113
43. Doss S, Schadt EE, Drake TA, Lusis AJ (2005) Cis-acting expression quantitative trait loci in mice. Genome Res 15:681–691

44. Wainwright M, Ravikumar P, Lafferty J (2007) High-dimensional graphical model selection using $l*1$-regularized logistic regression. In: Advances in neural information processing systems vol 19. p 1465
45. Choi M, Tan V, Anandkumar A, Willsky A (2011) Learning latent tree graphical models. J Mach Learn Res 12:1729–1770
46. Srebro N (2001) Maximum likelihood bounded tree-width markov networks. In: Proceedings of the 17th conference on uncertainty in artificial intelligence. Morgan Kaufmann Publishers Inc, pp 504–511
47. Friedman N, Nachman I (2000) Gaussian process networks. In: Proceedings of the 16th conference on uncertainty in artificial intelligence. Morgan Kaufmann Publishers Inc, pp 211–219
48. Tu Z, Wang L, Arbeitman MN, Chen T, Sun F (2006) An integrative approach for causal gene identification and gene regulatory pathway inference. Bioinformatics 22:e489–e496
49. Lee I, Date SV, Adai AT, Marcotte EM (2004) A probabilistic functional network of yeast genes. Science 306:1555–1558
50. Ernst J, Vainas O, Harbison CT, Simon I, Bar-Joseph Z (2007) Reconstructing dynamic regulatory maps. Mol Syst Biol 3
51. Deng M, Chen T, Sun F (2004) An integrated probabilistic model for functional prediction of proteins. J Comput Biol 11:463–475
52. Ucar D, Beyer A, Parthasarathy S, Workman CT (2009) Predicting functionality of protein-DNA interactions by integrating diverse evidence. Bioinformatics 25:i137–144
53. Ernst J, Beg QK, Kay KA, Balázsi G, Oltvai ZN et al (2008) A semi-supervised method for predicting transcription factor-gene interactions in Escherichia coli. PLoS Comput Biol 4:e1000044
54. Hwang D, Rust AG, Ramsey S, Smith JJ, Leslie DM et al (2005) A data integration methodology for systems biology. Proc Natl Acad Sci U S A 102:17296
55. modENCODE Consortium, Roy S, Ernst J, Kharchenko PV, Kheradpour P et al (2010) Identification of functional elements and regulatory circuits by Drosophila modENCODE. Science (New York) 330:1787–1797
56. Szklarczyk D, Franceschini A, Kuhn M, Simonovic M, Roth A et al (2010) The STRING database in 2011: functional interaction networks of proteins, globally integrated and scored. Nucleic Acid Res 39:D561–D568
57. Davis DA, Chawla NV (2011) Exploring and exploiting disease interactions from multi-relational gene and phenotype networks. PLoS ONE 6:e22670
58. Segal MR, Dahlquist KD, Conklin BR (2003) Regression approaches for microarray data analysis. J Comput Biol 10:961–980
59. Kim H, Hu W, Kluger Y (2006) Unraveling condition specific gene transcriptional regulatory networks in Saccharomyces cerevisiae. BMC Bioinform 7:165
60. Gao F, Foat B, Bussemaker H (2004) Defining transcriptional networks through integrative modeling of mRNA expression and transcription factor binding data. BMC Bioinform 5:31
61. Luscombe NM, Madan Babu M et al (2004) Genomic analysis of regulatory network dynamics reveals large topological changes. Nature 431:308–312
62. Tanay A, Sharan R, Kupiec M, Shamir R (2004) Revealing modularity and organization in the yeast molecular network by integrated analysis of highly heterogeneous genomewide data. Proc Natl Acad Sci U S A 101:2981–2986
63. Lemmens K, De Bie T, Dhollander T, De Keersmaecker S, Thijs I et al (2009) DISTILLER: a data integration framework to reveal condition dependency of complex regulons in Escherichia coli. Genome Biol 10:R27
64. Jeong H, Tombor B, Albert R, Oltvai ZN, Barabási AL (2000) The large-scale organization of metabolic networks. Nature 407:651–654
65. van Noort V, Snel B, Huynen MA (2004) The yeast coexpression network has a small-world, scale-free architecture and can be explained by a simple model. EMBO Rep 5:280–284

66. Yip A, Horvath S (2007) Gene network interconnectedness and the generalized topological overlap measure. BMC Bioinform 8:22
67. Clauset A, Shalizi CR, Newman MEJ (2009) Power-law distributions in empirical data. SIAM Rev 51:661
68. Marbach D, Prill RJ, Schaffter T, Mattiussi C, Floreano D et al (2010) Revealing strengths and weaknesses of methods for gene network inference. Proc Natl Acad Sci 107:6286–6291

Chapter 18
Direct Computer Mapping Based Modeling of a Multiscale Process Involving p53/miR-34a Signaling

Béla Csukás, Mónika Varga and Aleš Prokop

Abstract We studied a simplified multiscale biosystem with a new modeling and simulation methodology. The biosystem was a consciously, but arbitrarily selected multiscale part of the p53/miR-34a related signaling process that has an important role in tumor resistance, in cancer diagnostics, as well as in the therapy of various tumors. The multiscale model covered a vertical slice of the system from the change of a pathologic stage to the detailed dynamic molecular processes and vice versa. We employed the Direct Computer Mapping of process models for dynamic simulation of this typical multiscale, hybrid biosystem. The major advantage was the unified representation of the various quantitative and qualitative sub-models, as well as the easy combination of these various models within the unified simulating environment. Regardless to the limited number of components and interactions, the investigated fictitious illustrative example demonstrated many important and interesting features of the multiscale, hybrid biosystems. The model illustrates how the typical properties of the low level molecular events project onto the state properties of the higher scales. These properties (often called emergent properties) determine typical scenarios of lower scale states and actions. The simplified example, extracted from the independently developed, but coherent references, described some essential features about the modeled biological processes. We simulated the natural functioning of the p53/miR-34a signaling for the tumor suppression, as well as the various malfunctions of the system, resulting from tumor development. The simulation of addition of ectopic miR-34a demonstrates possible therapeutic intervention.Considering the uncertainties coming from the

B. Csukás (✉) · M. Varga
Research Group on Process Network Engineering, Kaposvár University, Kaposvár, Hungary
e-mail: csukas.bela@ke.hu

M. Varga
e-mail: varga.monika@ke.hu

A. Prokop (✉)
Chemical and Biomolecular Engineering Vanderbilt University, Nashville TN 37235, USA
e-mail: ales.prokop@vanderbilt.edu

A. Prokop and B. Csukás (eds.), *Systems Biology*,
DOI: 10.1007/978-94-007-6803-1_18,

very limited set of modeled components and interactions, as well as from the roughly estimated numerical parameters, constraints and initial values, the simplified model worked feasibly. Probably, this came from the more or less correct consideration of the feedback loops in the compact low level quantitative model. The studied process provides a good example for combining conservational (material balance based) and informational (sign based) models. This helps to understand the relativistic notion of informational process, as a well defined sub-process of the conservational process. This sub-process consumes and produces less amount of conservational measures than its complement, but effects more on its complement than vice versa. Conscious consideration of conservation based informational processes is an important lesson from biosystem modeling for computational tools and from computer simulations for biosystem research.

Keywords Biosystem modeling · Biosystem simulation · Hybrid process · Multiscale process · p53 · miR-34a · Signaling · Tumor development · Tumor suppression · Direct Computer Mapping · Conservational process · Informational process

Acronyms

AKT	Also known as Protein Kinase B (PKB), is a serine/threonine-specific protein kinase that is involved in cellular survival pathways, by inhibiting apoptotic processes
ATM	Ataxia telangiectasia mutated (ATM) is a protein kinase, involved in cellular survival pathways, by inhibiting apoptotic processes
ATR	Ataxia telangiectasia protein kinase that regulates phosphorylation of serine in DNA-damaged cells
B99	Provisionally named protein, whch transcriptional induction is dependent on wt p53 function after DNA damage
BAX	Bcl-2–associated X protein of the Bcl-2 gene family that promotes apoptosis by competing with Bcl-2
BCL2	B-cell lymphoma 2
CBP	CREB-binding protein, also known as CREBBP or CBP, is a protein that in humans is encoded by the CREBBP gene
CDC7	Cell division cycle 7-related protein kinase is an enzyme that in humans is encoded by the CDC7 gene
CDK 4/6	Cyclin-dependent kinase 4/6
Cyclin E2	A protein, binding to G1 phase Cdk2, which is required for the transition from G1 to S phase of the cell cycle that determines cell division.
Cyclin G1	Cdk-G1 cyclin complex begins to induce the initial processes of DNA replication
CDKCYC	An abbreviation referring to CDK4/6+CylinE2
c-Myc	

	Transcription factor that activates expression of many genes through binding on consensus sequences
Cop1	Protein believed to facilitate p53 degradation via the ubiquitin-proteasome pathway (similarly to MDM2)
CUL5	Cullin-5 is a protein, inhibiting cellular proliferation (encoded by the CUL5 gene)
DAE	Differential Algebraic Equation
DBC1	Deleted Gene in Breast Cancer 1
DCM	Direct Computer Mapping
DEVS	Discrete Event Systems
DNA	Deoxiribonucleonic Acid
E2F3	E2F transcription factor 3
EGF pathway	Epidemial growth factor pathway
EGFR	Epidermal growth factor receptor
EMT	Epithelial-mesenchymal transition
GADD45	Growth Arrest and DNA Damage or GADD45 genes are implicated as stress sensors against genotoxic/physiological stress, and modulate tumor formation
HDAC	Histone deacetylase inhibitors that can induce p21 (WAF1) expression, a regulator of p53's tumor suppressor activity
HER-2	Human Epidermal Growth Factor Receptor 2, also known as ErbB-2 or p185 is a protein encoded by the ERBB2 gene
HIF-1 beta	Hypoxia-inducible factor that respond to changes in available oxygen in the cellular environment, specifically, to decreases in oxygen, or hypoxia
HMGA2	High-mobility group of proteins, containing structural DNA-binding domains and functioning as a transcriptional regulating factor
ICAMs	Intercellular adhesive molecules
IGF-1	Insulin-like growth factor
IPDAE	Integro Partial Differential Algebraic Equations
Killer/DR5	Alternative abbreviation for tumor necrosis factor receptor superfamily, member 10b, official symbol TNFRSF10B
MAD2L1	MAD2L1 is a component of the mitotic spindle assembly checkpoint that prevents the onset of anaphase until all chromosomes are properly aligned at the metaphase plate
MDA-MB231	Human breast carcinoma cell line.
MDM2	Murine Double Minute
miR	Micro RNAs, small, non-coding RNA molecules
miR34a	A specific miR, participating in p53 controlled signaling processes
miR-34/a/b/c	A specific family if miRs, participating in p53 controlled signaling processes
MMPs	Matrix metallopeptidases

Notch	A family of transmembrane proteins, involved in lateral inhibition
Noxa	Phorbol-12-myristate-13-acetate-induced protein 1 encoded by the PMAIP1 gene, is a pro-apoptotic member of the Bcl-2 protein family
ODE	Ordinary Differential Equation
p110 (PI3K)	PI3K (p110 alpha and p110 beta) have differential effects on Akt activation and protection against oxidative stress-induced apoptosis in myoblasts
p21	Cyclin-dependent kinase inhibitor 1, encoded by the CDKN1A gene
p300	E1A binding protein p300 also known as EP300 or p300 is a protein that, in humans, is encoded by the EP300 gene
p53	Tumor suppressor protein, in humans it is encoded by the TP53 gene
p53Ac	Acetilated p53
p53mut	Non-functioning, mutant p53
p85α	Phosphatidylinositide 3-kinase (PI3K), involved in cell growth, proliferation, differentiation, etc.
PDE	Partial Differential Equation
PIG3	p53-inducible gene 3 (PIG3) identified in a screen for genes induced by p53 before the onset of apoptosis
Pirh2	Protein believed to facilitate p53 degradation via the ubiquitin-proteasome pathway (similarly to MDM2)
PP2A	Protein phosphatase 2 is an enzyme that in humans is encoded by the PPP2CA gene
PUMA	p53 upregulated modulator of apoptosis, also known as Bcl-2-binding component 3 (BBC3), is a pro-apoptotic protein encoded by the BBC3 gene
Ras	Family of proteins (discovered first as 'rat sarcoma protein', switching signaling that ultimately turn on genes involved in cell growth, differentiation and survival
RNA	Ribonucleonic acid
SIRT1	Silent Information Regulator 1
STN	State Transition Net
SySML	Systems Modeling Language
TGFβ	Transforming growth factor, beta receptor
UML	Unified Modeling Language
Wip1	A human protein phosphatase that is induced in response to ionizing radiation in a p53-dependent manner
Wt-p53	Wild type p53
XXX	A fictitious component, designating the DNA damage (like DBC1 in breast cancer)

18.1 Introduction

18.1.1 Challenge of Multiscale Process Modeling in Systems Biology

Computer assisted methods of process simulation and of simulation based process design and control had been developed before the up-to-date biosystem (biological) engineering appeared. The complexity of the multi-scale, hybrid bioengineering systems requires new modeling and computational methodologies.

The general formal models of the process systems had defined before the powerful Information Technology came. According to Kalman's definition, the State Space Model can be described by the state and output functions in the continuous time [1]. The abstract automaton representation of the discrete processes can be described by similar functions in the discrete time. These models emphasize the functionalities and do not deal with the structures behind.

In contrary, the net models focus on the description of process structure. This appeared first in the Petri Net [2], followed by the various State Transition Nets, until the higher order and Quantitative Petri Nets [3]. All of them belong to the General Net Theory [4].

The state-of-art in process modeling was analyzed by Marquardt [5], who reviewed the methodologies and tools, developed for simulation based problem solving. The significant evolution of process modeling methodologies has been motivated by the process industries [6].

The conventional approach focuses on the functioning, that can be described by a set of Ordinary Differential Equations (ODE's), Partial Differential Equations (PDE's), Differential Algebraic Equatios (DAE's) etc. [7]. This constructs do not distinguish between the additive and over-writeable semantics of the variables.

Recently the agent based and the model-driven approaches came into the limelight of modeling and simulation. There are two different approaches for model driven simulation. The general purpose frameworks, like UML2 [E1], SysML [E2], and Modelica [E3] offer a language, as well as a toolkit for any kind of computer models, on the one hand. On the other side, there are various domain specific model-driven solutions for the given fields of applications (e.g. for maritime surveillance system [8], for industrial control applications [9], etc.). Agent based approach offers another methodology to generate freely programmable specific applications [10]. However, as a golden mean, we need intermediate solutions that are not applicable for every simulation model, rather for a well defined but broad enough set of models.

The term "multiscale" means that the complex model is built from various parts with different spatial and/or temporal scales, while the more or less detailed parts can belong to different disciplines. The need for multiscale models appeared in many fields, e.g. in material science [7], in the computational system's biology [11, 12] and in process and product engineering [51].

Multiscale methodology usually means linking together the quite different modeling and computational methodologies, used for the description of the studied parts and scales. Sometimes "multiscale" is confused with "hierarchical", where the upper level models are decomposed into the more detailed lower levels. However, according to our understanding, in multiscale process modeling there is rather an event driven system in the organization of the complex simulation problem.

Nowadays, the paradigm of hierarchical modeling has been replaced by the idea of multiscale modeling. Comprehensive multilevel modeling of hierarchical systems underestimated the complexity of real world processes, as well as overestimated the capability of computers and computer modelers. Probably, the reason was that the respective ideas started from various isolated fields of natural science and engineering. For example, chemical engineering served well for medium complexity, where the processes are complex enough to form a tractable hierarchical systems, but simple enough to solve the problems by the available computational tools.

During the last decade, computational methods of problem solving have been more and more involved in the solution of biological and complex economical systems, however use of hierarchical approach became difficult. Hierarchy is an excellent methodology, applied originally for the recognition. However, because of lack of other methodologies, people started to use hierarchical control and design. The basic problem with it is that the space of problem solving is open, because each level employs the objectives, coming from outside, from the upper level.

On the other hand, very complex biosystem processes and global processes produced dramatic, but useful lessons for computer modelers and computational tool users. An interesting element of these lessons is that the model based control and design of global processes can and must learn from the organization and behavior of biological systems (e.g. the importance of the local solutions, cooperative feedback between the functionally connected neighbors, bi-directional multiscale feedback loops between the bottom-up and top-down behavior, etc.). It is less recognized that the study and understanding of biosystems can also learn from the experiences coming from the successes and failures of human made "artificial" processes [13].

Complex biosystems (e.g. from the cellular interactions to the medicated body) and complex global processes (e.g. agrifood process networks from the fork to the table) nowadays requires new multiscale methodologies. In the field of biology, new approaches are described, for example, in Meier-Schellersheim [12]. The example treated in our chapter attempts to develop a new approach for multiscale simulation in biology.

18.1.2 Importance of p53/miR-34a System in Natural and Therapeutic Tumor Suppression

Solving problems in system biology requires various approaches ranging from molecular processes up to the macro-organism, thus requires special attitude to handle these problems mathematically. Oncologic diseases are typical examples. Although there is an obvious progress in knowledge and treatment, the numbers of positively diagnosed patients are growing. As the oncologic diseases do not represent a single illness but a heterogeneous class of disorders with various causality and consequences, the quest for common attributes or effective prevention and therapy is a highly complicated task. Recent empiric procedures are frequently failing and leading only to suppression of symptoms. Similarly, our knowledge on prevention is coming from epidemiological studies based on previous observations. Meta-analysis of available data can help to identify some unexpected and therefore unattended interactions. Systemic analysis using simulated experiments and subsequent experimental (wet lab) validation may serve as a tool to search for complex therapeutic procedures. These approaches require input to biomedical research from bioinformatics and systems biology. Such approaches offer tools for mathematical description and subsequent modeling of complex system's behavior, such as complex cellular processes. Understanding of malignant processes and their modeling will open new possibilities for more exact prediction of preventive or therapeutic approaches.

p53 related signaling processes have a keynote role in development, suppression and therapy of tumors. A qualitative information of p53 signaling pathway is available in standard repositories (Applied Biosystems/Life technologies Corp., Biocarta LLC, SA Biosciences/Qiagen, KEGG, etc.), listing about 10–20 reactants and compartmentalization between cytoplasm and nucleus. None of them features acetylated reactant of p53Ac [14] necessary for the activation (only Applied Biosystems lists the acetylation in their commentary). References on quantitative simulation model are quite rare [15–17], and deal with a small sub-process in detail. Likewise, no acetylation is considered. The basic p53/MDM2 feedback loop, however, has been studied quite often, as a simple dynamic system (e.g. [18]).

Additional complication comes from the post-transcriptional control. The p53 mainly exerts its function through transcription regulation of its target genes to initiate various cellular responses. To maintain its proper function, p53 is tightly regulated by a wide variety of regulators in cells. Thus, p53, its regulators and regulated genes form a complex p53 related network which is composed of hundreds of genes and their products. Among the regulators, microRNAs are a class of endogenously expressed, small non-coding RNA molecules which play a key role in regulation of gene expression at the post-transcriptional level [19]. Among them miR-34a plays an important role in some cancer, particularly in colon cancer, lung cancer, chronic lymphatic leukemia and others. Yamakuchi and Lowenstein [20] outlined several outcomes (gene targets) of p53 activation, typically denoted as cellular responses (emergent properties, systemic properties).

Among them are cell cycle, sustained proliferation, cell cycle arrest, senescence, apoptosis, inhibition of angiogenesis, etc. The modeling and simulation of coupled p53/miRs is very rare. Lai et al. [21] outlined a combined p53/miR-34a model as an extension of the basic p53/MDM2 feedback loop (this model includes also acetylation). It is to be noted that modeling of p53 signaling or coupled p53/miR-34a loop would only represent a mere mathematical exercise, without the inclusion of the gene targets in the model. In addition, understanding the connection between the systemic (emergent) properties and p53/miR-34a control ought to represent a new hybrid approach in modeling of signaling pathways.

The human miR-34a was discovered computationally [22] and later verified experimentally [23]. This miR has recently been implicated in cancer, particularly with its expression relating to TP53 status [24]. The expression of such miR has also been confirmed in embryonic stem cells [25].

As a target of miR-34a the Silent Information Regulator 1 (SIRT1) gene is showed [26]. The miR-34a inhibition of SIRT1 leads to an increase in acetylated p53 and in expression of p21 and PUMA, transcriptional targets of p53 that regulate the cell cycle and apoptosis, respectively. Furthermore, miR-34a suppression of SIRT1 ultimately leads to apoptosis in wild-type human colon cancer cells but not in human colon cancer cells lacking p53. Finally, miR-34a itself is a transcriptional target of p53, suggesting a positive feedback loop between p53 and miR-34a. Thus, miR-34a functions as a tumor suppressor, in part, through a SIRT1/p53 pathway [27].

miR-34a inhibits human p53-mutant gastric cancer tumor spheres. In p53-deficient human gastric cancer cells, restoration of functional miR-34a inhibits cell growth and induces chemosensitization and apoptosis, indicating that miR-34a may restore p53 function. Restoration of miR-34a inhibits tumor sphere formation and growth, which has been reported to be correlated to the self-renewal of cancer stem cells. The self-renewal appears to be related to the direct modulation of downstream targets BCL2, Notch, and HMGA2, indicating that miR-34a may be involved in gastric cancer stem cell self-renewal/differentiation decision-making [26, 27].

Cancer is usually caused by multiple mutations and alterations of multiple signaling pathways which pose an extra challenge when defining the mechanisms underlining cancer therapy. Development of drug combination therapies for cancer can lead to more effective therapies to overcome drug resistance and to achieve maximal drug efficacy. Application of single-drug therapies are failing frequently due to the cell redundancy, drug-specific or multi-drug resistance formation. Concurrent application of two or more drugs helps to suppress resistance; however, it is often based on empiric knowledge from medical praxis. Mathematical modeling of signaling pathways and tumor cell behavior under in silico conditions may help to rationalize expenses needed for experimental verification of hypotheses.

Historically, [28] introduced a direct search method to optimize cancer chemotherapy regimens while [29, 30] has developed a qualitative conceptual framework for treating multiple genome abnormalities (blockage of multiple targets).

Al-Shyoukh et al. [31] established and validated a data-driven mathematical approach to systematically characterize signal-response relationships. Their results demonstrate how mathematical learning algorithms can enable systematic characterization of multi-signal induced biological activities. The proposed approach enables identification of input combinations that can result in desired biological responses. In retrospect, the results show that, unlike a single drug, a properly chosen combination of drugs can lead to a significant difference in the responses of different cell types, increasing the differential targeting of certain combinations. The successful validation of identified combinations demonstrates the power of this approach. Moreover, the approach enables examining the efficacy of all lower order mixtures of the tested signals. The approach also enables identification of system-level signaling interactions between the applied signals.

References on multiple targeting of signaling pathways are quite rare. Recently, a method based on stepwise direct search for identifying optimal combination of drugs for pain treatment has been introduced Curatolo and Sveticic [32]. Iadevaia et al. [33] analyzed the insulin-like growth factor (IGF-1) signaling network in the MDA-MB231 breast cancer cell line and the modeling predictions showed that optimal drug combinations inhibited cell signaling and proliferation. Bloom and Kloog [34] have considered targeting Ras pathway, while [35] described the rationale and results of clinical trials using biologically targeted agents in HER2-positive breast cancer patients. Single drugs that hit multiple targets and cocktails of biologically targeted agents have been also considered. Concluding, the combinatorial drug treatment may offer huge improvement in overall response.

A transfer of know-how, i.e. understanding of mathematical modeling and simulation principles by experimental biologists and vice versa, understanding of experimental procedures and real problems of experimental biology by bioinformatics is the main goal of future multi-disciplinary cooperation.

18.2 Example for a Hybrid Multiscale Biosystem

With the knowledge of the previous review we had a closer look at the p53/miR-34a interactions and tried to find an example for studying an oversimplified, but typical multiscale hybrid model.

18.2.1 Elements of the p53/miR-34a Related Puzzle

A wide variety of intracellular and extracellular stress signals are detected by the cell and communicate with the p53 protein by numerous mediators [19]. Figure 18.1 shows representative mediators. Stress signals promote the activation of p53, but it is mediated by MDM2 protein. Depending on the cell type, environmental context, as well as on the type and/or degree of stress, activated p53 selectively transcribes a group of its target genes (Fig. 18.1 shows representative examples) and initiates

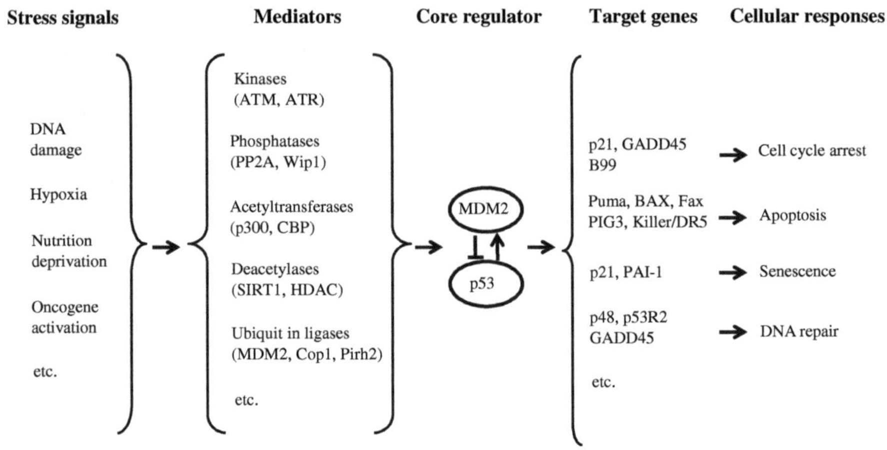

Fig. 18.1 p53 regulated mediators and target genes (according to [19])

various cellular responses to exert its function in tumor suppression. The scheme shows clearly the p53/MDM2 couple, as the "heart" of system.

Possible role of miRNAs (including miR-34a) has been shown in the same paper [19], as it is illustrated in Fig. 18.2.

According to the authors, the p53 induces the expression of a set of miRNAs, including miR-34/a/b/c, miR-145, miR-107, miR-192 and miR-215, which can all contribute to the role of p53 in tumor suppression as a new group of p53 target genes. miR-34a/b/c down-regulates CDK4 and CDK6 to induce cell cycle arrest, and down-regulates BCL2 to promote apoptosis. miR-145 down-regulates c-Myc to reduce cell proliferation. miR-192 and miR-215 down-regulate a group of genes which regulate DNA synthesis and cell cycle checkpoints, including CDC7, MAD2L1 and CUL5, to induce cell cycle arrest and reduce tumor cell growth. miR-107 down-regulates HIF-1 beta to negatively regulate hypoxia signaling and suppress angiogenesis. Consequently miR-34/a/b/c plays in intermediate role in functioning of p53 on CDK4, CDK6, Cyclin E2 and BCL2.

However, the picture is a little bit more complicated if we have a look at the feedback loops of miRNAs, inhibiting p53 (see Fig. 18.3, by [19]).

According to Feng et al. miR-125b and miR-504 directly down-regulate p53 protein levels and functions in apoptosis and cell cycle arrest through their direct binding to p53 3′-UTR. miR-34a up-regulates p53 activity and function by down-regulating SIRT1, which is a negative regulator of p53 through deacetylating acetylated p53 (p53Ac). miR-122 enhances p53 activity through its down-regulation of Cyclin G1, which forms a complex with PP2A phosphatase and enhances MDM2 activity to inhibit p53. miR-29 down-regulates p85a, a regulatory subunit of PI3 K, and thereby enhances p53 activity through the negative loop between PI3 K-AKT-MDM2 and p53.

Considering the single miR-34a, some essential features are summarized in Fig. 18.4. [20].

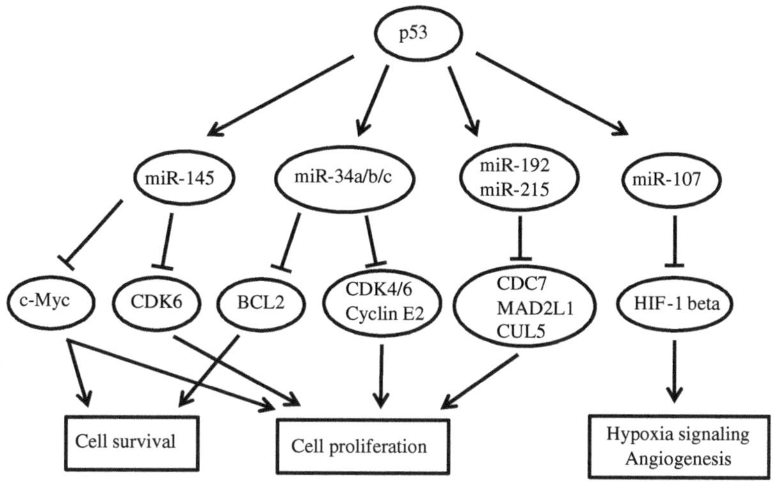

Fig. 18.2 Transcription regulation of specific miRNAs by p53 (according to [19])

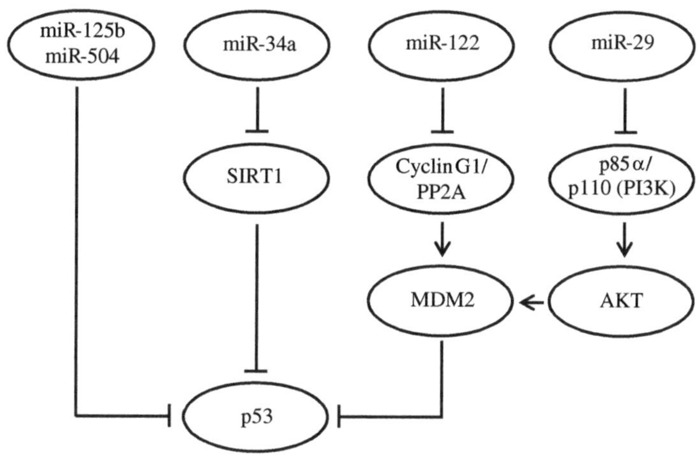

Fig. 18.3 Multiple miRNAs regulate the activity and function of p53 (according to [19])

According to this scheme p53 induces miR-34a expression, which increases p53 acetylation by suppressing SIRT1 expression. Resultant increase of p53 activity prolongs miR-34a expression. miR-34a induces apoptosis, cell growth arrest and senescence.

All of the schemes contain useful qualitative information about the p53/miR-34a system at a medium scale of processes. Considering the complexity of the system, the above seems to be too complicated for a simplified rule based interpretation on the one hand, as well as too ill-defined for a sophisticated quantitative analysis. In addition we need more details for linking the investigated system to the higher scale events and to lower scale dynamics.

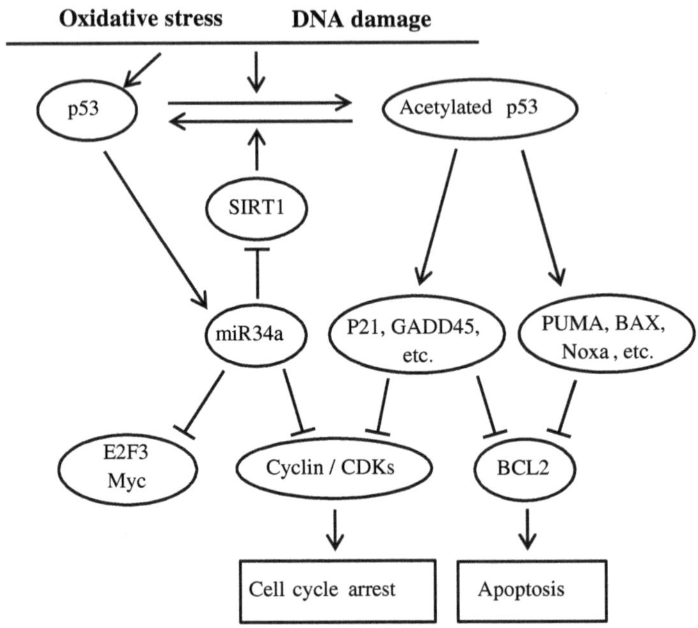

Fig. 18.4 Schematic representation of p53/miR-34a feedback loop (based on [20])

18.2.2 Three Independently Developed, Coherent Parts for a Multiscale View

We tried to find a more or less coherent set of literature sources that helps to narrow the horizontal complexity, but to traverse low, middle and high scale processes vertically. Having studied many related papers we have found three, more or less coherent key papers that make possible to study a p53/miR-34a related example, as well as the application of our hybrid, multiscale modeling methodology (Direct Computer Mapping of process models). In the following we introduce the three key papers, applied for our trial to analyze a very simple multiscale model.

18.2.2.1 High Scale: Role of p53/miR-34a System in Given Stage of a Pathological Process

Many years before discovering the significance of p53/miR-34a related signaling, [36] described the multistep process of colorectal tumorigenesis, emphasizing the possible identification of the underlying molecular events. Based on the overview of Vogelstein's model, a decade later Slabý et al. [50] focused on the role of microRNAs in colorectal cancer pathogenesis, as a possible translation of molecular biology into clinical application.

In the subsequent stages of “Normal epithelium”, “Early adenoma”, “Intermediate adenoma”, “Late adenoma”, “Colorectal cancer” and “Metastasis” let us focus on the 4th stage called “Loss of p53 function”, illustrated in Fig. 18.5.

According to Slabý et al., in the 4th stage, following the late adenoma, the choice between survival or colorectal carcinoma is determined by the p53/miR-34/a/b/c system. SIRT1, as well as CDK4, CDK6, Cyclin E2 and BCL2 are involved in the basic loop of this stage. This shows a big picture about the involvement of miR-34/a/b/c in the development of colorectal cancer. In the stage called “Loss of p53 function” the question of “(cell cycle arrest and/or apoptosis) or carcinoma” is decided. The cited keynote actors are: p53, miR-34/a/b/c, SIRT1, CDK4, CDK6, CyclinE2 and BCL2.

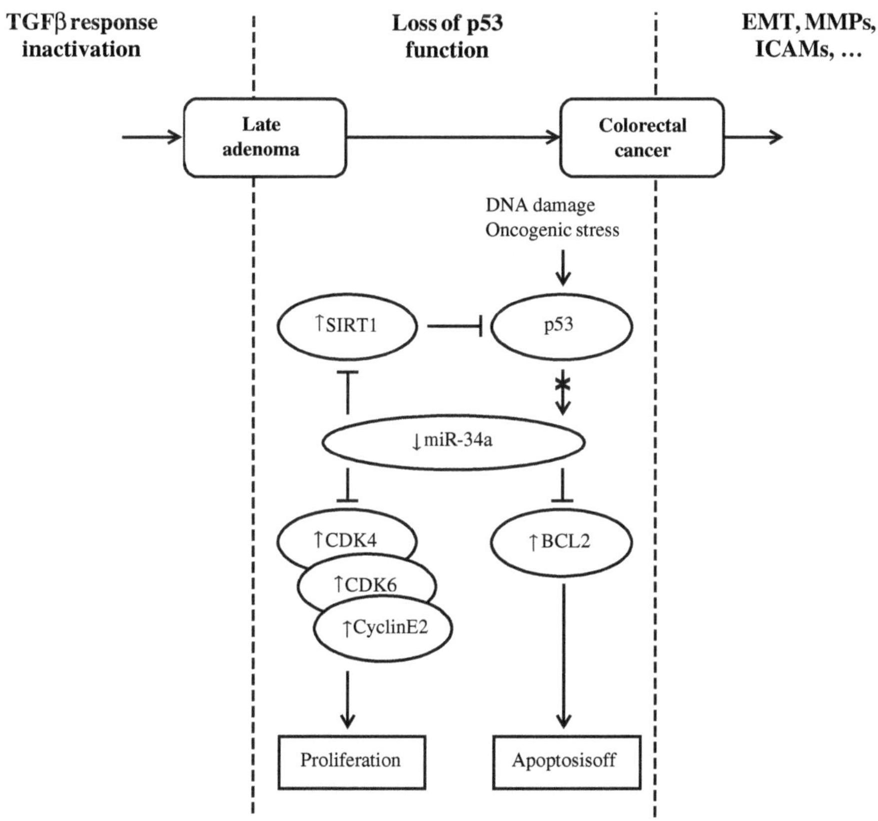

Fig. 18.5 MicroRNAs' involvement in 4th stage of colorectal cancer pathogenesis (based on Slabý et al. [50]). SIRT1 = silent information regulator 1, CDK4,6 = cyclin-dependent kinase 4,6, Cyclin E2 associates with Cdk2 in a functional kinase complex exhibiting catalytic control over cell cycle and the G_1/S transition, BCL2 = B cell lymphoma 2

18.2.2.2 Low Scale: Description of the Feedback Loops in the p53/miR-34a Interactions

In another context of breast cancer [37] described a possible, detailed low level dynamic model of miR-34a/p53 feedback structure, while p53 was activated by acetylation (see Fig. 18.6).

The authors studied a signaling module composed by p53, SIRT1 and miR-34a, based on the integration of experimental evidence with quantitative mathematical modeling. In the ODE based model the parameters were estimated heuristically, considering experimental data of [20, 26, 38]. In the model DNA damage can affect directly the p53 synthesis and activation, as well as indirectly through the DBC1 that inhibits the effect of SIRT1 on the deactivation of p53.

They modeled four different silencing mechanism of miR-34a affecting on SIRT1, and using numerical computations they compared the strength of the SIRT1′s two negative regulators (miR-34a and DBC1). Based on the analysis they concluded that miR-34a silences SIRT1 through translational repression, but DBC1 is a more efficient negative regulator of SIRT1 than miR-43a.

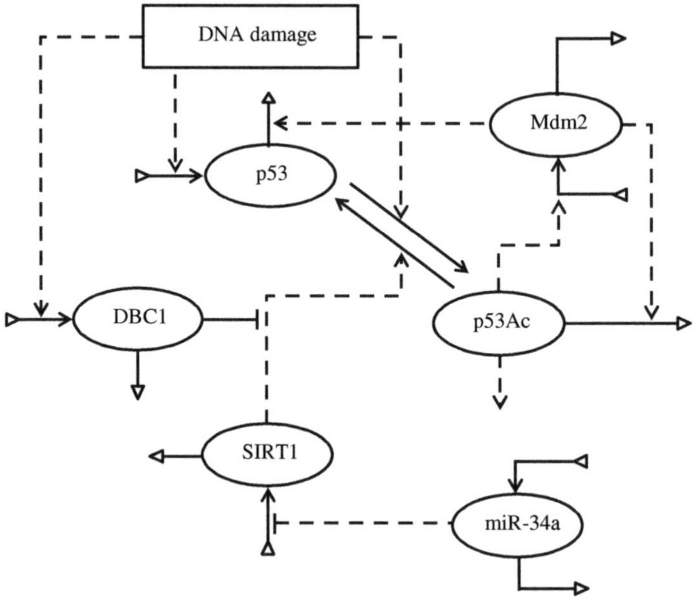

Fig. 18.6 Scheme of the detailed dynamic model of miR-34a/p53 feedback structure (based on [37]). *MDM2* murine double minute, p53 = deacetylated (inactive) p53, p53Ac = acetylated (active) p53, *SIRT*1 Silent Information Regulator 1, *DBC*1 Deleted Gene in Breast Cancer 1

18.2.2.3 Middle Scale: Qualitative Interpretation of the Cellular Scenarios in Development or Suppression of Tumor

Wong et al. [39] published an excellent review comparing the control schemes, corresponding to the various scenarios of the p53/miR-34 network, as follows

- regulated apoptosis and cell cycle arrest, resulting tumor suppression;
- abnormal regulation causing cell proliferation, resulting tumor development; and
- possible treatment of the dysregulated case by ectopic miR-34a expression.

Different scenarios for the signaling system involve SIRT1, CDK4, CDK6, Cyclin E2 and BCL2. Interestingly, the findings are coherent with the above described high and low scale models.

Figure 18.7 summarizes the molecular mechanisms in the p53/miR-34 network, involved in regulating cell apoptosis. p53 activates miR-34 after DNA damage and/or cellular stress, which subsequently inhibits expression of anti-apoptotic genes and results in cell apoptosis and tumor suppression. Stars indicate that as a consequence of the inhibition of anti-apoptotic cell cycle promoting proteins by miR-34a, apoptosis and cell cycle arrest can start. We can see also the regulation of SIRT1 by miR-34a as part of a positive feedback loop that leads to further activation of p53 once it has been activated. Abbreviations of anti-apoptotic proteins are the same as before, and E2F3 = E2F transcription factor 3. SIRT1 in previous works appeared as a harmful agent promoting inactivation of p53. Here it seems to have a direct effect on apoptosis and on cell cycle arrest.

Figure 18.8 explains the background of the "Loss of p53 function". It illustrates the abnormal regulation of the p53/miR-34 network causing cell proliferation and tumorigenesis. The function of miR-34 is lower in the presence of mutant p53 and/or miR-34 dysregulations. Over-expression of anti-apoptotic genes and proteins result in cell proliferation and tumor development. Stars indicate the lack of the respective promotion and inhibitions, caused by the abnormality in the downstream pathway.

Mutant p53 is not able to substitute for the role of p53. miR-34 dysregulation means the lack of the appropriately functioning miR-34.

In Fig. 18.9 the "miR-34 therapy" case is shown. Obviously this improves the failures caused by the dysregulated miR-34 and/or mutant p53. The function of miR-34 is reduced in the presence of mutant p53 and/or miR-34 dysregulation. Delivery of ectopic miR-34 recovers its function and results in cell apoptosis and tumor suppression via inhibition of anti-apoptotic genes. Stars indicate the lack of normal miR-34a promotion as well as the consequence of ectopic miR-34a, starting apoptosis and cell cycle arrest.

Stars reflect to abnormal transcriptional activation, as well as to the downregulation of anti-apoptotic proteins by miR-34. Ectopic expression is the expression of a gene in an abnormal place in an organism. This can be caused by a disease, or it can be artificially produced as a way to help determine what the function of that gene is, e.g. via introducing of a gene into the target organism (transient or stable transfection).

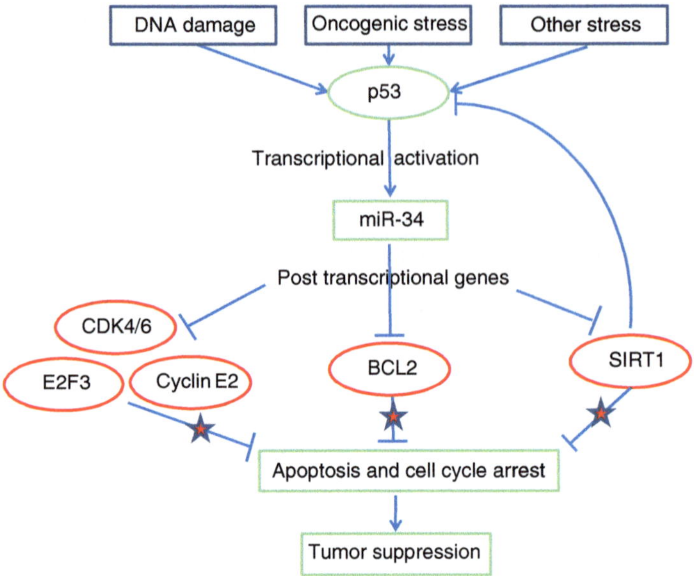

Fig. 18.7 p53/miR-34 network in regulating cell apoptosis (based on [39]). Stars indicate the lack of respective inhibitions

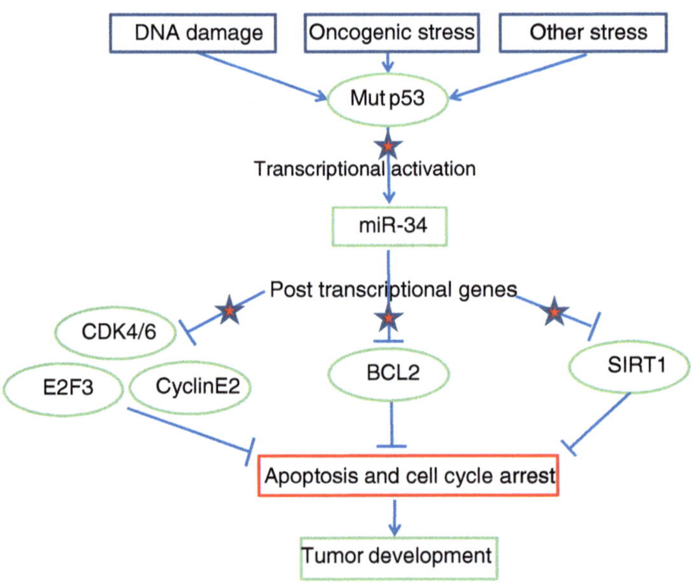

Fig. 18.8 Abnormal regulation of the p53-miR-34 network (based on [39]). *Stars* indicate the lack of respective promotion and inhibitions

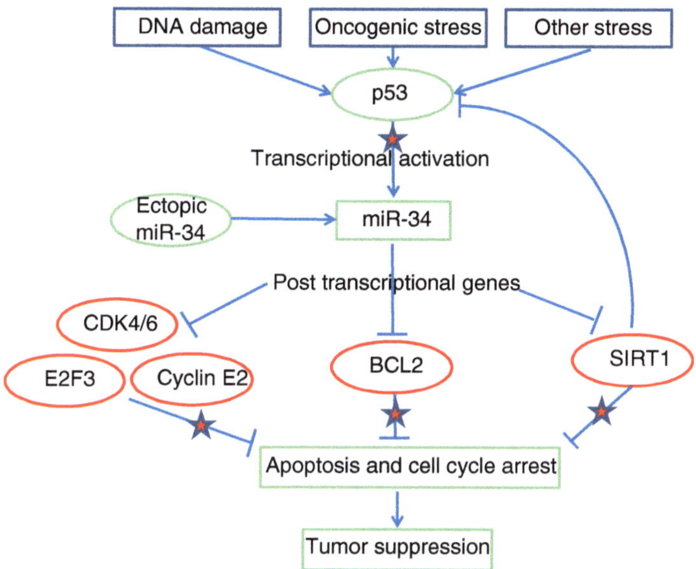

Fig. 18.9 Treatment of p53-mutant or miR34-dysfunctional cancer by ectopic miR-34 (based on [39]). *Stars* indicate the lack of respective promotion and inhibitions

18.3 New Tool for Multiscale Process Modeling and Simulation Methodology

Multiscale, hybrid processes of biosystems and of human-built process networks contain more complex elements and structures compared to the well established mathematical constructs. However these complex structures and functionalities might be mapped onto quite uniform elements of executable dynamic models, associated with local programs, while the simulation and the model based problem solving can be organized by a general purpose kernel program.

18.3.1 Direct Computer Mapping of Multiscale, Hybrid Processes

In our approach, called Direct Computer Mapping of process models [40, 41], the natural building blocks of the elementary states, actions and connections are mapped onto the elements of an executable code, directly.

The idea of Direct Computer Mapping (DCM) has evolved over last three decades. The basic principle of the method [40, 42] is that the simple state and action elements of the real world processes have to be mapped directly onto an executable code without interpreting them in the language of any particular mathematical constructs like ODE, PDE, DAE, IPDAE, STN, DEVS, etc. Hybrid processes are built from more complex elements and structures, than the well established mathematical constructs. However, these complex structures and functionalities might be mapped onto quite uniform elements of executable dynamic databases, associated with local (e.g. declarative) programs, while the simulation and the model based problem solving can be organized by a general purpose kernel program [43, 44, 45–47].

DCM contains a limited number of uniform computational building blocks for various processes and for different spatial and temporal scales. These general building blocks can be used for the flexible description of state and transition elements, as well as reading and modifying connections of various conservation based balance and rule based informational systems. The building elements can be associated with dedicated programs or program prototypes; accordingly, the functioning is embedded in the structure, flexibly. A general kernel, like an operational system, can execute all of the process models. The execution supports the spatial compartmentalization and the management of time-driven events, at different time horizons. The set of building blocks and execution algorithms can easily be adopted for building models with multiple spatial and temporal scales.

The key issue of multiscale paradigm is that it is impossible, and even not necessary to describe and calculate the whole system of multiscale processes simultaneously, in detail. Rather, depending on the investigated problem, the various subsets of processes must be calculated with various and changing detailness, that is determined by the environment of a given subprocess, e.g. by the signals coming from the lower or upper scales. The solution is supported by the generic capabilities of the brief programs, associated with the building blocks. It means that depending on the environmental situation, the building element can activate and/or generate a more detailed model for the given part.

The building and execution of effectively scalable process models is supported by the keynote feature that in our toolkit the structural skeleton of the quantitative and qualitative models is the same. It makes possible to change between quantitative 'a priori' models and qualitative rule based models automatically. It also helps at the stepwise discovery of the models (i.e. the identification of the structure and parameters). Accordingly, the qualitative expert knowledge can be transformed into quantitative models, based on the data, coming from the real world process (e.g. measurements). This is supported by the pair-wise joint set of qualitative and quantitative primitives.

An interesting case for using mixed quantitative and qualitative models is, when our model or a submodel runs at a given spatial and temporal scale, while the neighboring upper and/or lower level models are described only qualitatively. Depending on the experts' knowledge, we can change between qualitative and quantitative descriptions.

The robust solver, coming with the DCM, can be combined effectively by a robust multi-objective, discrete/continuous genetic algorithm. The evaluation feedback between dynamic simulation and genetic algorithm helps in identification and also in various problem solving activities (optimal control and/or design). Sometimes, genetic coding can utilize the knowledge, embedded in the property relationship lattice, too.

18.3.2 Declarative Syntax of Process Systems

The declarative description of the process model nets is summarized in Fig. 18.10. The architectural relations are symbolized by lines between the corresponding elements. Directed dashed lines help to understand the connection (unification) between the state or action elements and the associated program clauses. The coordination of the state and action elements with the reading and modifying connections is illustrated by directed bold lines in this Figure.

The list of elementary data (symbols, numbers) are contained in the d(.) functors, determining a Value by the triplet of the respective name, data list, and dimension. The modifiable content of the higher order constructs is determined by the list of these d(.) functors, called ValueList. All of the inputs, parameters and outputs, as well as all of the connections contain a ValueList.

The syntax of the state(.) and action(.) predicates are identical. The changeable content of these building elements is determined by the inputs (InpList), parameters (ParList) and outputs (OutList). InpList, ParList and OutList are dedicated lists, built from the i(.), c(.) and o(.) functors, respectively. Considering the net connections for the data flow, every functor corresponds to a so-called slot, characterized by slot name, slot type and ValueList.

The structurally and functionally different, multiple connections, interpreted between the state and action elements, connect the input and output slots, as it can be seen from the declaration of the connections. It is to be noted that the parameter slots can also be modified by special connections (e.g. coming from collaborating applications in the verification of the model); however this is not seen in Fig. 18.10.

Besides the input, parameter and output slots, the state and action elements are described by the following arguments: existence flag, identifying name, identifying scale coordinate, name of the corresponding program and the so-called Timings.

The spatial scale coordinate is defined by a list of integers that determines the spatial place (i.e. the functional and/or geometrical compartment) of the given element. It is to be noted that the elements are determined by the identifying name (StateName or ActName) and by its spatial scale (StateCoord, ActionCoord), together.

ProgramName identifies the program clause, associated with the given state or action element. The cardinality of programs is usually significantly less, than the number of elements. The program clauses can be defined with the so-called prototype elements, and they are saved in a dedicated partition or file. The syntax of the respective program clauses corresponds to the unification scheme (symbolized by dashed lines in Fig. 18.10). Actually, the StateProgramCode or ActionProgramCode clause bodies bind the free variables of OutList with the knowledge of the bound variables of InpList and ParList.

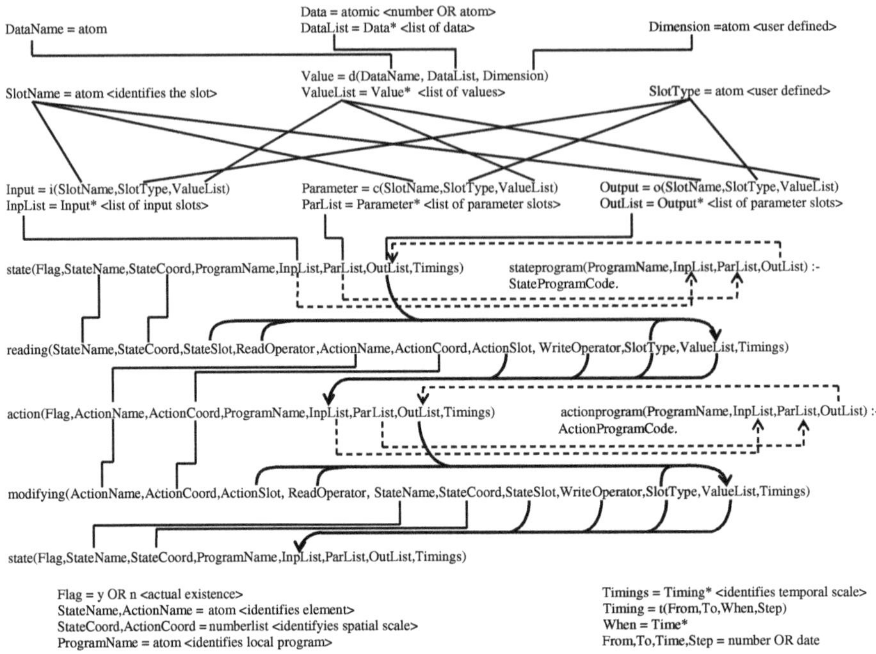

Fig. 18.10 Declarative syntax for the model-driven generation of simulation program

The syntax of the reading and modifying connections are also identical. The arguments of both predicates determine the identifying names, coordinates and slots for the sending and receiving sides of the connections. In addition, the connections declare the reading and writing operators for the data coming from and going to the specified slots, respectively. There are predefined, but extendable sets of reading and writing operators. For example writing operators can prescribe increases and decreases for the additive measures, removing or extending of d(.) elements from or to the respective lists, as well as overwriting of the signs (e.g. in case of rule based, qualitative models). The actual content is carried by the ValueList of the connections. The connections have also Timings, effecting on the execution of the model.

18.3.3 Model Driven Generation of the Executable Code

The above described declarative process architecture supports the model-driven, automatic generation of the executable code. This can be based on the graphical representation, extended with guarded data input, while the local programs can also be edited and tested during the declaration of the state and action prototypes.

The model generation is based on the description of the state and action prototypes. These prototypes determine also the number and type of the input, parameter and output slots, as well as the local program prototypes. All of the remaining state and action elements can be derived as the copies of the previously defined prototypes.

The reading and modifying connections can be generated automatically, by drawing edges between the respective sending and receiving slots.

The temporal behavior of the states, transitions and connections is described by the above mentioned Timings. It is a list of functors t(.), that makes possible to declare optionally different time scales and timings for the individual elements and connections in form of time periods, discrete times and time steps.

The event driven operation of the multiscale processes can be controlled by the informational connections, reading and modifying the respective state and transition slots.

Having finished the declaration of the model, filled with the appropriate initial data and parameters, the general purpose kernel program automatically generates the input files consisting of the executable state(.), action(.), reading(.) and modifying(.) facts, as well as of the stateprogram(.) and actionprogram(.) clauses.

18.3.4 Model Driven Process Simulation

The model-driven execution of the process simulation consists of six cyclically repeated, consecutive steps, as it can be seen in Fig. 18.11. The unification of the identifiers is signed by bold italics argument names. The identifiers and variables, determining the data flow, are symbolized by bold, underlined names in this Figure. The steps of the model-driven dynamic simulation are the followings:

1. The modifying connections change the content of the input slots of the state elements, according to the respective WriteOprator.
2. The state elements execute the associated program prototypes, which determine the new outputs of the states.
3. The reading connections read the content of the various state output slots, according to the prescribed ReadOperator.
4. The reading connections change the content of the input slots of the action elements, according to the respective WriteOperator.
5. The action elements execute the associated program prototypes, which determine the new outputs of the actions.
6. The modifying connections read the content of the various action output slots, according to the prescribed ReadOperator.

During the model-driven simulation, the prescribed spatial and temporal scales are taken into consideration, automatically. The necessary output reporting abilities may be embedded in the description of the model.

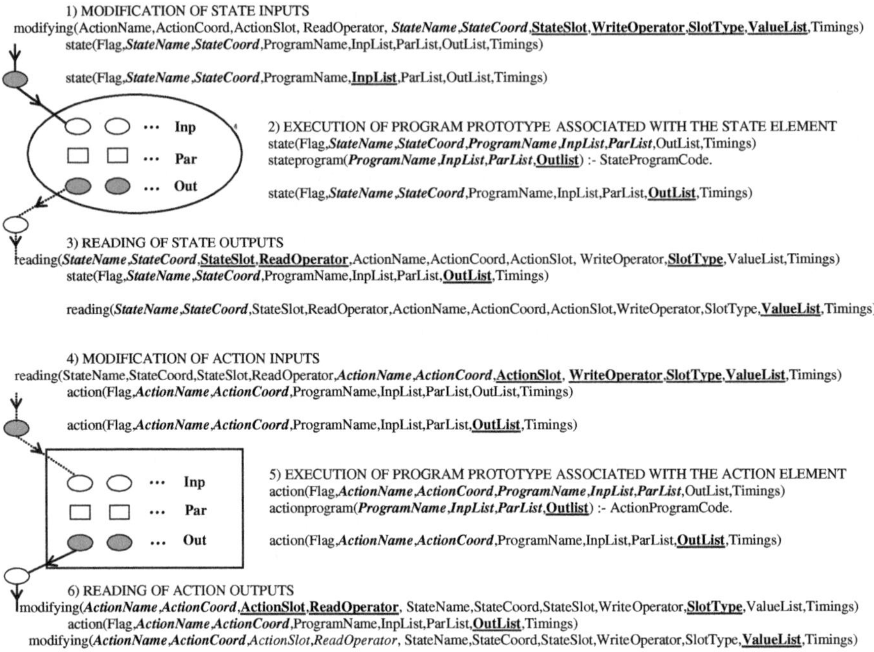

Fig. 18.11 Illustration of the model-driven execution of process simulation programs

Having finished the given full (or partial) simulations, the program automatically saves the complete final state of the process, which, having supplied by the additional parts (usually by additional connections) of the model, makes possible the stepwise continuation of the dynamic simulation.

18.3.5 Spatial and Temporal Multiscale Features

The state and transition elements are prepared for the declaration of the spatial scale in the form of a list of integer coordinates (see variable Coord). The list of integers (see Fig. 18.12) determines the spatial place (i.e. the functional and geometrical compartment) of the given element. The number of the ordered scales is optional, while [...,I] contains [...,I,J,...]. It is to be noted that the elements are determined by the identifying name (StateName or ActName) and Coord together.

This makes possible to use either the same identifying names with different coordinates, or specific identifying names for the various compartments, according to the user's convenience. The state and connection elements can be saved in individual, dedicated files. The only restriction is that the elements with same Coord must be together in the same dynamic database. In a marginal case all of the elements can be saved together in a single dynamic database, however it is usually less effective. The universal complement of the multiscale model is symbolized by Coord = [].

In the connections the input and output locations are taken into consideration by the identifying names and by the respective coordinates, together. The unification is based on both of them, as well as on the name of the respective slot. The connection may be local within a given sub-model, inter-compartmental between the compartments, as well as environmental between the compartments and the universal complement. A possible convention is that the connections must be stored with the sending compartment (except of the environmental input connections that must be present in the receiving compartment). Many possible cases are shown in Fig. 18.12, where the sub-models are symbolized with rectangles, and the connections belong to that compartment, where the bigger dotted end of the given edge is.

Temporal behavior of the states, transitions and connections can be described by the above mentioned list of functors

Timings = Timing*

Timing = t(From, To, [When1, When2, …, WhenM], Step)

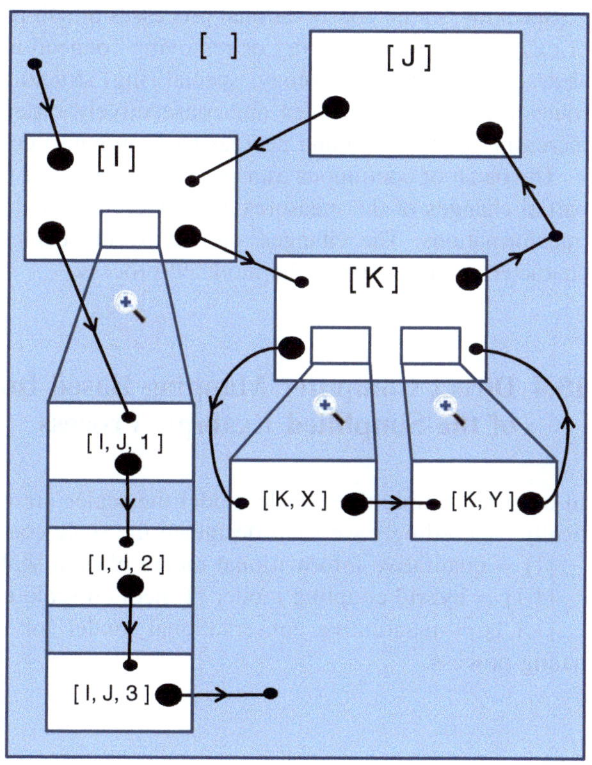

Fig. 18.12 Illustration of spatial scales and connections

This makes possible to declare optionally different timings for the individual elements or connections in the subsequent periods [From, To]. Timing may prescribe the execution in a given interval according to the prescribed time step, (Step) or at the prescribed times (When1, When2,…,etc.).

The event driven operation of the multiscale processes can be controlled by signaling connections, reading and modifying the respective states and transitions.

18.3.6 Holistic net and Network Properties

The execution capabilities of the general kernel engine support also the investigation of many tasks, including various sensitivity and flux analyses.

In the dynamic net structure of the process models the reading and modifying (overwriting, increasing, decreasing, etc.) connections determine the network (in abstract algebraic terminology: special ring) structure of the so-called influence routes. The influence routes are the consecutively ordered alternating series of the reading and modifying connections.

The influence routes make possible the analysis of the various kinds of structural and functional sensitivities (e.g. observability and controllability) of the modeled processes.

Specially, in the conservational processes or sub-processes the set of increasing or extending, and decreasing or removing connections determine the network (in abstract algebraic terminology: special ring) structure of the balance routes (flux routes). The balance routes are consecutively ordered alternating series of the increasing (extending) and decreasing (removing) connections.

The batch or continuous transports, carried by the balance routes, determine the partial changes of the measures along the connected series of transportations and transformations. The changes, carried by the balance routes, make possible to characterize the fluxes of the modeled process.

18.4 Direct Computer Mapping Based Implementation of the Simplified Example Process

In our very simple three-scale model the scales are embedded in each other vertically. Accordingly we shall designate the scale coordinates, as follows:

[1] = qualitative informational model of high scale pathological process;

[1,1] = hybrid coupling model of medium scale cellular events;

[1,1,1] = quantitative conservational model for low scale p53/miR-34a signaling process.

18.4.1 *Quantitative Conservational Model for Low Scale Core Processes of p53/miR-34a Signaling*

The quantitative balance model of the low scale core processes, describing the p53/miR-34a related control loops were prepared starting from the structure of [37], shown in Sect. 2.2.2. This structure was extended by the minimal set of signaling proteins, necessary for the simplified consideration of the main middle scale signaling functionalities behind the high scale pathological process. The state elements of the low scale model are summarized in Table 18.1.

Table 18.1 State elements of the low scale conservational model

Scale	Original name	Passive name	Type	Character	Initial value	Program
[1,1,1]	p53	p53	cons	permanantly existing	>0	measure
[1,1,1]	p53Ac	p53Ac	cons	ad hoc appears	0	measure
[1,1,1]	MDM2	mdm2	cons	ad hoc appears?	0	measure
[1,1,1]	miR-34a	miR34a	cons	ad hoc appears?	0	measure
[1,1,1]	mutant p53	p53_mut	cons	by mistake appears	0	measure
[1,1,1]	CDK4/ 6 + CylinE2	cdkcyc	cons	permanantly existing	>0	measure
[1,1,1]	BCL2	bcl2	cons	permanantly existing	>0	measure
[1,1,1]	SIRT1	sirt1	cons	permanantly existing?	>0	measure
[1,1,1]	XXX (like DBC1)	xxx	cons	ad hoc appears	0	measure
[1,1,1]	Resource	inp	cons	shows the overall consumption	>0	measure
[1,1,1]	Waste	out	cons	shows the overall wastes	0	measure

The modeled components were the followings:

p53	= inactive p53,
p53Ac	= active (acetylated) p53,
MDM2	= murine double minute,
miR-34a	= investigated miRNA,
mutant p53	= non-functioning p53,
CDK4/6 + CyclinE2	= lumped components, activating cell cycle arrest,
BCL2	= component, activating apoptosis,
XXX	= a fictitious component, designating the DNA damage (like DBC1 in breast cancer),
Resource	= finite pool of building elements for synthesis,
Waste	= pool of decomposed products.

In Table 18.1 the original name of each component is followed by the simplified name used in the computational databases. All of the state elements are of conservational types, i.e. atom conservation based additive measures. The character of the components describes the behavior of the given component, during the simulation that is determined by the initial and boundary conditions, as well by the applied model equations. Initial values show that we start from the possible zero initial conditions, as much as possible. Only the components with no modeled expression (synthesis) and p53 have non-zero initial values. In some cases, signed with question marks in the column of character, we could not decide about the initial conditions clearly. All of the state elements use the same prototype program called 'measure'.

The applied program prototype of measure, written in the

```
stateprogram(ProgramName,ParList,InpList,OutList):—StateProgramCode
```

syntax (see Fig. 18.11) is the following:

```
stateprogram(y,measures,[],[i(comp,dl,Extensive)],[o(conc,dl,Intensive)]) :-
  calculate_intensive(Extensive,Intensive),!.
  calculate_intensive([d(Basis,[M],BDim)|EL],[d(Basis,[M],BDim)|IL]) :-
   intensive(EL,M,BDim,[],IL).
     intensive([],_,_,RIL,IL) :-
    reverse(RIL,IL),!.
   intensive([d(Name,[Ext],EDim)|Other],M,BDim,Old,Result) :-
    atom_concat(EDim,'_',UDim),
    atom_concat(UDim,BDim,IDim),
    Int is Ext/M,!,
    intensive(Other,M,BDim,[d(Name,[Int],IDim)|Old],Result).
```

In this prototype the input slot accepts extensive quantities beginning with the respective reference measure (mass or volume), followed by the molar (chemical) quantity of the given biochemical component. In the simplified biological models it is usually difficult to identify the reference measure of the compartments, as well as the molar amounts are also uncertain. Considering this, similarly to other authors, we used a hypothetic reference unit of 1, and estimated amounts. The local program of prototype 'measure' is prepared for the calculation of concentrations from the extensive amounts. Assuming 1 for the reference unit this calculation is meaningless, accordingly we shall speak about simulated units (SU) in the interpretation of the results. Nevertheless in the possible further development, with the exact knowledge of the compartments and amounts we can do more realistic calculations for multiple, connected compartments ([1,1,1], [1,1,2], [1,1,3],…,etc.).

The action elements of the low scale model are summarized in Table 18.2. It is to be noted that two action elements have input data from the middle scale signaling model.

In Table 18.2 the original name of the actions (transformations) is followed by the simplified name used in the computational databases. The 19 action elements belong to 9 prototypes, as it signed in column Type. The action elements are characterized by the following properties:

Type	= name of the prototype transition;
Input data	= list of input conditions and concentrations, necessary for the calculations, where (State = Sign) means that the action takes place if the given State contains the Sign and (Conc, Component) means that calculation needs the actual concentration of the Component;
Parameters	= one or more numerical data for the calculation of the given action;
Output signs	= qualitative states, modified by the action directly;
Output decreases	= components, decreased by the action;
Output increases	= components, increased by the action;
Program	= identifier of the local program, associated with the prototype

The applied program prototypes will be described by their associated local programs written in the

```
actionprogram(ProgramName,ParList,InpList,OutList) :- ActionProgramCode
```

syntax (see Fig. 18.11) as follows:

Activation: if the signaling status from the middle scale is 'active', then calculates the first order transformation of p53 to p53Ac, i.e.:

```
actionprogram (y,activation,
[c(kin,dl,[d(k,[K],nd)])],
[i(status,dl,[d(sign,[Status],nd)]),i(p53,dl,[d(conc,[C],su)])],
[o(p53,dl,[d(comp,[Dec],su)]),o(p53Ac,dl,[d(comp,[Inc],su)])]) :-
g(dt,DT),
Status = active,
Inc is K*C*DT,
Dec is (−1)*Inc,!.
```

Inactivation: considering the SIRT1 determined second order kinetics, inhibited by XXX it deacetylates p53Ac, while the effect of XXX is controlled according to parameter A, i.e.:

```
actionprogram (y,inactivation,
[c(kin,dl,[d(k,[K],nd)]),c(alfa,dl,[d(alfa,[A],nd)])],
[i(p53Ac,dl,[d(conc,
[Cp53Ac],su)]),i(sirt1,dl,[d(conc,[CSirt1],su)]),i(xxx,dl,[d(conc,[Cxxx],su)])],
[o(p53Ac,dl,[d(comp,[Dec],su)]),o(p53,dl,[d(comp,[Inc],su)])]) :-
g(dt,DT),
Inc is K*Cp53Ac*CSirt1/(1 + A*Cxxx)*DT,
Dec is (-1)*Inc,!.
```

Table 18.2 Action elements of the low scale conservational model

Scale	Original name	Active name	Type	Input (conditions and/or rate determining concentrations)	Parameters	Output signs	Output decreases	Output increases	Program
[1,1,1]	p53 activation	proc1	activation	[(status = active), (conc,p53)]	k1	n/a	p53	p53Ac	activation
[1,1,1]	p53 inactivation	proc2	inactivatio	[(conc,p53Ac), (conc,si rt1)],	k2	n/a	p53Ac	p53	inactivation
[1,1,1]	p53 synthesis	proc3	mutated synthesis	[]	k3	n/a	inp	p53,p53mut	mutsynth
[1,1,1]	p53 degradation	proc4	promoted degradatio	[(conc,p53), (conc,mdcm2)]	k4	n/a	p53	out	promdegr
[1,1,1]	p53mut degradation	proc5	promoted degradatio	[(conc,p53mut), (conc,mdcm2)]	k5,mut	n/a	p53mut	out	promdegr
[1,1,1]	p53Ac degradation	proc6	promoted degradatio	[(conc,p53Ac), (conc,mdcm2)]	k6	n/a	p53Ac	out	promdegr
[1,1,1]	MDM2 synthesis	proc7	promoted synthesis	[(conc,p53Ac)]	k7	n/a	inp	mdm2	promsynth
[1,1,1]	MDM2 degradation	proc8	degradatio	[(conc,mdm2)]	k8	n/a	mdm2	out	degr
[1,1,1]	miR34a synthesis	proc9	promoted synthesis	[(conc,p53Ac)]	k9	n/a	inp	miR34a	promsynth
[1,1,1]	miR34a degradation	proc10	degradatio	[(conc,miR34a)]	k10	n/a	miR34a	out	degr
[1,1,1]	SIRT1 synthesis	proc11	inhibited synthesis	[(conc,miR34a)]	k11	n/a	inp	sirt1	inhibsynth
[1,1,1]	SIRT1 degradation	proc12	degradatio	[(conc,sirt1)]	k12	n/a	sirt1	out	degr
[1,1,1]	xxx synthesis	proc13	conditional synthesis	[(status = active)]	k13	n/a	inp	xxx	condsynth
[1,1,1]	xxx degradation	proc14	degradatio	[(conc,xxx)]	k14	n/a	xxx	out	degr
[1,1,1]	cdkcyc synthesis	proc15	inhibited synthesis	[(conc,miR34a)]	k15	n/a	inp	cdkcyc	inhibsynth
[1,1,1]	cdkcyc degradation	proc16	degradatio	[(conc,cdfcyc)]	k16	n/a	cdkcyc	out	degr
[1,1,1]	BCL2 synthesis	proc17	inhibited synthesis	[(conc,miR34a)]	k17	n/a	inp	bcl2	inhibsynth
[1,1,1]	BCL2 degradation	proc18	degradatio	[(conc,bcl2)]	k18	n/a	bcl2	out	degr
[1,1,1]	ectopic miR34a	proc19	inlet	[]	timed amount	n/a	inp	miR34a	inlet

Promoted synthesis: depending on the CP concentration of promoting component 'prom', it decreases the amount of component 'dec' and increases the amount of component 'inc', i.e.:

```
actionprogram (y,promsynth,
[c(kin,dl,[d(k,[K],nd)])],
[i(prom,dl,[d(conc,[CP],su)])],
[o(dec,dl,[d(comp,[Dec],su)]),o(inc,dl,[d(comp,[Inc],su)])]) :-
g(dt,DT),
Inc is K*CP*DT,
Dec is (−1)*Inc, !.
```

Inhibited synthesis: depending on the CI concentration of inhibiting component 'inhib' and on parameter A, it decreases the amount of component 'dec' and increases the amount of component 'inc', i.e.:

```
actionprogram (y,inhibsynth,
[c(kin,dl,[d(k,[K],nd)]),c(alfa,dl,[d(alfa,[A],nd)])],
[i(inhib,dl,[d(conc,[CI],su)])],
[o(dec,dl,[d(comp,[Dec],su)]),o(inc,dl,[d(comp,[Inc],su)])]) :-
g(dt,DT),
Inc is K/(1 + A*CI)*DT,
Dec is (−1)*Inc,!.
```

Conditional synthesis: if the signaling status from the middle scale is 'active', then calculates the first order transformation of 'dec' to 'inc', i.e.:

```
actionprogram (y,condsynth,
[c(kin,dl,[d(k,[K],nd)])],
[i(status,dl,[d(sign,[Status],nd)])],
[o(dec,dl,[d(comp,[Dec],su)]),o(inc,dl,[d(comp,[Inc],su)])]) :-
g(dt,DT),
Status = active,
Inc is K*DT,
Dec is (−1)*Inc,!.
```

Mutated synthesis: with the knowledge of the mutation ratio MF, synthesizes p53 (inc1) and p53mut (inc2) from the resource pool ('dec'), while in normal (healthy) synthesis MF = 0), i.e.:

```
actionprogram (y,mutsynth,
[c(kin,dl,[d(k,[K],nd)]),c(mut,dl,[d(m,[MF],nd)])],
[],
[o(dec,dl,[d(comp,[Dec],su)]),o(inc1,dl,[d(comp,[Incp53],su)]),o(inc2,dl,[d(comp,
[Incp53mut],su)])]) :-
g(dt,DT),
```

```
Incp53 is K*(1-MF)*DT,
Incp53mut is K*MF*DT,
Dec is (-1)*K*DT,!.
```

Degradation: calculates the first order decomposition of 'dec', resulting waste pool of 'inc', i.e.:

```
actionprogram (y,degradation,
[c(kin,dl,[d(k,[K],nd)])],
[i(dec,dl,[d(conc,[C],su)])],
[o(dec,dl,[d(comp,[Dec],su)]),o(inc,dl,[d(comp,[Inc],su)])]) :-
g(dt,DT),
Inc is K*C*DT,
Dec is (−1)*K*C*DT,!.
```

Promoted degradation: depending on the concentration of promoter 'prom', calculates decomposition of 'dec', resulting waste pool of 'inc', i.e.:

```
actionprogram (y,promdegr,
[c(kin,dl,[d(k,[K],nd)])],
[i(dec,dl,[d(conc,[C],su)]),i(prom,dl,[d(conc,[CP],su)])],
[o(dec,dl,[d(comp,[Dec],su)]),o(inc,dl,[d(comp,[Inc],su)])]) :-
g(dt,DT),
Inc is K*C*(1 + CP)*DT,
Dec is (−1)*Inc, !.
```

Inlet: transports an amount M of a given component 'inc' to the investigated model, i.e.:

```
actionprogram (y,inlet,
[],
[i(transport,dl,[d(conc,[M],su)])],
[o(inc,dl,[d(comp,[Inc],su)])]) :-
Inc is M,!.
```

The structure of the low scale dynamic model is illustrated in Fig. 18.13. The 'act' elements indicate that the low scale model forwards data to the middle scale signaling model (see later).

In the graphical representation of the process model net structures we use two kinds of nodes and three kinds of edges, as follows:

- ellipse = state element,
- rectangle = action element,
- dotted line = reading of signs or concentrations from state to action,

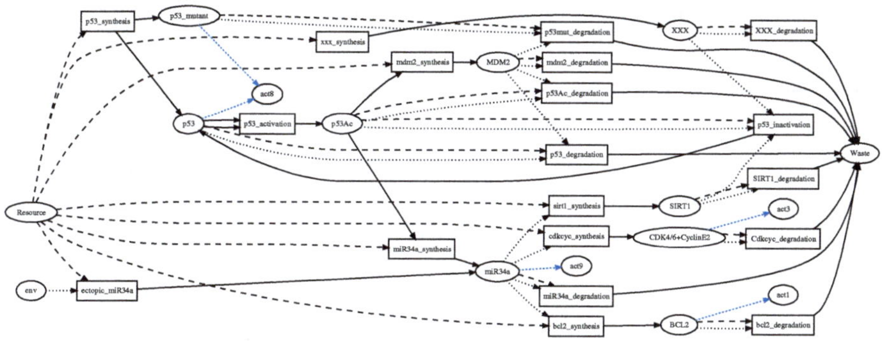

Fig. 18.13 Process model net structure of the low scale dynamic model. The *ellipses* and *rectangles* correspond to the state and action elements, listed in Tables 18.1 and 18.2, respectively

- solid line = increase of measures or overwriting the signs from action to state,
- dashed line = decrease of measures from action to state.

18.4.2 Qualitative Informational Model of High Scale Pathologic Properties

The high scale pathologic process can be described by an informational model consisting from sign representing state and rule representing action elements. This can be solved with same programmable elements and connections, only the operators and the local programs differ from the previously investigated conservational model.

The state elements of the high scale model [1] are summarized in Table 18.3. The three state elements have various, pre-defined possible values, as well as an initial value, as follows:

Pathogenic status (p_s): determines the actual state of the illness in the sense of the Vogelstein's model, described in Sect. 2.2.1. The possible values might be 'late_adenoma', 'tumor_suppression', 'tumor_development' and 'carcinoma'. Our simulation starts form 'late_adenoma'.

Oncogenic_status (o_s): contains information whether there exists any oncogenic stress or not. Accordingly it may have 'no' or 'yes' values, while we presume 'no' as an initial condition. This sign expresses the oncogenic boundary conditions for the investigated three scale process.

Signaling status (s_s): summarizes the state of the middle scale signaling scenarios. The possible values of the signaling status ('no', 'normal_p53_function', 'loss_p53_function' and 'restore_p53_function') can be understood as emergent properties, resulted from the underlying signaling processes. The initial value of

Table 18.3 State elements of the high scale pathologic model

Scale	Original name	Passive name	Type	Possible values	Initial value	Program
[1]	Pathogenic status	p_s	info	[late_adenoma, carcinoma, tumor_suppression, tumor development]	late_adenoma	sign
[1]	Signaling status	s_s	info	[no,normal_p53_function, loss_p53_fuction, restore_p53_function]	no	sign
[1]	Oncogenic status	o_s	info	[no,yes]	no	sign

'no' is in agreement of the temporary lack of oncogenic stress. The actual signaling status (highlighted by grey) comes from the middle scale signaling model.

All of the state elements belong to the same type (info), associated with the same program (sign). Accordingly the over-writeable (optionally multiple) content of the input slot is copied and stored in the output slot in each step of execution. The program of sign meta-prototype is the following:

```
stateprogram(y,sign,

[],
[i(Inp,dl,[d(sign,[InpSign],nd)])],
[o(Out,dl,[d(sign,[OutSign],nd)])]) :-

Out = Inp, OutSign = InpSign,!.
```

The action elements of the high scale model are summarized in Table 18.4.

The structure of the high scale informational model is illustrated in Fig. 18.14. The 'act' elements indicate that the low scale model forwards data to the middle scale signaling model (see later). The state element, determining the 'status' of the middle scale model is determined by two rules of the high scale system.

The functioning of the action elements correspond to the usual condition/action rules. The content of input slots corresponds to the various sets of conditions, initiating the firing. Two rules have output sign to the middle scale signaling model (highlighted by grey in Table 18.4). There are no measure decreases or increases associated with the rules. All of the action elements belong to the same type (rule), associated with the same program (rule). The program of rule meta-prototype is declared as follows:

```
actionprogram(y,rule,Param,Condition,Action) :-
conditions(Condition,Param,Consequence,0,0),
action(Consequence,[],Action),!.
v(y,rule,Param,Condition,[]) :-
conditions(Condition,Param,_,0,N),N > 0,!.
conditions([],Consequence,Consequence,Ack,Ack) :- !.
```

Table 18.4 Action elements of the high scale pathologic model

Scale	Original name	Active name	Type	Input (conditions and/or rate determining concentrations)	Parameters	Output signs	Output decrease	Output increase	Program
[1]	Permanent late	rule1	rule	[(p_s = late_adenoma), (o_s = no), (s_s = no)]	[]	[(p_s = late_adenoma)]	n/a	n/a	rule
[1]	Normal p53 functioning	rule2	rule	[(p_s = late_adenoma), (o_s = yes), (s_s = normal_p53_fuctioning)]	[]	[(p_s = tumor_supression), (o_s = no)]	n/a	n/a	rule
[1]	Loss of p53 functioning	rule3	rule	[(p_s = late_adenoma), (o_s = yes), (s_s = loss_p53_fuction)]	[]	[(p_s = tumor_development)]	n/a	n/a	rule
[1]	Restore of p53 functioning1	rule41	rule	[(p_s = tumor_development), (o_s = yes), (s_s = restore_p53_function)]	[]	[(p_s = tumor_supression), (o_s = no)]	n/a	n/a	rule
[1]	Restore of p53 functioning2	rule42	rule	[(p_s = tumor_development), (o_s = yes), (s_s = normal_p53_function)]	[]	[(p_s = tumor_supression), (o_s = no)]	n/a	n/a	rule
[1]	Development of	rule5	rule	[(p_s = tumor_development), (o_s = yes) (s_s = no)]	[]	[(p_s = carcinoma)]	n/a	n/a	rule
[1]	Activation of p53 signaling	rule61	rule	[(o_s = yes)]	[]	[(status = active)]	n/a	n/a	rule
[1]	Activation of p53 signaling	rule62	rule	[(o_s = no)]	[]	[(status = inactive)]	n/a	n/a	rule
[1]	Oncogenic stress	rule7	rule	[]	[]	[(o_s = yes)]	n/a	n/a	rule

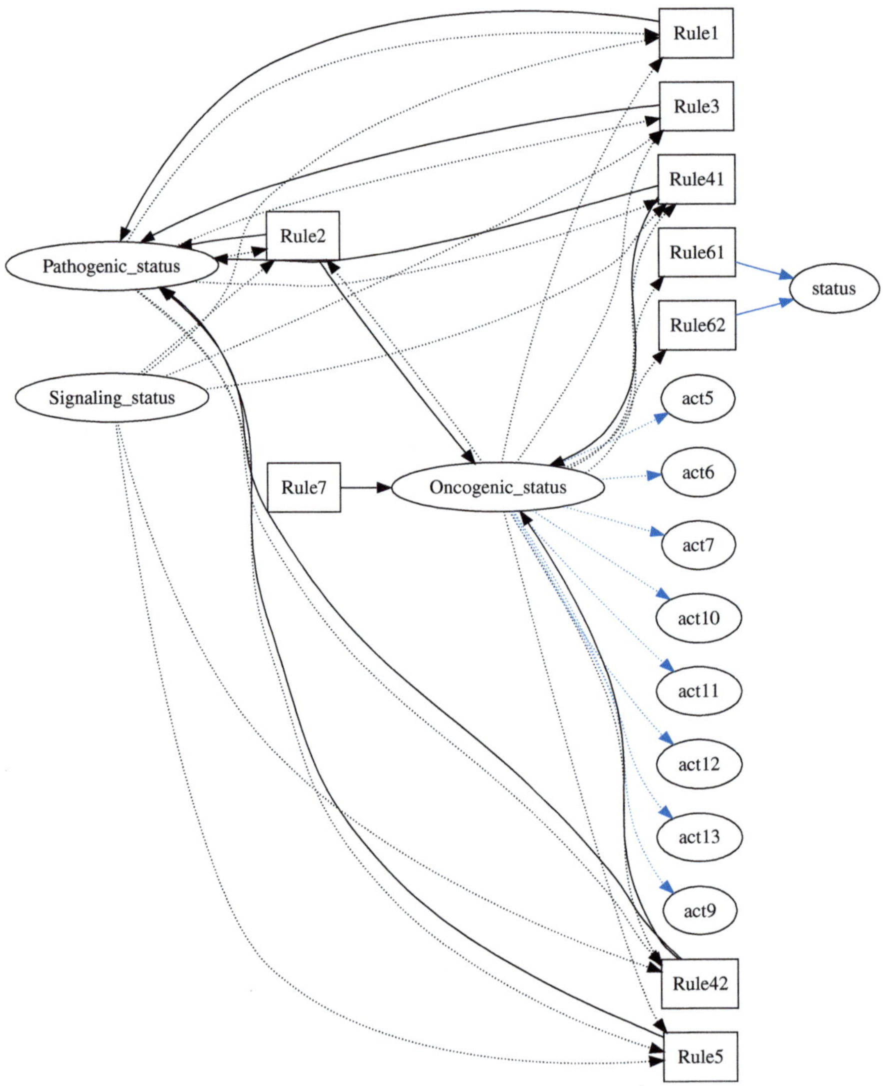

Fig. 18.14 Process model net structure of the high scale pathologic model. The *ellipses* and *rectangles* correspond to the state and action elements, listed in Tables 18.3 and 18.4, respectively

```
conditions([i(_,dl,[d(sign,[IContent],nd)])|OtherConditions],
[c(_,dl,[d(sign,[PContent],nd)])|OtherParam],RConsequence,Ack,RAck):-
identical(IContent,PContent,Check),
NAck is Ack + Check,!,
conditions(OtherConditions,OtherParam,RConsequence,NAck,RAck).
identical(IContent,PContent,0) :- IContent = PContent,!.
identical(_,_,1) :- !.
action([],Action,Action) :- !.
```

```
action([c(Info,dl,[d(sign,[PContent],nd)])|OtherConsequence],Action,RAction) :- !,
action(OtherConsequence,[o(Info,dl,[d(sign,[PContent],nd)])|Action],RAction).
```

This declarative program describes that if all of the input conditions are fulfilled, then all of the consequential actions have to be executed.

18.4.3 *Hybrid Coupling Model of Medium Scale Emergent Properties*

The middle scale signaling process is described by an informational model consisting from sign representing state and rule representing action elements.

The state elements of the middle scale model [1,1] are summarized in Table 18.5. The three state elements have various, pre-defined possible values, as well as an initial value, as follows:

Cell Cycle (cellcycle): determines the actual state of the cell cycle (proliferation), that may be 'functioning' or 'arrested'. The multiscale process model, starting from 'late adenoma' with 'no' oncogenic stress is characterized by the initial value of 'functioning'.

Apoptosis (apopt): characterizes the actual state of the apoptotic pathways, that may be 'off' or 'on'. The multiscale process model, starting from 'late adenoma' with 'no' oncogenic stress is characterized by the initial value of 'off'.

p53/miR-34a status (status): defines the actual state of the p53/miR-34a related signaling system, that may be 'inactive', 'active' or 'wrong'. The multiscale process model, starting from 'late adenoma' with 'no' oncogenic stress is characterized by the initial value of 'inactive'.

All of the state elements belong to the same type (info), associated with the same program (sign). The program of sign meta-prototype is the same as for the high scale pathologic model.

The action elements of the middle scale model are summarized in Table 18.6.

The functioning of the action elements correspond also to the usual condition/action rules. If the content of input slots corresponds to the various sets of conditions, then the rule fires, according to the output slots. There are no measure decreases or increases associated with the rules.

Table 18.5 State elements of the middle scale signaling model

Scale	Original name	Passive name	Type	Possible values	Initial value	Program
[1,1]	Cell Cycle	cellcyle	info	[functioning, arrested]	functioning	sign
[1,1]	Apoptosis	apopt	info	[off, on]	off	sign
[1,1]	p53 miR34a status	status	info	[inactive, active, wrong]	inactive	sign

Table 18.6 Action elements of the middle scale signaling model

Scale	Original name		Type	Input (conditions and/or rate determining concentrations)	Parameters	Output signs	Output decreases	Output increases	Program
[1,1]	Cell Cycle Arrest	act1	cond	[(cellcycle = functioning), (conc(bcl2) <=limit)]	limit	[(cell_cycle = arrested)]	n/a	n/a	cond
[1,1]	Cell Cycle Restart	act2	cond	[(cellcycle = arrested), (conc(bcl2) > limit)]	limit	[(cell_cycle = functioning)]	n/a	n/a	cond
[1,1]	Apoptosis Switch On	act3	cond	[(apoptosis = off), (conc(cdkcyc) <=limit)]	limit	[(apoptosis = on)]	n/a	n/a	cond
[1,1]	Apoptosis Switch Off	act4	cond	[(apoptosis = on), (conc(cdkcyc) > limit)]	limit	[(apoptosis = off)]	n/a	n/a	cond
[1,1]	Normal p53 Function1	act5	rule	[(o_s = yes), (status = active), (cell_cycle = arrested), (apoptosis = off)]	[]	[(s_s = normal_p53_function)]	n/a	n/a	rule
[1,1]	Normal p53 Function2	act6	rule	[(o_s = yes), (status = active), (cell_cycle = functioning), (apoptosis = on)]	[]	[(s_s = normal_p53_function)]	n/a	n/a	rule
[1,1]	Normal p53 Function3	act7	rule	[(o_s = yes), (status = active), (cell_cycle = arrested), (apoptosis = on)]	[]	[(s_s = normal_p53_function)]	n/a	n/a	rule
[1,1]	p53 Mutation	act8	ratio	[(conc,p53_mut), conc,p53)]	limit_ratio	[(status = wrong)]	n/a	n/a	ratio
[1,1]	miR34a Dysregulation	act9	cond	[(o_s = yes), (conc(miR34a) <=limit)]	limit	[(status = wrong)]	n/a	n/a	cond
[1,1]	Loss of p53 Function1	act10	rule	[(o_s = yes), (cell_cycle = functioning), (apoptosis = off)]	[]	[(s_s = loss_p53_function)]	n/a	n/a	rule
[1,1]	Restore p53 Function1	act11	rule	[(o_s = yes), (status = wrong), (cell_cycle = arrested), (apoptosis = off)]	[]	[(s_s = restore_p53_function), (status = active)]	n/a	n/a	rule
[1,1]	Restore p53 Function2	act12	rule	[(o_s = yes), (status = wrong), (cell_cycle = functioning), (apoptosis = on)]	[]	[(s_s = restore_p53_function), (status = active)]	n/a	n/a	rule
[1,1]	Restore p53 Function3	act13	rule	[(o_s = yes), (status = wrong), (cell_cycle = arrested), (apoptosis = on)]	[]	[(s_s = restore_p53_function), (status = active)]	n/a	n/a	rule

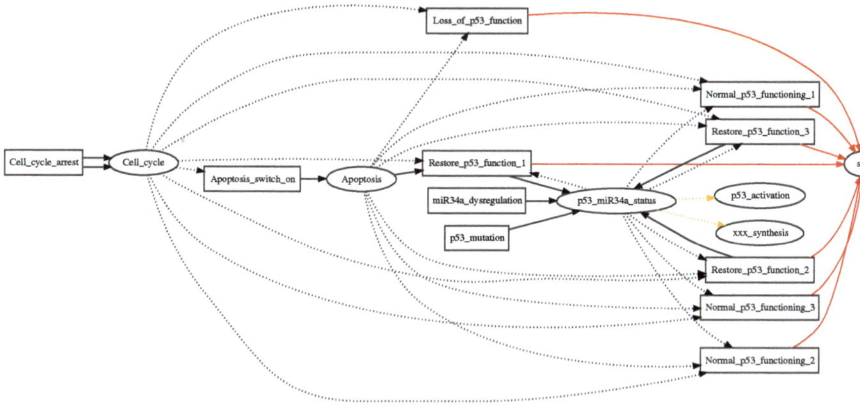

Fig. 18.15 Structure of the middle scale informational process. The *ellipses* and *rectangles* correspond to the state and action elements, listed in Tables 18.5 and 18.6, respectively

Part of the action elements belong to the type 'rule', with the program 'rule, introduced in high scale model. Another part of action elements is characterized by the 'cond' type, using the 'cond' program that checks for one qualitative sign and for one limit concentration, coming from the low scale model (highlighted grey). The actual limit values are declared as local parameters of the given actions.

Action prototype 'ratio' determines the ratio of two components (in our case the ratio of the normal and mutant p53, as follows:

```
actionprogram(y,ratio,
[c(limit_ratio,dl,[d(lr,[LR],nd)])],
[i(conc1,dl,[d(tomeg,[C1],su)]),i(conc2,dl,[d(tomeg,[C2],su)])],
[o(status,dl,[d(sign,[Result],nd)])]) :-
C2/(C1 + C2) > LR,
Result = wrong,!.
```

The structure of the middle scale model is illustrated in Fig. 18.15.

18.5 Simulation Based Reproduction of the Possible Scenarios in the Example Process

18.5.1 Generation and Simulation of the Multiscale Process Model

The generation and simulation of the multiscale process model is controlled by a supervisory file that determines

- the name and the multiscale coordinate of the individual models,
- the global time step of the system, and
- the simulation time (or the absolute starting and ending time).

The individual models are described in an extended Graphviz interface [48]. The general purpose interpreting program reads these files and prepares the User and Expert files of the given models. The User file will contain the declaration of the state(.) and action(.) elements, as well as the reading(.) and modifying(.) connections in the form introduced in part 18.3. The individual programs can run according to their optionally less time scale (with more frequent time steps), as well as their execution can be limited by time or by event driven control. The Expert file will contain the stateprogram(.) and actionprogram(.) clauses, declaring the individual local programs, associated with the various prototypes (remember the example programs, shown in part 18.4).

The simulation runs in a pseudo-parallel organization that supports the optional real parallelism. The kernel opens the models one after the other and executes the elements until the sum of the local time steps correspond to the global one. The connection between the scales is organized by special reading(.) and modifying(.) elements that are declared in the sending model or in the universal complement, in the sense of Fig. 18.12. These scale connecting elements get value in the sending model, and then they are copied to the receiving model to forward the data to the addressed element in the other model.

There might be pure conservational, pure informational and mixed models. Moreover the state and action prototypes can contain both conservational and informational input or output slots. It gives a flexible environment for the combined quantitative (e.g. conservation law based) and qualitative (e.g. rule based) simulation. It is to be noted that depending on the state and action elements, as well as on their local programs any other kind of modeling (e.g. stochastic, fuzzy, etc.) can be implemented.

Next we shall illustrate the multiscale simulation with a set of case studies. We emphasize that these case studies are based on the realistic knowledge, summarized in Part 18.4, however they are only oversimplified illustrations with roughly estimated data.

18.5.2 Tumor Suppression Resulted by the Normal Functioning of p53/miR34a System

The normal functioning of the p53/miR34a signaling is illustrated in Fig. 18.16. In this (and in the following) Figures the results obtained from the low, middle and high scale models are shown above each other, while vertical arrows emphasize the interaction between the scales. The abscissa shows the approximate time in hours. The changes of the signs in the state elements of the upper and middle scale models are illustrated by the vertical stepping of horizontal lines with. The low

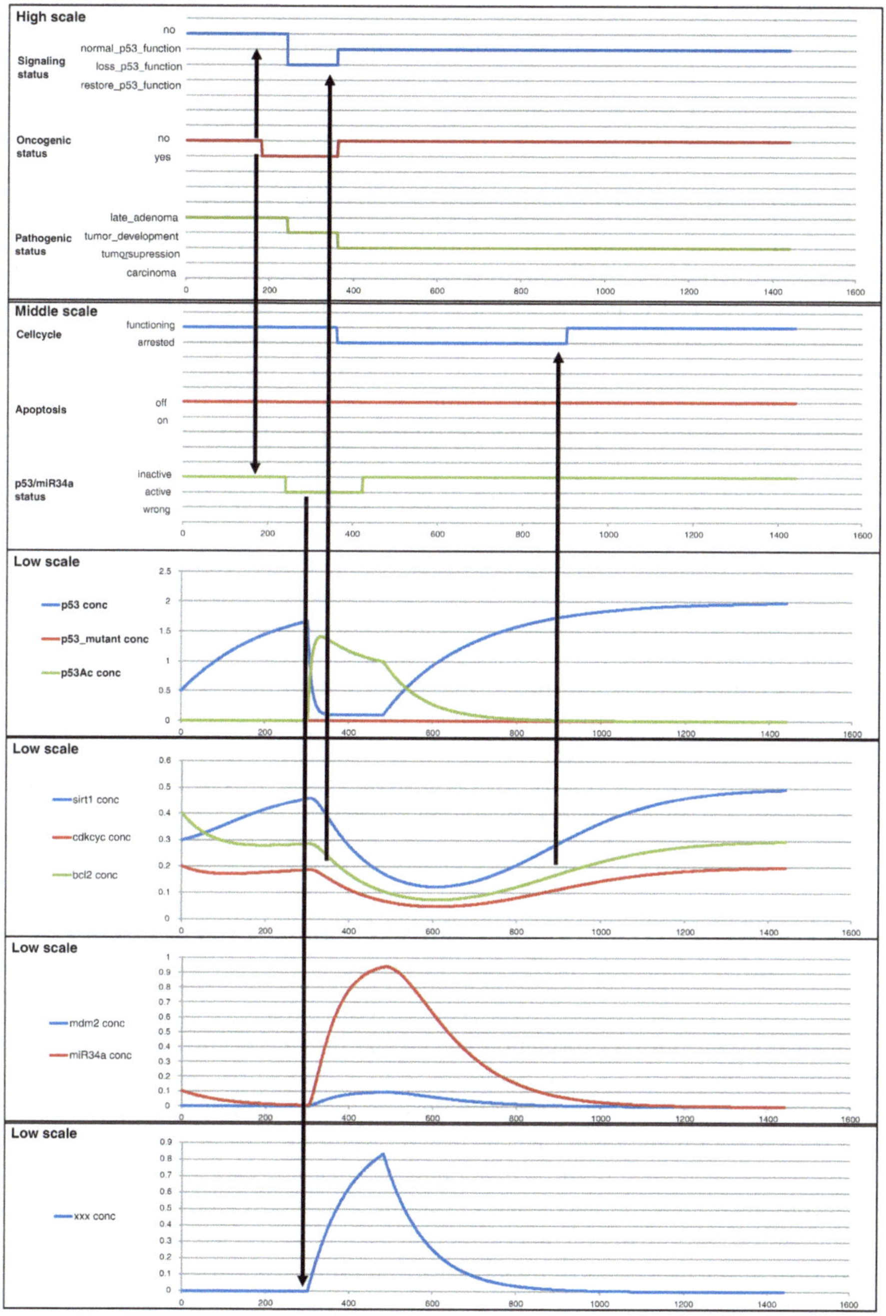

Fig. 18.16 Tumor suppression with normal p53/miR-34a functioning. Diagrams show the change of signs and concentrations in time. The *vertical arrows* emphasize the interactions between the scales

scale results are shown in four diagrams, showing the concentration of the typical groups of components in arbitrary simulation unit (SU).

All case studies start from the pathogenic status of 'late adenoma' from oncogenic status 'no' and with signaling status 'no'. Cell cycle is functioning and apoptosis is switched off by default. There are identical non zero initial conditions for p53 (or mutant p53), miR-34a, cdkcyc (CDK4/6 + CylinE2), bcl2 (BCL2) and sirt1 (SIRT1). Except of the mentioned changes, all of the kinetic parameters, concentration limits and other constants are identical in the case studies.

In case of normal p53/miR34a functioning, the oncogenic stress initiates tumor development and loss of p53 functioning. However, it simultaneously promotes the formation and activation (acetylation) of p53. If the stress is associated with a DNA damage, then the specific component xxx (like DBC1) may appear. Activated p53 promotes miR-34a formation that inhibits the expression of SIRT1. SIRT1 can accelerate deactivation of p53Ac, but the optionally present xxx inhibits this effect. On the other hand activated p53Ac promotes the expression of MDM2 that accelerates the decomposition of p53 and of p53Ac. These apparently contradictory influences help to avoid any kind of over-reaction. Every effect prepares its antagonistic one to keep the balance, suppressing the unnecessarily great changes.

Regardless to this equilibrium, miR-34a inhibits the expression of the anti-apoptotic and proliferation supporting proteins, accordingly cell cycle arrest and apoptosis will be promoted. In our case, because of the actually used arbitrary initial concentrations and constrains, first cell cycle will be arrested, accordingly normal p53 functioning will be reinstalled, and oncogenic stress will be survived. After this transient, the dangerous state is avoided and the processes are going toward a steady state, as well as cell cycle starts again. (It is to be noted that if it does not succeed, than apoptosis would be initiated.)

18.5.3 miR-34a Dysregulation Caused Tumor Development and its Ectopic miR-34a Therapy

Dysregulation of miR-34a formation means that the rate of miR-34a synthesis is decreased in the cell. This case is illustrated by a simulation, where the only difference from the previous run was a decrease in the kinetics of miR-34a synthesis with one order of magnitude. The calculated results are seen in Fig. 18.17. As the Figure shows, regardless to the activation of p53, the miR-34a level decreases.

Accordingly, after the oncogenic effect, the tumor development will be stabilized, resulting the loss of p53 function. Caused by the decreased miR-34a level, the expression of the post transcriptional genes, inhibiting the cell cycle arrest and the apoptosis will not be blocked, consequently proliferation and tumor development continues.

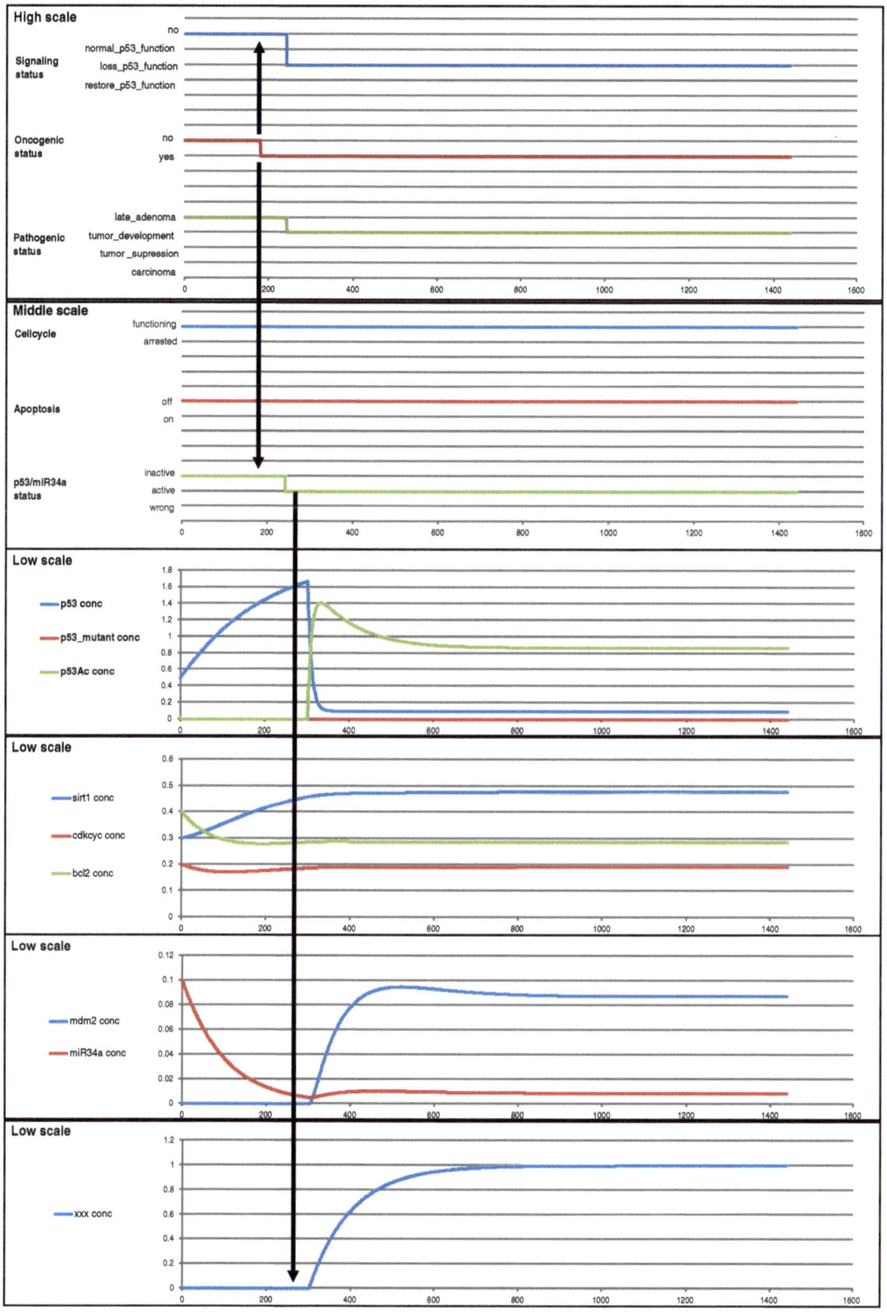

Fig. 18.17 miR-34a dysregulation caused tumor development. Diagrams show the change of signs and concentrations in time. The *vertical arrows* emphasize the interactions between the scales

Higher p53Ac level promotes the MDM2 expression, but finally all of the concentrations tend toward a modified steady state value, characterizing the permanent tumor development, going toward the next stage of carcinoma.

The fatal process can be avoided by a treatment with ectopic miR-34a. In the simulation, illustrated in Fig. 18.18, at the 720th hour we started an miR34a treatment, that increased the miR-34a concentration. Consequently, the expression of the post transcriptional genes, inhibiting the cell cycle arrest and the apoptosis will be blocked again, and the cell cycle will be arrested. This stops the proliferation, resulting in the tumor suppression, corresponding to the normal p53 function.

Having survived the oncogenic stress, the activation of the p53 decreases, causing a further decrease in the MDM2 and miR-34a synthesis. The lower MDM2 concentration decreases the degradation of p53 and p53Ac. Moreover the lower miR-34a concentration decreases the inhibition of the SIRT1 synthesis that has a positive feedback on p53 deactivation. Also the concentration of the anti-apoptotic and cell cycle motivating components is getting to increase.

18.5.4 Mutant p53 Caused Tumor Development and its Ectopic miR-34a Therapy

Another kind of miR_34a dysregulation is caused by the formation of mutant p53. The mutant p53 cannot be activated, accordingly it is not able to promote the miR-34a synthesis, accordingly the concentration of miR-34a decreases in the cell. This case is illustrated by a simulation, where the only difference from the Fig. 18.16 run is the formation of mutant p53 instead of the usual one. The calculated results are shown in Fig. 18.19. As the Figure shows, the miR-34a level converges to zero.

Accordingly, after the oncogenic effect, the tumor development will be stabilized, resulting the loss of p53 function. Again, caused by the decreased miR-34a level, the expression of the post transcriptional genes, inhibiting the cell cycle arrest and the apoptosis will not be blocked, consequently proliferation and tumor development continues.

Lack of p53Ac level blocks also the MDM2 expression, but finally all of the concentrations tend toward a modified steady state value, characterizing the permanent tumor development, going toward the next stage of carcinoma.

This fatal process can also be avoided by a treatment with ectopic miR-34a. In the simulation, illustrated in Fig. 18.20, at the 720th hour we started an miR34a treatment, that increased the miR-34a concentration. Consequently, the expression of the post transcriptional genes, inhibiting the cell cycle arrest and the apoptosis will be blocked again, and first the cell cycle will be arrested. This stops the proliferation that results in the tumor suppression, corresponding to the normal p53 function (as seen experimentally, e.g. in [49]).

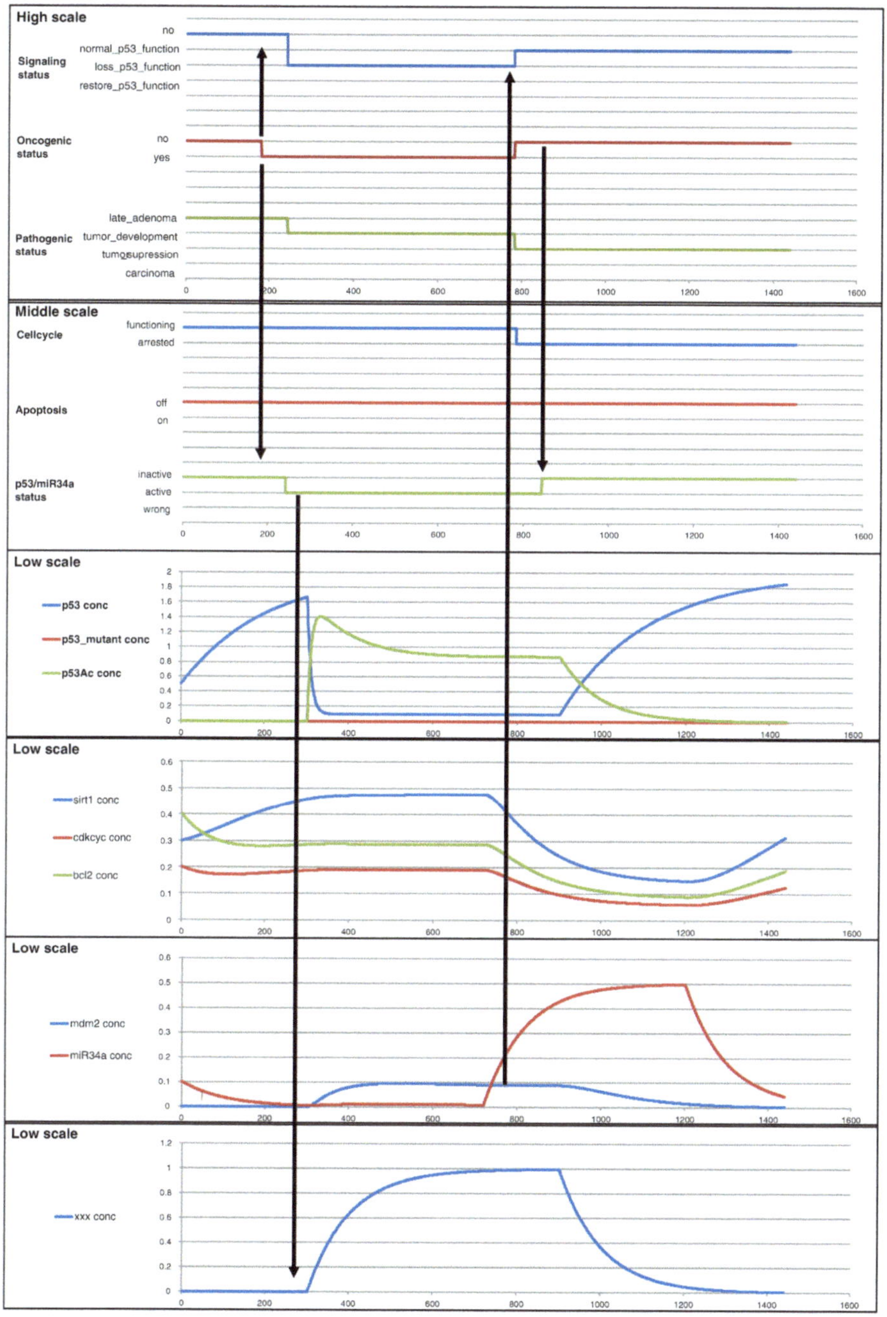

Fig. 18.18 Ectopic miR-34a therapy of miR-34a dysregulation caused tumor. Diagrams show the change of signs and concentrations in time. The *vertical arrows* emphasize the interactions between the scales

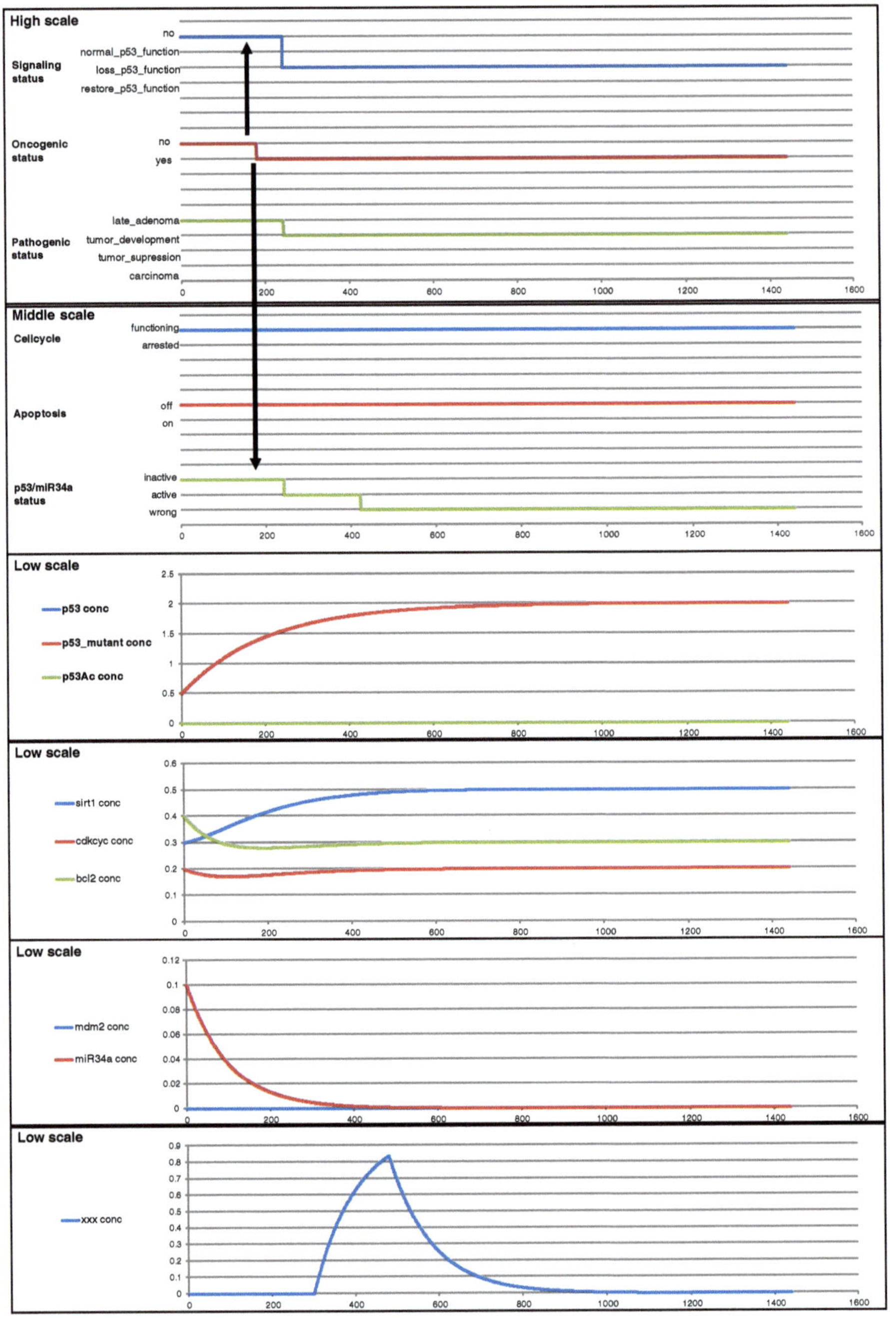

Fig. 18.19 Mutant p53 caused tumor development. Diagrams show the change of signs and concentrations in time. The *vertical arrows* emphasize the interactions between the scales

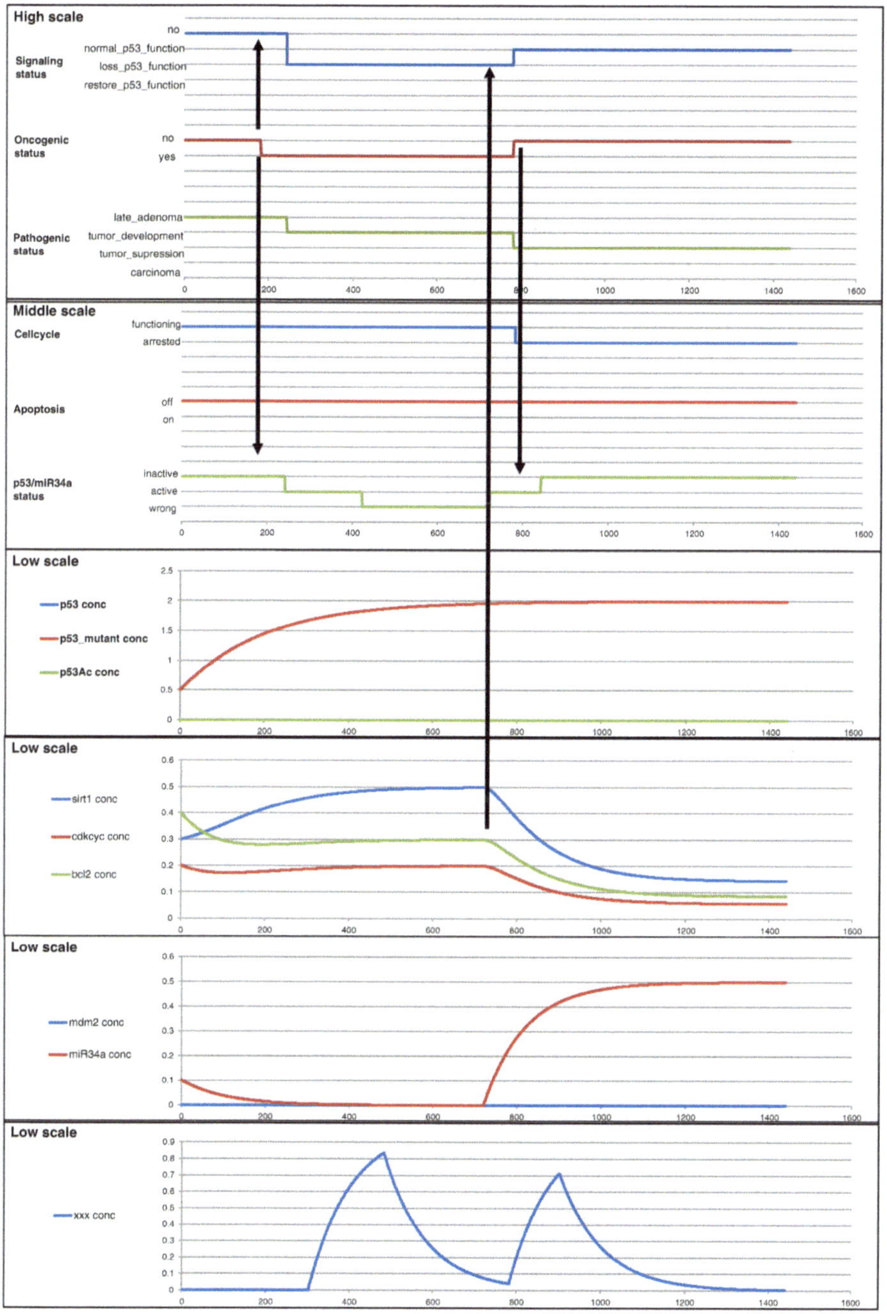

Fig. 18.20 Ectopic miR-34a therapy mutant p53 caused tumor. Diagrams show the change of signs and concentrations in time. The *vertical arrows* emphasize the interactions between the scales

18.6 Concluding Remarks on the Multiscale Simulation of a Simplified Example

We studied a simplified multiscale biosystem with a new modeling and simulation methodology. The biosystem was a consciously, but arbitrarily selected multiscale part of the p53/miR-34a related signaling process that has an important role in the tumor resistance, diagnostics, as well as in the therapy of various tumors.

The multiscale model covered the system from the change of a pathologic stage to the detailed dynamic molecular processes and vice versa, however the set of the considered components and interactions were extraordinarily limited, focusing on a heuristically selected, important subsystem. This simplification makes possible to provide an overview, and the critical evaluation of the difficulties and possibilities.

First of all we can conclude that Direct Computer Mapping of process models can be applied for modeling and simulation of a typical multiscale, hybrid biosystem. The major advantage is the unified representation of the various quantitative and qualitative sub-models, as well as the easy combination of these various models in a unified simulation environment. Another important feature is that the model can be extended with new sub-models, as well as with new elements (components, transformations, transportation) in the sub-models. Traversing of scales can be solved by declaring connections between the sub-models with the same syntax and semantics as within the sub-models, only the sending and the receiving coordinates are different. The declarative approach uses local variables and local programs for building elements that supports the free change (extension, removal, modification) of the multiscale, hybrid model, without considering the limitations of any mathematical construct or modeling language.

Regardless to the limited number of components and interactions, the investigated example demonstrates many important and interesting features of the multiscale, hybrid biosystems. Especially, one can see how the typical scenarios of the low level molecular events project onto the state properties in the higher scales. These properties (often called emergent properties) determine typical scenarios of lower scale states and actions. These properties express rather process based heuristics, than the sometimes "mystically" denoted emergence. In the studied simple model, the coupling of qualitative and quantitative information is viable, e.g. in the form of numerical relations, like constraints in the middle scale. The simplified example, extracted from the independently developed, but coherent references, describes some essential features about the modeled biological processes. We could simulate the natural functioning of the p53/miR-34a signaling for the tumor suppression, as well as the various malfunctions of the system. The simulation of adding ectopic miR-34a proved to be a possible therapeutic intervention.

Important findings of the investigated example are a new p53 model, featuring the miRNA control and inclusion of systemic (emergent) properties as cellular outputs and simulation of this model via new multiscale methodology. This example aims to challenge the "one-target, one-disease" tradition and seeks to develop multiple target strategies during the primary discovery steps by means of systems biology of a particular disease signaling pathway. Detailed signaling pathway description, including multiscale scenario, would then allow an interrogation of such pathway in silico, allowing for multi-hit scenario at drug discovery. This is an ultimate goal of the research.

In principle, the proposed approach could be applied to any biological system of any complexity. The simulation tool allows employing any kind of detail, is very robust and could handle even much larger systems with many reactants. Considering that the knowledge of signaling pathways would increase considerably in a near time, one would be able to simulate such detailed process system (for some signaling pathways detailed reaction is already available, e.g. for EGF).

We emphasize that considerable uncertainty may come from the arbitrary selection of the modeled part within a large biosystem. All of the biosystem models are well defined sets of state, action and connection elements. Other connections link the model with its (actually) non-modeled environment. The investigated subsystem (called also as universe of discourse) has to be separated from the non-modeled subsystem, unambiguously. State elements effect on action elements and, action elements effect on state elements, vice versa. Also the environment effects the studied model, as well as the given model effects on the environment. A feasible limitation is that we locate the actions to the studied model, while the environment can interact with the states. This separation means neither a geometrical border, nor the restriction to any spatial or temporal scale. Accordingly, the notion of environment refers to everything, outside of the consciously and/or arbitrarily selected process model.

All of the biosystem models are open, i.e. they cannot be isolated from their environment. However the feasible and useful biological models must be closed enough, with respect to the given study. One of the most important challenges in modeling is the appropriate and adequate outlining of the state and action elements for a given model.

The uncertainty of our model is quite clear in the limelight of the complexity shown in Figs. 18.2 and 18.3, as well as looking at the big picture in Fig. 18.1. Considering this, it seems to be a surprise that our simplified model works, at all. Probably this comes from the more or less correct consideration of the feedback loops at the low level quantitative model. Nevertheless, in a real application we have to make a systematic sensitivity analysis for the hidden increases and decreases of each component.

Another kind of uncertainty comes from the roughly estimated numerical parameters, constraints and initial values. However, it is to be noted that according to our experiences, a quantitative process with the modeled mutual feedback loops is less sensitive for the numerical values. Probably the robustness (and sustainability) of the biological processes is the consequence of the mutual (cooperative) feedback between the functionally connected neighbors.

Another important experience is that in the model development we had much more problems with the qualitative signs and rules. Sometimes it is difficult to avoid the deadlock and the hazard situations, appearing in the pure rule based parts. However the systematic analysis of the lower scale conservational sub-model helps to improve the rule based scales, too.

The studied process gives a good example for combining conservational (material balance based) and informational (sign based) models. In the lowest level of our multiscale hybrid model biochemical and physical processes with bio-chemical and physical entities take place, e.g.

- a protein comes from the environment;
- a protein goes out to the environment;
- a protein is synthesized;
- a protein is degraded;
- a protein regulates up (promotes) the synthesis of another one;
- a protein regulates down (inhibits) the synthesis of another one;
- proteins travel between and across the compartments;
- proteins react with each other (association, dissociation, etc.);
- proteins are activated and deactivated; etc.

All of the state elements are represented by various physical (or chemical) entities, as well as they represent some information by their existence. For example the proteins are built from a given amount of specifically sequenced amino acids, and these amino acids are built finally from specifically arranged atoms. However these proteins contain information about the possible actions, promoted or inhibited by them.

All of the action elements are represented by a set of physical (or chemical) changes, determined by the coordinated decreases or increases in the associated entities. For example, in a binding process, the amounts of free and bound component are decreased and increased, respectively, while the free binding site will be occupied. Nevertheless the binding process determines a rule in the language of signs (information), e.g. the bound protein switches off or on another action.

Finally, every component can behave either as a "conservational law based matter" or as an "abstract sign expressing information":

- DNA behaves as information, coding all of the proteins, but in replication it needs nucleotides;
- tRNAs, mRNAs behave as information, coding specific proteins, but they are built from nucleotides;
- miRs promote and inhibit various coding processes, but they are built from nucleotides;
- various proteins (enzymes) promote and inhibit various transformations and transportations, but they are built from amino acids;
- metabolic processes have rather conservational than informational character, etc.

Going down in the above list, the components and the processes seem to contain less information. Obviously, we can use a relativistic notion of informational process, as a well defined sub-process of the conservational process that consumes and produces less amount of conservation law based measures, than its complement, but effects more on its complement, than vice versa. Accordingly in biosystems all of the informational processes are conservation based information processes. In addition there is a finite pool of nucleotides, amino acids, proteins, metabolites etc. in the cell, while the over-expressions are limited by the conservation of the given building elements.

Conscious consideration of conservation based informational processes is an important lesson, learnt from biosystem modeling for computational tools and from computer simulations for biosystem research. Direct Computer Mapping makes possible the unified knowledge representation. In the further developments we must focus on the automatic model discovery, based on the common use of various knowledge elements. On the other hand experimental study of biosystems at various scales ought to be combined with multiscale modeling and simulation tools, based on the conscious use of conservation based information processes.

Acknowledgments We acknowledge Jaromir Mikeš, P.J. Šafárik University in Košice, Slovakia and Marek Mráz, Masaryk University Brno, Czech Republic and University of California, San Diego, USA for providing feedback and information on function of miRNAs.

The work was supported by Czech-Hungarian Bilateral project (#TÉT_10-1-2011-0024) and by Baross Gábor R&D Program (#REG-DD-09-2-2009-0101).

References

1. Kalman R, Falb P, Arbib M (1969) Topics in mathematical system theory. McGraw Hill, New York
2. Petri CA (1962) Kommunikation mit Automaten (Communication with Automatons), Schriften des Institut für Instrumentelle Mathematik, 2, Bonn
3. Chen M, Hofestädt R (2003) Quantitative Petri net model of gene regulated metabolic networks in the cell. In Silico Biol 3(0029):347–365
4. Brauer W (ed) (1980) Net theory and applications. Springer lecture notes in computer science 84, Springer, Berlin
5. Marquardt W (1996) Trends in Computer-aided Process Modeling. Comput Chem Eng 20 (6/7):591–609
6. Yang A, Braunschweig B, Fraga ES, Guessoum Z, Marquardt W, Nadjemi O, Paen D, Pinol D, Roux P, Sama S, Serra M, Stalker I (2008) A multi-agent system to facilitate component-based process modeling and design. Comput Chem Eng 32(10):2290–2305
7. Luscher AJ, McDowell DL, Bronkhorst CA (2010) A second gradient theoretical framework for hierarchical multiscale modeling of material. International Journal of Plasticity 26:1248-1275
8. Monperrus M, Jaozafy F, Marchalot G, Champeau J, Hoeltzener B, Jézéquel JM (2008) Model-driven simulation of a maritime surveillance system. In: Proceedings of the 4th European conference on model driven artchitecture. ECMDA-FA June 9–13, Berlin, Germany
9. Hästbacka D, Vepsäläinen T, Kuikka S (2011) Model-driven development of industrial process control applications. J Syst Softw 84:1100–1113
10. Fuentes-Fernández R, Galán JM, Hassan S, López-Paredes A, Pavón J (2010) Application of model driven techniques for agent-based simulation. In: Advances in practical applications of agents and multiagent systems. Advances in intelligent and soft computing, Springer 70/2010, pp 81–90

11. Dallon JC (2010) Multiscale modeling of cellular systems in biology. Curr Opin Colloid Interface Sci 15:24–31
12. Meier-Schellersheim M, Fraser IDC, Klauschen F (2009) Multiscale modeling for biologists. Wiley Interdisc Rev Syst Biol Med 1(1):4–14
13. Varga M, Csukás B (2011a) Development of sustainable agrifood interoperability—what can we learn from natural processes? In: Herdon M, Rózsa T, Szilágyi R (szerk.) agricultural informatics conference 2011: innovative information technologies in agriculture, Debrecen: Magyar Agrárinformatikai Szövetség, pp 64–72
14. Kruse JP, Gu W (2009) Modes of p53 regulation. Cell 137(4):609–622
15. Barjis I, Samarrai K, Samarrai R, Uzun O (2010) Modeling of p53 signaling pathway regulation. In: Proceedings of summerSim'10, 2010 summer simulation multiconference, Society for computer simulation international, San Diego, pp 506–513
16. Kim DH, Rho K, Kim S (2009) A theoretical model for p53 dynamics: identifying optimal therapeutic strategy for its activation and stabilization. Cell Cycle 8(22):3707–3716
17. Kim S, Aladjem MI, McFadden GB, Kohn KW (2010) Predicted functions of MdmX in fine-tuning the response of p53 to DNA damage. PLoS Comput Biol 6(2):e1000665
18. Geva-Zatorsky N, Dekel E, Batchelor E, Lahav G, Alon U (2010) Fourier analysis and systems identification of the p53 feedback loop. Proc Natl Acad Sci USA 107(30):13550–13555
19. Feng Z, Zhang C, Wu R, Hu W (2011) Tumor suppressor p53 meets microRNAs. J Mol Cell Biol 3:44–50
20. Yamakuchi M, Lowenstein CJ (2009) MiR-34, SIRT1 and p53 the feedback loop. Cell Cycle 8(5):712–715
21. Lai X, Schmitz U, Gupta S, Kunz M, Wolkenhauer O, Vera J (2011) On the regulation of microRNA target hubs: A systems biology perspective, 12th international conference on systems biology (ICSB), Heidelberg/Mainheim, Germany, 28 Aug-1 Sept
22. Lim LP, Glasner ME, Yekta S, Burge CB, Bartel DP (2003) Vertebrate microRNA genes. Science 299(5612):1540–x
23. Dostie J, Mourelatos Z, Yang M, Sharma A, Dreyfuss G (2003) Numerous microRNPs in neuronal cells containing novel microRNAs. RNA 9(2):180–186
24. Mraz M, Malinova K, Kotaskova J, Pavlova S, Tichy B, Malcikova J, Kozubik S, Smardova J et al (2009) MiR-34a, miR-29c and miR-17-5p are downregulated in CLL patients with TP53 abnormalities. Leukemia 23 (6):1159–1163
25. He L, He X, Lim LP et al (2007) A microRNA component of the p53 tumour suppressor network. Nature 447(7148):1130–1134
26. Yamakuchi M, Ferlito M, Lowenstein CJ (2008) miR-34a repression of SIRT1 regulates apoptosis. Proc Natl Acad Sci US 105(36):13421–13426
27. Ji Q, Hao X, Meng Y, Zhang M, Desano J, Fan D et al (2008) Restoration of tumor suppressor miR-34 inhibits human p53-mutant gastric cancer tumorspheres. BMC Cancer 8:266
28. Berenbaum M (1990) Direct search methods in the optimisation of cancer chemotherapy regimens. Br J Cancer 61:101–109
29. Workman P (2003) Strategies for treating cancers caused by multiple genome abnormalities: from concepts to cures? Curr Opin Investig Drugs 4(12):1410–1415
30. Workman P (2007) Drugging the cancer genome: new challenges of infrequent and combinatorial targets. Curr Opin Investig Drugs 8(6):445–446
31. Al-Shyoukh I, Yu F, Feng J, Yan K, Dubinett S, Ho CM, Shamma JS, Sun R (2011) Systematic quantitative characterization of cellular responses induced by multiple signals. BMC Syst Biol 30(5):88
32. Curatolo M, Sveticic G (2002) Drug combinations in pain treatment: a review of the published evidence and a method for finding the optimal combination. Best Pract Res Clin Anaesthesiol 16(4):507–519

33. Iadevaia S, Lu Y, Morales FC, Mills GB, Ram PT (2010) Identification of optimal drug combinations targeting cellular networks: Integrating phospho-proteomics and computational network analysis. Cancer Res 70(17):6704–6714
34. Blum R, Kloog Y (2005) Tailoring Ras-pathway—inhibitor combinations for cancer therapy. Drug Resist Update 8(6):369–380
35. Geuna E, Milani A, Redana S, Rossi V, Valabrega G, Aglietta M, Montemurro F (2011) Hitting multiple targets in HER2-positive breast cancer: proof of principle or therapeutic opportunity? Expert Opin Pharmacother 12(4):549–565
36. Fearon ER, Vogelstein B (1990) A genetic model for colorectal tumorigenesis. Cell 61:759–767
37. Lai X, Vera J, Wolkenhauer O (2010) Modeling miRNA regulation in signaling networks: miR-34a regulation of the p53/Sirt1 module. Nat Proc. doi:10.1038/npre.2010.5123.1
38. Suzuki HI, Yamagata K, Sugimoto K, Iwamoto T, Kato S, Miyazono K (2009) Modulation of microRNA processing by p53. Nature 460:529–533
39. Wong MYW, Yu Y, Walsh WR, Yang LY (2011) Micro-RNA family treatment of cancers with mutant or wild-type p53 (Review). Int J Oncol 38:1189–1195
40. Csukás B (1998) Simulation by direct mapping of the structural models onto executable programs. In AIChE annual meeting, miami, Paper 239/9
41. Csukás B, Varga M, Balogh S (2011a) Direct computer mapping of executable multiscale hybrid process architectures. In Proceedings of summer simulation multiconference'2011, Den Haag, pp 87–95
42. Csukás B (2000, 2005) Conservation, information, evolution—foundations of the process engineering. In: Környezet-tudomány'2000 Somogyi F (ed) Veszprémi Egyetemi Kiadó, 55–86. Copied in: A tudomány egésze. A magyar tudomány tudománypedagógiai szemléje Zsolnai J (ed) Nemzeti Tankönyvkiadó, Budapest (in Hungarian)
43. Csukás B, Balogh S, Kováts S, Aranyi A, Kocsis Z, Bartha L (1999) Process design by controlled simulation of the executable structural models. Comput Chem Eng 23:569–572
44. Csukás B, Varga M, Prokop A, Balogh S (2011b) Simulation based analysis of nanocarrier internalization—exciting challenges with a new computational tool. In: Intracellular delivery: fundamentals and applications. series: fundamental biomedical technologies, Prokop A (ed), Springer, vol 5, Part 1, pp 125–154. ISBN 978-94-007-1247-8
45. Temesvári K, Aranyi A, Balogh S, Csukás B (2004) Simulated moving bed separation of a two components steroid mixture. Chromatographia 60:189–199
46. Varga M, Balogh S, Csukás B (2010) Sector spanning agrifood process transparency with direct computer mapping. Agric Inform 1(2):73–83
47. Varga M, Csukás B, Balogh S (2011b) Dynamic model based methodology for agrifood process network interoperability. In: Proceedings of World computer congress on computers in agriculture, Prague, 309–323
48. Varga M (2009) Melléktermékeket hasznosító complex körfolyamat gazdasági optimalizálása (Economic optimization of sustainable complex processes). PhD Theses, Kaposvár University (in Hungarian)
49. Kumar B, Yadav A, Lang J, Teknos TN, Kumar P (2012) Dysregulation of microRNA-34a expression in head and neck squamous cell carcinoma promotes tumor growth and tumor angiogenesis. PLoS ONE 7(5):e37601
50. Slaby O, Svoboda M, Michalek J, Vyzula R (2009) MicroRNAs in colorectal cancer: translation of molecular biology into clinical application. Molecular Cancer 8(102):1–13
51. Yang A, Zhao Y (2009) From a generic paradigm to a generic tool set: exploring computer-aided multiscale modeling. Comput Aided Chem Eng 27:189–194

E-references

52. UML Forum. UML FAQ. http://www.uml-forum.com/. Accessed 10 May 2012
53. SysML Forum. Web community for Systems Modeling Language. http://www.sysmlforum.com/. Accessed 10 May 2012
54. Modelica. Homepage of Modelica Association. https://modelica.org/. Accessed 10 May 2012

Erratum to: On Different Aspects of Network Analysis in Systems Biology

Amphun Chaiboonchoe, Wiktor Jurkowski, Johann Pellet, Enrico Glaab, Alexey Kolodkin, Antonio Rausel, Antony Le Béchec, Stéphane Ballereau, Laurene Meyniel, Isaac Crespo, Hassan Ahmed, Vitaly Volpert, Vincent Lotteau, Nitin Baliga, Leroy Hood, Antonio del Sol, Rudi Balling and Charles Auffray

Erratum to: Chapter 6 in: A. Prokop and B. Csukás (eds.), *Systems Biology*, DOI 10.1007/978-94-007-6803-1_6

The spelling of the author name "Antonio Raussel" was incorrect. The name should read as "Antonio Rausell" in the Table of Contents, List of Contributors, and in Chap. 6.

The online version of the original chapter can be found under 10.1007/978-94-007-6803-1_6.

A. Chaiboonchoe · J. Pellet · S. Ballereau · L. Meyniel · H. Ahmed · V. Volpert
V. Lotteau · C. Auffray (✉)
European Institute for Systems Biology and Medicine, CNRS-UCBL-ENS,
Université de Lyon, 50 Avenue Tony Garnier, 69366 Lyon cedex 07, France
e-mail: cauffray@eisbm.org, achaiboonchoe@eisbm.org, jpellet@eisbm.org,
sballereau@eisbm.org, laurene.meyniel@inserm.fr, hahmed@eisbm.org,
vvolpert@eisbm.org, vlotteau@eisbm.org

W. Jurkowski · E. Glaab · A. Kolodkin · A. Rausel · A. Le Béchec · I. Crespo
A. d. Sol · R. Balling
Luxembourg Centre for Systems Biomedicine, University of Luxembourg, 7,
Avenue des Hauts-Fourneaux 4362 Esch-sur-Alzette, Luxembourg
e-mail: wiktor.jurkowski@uni.lu, enrico.glaab@uni.lu, alexey.kolodkin@uni.lu,
antonio.rausel@uni.lu, anthony.lebechec@uni.lu, isaac.crespo@uni.lu,
antonio.delsol@uni.lu, rudi.balling@uni.lu

A. Kolodkin · N. Baliga · L. Hood
Institute for Systems Biology, 401 Terry Avenue North, Seattle, WA 98109-5234, USA
e-mail: nbaliga@systemsbiology.org, lhood@systemsbiology.org

A. Prokop and B. Csukás (eds.), *Systems Biology*,
DOI: 10.1007/978-94-007-6803-1_19,

Index

A. Prokop and B. Csukás (eds.), *Systems Biology*,
DOI: 10.1007/978-94-007-6803-1,

MIX
Papier aus verantwortungsvollen Quellen
Paper from responsible sources
FSC® C105338

If you have any concerns about our products,
you can contact us on
ProductSafety@springernature.com

In case Publisher is established outside the EU,
the EU authorized representative is:
Springer Nature Customer Service Center GmbH
Europaplatz 3, 69115 Heidelberg, Germany

Printed by Libri Plureos GmbH
in Hamburg, Germany